CLINICAL NEUROENDOCRINOLOGY

HANDBOOK OF CLINICAL NEUROLOGY

Series Editors

MICHAEL J. AMINOFF, FRANÇOIS BOLLER, AND DICK F. SWAAB

VOLUME 124

ELSEVIER

EDINBURGH LONDON NEW YORK OXFORD PHILADELPHIA
ST LOUIS SYDNEY TORONTO 2014

CLINICAL NEUROENDOCRINOLOGY

Series Editors

MICHAEL J. AMINOFF, FRANÇOIS BOLLER, AND DICK F. SWAAB

Volume Editors

ERIC FLIERS, MÁRTA KORBONITS, AND JOHANNES A. ROMIJN

VOLUME 124

3rd Series

ELSEVIER

EDINBURGH LONDON NEW YORK OXFORD PHILADELPHIA
ST LOUIS SYDNEY TORONTO 2014

ELSEVIER B.V.
Radarweg 29, PO Box 211, 1000 AE Amsterdam, Netherlands
The Boulevard, Langford Lane, Kidlington, Oxford OX5 1GB, UK
225 Wyman Street, Waltham, MA 02451, USA

ISBN: 9780444596024

British Library Cataloguing in Publication Data
A catalogue record for this book is available from the British Library

Library of Congress Cataloging in Publication Data
A catalog record for this book is available from the Library of Congress

Notices
Knowledge and best practice in this field are constantly changing. As new research and experience broaden our understanding, changes in research methods, professional practices, or medical treatment may become necessary.

Practitioners and researchers must always rely on their own experience and knowledge in evaluating and using any information, methods, compounds, or experiments described herein. In using such information or methods they should be mindful of their own safety and the safety of others, including parties for whom they have a professional responsibility.

With respect to any drug or pharmaceutical products identified, readers are advised to check the most current information provided (i) on procedures featured or (ii) by the manufacturer of each product to be administered, to verify the recommended dose or formula, the method and duration of administration, and contraindications. It is the responsibility of practitioners, relying on their own experience and knowledge of their patients, to make diagnoses, to determine dosages and the best treatment for each individual patient, and to take all appropriate safety precautions.

To the fullest extent of the law, neither the Publisher nor the authors, contributors, or editors, assume any liability for any injury and/or damage to persons or property as a matter of products liability, negligence or otherwise, or from any use or operation of any methods, products, instructions, or ideas contained in the material herein.

The Publisher

Printed in United States of America
Transferred to Digital Printing, 2015
Commissioning Editor: Mica Haley
Development Editor: Michael Parkinson
Project Manager: Sujatha Thirugnana Sambandam
Designer/Design Direction: Alan Studholme

The Publisher's policy is to use **paper manufactured from sustainable forests**

Handbook of Clinical Neurology 3rd Series

Available titles

Vol. 79, The human hypothalamus: basic and clinical aspects, Part I, D.F. Swaab ISBN 9780444513571
Vol. 80, The human hypothalamus: basic and clinical aspects, Part II, D.F. Swaab ISBN 9780444514905
Vol. 81, Pain, F. Cervero and T.S. Jensen, eds. ISBN 9780444519016
Vol. 82, Motor neurone disorders and related diseases, A.A. Eisen and P.J. Shaw, eds. ISBN 9780444518941
Vol. 83, Parkinson's disease and related disorders, Part I, W.C. Koller and E. Melamed, eds. ISBN 9780444519009
Vol. 84, Parkinson's disease and related disorders, Part II, W.C. Koller and E. Melamed, eds. ISBN 9780444528933
Vol. 85, HIV/AIDS and the nervous system, P. Portegies and J. Berger, eds. ISBN 9780444520104
Vol. 86, Myopathies, F.L. Mastaglia and D. Hilton Jones, eds. ISBN 9780444518996
Vol. 87, Malformations of the nervous system, H.B. Sarnat and P. Curatolo, eds. ISBN 9780444518965
Vol. 88, Neuropsychology and behavioural neurology, G. Goldenberg and B.C. Miller, eds. ISBN 9780444518972
Vol. 89, Dementias, C. Duyckaerts and I. Litvan, eds. ISBN 9780444518989
Vol. 90, Disorders of consciousness, G.B. Young and E.F.M. Wijdicks, eds. ISBN 9780444518958
Vol. 91, Neuromuscular junction disorders, A.G. Engel, ed. ISBN 9780444520081
Vol. 92, Stroke – Part I: Basic and epidemiological aspects, M. Fisher, ed. ISBN 9780444520036
Vol. 93, Stroke – Part II: Clinical manifestations and pathogenesis, M. Fisher, ed. ISBN 9780444520043
Vol. 94, Stroke – Part III: Investigations and management, M. Fisher, ed. ISBN 9780444520050
Vol. 95, History of neurology, S. Finger, F. Boller and K.L. Tyler, eds. ISBN 9780444520081
Vol. 96, Bacterial infections of the central nervous system, K.L. Roos and A.R. Tunkel, eds. ISBN 9780444520159
Vol. 97, Headache, G. Nappi and M.A. Moskowitz, eds. ISBN 9780444521392
Vol. 98, Sleep disorders Part I, P. Montagna and S. Chokroverty, eds. ISBN 9780444520067
Vol. 99, Sleep disorders Part II, P. Montagna and S. Chokroverty, eds. ISBN 9780444520074
Vol. 100, Hyperkinetic movement disorders, W.J. Weiner and E. Tolosa, eds. ISBN 9780444520142
Vol. 101, Muscular dystrophies, A. Amato and R.C. Griggs, eds. ISBN 9780080450315
Vol. 102, Neuro-ophthalmology, C. Kennard and R.J. Leigh, eds. ISBN 9780444529039
Vol. 103, Ataxic disorders, S.H. Subramony and A. Durr, eds. ISBN 9780444518927
Vol. 104, Neuro-oncology Part I, W. Grisold and R. Sofietti, eds. ISBN 9780444521385
Vol. 105, Neuro-oncology Part II, W. Grisold and R. Sofietti, eds. ISBN 9780444535023
Vol. 106, Neurobiology of psychiatric disorders, T. Schlaepfer and C.B. Nemeroff, eds. ISBN 9780444520029
Vol. 107, Epilepsy Part I, H. Stefan and W.H. Theodore, eds. ISBN 9780444528988
Vol. 108, Epilepsy Part II, H. Stefan and W.H. Theodore, eds. ISBN 9780444528995
Vol. 109, Spinal cord injury, J. Verhaagen and J.W. McDonald III, eds. ISBN 9780444521378
Vol. 110, Neurological rehabilitation, M. Barnes and D.C. Good, eds. ISBN 9780444529015
Vol. 111, Pediatric neurology Part I, O. Dulac, M. Lassonde and H.B. Sarnat, eds. ISBN 9780444528919
Vol. 112, Pediatric neurology Part II, O. Dulac, M. Lassonde and H.B. Sarnat, eds. ISBN 9780444529107
Vol. 113, Pediatric neurology Part III, O. Dulac, M. Lassonde and H.B. Sarnat, eds. ISBN 9780444595652
Vol. 114, Neuroparasitology and tropical neurology, H.H. Garcia, H.B. Tanowitz and O.H. Del Brutto, eds. ISBN 9780444534903
Vol. 115, Peripheral nerve disorders, G. Said and C. Krarup, eds. ISBN 9780444529022
Vol. 116, Brain stimulation, A.M. Lozano and M. Hallett, eds. ISBN 9780444534972
Vol. 117, Autonomic nervous system, R.M. Buijs and D.F. Swaab, eds. ISBN 9780444534910
Vol. 118, Ethical and legal issues in neurology, J.L. Bernat and H.R. Beresford, eds. ISBN 9780444535016
Vol. 119, Neurologic aspects of systemic disease Part I, J. Biller and J.M. Ferro, eds. ISBN 9780702040863
Vol. 120, Neurologic aspects of systemic disease Part II, J. Biller and J.M. Ferro, eds. ISBN 9780702040870
Vol. 121, Neurologic aspects of systemic disease Part III, J. Biller and J.M. Ferro, eds. ISBN 9780702040887
Vol. 122, Multiple sclerosis and related disorders, D.S. Goodin, ed. ISBN 9780444520012
Vol. 123, Neurovirology, A.C. Tselis and J. Booss, eds. ISBN 9780444534880

Foreword

The Handbook of Clinical Neurology (HCN), founded by Pierre Vinken and George Bruyn in 1968, is a prestigious, multivolume reference work on disorders of the nervous system. Such an endeavor requires regular updates. Rapid advances in neurology and its closer relationship with an increasing number of other medical disciplines have led us also to add new topics as separate volumes since we took over in 2003 as editors of the current third series. The present title, Clinical Neuroendocrinology, is a novel title for the series. It includes a new range of topics for the HCN series as well as an update and expansion of topics discussed previously. It focuses on the pathophysiology, diagnosis, and treatment of diseases of the hypothalamus and pituitary gland. Some topics, such as pituitary and hypothalamic tumors, were dealt with in earlier volumes, but the information on genetic causes and treatment has changed tremendously and has been incorporated into the present volume. Various other topics, such as neuroendocrine mechanisms in athletes and long-term effects of treatment of pituitary adenomas, have never been included in the HCN series. Clinical neuroendocrinology of necessity involves the integration of a large number of medical disciplines with neurology, such as internal medicine, pediatrics, neurosurgery, neuroradiology, clinical genetics, and radiotherapy, as is evident from the present volume. The behavioral consequences of disorders and therapies, uncertainties in therapy, controversies and recent novel insights from research also receive attention here.

We were extremely pleased to have as volume editors three internationally renowned experts in clinical neuroendocrinology. Eric Fliers is Professor of Endocrinology at the Academic Medical Center of the University of Amsterdam, where he serves as Head of the Department of Endocrinology and Metabolism. He is the current chair of the Dutch Endocrine Society. He has his roots in research on the neuroendocrine nuclei in the postmortem human hypothalamus, and was involved in the foundation of the Netherlands Brain Bank. His current research is focused on neuroendocrine aspects of the hypothalamus–pituitary–thyroid axis. Márta Korbonits is Professor of Endocrinology and Metabolism at Barts and the London School of Medicine and Dentistry, Queen Mary University of London, where she heads the Centre for Endocrinology. With the help of the Familial Isolated Pituitary Adenomas (FIPA) consortium, she is currently defining the clinical characteristics of patients with familial pituitary adenoma syndromes and uncovering novel genetic variants causing these conditions. Johannes A. Romijn is Professor of Medicine at the Academic Medical Center of the University of Amsterdam and serves as chair of the Department of Medicine. He is the Editor-in-Chief of the European Journal of Endocrinology. His research has a main focus on neuroendocrine regulation of metabolism.

The three volume editors have assembled a truly international group of authors with acknowledged expertise to contribute to this volume and have produced with them an authoritative, comprehensive, and up-to-date account of clinical neuroendocrinology. Its availability electronically on Elsevier's Science Direct site as well as in print format should ensure its ready accessibility and facilitate searches for specific information. We are grateful to the volume editors and to all the contributors for their efforts in creating such an invaluable resource. As series editors we read and commented on each of the chapters with great interest. We are therefore confident that both clinicians and researchers in many different medical disciplines will find much in this volume to appeal to them.

As always it is a pleasure to thank Elsevier, our publishers – and in particular Michael Parkinson in Lochcarron, and Mica Haley and Kristi Anderson in San Diego – for their unfailing and expert assistance in the development and production of this volume.

Michael J. Aminoff
François Boller
Dick F. Swaab

Preface

Clinical neuroendocrinology focuses on the pathophysiology, diagnosis, and treatment of diseases of the hypothalamus and pituitary gland. Over the past years, much progress has been made in this field, sparked by several major breakthroughs including the discovery of leptin and the unraveling of its role in body weight regulation and metabolism; the identification of novel genetic causes of pituitary tumors, central hypothyroidism and hypogonadism; the recognition of oxytocin and vasopressin's roles in social behaviors; and the discovery of medicines effectively reducing the excessive endocrine activity of selected pituitary tumors.

The present volume aims to inform a broad readership of medical specialists involved in clinical neuroendocrinology about recent developments and state-of-the art knowledge in the field.

The first section of the volume focuses on major aspects of hypothalamic function and pathophysiology. The hypothalamus participates in complex and divergent pathophysiologic conditions such as disorders of water balance, psychiatric conditions, sleep disturbances, obesity, and critical illness. Sometimes, disturbances within the hypothalamus are the initiator of disease, as in rare forms of obesity, but more frequently diseases elsewhere in the body induce secondary pathophysiologic conditions in the hypothalamus, as in the syndrome of inappropriate antidiuretic hormone secretion (SIADH), sleep deprivation and critical illness. Hypothalamic neuropeptides such as vasopressin, oxytocin, and corticotropin-releasing hormone (CRH) are discussed in the context of neuroendocrine regulation, stress and water metabolism, and also as mediators of emotional and social behaviors. Usually, these different fields of hypothalamic function are not described in parallel.

The second section of the volume focuses on new developments in disorders of the pituitary gland. This section includes detailed descriptions of the pathophysiology, diagnosis, and treatment of different pituitary diseases, including active and inactive adenomas, Rathke's cleft cysts, craniopharyngioma, and unusual forms of hypopituitarism. Medical, surgical, and radiotherapeutic regimens are discussed in detail.

The third section of the volume is devoted to controversial issues and hot topics in clinical neuroendocrinology, including Nelson syndrome, familial pituitary tumors, and autoimmune hypophysitis. There are many uncertainties in endocrine substitution therapy for pituitary insufficiency. In this section these uncertainties regarding central adrenal insufficiency, hypothyroidism, and GH deficiency are discussed in different chapters. In addition, the irreversible consequences of pituitary tumors and/or their treatment on quality of life and neuropsychologic function are reviewed. The pituitary disease may be cured or controlled but it seems that patients may still suffer from a debilitating chronic syndrome. Finally, this part of the book includes a chapter on the neuroendocrine mechanisms involved with adaptation to physical exercise.

We are confident that the present volume covers most aspects of neuroendocrinology that are relevant for clinicians involved in the care of patients with neuroendocrine disease, and we expect that it will be of interest for internists, pediatricians, neurologists, neurosurgeons, neuroradiologists, clinical geneticists, and radiotherapists active in this field.

Eric Fliers
Márta Korbonits
Johannes A. Romijn

Contributors

N. Alband
Department of Endocrinology, Barts and the London School of Medicine, Queen Mary University of London, London, UK

O. Ansorge
Department of Neuropathology, Nuffield Department of Clinical Neurosciences, University of Oxford, John Radcliffe Hospital, Oxford, UK

E. Adams
Department of Endocrinology, Oxford Centre for Diabetes, Endocrinology and Metabolism, Churchill Hospital, Oxford, UK

R. Bachner-Melman
School of Social and Community Sciences, Ruppin Academic Center, Emek Hefer and Department of Psychology, Hebrew University of Jerusalem, Jerusalem, Israel

T.M. Barber
Division of Metabolic and Vascular Health, Warwick Medical School, University of Warwick, Coventry, UK

B.M.K. Biller
Neuroendocrine Unit, Massachusetts General Hospital and Harvard Medical School, Boston, MA, USA

A. Boelen
Department of Endocrinology and Metabolism, Academic Medical Center, University of Amsterdam, Amsterdam, The Netherlands

M. Bonomi
Department of Clinical Sciences and Community Health, University of Milan, and Division of Endocrine and Metabolic Diseases, San Luca Hospital, Istituto Auxologico Italiano, Milan, Italy

M. Buchfelder
Department of Neurosurgery, University of Erlangen-Nürnberg, Erlangen, Germany

P. Cappabianca
Department of Neurological Sciences, Division of Neurosurgery, Università degli Studi di Napoli Federico II, Naples, Italy

L.M. Cavallo
Department of Neurological Sciences, Division of Neurosurgery, Università degli Studi di Napoli Federico II, Naples, Italy

P. Chanson
Department of Endocrinology and Disorders of Reproduction, Hôpital Bicêtre; Reference Center for Rare Endocrine Disorders of Growth; Faculty of Medicine, Université Paris-Sud 11 and INSERM U693, Le Kremlin-Bicêtre, Paris, France

K. Cheer
Department of Endocrinology, Christie Hospital NHS Foundation Trust, Manchester, UK

A.P. Coll
University of Cambridge Metabolic Research Laboratories, MRC Metabolic Diseases Unit, Wellcome Trust–MRC Institute of Metabolic Science, Addenbrooke's Hospital, Cambridge, UK

M.T. Dattani
Developmental Endocrinology Research Group, Clinical and Molecular Genetics Unit, University College London–Institute of Child Health, London, UK

T.G. Dinan
Department of Psychiatry, University College Cork, Cork, Ireland

E. Donga
Department of Endocrinology, Leiden University Medical Center, Leiden, The Netherlands

R.P. Ebstein
Department of Psychology, National University of Singapore, Singapore

E.-M. Erfurth
Department of Endocrinology, Lund University, Lund, Sweden

E. Fliers
Department of Endocrinology and Metabolism, Academic Medical Center, University of Amsterdam, Amsterdam, The Netherlands

P. Kamenicky
Department of Endocrinology and Disorders of Reproduction, Hôpital Bicêtre; Reference Center for Rare Endocrine Disorders of Growth; Faculty of Medicine, Université Paris-Sud 11 and INSERM U693, Le Kremlin-Bicêtre, Paris, France

N. Karavitaki
Oxford Centre for Diabetes, Endocrinology and Metabolism, Churchill Hospital, Oxford, UK

M. Korbonits
Department of Endocrinology, Barts and the London School of Medicine, Queen Mary University of London, London, UK

L. Langouche
Laboratory and Department of Intensive Care Medicine, University of Leuven, Leuven, Belgium

R. Larder
University of Cambridge Metabolic Research Laboratories, MRC Metabolic Diseases Unit, Wellcome Trust–MRC Institute of Metabolic Science, Addenbrooke's Hospital, Cambridge, UK

S. Larkin
Department of Neuropathology, Nuffield Department of Clinical Neurosciences, University of Oxford, John Radcliffe Hospital, Oxford, UK

C.T. Lim
University of Cambridge Metabolic Research Laboratories, MRC Metabolic Diseases Unit, Wellcome Trust–MRC Institute of Metabolic Science, Addenbrooke's Hospital, Cambridge, UK

J.S. Loeffler
Department of Radiation Oncology, Massachusetts General Hospital, Boston, MA, USA

M.J. McCabe
Developmental Endocrinology Research Group, Clinical and Molecular Genetics Unit, University College London–Institute of Child Health, London, UK

M. Misra
Pediatric Endocrine and Neuroendocrine Units, Massachusetts General Hospital and Harvard Medical School, Boston, MA, USA

M.E. Molitch
Division of Endocrinology, Metabolism and Molecular Medicine, Northwestern University Feinberg School of Medicine, Chicago, IL, USA

H.L. Müller
Department of Pediatrics, Klinikum Oldenburg, Medical Campus University Oldenburg, Oldenburg, Germany

M. Naughton
Department of Psychiatry, University College Cork, Cork, Ireland

S.J.C.M.M. Neggers
Section of Endocrinology, Department of Medicine, Erasmus University Medical Center, Rotterdam, The Netherlands

K.S. Oh
Department of Radiation Oncology, Massachusetts General Hospital, Boston, MA, USA

I. Pashtan
Harvard Radiation Oncology Program, Boston, MA, USA

S. Pekic
Faculty of Medicine, University of Belgrade, and Clinic of Endocrinology, Diabetes and Metabolic Diseases, University Clinical Center Belgrade, Belgrade, Serbia

A.M. Pereira
Department of Endocrinology and Center for Endocrine Tumors, Leiden University Medical Center, Leiden, The Netherlands

L. Persani
Department of Clinical Sciences and Community Health, University of Milan, and Division of Endocrine and Metabolic Diseases, San Luca Hospital, Istituto Auxologico Italiano, Milan, Italy

F. Petraglia
Department of Molecular and Developmental Medicine, University of Siena, Siena, Italy

V. Popovic
Faculty of Medicine, University of Belgrade, and Clinic of Endocrinology, Diabetes and Metabolic Diseases, University Clinical Center Belgrade, Belgrade, Serbia

J.A. Romijn
Department of Medicine, Academic Medical Center, University of Amsterdam, Amsterdam, The Netherlands

S. Salenave
Department of Endocrinology and Disorders of Reproduction, Hôpital Bicêtre and Reference Center for Rare Endocrine Disorders of Growth, Le Kremlin-Bicêtre, Paris, France

S. Schlaffer
Department of Neurosurgery, University of Erlangen-Nürnberg, Erlangen, Germany

L.V. Scott
Department of Psychiatry, University College Cork, Cork, Ireland

D. Solari
Department of Neurological Sciences, Division of Neurosurgery, Università degli Studi di Napoli Federico II, Naples, Italy

F.M. Swords
Norwich Medical School and Directorate of Endocrinology, Norfolk and Norwich University Hospital NHS Foundation Trust, Norwich, UK

Y. Takahashi
Division of Diabetes and Endocrinology, Department of Internal Medicine, Kobe University Graduate School of Medicine, Kobe, Japan

P.J. Trainer
Department of Endocrinology, Christie Hospital NHS Foundation Trust, Manchester, UK

N.A. Tritos
Neuroendocrine Unit, Massachusetts General Hospital and Harvard Medical School, Boston, MA, USA

A.S.P. van Trotsenburg
Department of Paediatric Endocrinology, Academic Medical Center, University of Amsterdam, Amsterdam, The Netherlands

G. Van den Berghe
Laboratory and Department of Intensive Care Medicine, University of Leuven, Leuven, Belgium

A.J. van der Lely
Section of Endocrinology, Department of Medicine, Erasmus University Medical Center, Rotterdam, The Netherlands

J.G. Verbalis
Georgetown University, Washington, DC, USA

C. Voltolini
Department of Molecular and Developmental Medicine, University of Siena, Siena, Italy

J.A.H. Wass
Department of Endocrinology, Oxford Centre for Diabetes, Endocrinology and Metabolism, Churchill Hospital, Oxford, UK

Contents

SECTION 2 Disorders of the pituitary gland

Diagnostic approach

Pituitary pathologies

Therapeutic approach to pituitary tumors

SECTION 3 Controversial issues and hot topics

Section 1

Clinical aspects of hypothalamic function

Handbook of Clinical Neurology, Vol. 124 (3rd series)
Clinical Neuroendocrinology
E. Fliers, M. Korbonits, and J.A. Romijn, Editors

Chapter 1

Genetic aspects of hypothalamic and pituitary gland development

MARK J. McCABE AND MEHUL T. DATTANI*
Developmental Endocrinology Research Group, Clinical and Molecular Genetics Unit, University College London–Institute of Child Health, London, UK

INTRODUCTION

The primordial central nervous system develops during the third week of human gestation during neurulation, a process which gives rise to the neural plate with subsequent derivations into the spinal cord and brain. Fate map studies, which aim to follow the development of cells or tissues from early stages of embryogenesis, have shown that the pituitary, hypothalamus, optic nerves, and forebrain each develop from the anterior neural plate (Schlosser, 2006). Complex interactions of spatiotemporally regulated signaling molecules and transcription factors are critically important for their successful development.

The pituitary gland is a midline structure located in the sella turcica recess of the sphenoid bone at the base of the brain. It is composed of three lobes which have dual embryonic ectodermal origins, the oral ectoderm giving rise to the hormone-secreting anterior and intermediate lobes and the overlying neural ectoderm giving rise to the posterior lobe (Cohen, 2012). The posterior lobe is the only neural component of the pituitary gland and provides a direct link to the hypothalamus, which is also derived from the neural ectoderm. Maintained apposition and interactions between these two ectodermal layers is crucial for normal pituitary development. Insults to this developmental process can result in the loss or reduction of pituitary hormone-secreting cells resulting in congenital hypopituitarism, with phenotypes ranging from multiple pituitary hormone deficiencies (combined/multiple pituitary hormone deficiency (CPHD/MPHD)) to deficiencies in single hormones only, the most common isolated hormone deficiency being attributed to growth hormone (Alatzoglou and Dattani, 2009). Given its midline location, and that the pituitary gland is derived from the same region of the neural plate as the hypothalamus, optic nerves, and forebrain as described above, hypopituitarism is often associated with craniofacial/midline disorders affecting these structures also. Such disorders are characteristically heterogeneous but range from incompatibility with life, to holoprosencephaly (HPE) and cleft palate and septo-optic dysplasia (SOD), which will be described later (McCabe et al., 2011a).

This chapter will review the molecular basis underlying the development of the hypothalamo-pituitary axis and will detail how known defects in many of the required genes can lead to HPE and SOD as well as isolated CPHD/MPHD. Furthermore, this chapter will discuss the increasing evidence of overlapping genotypes between congenital hypopituitarism and Kallmann syndrome (KS), defined as the combination of hypogonadotropic hypogonadism (HH) and anosmia.

DEVELOPMENT OF THE HYPOTHALAMO-PITUITARY AXIS

Morphology

As mentioned briefly in the introduction, the three lobes of the pituitary are derived from two adjacent ectodermal layers. The primordium of the anterior lobe is termed Rathke's pouch (RP), and this structure develops through the dorsal invagination of the oral ectoderm toward the overlying neuroectoderm containing the primordium of the hypothalamus, the ventral diencephalon (VD). The invagination of RP involves tight regulation of cellular proliferation and subsequent differentiation events to give rise to five highly differentiated cell types secreting a total of six different hormones: (1) corticotrophs produce adrenocorticotropic hormone (ACTH), (2) thyrotrophs produce thyrotropin

*Correspondence to: Professor Mehul T. Dattani, UCL-Institute of Child Health, 30 Guilford Street, London, WC1N 1EH, UK. Tel: +44-207-905-2657, Fax: +44-207-404-619, E-mail: m.dattani@ucl.ac.uk

or thyroid-stimulating hormone (TSH), (3) somatotrophs produce growth hormone, (4) lactotrophs (which are derived from the same precursor cells as the somatotrophs; termed somatomammotrophs) produce prolactin, and (5) gonadotrophs produce follicle-stimulating hormone (FSH) and luteinizing hormone (LH) (Cohen, 2012). This invagination event also leads to the formation of the intermediate lobe, and this contains the melanotrophs which secrete pro-opiomelanocortin (POMC). POMC is a major precursor protein to endorphins, melanocyte-stimulating hormone (MSH), and ACTH (Alatzoglou and Dattani, 2009). Humans contain only a vestigial intermediate lobe and as such do not secrete large amounts of POMC-derived hormones. Once secreted, each of the hormones targets distant tissues and organs throughout the body.

As RP invaginates, part of the VD evaginates ventrally to form the infundibulum and later the posterior pituitary lobe and pituitary stalk. Throughout development there is a close association between the infundibulum and RP and the interactions and apposition between these structures must be maintained for successful organogenesis. The pituitary stalk acts as a physical connection between the pituitary gland and brain and contains the hypophyseal (hypothalamo-pituitary) portal system, as well as the neuronal connections traversing across the hypothalamic median eminence. These neurons originate from the supraoptic, suprachiasmatic, and paraventricular nuclei which are large hypothalamic magnocellular bodies located within the periventricular region of the hypothalamus (Szarek et al., 2010). The supraoptic and suprachiasmatic nuclei release arginine vasopressin while the paraventricular nuclei release oxytocin (Kelberman et al., 2009). Within the median eminence itself at the base of the hypothalamus is the capillary bed, into which the widely dispersed hypothalamic parvocellular neurons secrete hypophysiotropic hormones. These stimulate the release of the seven anterior/intermediate pituitary lobe hormones described above via the hypophyseal portal system. Interestingly, the parvocellular neurons also secrete oxytocin and arginine vasopressin, although at much lower concentrations than the magnocellular neurons, with the parvocellular-derived arginine vasopressin being implicated in acting synergistically with corticotropin-releasing hormone in regulating ACTH release. It is therefore evident that it is the hypothalamus that is the central mediator of growth, reproduction, and homeostasis, acting through the pituitary gland (Kelberman et al., 2009).

The anatomy of the developed hypothalamus is well understood. It extends from the anteriorly located optic chiasm to the posteriorly located mammillary body and is organized into distinct rostral to caudal regions: preoptic, anterior, tuberal, and mammillary (Szarek et al., 2010). The organ is also subdivided into three medial to lateral regions: periventricular, medial, and lateral (Szarek et al., 2010). The periventricular region was described above, but contained within the medial region is the medial preoptic nucleus, the anterior hypothalamus, the dorsomedial nucleus, the ventromedial nucleus, and the mammillary nuclei (Szarek et al., 2010). The lateral zone consists of the preoptic area and hypothalamic area.

Interestingly, however, deciphering hypothalamic development during embryogenesis has proved problematic, perhaps due to its anatomic complexity and highly diverse collection of cell groups and neuronal subtypes for which there is a dearth of literature defining the genetics and signaling and marker molecules involved in their delineation and identification (Blackshaw et al., 2010). Furthermore, genetic expression studies within the hypothalamus have knock-on effects on multiple neuronal subtypes and downstream physiologic processes. However, studies are slowly elucidating hypothalamic development. Structural organization of the developing human hypothalamus was nicely assessed by immunohistochemistry in more than 30 brains over the entire course of gestation, and provided evidence for architectural homologies between species, particularly that of the better characterized rat (Koutcherov et al., 2002, 2003). In addition, one recent study successfully labeled each major hypothalamic nucleus over the entire course of neurogenesis (Shimogori et al., 2010).

Timeline of hypothalamo-pituitary organogenesis

As described earlier, the hypothalamus and the pituitary are derived from the anterior neural plate. Their development is highly conserved across vertebrates including humans, and as such, this chapter will outline their development in the mouse, a model which is well characterized.

The growth and expansion of the brain during embryogenesis causes the embryonic head to bend anteriorly, resulting in the ventral displacement of the heart with the subsequent formation of a depression between the heart and brain, termed the stomodeum, or oral ectoderm. At embryonic (E) day 8.5, the upper edge of the stomodeum thickens, which signals the onset of pituitary organogenesis (Fig. 1.1). By E9.5, 1 day later, this thickened section of the stomodeum invaginates to form RP (Takuma et al., 1998; Rizzoti and Lovell-Badge, 2005), just prior to the commencement of hypothalamic neurogenesis at E10 (Shimogori et al., 2010). At E10.5, the infundibulum evaginates from the VD, which becomes morphologically evident in the neural ectoderm at E9.5 (Szarek et al., 2010), to come into contact with RP. This

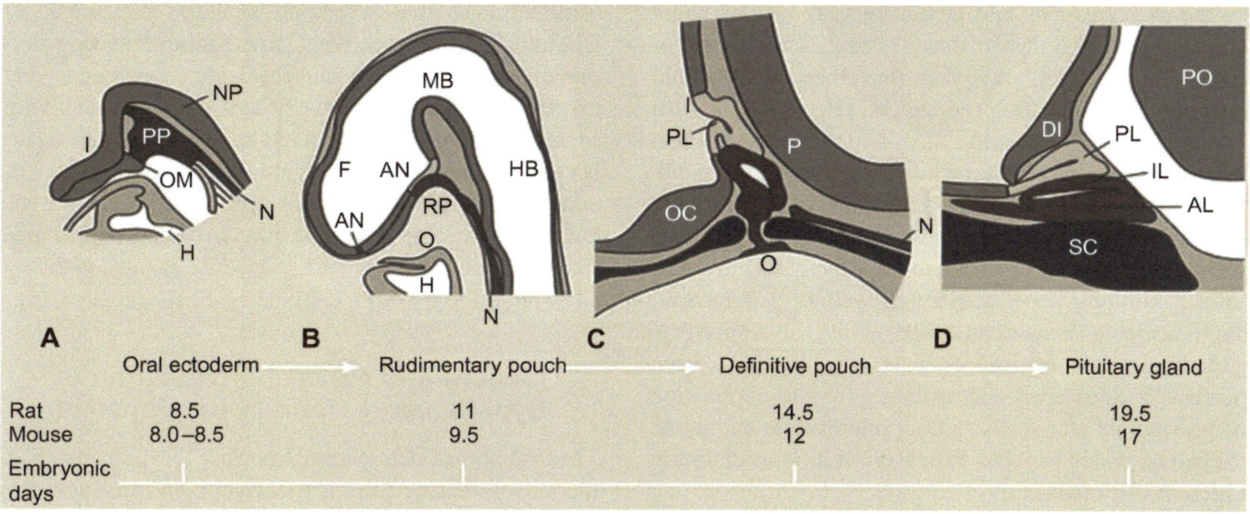

Fig. 1.1. Development of the murine pituitary. Thickening of the stomodeum (oral ectoderm) at E8.5 marks the onset of pituitary organogenesis (**A**), which is followed 1 day later by the invagination of a rudimentary pouch toward the overlying ventral diencephalon (**B**). The definitive Rathke's pouch is formed as its connection to the oral ectoderm is severed and the infundibulum evaginates from the ventral diencephalon to form the posterior pituitary (**C**). Progenitors of the hormone-secreting cell types then proliferate and terminally differentiate to produce the mature pituitary gland consisting of the anterior lobe, intermediate lobe and posterior lobe (**D**). E, embryonic day; I, infundibulum; NP, neural plate; N, notochord; PP, pituitary placode; OM, oral membrane; H, heart; F, forebrain; MB, midbrain; HB, hindbrain; RP, Rathke's pouch; AN, anterior neural pore; O, oral cavity; PL, posterior lobe; OC, optic chiasm; P, pontine flexure; PO, pons; IL, intermediate lobe; AL, anterior lobe; DI, diencephalon; SC, sphenoid cartilage (Sheng and Westphal, 1999).

contact is essential for successful organogenesis of the pituitary gland and is maintained throughout this process. At this moment, the connection between RP and the oral ectoderm is severed, leaving a fully developed pouch. Within this pouch, from E12.5 to E17.5, the progenitor cells of the five hormone-secreting cell types proliferate and terminally differentiate ventrally (Dasen and Rosenfeld, 1999; Ward et al., 2006). At a similar stage in development (E12–E14), the bulk of the hypothalamic neurons are born, concomitant with the highest expression levels of genes important for the regional patterning of hypothalamic progenitor cells, such as *Sim1, Sim2, Arx,* and *Nr5a1* (Shimogori et al., 2010). Neurogenesis is complete by E16 although expression of hypothalamic terminal differentiation markers peak postnatally (Shimogori et al., 2010).

In the pituitary, the first hormone-secreting cell types are the prospective thyrotrophs, detected at E11.5, which express transcription factor islet-1 (*Isl1*) and α-glycoprotein subunit (α-GSU) in the most ventral aspect of RP (Kelberman et al., 2009). After expressing thyroid-stimulating hormone subunit β (*Tshβ*) at E12.5, these cells are terminally differentiated although they disappear shortly after birth. At E12.5, the corticotrophs begin differentiating just dorsal to the intermediary thyrotrophs above. These express *Pomc*, as do the melanotrophs from E14.5 in the intermediate lobe. About this time, the definitive thyrotrophs are detected in RP (Kelberman et al., 2009). Detection of growth hormone and prolactin signals the commencement at E15.5 of somatotroph and lactotroph differentiation, respectively. The somatotrophs proliferate throughout the central and lateral aspects of the lobe while the lactotrophs remain restricted to the medial zone, adjacent to the intermediate lobe (Kelberman et al., 2009). Finally at E16.5, the gonadotrophs are detected with expression of Fsh subunit β (*Fshβ*), followed by Lh subunit β (*Lhβ*) 1 day later (Kelberman et al., 2009).

Recent work has determined that the spatial distribution of the hormone-secreting cell types within the pituitary is by no means random, and in fact the cells maintain same cell-type networks which facilitate a coordinated physiologic response to stimuli (Bonnefont et al., 2005; Budry et al., 2011; Hodson et al., 2012).

Perhaps one of the most significant advances of the past year was the production of a functional, three-dimensional anterior pituitary gland *in vitro* (Suga et al., 2011). The authors were able to reproduce a dual ectodermal cell layer with RP invagination. Through various culture conditions and application of induction factors, the authors were able to stimulate the differentiation of several anterior pituitary hormone cell types, most successfully the corticotrophs, followed by the somatotrophs and lactotrophs, as well as some gonadotrophs and thyrotrophs. Focusing on the corticotrophs, the authors revealed that these cells could respond to stimuli to produce ACTH, which in turn was suppressed by the addition of hydrocortisone. When these cells were transplanted into the kidneys of hypophysectomized mice, blood concentrations of ACTH were high following corticotropin-releasing hormone treatment and this was concomitant with increased locomotor and running-wheel activity, as well as an improved survival curve (Suga et al., 2011). While the laboratory-generated anterior pituitary lobe may not have been perfect, it does open new avenues for the application of pluripotent stem cells to treat nondiabetic endocrine disorders, which has thus far received little attention in regenerative medicine.

The beginning of hypothalamic hypophysiotropic hormone expression has been better characterized in the rat (Szarek et al., 2010), but in keeping with the murine model, gonadotropin-releasing hormone is expressed from E10.5 (produced from the only neuronal subset to originate outside of the hypothalamus) at the same time as the lactotroph inhibitor dopamine. At E13 and E13.5, respectively, thyrotropin-releasing hormone and corticotropin-releasing hormone are expressed. The production of the somatotroph inhibitor somatostatin has been detected at E17, although no earlier stages have been investigated, whereas the timing of expression of the somatotroph stimulator growth hormone-releasing hormone has not been determined (Szarek et al., 2010).

Genetic and molecular regulation of hypothalamo-pituitary development

The first part of this section describes the molecular regulation of murine pituitary development and how this fits with the timeline described above (Fig. 1.2). Human phenotypes derived from mutations in these same genes will also be described where known, and it will become apparent that there is considerable variation in penetrance and phenotypes. This section will then conclude with what is known about hypothalamic development.

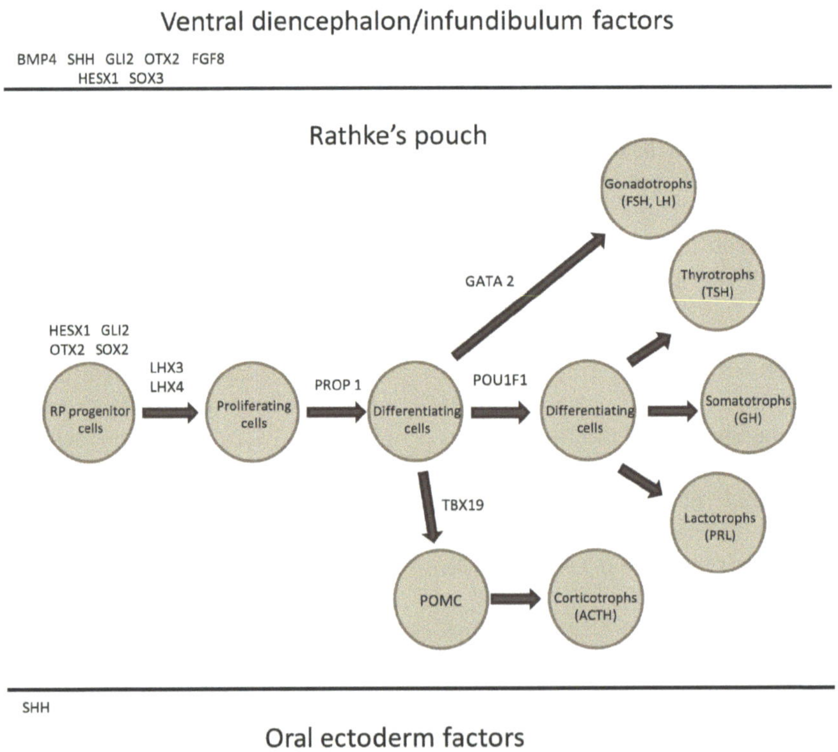

Fig. 1.2. Schematic representation of Rathke's pouch (RP) development from the earliest progenitor cells through to the five hormone secreting cell types. The transcription factors involved in regulating this process are expressed from the (**A**) ventral diencephalon or infundibulum (top), (**B**) oral ectoderm (bottom) or (**C**) from within the pouch itself (center). Note this figure is a highly simplified version of a much more complex process with the factors listed herein being those described within the text and not intended to be an exhaustive list. ACTH, adrenocorticotropic hormone; FSH, follicle-stimulating hormone; GH, growth hormone; LH, luteinizing hormone; POMC, pro-opiomelanocortin; PRL, prolactin; TSH, thyroid-stimulating hormone. (Reproduced from Bancalari et al., 2012; © Karger Publishers, Basel, Switzerland.)

Factors involved in the early formation of the pituitary

Bone morphogenetic protein 4 and the sonic hedgehog pathway

Bone morphogenetic protein 4 (BMP4) is the earliest signaling molecule detected in the prospective infundibulum at E8.5, occurring at the same time as the thickening of the stomodeum for RP formation. It is therefore presumed to be important for initial RP induction and with its expression continuing through to E14.5, it may play a role in RP maintenance as well (Kelberman et al., 2009). Loss of *Bmp4* through deletion usually results in early embryonic lethality (Kelberman et al., 2009), though those mice that do survive through to E10 present with no RP formation or thickening of the stomodeum (Kelberman et al., 2009).

In humans, *BMP4* expression is localized to the optic vesicle and optic cup, consistent across species prior to lens formation. In the brain, *BMP4* is expressed in the ventral diencephalon which is consistent with its known role in pituitary development, and is also expressed in the medial ganglionic eminence. Various mutations have been detected in *BMP4*, including missense (heterozygous, compound heterozygous, and homozygous), nonsense, chromosomal deletions, and frameshift variations, in patients presenting with various phenotypes including anophthalmia/microphthalmia, congenital glaucoma and sclerocornea, lateral ventricular dilatation, hypoplastic corpus callosum, hydrocephaly, cleft lip and palate, developmental and growth delay, and postaxial polydactyly, syndactyly, and brachydactyly (Bakrania et al., 2008; Reis et al., 2011). Very few clinical data exist with respect to patients presenting with pituitary phenotypes, the only case being a recent report of a chromosomal deletion of 14q22-q23 which encompassed *BMP4*, *OTX2,* and *SIX6* (Reis et al., 2011). Given that each of these three proteins is expressed in areas critical for successful pituitary development, it was not clear which protein was contributing to the phenotype, if not all. Further clarification of the pituitary phenotype in cases of isolated *BMP4* mutations is therefore required.

Along with fibroblast growth factor 8 (FGF8; described later), BMP4 appears to be involved in the spatially restricted distribution of sonic hedgehog (SHH) in the oral ectoderm. In mice, *Shh* is extensively expressed throughout the VD until E14.5 and the oral ectoderm except for RP until E12.5. Shh is a member of the hedgehog family of morphogens. Shh triggers the dissociation of transmembrane receptor Patched from its co-receptor Smoothened which results in the activation of the GLI family of transcription factors. These are involved in activating or repressing target genes (Kelberman et al., 2009). Patched and the GLI family of transcription factors are expressed within RP (with the latter also expressed in the VD) suggesting that Shh mediates pituitary development via these pathways. Shh knockout mice exhibit severe phenotypes of cyclopia and loss of brain midline structures, such that the pituitary phenotype cannot be assessed. However, it has been observed in mice transgenic for an Shh inhibitor that RP does develop, probably due to the normal expression of *Bmp4* and *Fgf8,* but it is severely hypoplastic. Humans presenting with mutations in *SHH* exhibit holoprosencephaly (HPE), defined as the failure of the brain to divide into two cerebral hemispheres. This is often accompanied by craniofacial malformations and most commonly, developmental delay. Approximately 20 mutations in *SHH* have been identified in patients with HPE, with at least eight mutant proteins impairing the production of the active SHH N-terminal (Traiffort et al., 2004). As per *BMP4*, mutations in *SHH* have not been detected in patients with hypopituitarism, either in isolation or in association with other disorders. Unsurprisingly, patients with *GLI2* mutations also present with HPE in association with craniofacial/midline abnormalities, such as a single central incisor, single nares, optic nerve hypoplasia, partial agenesis of the corpus callosum, and pituitary gland dysfunction, as well as postaxial polydactyly. There have been 17 heterozygous missense, nonsense, and frameshift *GLI2* mutations identified to date in patients with variable phenotypes such as HPE and craniofacial abnormalities, polydactyly, and HH (Table 1.1), which are often associated with panhypopituitarism or isolated growth hormone deficiency (IGHD). Mutations can also present in patients with hypopituitarism in the absence of HPE (França et al., 2010). *Gli2* knockout mice are perinatal lethal with severe cartilage and bone developmental abnormalities (Zhao et al., 2006).

FGF8

FGF8 is a member of the large 25-member fibroblast growth factor family. Infundibular expression is detected by E9.5 and its role in mediating the expansion of RP is through inducing expression of *Lhx3* and *Lhx4* (McCabe et al., 2011b). *Fgf8* hypomorphic mice present with a markedly hypoplastic anterior pituitary gland and in some cases, the posterior pituitary can be absent (McCabe et al., 2011b). These were associated with midline defects indicative of HPE. Hypothalamic arginine vasopressin and oxytocin neurons are reduced in number and this is coincident with a recent report of a patient presenting with diabetes insipidus in association with the first reported autosomal recessive case of HPE (Table 1.1) (McCabe et al., 2011b) attributed to *FGF8.* Generally, however, mutations in *FGF8* are autosomal

Table 1.1

Mutational characteristics of transcription factors mediating pituitary development

Gene	Inheritance	Types	Endocrinopathy	Phenotype
GLI2	Haploinsufficiency	Missense, frameshift	CPHD (GH, TSH, LH, FSH, ACTH)	HPE, craniofacial abnormalities, polydactyly, HH, partial ACC
FGF8	AR, AD	Missense, chromosome deletion	LH, FSH, DI	HH, anosmia, HPE, Moebius syndrome, SOD
LHX3	AR	Missense, nonsense, frameshift, splicing	CPHD (GH, TSH, LH, FSH, PRL, ACTH)	Limited neck rotation, short cervical spine, sensorineural deafness
LHX4	AD	Missense, frameshift	CPHD (GH, TSH, ACTH), variable GnD	Cerebellar abnormalities
HESX1	AR, AD	Missense, frameshift	IGHD or CPHD	SOD
SOX2	AD	Missense, nonsense, frameshift	LH, FSH, variable GHD	Anophthalmia/microphthalmia, esophageal atresia, genital tract abnormalities, hypothalamic hamartoma, sensorineural hearing loss, diplegia
SOX3	XL	Variations in polyalanine tract length, chromosome duplication	IGHD or CPHD	Mental retardation, infundibular hypoplasia, EPP, midline abnormalities
OTX2	AD	Missense, nonsense, microdeletion	IGHD or CPHD	Anophthalmia/microphthalmia, coloboma, developmental delay
PROP1	AR	Missense, nonsense, frameshift, splicing	CPHD (GH, TSH, LH, FSH, PRL), evolving ACTH deficiencies	May show transient anterior pituitary hyperplasia
POU1F1	AR, AD	Missense, nonsense, frameshift, splicing	CPHD (GH, TSH, PRL)	Variable anterior pituitary hypoplasia
TBX19	AR	Missense, nonsense, frameshift, splicing	ACTH	Neonatal hypoglycaemia, neonatal cholestatic jaundice

ACC, agenesis of the corpus callosum; ACTH, adrenocorticotropic hormone; AD, autosomal dominant; AR, autosomal recessive; CPHD, combined pituitary hormone deficit; DI, diabetes insipidus; EPP, ectopic posterior pituitary; FSH, follicle stimulating hormone; GH, growth hormone; GHD, growth hormone deficiency; GnD, gonadotropin deficiency; HH, hypogonadotropic hypogonadism; HPE, holoprosencephaly; IGHD, isolated growth hormone deficiency; LH, luteinizing hormone; PRL, prolactin; SOD, septo-optic dysplasia; TSH, thyroid stimulating hormone; XL, X-linked.

dominant, and apart from one other heterozygous variant, which was detected in a child with SOD in conjunction with Moebius syndrome (characterized by malformation of the sixth and seventh craniofacial nerves) but a normal pituitary (McCabe et al., 2011b), all other *FGF8* mutations have previously been reported in association with KS, but not with hypopituitarism.

Lim homeodomain transcription factors

After induction by Fgf8, Lhx3 is expressed strongly and uniformly in RP from E9.5, and is also detected in the ventral hindbrain and spinal cord (Sheng et al., 1996). Expression persists throughout the adult pituitary as well, suggesting a maintenance role for Lhx3 in at least one of the anterior pituitary cell types (Cohen, 2012). While initial formation of RP does occur in $Lhx3^{-/-}$ mice, these animals die shortly after birth and present with defects in the differentiation of all hormone-secreting cell types. Similarly, in humans, homozygous or compound heterozygous carriers of *LHX3* mutations present with CPHD and cervical abnormalities with or without restricted neck rotation. Some patients also present with sensorineural hearing loss and mutations can also be frameshift or splicing anomalies (Table 1.1) (Kelberman et al., 2009). The anterior pituitary can present as hypoplastic or enlarged on magnetic resonance imaging (MRI) with a structurally normal posterior pituitary. Microadenoma is also occasionally observed. Recently, heterozygous carriers of a dominant negative *LHX3* mutation have been shown to manifest a phenotype characterized by limited rotation of the neck.

Lhx4 is closely related to *Lhx3* and is also expressed throughout RP from E9.5. Unlike the constant expression of the latter, however, *Lhx4* is restricted to the anterior lobe alone by E12.5 and is downregulated by E15.5 (Sheng et al., 1997). While *Lhx4* mutants die shortly after birth, they do develop an anterior pituitary gland containing each of the differentiated cell types at reduced numbers (in contrast to *Lhx3* mutants), but it is hypoplastic. Patients with heterozygous missense or frameshift mutations in *LHX4* have variable hormone phenotypes, including GHD and variable TSH, gonadotropin and ACTH deficiencies with a hypoplastic anterior pituitary, with or without an ectopic posterior pituitary (Table 1.1) (Cohen, 2012).

Lhx2 is another member of the same family and was only recently implicated in murine pituitary formation. Previously, its disruption in mice led to anophthalmia and malformation of the cerebral cortex, but it is also expressed in the VD, infundibulum, and posterior pituitary. Similarly to *Lhx4*, its disruption in mice led to disorganized RP morphology although all endocrine cell lineages were present (McCabe et al., 2011a). Human carriers of *LHX2* mutations with hypopituitarism have not been described, although recently one patient presenting with anophthalmia and a heterozygous p.P43A variant, predicted to be functionally deleterious, was reported. The paternal carrier was phenotypically normal (Desmaison et al., 2010), but as seen with other genes, there is considerable phenotypic variability and the proband's phenotype could possibly be caused by mutations in two or more genes, one of which may include *LHX2*; although it would be important to confirm that this is a loss of function allele.

Homeobox embryonic stem cell 1 (*HESX1*)

HESX1 is a member of the *paired-like* homeobox gene family, the members of which are defined as transcriptions factors containing a DNA-binding homeodomain. Early on in embryogenesis it is ubiquitously expressed in the anterior visceral endoderm and adjacent ectoderm in an area which will develop into the prosencephalon and forebrain. By E9.0, its expression is restricted to the primordial RP and VD (McCabe et al., 2011a). Acting as a transcriptional repressor, Hesx1 appears to regulate cellular proliferation and patterning within RP, before it gradually downregulates concomitantly with an increase in the expression of another *paired-like* homeobox gene, prophet of Pit1 (*Prop1*; discussed later) (Carvalho et al., 2010). *Hesx1* expression ceases by E13.5 and this gradual downregulation is an absolute necessity for successful pituitary development. *Hesx1* activation is dependent upon Lhx1 and Lhx3 via its 5′ promoter region. Later stages of expression appear to be dependent on 3′ elements (Kelberman et al., 2009).

Mutations, missense, or frameshift in *HESX1* may be recessive or dominant. Hormone anomalies in humans with *HESX1* mutations are quite variable, ranging from IGHD to CPHD; nonetheless, these are often associated with septo-optic dysplasia (SOD), which is characterized by optic nerve hypoplasia and/or brain midline abnormalities, such as agenesis of the corpus callosum and/or septum pellucidum. Pituitary phenotypes on MRI are highly variable across patients, with images suggesting an apparently structurally normal pituitary to much more severe presentations of anterior pituitary hypoplasia or aplasia and an undescended or ectopic posterior pituitary (Kelberman et al., 2009). Mutations in *HESX1* (Table 1.1) can also result in isolated hypopituitarism, which is suggestive of the pituitary gland being highly sensitive to HESX1 dosage (Kelberman et al., 2009).

SOX2 and SOX3

These two proteins are transcription factor members of the SRY-related high mobility group (HMG) box (SOX) family and are early markers of progenitor cells; their expression is downregulated as cells differentiate. Murine studies of *Sox2* with respect to pituitary development have proven partly problematic, in that complete knockouts lead to early embryonic lethality and hypomorphic models produce relatively mild brain phenotypes, possibly due to a redundancy with *Sox1* and *Sox3* (Jayakody et al., 2012). Recently, Jayakody et al. (2012) generated a selective *Sox2* knockout, in which murine cells which express *Hesx1*, i.e., Rathke's pouch, would not express *Sox2*. These embryos survived through to birth and presented with significant anterior pituitary hypoplasia detectable from E12.5–E14.5 with normal intermediate and posterior lobes. There was a marked reduction in the transcription factor, POU domain, class 1 (Pou1F1 or Pit1; described later), which is required for the differentiation of several hormone-secreting cell lineages. As a consequence, a marked reduction in the differentiation of somatotrophs and thyrotrophs was observed (Jayakody et al., 2012). This was probably due to the reduced proliferation of periluminal progenitors which reside along the cleft between the anterior and intermediate pituitary lobes, meaning that while gonadotroph and corticotroph progenitors could be born, there were simply not enough progenitors to give rise to the later thyrotrophs or somatotrophs (Jayakody et al., 2012). In human embryos, strong *SOX2* expression is detected from 4.5 to 9 weeks of development in RP and this is maintained throughout anterior pituitary development. While it is also detected in the diencephalon, it is not expressed in the infundibulum, or subsequent posterior pituitary. In humans, *de novo* autosomal dominant *SOX2* mutations have been described in association with complex phenotypes. Initially, *SOX2* mutations had not been associated with pituitary abnormalities, having been implicated most prominently in bilateral anophthalmia, severe microphthalmia, learning difficulties, esophageal atresia, and genital abnormalities (Table 1.1). However, after further careful phenotypic characterization, potential phenotypes associated with *SOX2* mutations were expanded to include anterior pituitary hypoplasia, HH, and variable GHD, often in association with hippocampal abnormalities, corpus callosum agenesis, esophageal atresia, hypothalamic hamartoma, and sensorineural hearing loss (McCabe et al., 2011a). The pituitary is usually small in the majority of patients as revealed by MRI, but occasionally it can appear enlarged and remains so for years. As described earlier in this chapter, the cells of RP proliferate ventrally, away from the lumen. However, in the periluminal area, the embryo maintains a proliferative zone, which persists through to adulthood, containing progenitors (Kelberman et al., 2009). While SOX2 is expressed throughout the anterior pituitary initially, its localization becomes restricted to this proliferative zone as endocrine cell differentiation proceeds. It is still detectable here in the mature gland (Kelberman et al., 2009).

The dosage of another SOX family member, SOX3, is crucial for normal hypothalamo-pituitary development; over- or underdosage of *SOX3* can lead to hypopituitarism or infundibular hypoplasia. Knockout mice present with variable phenotypes including poor growth, craniofacial defects and variable endocrine deficits. RP in hemizygote embryos presents with abnormal bifurcations which persist into adulthood. While expression of the protein in RP itself is not detected, expression is prolific in the VD, and the extra bifurcations in the mutant could be attributed to the concomitant expansion of the Fgf8 and Bmp4 expression domains (Rizzoti et al., 2004). In humans, *SOX3* is located on the X chromosome, and, as observed in mice, mutations are associated with variable phenotypes. MRI usually reveals a small anterior pituitary which is coincident with variable hormone phenotypes including IGHD or panhypopituitarism. The posterior pituitary is usually undescended/ectopic and dysgenesis of the corpus callosum can be observed. These phenotypes are often associated with variable developmental delay (Table 1.1) (Alatzoglou et al., 2011). The recent detection of a deletion within the polyalanine tract of *SOX3* was associated with hypopituitarism in a heterozygous female patient. Interestingly, this variant was associated with a gain of function, whereas previously described polyalanine tract expansions had been associated with loss of function (Woods et al., 2005; Alatzoglou et al., 2011).

Orthodentic homeobox 2 (OTX2)

The role of this transcription factor in hypothalamo-pituitary development remains largely unclear. In mice, its expression is restricted to the developing neural and sensory structures such as the brain, eyes, nose, and ears. Homozygous null mutant mice, which die midgestation due to impaired gastrulation, exhibit severe malformations of the forebrain structures as well as the eyes. Heterozygous mice present with variable eye phenotypes ranging from normal to severe (anophthalmia/microphthalmia) and brain structural abnormalities (holoprosencephaly or anencephaly), which is consistent with the expression pattern for *Otx2*.

However, recent studies have revealed that *Otx2* is also expressed in the VD by E10.5, where it may potentially regulate *Fgf8* or *Bmp* induction of RP formation (Mortensen et al., 2011). It is also expressed in RP at the same stage suggesting an intrinsic role in RP development also, and this is consistent with its proposed capability of activating *Hesx1* expression. While *Otx2* gene expression is only detectable through to E12.5 (in RP), protein expression persists through to E14.5. Two days later, Otx2/*Otx2* disappears altogether from the RP and VD. Its potential role in pituitary development was recently suggested in *Prop1*-mutant mice (Mortensen et al., 2011). In these animals, *Otx2* expression in RP persists through to E16.5, which is 4 days later than its usual disappearance from the RP, concomitant with the peak in *Prop1* expression, and 2 days later than any obvious pituitary morphology defects become apparent. This suggests that Prop1 may regulate the genetic expression of factors which suppress *Otx2* expression, thereby implicating a role for Otx2 in pituitary development. Further evidence of a role for Otx2 in hypothalamo-pituitary development was shown in GnRH-neuron-*Otx2* knockout mice which exhibited hypogonadotropic hypogonadism (Diaczok et al., 2011). These data are consistent with human *OTX2*-mutant phenotypes. These can encompass highly variable hypopituitary phenotypes ranging from IGHD to panhypopituitarism and HH, most commonly in association with severe eye abnormalities including anophthalmia and microphthalmia. To date, heterozygous missense, nonsense and frameshift *OTX2* mutations, of which 20 have been described, account for 2–3% of severe eye abnormalities (Table 1.1) (McCabe et al., 2011a).

Factors regulating cellular differentiation

PROP1 and *POU1F1/PIT1*

Like Hesx1, Prop1 is a *paired-like* homeobox transcription factor, but unlike Hesx1, it acts as a transcriptional activator to stimulate pituitary cell differentiation as its primary role. Hesx1 represses Prop1 expression from E9.0. It gradually becomes detectable in RP but its activity is still attenuated through the formation of heterodimers with Hesx1. By E12.5, it is believed that the levels of Prop1 are sufficient to displace Hesx1 to the point where homodimers between Prop1 proteins are formed. From this time point, Prop1 stimulates the differentiation and proliferation of each of the hormone-secreting cell lineages. Recessive mutations in *PROP1* are associated with GH, prolactin, and TSH deficiencies but can also manifest as deficiencies in gonadotropins and ACTH which may be present at a young age, or develop over time. While the posterior pituitary remains structurally normal, the anterior pituitary on MRI can appear either small or large, where it can wax and wane in size before eventual involution (Ward et al., 2005; Kelberman et al., 2009). Mutations in *PROP1* (Table 1.1) are the most prevalent cause of CPHD, accounting for up to 50% of familial cases, although the incidence of *PROP1* mutations is much lower in sporadic cases.

Currently, it is not entirely clear how Prop1 drives gonadotroph and corticotroph differentiation; however, with respect to the somatotroph, lactotroph, and thyrotroph lineages, Prop1 activates the POU domain class 1 (Pou1F1 or Pit1) transcription factor at E13.5 (Bodner et al., 1988). Consequently, mutations in *POU1F1* are associated with the hormone deficiencies corresponding to these three cell types (Table 1.1). Again the posterior pituitary is structurally normal on MRI, but the anterior pituitary can present as either normal or small in appearance. POU1F1 is thought to synergize with another transcription factor, PITX1, to augment prolactin gene expression and to a lesser extent, growth hormone gene expression (Kelberman et al., 2009). PITX1 in gonadotrophs drives *Lhβ*. Currently, however, mutations in *PITX1* have not been associated with hypopituitarism.

GATA2

GATA2 is one member of a six member transcription factor family. It has dual functions where it acts as (1) a stem cell maintenance factor in some tissues and (2) a promoter of cellular differentiation in other tissues (Morceau et al., 2004). After BMP2 induction at E10.5, *Gata2* is expressed in the ventral RP where it regulates α-GSU. α-GSU is the common subunit of Lh, Fsh and Tsh, thus Gata2 is a marker for the prospective, and then definitive, gonadotrophs and thyrotrophs, with expression maintained in the adult. Like Pitx1, Gata2 can synergize with Pou1F1 to induce *Tshβ* expression and it also synergizes with Nr5a1 to increase *Lhβ* expression (Kelberman et al., 2009). Despite these roles, no *GATA2* mutations have been described in humans in association with hypopituitarism.

TBX19

Previously known as TPIT, TBX19 is a member of the T-box family of transcription factors. It is exclusively expressed in the developing pituitary, first at E12.5 in *Pomc*-positive cells and then in the corticotrophs and melanotrophs where it is maintained in the adult gland (Lamolet et al., 2001). The transcription of *Pomc* requires cooperation between Tbx19 and Pitx1, whereby they bind to contiguous sites within the same regulatory element. This dependence upon Tbx19 is evident in *Tbx19* knockout mice which almost completely lack *Pomc*-expressing cells, resulting in severe ACTH and glucocorticoid deficiencies (Kelberman et al., 2009). Conversely, Tbx19 antagonizes the gene (nuclear receptor subfamily 5, group A, member 1 (*Nr5a1*)) encoding the transcription factor steroidogenic factor 1 (Sf1), which is important for gonadotroph development. Thus, Tbx19 is essential for normal corticotroph and melanotroph fate determination while actively suppressing gonadotroph differentiation. In humans, autosomal recessive mutations in *TBX19* are the most common cause of isolated ACTH deficiency and the patients present with life-threatening hypoglycemia in the neonatal period (Table 1.1) (Kelberman et al., 2009).

Factors involved in hypothalamic formation

As described previously, our understanding of the pathways underlying hypothalamic development is relatively rudimentary. However, recent studies have improved our knowledge of this process somewhat. This section will summarize recent data concerning hypothalamic development. It will become apparent that genetic pathways involved in hypothalamic formation can overlap with those involved in pituitary organogenesis. What is interesting is that while several murine mutants have been generated lacking genes important for hypothalamic development, only one has thus far been implicated in a human hypothalmo-pituitary disorder (Webb et al., 2013). The remaining genes however, provide a source of potential candidates for relevant phenotypes.

Studies have revealed that major signaling pathways, such as Wnt, Shh, bone morphogenetic proteins, and the Lhx transcription factors, have key roles in hypothalamic induction and patterning (Blackshaw et al., 2010). *Shh* murine mutants develop HPE which is coincident with the failure of hypothalamic anlagen induction, and its overexpression in zebrafish leads to ectopic expression of hypothalamic markers (Szarek et al., 2010). Bone morphogenetic proteins appear to negatively regulate *Shh* expression, such that Shh appears to be important for the proliferation and patterning of anterior hypothalamic neural progenitors while Bmp7 appears to be important for the same event at the posteroventral aspects (Blackshaw et al., 2010). Indeed, expanded Shh signaling disrupts posterior patterning (Blackshaw et al., 2010). The downregulation of Shh signaling by bone morphogenetic proteins is essential for the upregulation of *Fgf8* and *Fgf10*, which may also have roles in hypothalamic patterning, although currently this remains to be fully elucidated. Wnt signaling follows a posterior-anterior expression gradient with the pathway appearing to be important for the development of ventral hypothalamic identity (Blackshaw et al., 2010). These initial patterning pathways appear to lead to specific regions along the primordial hypothalamic anlagen, which are subject to specific transcription factor "codes" (Blackshaw et al., 2010).

Transcription factors belonging to the basic helix-loop-helix (bHLH) family appear to play major roles in hypothalamic neurogenesis (Szarek et al., 2010). One such factor, *Sim1*, is expressed in paraventricular nuclei (PVN), supraoptic nuclei (SON), and anterior periventricular nuclei (aPeVN) from E10.5 through to postnatal life (Szarek et al., 2010). Mice that are homozygous mutants for *Sim1* die postnatally and lack almost all of the neurons of the PVN and SON, including those which express thyrotropin-releasing hormone and corticotropin-releasing hormone (Szarek et al., 2010). Somatostatin-secreting neurons in the aPeVN are also largely absent. Consolidating this role for *Sim1* in the development of these neurons, murine mutants for its dimerization partner *Arnt2* display similar phenotypes. The expression of the gene *Brn2*, which encodes a POU domain transcription factor, is absent from their target neurons. Any direct regulatory mechanisms of this gene by *Sim1* and *Arnt2* are yet to be proven, but *Brn2* is required for the differentiation of the corticotropin-releasing hormone, oxytocin (OT) and arginine vasopressin (AVP) secreting neurons of the PVN and SON (Szarek et al., 2010). Interestingly, *ARNT2* was recently the first gene implicated in a human hypothalamic disorder, previously undescribed. A novel, homozygous frameshift mutation was detected in 6 members of a highly consanguineous family and was associated with secondary microcephaly with fronto-temporal lobe hypoplasia, multiple pituitary hormone deficiency, seizures, severe visual impairment and abnormalities of the kidneys and urinary tract (Webb et al., 2013).

Orthopedia (*Otp*) is a homeobox transcription factor, which like *Sim1* and *Arnt2* appears to be important for the development of the PVN, SON, aPeVN, and the arcuate nucleus (ARC). Murine mutants, which die soon after birth, display reduced formation of these nuclei, which may be due to the postulated role of *Otp* in regulating cellular proliferation (Szarek et al., 2010). Interestingly, *Otp* mutants also lack *Brn2* expression; however, the relationship between *Otp* and the *Sim1/Arnt2* pathways in the development of these neurons remains to be established (Caqueret et al., 2006; Szarek et al., 2010).

Nkx2.1 is a transcription factor which is induced in early central nervous system development. *Nkx2.1* mutant mice die at birth and present with severe abnormalities in the ventral hypothalamus, including agenesis of the ARC and ventromedial nucleus (VMH; linked to innate behavioral responses including feeding, fear, thermoregulation, and sexual activity). Interestingly, RP fails to form in the same mutant mice (Szarek et al., 2010). *Nr5a1*, which encodes Sf1, as mentioned previously, and is classically associated with normal gonadal and adrenal development, also appears to be important for VMH development. *Nr5a1* mutant mice lack normal survival and migration of these neuronal precursors from the ventricular zone (Blackshaw et al., 2010; Szarek et al., 2010). *Nr5a1* and *Nkx2.1* expression is nonoverlapping, and it appears that Sf1 represses *Nkx2.1* expression *in vitro* (Blackshaw et al., 2010; Szarek et al., 2010).

Homeobox genes *Hmx2* and *Hmx3* appear to be important for the differentiation of GHRH-secreting neurons in the ARC, and are subsequently expressed in the ventral hypothalamus from E10.5. While the overall number of neurons present in the ARC does not appear to change, mutant mice exhibit dwarfism and a reduced number of GHRH-secreting neurons. In the same animals, expression of *Gsh1*, which is required for *Ghrh* expression, is absent (Szarek et al., 2010). *Mash1* is another transcription factor which is required for the expression of *Gsh1*. It is expressed in the ventral retrochiasmatic neuroepithelium from E10.5 to E12.5 and is required for neurogenesis and subtype specification in many regions of the central nervous system (Szarek et al., 2010). *Mash1* knockout mice exhibit hypoplasia of the ARC and VMH, yet this phenotype can be rescued by the knockin *Ngn2* mutant, which, like *Mash1*, is another member of the bHLH transcription factor family (Szarek et al., 2010).

The development of the suprachiasmatic nuclei (SCN) may, to some extent, be dependent on the Lim homeodomain transcription factor family (Blackshaw et al., 2010). *Lhx1* and *Lhx8* are expressed in the anterior hypothalamic neuroepithelium from E12.5. Upon the completion of neurogenesis at E16.5, *Lhx1* expression is restricted to the SCN. *Lhx1* mutant mice display severe disruption of SCN neuronal termination which is coincident with circadian rhythm behavioral problems (Blackshaw et al., 2010).

Sox3, which was described above with respect to pituitary development, may also be important for hypothalamic development. It is expressed throughout the ventral diencephalon and from E12.5 is restricted to nuclei containing parvicellular neuronal subtypes (Szarek et al., 2010). Thus, the pathophysiologic basis of pituitary hormone deficiencies in murine models may be due to the failure of parvicellular neuronal development. *Sox3* is also expressed in the median eminence, which, as described earlier, is where all of the hypophysiotropic hormones come together before traversing the pituitary stalk into the pituitary gland. Defects in morphology here, potentially due to *Sox3* disruption, may also account for pituitary hormone deficiencies.

CONGENITAL HYPOPITUITARISM AND ASSOCIATED DEFECTS

As already mentioned, being a midline structure, defects during pituitary development can have knock-on effects on the development of other associated structures, and these can consequently manifest as highly variable phenotypes including HPE and SOD. The list of genes associated with hypothalamo-pituitary development in the previous section was not exhaustive, but when all genes known to regulate the development of this axis are taken into account, they still only account for a very low proportion (approximately 5–10% of patients studied) of patients with hypopituitarism with/without craniofacial/midline disorders. This indicates that many more causative genes or other etiologic factors remain to be identified.

The most common forebrain anomaly in humans is HPE, which occurs at an incidence of 1:10 000–20 000 in the general population but as high as 1:250 in conceptuses (McCabe et al., 2011a). It is defined by varying degrees of separation of the cerebral hemispheres and ventricles (McCabe et al., 2011a). The corpus callosum is lost when the frontal and parietal lobes fail to divide posteriorly (Fig. 1.3). Facial features associated with the condition are classically midline and include cyclopia, anophthalmia, midfacial hypoplasia, hypotelorism, cleft lip and/or palate, and a single central incisor. There is an increasing number of genetic factors being implicated in the condition, and these include members of the sonic hedgehog signaling pathway such as *SHH*, *ZIC2*, *TGIF1*, *PTCH1*, *GLI2*, *DISP1*, *TDGF1*, *GAS1*, *EYA4*, and *FOXH1* (McCabe et al., 2011b). Recently, a mutation in *FGF8* provided the first genetic cause of autosomal recessive HPE (McCabe et al., 2011b).

SOD is a complex and highly variable disorder which is diagnosed when any two of the following three phenotypes are present: (1) optic nerve hypoplasia, (2) midline neuroradiologic abnormalities such as agenesis of the corpus callosum and septum pellucidum, and (3) anterior pituitary hypoplasia with consequent hypopituitarism (McCabe et al., 2011a). Despite SOD being largely sporadic and inclusive of possible environmental induced pathologies through smoking and drug/alcohol abuse, the number of genetic factors implicated in this condition is also increasing and currently includes *HESX1*, *OTX2*, *SOX2* and *SOX3* (Table 1.1). These genes, and those listed for HPE, above, are all expressed very early in forebrain and pituitary development and so it is not surprising that mutations affecting these genes can induce the disorders described herein.

OVERLAP BETWEEN CONGENITAL HYPOPITUITARISM AND MIDLINE DEFECTS WITH KALLMANN SYNDROME

Up until recently, these two groups of disorders were considered quite separate; however, it has always been acknowledged that they share overlapping phenotypes such as craniofacial defects (high-arched palate, cleft lip/palate), dental agenesis, and sensorineural hearing loss (McCabe et al., 2011a). A number of genes have been implicated in the etiology of Kallmann syndrome (KS) including *KAL1*, *FGFR1*, *FGF8*, *PROKR2*, and *PROK2*. Recently, chromodomain helicase DNA binding protein-7 (CHD7), bromodomain and WD repeat-containing protein 2 (WDR11), negative elongation factor (NELF), and semaphorin 3A (SEMA3A) have been added to this list (Kim and Layman, 2011; Raivio et al., 2012; Young et al., 2012).

The first five original genes listed above are all implicated in the maturation and/or migration of the gonadotropin-releasing hormone neurons to the hypothalamus, which, as mentioned previously, are the only hypothalamic neurons to originate from outside the organ. In addition, these genes are involved in the development of the olfactory complex, meaning that the association of KS

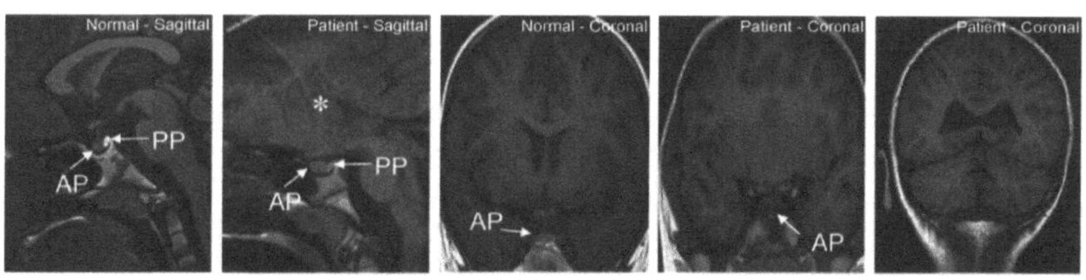

Fig. 1.3. Sagittal and coronal MRI scans of a normal control subject and a patient with the craniofacial/midline disorder holoprosencephaly. The normal scan shows an intact corpus callosum above a healthy anterior pituitary (AP) and posterior pituitary (PP) (arrows) as well as normal division of the brain into its cerebral hemispheres. In the patient, who has autosomal recessive holoprosencephaly due to a missense mutation in *FGF8*, the corpus callosum is absent (*) and the anterior pituitary is enlarged. The failure of the brain to divide into its cerebral hemispheres in the patient is clearly evident in the coronal views, particularly in the far right image, where, when compared to the normal coronal scan, the corpus callosum is clearly missing (arrowhead). This figure is an example and is not completely representative due to high variations in phenotypes seen across patients. The pituitary phenotype can include anterior pituitary hypoplasia and an ectopic posterior pituitary, neither of which are shown here.

with anosmia is what differentiates the condition from HH. Consistent with the known overlap in phenotypes between KS and the hypopituitarism spectrum is the recent detection of variations in *FGFR1*, *FGF8*, and *PROKR2* in cases of CPHD, SOD, and HPE (McCabe et al., 2011b; Raivio et al., 2012). Thus, as with other genes classically associated with SOD, such as *HESX1*, sequence variants in classic "KS" genes may be associated with highly variable phenotypes. This suggests that the etiology of the spectrum of hypopituitary disorders may, at times, be digenic or oligogenic, a phenomenon which is well established in the etiology of KS.

Thus the molecular basis of both of these groups of disorders is very complex, and it could involve multiple genetic factors as well as environmental involvement. Such complexity is consistent with the small percentage of patients with hypopituitarism to have been assigned a genetic cause, as mentioned above. Probing further into the genetics of KS may provide an avenue of identifying new candidates for hypopituitarism, but the scope and ability to identify a whole new range of genetic targets has opened up with the recent advent of homozygosity mapping and whole exome sequencing.

CONCLUSION AND FUTURE DIRECTIONS

The development of the hypothalamo-pituitary axis is a highly complex and organized process that involves multiple signaling molecules and transcription factors arranged in a spatiotemporal fashion. In this chapter, we have provided an overview of hypothalamo-pituitary development and some of the molecules involved in this process. As demonstrated herein, mutations in genes encoding such molecules can lead to congenital hypopituitarism in association with HPE and SOD. However, the percentage of patients with disorders such as HPE and SOD in whom a genetic cause has been described is small, and while recent evidence of a genotypic overlap with KS offers a new avenue of potential gene candidates, the advent of homozygosity mapping and whole exome sequence appears to be potentially more fruitful. Furthermore, the development of the pituitary gland is tightly entwined with that of the hypothalamus, for which our understanding is relatively rudimentary, due to its anatomic complexity of highly diverse cell groups and neuronal subgroups. In addition, only one study has thus far identified a genetic cause of specific hypothalamic defects in humans. Given the integral association between the hypothalamus and pituitary gland, future work using the aid of newly acquired technologies could also provide new genetic candidates for hypothalamic disorders and how these may impact upon human phenotypes, including those associated with the etiology of hypopituitarism. From this, we would gain a better understanding of hypothalamo-pituitary development in humans.

REFERENCES

Alatzoglou KS, Dattani MT (2009). Genetic forms of hypopituitarism and their manifestation in the neonatal period. Early Hum Dev 85: 705–712.

Alatzoglou KS, Kelberman D, Cowell CT et al. (2011). Increased transactivation associated with SOX3 polyalanin tract deletion in a patient with hypopituitarism. J Clin Endocrinol Metab 96: E658–E690.

Bakrania P, Efthymiou M, Klein JC et al. (2008). Mutations in *BMP4* cause eye, brain, and digit development anomalies: overlap between the *BMP4* and hedgehog signalling pathways. Am J Hum Genet 82: 304–391.

Bancalari RE, Gregory C, McCabe MJ et al. (2012). Pituitary Gland Development: An Update, Karger Publishers, Basel.

Blackshaw S, Scholpp S, Placzek M et al. (2010). Molecular pathways controlling development of thalamus and hypothalamus: from neural specification to circuit formation. J Neurosci 30: 14925–14930.

Bodner M, Castrillo JL, Theill LE et al. (1988). The pituitary specific transcription factor GHF-1 is a homeobox-containing protein. Cell 55: 505–518.

Bonnefont X, Lacampagne A, Sanchez-Hormigo A et al. (2005). Revealing the large-scale network organization of growth hormone-secreting cells. Proc Natl Acad Sci U S A 102: 16880–16885.

Budry L, Lafont C, El Yandouzi T et al. (2011). Related pituitary cell lineages develop into interdigitated 3D cell networks. Proc Natl Acad Sci U S A 108: 12515–12520.

Caqueret A, Boucher F, Michaud JL et al. (2006). Laminar organization of the early developing anterior hypothalamus. Dev Biol 298: 95–106.

Carvalho LR, Brinkmeier ML, Castinetti F et al. (2010). Corepressors TLE1 and TLE3 interact with HESX1 and PROP1. Mol Endocrinol 24: 754–765.

Cohen LE (2012). Genetic disorders of the pituitary. Curr Opin Endocrinol Diabetes Obes 19: 33–39.

Dasen J, Rosenfeld M (1999). Signaling mechanisms in pituitary morphogenesis and cell fate determination. Curr Opin Cell Biol 11: 669–677.

Desmaison A, Vigouroux A, Rieubland C et al. (2010). Mutations in the *LHX2* gene are not a frequent cause of micro/anophthalmia. Mol Vis 16: 2847–2849.

Diaczok D, Divall S, Matuso I et al. (2011). Deletion of Otx2 in GnRH neurons results in a mouse model of hypogonadotrophic hypogonadism. Mol Endocrinol 25: 833–846.

França MM, Jorge AAL, Carvahlo LRS et al. (2010). Novel heterozygous nonsense GLI2 mutations in patients with hypopituitarism and ectopic posterior pituitary lobe without holoprosencephaly. J Clin Endocr Metab 95: E384–E391.

Hodson D, Schaeffer M, Romanò N et al. (2012). Existence of long-lasting experience-dependent plasticity in endocrine cell networks. Nat Commun 3: 1–10.

Jayakody SA, Andoniadou CL, Gaston-Massuet C et al. (2012). SOX2 regulates the hypothalamic-pituitary axis at multiple levels. J Clin Invest 122: 3635–3646.

Kelberman D, Rizzoti K, Lovell-Badge R et al. (2009). Genetic regulation of pituitary gland development in human and mouse. Endocr Rev 30: 790–829.

Kim HG, Layman LC (2011). The role of CHD7 and the newly identified WDR11 gene in patients with idiopathic hypogonadotrophic hypogonadism and Kallmann syndrome. Mol Cell Endocrinol 346: 74–83.

Koutcherov Y, Mai JK, Ashwell KW et al. (2002). Organization of the human hypothalamus in fetal development. J Comp Neurol 446: 301–324.

Koutcherov Y, Mai JK, Paxinos G (2003). Hypothalamus of the human fetus. J Chem Neuroanat 26: 253–270.

Lamolet B, Pulichino AM, Lamonerie T (2001). A pituitary cell-restricted T box factor, Tpit, activates POMC transcription in cooperation with Pitx homeoproteins. Cell 104: 849–859.

McCabe MJ, Alatzoglou KS, Dattani MT (2011a). Septo-optic dysplasia and other midline defects: the role of transcription factors: *HESX1* and beyond. Best Pract Res Clin Endocrinol Metabol 25: 115–124.

McCabe MJ, Gaston-Massuet C, Tziaferi V et al. (2011b). Novel *FGF8* mutations associated with recessive holoprosencephaly, craniofacial defects and hypothalamo-pituitary dysfunction. J Clin Endocr Metab 96: E1709–E1718.

Morceau F, Schnekenburger M, Dicato M et al. (2004). GATA-1: friends, brothers, and coworkers. Ann N Y Acad Sci 1030: 537–554.

Mortensen AH, MacDonald JW, Gosh D et al. (2011). Candidate genes for panhypopituitarism identified by gene expression profiling. Physiol Genomics 43: 1105–1116.

Raivio T, Avbelj M, McCabe MJ et al. (2012). Genetic overlap in Kallmann syndrome, combined pituitary hormone deficiency and septo-optic dysplasia. J Clin Endocrinol Metab 97: E694–E699.

Reis LM, Tyler RC, Schilter KF et al. (2011). *BMP4* loss-of-function mutations in developmental eye disorders including SHORT syndrome. Hum Genet 130: 495–504.

Rizzoti K, Lovell-Badge R (2005). Early development of the pituitary gland: induction and shaping of Rathke's pouch. Rev Endocr Metab Disord 6: 161–172.

Rizzoti K, Brunelli S, Carmignac D et al. (2004). SOX3 is required during the formation of the hypothalamo-pituitary axis. Nat Genet 36: 247–255.

Schlosser G (2006). Induction and specification of cranial placodes. Dev Biol 294: 303–351.

Sheng HZ, Westphal H (1999). Early steps in pituitary organogenesis. Trends Genet 15: 236–240.

Sheng HZ, Zhadanov AB, Mosinger Jr B et al. (1996). Specification of pituitary cell lineages by the LIM homeobox gene *Lhx3*. Science 272: 1004–1007.

Sheng HZ, Moriyama K, Yamashita T et al. (1997). Multistep control of pituitary organogenesis. Science 278: 1809–1812.

Shimogori T, Lee DA, Miranda-Angulo A et al. (2010). A genomic atlas of mouse hypothalamic development. Nat Neurosci 13: 767–775.

Suga H, Kadoshima T, Minaguchi M et al. (2011). Self-formation of functional adenohypophysis in three-dimensional culture. Nature 480: 57–62.

Szarek E, Cheah PS, Schwartz J et al. (2010). Molecular genetics of the developing neuroendocrine hypothalamus. Mol Cell Endocrinol 323: 115–123.

Takuma N, Shen HZ, Furuta Y et al. (1998). Formation of Rathke's pouch requires dual induction from the diencephalon. Development 125: 4835–4840.

Traiffort E, Dubourg C, Faure H et al. (2004). Functional characterization of sonic hedgehog mutations associated with holoprosencephaly. J Biol Chem 279: 42889–42897.

Ward RD, Raetzman LT, Suh H et al. (2005). Role of PROP1 in pituitary gland growth. Mol Endocrinol 19: 698–710.

Ward RD, Stone BM, Raetzman LT et al. (2006). Cell proliferation and vascularization in mouse models of pituitary hormone deficiency. Mol Endocrinol 20: 1378–1390.

Webb EA, Al Mutair A, Kelberman D et al. (2013). *ARNT2* mutation causes hypopituitarism, post-natal microcephaly, visual and renal anomalies. Brain 136: 3096–3105.

Woods KS, Cundall M, Turton J et al. (2005). Over- and under-dosage of SOX3 is associated with infundibular hypoplasia and hypopituitarism. Am J Hum Genet 76: 833–849.

Young J, Metay C, Bouligand J et al. (2012). *SEMA3A* deletion in a family with Kallmann syndrome validates the role of semaphoring 3A in human puberty and olfactory system development. Hum Reprod 27: 1460–1465.

Zhao M, Qiao M, Harris SE et al. (2006). The zinc finger transcription factor Gli2 mediates bone morphogenetic protein 2 expression in osteoblasts in response to hedgehog signalling. Mol Cell Biol 26: 6197–6208.

Handbook of Clinical Neurology, Vol. 124 (3rd series)
Clinical Neuroendocrinology
E. Fliers, M. Korbonits, and J.A. Romijn, Editors

Chapter 2

Neuroendocrinology of pregnancy and parturition

CHIARA VOLTOLINI AND FELICE PETRAGLIA*
Department of Molecular and Developmental Medicine, University of Siena, Siena, Italy

INTRODUCTION

A successful pregnancy requires a series of maternal adaptive mechanisms driven by the brain. In this context, the neuroendocrine system is active in modulating and orchestrating important biologic functions through a complex system of hormonal regulation. These neuroendocrine changes are critical for the maintenance of a pregnancy and the timing of parturition, for fetal growth and development, and for maternal/fetal protection from adverse programming (Brunton and Russell, 2008a).

Neuroendocrine circuits can have powerful inhibitory as well as excitatory roles. Hormones induce a cascade of changes in the brain that affect several aspects of maternal physiology by: (1) reducing stress reactions and evoking maternal behavior; (2) preventing further ovulation and the risk of another, competing pregnancy; (3) increasing appetite, expanding blood volume, and modifying breathing; (4) preparing the neuroendocrine circuits that drive the birth process and stimulating the mammary glands to produce milk (Russell, 2000).

The neuroendocrine system in pregnancy is not limited to the maternal brain, but includes both the fetal brain and the placenta, itself a neuroendocrine organ controlling some of the adaptive mechanisms during pregnancy by autocrine, paracrine, and endocrine pathways. Therefore, the concept of neuroendocrinology during pregnancy becomes broader and more complex than in the nonpregnant situation (Douglas and Ludwig, 2008).

Brain and placenta are both central organs in the presence of stress, and maternal, fetal, and placental hypothalamus–pituitary–adrenal (HPA) axes play a significant role activating a series of responses that contribute to the maintenance of physiologic conditions while avoiding adverse effects of stress on both mother and offspring (Petraglia et al., 1996a; Brunton et al., 2008a).

The present chapter focuses on the role of the placenta as a primary source of hypothalamus-like and other neuroactive hormones in the physiology and pathology of pregnancy.

THE HYPOTHALAMUS–PITUITARY–TARGET GLAND AXES DURING PREGNANCY

During pregnancy, the maternal brain is exposed to a concentration of hormones never previously experienced; these are either proteins and peptides with restricted access to the brain, or steroids, which enter the brain from the circulation or are generated directly within the brain. As a result, the control of neuroendocrine systems and of other major physiologic systems is dramatically altered, with up- or downregulation of specific central pathways, showing that brain is not simply a passive receiver of endocrine signals or of altered expression of hormone receptors (Wagner and Morrell, 1996; Francis et al., 2002; Mann and Babb, 2005).

Adaptation of the maternal brain during pregnancy serves to optimize pregnancy outcome, ensuring an appropriate internal environment for the pregnant state. In fact, adaptive mechanisms are important in providing essential and optimal conditions for a maintained and successful pregnancy, for parturition and lactation, and for protecting the developing fetus from events that may have adverse programming effects on intrauterine and postnatal life. In particular, the maternal brain is prepared to support the birth process and ensure the immediate expression of maternal nurturing behavior and lactation postpartum. During pregnancy, the action of many neurohormones on the maternal brain is temporary, since they are aimed at adjusting the functioning of systems to maintain pregnancy and largely reversed

*Correspondence to: Felice Petraglia, Department of Obstetrics and Gynecology, Policlinico Santa Maria alle Scotte, Viale Bracci, 53100 Siena, Italy. Tel: +39-0577-586602, Fax: +39-0577-233454, E-mail: felice.petraglia@unisi.it

after birth; however, other neurohormone actions are longer lasting, leading to permanent behavioral changes (Fleming and Sarker, 1990). The activities of multiple neuroendocrine systems are modified in pregnancy to achieve these requirements (Brunton et al., 2008b).

In this context, the human placenta plays a primary role. Indeed, this organ is capable of synthesizing and releasing several neuroactive factors, such as hypothalamus-like and pituitary-like hormones. The neurohormones produced by placental tissue act locally in modulating the release of the pituitary-like hormones, resembling the organization of the hypothalamus–pituitary–target gland axes; moreover, they are chemically identical to, and have the same biologic activities as, their neuronal counterparts, suggesting also a neuroectodermal embryologic origin of placental trophoblast (Petraglia et al., 1990a). Therefore, hypothalamic functions during pregnancy are integrated by the placenta. The effect of these placental neuroendocrine factors also impacts on the fetal brain and neuroendocrine functions. For this reason, pregnancy is a highly complex neuroendocrine condition with three neuroendocrine systems working at the same time for an optimal outcome.

Activity of hypothalamus–pituitary–adrenal axis

The hypothalamus–pituitary–adrenal (HPA) axis plays a key role in the neuroendocrine response to stress via cortisol secretion, acting to restore homeostasis following stressful events for survival (Brunton and Russell, 2010). Corticotropin-releasing hormone (CRH) represents the main regulator of the axis: released from hypothalamus, this neurohormone stimulates adrenocorticotropic hormone (ACTH) release from the anterior pituitary, and consequently glucocorticoid secretion from the adrenal cortex which acts to facilitate appropriate stress responses by promoting energy mobilization, cardiovascular responses, and immune responses (Fig. 2.1). Negative feedback action inhibits the effectors of the stress response when the stress no longer poses a threat (Vale et al., 1981; Besedovsky and del Rey, 1992; Dallman et al., 1992). The maternal–placental–fetal HPA axes play a fundamental role during pregnancy and parturition.

Maternal hypothalamus–pituitary–adrenal axis

The secretion of maternal HPA axis hormones increases throughout pregnancy and is related to placental function, as circulating maternal CRH originates almost entirely from the placenta (Petraglia et al., 1996a). The responsiveness of the HPA axis to psychological and physical stressors is partially reduced throughout pregnancy, and characterized by lower ACTH and corticosterone secretion, as supported by the observation in rats of reduced CRH and vasopressin mRNA expression in the hypothalamic parvocellular paraventricular nucleus (pPVN). This adaptation may serve to neutralize the impact of stress by reducing stress-induced fetal exposure to maternal glucocorticoid, thus minimizing the risk of unfavorable glucocorticoid programming (Welberg and Seckl, 2001) while at the same time promoting the anabolic adaptations in the mother necessary for successful pregnancy (Herrera, 2000).

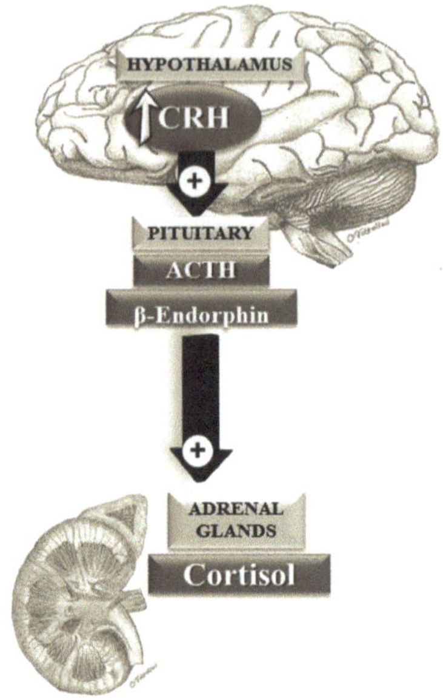

Fig. 2.1. Maternal changes of stress hormones during pregnancy. The hypothalamic–pituitary–adrenal (HPA) axis plays a key role in the neuroendocrine response to stress, and the secretion of maternal HPA axis hormones increases throughout pregnancy. Corticotropin-releasing hormone (CRH) represents the main regulator of the axis as, when released from the hypothalamus, it stimulates adrenocorticotropic hormone (ACTH) release from the anterior pituitary, and consequently cortisol secretion from the adrenal cortex which acts to facilitate appropriate stress responses by promoting energy mobilization, cardiovascular responses, and immune responses.

Unlike CRH, it is uncertain whether maternal plasma ACTH originates from the maternal pituitary and/or from the placenta. Maternal ACTH and cortisol, but not CRH, undergo a typical circadian rhythmicity in maternal circulation, suggesting that CRH is mostly of placental origin (Waddell, 1993). Moreover, hypersecretion of CRH and cortisol in pregnancy is not glucocorticoid-sensitive, showing a distinct regulation of hypothalamic and placental CRH (Tropper et al., 1987). From term pregnancy towards labor, circulating CRH increases and is bioactive on pituitary and placental ACTH release at the time of labor stress. Considering also the increased maternal levels of neuropeptide Y (NPY), β-endorphin (β-END), and oxytocin (OT) at delivery, the activation of the stress-related pathways and of neurobehavioral changes (mood and pain) at parturition act simultaneously (Petraglia et al., 1990b).

Early and mid-pregnancy

Animal studies show HPA axis activity at the circadian nadir to be similar to that observed out of pregnancy. However, compared with prepregnancy, basal HPA axis activity is strikingly altered. In rats, circulating peak levels of ACTH and corticosterone are reduced by day 2 compared with diestrus (Atkinson and Waddell, 1995). As pregnancy proceeds, the circadian pattern remains suppressed and corticosterone levels decline further until mid-pregnancy (day 10), when they begin to increase again (Atkinson and Waddell, 1995; Mizoguchi et al., 1997). The suppression of the diurnal increase in ACTH secretion is more striking and is sustained throughout pregnancy (Atkinson and Waddell, 1995). In women, early pregnancy salivary cortisol levels are also lower than in late pregnancy, and differences between nadir and peak (early light phase in humans) become greater as cortisol levels increase throughout pregnancy (Obel et al., 2005; Harville et al., 2007). Interestingly, in early pregnancy high glucocorticoid secretion is associated with miscarriage in women at 1–3 weeks postconception compared with women with ongoing gestation (Nepomnaschy et al., 2006), suggesting that maternal stress in the first trimester is associated with a higher risk for spontaneous abortion.

The HPA axis responses to stressors in rodents are not different from nonpregnant animals during early pregnancy. At the time of implantation, rodents exhibit increased ACTH secretion, even though the risk of pregnancy failure is high due to inhibition of progesterone secretion and action (Joachim et al., 2003; Nakamura et al., 2008). Furthermore, in the first half of pregnancy, HPA axis secretory responses to acute physical and emotional stressors remain similar to those of nonpregnant females (Nakamura et al., 1997; Neumann et al., 1998). In contrast, in women the experience of "chronic stressful life events" during early pregnancy blunts salivary cortisol levels in morning samples, without affecting evening levels (Obel et al., 2005).

Late pregnancy

In rodents, late pregnancy is associated with reduced basal activity of the HPA axis. In the pPVN, expression of CRH and vasopressin mRNA is decreased (Johnstone et al., 2000); similarly, in the anterior pituitary, mRNA expression for pro-opiomelanocortin (POMC) is reduced (Ma et al., 2005). Despite suppression of the normal circadian rhythm in ACTH secretion in the second half of pregnancy, the circadian variation in corticosterone secretion is maintained, implying an alteration of the plasma ACTH–corticosterone relationship. This may reflect increased sensitivity of the adrenal gland to ACTH by the action of increased circulating estrogen (Carr et al., 1981). Daily integrated circulating corticosterone levels increase progressively from mid-gestation until term in the rat (Atkinson and Waddell, 1995). However, physiologically available corticosterone is reduced due to greatly elevated levels of circulating corticosterone-binding globulin in late pregnancy (Douglas et al., 2003).

The responsiveness of the HPA axis to stressors is progressively attenuated during late pregnancy, in women (de Weerth and Buitelaar, 2005) as in other species. In rats, HPA axis responses to a wide range of stressors in late pregnancy are suppressed, as indicated by reduced ACTH and corticosterone responses (Brunton et al., 2005, 2006; Douglas et al., 2005). Similarly, in late pregnant women, exogenously administered CRH fails to increase ACTH or corticosterone secretion (Schulte et al., 1990) and suppressed salivary cortisol responses are observed following exposure to the cold pressor test (Kammerer et al., 2002).

Parturition

In women, circulating levels of CRH, ACTH and cortisol are increased at term and during labor (Petraglia et al., 1990d; Ochedalski et al., 2001), although they are not necessarily indicative of maternal HPA axis activation. The human placenta and endometrium synthesize and secrete CRH which drives fetal ACTH and cortisol secretion increasingly towards term, whereas the placenta is also a source of ACTH (Petraglia et al., 1996a). Placental CRH has complex effects including a role in the onset of labor (Vitoratos et al., 2006), resembling the timer of a biologic clock counting from the early stages of gestation and signaling the timing of labor onset (McLean et al., 1995; Grammatopoulos, 2008). Therefore, although the process of parturition represents one of the most prominent life stressors, placental rather than HPA axis mechanisms are more likely to be activated.

FETAL HYPOTHALAMUS–PITUITARY–ADRENAL AXIS

The actions of the fetal HPA axis, together with those of the placenta, are essential in fetal development, maturation and homeostasis, and eventually in preparation for the survival of the neonate (Korebrits et al., 1998; Mastorakos and Ilias, 2003).

Fetal HPA axis activity starts at mid-gestation, after the maturation of the fetal pituitary (Gitau et al., 2001). Interestingly, fetal stress responses are independent from those of the mother: the fetal hypothalamus releases CRH, together with placental CRH, inducing fetal pituitary ACTH secretion. ACTH in turn controls adrenocortical functional development, including angiogenesis and expression of steroidogenetic enzymes (Mesiano and Jaffe, 1997).

In the fetal adrenal gland, the fetal zone represents the principal localization of dehydroepiandrosterone sulfate (DHEAS) synthesis (Pepe and Albrecht, 1995), constituting the substrate for placental estrogen synthesis; thus, CRH may stimulate adrenal steroidogenesis, thereby providing the substrate for the placental production of estrogens, which favor parturition by inducing contraction (Grammatopoulos and Hillhouse, 1999). The fetal adrenal gland also contains the definitive zone, representing the main site of mineralocorticoid synthesis, and the transitional zone where cortisol is synthesized *de novo* after the 28th week of pregnancy (Mesiano and Jaffe, 1997). Although fetal stress responses are independent, it is important to know that about one-third of variations in fetal cortisol levels are attributable to maternal cortisol levels.

Parturition is a very stressful condition for the fetus and an adequate adrenocortical secretion of steroids (mainly cortisol) enables its adaptation to extrauterine life. Stimulation of the fetal pituitary by CRH increases ACTH production and, consequently, the synthesis of cortisol by the fetal adrenal gland and maturation of fetal lungs. In turn, the rising fetal cortisol concentrations further stimulate placental CRH production. The maturation of the fetal lungs as a result of increasing cortisol concentrations represents a fundamental aspect of fetal adaptive mechanisms to extrauterine life activated by the stress of delivery. Moreover, fetal lung maturation is associated with increased production of surfactant protein A and phospholipids, both proinflammatory agents, that may stimulate myometrial contractility through increased production of prostaglandins by fetal membranes and the myometrium itself (Smith, 2007).

The fetal zone of the adrenal glands involutes rapidly after delivery of the placenta, indicating that placental factors, such as CRH, maintain its function; therefore after parturition, the adrenal cortex undergoes radical remodeling although normally there is no concomitant clinically evident adrenocortical deficiency in infants born at term (Mesiano and Jaffe, 1997).

Activity of hypothalamus–pituitary–gonadal axes

The hypothalamus–pituitary–gonadal axis is the major system controlling reproductive functions in vertebrates. The main regulator is represented by gonadotropin-releasing hormone (GnRH) produced by the hypothalamus that in the female acts on the anterior pituitary to induce the synthesis and secretion of luteinizing hormone (LH) and follicle-stimulating hormone (FSH), in turn exerting their actions on the ovaries (Stojilkovic et al., 1994; Sasaki and Norwitz, 2011).

From early pregnancy, low levels of LH and FSH reflect decreased functioning of the hypothalamus–pituitary–gonadal axis. Reduction of maternal LH and FSH levels may depend also on the rise of maternal serum inhibin levels during the first weeks of gestation. Moreover, secretion of maternal progesterone switches from gonads to placenta within 10–12 weeks of gestation, limiting the role of ovarian functioning in pregnancy (Petraglia et al., 1996a).

However, similarly to CRH, bioactive GnRH is also produced by human placenta. In relation to gestational age, its concentration peaks in early placental specimens (8 weeks of gestation) and decreases in samples collected in subsequent periods of gestation (Miyake et al., 1982); in parallel, maternal circulating GnRH levels are high in the first half of pregnancy and decline later on (Siler-Khodr et al., 1984). Interestingly, placental and circulating GnRH levels increase during the first trimester, probably modulating human chorionic gonadotropin (hCG) production by trophoblast cells (Lee et al., 2010).

Activity of the other neuroendocrine axes

Hypothalamus–prolactin axis

Prolactin (PRL) is produced by lactotroph cells in the anterior pituitary gland under the inhibitory control of dopamine (DA), released from mediobasal hypothalamic neurons. Interestingly, PRL itself feeds back on the hypothalamus to regulate its own secretion. Suckling of the nipple during breastfeeding represents the prototypical stimulus for PRL secretion, acting directly on the pituitary gland and hypothalamus, while the estradiol-induced PRL secretion is entrained to the light/dark cycle (Grattan and Kokay, 2008).

There are several adaptations to maintain high levels of lactogenic hormones throughout pregnancy and lactation. This is achieved by different mechanisms in different species, but is likely to involve marked changes to the feedback inhibitory systems (Larsen and Grattan, 2012). High circulating PRL levels are essential for maintaining pregnancy by providing luteotropic support to the corpus luteum, thereby stimulating progesterone secretion in early pregnancy (Freeman et al., 2000). Moreover, PRL has an immunomodulatory function, by regulating cytokine profile and enhancing IL-12 and IL-10 production in a stimulus-specific manner (Matalka and Ali, 2005). The increasing PRL levels throughout gestation are linked to increasing estrogen levels, which also rise from early pregnancy; moreover, PRL receptor expression is upregulated in the hypothalamus during pregnancy (Pi and Grattan, 1999), suggesting that the hypothalamus is highly sensitive to PRL at this time. In humans, the placenta produces an increasing amount of placental lactogen (hPL) throughout pregnancy (Ben-Jonathan et al., 2008), which is a molecule structurally similar to PRL that can bind to PRL receptors. Differently from PRL, hPL is not subjected to hypothalamic regulation by DA, hence providing a source of lactogenic hormones that is not subject to the normal negative feedback regulation (Grattan and Kokay, 2008). The amount of hPL peaks at mid-gestation and appears to be predominantly secreted into the amniotic fluid and in fetal and maternal body fluids, including cerebrospinal fluid. Therefore, during human pregnancy a large increase in both pituitary-produced PRL and placental hPL have been recognized (Larsen and Grattan, 2012). Elevated lactogen hormones contribute to regulating several functions: (1) establishment of maternal behavior (Bridges et al., 1985); (2) food intake, by increasing orexigenic drive and establishing leptin resistance in the mother (Naef and Woodside, 2007); (3) suppression of stress responses during late pregnancy and lactation (Torner and Neumann, 2002), to minimize the risk of adverse fetal programming from glucocorticoids (Brunton and Russell, 2008b); (4) prevention of stress-induced hyperthermia (Drago and Amir, 1984); (5) regulation of oxytocin neurons during parturition and lactation (Russell et al., 2001), hence contributing to myometrial contractility and milk ejection; (6) maternal recognition of the offspring, possibly by PRL-induced neurogenesis (Shingo et al., 2003).

Hypothalamus–growth hormone axis

Synthesis and secretion of pituitary growth hormone (GH), which is involved in several functions such as promoting development and metabolism, is regulated by two hypothalamic-releasing factors: GH-releasing hormone (GHRH) and somatostatin (SST) (Sassolas, 2000).

The regulation of the somatotropic axis is critical in pregnancy. It is well established that in humans circulating concentrations of GH-like immunoreactivity dramatically increase during pregnancy (Handwerger and Freemark, 2000), although this is not associated with increased transcription of the pituitary gene. In fact, it rather reflects placental production of a bioactive GH variant (Frankenne et al., 1990). During early pregnancy, pituitary GH is the only GH measurable in maternal serum and it is secreted in a highly pulsatile pattern; conversely, in late pregnancy, 85% of circulating levels of bioactive GH derive from placenta, whose secretion is not episodic, with maternal serum concentrations remaining relatively constant during a 24 hour period (Alsat et al., 1998). Moreover, unlike pituitary GH, the placental variant is unresponsive to hypothalamic GHRH (de Zegher et al., 1990), which is nevertheless produced by placental tissue as well (Berry et al., 1992). These considerations suggest that placental GH secretion is constitutively at its maximal level (de Zegher et al., 1990) and/or that placental GHRH exerts roles unrelated to GH secretion, such as the regulation of fetal pituitary GH secretion (Nogués et al., 1997). Glucose appears to be the primary modulator of placental GH secretion, as demonstrated by *in vitro* studies (Patel et al., 1995) and by the fact that hyperglycemia in diabetic pregnant women is associated with reduced circulating GH concentrations (McIntyre et al., 2000). In addition, hypoglycemia induces placental GH synthesis with subsequent increase of maternal blood glucose levels, suggesting that placental GH may thus protect the fetus from nutrient deficiency (Alsat et al., 1998). Placental GH stimulates insulin-like growth factor 1 (IGF-1) and its binding protein (IGFBP-3), reducing plasma clearance of IGF-1 and resulting in negative feedback suppression of maternal GH secretion (Daughaday et al., 1990). Interestingly, high circulating levels of placental GH and IGF-1 may explain the biochemical "acromegalic-like" state of the third trimester of pregnancy. Concerning parturition, increased GH secretion clearly occurs without a clear physiologic significance: it may occur secondary to other endocrine changes or reflect a nonspecific response to the stress involved (Hull and Harvey, 2002).

Hypothalamus–pituitary–thyroid axis

The neuroendocrine system controlling thyroid function is regulated by the hypothalamic thyrotropin-releasing hormone (TRH) that in turn stimulates the release of thyroid-stimulating hormone (TSH) by the anterior pituitary, representing the major effector on thyroid function.

The responsiveness of the maternal hypothalamus–pituitary–thyroid axis can be considered to function normally in pregnancy (Glinoer, 1997). Interestingly, it is well known that human placenta contains immunoreactivity for TRH, although its bioactivity is still controversial. A major role is played by placental hCG: it bears a structural resemblance to TSH and is able to bind TSH receptors with subsequent thyrotropic activity (Mori et al., 1988), demonstrating its role as a thyroid regulator in normal pregnancy (Glinoer et al., 1990). To date, there has been a great deal of compelling evidence indicating a transient fall in maternal serum TSH near the end of the first trimester in normal pregnancy, and that this partial TSH suppression is associated with the elevation in circulating hCG. Thus, the lowering of TSH corresponds to a transient and partial blunting of the pituitary–thyroid axis associated with an increased hormonal output from the thyroid, which may sometimes even induce a mild thyrotoxicosis during pregnancy. Even though maternal and fetal thyroid functions are autonomously regulated, they are not completely independent of one another. Despite the fact that the placenta acts as a barrier between maternal and fetal hypothalamus–pituitary–thyroid axes, there is evidence of at least some transplacental passage of maternal thyroid hormones. This is probably important in the early stages of fetal development, as it protects the fetus from adverse neurodevelopmental programming determined by a shortage of thyroid hormone (Williams, 2008).

THE PLACENTA: A NEUROENDOCRINE ORGAN

In the past, the placenta was believed to be a largely passive organ mainly responsible for delivering nutrients to the fetus. With progress in obstetric research, this concept has gradually shifted to one that recognizes the placenta as a transient neuroendocrine organ and a central regulator of maternal–placental–fetal physiology. In this revised concept, the placenta ensures appropriate physiologic milieus for normal growth and development of fetal, placental, and maternal tissues necessary for a successful pregnancy (Petraglia et al., 2010). Indeed, the placenta represents a very metabolically active organ during pregnancy. It is a source of a large number of "information" molecules that, when released, can exert their biologic effects on the placenta itself but can also enter the maternal and fetal circulation, thus acting as autocrine, paracrine, and endocrine factors (Petraglia et al., 1996a).

The placenta performs a primary role in the regulation of maternal biologic functions that, before pregnancy, are mainly coordinated by the central nervous system, such as energy mobilization, cardiovascular responses, metabolic and immune functions, mood and behavior. During intrauterine life it also provides hormones involved in fetal growth, development, and maturation (Wood, 2005; O'Donnell et al., 2009; Brunton and Russell, 2011).

Placental neuroendocrine activity is influenced by the anatomic characteristics of this organ. As pregnancy advances, placental structure progressively changes: in the first trimester the relative number of trophoblasts increases and the feto-maternal exchange begins to be dominated by the secretory function of the placenta, while throughout the second and third trimester the structure of the placenta further adapts, with the villi consisting mainly of fetal capillaries supported by sparse stroma. This structural arrangements facilitates the transport of compounds including hormones and their precursors across the feto-maternal interface (Feldt-Rasmussen and Mathiesen, 2011).

As stated above, among the various placental hormones several neuroactive factors have been recognized, including hypothalamic and pituitary hormones, growth factors, steroid hormones, vasoactive peptides, metabolic hormones, and many others (Table 2.1), that

Table 2.1

Placental neuroactive factors

Hypothalamus-like hormones	Corticotropin-releasing hormone (CRH)
	Urocortins (Ucn1, Ucn2, Ucn3)
	Gonadotropin-releasing hormone (GnRH)
	Growth hormone-releasing hormone (GHRH)
	Somatostatin (SST)
	Thyrotropin-releasing hormone (TRH)
	Oxytocin (OT)
Pituitary-like hormones	Human chorionic gonadotropin (hCG)
	Adrenocorticotropic hormone (ACTH)
	Opioids (β-endorphin)
	Placental growth hormone (PGH)
	Human placental lactogen (hPL)
Other neurohormones	Neuropeptide Y (NPY)
	Chromogranin A (CgA)
	Neurokinin B (NKB)
	Monoamines
Growth hormones	Transforming growth factor-β (TGF-β)
	Activin A
	Follistatin
	Follistatin-related gene (FLRG)
	Inhibin A and B
	Insulin-like growth factors (IGFs)
	IGF-binding proteins (IGF-BPs)
	Fibroblast growth factor (FGF)
	Epidermal growth factor (EGF)
	Vascular endothelial growth factor (VEGF)
Steroid hormones	Progesterone
	Estrogens
Vasoactive peptides	Adrenomedullin (ADM)
	Endothelins (ETs)
	Calcitonin gene-related peptide (CGRP)
	Parathyroid hormone-related peptide (PTH-rp)
Metabolic hormones	Leptin
	Ghrelin

are also produced by and are active in the brain, suggesting that placenta performs regulatory functions that are similar or analogous to ones ascribed to the brain (Petraglia et al., 1990a; Reis et al., 2001). The network of peptide signaling substances expressed in human placenta is highly specialized in regulating several functions: (1) implantation, through the regulation of local immune processes, mechanisms of invasion of maternal decidua, and trophoblast differentiation; (2) regulation of maternal and fetal endocrine glands; (3) modulation of maternal mood and behavior; (4) regulation of fetal growth, through the transfer of gas and nutrients such as oxygen, glucose, and amino acids; (5) timing of delivery, through the regulation of maternal contractility.

However, neurohormonal changes in pregnancy are associated not just with gestational physiology but also with pathologic conditions potentially threatening maternal and fetal well-being, as expression of the onset of pathologic mechanisms or even activation of adaptive responses by the feto-placental unit (Petraglia and Florio, 2001; Torricelli et al., 2012).

STRESS-RELATED HORMONES: IMPLICATIONS IN PHYSIOLOGIC PREGNANCY AND PARTURITION AND IN OBSTETRIC COMPLICATIONS

Corticotropin-releasing hormone family

The CRH family includes CRH and urocortins (Ucns), neuropeptides that are known as the main regulators of adaptive responses to stress (Petraglia et al., 2010). In mammals, the CRH/Ucn family consists of at least four ligands: CRH, Ucn (Vaughan et al., 1995), Ucn2 and Ucn3 (Hsu and Hsueh, 2001). All these peptides, which are produced mainly by the hypothalamus but also by several organs, though in small amounts, have been found in the human placenta and in fetal membranes (Petraglia et al., 1996b; Imperatore et al., 2006; Vitoratos et al., 2006), which, during pregnancy, may even represent their major source. As several studies have confirmed, the CRH/Ucn peptides are implicated as important neuroendocrine mediators in the physiology of early and late pregnancy and in the mechanisms of parturition (Zoumakis et al., 2009).

Corticotropin-releasing hormone

Placental expression and regulation

Placental CRH peptide and mRNA expression are higher at term than in early gestation (Saijonmaa et al., 1988; Grammatopoulos, 2008). Placental CRH mRNA is located in the cytotrophoblast, syncytiotrophoblast, and intermediate trophoblast at term (Petraglia et al., 1987; Riley et al., 1991); moreover, CRH mRNA is also expressed by the subepithelial layer of the amnion, the reticular layer of the chorion, the decidual stromal cells, and human umbilical vein endothelial cells (Petraglia et al., 1987; Simoncini et al., 1999).

Placental CRH is identical to the hypothalamic isoform in structure, immunoreactivity, and bioactivity (Chan et al., 1988; Sasaki et al., 1988). In nonpregnant women, plasma CRH levels are very low (around 15 pg/mL) or undetectable (Sasaki et al., 1987). Human placenta expresses large amounts of CRH (>1000 times higher than in myometrium and choriodecidua), resulting in high CRH levels in maternal serum during pregnancy (Riley et al., 1991). Maternal serum CRH levels increase exponentially from early pregnancy to approximately 800 pg/mL during the third trimester and peak (2000–3000 pg/mL) during labor. The peptide becomes undetectable within 24 hours after delivery (Goland et al., 1986; Economides et al., 1987). CRH is also measurable in amniotic fluid (Maser-Gluth et al., 1987) and cord serum (Nagashima et al., 1987).

Mechanisms regulating placental CRH release are almost identical to those regulating hypothalamic release. Prostaglandins (PGs), neurotransmitters (norepinephrine, acetylcholine), neuropeptides (angiotensin II, arginine vasopressin) and cytokines (IL-1) all stimulate CRH secretion from cultured placental cells *in vitro* (Petraglia et al., 1989, 1990c). On the other hand, progesterone (P4), nitric oxide (Smith, 2007) and estrogens (Ni et al., 2002) decrease placental CRH production. Interestingly, in contrast to the hypothalamus, placental CRH release is positively upregulated by glucocorticoids (Robinson et al., 1988). This positive feedforward system is a unique feature of placental tissue during pregnancy, suggesting that the demand for CRH biologic functions increases as pregnancy progresses towards term (Petraglia et al., 2010).

Physiologic pregnancy and parturition

During human pregnancy, CRH targets multiple fetal–maternal tissues, including the placenta, fetal membranes, myometrial smooth muscle, and fetal adrenals (Grammatopoulos, 2007; Smith, 2007). CRH has been shown to be involved in several biologic functions, including regulation of hormonal release, vascular tonus, immune functions and myometrial contractility (Petraglia et al., 2010) (Fig. 2.2).

Regarding endocrine functions, in the placenta and in fetal membranes CRH regulates: (1) human trophoblast cell growth and invasion through regulation of carcinoembryonic antigen-related cell adhesion molecule-1 (CEACAM-1); (2) tissue remodeling through modulation of secretion of the matrix-degrading protease matrix

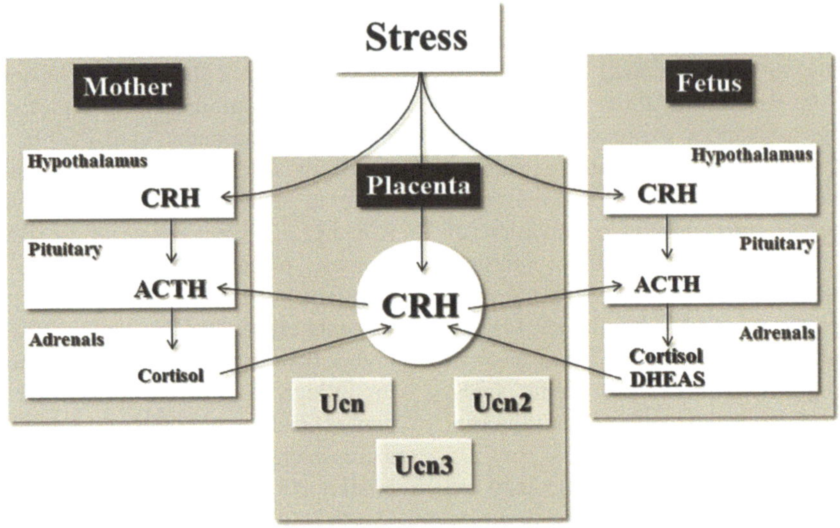

Fig. 2.2. Stress and maternal–placental–fetal HPA axes. During pregnancy maternal and fetal hypothalamic–pituitary–adrenal (HPA) axes are integrated by the placenta, and the three neuroendocrine systems work at the same time for an optimal outcome. Placental corticotropin-releasing hormone (CRH) targets multiple fetal–maternal tissues, including placenta, fetal membranes, myometrial smooth muscle, and fetal adrenals. In fact CRH regulates several biologic functions, including hormonal release, vascular tonus, immune functions, and myometrial contractility; moreover, it also stimulates cortisol and dehydroepiandrosterone sulphate (DHEAS) production from fetal adrenal gland at the end of pregnancy, suggesting a role in fetal lung maturation and adaptive mechanisms in response to the stress of parturition. Urocortins belong to the CRH superfamily and have recently been implicated as novel neuroendocrine mediators in physiologic and pathologic pregnancy and parturition, exerting complementary or sometimes contrasting actions to fine-tune CRH biologic effects.

metalloproteinase-9 (MMP-9); (3) placental vascular tone through activation of the nitric oxide pathway; and (4) ACTH, PG generation, and bioavailability (Clifton et al., 1994; Li and Challis, 2005; Bamberger et al., 2006). CRH also plays a pivotal role in regulating estrogens and progesterone production in the third trimester (Rainey et al., 2004; Kalantaridou et al., 2010), corroborating the hypothesis that placental CRH production is linked to the length of gestation in humans. Interestingly, it stimulates cortisol and DHEAS production from fetal adrenals at the end of pregnancy, suggesting a role for CRH in fetal lung maturation and adaptive mechanisms in response to the stress of parturition (Pepe and Albrecht, 1995; Sirianni et al., 2005) (Fig. 2.3).

CRH is also a potent vasoactive molecule acting on human vascular endothelium as well as smooth muscle and the feto-placental circulation (Dashwood et al., 1987), and of great physiologic importance during human pregnancy. CRH is a potent relaxant of the uterine arteries and, when administered chronically in pregnant rats, it causes a decrease in blood pressure (Jain et al., 1998). CRH-induced vasorelaxation is a specific receptor-operated, endothelium-dependent effect mediated by the nitric oxide–cyclic guanosine monophosphate (cGMP) pathway (Clifton et al., 1995; Jain et al., 1998), but it may be also an endothelium-independent effect through a direct action on the vascular smooth muscle that is not mediated by cAMP, K^+ channels, or Ca^{++} channels.

Regarding immune functions, it is now established that CRH may influence embryo implantation (Makrigiannakis et al., 2004); in a process involving a series of immune and nonimmune cells, invasive trophoblast cells, and maternal endometrium, an intricate network of locally acting peptides regulates the complex processes of trophoblast invasion and placentation (Anin et al., 2004; van den Brûle et al., 2005). The intrauterine Fas/Fas ligand (FasL) system has been considered primarily as a mechanism used by trophoblasts to escape maternal immune attack and interestingly, CRH has been shown to regulate FasL expression in several cells, including trophoblast (Dermitzaki et al., 2002). At the maternal–fetal interface, trophoblast cells are important in producing significant proinflammatory cytokines in response to physiologic and pathologic conditions (Griesinger et al., 2001; Abrahams and Mor, 2005). High levels of CRH have been shown to be released locally at the site of inflammation and it has been demonstrated to serve a proinflammatory function by activating the immune/inflammatory response, including: (1) decidualization by regulating endometrial PGs and proinflammatory cytokines IL-1 and IL-6 in cultured human endometrial cells (Zoumakis et al., 2000); (2) regulation of proinflammatory cytokine expression in myometrial

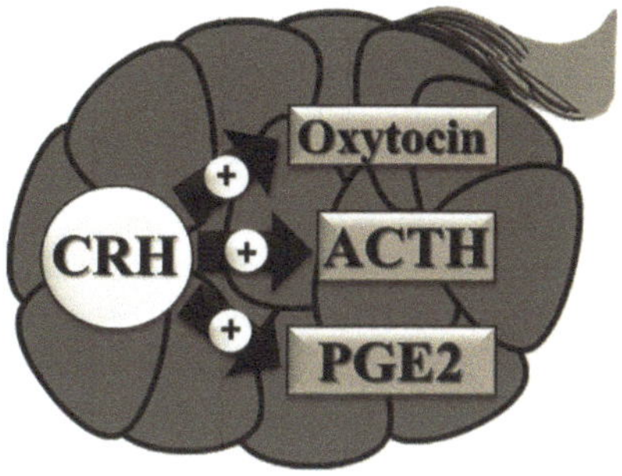

Fig. 2.3. Corticotropin-releasing hormone (CRH) and regulation of endocrine function during pregnancy. In placenta and fetal membranes CRH regulates several endocrine/paracrine/autocrine functions that contribute to maintain maternal and fetal adaptation during pregnancy and parturition. In particular CRH, maternal levels of which rise significantly at the time of labor, both term and preterm, has been proposed to constitute a biologic clock triggering the onset of parturition. In fact, evidence from the literature attests that placental CRH stimulates the local production of adrenocorticotropic hormone (ACTH), prostaglandins, and oxytocin through the binding of specific receptors (CRH-R1, CRH-R2) that are localized in fetoplacental tissues. CRH also regulates estrogens and progesterone production in the third trimester, corroborating the hypothesis that its production by the placenta is linked to the length of gestation. Finally, CRH stimulates cortisol and dehydroepiandrosterone sulphate (DHEAS) production from fetal adrenals at the end of pregnancy, suggesting a role for this neurohormone in fetal lung maturation and adaptive mechanisms in response to the stress of parturition.

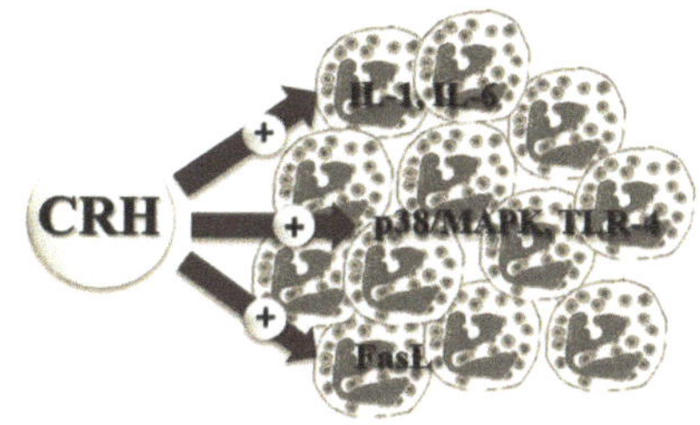

Fig. 2.4. Corticotropin-releasing hormone (CRH) and regulation of immune function during pregnancy. Trophoblast cells are important in producing significant proinflammatory cytokines in response to physiologic and pathologic conditions. CRH has been shown to be released in high levels locally at the sites of inflammation and to serve a proinflammatory function by activating the immune/inflammatory response, including regulation of: (1) prostaglandins and proinflammatory cytokines in trophoblast cells; (2) proinflammatory cytokine expression in myometrial and trophoblast cells, associated with selected activation of p38/MAPK signaling and Toll-like receptor (TLR)-4 expression; and (3) FasL expression in several cells, including trophoblast.

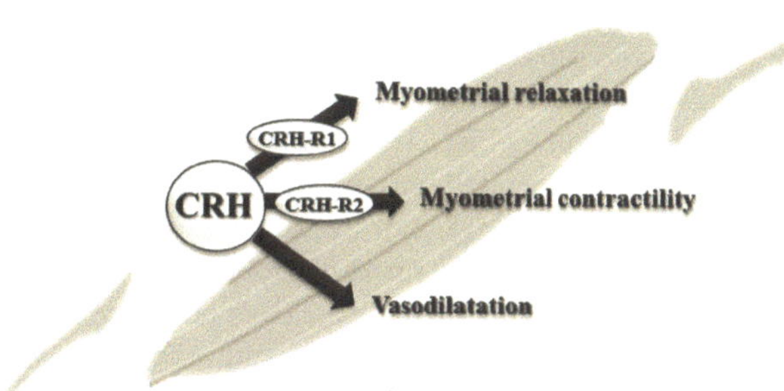

Fig. 2.5. Corticotropin-releasing hormone (CRH) and regulation of myometrial contractility and vascular tonus during pregnancy. CRH is involved in both relaxation and contraction of myometrium and this has been demonstrated to be likely dependent on different patterns of expression and biologic effects of CRH receptors (CRH-Rs). CRH-R1 contributes to maintenance of myometrial relaxation during pregnancy through activation of the adenylyl cyclase/cAMP pathway; while at term CRH binding of CRH-R2 stimulates the phospholipase C/inositol triphosphate, ERK1/2, and RhoA pathways and increases myosin light chain (MLC20) phosphorylation, promoting myometrial contractility. CRH is also a potent vasoactive molecule: CRH-induced vasorelaxation is a specific, receptor-operated, endothelium-dependent effect mediated by the nitric oxide-cGMP pathway, but it may be also an endothelium-independent effect through a direct action on the vascular smooth muscle.

and trophoblast cells, associated with selected activation of p38/MAPK signaling and Toll-like receptor (TLR)-4 expression (Guleria and Pollard, 2000; Kumazaki et al., 2004; Tsatsanis et al., 2006); and (3) suppression of endotoxin-induced proinflammatory cytokine production and survival of mice subjected to lipopolysaccharide (LPS)-induced septic shock by blockage of CRH receptors (Agelaki et al., 2002) (Fig. 2.4).

CRH also regulates myometrial contractility, exerting diverse roles at different stages of gestation. In fact, CRH is involved in both relaxation and contraction of myometrium and this has been demonstrated to be likely dependent on different patterns of expression and biologic effects of CRH receptors (CRH-Rs) (Petraglia et al., 2010). CRH-R1 contributes to the maintenance of myometrial relaxation during pregnancy through activation of the adenylyl cyclase/cAMP pathway (Grammatopoulos et al., 1996; Mignot et al., 2005). In contrast, at term CRH binding induces phosphorylation of CRH-R2 variants, with subsequent stimulation of the phospholipase C/inositol triphosphate, ERK1/2, and RhoA pathways and increase of myosin light chain (MLC20) phosphorylation, promoting myometrial contractility (Karteris et al., 2004) (Fig. 2.5). However, the hypothesis of clearly distinct roles for CRH-R1 (prorelaxation) and CRH-R2 (procontractile) has been challenged by the finding of a region-specific change in CRH receptor subtypes in the uterus identifying CRH-R2 as one of the fundal genes significantly increased during labor (Stevens et al., 1998; Markovic et al.,

2007) while CRH-R1 has been shown to be expressed in the lower segment of the uterus and upregulated, rather than downregulated, with the onset of labor (Markovic et al., 2007). Putative dynamic and changing alternative splicing of CRH-Rs within the myometrium during pregnancy and labor may in part explain this phenomenon.

Implications for maternal/fetal adverse programming

CRH has been found to be abnormally regulated in the presence of several obstetric complications.

Preterm birth (PTB) is a syndrome that may be initiated by several causes and evidence supports the involvement of the placenta in the mechanisms activating premature onset of labor. In fact, CRH mRNA expression is significantly higher in preterm than in term placentas (Torricelli et al., 2007), and maternal serum CRH concentrations in patients with subsequent PTB parallels the CRH curve of normal pregnancy, but the level is significantly displaced upward (Wolfe et al., 1988). At mid-gestation, maternal CRH levels are higher in women who subsequently have spontaneous PTB than in those delivering at term (Hobel et al., 1999). Moreover, maternal circulating CRH levels have been found to be higher in PTB associated with microbial invasion of the amniotic cavity (MIAC) than in the absence of MIAC (Warren et al., 1992; Petraglia et al., 1995). Interestingly, recent data from our group indicates that chorioamnionitis associated with PTB activates placental CRH pathway *in vivo* (Torricelli et al., 2011). In fact, trophoblast samples collected from PTB associated with chorioamnionitis show upregulation of CRH in comparison to PTB not associated with chorioamnionitis. These changes have been confirmed *in vitro* by treating placental trophoblast with LPS, suggesting CRH's potential importance in infection-mediated PTB (Torricelli et al., 2011).

Interestingly, maternal CRH levels have been reported to be lower in women delivering post-term than in those delivering at term, corroborating the important role of this neurohormone in regulating the placental "clock" of human pregnancy and determining its length (Torricelli et al., 2006).

Pre-eclampsia (PE) and pregnancy-induced hypertension (PIH) represent other obstetric complications in which a deregulation of CRH has been found. Maternal plasma and cord blood CRH levels are higher in women affected by pregnancy-induced hypertension (PIH) and PE compared to healthy women (Laatikainen et al., 1991; Liapi et al., 1996; Petraglia et al., 1996c; Farina et al., 2004; Florio et al., 2004a). Moreover, both placental CRH release into fetal plasma and CRH peptide content are higher in PE than in uncomplicated pregnancy (Goland et al., 1995). Additional data confirm that pregnancies complicated by PE and intrauterine growth restriction (IUGR) are associated with abnormal placental vascular resistance and abnormally high umbilical vein CRH levels (Trudinger et al., 1985; Goland et al., 1993), reinforcing the concept of the importance of CRH in the control of human feto-placental circulation.

In miscarriage, CRH peptide has been found to be more concentrated in the placenta and the product of conceptus in spontaneous abortion than in elective abortion of the same gestational age (Madhappan et al., 2003). A possible role for CRH in mediating trophoblast apoptosis through the induction of lymphocytic FasL has been hypothesized (Minas et al., 2007).

Urocortins

Ucn, Ucn2 and Ucn3 were identified as mammalian CRH-like paralogs a few years after CRH isolation. Ucn was identified in 1995 and was shown to share a similar primary structure and bioactivity with CRH (Skelton et al., 2000). Ucn has high affinity for every CRH-binding site identified to date, in particular CRH receptor type 1 (CRH-R1) and type 2 (CRH-R2) (Hillhouse et al., 2002). Although Ucn activates both CRH receptors, the lack of a pervasive Ucn projection to CRH-R2-expressing cells (Bittencourt et al., 1999) and the absence of CRH/Ucn projections to brain anxiety centers (Weninger et al., 1999) pointed to the existence of additional CRH-related peptides. This was confirmed when Ucn2 (also named stresscopin-related peptide) and Ucn3 (also named stresscopin) were isolated (Hsu and Hsueh, 2001).

Placental expression and regulation

In gestational tissues, syncytiotrophoblast and – rarely – cytotrophoblast or mesenchymal cells of placental villi show immunoreactivity for Ucn. Ucn mRNA and peptide are also expressed by fetal membranes, collected in the first and third trimesters, amnion epithelial cells, the subepithelial layer of the amnion and the reticular layer of the chorion (Petraglia et al., 1996b; Clifton et al., 2000; Gu et al., 2001).

Similarly, Ucn2 and Ucn3 mRNA have been detected widely in brain, colon, small intestine, muscle, stomach, thyroid, adrenal, pancreas, heart, adrenal, peripheral blood cells, and spleen and placenta (Petraglia et al., 2010). In addition, immunohistochemical analysis using a polyclonal antibody for Ucn2 and Ucn3 detected specific protein signals in the same tissues. Human trophoblasts, fetal membranes, and maternal decidua express mRNA and immunoreactive Ucn2 and Ucn3 throughout gestation (Imperatore et al., 2006). Their localization shows some differences with Ucn and CRH (Challis et al., 2001; Florio et al., 2004b). Ucn2 and Ucn3 are localized in syncytiotrophoblast and extravillous trophoblast cells, but Ucn2 is localized to blood vessel endothelial cells, leading to the suggestion of a role of Ucn2 in regulating the placental vascular endothelial behavior. With respect to the fetal membranes, Ucn2 is distributed only in amnion, while Ucn3 is found in both amnion and chorionic cells (Imperatore et al., 2006).

Although previous studies demonstrated that Ucn production is increased in hypoxic conditions (Ikeda et al., 1998), Ucn secretion and mRNA expression are not affected by hypoxia in human gestational trophoblasts *in vitro*, suggesting that the mode of placental regulation of Ucn may be unique to pregnancy (Choy et al., 2004). As to placental Ucn2 and Ucn3 regulation, few data are available. Recently, it has been demonstrated that Ucn2 affects placental conversion of fetal adrenal C19 steroid precursors into estradiol (E2) and appears to do so by increasing P450 aromatase expression via CRH-R2, thus suggesting a putative novel role for CRH family peptides in the sequence of steps involved in estrogen biosynthesis. If this is confirmed, it may have important implications for the physiology of pregnancy and parturition (Imperatore et al., 2009). Moreover, Ucn2 and Ucn3 expression in first trimester and term trophoblasts is increased in a hypoxic environment through a hypoxia-inducible factor 1 (HIF-1)-dependent process (Imperatore et al., 2010).

Outside the brain, Ucn localizes to several tissues and organs, including adipose tissue (Seres et al., 2004), heart (Kimura et al., 2002), immunologic tissue (Uzuki et al., 2001), skin (Slominski and Wortsman, 2000), small intestine, and colon (Muramatsu et al., 2000).

Physiologic pregnancy and parturition

Ucns bind to CRH-Rs with different affinity, thus exerting complementary or sometimes contrasting actions to fine-tune CRH biologic effects.

Ucn is a potential modulator of neuroendocrine activity and stress-related behavior, and similarly to CRH, is involved in important biologic functions such as regulation of hormones, vascular tonus, immune functions, and myometrial contractility (Petraglia et al., 2010). Ucn, which is able to activate the HPA axis equipotently or more potently than CRH, stimulates placental ACTH secretion as *in vitro* data have demonstrated, without any significant difference from CRH-induced ACTH release (Petraglia et al., 1999). Moreover, similarly to CRH, Ucn stimulates prostaglandin (PG) E2 release from human placental explants at term in a dose-dependent manner (Muramatsu et al., 2001), as well as placental secretion of activin A (Florio et al., 2002a), a member of the transforming growth factor-β (TGF-β) superfamily that is abundantly produced and secreted by the human placenta (Florio et al., 2002a). It also stimulates oxytocin release by placental cells in culture (Florio et al., 1996). Ucn2 and Ucn3 modulate HPA axis activity at the hypothalamic level in a paracrine or autocrine fashion, but unlike CRH and Ucn, peripheral Ucn2 or Ucn3 administration does not increase either pituitary (Tanaka et al., 2003; Venihaki et al., 2004) or placental ACTH secretion (Imperatore et al., 2006).

Placental Ucn also appears to act as a relaxant on the utero-placental vasculature via activation of the nitric oxide-cGMP system (Jones and Challis, 1990; Clifton et al., 1995), an effect mediated by binding to CRH-R2. Moreover, Ucn maximizes the release of products such as ACTH or PGs *in vivo* by causing vasodilatation of placental vascular tissue (Clifton et al., 1996; Petraglia et al., 1999).

Regarding immune functions, it is well established that Ucn serves anti-inflammatory actions in several cells and tissues (Poliak et al., 1997; Agnello et al., 1998; Wang et al., 2007). Interestingly Ucn treatment modulates lipopolysaccharide (LPS)-induced TNF-α secretion and IL-4 and IL-10 release in trophoblast cultured cells, suggesting an immunomodulatory role of this neuropeptide in placenta in the presence of infective stimuli (Torricelli et al., 2009a). A few data are available on the involvement of Ucn2 and Ucn3 in the regulation of immune functions. A recent *in vitro* study from our group has shown that Ucn2 increases, while Ucn3 decreases, placental TNF-α and IL-10 secretion after LPS-treatment, suggesting a pro- and an anti-inflammatory action for Ucn2 and Ucn3, respectively, in gestational tissues (Novembri et al., 2011).

Human myometrium expresses Ucn (Clifton et al., 2000) and a twofold increase of contractility is observed when Ucn is added after PGF2 administration (Petraglia et al., 1999). Ucn activates diverse intracellular signaling pathways that contribute to the activation of myometrial contractility (Aggelidou et al., 2002), such as p42/p44 MAPK (Nohara et al., 1996; Ohmichi et al., 1997). Moreover, Ucn stimulates MMP-9 protein level in the culture medium of chorionic trophoblast, syncytiotrophoblast, and amniotic epithelial cells (Li and Challis, 2005), suggesting a local role in tissue remodeling and cervical ripening at the time of labor. A role for Ucn2 in the control of myometrial contractility during human pregnancy has been demonstrated, involving binding to CRH-R2 and sequential activation of PKC, leading to MLC20 phosphorylation (Tropper et al., 1992; Karteris et al., 2004) and enabling actin–myosin interaction and cell contraction. Moreover, neither Ucn2 nor Ucn3 affects mRNA expression of NO synthase isoforms and protein expression or activity of soluble guanylate cyclase (Wolfe et al., 1988), with stimulatory effect on the human myometrium contractility. Finally, a role for Ucns in the local regulation of PGs has been hypothesized, as Ucn increases PG release while Ucn2 and Ucn3 inhibit cyclooxygenase (COX)-2 and 15-hydroprostaglandin dehydrogenase (PGDH) expression in cultured placental cells (Gao et al., 2008).

Implications for maternal/fetal adverse programming

Ucn has been extensively studied in several obstetric complications such as PTB, PE and hypertensive disorders, miscarriage, post-term pregnancy, and Down syndrome (Torricelli et al., 2009b). However, there are fewer data on Ucn2 and Ucn3.

Maternal plasma and cord blood Ucn levels are higher in women delivering preterm compared to those delivering at term, while Ucn mRNA expression does not change between term and preterm placenta. These data, together with the finding that Ucn levels in arterial cord blood are higher than in venous cord blood and in maternal plasma, suggest a fetal rather than an exclusively placental source of the peptide at preterm parturition (Florio et al., 2005). Moreover, recent data from our group showed that chorioamnionitis associated with PTB activates placental Ucn pathways *in vivo* (Torricelli et al., 2011). In fact, trophoblast samples collected from PTB associated with chorioamnionitis show upregulation of Ucn2 and downregulation of Ucn and Ucn3 in comparison with preterm deliveries not associated with chorioamnionitis. These changes have been confirmed *in vitro* by treating placental trophoblast with LPS, suggesting their potential importance in infection-mediated PTB (Torricelli et al., 2011). As regards post-term pregnancy, maternal blood Ucn does not change in post-term compared to term laboring women. However, higher Ucn levels have been found in women undergoing induction of labor for post-term pregnancy and responding within 12 hours, compared to induced and nonresponding women, reinforcing the hypothesis of a major fetal contribution for this neurohormone in the mechanisms of physiologic and pathologic labor (Torricelli et al., 2006).

Similarly to PTB, maternal plasma and cord blood Ucn levels are higher in women affected by PIH, PE, or PE associated with IUGR when compared with healthy women, while Ucn mRNA expression does not change between pathologic and normal placenta; these data, together with the finding that Ucn levels in arterial cord blood are higher than in venous cord blood and in maternal plasma, again suggest a fetal major source of the peptide in pregnancy-related hypertensive disorders (Florio et al., 2006). Recently, placental mRNA expression of Ucn2 and Ucn3 has been evaluated in relation to PE (Imperatore et al., 2010): all PE placentas appeared to express significantly higher Ucn2 and Ucn3 mRNA compared to controls. Interestingly, early PE samples show stronger immunoreactivity for Ucn2 than for Ucn3, while, vice versa, Ucn3 immunostaining was stronger in late PE samples. Moreover, Ucn2 transcript levels have been shown to increase in placental explants exposed to *in vitro* hypoxia reoxygenation, suggesting that increased placental expression of the peptides may reflect a response to the oxidative stress as well as involvement in the pathogenesis of PE. Interestingly, Ucn has been found to contribute to the pathogenesis of IUGR possibly through negative regulation of placental system A activity, which represents a placental amino acid transporter whose normal activity is fundamental for maintaining fetal growth (Giovannelli et al., 2011).

Ucn peptide has been found to be more concentrated in the product of conceptus in spontaneous abortion than in elective abortion of the same gestational age (Madhappan et al., 2003).

Oxytocin

OT is synthesized by neurons of the supraoptic nucleus and PVN in the hypothalamus and secreted by the posterior pituitary; its major target organs are the pregnant uterus and mammary glands, as it regulates myometrial contractility and milk ejection (Brownstein et al., 1980; Rose et al., 1996). OT and its carrier molecule neurophysin are stored in PVN axon terminals until neural inputs elicit their release (Renaud and Bourque, 1991). It is a classic example of a peptide hormone, produced by hypothalamic neurons and secreted into the general circulation, in particular at the time of parturition and during lactation. Although placental expression and secretion of OT have been shown (Sakai et al., 1993), their contribution to the mechanisms of parturition remains unknown at this stage.

Human decidua expresses greater levels of OT mRNA than amnion, chorion, and trophoblast (Chibbar et al., 1993). Because the placental content of OT is approximately five times greater than in the posterior pituitary lobe, it has been speculated that the placenta might be the main source of OT during pregnancy (Nakazawa et al., 1984). Placental OT secretion is increased by several paracrine factors such as CRH, activin A, and PGs operating within human intrauterine tissues (Florio et al., 1996; Mitchell et al., 1998).

At the onset of labor, the uterine sensitivity to OT markedly increases in association with both an upregulation of OT receptor mRNA levels and a strong increase in the density of myometrial OT receptors, reaching a peak during early labor (Havelock et al., 2005). Since OT infusion can initiate labor and an OT antagonist is effective in delaying threatened preterm labor in several species, an important role for OT in driving parturition is suggested (Chard, 1989; Williams et al., 1998).

OT is responsible for the milk-ejecting activity of the breast gland, and a complementary role in regulating milk production through control of prolactin secretion has been considered (Brunton and Russell, 2010).

Finally, OT within the brain plays a pivotal role in the mutual mother–infant bonding process, although the underlying mechanisms in the brains of infants has not yet been clarified. Central OT has been suggested to play a role in: (1) an affective "calm" response displayed by infants during social contact; (2) paw sucking in neonatal rats; (3) an infant's physical development and subsequent development of adequate attachment behaviors (Nagasawa et al., 2012).

The role of stress in maternal and fetal adverse programming

Stress is a state of threatened homeostasis. Potential threats to pregnancy homeostasis due to stressors occurring from implantation and early pregnancy towards parturition may predispose to increased circulating levels of HPA hormones in both maternal and fetal circulation (Florio et al., 2002b; Vrekoussis et al., 2010).

In the event of acute or chronic metabolic, physical, or infectious stressful conditions, the placenta takes part in a stress syndrome by releasing neurohormones which may help to adequately influence uterine perfusion, maternal metabolism, fluid balance, and possibly uterine contractility, thereby protecting the mother and the fetus from a hostile environment; but it may also itself contribute to the activation of pathologic mechanisms leading to obstetric disorders, such as miscarriage, PTB, PE, and IUGR (Florio et al., 2002b).

In fact, abnormal trophoblast invasion, deficient remodeling of spiral arteries with high-resistance placental vessels, and subsequent placental dysfunction may reflect a state of threatened homeostasis for the feto-placental unit that responds to intrauterine stressors by increasing CRH levels and that may predispose to the development of obstetric hypoxic disorders, such as PE and/or IUGR. Similarly, in the presence of stressful pathologic conditions such as infections, increased placental CRH release may influence premature onset of labor with subsequent PTB, allowing the feto-placental unit to escape an adverse environment (Florio et al., 2002b; Petraglia et al., 2010).

Several studies have confirmed the association between prenatal maternal stress and negative birth outcomes, as well as long-term consequences on the disease risk of the offspring. Increased maternal glucocorticoid levels and maternal exposure to glucocorticoids have been associated with IUGR, insulin resistance, and lifelong deregulation of the HPA axis (Wadhwa, 2005). The effect on the fetus of maternal glucocorticoid secretion is regulated by the placental enzyme 11β-hydroxysteroid-dehydrogenase-type 2 (11β-HSD2), which protects the developing fetus by converting cortisol to the inactive form, cortisone; therefore, increased maternal glucocorticoid levels or decreased 11β-HSD2 activity can lead to increased exposure (Benediktsson et al., 1997). Moreover, several experimental and human studies have demonstrated that a wide spectrum of stressful prenatal stimuli increase the risk for adverse metabolic and neurodevelopmental abnormalities in the offspring, such as the development of insulin resistance and type 2 diabetes, cardiovascular disease, and impaired cognitive development, behavioral problems, autism, and schizophrenia (O'Donnell et al., 2009).

Similarly, the fetus might respond to adverse intrauterine circumstances through increased CRH production and consequent cortisol release. Inappropriate elevations of fetal cortisol may impair fetal growth and predispose to later life diseases such as cardiovascular disease and insulin resistance, a phenomenon called "fetal programming" to which the HPA axis may be highly susceptible during fetal development on the basis of the timing and intensity of the adverse intrauterine stressors (Matthews, 2002). Moreover, experimental and human studies indicate that early life stress induces persistent alterations of gene expression due to epigenetic changes that may lead to several metabolic and neurodevelopmental disorders (Phillips et al., 1998).

CONCLUSIONS

Pregnancy is a complex physiologic condition which also involves the neuroendocrine system. Maternal changes in the hypothalamus–pituitary–adrenal, –gonadal, –GH and –thyroid axes are well-established and are considered critical for the maternal homeostasis necessary to maintain an ongoing pregnancy.

A major support for maternal neuroendocrine changes derives from the placenta. It is now accepted that the placenta should be considered a neuroendocrine organ rich in neurohormones, neuropeptides, and neurosteroids. Their expression and secretion from trophoblasts suggest an involvement in several physiologic mechanisms related to pregnancy maintenance and development.

Stress-related hormones are the most deeply studied for their putative role at parturition. CRH, Ucns, and oxytocin are key placental neuroendocrine factors which mediate both endocrine (metabolism, immune function, cardiovascular changes) and paracrine (uterine contractility, local hormone production) mechanisms involved in term and preterm birth.

However, hormonal changes are not necessarily beneficial, as they may predispose mother and fetus to potentially life-threatening disorders (Brunton and Russell, 2008a, 2010; Petraglia et al., 2010). The excessive/reduced release of some placental neurohormones in association with gestational disease may be part of an adaptive response of the feto-placental unit to adverse environmental conditions (hypertension, hypoxia, infection). In the setting of severe maternal and/or fetal stress elicited by pathologic conditions, placental neurohormones coordinate adaptive changes in uterine perfusion, maternal metabolism, fluid balance, and possibly uterine contractility, contributing to the development of gestational disorders.

References

Abrahams VM, Mor G (2005). Toll-like receptors and their role in the trophoblast. Placenta 26: 540–547.

Agelaki S, Tsatsanis C, Gravanis A et al. (2002). Corticotropin-releasing hormone augments proinflammatory cytokine production from macrophages in vitro and in lipopolysaccharide-induced endotoxin shock in mice. Infect Immun 70: 6068–6074.

Aggelidou E, Hillhouse EW, Grammatopoulos DK (2002). Up-regulation of nitric oxide synthase and modulation of the guanylate cyclase activity by corticotropin-releasing hormone but not urocortin II or urocortin III in cultured human pregnant myometrial cells. Proc Natl Acad Sci U S A 99: 3300–3305.

Agnello D, Bertini R, Sacco S et al. (1998). Corticosteroid-independent inhibition of tumor necrosis factor production by the neuropeptide urocortin. Am J Physiol 275: E757–E762.

Alsat E, Guibourdenche J, Couturier A et al. (1998). Physiological role of human placental growth hormone. Mol Cell Endocrinol 140: 121–127.

Anin SA, Vince G, Quenby S (2004). Trophoblast invasion. Hum Fertil (Camb) 7: 169–174.

Atkinson HC, Waddell BJ (1995). The hypothalamic-pituitary-adrenal axis in rat pregnancy and lactation: circadian variation and interrelationship of plasma adrenocorticotropin and corticosterone. Endocrinology 136: 512–520.

Bamberger AM, Minas V, Kalantaridou SN et al. (2006). Corticotropin-releasing hormone modulates human trophoblast invasion through carcinoembryonic antigen-related cell adhesion molecule-1 regulation. Am J Pathol 168: 141–150.

Benediktsson R, Calder AA, Edwards CR et al. (1997). Placental 11 beta-hydroxysteroid dehydrogenase: a key regulator of fetal glucocorticoid exposure. Clin Endocrinol (Oxf) 46: 161–166.

Ben-Jonathan N, LaPensee CR, LaPensee EW (2008). What can we learn from rodents about prolactin in humans? Endocr Rev 29: 1–41.

Berry SA, Srivastava CH, Rubin LR et al. (1992). Growth hormone-releasing hormone-like messenger ribonucleic acid and immunoreactive peptide are present in human testis and placenta. J Clin Endocrinol Metab 75: 281–284.

Besedovsky HO, del Rey A (1992). Immune-neuroendocrine circuits: integrative role of cytokines. Front Neuroendocrinol 13: 61–94.

Bittencourt JC, Vaughan J, Arias C et al. (1999). Urocortin expression in rat brain: evidence against a pervasive relationship of urocortin-containing projections with targets bearing type 2 CRF receptors. J Comp Neurol 415: 285–312.

Bridges RS, DiBiase R, Loundes DD et al. (1985). Prolactin stimulation of maternal behavior in female rats. Science 227: 782–784.

Brownstein MJ, Russell JT, Gainer H (1980). Synthesis, transport, and release of posterior pituitary hormones. Science 207: 373–378.

Brunton PJ, Russell JA (2008a). The expectant brain: adapting for motherhood. Nat Rev Neurosci 9: 11–25.

Brunton PJ, Russell JA (2008b). Attenuated hypothalamo-pituitary-adrenal axis responses to immune challenge during pregnancy: the neurosteroid opioid connection. J Physiol 586: 369–375.

Brunton PJ, Russell JA (2010). Endocrine induced changes in brain function during pregnancy. Brain Res 1364: 198–215.

Brunton PJ, Russell JA (2011). Neuroendocrine control of maternal stress responses and fetal programming by stress in pregnancy. Prog Neuropsychopharmacol Biol Psychiatry 35: 1178–1191.

Brunton PJ, Meddle SL, Ma S et al. (2005). Endogenous opioids and attenuated hypothalamic-pituitary-adrenal axis responses to immune challenge in pregnant rats. J Neurosci 25: 5117–5126.

Brunton PJ, Sabatier N, Leng G et al. (2006). Suppressed oxytocin neuron responses to immune challenge in late pregnant rats: a role for endogenous opioids. Eur J Neurosci 23: 1241–1247.

Brunton PJ, Russell JA, Douglas AJ (2008a). Adaptive responses of the maternal hypothalamic-pituitary-adrenal axis during pregnancy and lactation. J Neuroendocrinol 20: 764–776.

Brunton PJ, Arunachalam S, Russel JA (2008b). Control of neurohypophysial hormone secretion, blood osmolality and volume in pregnancy. J Physiol Pharmacol 59: 27–45.

Carr BR, Parker Jr CR, Madden JD et al. (1981). Maternal plasma adrenocorticotropin and cortisol relationships throughout human pregnancy. Am J Obstet Gynecol 139: 416–422.

Challis JR, Sloboda D, Matthews SG et al. (2001). The fetal placental hypothalamic-pituitary-adrenal (HPA) axis, parturition and post natal health. Mol Cell Endocrinol 185: 135–144.

Chan EC, Thomson M, Madsen G et al. (1988). Differential processing of corticotrophin-releasing hormone by the human placenta and hypothalamus. Biochem Biophys Res Commun 153: 1229–1235.

Chard T (1989). Fetal and maternal oxytocin in human parturition. Am J Perinatol 6: 145–152.

Chibbar R, Miller FD, Mitchell BF (1993). Synthesis of oxytocin in amnion, chorion, and decidua may influence the timing of human parturition. J Clin Invest 91: 185–192.

Choy MY, Leung TN, Leung PS et al. (2004). Effect of hypoxia on urocortin production in human gestational trophoblasts in vitro. Am J Reprod Immunol 52: 118–123.

Clifton VL, Read MA, Leitch IM et al. (1994). Corticotropin-releasing hormone-induced vasodilatation in the human fetal placental circulation. J Clin Endocrinol Metab 79: 666–669.

Clifton VL, Read MA, Leitch IM et al. (1995). Corticotropin-releasing hormone-induced vasodilatation in the human fetal-placental circulation: involvement of the nitric oxide-cyclic guanosine 3′,5′-monophosphate-mediated pathway. J Clin Endocrinol Metab 80: 2888–2893.

Clifton VL, Read MA, Boura AL et al. (1996). Adrenocorticotropin causes vasodilatation in the human fetal-placental circulation. J Clin Endocrinol Metab 81: 1406–1410.

Clifton VL, Gu Q, Murphy VE et al. (2000). Localization and characterization of urocortin during human pregnancy. Placenta 21: 782–788.

Dallman MF, Akana SF, Scribner KA et al. (1992). Stress, feedback and facilitation in the hypothalamo-pituitary-adrenal axis. J Neuroendocrinol 4: 517–526.

Dashwood MR, Andrews HE, Wei ET (1987). Binding of [125I]Tyr-corticotropin-releasing factor to rabbit aorta is reduced by removal of the endothelium. Eur J Pharmacol 135: 111–112.

Daughaday WH, Trivedi B, Winn HN et al. (1990). Hypersomatotropism in pregnant women, as measured by a human liver radioreceptor assay. J Clin Endocrinol Metab 70: 215–221.

de Weerth C, Buitelaar JK (2005). Physiological stress reactivity in human pregnancy – a review. Neurosci Biobehav Rev 29: 295–312.

de Zegher F, Vanderschueren-Lodeweyckx M, Spitz B et al. (1990). Perinatal growth hormone (GH) physiology: effect of GH-releasing factor on maternal and fetal secretion of pituitary and placental GH. J Clin Endocrinol Metab 71: 520–522.

Dermitzaki E, Tsatsanis C, Gravanis A et al. (2002). Corticotropin-releasing hormone induces Fas ligand production and apoptosis in PC12 cells via activation of p38 mitogen-activated protein kinase. J Biol Chem 277: 12280–12287.

Douglas AJ, Ludwig M (2008). The role of neuroendocrinology in the perinatal period. J Neuroendocrinol 20: 417–418.

Douglas AJ, Brunton PJ, Bosch OJ et al. (2003). Neuroendocrine responses to stress in mice: hyporesponsiveness in pregnancy and parturition. Endocrinology 144: 5268–5276.

Douglas AJ, Meddle SL, Toschi N et al. (2005). Reduced activity of the noradrenergic system in the paraventricular nucleus at the end of pregnancy: implications for stress hyporesponsiveness. J Neuroendocrinol 17: 40–48.

Drago F, Amir S (1984). Effects of hyperprolactinaemia on core temperature of the rat. Brain Res Bull 12: 355–358.

Economides D, Linton E, Nicolaides K et al. (1987). Relationship between maternal and fetal corticotrophin-releasing hormone-41 and ACTH levels in human mid-trimester pregnancy. J Endocrinol 114: 497–501.

Farina A, Chan CW, Chiu RW et al. (2004). Circulating corticotropin-releasing hormone mRNA in maternal plasma: relationship with gestational age and severity of preeclampsia. Clin Chem 50: 1851–1854.

Feldt-Rasmussen U, Mathiesen ER (2011). Endocrine disorders in pregnancy: physiological and hormonal aspects of pregnancy. Best Pract Res Clin Endocrinol Metab 25: 875–884.

Fleming AS, Sarker J (1990). Experience-hormone interactions and maternal behavior in rats. Physiol Behav 47: 1165–1173.

Florio P, Lombardo M, Gallo R et al. (1996). Activin A, corticotropin-releasing factor and prostaglandin F2 alpha increase immunoreactive oxytocin release from cultured human placental cells. Placenta 17: 307–311.

Florio P, Arcuri F, Ciarmela P et al. (2002a). Identification of urocortin mRNA and peptide in the human endometrium. J Endocrinol 173: R9–R14.

Florio P, Severi FM, Ciarmela P et al. (2002b). Placental stress factors and maternal-fetal adaptive response: the corticotropin-releasing factor family. Endocrine 19: 91–102.

Florio P, Imperatore A, Sanseverino F et al. (2004a). The measurement of maternal plasma corticotropin-releasing factor (CRF) and CRF-binding protein improves the early prediction of preeclampsia. J Clin Endocrinol Metab 89: 4673–4677.

Florio P, Vale W, Petraglia F (2004b). Urocortins in human reproduction. Peptides 25: 1751–1757.

Florio P, Torricelli M, Galleri L et al. (2005). High fetal urocortin levels at term and preterm labor. J Clin Endocrinol Metab 90: 5361–5365.

Florio P, Torricelli M, De Falco G et al. (2006). High maternal and fetal plasma urocortin levels in pregnancies complicated by hypertension. J Hypertens 24: 1831–1840.

Francis K, Meddle SL, Bishop VR et al. (2002). Progesterone receptor expression in the pregnant and parturient rat hypothalamus and brainstem. Brain Res 927: 18–26.

Frankenne F, Scippo ML, Van Beeumen J et al. (1990). Identification of placental human growth hormone as the growth hormone-V gene expression product. J Clin Endocrinol Metab 71: 15–18.

Freeman ME, Kanyicska B, Lerant A et al. (2000). Prolactin: structure, function, and regulation of secretion. Physiol Rev 80: 1523–1631.

Gao L, Lu C, Xu C et al. (2008). Differential regulation of prostaglandin production mediated by corticotropin-releasing hormone receptor type 1 and type 2 in cultured human placental trophoblasts. Endocrinology 149: 2866–2876.

Giovannelli A, Greenwood SL, Desforges M et al. (2011). Corticotrophin-releasing factor and urocortin inhibit system A activity in term human placental villous explants. Placenta 32: 99–101.

Gitau R, Fisk NM, Teixeira JM et al. (2001). Fetal hypothalamic-pituitary-adrenal stress responses to invasive procedures are independent of maternal responses. J Clin Endocrinol Metab 86: 104–109.

Glinoer D (1997). The regulation of thyroid function in pregnancy: pathways of endocrine adaptation from physiology to pathology. Endocr Rev 18: 404–433.

Glinoer D, de Nayer P, Bourdoux P et al. (1990). Regulation of maternal thyroid during pregnancy. J Clin Endocrinol Metab 71: 276–287.

Goland RS, Wardlaw SL, Stark RI et al. (1986). High levels of corticotropin-releasing hormone immunoactivity in maternal and fetal plasma during pregnancy. J Clin Endocrinol Metab 63: 1199–1203.

Goland RS, Jozak S, Warren WB et al. (1993). Elevated levels of umbilical cord plasma corticotropin-releasing hormone in growth-retarded fetuses. J Clin Endocrinol Metab 77: 1174–1179.

Goland RS, Conwell IM, Jozak S (1995). The effect of pre-eclampsia on human placental corticotrophin-releasing hormone content and processing. Placenta 16: 375–382.

Grammatopoulos DK (2007). The role of CRH receptors and their agonists in myometrial contractility and quiescence during pregnancy and labour. Front Biosci 12: 561–571.

Grammatopoulos DK (2008). Placental corticotrophin-releasing hormone and its receptors in human pregnancy and labour: still a scientific enigma. J Neuroendocrinol 20: 432–438.

Grammatopoulos DK, Hillhouse EW (1999). Role of corticotropin-releasing hormone in onset of labour. Lancet 354: 1546–1549.

Grammatopoulos D, Stirrat GM, Williams SA et al. (1996). The biological activity of the corticotropin-releasing hormone receptor-adenylate cyclase complex in human myometrium is reduced at the end of pregnancy. J Clin Endocrinol Metab 81: 745–751.

Grattan DR, Kokay IC (2008). Prolactin: a pleiotropic neuroendocrine hormone. J Neuroendocrinol 20: 752–763.

Griesinger G, Saleh L, Bauer S et al. (2001). Production of pro- and anti-inflammatory cytokines of human placental trophoblasts in response to pathogenic bacteria. J Soc Gynecol Investig 8: 334–340.

Gu Q, Clifton VL, Schwartz J et al. (2001). Characterization of urocortin in human pregnancy. Chin Med J (Engl) 114: 618–622.

Guleria I, Pollard JW (2000). The trophoblast is a component of the innate immune system during pregnancy. Nat Med 6: 589–593.

Handwerger S, Freemark M (2000). The roles of placental growth hormone and placental lactogen in the regulation of human fetal growth and development. J Pediatr Endocrinol Metab 13: 343–356.

Harville EW, Savitz DA, Dole N et al. (2007). Patterns of salivary cortisol secretion in pregnancy and implications for assessment protocols. Biol Psychol 74: 85–91.

Havelock JC, Keller P, Muleba N et al. (2005). Human myometrial gene expression before and during parturition. Biol Reprod 72: 707–719.

Herrera E (2000). Metabolic adaptations in pregnancy and their implications for the availability of substrates to the fetus. Eur J Clin Nutr 54: S47–S51.

Hillhouse EW, Randeva H, Ladds G et al. (2002). Corticotropin-releasing hormone receptors. Biochem Soc Trans 30: 428–432.

Hobel CJ, Dunkel-Schetter C, Roesch SC et al. (1999). Maternal plasma corticotropin-releasing hormone associated with stress at 20 weeks' gestation in pregnancies ending in preterm delivery. Am J Obstet Gynecol 180: S257–S263.

Hsu SY, Hsueh AJ (2001). Human stresscopin and stresscopin-related peptide are selective ligands for the type 2 corticotropin-releasing hormone receptor. Nat Med 7: 605–611.

Hull KL, Harvey S (2002). GH as a co-gonadotropin: the relevance of correlative changes in GH secretion and reproductive state. J Endocrinol 172: 1–19.

Ikeda K, Tojo K, Sato S et al. (1998). Urocortin, a newly identified corticotropin-releasing factor-related mammalian peptide, stimulates atrial natriuretic peptide and brain natriuretic peptide secretions from neonatal rat cardiomyocytes. Biochem Biophys Res Commun 250: 298–304.

Imperatore A, Florio P, Torres PB et al. (2006). Urocortin 2 and urocortin 3 are expressed by the human placenta, deciduas, and fetal membranes. Am J Obstet Gynecol 195: 288–295.

Imperatore A, Li W, Petraglia F (2009). Urocortin 2 stimulates estradiol secretion from cultured human placental cells: an effect mediated by the type 2 corticotrophin-releasing hormone (CRH) receptor. Reprod Sci 16: 551–558.

Imperatore A, Rolfo A, Petraglia F et al. (2010). Hypoxia and preeclampsia: increased expression of urocortin 2 and urocortin 3. Reprod Sci 17: 833–843.

Jain V, Shi SQ, Vedernikov YP et al. (1998). In vivo effects of corticotropin-releasing factor in pregnant rats. Am J Obstet Gynecol 178: 186–191.

Joachim R, Zenclussen AC, Polgar B et al. (2003). The progesterone derivative dydrogesterone abrogates murine stress-triggered abortion by inducing a Th2 biased local immune response. Steroids 68: 931–940.

Johnstone HA, Wigger A, Douglas AJ et al. (2000). Attenuation of hypothalamic-pituitary-adrenal axis stress responses in late pregnancy: changes in feedforward and feedback mechanisms. J Neuroendocrinol 12: 811–822.

Jones SA, Challis JR (1990). Effects of corticotropin-releasing hormone and adrenocorticotropin on prostaglandin output by human placenta and fetal membranes. Gynecol Obstet Invest 29: 165–168.

Kalantaridou SN, Zoumakis E, Makrigiannakis A et al. (2010). Corticotropin-releasing hormone, stress and human reproduction: an update. J Reprod Immunol 85: 33–39.

Kammerer M, Adams D, Castelberg Bv et al. (2002). Pregnant women become insensitive to cold stress. BMC Pregnancy Childbirth 2: 8.

Karteris E, Hillhouse EW, Grammatopoulos D (2004). Urocortin II is expressed in human pregnant myometrial cells and regulates myosin light chain phosphorylation: potential role of the type-2 corticotropin-releasing hormone receptor in the control of myometrial contractility. Endocrinology 145: 890–900.

Kimura Y, Takahashi K, Totsune K et al. (2002). Expression of urocortin and corticotropin-releasing factor receptor subtypes in the human heart. J Clin Endocrinol Metab 87: 340–346.

Korebrits C, Ramirez MM, Watson L et al. (1998). Maternal corticotropin-releasing hormone is increased with impending preterm birth. J Clin Endocrinol Metab 83: 1585–1591.

Kumazaki K, Nakayama M, Yanagihara I et al. (2004). Immunohistochemical distribution of Toll-like receptor 4 in term and preterm human placentas from normal and complicated pregnancy including chorioamnionitis. Hum Pathol 35: 47–54.

Laatikainen T, Virtanen T, Kaaja R et al. (1991). Corticotropin-releasing hormone in maternal and cord plasma in pre-eclampsia. Eur J Obstet Gynecol Reprod Biol 39: 19–24.

Larsen CM, Grattan DR (2012). Prolactin, neurogenesis, and maternal behaviors. Brain Behav Immun 26: 201–209.

Lee HJ, Snegovskikh VV, Park JS et al. (2010). Role of GnRH-GnRH receptor signaling at the maternal-fetal interface. Fertil Steril 94: 2680–2687.

Li W, Challis JR (2005). Corticotropin-releasing hormone and urocortin induce secretion of matrix metalloproteinase-9 (MMP-9) without change in tissue inhibitors of MMP-1 by cultured cells from human placenta and fetal membranes. J Clin Endocrinol Metab 90: 6569–6574.

Liapi CA, Tsakalia DE, Panitsa-Faflia CC et al. (1996). Corticotropin-releasing-hormone levels in pregnancy-induced hypertension. Eur J Obstet Gynecol Reprod Biol 68: 109–114.

Ma S, Shipston MJ, Morilak D et al. (2005). Reduced hypothalamic vasopressin secretion underlies attenuated adrenocorticotropin stress responses in pregnant rats. Endocrinology 146: 1626–1637.

Madhappan B, Kempuraj D, Christodoulou S et al. (2003). High levels of intrauterine corticotropin-releasing hormone, urocortin, tryptase, and interleukin-8 in spontaneous abortions. Endocrinology 144: 2285–2290.

Makrigiannakis A, Zoumakis E, Kalantaridou S et al. (2004). Participation of maternal and fetal CRH in early phases of human implantation: the role of antalarmin. Curr Drug Targets Immune Endocr Metabol Disord 4: 75–78.

Mann PE, Babb JA (2005). Neural steroid hormone receptor gene expression in pregnant rats. Brain Res Mol Brain Res 142: 39–46.

Markovic D, Vatish M, Gu M et al. (2007). The onset of labor alters corticotropin-releasing hormone type 1 receptor variant expression in human myometrium: putative role of interleukin-1beta. Endocrinology 148: 3205–3213.

Maser-Gluth C, Lorenz U, Vecsei P (1987). In pregnancy, corticotropin-releasing-factor in maternal blood and amniotic fluid correlates with the gestational age. Horm Metab Res Suppl 16: 42–44.

Mastorakos G, Ilias I (2003). Maternal and fetal hypothalamic-pituitary-adrenal axes during pregnancy and postpartum. Ann N Y Acad Sci 997: 136–149.

Matalka KZ, Ali DA (2005). Stress-induced versus preovulatory and pregnancy hormonal levels in modulating cytokine production following whole blood stimulation. Neuroimmunomodulation 12: 366–374.

Matthews SG (2002). Early programming of the hypothalamo-pituitary-adrenal axis. Trends Endocrinol Metab 13: 373–380.

McIntyre HD, Serek R, Crane DI et al. (2000). Placental growth hormone (GH), GH-binding protein, and insulin-like growth factor axis in normal, growth-retarded, and diabetic pregnancies: correlations with fetal growth. J Clin Endocrinol Metab 85: 1143–1150.

McLean M, Bisits A, Davies J et al. (1995). A placental clock controlling the length of human pregnancy. Nat Med 1: 460–463.

Mesiano S, Jaffe RB (1997). Developmental and functional biology of the primate fetal adrenal cortex. Endocr Rev 18: 378–403.

Mignot TM, Paris B, Carbonne B et al. (2005). Corticotropin-releasing hormone effects on human pregnant vs. nonpregnant myometrium explants estimated from a mathematical model of uterine contraction. J Appl Physiol 99: 1157–1163.

Minas V, Jeschke U, Kalantaridou SN (2007). Abortion is associated with increased expression of FasL in decidual leukocytes and apoptosis of extravillous trophoblasts: a role for CRH and urocortin. Mol Hum Reprod 13: 663–673.

Mitchell BF, Fang X, Wong S (1998). Oxytocin: a paracrine hormone in the regulation of parturition? Rev Reprod 3: 113–122.

Miyake A, Sakumoto T, Aono T et al. (1982). Changes in luteinizing hormone-releasing hormone in human placenta throughout pregnancy. Obstet Gynecol 60: 444–449.

Mizoguchi Y, Yamaguchi H, Aoki F et al. (1997). Corticosterone is required for the prolactin receptor gene expression in the late pregnant mouse mammary gland. Mol Cell Endocrinol 132: 177–183.

Mori M, Amino N, Tamaki H et al. (1988). Morning sickness and thyroid function in normal pregnancy. Obstet Gynecol 72: 355–359.

Muramatsu Y, Fukushima K, Iino K et al. (2000). Urocortin and corticotropin-releasing factor receptor expression in the human colonic mucosa. Peptides 21: 1799–1809.

Muramatsu Y, Sugino N, Suzuki T et al. (2001). Urocortin and corticotropin-releasing factor receptor expression in normal cycling human ovaries. J Clin Endocrinol Metab 86: 1362–1369.

Naef L, Woodside B (2007). Prolactin / leptin interactions in the control of food intake in rats. Endocrinology 148: 5977–5983.

Nagasawa M, Okabe S, Mogi K et al. (2012). Oxytocin and mutual communication in mother-infant bonding. Front Hum Neurosci 6: 31.

Nagashima K, Yagi H, Yunoki H et al. (1987). Cord blood levels of corticotropin-releasing factor. Biol Neonate 51: 1–4.

Nakamura H, Seto T, Nagase H et al. (1997). Inhibitory effect of pregnancy on stress-induced immunosuppression through corticotropin releasing hormone (CRH) and dopaminergic systems. J Neuroimmunol 75: 1–8.

Nakamura K, Sheps S, Arck PC (2008). Stress and reproductive failure: past notions, present insights and future directions. J Assist Reprod Genet 25: 47–62.

Nakazawa K, Makino T, Iizuka R (1984). Immunohistochemical study on oxytocin-like substance in the human placenta. Endocrinol Jpn 31: 763–768.

Nepomnaschy PA, Welch KB, McConnell DS et al. (2006). Cortisol levels and very early pregnancy loss in humans. Proc Natl Acad Sci U S A 103: 3938–3942.

Neumann ID, Johnstone HA, Hatzinger M et al. (1998). Attenuated neuroendocrine responses to emotional and physical stressors in pregnant rats involve adenohypophysial changes. J Physiol 508: 289–300.

Ni X, Nicholson RC, King BR et al. (2002). Estrogen represses whereas the estrogen-antagonist ICI 182780 stimulates placental CRH gene expression. J Clin Endocrinol Metab 87: 3774–3778.

Nogués N, Del Río JA, Pérez-Riba M et al. (1997). Placenta-specific expression of the rat growth hormone-releasing hormone gene promoter in transgenic mice. Endocrinology 138: 3222–3227.

Nohara A, Ohmichi M, Koike K et al. (1996). The role of mitogen-activated protein kinase in oxytocin-induced contraction of uterine smooth muscle in pregnant rat. Biochem Biophys Res Commun 229: 938–944.

Novembri R, Torricelli M, Bloise E et al. (2011). Effects of urocortin 2 and urocortin 3 on IL-10 and TNF-α expression and secretion from human trophoblast explants. Placenta 32: 969–974.

O'Donnell K, O'Connor TG, Glover V (2009). Prenatal stress and neurodevelopment of the child: focus on the HPA axis and role of the placenta. Dev Neurosci 31: 285–292.

Obel C, Hedegaard M, Henriksen TB et al. (2005). Stress and salivary cortisol during pregnancy. Psychoneuroendocrinology 30: 647–656.

Ochedalski T, Zylińska K, Laudański T et al. (2001). Corticotrophin-releasing hormone and ACTH levels in maternal and fetal blood during spontaneous and oxytocin-induced labour. Eur J Endocrinol 144: 117–121.

Ohmichi M, Koike K, Kimura A et al. (1997). Role of mitogen-activated protein kinase pathway in prostaglandin F2alpha-induced rat puerperal uterine contraction. Endocrinology 138: 3103–3111.

Patel N, Alsat E, Igout A et al. (1995). Glucose inhibits human placental GH secretion, in vitro. J Clin Endocrinol Metab 80: 1743–1746.

Pepe GJ, Albrecht ED (1995). Actions of placental and fetal adrenal steroid hormones in primate pregnancy. Endocr Rev 16: 608–648.

Petraglia F, Florio P (2001). Stress hormones, human pregnancy and parturition. Stress 4: 217–218.

Petraglia F, Sawchenko PE, Rivier J et al. (1987). Evidence for local stimulation of ACTH secretion by corticotropin-releasing factor in human placenta. Nature 328: 717–719.

Petraglia F, Sutton S, Vale W (1989). Neurotransmitters and peptides modulate the release of immunoreactive corticotropin-releasing factor from cultured human placental cells. Am J Obstet Gynecol 160: 247–251.

Petraglia F, Volpe A, Genazzani AR et al. (1990a). Neuroendocrinology of the human placenta. Front Neuroendocrinol 11: 6–37.

Petraglia F, Calzà L, Garuti GC et al. (1990b). New aspects of placental endocrinology. J Endocrinol Invest 13: 353–371.

Petraglia F, Garuti GC, De Ramundo B et al. (1990c). Mechanism of action of interleukin-1 beta in increasing corticotropin-releasing factor and adrenocorticotropin hormone release from cultured human placental cells. Am J Obstet Gynecol 163: 1307–1312.

Petraglia F, Giardino L, Coukos G et al. (1990d). Corticotropin-releasing factor and parturition: plasma and amniotic fluid levels and placental binding sites. Obstet Gynecol 75: 784–789.

Petraglia F, Aguzzoli L, Florio P et al. (1995). Maternal plasma and placental immunoreactive corticotrophin-releasing factor concentrations in infection-associated term and pre-term delivery. Placenta 16: 157–164.

Petraglia F, Florio P, Nappi C et al. (1996a). Peptide signaling in human placenta and membranes: autocrine, paracrine, and endocrine mechanisms. Endocr Rev 17: 156–186.

Petraglia F, Florio P, Gallo R et al. (1996b). Human placenta and fetal membranes express human urocortin mRNA and peptide. J Clin Endocrinol Metab 81: 3807–3810.

Petraglia F, Florio P, Benedetto C et al. (1996c). High levels of corticotropin-releasing factor (CRF) are inversely correlated with low levels of maternal CRF-binding protein in pregnant women with pregnancy-induced hypertension. J Clin Endocrinol Metab 81: 852–856.

Petraglia F, Florio P, Benedetto C et al. (1999). Urocortin stimulates placental adrenocorticotropin and prostaglandin release and myometrial contractility in vitro. J Clin Endocrinol Metab 84: 1420–1423.

Petraglia F, Imperatore A, Challis JR (2010). Neuroendocrine mechanisms in pregnancy and parturition. Endocr Rev 31: 783–816.

Phillips DI, Barker DJ, Fall CH et al. (1998). Elevated plasma cortisol concentrations: a link between low birth weight and the insulin resistance syndrome? J Clin Endocrinol Metab 83: 757–760.

Pi XJ, Grattan DR (1999). Increased prolactin receptor immunoreactivity in the hypothalamus of lactating rats. J Neuroendocrinol 11: 693–705.

Poliak S, Mor F, Conlon P et al. (1997). Stress and autoimmunity: the neuropeptides corticotropin-releasing factor and urocortin suppress encephalomyelitis via effects on both the hypothalamic-pituitary-adrenal axis and the immune system. J Immunol 158: 5751–5756.

Rainey WE, Rehman KS, Carr BR (2004). The human fetal adrenal: making adrenal androgens for placental estrogens. Semin Reprod Med 22: 327–336.

Reis FM, Florio P, Cobellis L et al. (2001). Human placenta as a source of neuroendocrine factors. Biol Neonate 79: 150–156.

Renaud LP, Bourque CW (1991). Neurophysiology and neuropharmacology of hypothalamic magnocellular neurons secreting vasopressin and oxytocin. Prog Neurobiol 36: 131–169.

Riley SC, Walton JC, Herlick JM et al. (1991). The localization and distribution of corticotropin-releasing hormone in the human placenta and fetal membranes throughout gestation. J Clin Endocrinol Metab 72: 1001–1007.

Robinson BG, Emanuel RL, Frim DM et al. (1988). Glucocorticoid stimulates expression of corticotropin-releasing hormone gene in human placenta. Proc Natl Acad Sci U S A 85: 5244–5248.

Rose JP, Wu CK, Hsiao CD et al. (1996). Crystal structure of the neurophysin-oxytocin complex. Nat Struct Biol 3: 163–169.

Russell JA (2000). Neuroendocrinology briefings 12. The maternal brain. J Neuroendocrinol 12: 1143–1144.

Russell JA, Douglas AJ, Ingram CD (2001). Brain preparations for maternity – adaptive changes in behavioral and neuroendocrine systems during pregnancy and lactation. An overview. Prog Brain Res 133: 1–38.

Saijonmaa O, Laatikainen T, Wahlström T (1988). Corticotrophin-releasing factor in human placenta: localization, concentration and release in vitro. Placenta 9: 373–385.

Sakai A, Makino T, Kato Y et al. (1993). A uterine muscle contractile substance obtained from a human placental cDNA library. Gynecol Endocrinol 7: 77–82.

Sasaki K, Norwitz ER (2011). Gonadotropin-releasing hormone/gonadotropin-releasing hormone receptor signaling in the placenta. Curr Opin Endocrinol Diabetes Obes 18: 401–408.

Sasaki A, Shinkawa O, Margioris AN et al. (1987). Immunoreactive corticotropin-releasing hormone in human plasma during pregnancy, labor, and delivery. J Clin Endocrinol Metab 64: 224–229.

Sasaki A, Tempst P, Liotta AS et al. (1988). Isolation and characterization of a corticotropin-releasing hormone-like peptide from human placenta. J Clin Endocrinol Metab 67: 768–773.

Sassolas G (2000). Growth hormone-releasing hormone: past and present. Horm Res 53: 88–92.

Schulte HM, Weisner D, Allolio B (1990). The corticotrophin releasing hormone test in late pregnancy: lack of adrenocorticotrophin and cortisol response. Clin Endocrinol (Oxf) 33: 99–106.

Seres J, Bornstein SR, Seres P et al. (2004). Corticotropin-releasing hormone system in human adipose tissue. J Clin Endocrinol Metab 89: 965–970.

Shingo T, Gregg C, Enwere E et al. (2003). Pregnancy-stimulated neurogenesis in the adult female forebrain mediated by prolactin. Science 299: 117–120.

Siler-Khodr TM, Khodr GS, Valenzuela G (1984). Immunoreactive gonadotropin-releasing hormone level in maternal circulation throughout pregnancy. Am J Obstet Gynecol 150: 376–379.

Simoncini T, Apa R, Reis FM et al. (1999). Human umbilical vein endothelial cells: a new source and potential target for corticotropin-releasing factor. J Clin Endocrinol Metab 84: 2802–2806.

Sirianni R, Rehman KS, Carr BR et al. (2005). Corticotropin-releasing hormone directly stimulates cortisol and the cortisol biosynthetic pathway in human fetal adrenal cells. J Clin Endocrinol Metab 90: 279–285.

Skelton KH, Owens MJ, Nemeroff CB (2000). The neurobiology of urocortin. Regul Pept 93: 85–92.

Slominski A, Wortsman J (2000). Neuroendocrinology of the skin. Endocr Rev 21: 457–487.

Smith R (2007). Parturition. N Engl J Med 356: 271–283.

Stevens MY, Challis JR, Lye SJ (1998). Corticotropin-releasing hormone receptor subtype 1 is significantly up-regulated at the time of labor in the human myometrium. J Clin Endocrinol Metab 83: 4107–4115.

Stojilkovic SS, Reinhart J, Catt KJ (1994). Gonadotropin-releasing hormone receptors: structure and signal transduction pathways. Endocr Rev 15: 462–499.

Tanaka Y, Makino S, Noguchi T et al. (2003). Effect of stress and adrenalectomy on urocortin II mRNA expression in the hypothalamic paraventricular nucleus of the rat. Neuroendocrinology 78: 1–11.

Torner L, Neumann ID (2002). The brain prolactin system: involvement in stress response adaptations in lactation. Stress 5: 249–257.

Torricelli M, Ignacchiti E, Giovannelli A et al. (2006). Maternal plasma corticotrophin-releasing factor and urocortin levels in post-term pregnancies. Eur J Endocrinol 154: 281–285.

Torricelli M, Giovannelli A, Leucci E et al. (2007). Labor (term and preterm) is associated with changes in the placental mRNA expression of corticotrophin-releasing factor. Reprod Sci 14: 241–245.

Torricelli M, Voltolini C, Bloise E et al. (2009a). Urocortin increases IL-4 and IL-10 secretion and reverses LPS-induced TNF-alpha release from human trophoblast primary cells. Am J Reprod Immunol 62: 224–231.

Torricelli M, Voltolini C, Biliotti G et al. (2009b). Urocortin in amniotic fluid and Down syndrome. Prenat Diagn 29: 806–807.

Torricelli M, Novembri R, Bloise E et al. (2011). Changes in placental CRH, urocortins, and CRH-receptor mRNA expression associated with preterm delivery and chorioamnionitis. J Clin Endocrinol Metab 96: 534–540.

Torricelli M, Voltolini C, De Bonis M et al. (2012). The identification of high risk pregnancy: a new challenge in obstetrics. J Matern Fetal Neonatal Med 25: 2–5.

Tropper PJ, Goland RS, Wardlaw SL et al. (1987). Effects of betamethasone on maternal plasma corticotropin releasing factor, ACTH and cortisol during pregnancy. J Perinat Med 15: 221–225.

Tropper PJ, Warren WB, Jozak SM et al. (1992). Corticotropin releasing hormone concentrations in umbilical cord blood of preterm fetuses. J Dev Physiol 18: 81–85.

Trudinger BJ, Giles WB, Cook CM et al. (1985). Fetal umbilical artery flow velocity waveforms and placental resistance: clinical significance. Br J Obstet Gynaecol 92: 23–30.

Tsatsanis C, Androulidaki A, Alissafi T et al. (2006). Corticotropin-releasing factor and the urocortins induce the expression of TLR4 in macrophages via activation of the transcription factors PU.1 and AP-1. J Immunol 176: 1869–1877.

Uzuki M, Sasano H, Muramatsu Y et al. (2001). Urocortin in the synovial tissue of patients with rheumatoid arthritis. Clin Sci (Lond) 100: 577–589.

Vale W, Spiess J, Rivier C et al. (1981). Characterization of a 41-residue ovine hypothalamic peptide that stimulates secretion of corticotropin and beta-endorphin. Science 213: 1394–1397.

van den Brûle F, Berndt S, Simon N et al. (2005). Trophoblast invasion and placentation: molecular mechanisms and regulation. Chem Immunol Allergy 88: 163–180.

Vaughan J, Donaldson C, Bittencourt J et al. (1995). Urocortin, a mammalian neuropeptide related to fish urotensin I and to corticotropin-releasing factor. Nature 378: 287–292.

Venihaki M, Sakihara S, Subramanian S et al. (2004). Urocortin III, a brain neuropeptide of the corticotropin-releasing hormone family: modulation by stress and attenuation of some anxiety-like behaviours. J Neuroendocrinol 16: 411–422.

Vitoratos N, Papatheodorou DC, Kalantaridou SN et al. (2006). "Reproductive" corticotropin-releasing hormone. Ann N Y Acad Sci 1092: 310–318.

Vrekoussis T, Kalantaridou SN, Mastorakos G et al. (2010). The role of stress in female reproduction and pregnancy: an update. Ann N Y Acad Sci 1205: 69–75.

Waddell BJ (1993). The placenta as hypothalamus and pituitary: possible impact on maternal and fetal adrenal function. Reprod Fertil Dev 5: 479–497.

Wadhwa PD (2005). Psychoneuroendocrine processes in human pregnancy influence fetal development and health. Psychoneuroendocrinology 30: 724–743.

Wagner CK, Morrell JI (1996). Levels of estrogen receptor immunoreactivity are altered in behaviorally-relevant brain regions in female rats during pregnancy. Brain Res Mol Brain Res 42: 328–336.

Wang MJ, Lin SZ, Kuo JS et al. (2007). Urocortin modulates inflammatory response and neurotoxicity induced by microglial activation. J Immunol 179: 6204–6214.

Warren WB, Patrick SL, Goland RS (1992). Elevated maternal plasma corticotropin-releasing hormone levels in pregnancies complicated by preterm labor. Am J Obstet Gynecol 166: 1198–1207.

Welberg LA, Seckl JR (2001). Prenatal stress, glucocorticoids and the programming of the brain. J Neuroendocrinol 13: 113–128.

Weninger SC, Dunn AJ, Muglia LJ et al. (1999). Stress-induced behaviors require the corticotropin-releasing hormone (CRH) receptor, but not CRH. Proc Natl Acad Sci U S A 96: 8283–8288.

Williams GR (2008). Neurodevelopmental and neurophysiological actions of thyroid hormone. J Neuroendocrinol 20: 784–794.

Williams PD, Bock MG, Evans BE et al. (1998). Progress in the development of oxytocin antagonists for use in preterm labor. Adv Exp Med Biol 449: 473–479.

Wolfe CD, Patel SP, Linton EA et al. (1988). Plasma corticotrophin-releasing factor (CRF) in abnormal pregnancy. Br J Obstet Gynaecol 95: 1003–1006.

Wood CE (2005). Estrogen/hypothalamus-pituitary-adrenal axis interactions in the fetus: the interplay between placenta and fetal brain. J Soc Gynecol Investig 12: 67–76.

Zoumakis E, Margioris AN, Stournaras C et al. (2000). Corticotrophin-releasing hormone (CRH) interacts with inflammatory prostaglandins and interleukins and affects the decidualization of human endometrial stroma. Mol Hum Reprod 6: 344–351.

Zoumakis E, Kalantaridou SN, Makrigiannakis A (2009). CRH-like peptides in human reproduction. Curr Med Chem 16: 4230–4235.

Handbook of Clinical Neurology, Vol. 124 (3rd series)
Clinical Neuroendocrinology
E. Fliers, M. Korbonits, and J.A. Romijn, Editors

Chapter 3

Disorders of water metabolism: diabetes insipidus and the syndrome of inappropriate antidiuretic hormone secretion

JOSEPH G. VERBALIS*
Georgetown University, Washington, DC, USA

Disorders of body fluids are among the most commonly encountered problems in the practice of clinical medicine. This is in large part because many different disease states can potentially disrupt the finely balanced mechanisms that control the intake and output of water and solute. Since body water is the primary determinant of the osmolality of the extracellular fluid (ECF), disorders of body water homeostasis can be broadly divided into *hypo-osmolar disorders*, in which there is an excess of body water relative to body solute, and *hyperosmolar disorders*, in which there is a deficiency of body water relative to body solute. Because sodium is the main constituent of plasma osmolality, these disorders are typically characterized by *hyponatremia* and *hypernatremia*, respectively. Before discussing these disorders, this chapter will briefly review the regulatory mechanisms underlying water and sodium metabolism, the two major determinants of body fluid homeostasis.

WATER METABOLISM

Water metabolism represents a balance between the intake and excretion of water. Each side of this balance equation can be considered to consist of a *"regulated"* and an *"unregulated"* component, the magnitudes of which can vary quite markedly under different physiologic and pathophysiologic conditions. The unregulated component of water intake consists of the intrinsic water content of ingested foods, the consumption of beverages primarily for reasons of palatability or desired secondary effects (e.g., caffeine), or for social or habitual reasons (e.g., alcoholic beverages), whereas the regulated component of water intake consists of fluids consumed in response to a perceived sensation of thirst. Similarly, the unregulated component of water excretion occurs via insensible water losses from a variety of sources (cutaneous losses from sweating, evaporative losses in exhaled air, gastrointestinal losses) as well as the obligate amount of water that the kidneys must excrete to eliminate solutes generated by body metabolism, whereas the regulated component of water excretion is composed of the renal excretion of free water in excess of the obligate amount necessary to excrete metabolic solutes (Verbalis, 1997b). In effect, the regulated components are those that act to maintain water balance by compensating for whatever perturbations result from unregulated water losses or gains. Within this framework, it is clear that the two major mechanisms responsible for regulating water metabolism are thirst and pituitary secretion of the hormone vasopressin.

Thirst

Thirst is the body's defense mechanism to increase water consumption in response to perceived deficits of body fluids. Thirst can be stimulated in animals and man either by intracellular dehydration caused by increases in the effective osmolality of the ECF, or by intravascular hypovolemia caused by losses of ECF. Substantial evidence to date has supported mediation of the former by osmoreceptors located in the anterior hypothalamus of the brain, whereas the latter appears to be stimulated primarily via activation of low- and/or high-pressure baroreceptors, with a likely contribution from circulating angiotensin II during more severe degrees of intravascular hypovolemia and hypotension (Fitzsimons, 1992; Stricker and Verbalis, 2013). Controlled studies in animals have consistently reported thresholds for osmotically induced drinking ranging from 1% to 4% increases in plasma osmolality above basal levels; analogous studies in humans using quantitative estimates of subjective symptoms of thirst have confirmed that increases in plasma

*Correspondence to: Dr. Joseph G. Verbalis, 232 Building D, Georgetown University, 4000 Reservoir Rd. NW, Washington, DC 20007, USA. Tel: +1-202-687-2818, Fax: +1-202-687-2040, E-mail: verbalis@georgetown.edu

osmolality of similar magnitudes are necessary to produce an unequivocal sensation described as "thirst" (Robertson, 1983; Thompson et al., 1986).

Conversely, the threshold for producing hypovolemic, or extracellular, thirst is significantly greater in both animals and humans. Studies in several species have shown that sustained decreases in plasma volume or blood pressure of at least 4–8%, and in some species 10–15%, are necessary to consistently stimulate drinking. In humans, it has been difficult to demonstrate any effects of mild to moderate hypovolemia to stimulate thirst independently of osmotic changes occurring with dehydration. This blunted sensitivity to changes in extracellular fluid volume or blood pressure in humans probably represents an adaptation that occurred as a result of the erect posture of primates, which predisposes them to wider fluctuations in blood and atrial filling pressures as a result of orthostatic pooling of blood in the lower body; stimulation of thirst (and vasopressin secretion) by such transient postural changes in blood pressure might lead to overdrinking and inappropriate antidiuresis in situations where the ECF volume was actually normal but only transiently maldistributed. Consistent with a blunted response to baroreceptor activation, studies have also shown that systemic infusion of angiotensin II to pharmacologic levels is a much less potent stimulus to thirst in humans than in animals (Phillips et al., 1985). Nonetheless, this response is not completely absent in humans, as demonstrated by rare cases of polydipsia in patients with pathologic causes of hyper-reninemia.

Although osmotic changes clearly are more effective stimulants of thirst than are volume changes in humans, it is not clear whether relatively small changes in plasma osmolality are responsible for day-to-day fluid intakes. Most humans consume the majority of their ingested water as a result of the unregulated components of fluid intake discussed previously, and generally ingest volumes in excess of what can be considered to be actual "need" (de Castro, 1988). Consistent with this observation is the fact that under most conditions plasma osmolalities in man remain within 1–2% of basal levels, and these relatively small changes in plasma osmolality are generally below the threshold levels that have been found to stimulate thirst in most individuals. This suggests that despite the obvious vital importance of thirst during pathologic situations of hyperosmolality and hypovolemia, under normal physiologic conditions water balance in man is accomplished more by regulated free water excretion than by regulated water intake (Verbalis, 1997b).

Vasopressin secretion

The primary physiologic action of *arginine vasopressin (AVP)* is its function as a water-retaining hormone. The central sensing system (*osmostat*) for control of release of AVP is located in the hypothalamus anterior to the third ventricle that also includes the circumventricular organ, the *organum vasculosum of the lamina terminalis (OVLT)*. The osmostat controls release of AVP to cause water retention, and also stimulates thirst to cause water repletion (Stricker and Verbalis, 2013). Osmotic regulation of AVP release and thirst are usually closely coupled, but experimental lesions and some pathologic situations in humans demonstrate that each can be regulated independently (Baylis and Thompson, 1988). The primary extracellular osmolyte to which the osmoreceptor responds is sodium. Under normal physiologic conditions, glucose and urea readily traverse neuron cell membranes and do not stimulate release of AVP. Basal osmolality in normal subjects lies between 280 and 295 mOsm/kg H_2O, but for each individual osmolality is maintained within narrow ranges. Increases in plasma osmolality of as little as 1–2% will stimulate the osmoreceptors to release AVP. Basal plasma levels of AVP are 0.5–2 pg/mL, which are sufficient to maintain urine osmolality above plasma osmolality and urine volume in the range of 2–3 L/day. When AVP levels are suppressed below 0.5 pg/mL, maximum urine osmolality decreases to less than 100 mOsm/kg H_2O and a free water diuresis ensues to levels approaching 800–1000 mL/hour (18–24 L/day). Increases in plasma osmolality cause a linear increase in plasma AVP and a corresponding linear increase in urine osmolality (Robertson, 1976). At a plasma osmolality of approximately 295 mOsm/kg H_2O, urine osmolality is maximally concentrated to 1000–1200 mOsm/kg H_2O. Thus, the entire physiologic range of urine concentration is accomplished by relatively small changes in plasma AVP of 0–5 pg/mL (Robinson and Verbalis, 2011).

AVP secretion is also stimulated by low blood volume and pressure. High-pressure baroreceptors are located in the aorta and carotid sinus, and low-pressure baroreceptors are located in the right and left atria. Stimuli for pressure and volume receptors are carried via the glossopharyngeal (ninth) and vagal (tenth) cranial nerves to the *nucleus tractus solitarius* in the brainstem. Subsequent secondary and tertiary projections converge on the magnocellular neurons, where they provide inhibitory as well as excitatory inputs. Decreases in blood pressure or vascular volume stimulate AVP release, whereas situations that increase blood volume or left atrial pressure (e.g., negative-pressure breathing) decrease secretion of AVP. The release of AVP in response to changes in volume or pressure is less sensitive than the release in response to osmoreceptors, and generally a 10–15% reduction in blood volume or pressure is needed to stimulate release of AVP. However, once arterial pressure falls below this threshold, the stimulated response is exponential and plasma levels of AVP achieved that are markedly greater than those achieved by osmotic stimulation (Robertson, 1976). Other nonosmotic stimuli, such as nausea and intestinal traction, also act through similar nonosmotic neural pathways to release AVP.

Vasopressin actions

Three known receptor subtypes mediate the actions of AVP. They all are classic G protein-coupled receptors with seven transmembrane domains, and are classified according to the second messenger system to which they are coupled (Thibonnier et al., 1998). The AVP V1a (V1aR) and V1b (V1bR) receptors are linked to the phosphoinositol signaling pathway via Gaq/11 GTP binding proteins that activate phospholipase C activity, with intracellular calcium acting as the second messenger. V1aR are present on vascular smooth muscle cells, hepatocytes, and platelets, and mediate the well-known pressor effects of AVP on peripheral resistance and blood pressure. V1bR are found predominately on corticotroph cells of the anterior pituitary, where they mediate corticotropin (ACTH) release in concert with the well-known effects of corticotropin-releasing hormone (CRH). V2R, or antidiuretic receptors, are mainly localized in the collecting duct cells of the kidney where they regulate water excretion. V2R are G protein-coupled receptors that activate adenylyl cyclase with subsequent increased intracellular cyclic AMP levels upon ligand activation. The increased cAMP initiates the movement of *aquaporin-2 (AQP2)* water channels from the cytoplasm to the apical (luminal) membrane of the collecting duct cells. Once inserted into the apical membrane, these channels allow facilitated rapid transport of water from the collecting duct lumen into the cell along osmotic gradients (Knepper, 1997). The water then exits the cell through the basolateral membrane and into the kidney medullary circulation via aquaporin-3 and aquaporin-4 water channels, which are constitutively present in the basolateral membrane. This entire process is termed *antidiuresis*. In the absence of AVP, the AQP2 channels are re-internalized from the apical membrane into subapical vesicles. This prevents active reabsorption of water from the collecting duct lumen, resulting in *diuresis*. In addition to this rapid "shuttling" of the AQP2 water channels to regulate water reabsorption on a minute-to-minute basis, AVP also acts via V2R to regulate long-term stores of AQP2; i.e., increased AVP stimulates AQP2 synthesis and the absence of AVP results in decreased AQP2 synthesis (Knepper, 1998). The hypertonic medullary interstitium determines the maximum concentration of the final urine, which is isotonic with the inner medulla of the kidney under conditions of maximal antidiuresis (Knepper, 1997; Nielsen et al., 2002).

Integration of thirst and AVP secretion

A synthesis of what is presently known about the regulation of thirst and AVP secretion in man leads to a relatively simple but elegant system to maintain water balance (Verbalis, 1997b; Stricker and Verbalis, 2013). Under normal physiologic conditions, the sensitivity of the osmoregulatory system for AVP secretion accounts for maintenance of plasma osmolality within narrow limits by adjusting renal water excretion to small changes in osmolality. Stimulated thirst does not represent a major regulatory mechanism under these conditions, and unregulated fluid ingestion supplies adequate water in excess of true "need," which is then excreted in relation to osmoregulated pituitary AVP secretion. However, when unregulated water intake cannot adequately supply body needs in the presence of plasma AVP levels sufficient to produce maximal antidiuresis, then plasma osmolality rises to levels that stimulate thirst and produce water intake proportional to the elevation of osmolality above this threshold. In such a system thirst essentially represents a backup mechanism called into play when pituitary and renal mechanisms prove insufficient to maintain plasma osmolality within a few percent of basal levels. This arrangement has the advantage of freeing man from frequent episodes of thirst that would require a diversion of activities toward behavior oriented to seeking water when water deficiency is sufficiently mild to be compensated for by renal water conservation, but would stimulate water ingestion once water deficiency reaches potentially harmful levels. Stimulation of AVP secretion at plasma osmolalities below the threshold for subjective thirst acts to maintain an excess of body water sufficient to eliminate the need to drink whenever slight elevations in plasma osmolality occur. This system of differential effective thresholds for thirst and AVP secretion nicely complements many studies that have demonstrated excess unregulated, or "need-free", drinking in both man and animals (Fitzsimons, 1992).

SODIUM METABOLISM

Maintenance of sodium homeostasis requires a balance between intake and excretion of Na^+. As in the case of water metabolism, it is possible to define regulated and unregulated components of both Na^+ intake and Na^+ excretion. Unlike water intake, however, there is little evidence in humans to support a significant role for regulated Na^+ intake, with the possible exception of some pathologic conditions. Consequently, there is an even greater dependence on mechanisms for regulated renal excretion of sodium than is the case for excretion of water (Verbalis, 1997a). Whether for this reason or not, the mechanisms for renal excretion of sodium are more numerous and substantially more complex than the relatively simple, albeit quite efficient, system for AVP-regulated excretion of water.

Salt appetite

The only solute for which any specific appetite has been clearly demonstrated in man is sodium (as with animals, this is generally expressed as an appetite for the chloride salt of sodium, so it is usually called *salt appetite*). Because of the importance of Na^+ for ensuring maintenance of the ECF volume, which in turn directly supports blood volume and pressure, its uniqueness insofar as meriting a specific mechanism for regulated intake seems appropriate. However, despite abundant evidence in many different species demonstrating a salt appetite that is proportionately related to Na^+ losses (Denton, 1982), there is only one pathologic condition in which a specific stimulated sodium appetite has been unequivocally observed in humans, namely Addison's disease caused by adrenal insufficiency. Almost since the initial discovery of this disorder, salt craving has remained one of the well-known manifestations of Addison's disease (Wilkins and Richter, 1940). A robust salt appetite also occurs prominently in adrenalectomized animals, and appears to be related in part to the high plasma levels of adrenocorticotropic hormone (ACTH) produced as a result of the loss of cortisol feedback on the pituitary. However, despite the presence of Na^+ deficiency in most patients with untreated Addison's disease, only 15–20% of such patients manifest salt-seeking behavior (Orth and Kovacs, 1998). Even more striking is the apparent absence of salt appetite during a variety of other disorders causing severe Na^+ and ECF volume depletion in humans (e.g., patients with hemorrhagic blood loss, diuretic-induced hypovolemia, or hypotension of any etiology become thirsty when intravascular deficits are marked, but almost never express a pronounced desire for salty foods or fluids).

Renal sodium excretion

Although specific mechanisms exist for regulated renal excretion of all major electrolytes, none is as numerous or as complex as those controlling Na^+ excretion, which is not surprising in view of the fact that maintenance of ECF volume is crucial to normal health and function. The most important of these mechanisms are glomerular filtration rate, aldosterone secretion and renal effects, and intrarenal hemodynamic and peritubular factors. In view of the complexity of these mechanisms, the reader is referred to more complete reviews of this topic (Kirchner and Stein, 1994; Reeves and Andreoli, 2001).

HYPO-OSMOLALITY

Hypo-osmolality indicates excess water relative to solute in the ECF; because water moves freely between the ECF and the intracellular fluid (ICF), this also indicates an excess of total body water relative to total body solute. Imbalances between body water and solute can be generated either by *depletion* of body solute more than body water, or by *dilution* of body solute from increases in body water more than body solute. This is an oversimplification of complex physiology, and most hypo-osmolar states include components of both solute depletion and water retention. Nonetheless, this general concept has proven to be useful because it provides a simple framework for understanding the basic etiologies of hypo-osmolar disorders.

Differential diagnosis

Definitive identification of the etiology of hypo-osmolality is not always possible at the time of presentation, but categorization according to the patient's ECF volume status represents the first step in ascertaining the underlying cause of the disorder (Verbalis, 2012).

Decreased ECF volume (hypovolemia)

Clinically detectable hypovolemia indicates some degree of solute depletion. Even isotonic or hypotonic fluid losses can cause hypo-osmolality if water or hypotonic fluids are subsequently ingested or infused. A low urine sodium concentration (U_{Na}) suggests a nonrenal cause of solute depletion, whereas a high U_{Na} suggests renal causes of solute depletion. Diuretic use is the most common cause of hypovolemic hypo-osmolality. Most etiologies of solute losses causing hypovolemic hypo-osmolality will be clinically apparent, although some salt-wasting nephropathies and mineralocorticoid deficiency may be difficult to diagnose during early phases of these diseases.

Normal ECF volume (euvolemia)

Virtually any disorder causing hypo-osmolality can present with a volume status that appears normal by standard methods of clinical evaluation. Because clinical assessment of volume status is not very sensitive, the presence of normal or low blood urea nitrogen and uric acid concentrations are helpful laboratory correlates of relatively normal ECF volume. In these cases, a low U_{Na} (<30 mmol/L) suggests depletional hypo-osmolality secondary to ECF losses with subsequent volume replacement by water or other hypotonic fluids. Such patients may appear euvolemic by the usual clinical parameters used to assess hydration status. A high

U_{Na} ($\geq$ 30 mmol/L) generally indicates a dilutional hypo-osmolality such as the syndrome of inappropriate antidiuretic hormone secretion (SIADH), the most common cause of euvolemic hypo-osmolality. The clinical criteria necessary to diagnose SIADH remain as initially defined by Bartter and Schwartz in 1967 (Bartter and Schwartz, 1967). Many different disorders are associated with SIADH; these can be divided into four major etiologic groups: tumors, CNS disorders, drug effects, and pulmonary diseases.

Increased ECF volume (hypervolemia)

Clinically detectable hypervolemia indicates whole body sodium excess, and hypo-osmolality in these patients suggests a relatively decreased intravascular volume and/or pressure leading to water retention as a result of elevated plasma AVP levels and decreased distal delivery of glomerular filtrate to the kidneys. Such patients usually have a low U_{Na} because of secondary hyperaldosteronism, but under certain conditions the U_{Na} may be elevated (e.g., diuretic therapy). Hyponatremia generally does not occur until relatively advanced stages of congestive heart failure, cirrhosis, or the nephrotic syndrome, by which time diagnosis is usually not difficult. Renal failure can also cause retention of both sodium and water.

Clinical manifestations

The clinical manifestations of hyponatremia are largely neurologic, and primarily reflect brain edema resulting from osmotic water shifts into the brain (Verbalis, 2010). These range from nonspecific symptoms such as headache and confusion to more severe manifestations such as decreased sensorium, coma, seizures, and death. Significant central nervous system (CNS) symptoms generally do not occur until the serum sodium concentration ([Na^+]) falls below 125 mmol/L, and the severity of symptoms can be roughly correlated with the degree of hypo-osmolality. Individual variability is marked, and for any patient the level of serum [Na^+] at which symptoms will appear cannot be accurately predicted. Several factors other than the severity of the hypo-osmolality also affect the degree of neurologic dysfunction. The most important is the time course over which hypo-osmolality develops. Rapid development of severe hypo-osmolality frequently causes marked neurologic symptoms, whereas gradual development over several days or weeks is often associated with relatively mild symptomatology despite profound degrees of hypo-osmolality. This is because the brain counteracts osmotic swelling by extruding extracellular and intracellular solutes, including potassium and a variety of small organic molecules (amino acids, polyols, and methylamines) called *organic osmolytes* (Gullans and Verbalis, 1993). Since this is a time-dependent process, rapid development of hypo-osmolality can result in brain edema before this adaptation occurs, but with slower development of the same degree of hypo-osmolality brain cells can lose solute sufficiently rapidly to prevent cell swelling, brain edema, and neurologic dysfunction (Verbalis, 2010). Underlying neurologic disease also affects the level of hypo-osmolality at which CNS symptoms appear; moderate hypo-osmolality is of little concern in an otherwise healthy patient, but can cause morbidity in a patient with an underlying seizure disorder. Non-neurologic metabolic disorders (hypoxia, hypercapnia, acidosis, hypercalcemia, etc.) similarly can affect the level of osmolality at which CNS symptoms occur.

Therapy

Correction of hyponatremia is associated with markedly improved neurologic outcomes in patients with severely symptomatic hyponatremia. In a retrospective review of patients who presented with severe neurologic symptoms and serum [Na^+] < 125 mmol/L, prompt therapy with isotonic or hypertonic saline resulted in a correction in the range of 20 mmol/L over several days and neurologic recovery in almost all cases. In contrast, in patients who were treated with fluid restriction alone, there was very little correction over the study period (< 5 mmol/L over 72 hours), and the neurologic outcomes were much worse, with most of these patients either dying or entering a persistently vegetative state (Ayus, 1986). Consequently, based on this and many similar retrospective analyses, prompt therapy to rapidly increase the serum [Na^+] represents the standard of care for treatment of patients presenting with severe life-threatening symptoms of hyponatremia.

Chronic hyponatremia is much less symptomatic as a result of the process of brain volume regulation. Because of this adaptation process, chronic hyponatremia is arguably a condition that clinicians feel they may not need to be as concerned about, which has been reinforced by the common usage of the descriptor *"asymptomatic hyponatremia"* for many such patients. However, it is clear that many such patients very often do have neurologic symptoms, even if milder and more subtle in nature, including headaches, nausea, mood disturbances, depression, difficulty concentrating, slowed reaction times, unstable gait, increased falls, confusion, and disorientation (Renneboog et al., 2006). Consequently, all patients with hyponatremia who manifest any neurologic symptoms that could possibly be related to the hyponatremia should be considered as potential candidates for treatment of the hyponatremia, regardless of the chronicity of the hyponatremia or the level of serum [Na^+].

Currently available therapies for treatment of hyponatremia

Conventional management strategies for hyponatremia range from saline infusion and fluid restriction to pharmacologic measures to adjust fluid balance. Consideration of treatment options should always include an evaluation of the benefits as well as the potential toxicities of any therapy, and must be individualized for each patient (Verbalis, 2013; Verbalis et al., 2013). It should always be remembered that sometimes simply stopping treatment with an agent that is associated with hyponatremia is sufficient to reverse a low serum [Na^+].

Isotonic saline

The treatment of choice for depletional hyponatremia (i.e., hypovolemic hyponatremia) is isotonic saline ([Na^+] = 154 mmol/L) to restore ECF volume and ensure adequate organ perfusion. This initial therapy is appropriate for patients who either have clinical signs of hypovolemia, or in whom a spot U_{Na} is < 30 mmol/L. However, this therapy is ineffective for dilutional hyponatremias such as SIADH (Schwartz et al., 1957), and continued inappropriate administration of isotonic saline to a euvolemic patient may worsen their hyponatremia (Steele et al., 1997) and/or cause fluid overload.

Hypertonic saline

Acute hyponatremia presenting with severe neurologic symptoms is life-threatening, and should be treated promptly with hypertonic solutions, typically 3% NaCl ([Na^+] = 513 mmol/L), as this represents the most reliable method to quickly raise the serum [Na^+]. A continuous infusion of hypertonic NaCl is usually utilized in inpatient settings. Various formulae have been suggested for calculating the initial rate of infusion of hypertonic solutions (Adrogue and Madias, 2000b), but until now there has been no consensus regarding optimal infusion rates of 3% NaCl. One of the simplest methods to estimate an initial 3% NaCl infusion rate utilizes the following relationship (Verbalis, 2013; Verbalis et al., 2013):

$$\begin{aligned}&\text{Patients weight (kg)}\\&\quad\times \text{desired correction rate (mEq/L/hour)}\\&\quad= \text{infusion rate of 3\% NaCl (mL/hour)}\end{aligned}$$

Depending on individual hospital policies, the administration of hypertonic solutions may require special considerations (e.g., placement in the intensive care unit, sign-off by a consultant, etc.).

An alternative option for more emergent situations is administration of a 100 mL bolus of 3% NaCl, repeated 1–2 times every 30 min if there is no clinical improvement, which has been recommended by a consensus conference organized to develop guidelines for prevention and treatment of exercise-induced hyponatremia, an acute and potentially lethal condition (Hew-Butler et al., 2008). Injecting this amount of hypertonic saline intravenously raises the serum [Na^+] by an average of 2–4 mmol/L, which is well below the recommended maximal daily rate of change of 10–12 mmol/24 hours or 18 mmol/48 hours (Sterns et al., 1994). Because the brain can only accommodate an average increase of approximately 7–8% in brain volume before herniation occurs, quickly increasing the serum [Na^+] by as little as 2–4 mmol/L in acute hyponatremia can effectively reduce brain swelling and intracranial pressure (Battison et al., 2005).

Many physicians are hesitant to use hypertonic saline in patients with chronic hyponatremia, because it can cause an overly rapid correction of serum sodium levels that can lead to the *osmotic demyelination syndrome* (ODS) (Sterns et al., 1986). Nonetheless, this remains the treatment of choice for patients with severe neurologic symptoms, even when the time course of the hyponatremia is nonacute or unknown.

Fluid restriction

For patients with chronic hyponatremia, fluid restriction has been the most popular and most widely accepted treatment. When SIADH is present, fluids should generally be limited to 500–1000 mL/24 hours. Because fluid restriction increases the serum [Na^+] largely by underreplacing the excretion of fluid by the kidneys, some have advocated an initial restriction to 500 mL less than the 24 hour urine output (Robertson, 2006). When instituting a fluid restriction, it is important for the nursing staff and the patient to understand that this includes *all* fluids that are consumed, not just water. Generally the water content of ingested food is not included in the restriction because this is balanced by insensible water losses (perspiration, exhaled air, feces, etc.), but caution should be exercised with foods that have high fluid concentrations (such as fruits and soups). Restricting fluid intake can be effective when properly applied and managed in selected patients, but serum [Na^+] is increased only slowly (1–2 mmol/L/day) even with severe restriction (Schwartz et al., 1957). In addition, this therapy is often poorly tolerated because of an associated increase in thirst leading to poor compliance with long-term therapy. However, it is economically favorable, and some patients do respond well to this option.

Fluid restriction should not be used with hypovolemic patients, and is particularly difficult to maintain in patients with very elevated urine osmolalities secondary to high AVP levels; in general, if the sum of urine Na^+

and K^+ exceeds the serum [Na^+], most patients will not respond to a fluid restriction since an electrolyte-free water clearance will be difficult to achieve (Furst et al., 2000; Berl, 2008; Decaux, 2009). In addition, fluid restriction is not practical for some patients, particularly patients in intensive care settings, who often require administration of significant volumes of fluids as part of their treatment.

Demeclocycline

Demeclocycline, a tetracycline antibiotic, inhibits adenylyl cyclase activation after AVP binds to the V2R in the kidney, and thus targets the underlying pathophysiology of SIADH. This therapy is typically used when patients find severe fluid restriction unacceptable and the underlying disorder cannot be corrected. However, demeclocycline is not approved by the US Food and Drug Administration (FDA) or the European Medicines Agency (EMA) to treat hyponatremia, and can cause nephrotoxicity in patients with heart failure and cirrhosis, although this is usually reversible (Singer and Rotenberg, 1973).

Mineralocorticoids

Administration of mineralocorticoids, such as fludrocortisone, has been shown to be useful in a small number of elderly patients (Ishikawa et al., 1996). However, the initial studies of SIADH did not show it to be of benefit in patients with SIADH, and it carries the risk of fluid overload and hypertension. Consequently, it is rarely used as primary therapy for hyponatremia.

Urea

Administration of urea has been successfully used to treat hyponatremia because it induces osmotic diuresis and augments free water excretion. Effective doses of urea for treatment of hyponatremia are 30–90 g daily in divided doses (Decaux and Genette, 1981). Unfortunately, its use is limited because there is no US Pharmacopeia (USP) formulation for urea, it is not currently approved by any regulatory agency for treatment of hyponatremia, and there are limited data to support its long-term use. Furthermore, urea is associated with poor palatability leading to patient compliance problems. However, patients with feeding tubes may be excellent candidates for urea therapy since palatability is not a concern, and the use of fluid restriction may be difficult in some patients with high obligate intake of fluids as part of their nutritional and medication therapy. Although mild azotemia can occur with urea therapy, this rarely reaches clinically significant levels.

Furosemide and NaCl

The use of furosemide (20–40 mg/day) coupled with a high salt intake (200 mmol/day), which represents an extension of the treatment of acute symptomatic hyponatremia to the chronic management of euvolemic hyponatremia, has also been reported to be successful in selected cases (Decaux et al., 1981). However, the long-term efficacy and safety of this approach is unknown.

Arginine vasopressin receptor (AVPR) antagonists

Clinicians have used all of the above conventional therapies for hyponatremia over the past decades. However, conventional therapies for hyponatremia, although effective in specific circumstances, are suboptimal for many different reasons, including variable efficacy, slow responses, intolerable side-effects, and serious toxicities. But perhaps the most striking deficiency of most conventional therapies is that, with the exception of demeclocycline, these therapies do not directly target the underlying cause of almost all dilutional hyponatremias, namely inappropriately elevated plasma AVP levels. A new class of pharmacologic agents, *vasopressin receptor antagonists*, also called *"vaptans"*, which directly block AVP-mediated receptor activation, have recently been developed for clinical use (Thibonnier et al., 2001; Greenberg and Verbalis, 2006).

Conivaptan and tolvaptan are competitive receptor antagonists of the AVP V2R and have been approved by the FDA for the treatment of euvolemic and hypervolemic hyponatremia and by the EMA for treatment of hyponatremia caused by SIADH. These agents compete with AVP/ADH for binding at its site of action in the kidney, thereby blocking the antidiuresis caused by elevated AVP levels and directly attacking the underlying pathophysiology of dilutional hyponatremia. AVPR antagonists produce electrolyte-free water excretion (called *"aquaresis"*) without affecting renal sodium and potassium excretion (Ohnishi et al., 1993). The overall result is a reduction in body water without natriuresis, which leads to an increase in the serum [Na^+]. One of the major benefits of this class of drugs is that serum [Na^+] is significantly increased by an average of 4–8 mmol/L within 24–48 hours (Schrier et al., 2006; Zeltser et al., 2007); this is considerably faster than the effects of fluid restriction, which can take many days. Also, compliance has not been shown to be problem for vaptans, whereas this is a major problem with attempted long-term use of fluid restriction.

Conivaptan is FDA approved for euvolemic and hypervolemic hyponatremia in hospitalized patients.

It is available only as an intravenous preparation, and is given as a 20 mg loading dose over 30 min, followed by a continuous infusion of 20 or 40 mg (Li-Ng and Verbalis, 2010). Generally, the 20 mg continuous infusion is used for the first 24 hours to gauge the initial response. If the correction of serum [Na^+] is felt to be inadequate (e.g., < 5 mmol/L), then the infusion rate can be increased to 40 mg/day. Therapy is limited to a maximum duration of 4 days because of drug-interaction effects with other agents metabolized by the CYP3A4 hepatic isoenzyme. Importantly, for conivaptan and all other vaptans, it is critical that the serum [Na^+] concentration is measured frequently during the active phase of correction of the hyponatremia (a minimum of every 6–8 hours for conivaptan, but more frequently in patients with risk factors for development of the osmotic demyelination syndrome (ODS), such as severely low serum [Na^+], malnutrition, alcoholism and hypokalemia (Verbalis et al., 2013). If the correction approaches 12 mmol/L in the first 24 hours, the infusion should be stopped and the patient monitored on a fluid restriction. If the correction exceeds 12 mmol/L, consideration should be given to administering sufficient water, either orally or as intravenous D_5W to bring the overall correction below 12 mEq/L. The maximum correction limit should be reduced to 8 mEq/L over the first 24 hours in patients with risk factors for development of ODS mentioned previously. The most common adverse effects include infusion-site reactions, which are generally mild and usually do not lead to treatment discontinuation, headache, thirst, and hypokalemia (Zeltser et al., 2007).

Tolvaptan, an oral AVPR antagonist, is FDA approved for treatment of dilutional hyponatremias. In contrast to conivaptan, oral administration allows it to be used for both short- and long-term treatment of hyponatremia (Schrier et al., 2006). Similar to conivaptan, tolvaptan treatment must be initiated in the hospital so that the rate of correction can be monitored carefully. Patients with a serum [Na^+] < 125 mmol/L are eligible for therapy with tolvaptan as primary therapy; if the serum [Na^+] is ≥ 125 mmol/L, tolvaptan therapy is only indicated if the patient has symptoms that could be attributable to the hyponatremia and the patient is resistant to attempts at fluid restriction (Otsuka Parmaceutical Co., 2009). The starting dose of tolvaptan is 15 mg on the first day, and the dose can be titrated to 30 mg and 60 mg at 24 hour intervals if the serum [Na^+] remains < 135 mEq/L or the increase in serum [Na^+] has been ≤ 5 mmol/L in the previous 24 hours. As with conivaptan, it is essential that the serum [Na^+] concentration is measured frequently during the active phase of correction of the hyponatremia (a minimum of every 6–8 hours, but more frequently in patients with risk factors for development of ODS). Limits for safe correction of hyponatremia and methods to compensate for overly rapid corrections are the same as described previously for conivaptan. One additional factor that helps to avoid overly rapid correction with tolvaptan is the recommendation that fluid restriction not be used during the active phase of correction, thereby allowing the patient's thirst to compensate for an overly vigorous aquaresis. Common side-effects include dry mouth, thirst, increased urinary frequency, dizziness, nausea and orthostatic hypotension, which were relatively similar between placebo and tolvaptan groups in clinical trials (Schrier et al., 2006; Otsuka Pharmaceutical Co., 2009).

Because inducing increased renal fluid excretion via either a diuresis or an aquaresis can cause or worsen hypotension in patients with hypovolemic hyponatremia, vaptans are contraindicated in this patient population (Verbalis et al., 2013). However, clinically significant hypotension was not observed in either the conivaptan or tolvaptan clinical trials in euvolemic and hypervolemic hyponatremic patients. Although vaptans are not contraindicated with decreased renal function, these agents generally will not be effective if the serum creatinine is > 3.0 mg/dL. By virtue of their solubility properties, the vaptans are readily absorbed through the gastrointestinal tract, and therefore likely cross the blood–brain barrier as well, but alterations of central nervous system (CNS) functions have not been observed to date with clinical use of vaptans. This may be due to the fact that most of the CNS effects of AVP have been attributed to V1aR, with no evidence to support the presence or biologic function of V2R in the CNS (Thibonnier et al., 1998); however, such effects have not been reported even with clinical use of conivaptan, a combined V1aR/V2R antagonist. Therefore, until more directed and sensitive studies are done to assess potential CNS effects of AVP receptor antagonists, it must be concluded that such effects are absent, or minimal.

Hyponatremia treatment guidelines

Although various authors have published recommendations for the treatment of hyponatremia (Adrogue and Madias, 2000b; Ellison and Berl, 2007; Verbalis et al., 2013; Sterns et al., 2009; Verbalis, 2009), no standardized treatment algorithms have yet been widely accepted. A synthesis of existing expert recommendations for treatment of hyponatremia is illustrated in Figure 3.1. This algorithm is based primarily on the symptomatology of hyponatremic patients, rather than the serum [Na^+] or the chronicity of the hyponatremia, the latter being often difficult to ascertain.

A careful neurologic history and assessment should always be done to identify potential causes of the

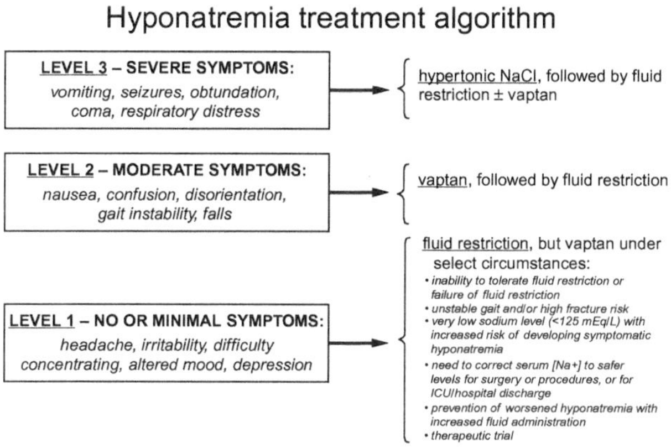

Fig. 3.1. Proposed treatment algorithm for hyponatremia based on degree of symptomatology.

patient's symptoms other than hyponatremia, although it will not always be possible to exclude an additive contribution from the hyponatremia to an underlying neurologic condition. In this algorithm, patients are divided into three groups based on their presenting symptoms: *severe symptoms* (seizures, coma, respiratory arrest, obtundation, and vomiting; these symptoms usually imply a more acute onset or worsening of hyponatremia requiring immediate active treatment); *moderate symptoms* (nausea, confusion, disorientation, and altered mental status; these symptoms may be either chronic or acute, but allow time to elaborate a more deliberate approach to treatment); *mild symptoms* (headache, irritability, inability to concentrate, altered mood, and depression, or a virtual absence of discernible symptoms; these symptoms usually indicate that the patient may have chronic or slowly evolving hyponatremia) (Verbalis, 2012).

Patients with severe symptoms should be treated with hypertonic saline as first-line therapy, followed by fluid restriction with or without AVPR antagonist therapy. Patients with moderate symptoms will benefit from a regimen of vaptan therapy or limited hypertonic saline administration followed by fluid restriction or long-term vaptan therapy. Although moderate neurologic symptoms can indicate that a patient is in an early stage of acute hyponatremia, they more often indicate a chronically hyponatremic state with sufficient brain volume adaptation to prevent marked symptomatology from cerebral edema. Regardless, close monitoring of these patients in a hospital setting is warranted until the symptoms improve or stabilize. Patients with no or minimal symptoms should be managed initially with fluid restriction, although treatment with vaptans may be appropriate for a wide range of specific clinical conditions, foremost of which is a failure to improve the serum $[Na^+]$ despite reasonable attempts at fluid restriction.

A special case is when spontaneous correction of hyponatremia occurs at an undesirably rapid rate as a result of the onset of a water diuresis. This can occur following cessation of desmopressin therapy in a patient who has become hyponatremic, replacement of glucocorticoids in a patient with adrenal insufficiency, replacement of solutes in a patient with diuretic-induced hyponatremia, or spontaneous resolution of transient SIADH. Brain damage from ODS can clearly ensue in this setting if the preceding period of hyponatremia has been of sufficient duration (usually ≥ 48 hours) to allow brain volume regulation to occur. If the previously discussed correction parameters have been exceeded and the correction is proceeding more rapidly than planned (usually because of continued excretion of hypotonic urine), the pathologic events leading to demyelination can be reversed by administration of hypotonic fluids and desmopressin. Efficacy of this approach is suggested both from animal studies (Soupart et al., 1994) and from case reports in humans (Goldszmidt and Iliescu, 2000; Sterns et al., 2009), even when patients are overtly symptomatic (Oya et al., 2001).

Although this algorithm is based on presenting symptoms at the time of initial evaluation, it should be remembered that in some cases patients initially exhibit more moderate symptoms because they are in the early stages of hyponatremia. In addition, some patients with minimal symptoms are prone to develop more symptomatic hyponatremia during periods of increased fluid ingestion. In support of this, approximately 70% of 31 patients presenting to a university hospital with symptomatic hyponatremia and a mean serum $[Na^+]$ of 119 mmol/L had pre-existing "asymptomatic" hyponatremia as the most common risk factor identified (Bissram et al., 2007). Consequently, therapy of hyponatremia should also be considered to prevent progression from lower to higher levels of symptomatic hyponatremia, particularly in patients with a past history of repeated presentations for symptomatic hyponatremia.

Monitoring the serum $[Na^+]$ in hyponatremic patients

The frequency of serum $[Na^+]$ monitoring is dependent on both the severity of the hyponatremia and the therapy chosen. In all hyponatremic patients neurologic symptomatology should be carefully assessed very early in the diagnostic evaluation to ascertain the symptomatic severity of the hyponatremia and to determine whether the patient requires more urgent therapy. All patients undergoing active treatment with hypertonic saline for symptomatic hyponatremia should have frequent monitoring of serum $[Na^+]$ and ECF volume status (every 2–4 hours) to ensure that the serum $[Na^+]$ does not exceeded the recommended levels during the active phase of correction (Verbalis et al., 2013), since overly rapid correction of serum sodium can cause damage to the myelin sheath of nerve cells, resulting in ODS (Sterns et al., 1986). Patients treated with vaptans for mild to moderate symptoms should have serum [Na^+] monitored every 6–8 hours during the active phase of correction, which will generally be the first 24–48 hours of therapy. Active treatment with hypertonic saline or vaptans should be stopped when the patient's symptoms are no longer present, a safe serum [Na^+] (usually > 120 mmol/L) has been achieved, or the rate of correction has reached 12 mmol/L within 24 hours or 18 mmol/L within 48 hours (Sterns et al., 1994; Verbalis et al., 2013). Importantly, ODS has not yet been reported either in clinical trials or with therapeutic use of any vaptan as monotherapy to date. In patients with a stable level of serum [Na^+] treated with fluid restriction or therapies other than hypertonic saline, measurement of serum [Na^+] daily is generally sufficient, since levels will not change that quickly in the absence of active therapy or large changes in fluid intake or administration.

Long-term treatment of chronic hyponatremia

Some patients will benefit from continued treatment of hyponatremia following discharge from the hospital. In many cases, this will consist of a continued fluid restriction. However, as discussed previously, long-term compliance with this therapy is poor due to the increased thirst that occurs with more severe degrees of fluid restriction. Thus, for selected patients who have responded to tolvaptan in the hospital, consideration should be given to continuing the treatment as an outpatient after discharge. In patients with established chronic hyponatremia, tolvaptan has been shown to be effective at maintaining a normal [Na^+] for as long as 3 years of continued daily therapy (Berl et al., 2010). However, many patients with inpatient hyponatremia have a transient form of SIADH, without need for long-term therapy. Selection of which patients with inpatient hyponatremia are candidates for long-term therapy should be based on the etiology of the SIADH. In all cases, consideration should be given to a trial of stopping the drug 2–4 weeks following discharge to see if hyponatremia is still present. A reasonable period of tolvaptan cessation to evaluate the presence of continued SIADH is 7 days, since this period was sufficient for demonstration of a recurrence of hyponatremia in the tolvaptan SALT trials (Schrier et al., 2006; Berl et al., 2010). Serum [Na^+] should be monitored every 2–3 days following cessation of tolvaptan so that the drug can be resumed as quickly as possible in those patients with recurrent hyponatremia, since the longer the patient is hyponatremic the greater the risk of subsequent ODS with overly rapid correction of the low serum [Na^+].

HYPEROSMOLALITY

Hyperosmolality indicates a deficiency of water relative to solute in the ECF. Because water moves freely between the ICF and ECF, this also indicates a deficiency of total body water relative to total body solute. Although hypernatremia can be caused by an excess of body sodium, the vast majority of cases are due to losses of body water in excess of body solutes, caused by either insufficient water intake or excessive water excretion. Consequently, most of the disorders causing hyperosmolality are those associated with inadequate water intake and/or deficient pituitary AVP secretion. Although hyperosmolality from inadequate water intake is seen frequently in clinical practice, this is usually not due to an underlying defect in thirst but rather results from a generalized incapacity to obtain and/or ingest fluids, often stemming from a depressed sensorium.

Etiologies and diagnosis

Evaluation of the patient's ECF volume status is important as a guide to fluid replacement therapy, but is not as useful for differential diagnosis since most hyperosmolar patients will manifest some degree of hypovolemia. Rather, assessment of urinary concentrating ability provides the most useful data with regard to the type of disorder present. Using this approach, disorders of hyperosmolality can be categorized as those in which renal water conservation mechanisms are intact but are unable to compensate for inadequately replaced losses of hypotonic fluids from other sources, or those in which renal concentrating defects are a contributing factor to the deficiency of body water (Verbalis, 2012).

Diabetes insipidus

Diabetes insipidus (DI) can result from either inadequate AVP secretion (*central or neurogenic DI*) or inadequate renal response to AVP (*nephrogenic DI*). Central DI is caused by a variety of acquired or congenital anatomic lesions that disrupt the neurohypophysis, including pituitary surgery, tumors, trauma, hemorrhage, thrombosis, infarction, or granulomatous disease (Robertson, 1995), as well as less commonly by genetic mutations of the AVP gene (Babey et al., 2011). Severe nephrogenic DI is most commonly congenital due to defects in the gene for the AVP V2R (X-linked recessive pattern of inheritance) or in the gene for the AQP2 water channel (autosomal recessive pattern of inheritance) (Fujiwara and Bichet, 2005), but relief of chronic urinary obstruction or therapy with drugs such as lithium can cause an acquired form sufficient to warrant treatment. Acquired nephrogenic DI can result from hypokalemia or hypercalcemia, but the mild concentrating defect generally does not by itself cause hypertonicity and responds to correction of the underlying disorder (Khanna, 2006). Regardless of the etiology of the DI, the end result is a water diuresis due to an inability to concentrate urine appropriately.

Because patients with DI do not have impaired urine Na^+ conservation, the ECF volume is generally not markedly decreased and regulatory mechanisms for maintenance of osmotic homeostasis are primarily activated: stimulation of thirst and pituitary AVP secretion (to whatever degree the neurohypophysis is still able to secrete AVP). In cases where AVP secretion is totally absent (complete DI), patients are dependent entirely on water intake for maintenance of water balance. However, in cases where some residual capacity to secrete AVP remains (partial DI), plasma osmolality can eventually reach levels that allow moderate degrees of urinary concentration. Although untreated DI can lead to both hyperosmolality and volume depletion, until the water losses become severe, volume depletion is minimized by osmotic shifts of water from the ICF into the more osmotically concentrated ECF (Robinson and Verbalis, 2011).

Osmoreceptor dysfunction

The primary osmoreceptors that control AVP secretion and thirst are located in the anterior hypothalamus, and lesions of this region in animals cause hyperosmolality through a combination of impaired thirst and osmotically stimulated AVP secretion (Johnson and Buggy, 1978). Initial reports in humans described this syndrome as "essential hypernatremia," and subsequent studies used the term "*adipsic hypernatremia*" in recognition of the profound thirst deficits found in most of the patients. All of these syndromes are now grouped together as *disorders of osmoreceptor function* (Baylis and Thompson, 1988). Most of the cases reported to date have represented various degrees of osmoreceptor destruction associated with different brain lesions (Baylis and Thompson, 2001). In contrast to lesions causing central DI, these lesions usually occur more rostrally in the hypothalamus. For all cases of osmoreceptor dysfunction it is important to remember that afferent pathways from the brainstem to the hypothalamus generally remain intact; therefore, these patients will usually have normal AVP and renal concentrating responses to baroreceptor-mediated stimuli such as hypovolemia and hypotension.

Differential diagnosis

Distinguishing between central and nephrogenic DI in a patient who is already hyperosmolar entails simply evaluating the response to a trial of AVP or desmopressin. Administration of AVP (5 units subcutaneously) or, preferably, the selective AVP V2R agonist desmopressin (2 μg subcutaneously or intravenously), should cause a significant increase in urine osmolality within 1–2 hours after injection in patients with central DI, indicating insufficient endogenous AVP secretion. An absent or suboptimal response suggests renal resistance to AVP effects and, therefore, nephrogenic DI. Although conceptually simple, interpretational difficulties often arise because the water diuresis produced by AVP deficiency causes a downregulation of AQP2 synthesis along with a wash-out of the renal medullary concentrating gradient, such that increases in urine osmolality in response to administered AVP or desmopressin are not as great as would be expected (see Table 3.1, the interpretation of urine concentration after AVP/desmopressin).

Because patients with DI generally have an intact thirst mechanism, such patients often do not present with hyperosmolality, but rather have normal plasma osmolality and serum sodium levels with polyuria and polydipsia (Robertson, 1995). In these cases, a fluid deprivation test should be performed in order to raise the serum osmolality and confirm the diagnosis of DI (see Table 3.1 for the procedure and for the interpretation of a fluid deprivation test).

When a diagnosis of central DI is made, magnetic resonance imaging (MRI) of the hypothalamus and neurohypophyseal tract is mandatory to rule out a neoplasm or granulomatous disease as an etiology. In individuals with a normal posterior pituitary, the presence of a pituitary "bright spot" (i.e., a hyperintense signal in the absence of contrast administration, representing the AVP-containing neurosecretory granules) is usually seen on T1-weighted noncontrast sagittal images. Conversely, in patients with central DI, the bright spot is usually absent. However, this test is not definitive, since the pituitary bright spot decreases with age and with disorders that cause dehydration, and it can be present in up to 5% of patients with DI due to the presence of pituitary oxytocin, which also leads to a hyperintense signal on T1-weighted imaging (Robinson and Verbalis, 2011).

Evaluation of anterior pituitary function should be performed in all patients with central DI, especially if glucocorticoid administration and/or replacement unmasks underlying DI. Adrenal insufficiency can cause hypersecretion of AVP, which may be due in part to reductions in systemic blood pressure and cardiac output caused by cortisol deficiency, thereby stimulating pituitary AVP release. Cortisol deficiency is also known to cause increased AVP release from the median eminence into the pituitary portal circulation in an attempt to increase ACTH secretion via effects at pituitary

Table 3.1

Fluid deprivation test for the diagnosis of diabetes insipidus

Procedure

1. Initiation of the deprivation period depends on the severity of the diabetes insipidus (DI). In routine cases the patient should not drink any fluids after dinner the day before the test. In cases with more severe polyuria and polydipsia this may be too long a period without fluids and the water deprivation should be begun early (e.g., 6 a.m.) on the morning of the test.
2. Obtain plasma and urine osmolality, serum electrolytes, and a plasma arginine vasopressin (AVP) level at the start of the test.
3. Measure urine volume and osmolality hourly or with each voided urine.
4. Stop the test when body weight decreases by $\geq 3\%$, the patient develops orthostatic blood pressure changes, the urine osmolality reaches a plateau (i.e., $<10\%$ change over two or three consecutive measurements), or the serum $[Na+] > 145$ mmol/L.
5. Obtain plasma and urine osmolality, serum electrolytes, and a plasma AVP level at the end of the test, when the plasma osmolality is elevated, preferably >300 mOsm/kg H_2O.
6. If the serum $[Na+] < 146$ mmol/L or the plasma osmolality <300 mOsm/kg H_2O when the test is stopped, then consider a short infusion of hypertonic saline (3% NaCl at a rate of 0.1 mL/kg/min for 1–2 h) to reach these endpoints.
7. If hypertonic saline infusion is not required to achieve hyperosmolality, administer AVP (5 U) or desmopressin (1 μg) subcutaneously and continue following urine osmolality and volume for an additional 2 h.

Interpretation

1. An unequivocal urine concentration after AVP/desmopressin ($>50\%$ increase) indicates central DI and an unequivocal absence of urine concentration ($<10\%$) strongly suggests nephrogenic DI or primary polydipsia.
2. Differentiating between nephrogenic DI and primary polydipsia, as well as for cases in which the increase in urine osmolality after AVP/desmopressin administration is more equivocal (e.g., 10–50%), is best done using the relation between plasma AVP levels and plasma osmolality obtained at the end of the dehydration period and/or hypertonic saline infusion and the relation between plasma AVP levels and urine osmolality under basal conditions (Robertson, 1995; Verbalis, 2012).

V1bR on corticotrophs, which may also contribute to increased systemic AVP levels (Oelkers, 1989).

The recent development of a commercial assay for the C-terminal glycoprotein of the AVP prohormone copeptin offers the possibility of a more stable and easier to measure marker of AVP secretion in response to induced dehydration and hypertonicity. Copeptin is secreted from the posterior pituitary in equimolar amounts as AVP, but is not degraded proteolytically as rapidly as AVP; in addition, its larger size renders measurement by two-site immunoassays possible, in contrast to AVP, which is too small for use of this technique (Morgenthaler et al., 2008). However, the clinical use of copeptin levels as a surrogate marker of AVP secretion during fluid deprivation tests will require standardization of normal and abnormal responses relative to plasma AVP levels.

Clinical manifestations

The characteristic clinical symptoms of DI are the polyuria and polydipsia that result from the underlying impairment of urinary concentrating mechanisms, which have already been covered in the previous section discussing pathophysiology of specific types of DI. Interestingly, patients with DI typically describe a craving for cold water, which appears to quench their thirst better. Patients with CDI also typically describe a precipitous onset of their polyuria and polydipsia, which simply reflects the fact that urinary concentration can be maintained fairly well until the number of AVP-producing neurons in the hypothalamus decreases to 10–15% of normal, after which plasma AVP levels decrease to the range where urine output increases dramatically (Heinbecker and White, 1941).

However, patients with DI, and particularly those with osmoreceptor dysfunction syndromes, can also present with varying degrees of hyperosmolality and dehydration depending on their overall hydration status. It is therefore important to be aware of the clinical manifestations of hyperosmolality as well. These can be divided into the signs and symptoms produced by dehydration, which are largely cardiovascular, and those caused by the hyperosmolality itself, which are predominantly neurologic and reflect brain dehydration as a result of osmotic water shifts out of the central nervous system (Verbalis, 2012). Cardiovascular manifestations of hypertonic dehydration include hypotension, azotemia, acute tubular necrosis secondary to renal hypoperfusion or rhabdomyolysis, and shock. Neurologic manifestations range from nonspecific symptoms such as irritability and cognitive dysfunction to more severe manifestations of *hypertonic encephalopathy* such as disorientation, decreased level of consciousness, obtundation, chorea, seizures, coma, focal neurologic deficits, subarachnoid hemorrhage, and cerebral infarction. The severity of symptoms can be roughly correlated with the degree of hyperosmolality, but individual variability is marked and for any single patient the level of serum [Na^+] at which symptoms will appear cannot be accurately predicted.

Similar to hypo-osmolar syndromes, the length of time over which hyperosmolality develops can markedly affect the clinical symptomatology. Rapid development of severe hyperosmolality is frequently associated with marked neurologic symptoms, whereas gradual development over several days or weeks generally causes milder symptoms. In this case, the brain counteracts osmotic shrinkage by increasing intracellular content of solutes. These include electrolytes such as potassium and a variety of *organic osmolytes*, which were previously called *idiogenic osmoles*; for the most part these are the same organic osmolytes that are lost from the brain during adaptation to hypo-osmolality (Gullans and Verbalis, 1993). The net effect of this process is to protect the brain against excessive shrinkage during sustained hyperosmolality. However, once the brain has adapted by increasing its solute content, rapid correction of the hyperosmolality can produce brain edema, since it takes a finite time (24–48 hours in animal studies) to dissipate the accumulated solutes, and until this process has been completed the brain will accumulate excess water as plasma osmolality is normalized (Verbalis, 2010). This effect is most often seen in dehydrated pediatric patients who can develop seizures with rapid rehydration, but it has been described only rarely in adults.

Therapy

The general goals of treatment of all hyperosmolar disorders are (1) correction of pre-existing water deficits, and (2) reduction in ongoing excessive urinary water losses. The specific therapy required varies with the clinical situation. Awake ambulatory patients with diabetes insipidus and normal thirst have little body water deficit but benefit from relief of the polyuria and polydipsia that disrupt normal activities. In contrast, comatose patients with or without diabetes insipidus are unable to drink in response to thirst, and in these patients progressive hypertonicity may be life-threatening. The established water deficit may be estimated using the following formula (Robinson and Verbalis, 1997):

$$\text{Water deficit} = 0.6 \times \text{premorbid weight} \times [1 - 140/\text{serum}\,[Na^+]\,(\text{mmol/L})]$$

This formula is dependent on several assumptions (total body water is approximately 60% of body weight, no body solute is lost as hypertonicity develops, and the premorbid [Na^+] is 140 mmol/L), but nonetheless provides a valid estimate of the approximate total body water deficit. To reduce the risk of central nervous system injury from protracted exposure to severe hypertonicity, the serum osmolality should be lowered to the range of 330 mOsm/kg H_2O within the first 24 hours of therapy. As noted previously, because the organic osmolytes accumulated in the brain during chronic hyperosmolality cannot be immediately dissipated, further correction to a normal osmolality should be spread over the subsequent 1–3 days to avoid producing cerebral edema during treatment, particularly in pediatric patients (Adrogue and Madias, 2000a).

It should be remembered that formulae which estimate body water deficits do not take ongoing water losses into account. Consequently, frequent serum electrolyte determinations must be made, and the administration rate of oral water or intravenous 5% dextrose in water should be adjusted accordingly. For example, the estimated water deficit of a 70 kg patient whose serum [Na^+] is 160 mmol/L would be 5.25 L. In such an individual, administration of water at a rate > 200 mL per hour would be required simply to correct the established deficit over 24 hours, but additional fluid would be needed to keep up with any ongoing losses in a patient with diabetes insipidus until a response to treatment has occurred.

A variety of antidiuretic agents have been used to treat central diabetes insipidus, but desmopressin is the treatment of choice for this disorder. Desmopresin was synthesized as a selective antagonist of AVP V2R, and it is particularly useful therapeutically because it has a much longer half-life than AVP and is devoid of the pressor activity of AVP at vascular V1aR (Robinson, 1976). Desmopressin is generally administered intranasally (5–20 μg every 8–24 hours), but can be given parenterally in acute situations (1–2 μg via the intravenous, intramuscular, or subcutaneous route). For both the intranasal and parenteral preparations, increasing the administered dose generally has the effect of prolonging the duration of antidiuresis rather than increasing its magnitude; consequently, altering the dose can be useful to reduce the required frequency of administration. Synthetic AVP (Pitressin) can also be used to treat central DI, but its use is limited by a much shorter half-life necessitating more frequent dosing or a continuous infusion, and the production of pressor effects due to vasoconstriction.

Nephrogenic diabetes insipidus is more difficult to treat since the kidney is resistant to all AVP-type agents. Limited responses can sometimes be achieved using thiazide diuretics (any drug of the thiazide class may be used with equal potential for benefit). Thiazides cause natriuresis by blocking sodium absorption in the cortical diluting site; when combined with dietary sodium restriction a modest hypovolemia results, which stimulates isotonic proximal tubular solute reabsorption and diminishes solute delivery to the distal parts of the nephron. Together, these effects diminish free water clearance independently of actions of AVP, thereby decreasing the polyuria of patients with nephrogenic DI (Sands and Bichet, 2006). Monitoring for hypokalemia is necessary and K^+ supplementation is occasionally required. Care must be exercised when treating patients taking lithium with diuretics, since the induced contraction of plasma volume may increase lithium concentrations by increasing proximal tubular absorption and worsen potential toxic effects of the therapy (Grunfeld and Rossier, 2009). Because prostaglandins increase renal medullary blood flow and diminish medullary solute concentration, effects that modestly decrease the interstitial gradient for water reabsorption, drugs that block renal prostaglandin synthesis (e.g., nonsteroidal anti-inflammatory agents) can increase non-AVP-mediated water reabsorption and impair urinary dilution, thereby reducing free water clearance and urine output. Although these agents are somewhat effective in central diabetes insipidus, their main usefulness is as adjunctive therapy in nephrogenic diabetes insipidus, in which more direct antidiuretic therapies are limited.

References

Adrogue HJ, Madias NE (2000a). Hypernatremia. N Engl J Med 342: 1493–1499.

Adrogue HJ, Madias NE (2000b). Hyponatremia. N Engl J Med 342: 1581–1589.

Ayus JC (1986). Diuretic-induced hyponatremia [editorial]. Arch Intern Med 146: 1295–1296.

Babey M, Kopp P, Robertson GL (2011). Familial forms of diabetes insipidus: clinical and molecular characteristics. Nat Rev Endocrinol 7: 701–714.

Bartter FC, Schwartz WB (1967). The syndrome of inappropriate secretion of antidiuretic hormone. Am J Med 42: 790–806.

Battison C, Andrews PJ, Graham C et al. (2005). Randomized controlled trial on the effect of a 20% mannitol solution and a 75% saline/6% dextran solution on increased intracranial pressure after brain injury. Crit Care Med 33: 196–202.

Baylis PH, Thompson CJ (1988). Osmoregulation of vasopressin secretion and thirst in health and disease. Clin Endocrinol (Oxf) 29: 549–576.

Baylis PH, Thompson CJ (2001). Diabetes insipidus and hyperosmolar syndromes. In: KL Becker (Ed.), Principles and Practice of Endocrinology and Metabolism. 3rd edn. Lippincott Williams Wilkins, Philadelphia, pp. 285–293.

Berl T (2008). Impact of solute intake on urine flow and water excretion. J Am Soc Nephrol 19: 1076–1078.

Berl T, Quittnat-Pelletier F, Verbalis JG et al. (2010). Oral tolvaptan is safe and effective in chronic hyponatremia. J Am Soc Nephrol 21: 705–712.

Bissram M, Scott FD, Liu L et al. (2007). Risk factors for symptomatic hyponatraemia: the role of pre-existing asymptomatic hyponatraemia. Intern Med J 37: 149–155.

de Castro JM (1988). A microregulatory analysis of spontaneous fluid intake in humans: evidence that the amount of liquid ingested and its timing is mainly governed by feeding. Physiol Behav 3: 705–714.

Decaux G (2009). The syndrome of inappropriate secretion of antidiuretic hormone (SIADH). Semin Nephrol 29: 239–256.

Decaux G, Genette F (1981). Urea for long-term treatment of syndrome of inappropriate secretion of antidiuretic hormone. Br Med J (Clin Res Ed) 283: 1081–1083.

Decaux G, Waterlot Y, Genette F et al. (1981). Treatment of the syndrome of inappropriate secretion of antidiuretic hormone with furosemide. N Engl J Med 304: 329–330.

Denton D (1982). The Hunger for Salt: An Anthropological Physiological and Medical Analysis. Springer-Verlag, Berlin.

Ellison DH, Berl T (2007). Clinical practice. The syndrome of inappropriate antidiuresis. N Engl J Med 356: 2064–2072.

Fitzsimons JT (1992). Physiology and pathophysiology of thirst and sodium appetite. In: DW Seldin, G Giebisch (Eds.), The Kidney. Physiology and Pathophysiology. Raven Press, New York, pp. 1615–1648.

Fujiwara TM, Bichet DG (2005). Molecular Biology of Hereditary Diabetes Insipidus. J Am Soc Nephrol 16: 2836–2846.

Furst H, Hallows KR, Post J et al. (2000). The urine/plasma electrolyte ratio: a predictive guide to water restriction. Am J Med Sci 319: 240–244.

Goldszmidt MA, Iliescu EA (2000). DDAVP to prevent rapid correction in hyponatremia. Clin Nephrol 53: 226–229.

Greenberg A, Verbalis JG (2006). Vasopressin receptor antagonists. Kidney Int 69: 2124–2130.

Grunfeld JP, Rossier BC (2009). Lithium nephrotoxicity revisited. Nat Rev Nephrol 5: 270–276.

Gullans SR, Verbalis JG (1993). Control of brain volume during hyperosmolar and hypoosmolar conditions. Annu Rev Med 44: 289–301.

Heinbecker P, White HL (1941). Hypothalamico-hypophyseal system and its relation to water balance in the dog. Am J Physiol 133: 582–593.

Hew-Butler T, Ayus JC, Kipps C et al. (2008). Statement of the Second International Exercise-Associated Hyponatremia Consensus Development Conference, New Zealand, 2007. Clin J Sport Med 18: 111–121.

Ishikawa S, Fujita N, Fujisawa G et al. (1996). Involvement of arginine vasopressin and renal sodium handling in pathogenesis of hyponatremia in elderly patients. Endocr J 43: 101–108.

Johnson AK, Buggy J (1978). Periventricular preoptic-hypothalamus is vital for thirst and normal water economy. Am J Physiol 234: R122–R129.

Khanna A (2006). Acquired nephrogenic diabetes insipidus. Semin Nephrol 26: 244–248.

Kirchner KA, Stein JH (1994). Sodium metabolism. In: RG Narins (Ed.), Clinical Disorders of Fluid and

Electrolyte Metabolsim. 5th edn. McGraw-Hill, New York, pp. 45–80.
Knepper MA (1997). Molecular physiology of urinary concentrating mechanism: regulation of aquaporin water channels by vasopressin. Am J Physiol 272: F3–F12.
Knepper MA (1998). Long-term regulation of urinary concentrating capacity. Am J Physiol 275: F332–F333.
Li-Ng M, Verbalis JG (2010). Conivaptan: evidence supporting its therapeutic use in hyponatremia. Core Evid 4: 83–92.
Morgenthaler NG, Struck J, Jochberger S et al. (2008). Copeptin: clinical use of a new biomarker. Trends Endocrinol Metab 19: 43–49.
Nielsen S, Frokiaer J, Marples D et al. (2002). Aquaporins in the kidney: from molecules to medicine. Physiol Rev 82: 205–244.
Oelkers W (1989). Hyponatremia and inappropriate secretion of vasopressin (antidiuretic hormone) in patients with hypopituitarism. N Engl J Med 321: 492–496.
Ohnishi A, Orita Y, Okahara R et al. (1993). Potent aquaretic agent. A novel nonpeptide selective vasopressin 2 antagonist (OPC-31260) in men. J Clin Invest 92: 2653–2659.
Orth DN, Kovacs WJ (1998). The adrenal cortex. In: JD Wilson, DW Foster, HM Kronenberg et al. (Eds.), Williams Textbook of Endocrinology. 9th edn. WB Saunders, Philadelphia, pp. 517–664.
Otsuka Pharmaceutical Co., Ltd., Tokyo (2009). Samsca (tolvaptan) prescribing information.
Oya S, Tsutsumi K, Ueki K et al. (2001). Reinduction of hyponatremia to treat central pontine myelinolysis. Neurology 57: 1931–1932.
Phillips PA, Rolls BJ, Ledingham JG et al. (1985). Angiotensin II-induced thirst and vasopressin release in man. Clin Sci (Lond) 68: 669–674.
Reeves WB, Andreoli TE (2001). Tubular sodium transport. In: RW Schrier (Ed.), Diseases of the Kidney and Urinary Tract. 7th edn. Lippincott Williams Wilkins, Philadelphia, pp. 135–175.
Renneboog B, Musch W, Vandemergel X et al. (2006). Mild chronic hyponatremia is associated with falls unsteadiness and attention deficits. Am J Med 119: 71.
Robertson GL (1976). The regulation of vasopressin function in health and disease. Recent Prog Horm Res 33: 333–385.
Robertson GL (1983). Thirst and vasopressin function in normal and disordered states of water balance. J Lab Clin Med 101: 351–371.
Robertson GL (1995). Diabetes insipidus. Endocrinol Metab Clin North Am 24: 549–572.
Robertson GL (2006). Regulation of arginine vasopressin in the syndrome of inappropriate antidiuresis. Am J Med 119 (7 Suppl 1): S36–S42.
Robinson AG (1976). DDAVP in the treatment of central diabetes insipidus. N Engl J Med 294: 507–511.
Robinson AG, Verbalis JG (1997). Diabetes insipidus. Curr Ther Endocrinol Metab 6: 1–7.
Robinson AG, Verbalis JG (2011). Posterior pituitary. In: S Melmed, KS Polonsky, P Reed Larsen et al. (Eds.), Williams Textbook of Endocrinology. 12th edn. WB Saunders, Philadelphia, pp. 291–323.
Sands JM, Bichet DG (2006). Nephrogenic diabetes insipidus. Ann Intern Med 144: 186–194.
Schrier RW, Gross P, Gheorghiade M et al. (2006). Tolvaptan a selective oral vasopressin V2-receptor antagonist for hyponatremia. N Engl J Med 355: 2099–2112.
Schwartz WB, Bennett S, Curelop S et al. (1957). A syndrome of renal sodium loss and hyponatremia probably resulting from inappropriate secretion of antidiuretic hormone. Am J Med 23: 529–542.
Singer I, Rotenberg D (1973). Demeclocycline-induced nephrogenic diabetes insipidus. In-vivo and in-vitro studies. Ann Intern Med 79: 679–683.
Soupart A, Penninckx R, Crenier L et al. (1994). Prevention of brain demyelination in rats after excessive correction of chronic hyponatremia by serum sodium lowering. Kidney Int 45: 193–200.
Steele A, Gowrishankar M, Abrahamson S et al. (1997). Postoperative hyponatremia despite near-isotonic saline infusion: a phenomenon of desalination. [see comments] Ann Intern Med 126: 20–25.
Sterns RH, Riggs JE, Schochet Jr SS (1986). Osmotic demyelination syndrome following correction of hyponatremia. N Engl J Med 314: 1535–1542.
Sterns RH, Cappuccio JD, Silver SM et al. (1994). Neurologic sequelae after treatment of severe hyponatremia: a multicenter perspective. J Am Soc Nephrol 4: 1522–1530.
Sterns RH, Nigwekar SU, Hix JK (2009). The treatment of hyponatremia. Semin Nephrol 29: 282–299.
Stricker EM, Verbalis JG (2013). Water and salt intake and body fluid homeostasis. In: LR Squire, D Berg, FE Bloom et al. (Eds.), Fundamental Neuroscience. 4th edn. Academic Press, Waltham MA, pp. 783–797.
Thibonnier M, Conarty DM, Preston JA et al. (1998). Molecular pharmacology of human vasopressin receptors. Adv Exp Med Biol 449: 251–276.
Thibonnier M, Coles P, Thibonnier A et al. (2001). The basic and clinical pharmacology of nonpeptide vasopressin receptor antagonists. Annu Rev Pharmacol Toxicol 41: 175–202.
Thompson CJ, Bland J, Burd J et al. (1986). The osmotic thresholds for thirst and vasopressin release are similar in healthy man. Clin Sci (Lond) 71: 651–656.
Verbalis JG (1997a). Body sodium and extracellular fluid volume. In: RL Jamison, R Wilkinson (Eds.), Nephrology. Chapman Hall Medical, London, pp. 95–101.
Verbalis JG (1997b). Body water and osmolality. In: RL Jamison, R Wilkinson (Eds.), Nephrology. Chapman Hall Medical, London, pp. 89–94.
Verbalis JG (2009). Hyponatremia and hypo-osmolar disorders. In: A Greenberg (Ed.), Primer on Kidney Diseases. 5th edn. Saunders Elsevier, Philadelphia, pp. 52–59.
Verbalis JG (2010). Brain volume regulation in response to changes in osmolality. Neuroscience 168: 862–870.
Verbalis JG (2012). Disorders of water balance. In: MW Taal, GM Chertow, PA Marsden et al. (Eds.), Brenner Rector's The Kidney. 9th edn. Elsevier, Philadelphia, pp. 540–594.
Verbalis JG (2013). The syndrome of inappropriate antidiuretic hormone secretion and other hypoosmolar

disorders. In: TM Coffman, RJ Falk, BA Molitoris et al. (Eds.), Schrier's Diseases of the Kidney. 9th edn. Lippincott Williams Wilkins, Philadelphia, pp. 2012–2054.

Verbalis JG, Goldsmith SR, Greenberg A et al. (2013). Diagnosis, evaluation, and treatment of hyponatremia: expert panel recommendations. Am J Med 126 (10 Suppl 1): S1–S42.

Wilkins L, Richter CP (1940). A great craving for salt by a child with cortico-adrenal insufficiency. J Am Med Assoc 114: 866–868.

Zeltser D, Rosansky S, van Rensburg H et al. (2007). Assessment of the efficacy and safety of intravenous conivaptan in euvolemic and hypervolemic hyponatremia. Am J Nephrol 27: 447–457.

Handbook of Clinical Neurology, Vol. 124 (3rd series)
Clinical Neuroendocrinology
E. Fliers, M. Korbonits, and J.A. Romijn, Editors

Chapter 4

The role of oxytocin and vasopressin in emotional and social behaviors

RACHEL BACHNER-MELMAN[1,2] AND RICHARD P. EBSTEIN[3]*
[1]*School of Social and Community Sciences, Ruppin Academic Center, Emek Hefer, Israel*
[2]*Department of Psychology, Hebrew University of Jerusalem, Jerusalem, Israel*
[3]*Department of Psychology, National University of Singapore, Singapore*

INTRODUCTION

The endocrine system and the plethora of signaling molecules that have developed over the course of vertebrate evolution figure prominently in mediating social and emotional behaviors. The neuropeptides oxytocin (OT) and arginine vasopressin (AVP) share a similar chemical structure, a long evolutionary history, and a role in molding social interactions across the vertebrates (Carter et al., 2008; Heinrichs and Domes, 2008). Converging lines of evidence have recently highlighted these nonapeptides as modulators of social and emotional behavior (Meyer-Lindenberg et al., 2011). In the brain, OT and AVP act as neurohormones or neuromodulators, and their receptors are distributed in various regions associated with the central nervous control of stress and anxiety (Landgraf and Neumann, 2004). OT and AVP modulate the integration of excitatory information for emotional aspects of the social brain and interact with the mesolimbic reward pathway to regulate dopaminergic circuits for reward (Ludwig and Leng, 2006).

Basic social behaviors are necessary to most animals, whether they tend to live alone or in groups. Animals need to cooperate with others to find food, create a nest or shelter to rear their young, and defend it from predators. They need to connect to a mate to reproduce and parent their offspring. Sociality and emotionality are mediated by multiple systems and use sensory, autonomic, emotional, and motor mechanisms to permit approach to others or prevent it and cause withdrawal behaviors. The endocrine system and the signaling molecules that have evolved in vertebrates, in particular OT and AVP, play a prominent part in mediating a number of social and affiliative behaviors in species ranging from fish to humans.

Even though we humans necessarily have limited understanding of and access to emotions in animal experience, emotional behaviors such as aggression, trust, bonding, attachment and maternal behaviors, stress, and altruism would appear to bear at least some relevance to human emotional experience. The identification of genes fostering emotional and social behaviors has become a fascinating area of research leveraging on the increasing power of cutting-edge genomic tools, and is of considerable relevance to human emotional behavior.

We will now examine the role of the nonapeptide hormones OT and AVP in relevant research with nonclinical populations, before examining their role in clinical populations.

NONCLINICAL POPULATIONS

Oxytocin

OT's significant role in human social and emotional behaviors has become increasingly clear over the past few years (Donaldson and Young, 2008; Heinrichs et al., 2009). Imaging studies reinforce the role of OT in influencing our social and emotional behavior, and there is strong evidence that it modulates activity in the amygdala and other brain regions (Baumgartner et al., 2008). OT receptors in the human brain are preferentially located in areas associated with emotions and social behaviors, especially the amygdala, hypothalamus, and anterior cingulate (Loup et al., 1991; Verbalis, 1999).

*Correspondence to: Richard P. Ebstein, Psychology Department, Faculty of Arts and Sciences, National University of Singapore, Block AS4, 9 Arts Link, Singapore 117570. Tel: +65-91140010, E-mail: rpebstein@gmail.com

Interestingly, there is considerable evidence that dendrites are "a major source of peptides released in the brain; this release is not specifically targeted at synapses, and the long half-life of peptides in the CNS and their abundance in the extracellular fluid mean that, after release, they can diffuse to distant targets. At their targets, peptides seem to be able to functionally reorganize neuronal networks, providing a substrate for prolonged behaviours" (see review by Ludwig and Leng, 2006). Hence distribution of oxytocin and vasopressin receptors in the brain is more informative than fibers regarding distribution of these neuropeptides.

The mediation of OT release in the brain by ADP-ribosyl cyclase and /or CD38 was discovered as a result of a seminal paper by Higashida and his group (Jin et al., 2007; see Fig. 4.1). CD38 is a multifunctional molecule (ectoenzyme) that has both enzymatic and receptor properties and plays a key role in a range of tissue processes such as proliferation, differentiation, migration, adhesion, and secretion. In the brain, CD38 is found in neurons and glial cells, shows intracellular or plasma membrane location, and is enriched in neuronal perikarya and dendrites (Mizuguchi et al., 1995; Ceni et al., 2006). It is critical for the release of OT but not AVP (Jin et al., 2007). OT plays a central role in many emotional behaviors (Table 4.1), such as those noted below.

Trust

Hormonal effects on human behavior have been especially noticeable with respect to trust and trustworthiness. It has been repeatedly shown that higher oxytocin levels are associated with trustworthy behavior (Heinrichs and Domes, 2008; Campbell, 2010) and that the intranasal administration of OT increases both trust (Kosfeld et al., 2005) and the perception of faces as being trustworthy (Theodoridou et al., 2009). Interestingly, Zak and his colleagues (2005) have shown that not only does oxytocin increase human trust in social interactions, but that trust has a positive feedback effect on OT levels. The latter are higher in people who receive money as a result of trusting in comparison to OT levels in people who receive money that was transferred unintentionally. OT therefore appears to increase trustees' oxytocin levels and make them even more trustworthy.

Mind reading

The ability to "read the mind" of other individuals, or infer their mental state by interpreting subtle social cues, is indispensable in human social interaction and central to emotional experience. Domes et al. (2007b) found that after the application of oxytocin, subjects made fewer errors in identifying facial emotional expression using the Reading the Mind in the Eyes Test (RMET). It appears that oxytocin improves emotion recognition specifically when a task is demanding, because it improved performance only on more difficult test items that generated less than 70% accuracy at baseline. In other studies, OT selectively improved the recognition of happy facial expressions but impaired the decoding of negative facial expressions (Di Simplicio et al., 2009; Marsh et al., 2010).

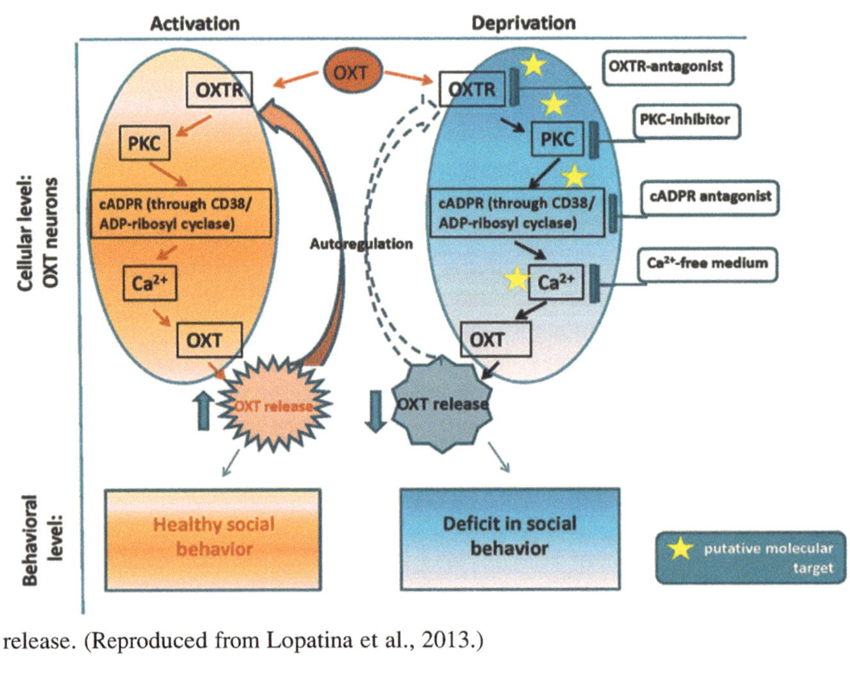

Fig. 4.1. Oxytocin release. (Reproduced from Lopatina et al., 2013.)

Table 4.1

Studies of the effects of oxytocin and arginine vasopressin on various emotional behaviors

Functional domain	Model paradigm	Effect of OT	Effect of AVP	Organism	References
Pair-bonding	Laboratory behavior	+	+	Prairie vole	(Carter et al., 1992; Winslow et al., 1993)
	Questionnaire Genetic	+		Human	(Walum et al., 2012)
	Questionnaire Genetic AVPR la		+	Human	(Walum et al., 2008)
	Plasma OT Distressed pair-bonding	+	+	Human	(Taylor et al., 2010)
Romantic attachment	Field study Plasma OT	+		Human	(Schneiderman et al., 2012)
Parenting	LAB Plasma OT Genetic	+		Human	(Bakermans-Kranenburg and van Ijzendoorn, 2008; Feldman et al., 2010, 2011, 2012; Gordon et al., 2010; Galbally et al., 2011; Feldman, 2012; Weisman et al., 2012)
Empathy	Questionnaire Genetic OXTR	+		Human	(Rodrigues et al., 2009; Wu et al., 2012)
	Sniffing OT	Context-dependent		Human	(Bartz et al., 2010)
	Sniffing AVP	Context dependent		Human	(Thompson et al., 2004, 2006; Uzefovsky et al., 2012)
Trust	LAB games Sniffing	↑		Human	(Zak, 2005; Kosfeld et al., 2005)
	LAB games Plasma OT	U shaped		Human	(Zhong et al., 2012)
	LAB games Genetic	+		Human	(Krueger et al., 2012)
Mind reading and facial processing	LAB Sniffing OT	+		Human	(Domes et al., 2007a, b; Guastella et al., 2008b)
Altruism	LAB games Sniffing Genetic	Overall OT enhances altruism but context dependent	No effect	Human	(Zak et al., 2007; Israel et al., 2009, 2012; De Dreu et al., 2010; Declerck et al., 2013)
	Lab games AVP AVPR la Genetic			Human	(Knafo et al., 2008a)
Moral judgment	Questionnaire Moral dilemmas Genetic OXTR	Intention Perspective taking		Human	(Walter et al., 2012)
			Clinical studies		
Autism	Genetic evidence AVPR la		+	Human	(Kim et al., 2002; Wassink et al., 2004; Yirmiya et al., 2006)
	Genetic evidence OXTR	+		Human	(Wu et al., 2005; Lerer et al., 2008; Gregory et al., 2009)
	CD38 Related to OT	+		Human	(Jin et al., 2007; Lerer et al., 2010; Ebstein et al., 2011; Riebold et al., 2011; Higashida et al., 2012)

Continued

Table 4.1

Continued

Functional domain	Model paradigm	Effect of OT	Effect of AVP	Organism	References
	Sniffing OT		+	human	(Andari et al., 2010; Guastella et al., 2010; Kuehn, 2011)
Schizophrenia	Plasma AVP		+	Human	(Rubin et al., 2013)
	Genetic evidence OXTR AVP	+	+	Human	(Souza et al., 2010; Teltsh et al., 2011; Montag et al., 2013)
	Drug target	+		Human	(Feifel, 2012)
	Plasma OT Trust game	+		Human	(Keri et al., 2009)
Anxiety and depression	Knockout mouse	+		Mouse	(Caldwell et al., 2008)
	Plasma OT Genetic OXTR	+		Human	(Parker et al., 2010; Weisman et al., 2013; Apter-Levy et al., 2013)
Social stress	Sniffing OT Genetic OXTR	+		Human	(Heinrichs et al., 2001, 2003; Chen et al., 2011; Kumsta and Heinrichs, 2013)
	Sniffing AVP		+	Human	(Shalev et al., 2011)

OT, oxytocin; AVP, arginine vasopressin; OXTR, OT receptor.

Empathy

A prerequisite for mental wellbeing and positive social interactions is empathy, the ability to share the other's feelings (Hein and Singer, 2008). The developmental emergence of empathy was studied by Knafo and colleagues (2008b), who examined the contribution of genes and environment to an empathy factor in 409 pairs of young twins. No genetic influences were found at 14 and 20 months, with strong shared environmental influences accounting for most of the variance. However, at 24 and 36 months, genetics accounted for 34–47% of the variance in the common empathy factor, while shared environment effects decreased from 0.69 at 14 months to 0 at 36 months. Overall, genetics accounted for both change and continuity in empathy, but their role changed as children grew up. In another twin study conducted with children 3.5 years of age, moderate heritabilities were estimated for individual differences in empathy, and the rest of the variance was accounted for by nonshared environment (Knafo et al., 2009).

In a recent study designed to examine the role of specific genes contributing to empathy (Rodrigues et al., 2009), the authors examined individual differences at a polymorphic site in the oxytocin receptor (OXTR) gene, rs53576, previously found to be associated with autism (Wu et al., 2005) and maternal sensitivity (Bakermans-Kranenburg and van Ijzendoorn, 2008). Compared with individuals homozygous for the G allele of rs53576 (GG), individuals with one or two copies of the A allele (AG/AA) exhibited lower behavioral and dispositional empathy.

Positive communication between couples

OT is involved in couple interaction and close relationships in humans. It increases gaze to the eye region of faces (Guastella et al., 2008a), which may be one mechanism that enables oxytocin to enhance emotion recognition, interpersonal communication, and social approach behavior.

Feldman and colleagues (Schneiderman et al., 2012) found plasma OT to be significantly higher in new lovers compared to singles. The author concluded that OT may play an important role during the first stages of romantic attachment, lending support to evolutionary models according to which parental and romantic attachment share underlying biobehavioral mechanisms (Feldman, 2012).

A support intervention (warm touch enhancement) enhanced salivary oxytocin in husbands and wives relative to controls (Holt-Lunstad et al., 2008). Gouin and colleagues (2010) admitted 37 couples for a 24 hour visit to a hospital research unit and created small blister wounds on their forearms before giving them a structured social support interaction task. Following discharge, blister sites were monitored daily and blood samples were collected. Higher oxytocin levels were associated with more positive communication behaviors during the structured interaction task. The blister wounds of individuals in the upper oxytocin quartile healed faster than those of participants in lower oxytocin quartiles. These results confirm and extend prior evidence implicating oxytocin in couples' positive and negative communication behaviors and provide further evidence of its role in wound healing, an important health outcome.

In a double-blind placebo-controlled design (Ditzen et al., 2009), 47 heterosexual couples received oxytocin or placebo intranasally before engaging in a couple conflict discussion in the laboratory. The conflict session was videotaped and coded for verbal and nonverbal interaction behavior, such as eye contact, nonverbal positive behavior, and self-disclosure. Salivary cortisol was repeatedly measured during the experiment. Oxytocin significantly increased positive communication in relation to negative communication behavior during the conflict discussion and significantly reduced salivary cortisol levels after the conflict compared with placebo. These results are in line with animal studies indicating that central oxytocin facilitates approach and pair-bonding behavior.

Generosity and altruism

OT has been implicated in a variety of prosocial processes, and much of this research has used social dilemma tasks. Israel and colleagues (2009) demonstrated association between 15 single-tagging SNPs across the OT receptor (OXTR) gene and altruism as defined in a dictator game (Hoffman et al., 1996) and a social values orientation paradigm (Van Lange, 1999).

Zak et al. (2007) tested whether OT prompts generosity between anonymous human strangers by using two decision-making tasks from experimental economics, the ultimatum game and the dictator game, after the administration of either intranasal oxytocin or placebo. In both games, participants were given 10 dollars and asked to share an amount of their choice with another individual whose identity was masked. In the ultimatum game, unlike in the dictator game, the person receiving the money could refuse the offer, in which case neither player would receive anything. OT raised generosity in the ultimatum game, but not the dictator game, by 80% over placebo. OT may therefore play a role in giving situations where there is a threat of punishment and an emotional identification with another person whose motives need to be understood.

Bonding and attachment

Intriguingly, OT plasma levels have been linked to individual patterns of maternal–fetal attachment (Levine et al., 2007). Salivary OT levels were found to be associated with bonding to parents and inversely related to psychological distress, in particular depressive symptoms (Gordon et al., 2008).

Oxytocin and mirror neurons

The mirror neuron system (MNS), originally discovered in monkeys (Rizzolatti et al., 1996), has evolved in humans into a broad neural system that enables the simulation and understanding of others' intentions, thoughts, and feelings (Gallese et al., 2004). The discovery that neural activity measured by electroencephalogram (EEG) oscillations in monkeys reflects both perception and execution of biologic motion led authors to tentatively trace the suppression of μ rhythms to a human mirror neuron system (for a review see Pineda, 2005). For example, μ rhythms, measured between 8 and 12 Hz over somatomotor regions, become desynchronized and less powerful when a person moves (Gastaut et al., 1952) or observes another person's actions (Muthukumaraswamy and Johnson, 2004; Muthukumaraswamy et al., 2004). Several recent studies have linked EEG μ suppression to higher social and emotional information processing including social skills (Oberman et al., 2007), theory of mind (Pineda and Hecht, 2009; Perry et al., 2010b), and empathy (Cheng et al., 2008a, b).

Perry et al. (2010a) investigated the effect of intranasally administered OT on the suppression of EEG rhythms in the low and high α/μ range, and in the β band (15–25 Hz). The administration of OT was found to enhance α/μ and β suppression. This study links two previously separate focuses of social and emotional functioning in humans: the nonapeptide hormone OT and mirror neuron systems. It shows that OT affects EEG rhythms in the α/μ and β ranges differentially in tasks of biologic motion and nonbiologic motion. OT therefore appears to modulate an EEG oscillation thought to partially mediate complex interpersonal emotional processes such as theory of mind and empathy. By showing that OT enhances μ suppression, the Perry et al. (2010a) study establishes a common groundwork that appears to link OT and mirror neurons.

Vasopressin

The actions of OT and vasopressin (AVP) are frequently in opposing directions (Carter et al., 2008). OT shows anxiolytic and antistress effects in both genders that seem to be localized within the central amygdala and the hypothalamic paraventricular nucleus (Neumann, 2008). In contrast, AVP is associated with arousal, vigilance, and defensive behaviors (Carter, 1998; Landgraf and Neumann, 2004).

The role of AVP in human behavior is still relatively poorly understood and the intranasal administration of AVP in studies with people has been scarce in comparison with the intranasal administration of OT. Relatively few investigations (Pietrowsky et al., 1996; Born et al., 2002; Thompson et al., 2006; Zink et al., 2011) have used intranasal AVP with the aim of helping to unravel the mode of AVP action on the human emotional and social brain. Androgens control AVP synthesis in several brain areas (De Vries et al., 1994), supporting a role for this hormone specifically in male emotions and social interactions. Below we examine evidence for a role for AVP in specific facets of human social and emotional behavior. The vasopressin system is illustrated in Figure 4.2.

Aggression

Animal models show a role for AVP in male aggressive behavior (Winslow et al., 1993). Research using the intranasal administration of AVP in rodents shows a sex-specific effect role for AVP, which promotes aggression in the face of territorial intrusions in males and influences mating (Winslow et al., 1993; Wang et al., 1994). Thompson and colleagues (2004, 2006) found that intranasal administration of AVP stimulated agonistic facial motor patterns in response to the faces of unfamiliar men and decreased perceptions of the friendliness of those faces. Interestingly, AVP had an opposite effect in women, increasing perceptions of the friendliness of those faces. These findings point to a major role of AVP in modulating the perception of emotions differentially for men and women.

Additional evidence for a role of AVP in human aggression is suggested by a study of cerebrospinal fluid levels of AVP that were positively correlated with aggressive behaviors in subjects suffering from personality disorders (Coccaro et al., 1998).

Psychosocial stress

AVP, which has been related to both social behavior and hypothalamus–pituitary–adrenal (HPA) axis regulation, seems to play a role in the modulation of stress in psychosocial situations (Goodson, 2008). Intranasal AVP may directly activate the HPA axis in the presence or absence of stressful cues, and stressors in general may

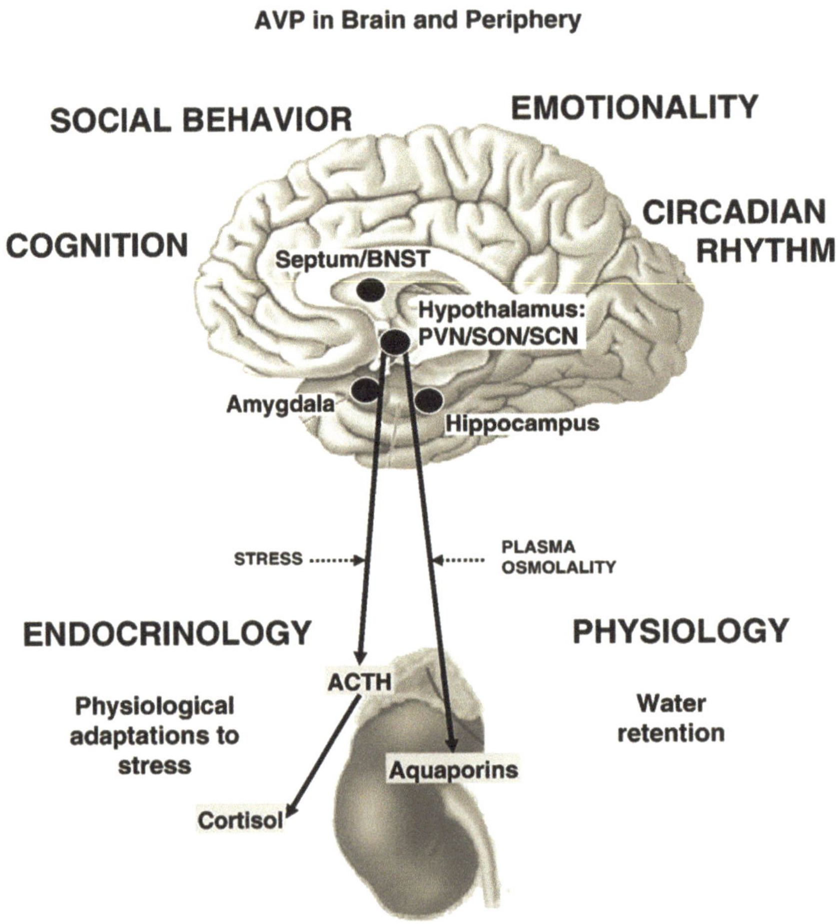

Fig. 4.2. The vasopressin system. (Reproduced from Frank and Landgraf, 2008.)

interact with AVP to trigger a rise in cortisol even in the absence of a social evaluative threat. Shalev et al. (2009) used the Trier Social Stress Test (TSST) (Kirschbaum et al., 1993), a paradigm effective in evaluating stress levels under controlled laboratory conditions, to address the importance of evaluative social stress. The TSST uses public speaking and mental arithmetic tasks to evoke a salivary cortisol response to stress. Individual responses are influenced by gender (Kirschbaum et al., 1995; Uhart et al., 2006), genes (Kumsta et al., 2007; Shalev et al., 2009), environmental stressors (Macmillan et al., 2009), and other factors (Kudielka et al., 2009). HPA axis reactivity was assessed by measuring salivary cortisol, and central nervous system reactivity was measured by monitoring blood pressure and pulse rate.

To test the hypothesis that the effect of intranasal AVP on the salivary cortisol response is contingent upon social contexts, Shalev et al. (2011) administered the full TSST to one group and devised experimental conditions aimed at isolating the social evaluative threat in three distinct control groups. The first was a "no task" group controlling for direct physiologic influences of AVP administration on HPA reactivity under a no stress condition. Participants were administered intranasal AVP or placebo while sitting alone in a controlled environment, in the absence of stressful stimuli. The second experiment was a "no audience" situation. Participants performed the TSST tasks, but no audience and cameras were present, so that they were experiencing low social evaluative threat but enough stress to trigger a cortisol response. The third experiment was designed to evoke physiologic stress (cortisol, blood pressure, and heart rate) but not a social stress response. Participants rode an exercise bike without an audience and without a camera. AVP augmented the response to social stress solely in the experimental condition of full TSST. No AVP-related augmentation was seen in any of the three control conditions. AVP therefore appears to increase sensitivity to the social milieu, specifically to the presence of observers. Stress in people is no doubt conferred through hedonic processes that relate to respect, self-esteem, acceptance, and positive social attention (Dickerson and Kemeny, 2004).

EMPATHY

The ability to empathize, or accurately perceive another's emotions and respond to them in an appropriate way, lies at the core of human experience and uniqueness. Bachner-Melman et al. (2005) found linkage between two microsatellite markers near the vasopressin receptor (AVPR1A) gene located on chromosome 12q14-15 (RS1 and RS3) and sibling relationships. Bisceglia and colleagues (2012) reported an association between RS3 and a composite of "maternal sensitivity". It is a multi-faceted concept involving at least two processes: cognitive empathy (CE) – the ability to perceive what the other is feeling – and emotional empathy – matching an emotional response to the feelings of the other (Davis, 1980). Facial expressions are salient representatives of our emotions and interpreting facial expression appears to be a robust indicator of CE (Baron-Cohen et al., 2001).

When Uzefovsky et al. (2012) examined the effect of intranasal AVP versus placebo on the ability of male participants to identify the emotions exhibited in RMET eye pictures, the AVP group made significantly more errors than the placebo group ($t(37) = 2.199$, $p = 0.034$, $D = 0.72$). Given the known gender-specific effect of AVP (Thompson et al., 2004, 2006), the perceived gender of the individuals photographed was examined in the context of emotion recognition (the RMET includes photographs of both men and women). When scores were stratified by the sex of the person in the photograph, a large AVP effect was observed ($t(37) = -2.77$, $p = 0.009$, $D = 0.91$), but only when males observed male photographs. No effect of AVP was observed when males observed female photographs. The RMET pictures were then sorted into negative versus positive emotions. The effect of AVP on reducing empathy was observed for negative ($t(37) = -2.38$, $p = 0.023$) but not positive emotions, which supports the results of Thompson et al. (2004) showing a role for AVP in inducing aggressive emotions selectively in men.

AVP therefore seems to diminish CE in males, but only specifically towards other males. Importantly, this finding echoes previous knowledge derived from animal models showing a role for AVP in aggression-related male social behavior (Winslow and Insel, 1993). In humans, physical aggressive behavior by males is more common than by females (Archer and Coyne, 2005) and it is more commonly directed towards other males. Feelings of aggression, which increase the probability of a physical attack, are accompanied by diminished ability to empathize with the possible victim (Preston and de Waal, 2002). All these claims point to a role for AVP in reducing the CE of males towards other males.

Empathy has also been hypothesized to underlie processes involved in prosocial behavior such as altruism (Batson et al., 1991).

ALTRUISM

Altruism, or the provision of benefit for others at the cost of one's own, is a prominent trait characteristic of our species. Altruistic behavior challenges evolutionary theory, in that natural selection favors prosocial traits over selfish ones. It poses not only an evolutionary but an economic paradox, seeming to contradict the principle of profit maximization. By all accounts, *Homo sapiens* shows extraordinarily altruistic tendencies. Numerous explanations have been proposed, including kin selection, reciprocal altruism, indirect reciprocity, and altruistic punishment (Sigmund and Hauert, 2002; Fehr and Rockenbach, 2004).

An important experimental approach in understanding human altruistic behavior and its biologic underpinnings has been the adoption of behavioral economics paradigms. Knafo et al. (2008a) used a molecular genetic approach combined with a classic behavioral economic paradigm, the dictator game. Participants "dictate" how much of a fixed sum of money they receive they will give to a passive recipient. The length of the arginine vasopressin 1a receptor promoter (AVPR1a) RS3 microsatellite predicted fund allocations. The length of the RS3 microsatellite was associated with greater giving behavior. Meyer-Lindenberg et al. (2009) also found that long AVPR1a alleles predicted greater amygdala activation during functional imaging employing an emotional face-matching paradigm.

CLINICAL POPULATIONS

Whereas the discussion of OT and AVP has so far focused on aspects of normal behavior and normal sociality, accumulating evidence suggests that OT and AVP also contribute to psychopathology. In particular, these two nonapeptides have been linked to autism spectrum disorder (ASD), a condition characterized by widespread anomalies in the domain of emotions.

Oxytocin and vasopressin in autism spectrum disorders

It was first suggested over a decade ago that OT and AVP might contribute to ASD (Modahl et al., 1992; Panksepp, 1993; Waterhouse et al., 1996; Freeman, 1997; Insel et al., 1999). The connection between these hormones and ASD is a good example of the relevance of animal models to human behavior. In 1987 Modahl (Modahl et al., 1998) found lower plasma OT levels in children with autism than normal children. OT increased with age in the normal children in this study, but not the autistic children. Whereas elevated OT was associated with higher scores on social and developmental measures for the normal children, it was associated with *lower* scores for the autistic children. These relationships were strongest in a subset of autistic children identified as "aloof".

Several studies of ASD have found abnormal μ suppression, hypothesized to involve mirror neurons and OT (Perry et al., 2010a), when ASD individuals viewed actions performed by others. When the participants performed the same actions themselves, μ suppression was normal (Martineau et al., 2004; Oberman et al., 2008).

Direct experimental evidence suggesting a role of AVP and OT in ASD was provided by association studies linking AVPR1a with idiopathic autism (Kim et al., 2002; Wassink et al., 2004; Yirmiya et al., 2006). The human AVP V1a (AVPR1a) receptor gene is relatively simple, containing two exons and one intron, located at 12q14-15 with three polymorphisms in the 5′ flanking region and one in the intron (Thibonnier et al., 1994; Thibonnier, 2004). The 5′ flanking region microsatellites RS1 and RS3 have received the most attention, with links to a diverse set of interpersonal skills from sibling relationships (Bachner-Melman et al., 2005) to musical ability (Granot et al., 2007).

Kim and colleagues (2002) genotyped two microsatellite polymorphisms (RS3 and RS1) from the 5′ flanking region of AVPR1A for 115 autism trios and found nominally significant transmission disequilibrium between autism and RS1 by a family-based association test that was not significant after Bonferroni correction. In a second study, by Wassink and colleagues (2004), association was observed with both promoter region markers, but only for children with normal language abilities. In a third study (Yirmiya et al., 2006), no significant association was found between RS1 and RS3 and autism, although significant transmission disequilibrium was observed between an intronic microsatellite (AVR) and the promoter region microsatellites. Haplotype analysis showed significant transmission disequilibrium in autism families when all three microsatellites (two promoter region markers and the intronic microsatellite) were analyzed. Association between AVPR1a and daily living and communication skills, as measured by the Vineland Adaptive Behavioral Scales (VABS) and ADOS-G, was also observed. To summarize, both translational research and more direct molecular genetic studies indicate a provisional role for the AVPR1a receptor in contributing to the etiology of ASD.

An interesting Swedish study strengthens the connection between AVPR1a and emotional behavior (Walum et al., 2008). The RS3 repeat polymorphism was found to be associated with traits reflecting the establishment of intimacy in males: pair-bonding including partner bonding, perceived marital problems, as well as marital status. The RS3 genotype of the males also affected marital quality as perceived by their spouses.

Intriguingly, the 334 bp risk allele that doubles the risk of marital crisis is the same allele that is overtransmitted in ASD (Kim et al., 2002). Yirmiya et al. (2006) took a closer look at the role of this allele in autism and also found overtransmission of this second most common allele (RS3 allele 5). This strengthens the notion that the second most common RS3 allele contributes risk for dysfunctional social behaviors. Furthermore, a recent imaging study by Meyer-Lindenberg's group (Meyer-Lindenberg et al., 2009) showed that the second most common 334 bp risk allele of RS3 (present in 21.3% of subjects) showed differential overactivation of the left and right sections of the amygdala.

Oxytocin and vasopressin in eating pathology

An association between the RS3 microsatellite polymorphism in the AVPR1A promoter region and pathologic dieting was observed by Bachner-Melman et al. (2004). The AVPR1A receptor may contribute to dieting behavior via the mechanism of stress-induced feeding problems, since vasopressin is involved in the leptin-induced activation of the hypothalamic pituitary axis (HPA) axis (Morimoto et al., 2000; Levine, 2001). Stressors and coping difficulties commonly precede the development of anorexia nervosa, so that the HPA axis no doubt plays a role in the etiology of this often fatal disorder (Horesh et al., 1996; Welch et al., 1997).

Evidence for an involvement of OT neurons in eating disorders is suggested by a report showing a robust reduction in the number of OT-expressing neurons of the Prader–Willi syndrome (PWS) (Swaab et al., 1995). Interestingly, in this syndrome the number of AVP-expressing neurons in the PVN did not change significantly. Although these results were in PWS patients, they nevertheless suggest that OT neurons of the PVN likely play a role in ingestive behavior as "satiety neurons" in the human hypothalamus.

Oxytocin and vasopressin in depression and anxiety

Earlier studies suggest a role of OT as well as AVP in depression (Swaab et al., 1995; Purba et al., 1996) based on an increase in the number of OT and AVP immunoreactive cells in postmortem brain tissue from depressed patients. A later investigation revealed a 60% increase of vasopressin mRNA expression in depressed compared with control subjects (Meynen et al., 2006). In the melancholic subgroup, AVP mRNA expression was significantly increased in both the supraoptic nucleus (SON) and the paraventricular nucleus (PVN) compared with control subjects.

More recent findings further support that functional irregularities in OT biology may play a role in the pathophysiology of depression and anxiety. OT mediates an antidepressant-like effect in male mice, which disappears in OTR knockout mice (Matsushita et al., 2010, 2012). In humans, plasma OT levels are associated with major depressive disorders (Frasch et al., 1995; Scantamburlo et al., 2007). The symptom severity of social anxiety, adjusted for age and gender in a healthy group of subjects, was found to be associated with higher plasma oxytocin levels (Hoge et al., 2008). An OT receptor polymorphism is associated with major depression and anxiety in adolescent girls (Thompson et al., 2011), and nasal administration of OT improves some symptoms of depression and anxiety (Scantamburlo et al., 2007; Neumann and Landgraf, 2012).

Methodologic issues in oxytocin and vasopressin research

Currently there are differences in opinion surrounding the measurement of oxytocin (Szeto et al., 2011) and particularly concerning the requirement of sample extraction. The commercially available oxytocin EIA kit from Enzo Life Sciences (formerly Assay Designs), which has been validated for linearity, cross-reactivity, matrix effects, accuracy, precision, and recovery (Carter et al., 2007), has been used in many studies. The experience of some investigators suggests that extraction of the samples leads to significant loss of measurable oxytocin. Importantly, the oxytocin data from nonextracted samples makes biologic sense as compared to those from extracted samples, which often gave rise to nondetectable levels. (For a fuller discussion see Zhong et al., 2012.)

Evidence that intranasally administered small peptides can permeate the blood–brain barrier was summarized in the study by Born and colleagues (2002). Moreover, intranasal administration of oxytocin is also detected in a substantial rise in plasma levels 30 minutes after administration (Gossen et al., 2012). Interestingly, there is a slight augmentation of endogenous testosterone levels 210 minutes after oxytocin. Generally, 24 IU of oxytocin are administered intranasally and the effect of the hormone examined 45 minutes later (Domes et al., 2007b).

CD38 and abnormal social and emotional behaviors

Accumulating evidence that OT plays a role in dysfunctional social relationships as well as normal ones makes CD38 a key mediator of OT brain release (Higashida et al., 2007; Jin et al., 2007), a potential focus of interest in normal human and abnormal emotional behaviors, for example in autism (Bartz and McInnes, 2007; Young, 2007). To summarize: two research groups have recently examined the role of CD38 in autism. Higashida's group (Munesue et al., 2010) analyzed 10 SNPs and mutations of CD38 by resequencing DNAs based on a case-control study of Japanese and Caucasian families with a child with ASD. CD38 SNPs rs6449197 and rs3796863 showed significant associations with a subset of ASD subjects

(IQ > 70; designated as high functioning autism (HFA)) in the Caucasian but not the Japanese family trios.

In another study of CD38 (Lerer et al., 2010), all tagging SNPs across the CD38 gene region were examined in 170 subjects with ASD. Individual SNPs and haplotypes were tested for association with ASD and CD38 gene expression was measured in lymphoblastoid cell lines (LBC) derived from ASD subjects and unaffected parents. When ASD subjects were divided into high and low functioning groups based on an IQ cutoff of 70, significant association was observed between low functioning ASD and three to seven haplotypes. The SNP (rs3796863) found to be significantly associated with ASD in the Munesue et al. (2010) study was located in all but one of the significant haplotypes (Lerer et al., 2010).

CD38 mRNA levels were then examined to determine whether CD38 expression in peripheral cells might be a biomarker for ASD. Main effects were observed for diagnosis and sex. A highly significant reduction in CD38 expression was observed in cells from the children with ASD compared to "unaffected" parents.

These results (Lerer et al., 2010) were partially replicated in a subsequent investigation (Riebold et al., 2011) conducted with the same sample with 38 additional cell lines, so that both parents were included. Cells in culture or frozen lines were thawed and cultured and their CD38 mRNA levels measured. CD38 expression in ASD patient lines was significantly lower than in those derived from the patients' parents. Although these results are not a fully independent replication, we believe they considerably strengthen the hypothesis that reduced CD38 transcription is a characteristic of peripheral lymphocyte cells derived from ASD subjects (Lerer et al., 2010).

Endophenotypes

The search for biologic endophenotypes of behavior allows across-syndrome cohort selection, including nonclinical subjects, based on trait homogeneity. This approach can loosen the ties between biologic research and clinical syndromes (Levy and Ebstein, 2009). Decomposing syndromic disorders into intermediate or endophenotypes can be seen as a necessary intermediate step in understanding the links between genetics and behavior (Gottesman and Gould, 2003). Examining the role of OT and AVP in characteristics of both nonclinical and clinical syndromes can be highly informative, for example for ASD. Studies in nonclinical cohorts have revealed the rich roles of OT and AVP in molding emotional behaviors in humans with implications for understanding deficits seen in autism. We believe these investigations of nonclinical cohorts have provided the scientific basis for clinical trials in several centers where OT is being tested as a novel therapeutic agent in ASD.

FUTURE DIRECTIONS

Knowledge about the connection between social behavior, emotions, and the human brain has recently taken huge strides by leveraging neuroimaging, neuroendocrinology, and more recently, neurogenetics. Neuroimaging studies have delineated the neural correlates of emotional responses, especially in social interactions (Adolphs, 2003). Whereas pheromones allow animals to connect and transfer social information (Dulac and Torello, 2003), humans process emotional behavior and messages to a large extent via visual and face perception, and a large body of research has developed in this realm (Tsao et al., 2003). Future studies could stratify responses to intranasal OT by genotype. Both the OXTR (Mizumoto et al., 1997; Kusui et al., 2001; Gregory et al., 2009) and AVPR1a genes are characterized by epigenetic markings that may well be a good focus for future investigation. Our understanding of variance in individual responses to nonapeptides could be improved by factoring in environmental variables using the methylation of CpG islands.

The discovery of primate mirror neurons system neuroscience of emotional behaviors has dramatically impacted neuroscience. The hormone OT may modulate the brain's mirror neuron system (Perry et al., 2010a). Imitation and mimicry are automatic behaviors that facilitate empathy (Iacoboni, 2009). The discovery that OT modulates μ suppression may provide insight into the mechanism by which OT enhances empathy and other-regarding emotional behaviors such as altruism and trust.

Mirror neurons discharge during the execution and the observation of actions, and so represent the overlap between perception and action. Broadly congruent mirror neuron cells fire during one's own and others' actions, and may prove to be a simple neural mechanism for recognizing and coding others' actions and for promoting cooperative behavior (Newman-Norlund et al., 2007).

Future research should further examine the role of OT in modulating mirror neuron function, since it appears to affect a wide range of human emotional and social behaviors. A neurogenetic strategy stratifying individual differences in OT modulation of μ suppression by genotype information would be valuable.

Studies involving the intranasal administration of AVP have lagged considerably behind those involving OT. This gap could be due to the fact that OT, unlike AVP, is available commercially as a nasal spray. Future research should aim to reduce this gap and more studies using intranasal AVP are needed in order to better understand the role of this hormone, coupled with OT, in shaping social and emotional behavior.

The work of Thompson and colleagues (2004, 2006) and Uzefovsky and colleagues (2012) points to a gender-specific effect for intranasal AVP on facial information processing, a precondition for empathy. The imaging strategy adopted by Meyer-Lindenberg and colleagues showed that human amygdala function is strongly associated with genetic variation in AVPR1a (Meyer-Lindenberg et al., 2009). The intranasal administration of this nonapeptide also seems to enhance HPA axis reactivity but only in social contexts, for example when there is a direct threat to the maintenance of our social selves (Shalev et al., 2011). Stress evoked by social threat is an integral part of emotional functioning and is related to self-esteem and in extreme forms, to poor mental health (e.g., social phobia).

Another area of interest for future research is the role of CD38 and retinoids. Provisional results point to a provisional role for CD38 in conferring vulnerability to ASD (Ebstein et al., 2009, 2011; Lerer et al., 2010; Riebold et al., 2011). These investigations should be extended to nonclinical populations. The induction by all-*trans*-retinoic acid of CD38 can be rescued *in vitro* by ATRA treatment, as can cell lines showing decreased CD38 transcription (Lerer et al., 2010; Riebold et al., 2011). Vitamin A may therefore have a modulatory role in these processes. The hypothesis that the intranasal administration of vitamin A drops mimics the effect of OT treatment could therefore be investigated.

Finally, deficits in emotional functioning characterize a number of psychopathologies such as autism, schizophrenia, personality disorders, and social phobias. Understanding the molecular architecture of the social brain from neural correlates to neurogenetics in nonclinical subjects will help unravel the complexity of psychopathologies characterized by social and emotional deficits. This connection goes in both directions, and an understanding of deficits in the social brain in psychopathology may also enhance our appreciation of normal empathy and mentalization processes.

References

Adolphs R (2003). Cognitive neuroscience of human social behaviour. Nat Rev Neurosci 4: 165–178.

Andari EJR, Duhamel T, Zalla E et al. (2010). Promoting social behavior with oxytocin in high-functioning autism spectrum disorders. Proc Natl Acad Sci U S A 107: 4389–4394.

Apter-Levy YM, Feldman A, Vakart RP et al. (2013). Impact of maternal depression across the first 6 years of life on the child's mental health, social engagement and empathy: the moderating role of oxytocin. Am J Psychiatry 170: 1161–1168.

Archer J, Coyne SM (2005). An integrated review of indirect relational and social aggression. Pers Soc Psychol Rev 9: 212–230.

Bachner-Melman R, Zohar AH, Elizur Y et al. (2004). Association between a vasopressin receptor AVPR1A promoter region microsatellite and eating behavior measured by a self-report questionnaire (Eating Attitudes Test) in a family-based study of a nonclinical population. Int J Eat Disord 36: 451–460.

Bachner-Melman R, Zohar AH, Bacon-Shnoor N et al. (2005). Linkage between vasopressin receptor AVPR1A promoter region microsatellites and measures of social behavior in humans. J Indiv Differ 26: 2–10.

Bakermans-Kranenburg MJ, van Ijzendoorn MH (2008). Oxytocin receptor (OXTR) and serotonin transporter (5-HTT) genes associated with observed parenting. Soc Cogn Affect Neurosci 3: 128–134.

Baron-Cohen S, Wheelwright S, Hill J et al. (2001). The Reading the Mind in the Eyes Test revised version: a study with normal adults and adults with Asperger syndrome or high-functioning autism. J Child Psychol Psychiatry 42: 241–251.

Bartz JA, McInnes LA (2007). CD38 regulates oxytocin secretion and complex social behavior. Bioessays 29: 837–841.

Bartz JA, Zaki J, Bolger N (2010). Oxytocin selectively improves empathic accuracy. Psychol Sci 21: 1426–1428.

Batson CD, Batson JG, Slingsby JK et al. (1991). Empathic joy and the empathy-altruism hypothesis. J Pers Soc Psychol 61: 413–426.

Baumgartner T, Heinrichs M, Vonlanthen A et al. (2008). Oxytocin shapes the neural circuitry of trust and trust adaptation in humans. Neuron 58: 639–650.

Bisceglia R, Jenkins JM, Wigg KG et al. (2012). Arginine vasopressin 1a receptor gene and maternal behavior: evidence of association and moderation. Genes Brain Behav 11: 262–268.

Born J, Lange T, Kern W et al. (2002). Sniffing neuropeptides: a transnasal approach to the human brain. Nat Neurosci 5: 514–516.

Caldwell HK, Stephens SL, Young 3rd WS (2008). Oxytocin as a natural antipsychotic: a study using oxytocin knockout mice. Mol Psychiatry 14: 190–196.

Campbell A (2010). Oxytocin and human social behavior. Pers Soc Psychol Rev 14: 281–295.

Carter CS (1998). Neuroendocrine perspectives on social attachment and love. Psychoneuroendocrinology 23: 779–818.

Carter CS, Williams JR, Witt DM et al. (1992). Oxytocin and social bonding. Ann N Y Acad Sci 652: 204–211.

Carter CS, Pournajafi-Nazarloo H, Kramer KM et al. (2007). Oxytocin: behavioral associations and potential as a salivary biomarker. Ann N Y Acad Sci 1098: 312–322.

Carter CS, Grippo AJ, Pournajafi-Nazarloo H et al. (2008). Oxytocin vasopressin and sociality. Prog Brain Res 170: 331–336.

Ceni C, Pochon N, Villaz M et al. (2006). The CD38-independent ADP-ribosyl cyclase from mouse brain synaptosomes: a comparative study of neonate and adult brain. Biochem J 395: 417–426.

Chen FS, Kumsta R, von Dawans B et al. (2011). Common oxytocin receptor gene (OXTR) polymorphism and social support interact to reduce stress in humans. Proc Natl Acad Sci U S A 108: 19937–19942.

Cheng Y, Lee PL, Yang CY et al. (2008a). Gender differences in the mu rhythm of the human mirror-neuron system. PLoS One 3: e2113.

Cheng Y, Yang CY, Lin CP et al. (2008b). The perception of pain in others suppresses somatosensory oscillations: a magnetoencephalography study. Neuroimage 40: 1833–1840.

Coccaro EF, Kavoussi RJ, Hauger RL et al. (1998). Cerebrospinal fluid vasopressin levels: correlates with aggression and serotonin function in personality-disordered subjects. Arch Gen Psychiatry 55: 708–714.

Davis MH (1980). A multidimensional approach to individual differences in empathy. JSAS Catalog of Selected Documents in Psychology 10: 85.

De Dreu CK, Greer LL, Handgraaf MJ et al. (2010). The neuropeptide oxytocin regulates parochial altruism in intergroup conflict among humans. Science 328: 1408–1411.

De Vries GJ, Wang Z, Bullock NA et al. (1994). Sex differences in the effects of testosterone and its metabolites on vasopressin messenger RNA levels in the bed nucleus of the stria terminalis of rats. J Neurosci 14: 1789–1794.

Declerck C, Boone C, Kiyonari T (2013). The effect of oxytocin on cooperation in a prisoner's dilemma depends on the social context and a person's social value orientation. Soc Cogn Affect Neurosci 9: 802–809. http://dx.doi.org/10.1093/scan/nst040.

Di Simplicio M, Massey-Chase R, Cowen PJ et al. (2009). Oxytocin enhances processing of positive versus negative emotional information in healthy male volunteers. J Psychopharmacol 23: 241–248.

Dickerson SS, Kemeny ME (2004). Acute stressors and cortisol responses: a theoretical integration and synthesis of laboratory research. Psychol Bull 130: 355–391.

Ditzen B, Schaer M, Gabriel B et al. (2009). Intranasal oxytocin increases positive communication and reduces cortisol levels during couple conflict. Biol Psychiatry 65: 728–731.

Domes G, Heinrichs M, Glascher J et al. (2007a). Oxytocin attenuates amygdala responses to emotional faces regardless of valence. Biol Psychiatry 62: 1187–1190.

Domes G, Heinrichs M, Michel A et al. (2007b). Oxytocin improves mind-reading in humans. Biol Psychiatry 61: 731–733.

Donaldson ZR, Young LJ (2008). Oxytocin vasopressin and the neurogenetics of sociality. Science 322: 900–904.

Dulac C, Torello AT (2003). Molecular detection of pheromone signals in mammals: from genes to behaviour. Nat Rev Neurosci 4: 551–562.

Ebstein RP, Israel S, Lerer E et al. (2009). Arginine vasopressin and oxytocin modulate human social behavior. Ann N Y Acad Sci 1167: 87–102.

Ebstein RP, Mankuta D, Yirmiya N et al. (2011). Are retinoids potential therapeutic agents in disorders of social cognition including autism? FEBS Lett 585: 1529–1536.

Fehr E, Rockenbach B (2004). Human altruism: economic neural and evolutionary perspectives. Curr Opin Neurobiol 14: 784–790.

Feifel D (2012). Oxytocin as a potential therapeutic target for schizophrenia and other neuropsychiatric conditions. Neuropsychopharmacology 37: 304–305.

Feldman R (2012). Oxytocin and social affiliation in humans. Horm Behav 61: 380–391.

Feldman R, Gordon I, Schneiderman I et al. (2010). Natural variations in maternal and paternal care are associated with systematic changes in oxytocin following parent-infant contact. Psychoneuroendocrinology 35: 1133–1141.

Feldman R, Gordon I, Zagoory-Sharon O (2011). Maternal and paternal plasma salivary and urinary oxytocin and parent-infant synchrony: considering stress and affiliation components of human bonding. Dev Sci 14: 752–761.

Feldman R, Zagoory-Sharon O, Weisman O et al. (2012). Sensitive parenting is associated with plasma oxytocin and polymorphisms in the OXTR and CD38 genes. Biol Psychiatry 72: 175–181.

Frank E, Landgraf R (2008). The vasopressin system – from antidiuresis to psychopathology. Eur J Pharmacol 583: 226–242.

Frasch A, Zetzsche T, Steiger A et al. (1995). Reduction of plasma oxytocin levels in patients suffering from major depression. Adv Exp Med Biol 395: 257–258.

Freeman WJ (1997). Neurohumoral brain dynamics of social group formation. Implications for autism. Ann N Y Acad Sci 807: 501–503.

Galbally M, Lewis AJ, Ijzendoorn M et al. (2011). The role of oxytocin in mother-infant relations: a systematic review of human studies. Harv Rev Psychiatry 19: 1–14.

Gallese V, Keysers C, Rizzolatti G (2004). A unifying view of the basis of social cognition. Trends Cogn Sci 8: 396–403.

Gastaut H, Terzian H, Gastaut Y (1952). Study of a little electroencephalographic activity: rolandic arched rhythm. Mars Med 89: 296–310.

Goodson JL (2008). Nonapeptides and the evolutionary patterning of sociality. Prog Brain Res 170: 3–15.

Gordon I, Zagoory-Sharona O, Leckman JF et al. (2010). Oxytocin and the development of parenting in humans. Biol Psychiatry 68: 377–382.

Gordon I, Zagoory-Sharon O, Schneiderman I et al. (2008). Oxytocin and cortisol in romantically unattached young adults: associations with bonding and psychological distress. Psychophysiology 45: 349–352.

Gossen A, Hahn A, Westphal L et al. (2012). Oxytocin plasma concentrations after single intranasal oxytocin administration – a study in healthy men. Neuropeptides 46: 211–215.

Gottesman II, Gould TD (2003). The endophenotype concept in psychiatry: etymology and strategic intentions. Am J Psychiatry 160: 636–645.

Gouin JP, Carter CS, Pournajafi-Nazarloo H et al. (2010). Marital behavior oxytocin vasopressin and wound healing. Psychoneuroendocrinology 35: 1082–1090.

Granot RY, Frankel Y, Gritsenko V et al. (2007). Provisional evidence that the arginine vasopressin 1a receptor gene is associated with musical memory. Evol Hum Behav 28: 313–318.

Gregory SG, Connelly JJ, Towers AJ et al. (2009). Genomic and epigenetic evidence for oxytocin receptor deficiency in autism. BMC Med 7: 62.

Guastella AJ, Mitchell PB, Mathews F (2008a). Oxytocin enhances the encoding of positive social memories in humans. Biol Psychiatry 64: 256–258.

Guastella AJ, Mitchell PB, Dadds MR (2008b). Oxytocin increases gaze to the eye region of human faces. Biol Psychiatry 63: 3–5.

Guastella AJ, Einfeld SL, Gray KM et al. (2010). Intranasal oxytocin improves emotion recognition for youth with autism spectrum disorders. Biol Psychiatry 67: 692–694.

Hein G, Singer T (2008). I feel how you feel but not always: the empathic brain and its modulation. Curr Opin Neurobiol 18: 153–158.

Heinrichs M, Domes G (2008). Neuropeptides and social behaviour: effects of oxytocin and vasopressin in humans. Prog Brain Res 170: 337–350.

Heinrichs M, Meinlschmidt G, Neumann I et al. (2001). Effects of suckling on hypothalamic–pituitary–adrenal axis responses to psychosocial stress in postpartum lactating women. J Clin Endocrinol Metab 86: 4798–4804.

Heinrichs M, Baumgartner T, Kirschbaum C et al. (2003). Social support and oxytocin interact to suppress cortisol and subjective responses to psychosocial stress. Biol Psychiatry 54: 1389–1398.

Heinrichs M, von Dawans B, Domes G (2009). Oxytocin vasopressin and human social behavior. Front Neuroendocrinol 30: 548–557.

Higashida H, Salmina AB, Olovyannikova RY et al. (2007). Cyclic ADP-ribose as a universal calcium signal molecule in the nervous system. Neurochem Int 51: 192–199.

Higashida H, Yokoyama S, Kikuchi M et al. (2012). CD38 and its role in oxytocin secretion and social behavior. Horm Behav 61: 351–358.

Hoffman E, McCabe K, Smith VL (1996). Social distance and other-regarding behavior in dictator games. Am Econ Rev 86: 653–660.

Hoge EA, Pollack MH, Kaufman RE et al. (2008). Oxytocin levels in social anxiety disorder. CNS Neurosci Ther 14: 165–170.

Holt-Lunstad J, Birmingham WA, Light KC (2008). Influence of a warm touch support enhancement intervention among married couples on ambulatory blood pressure oxytocin alpha amylase and cortisol. Psychosom Med 70: 976–985.

Horesh N, Apter A, Ishai J et al. (1996). Abnormal psychosocial situations and eating disorders in adolescence. J Am Acad Child Adolesc Psychiatry 35: 921–927.

Iacoboni M (2009). Imitation, empathy, and mirror neurons. Annu Rev Psychol 60: 653–670.

Insel TR, O'Brien DJ, Leckman JF (1999). Oxytocin vasopressin and autism: is there a connection? Biol Psychiatry 45: 145–157.

Israel S, Lerer E, Shalev I et al. (2009). The oxytocin receptor (OXTR) contributes to prosocial fund allocations in the dictator game and the social value orientations task. PLoS One 4: e5535.

Israel S, Weisel O, Ebstein RP et al. (2012). Oxytocin but not vasopressin increases both parochial and universal altruism. Psychoneuroendocrinology 37: 1341–1344.

Jin D, Liu HX, Hirai H et al. (2007). CD38 is critical for social behaviour by regulating oxytocin secretion. Nature 446: 41–45.

Keri S, Kiss I, Kelemen O (2009). Sharing secrets: oxytocin and trust in schizophrenia. Soc Neurosci 4: 287–293.

Kim SJ, Young LJ, Gonen D et al. (2002). Transmission disequilibrium testing of arginine vasopressin receptor 1A (AVPR1A) polymorphisms in autism. Mol Psychiatry 7: 503–507.

Kirschbaum C, Pirke KM, Hellhammer DH (1993). The "Trier Social Stress Test" – a tool for investigating psychobiological stress responses in a laboratory setting. Neuropsychobiology 28: 76–81.

Kirschbaum C, Klauer T, Filipp SH et al. (1995). Sex-specific effects of social support on cortisol and subjective responses to acute psychological stress. Psychosom Med 57: 23–31.

Knafo A, Israel S, Darvasi A et al. (2008a). Individual differences in allocation of funds in the dictator game associated with length of the arginine vasopressin 1a receptor RS3 promoter region and correlation between RS3 length and hippocampal mRNA. Genes Brain Behav 7: 266–275.

Knafo A, Zahn-Waxler C, Van Hulle C et al. (2008b). The developmental origins of a disposition toward empathy: genetic and environmental contributions. Emotion 8: 737–752.

Knafo A, Zahn-Waxler C, Davidov M et al. (2009). Empathy in early childhood: genetic, environmental and affective contributions. Ann N Y Acad Sci 1167: 103–114.

Kosfeld M, Heinrichs M, Zak PJ et al. (2005). Oxytocin increases trust in humans. Nature 435: 673–676.

Krueger F, Parasuraman R, Iyengar V et al. (2012). Oxytocin receptor genetic variation promotes human trust behavior. Front Hum Neurosci 6: 4.

Kudielka BM, Hellhammer DH, Wust S (2009). Why do we respond so differently? Reviewing determinants of human salivary cortisol responses to challenge. Psychoneuroendocrinology 34: 2–18.

Kuehn BM (2011). Scientists probe oxytocin therapy for social deficits in autism schizophrenia. JAMA 305: 659–661.

Kumsta R, Heinrichs M (2013). Oxytocin stress and social behavior: neurogenetics of the human oxytocin system. Curr Opin Neurobiol 23: 11–16.

Kumsta R, Entringer S, Koper JW et al. (2007). Sex specific associations between common glucocorticoid receptor gene variants and hypothalamus–pituitary–adrenal axis responses to psychosocial stress. Biol Psychiatry 62: 863–869.

Kusui C, Kimura T, Ogita K et al. (2001). DNA methylation of the human oxytocin receptor gene promoter regulates tissue-specific gene suppression. Biochem Biophys Res Commun 289: 681–686.

Landgraf R, Neumann ID (2004). Vasopressin and oxytocin release within the brain: a dynamic concept of multiple and variable modes of neuropeptide communication. Front Neuroendocrinol 25: 150–176.

Lerer E, Levi S, Salomon S et al. (2008). Association between the oxytocin receptor (OXTR) gene and autism: relationship to Vineland Adaptive Behavior Scales and cognition. Mol Psychiatry 13: 980–988.

Lerer E, Levi S, Israel S et al. (2010). Low CD38 expression in lymphoblastoid cells and haplotypes are both associated with autism in a family-based study. Autism Res 3: 293–302.

Levine S (2001). Primary social relationships influence the development of the hypothalamic–pituitary–adrenal axis in the rat. Physiol Behav 73: 255–260.

Levine A, Zagoory-Sharon O, Feldman R et al. (2007). Oxytocin during pregnancy and early postpartum: individual patterns and maternal-fetal attachment. Peptides 28: 1162–1169.

Levy Y, Ebstein RP (2009). Research review: crossing syndrome boundaries in the search for brain endophenotypes. J Child Psychol Psychiatry 50: 657–668.

Lopatina O, Inzhutova A, Salmina AB et al. (2013). The roles of oxytocin and CD38 in social or parental behaviors. Front Neurosci 6: 182.

Loup F, Tribollet E, Dubois-Dauphin M et al. (1991). Localization of high-affinity binding sites for oxytocin and vasopressin in the human brain. An autoradiographic study. Brain Res 555: 220–232.

Ludwig M, Leng G (2006). Dendritic peptide release and peptide-dependent behaviours. Nat Rev Neurosci 7: 126–136.

Macmillan HL, Georgiades K, Duku EK et al. (2009). Cortisol response to stress in female youths exposed to childhood maltreatment: results of the youth mood project. Biol Psychiatry 66: 62–68.

Marsh AA, Yu HH, Pine DS et al. (2010). Oxytocin improves specific recognition of positive facial expressions. Psychopharmacology (Berl) 209: 225–232.

Martineau J, Schmitz C, Assaiante C et al. (2004). Impairment of a cortical event-related desynchronisation during a bimanual load-lifting task in children with autistic disorder. Neurosci Lett 367: 298–303.

Matsushita H, Tomizawa K, Okimoto N et al. (2010). Oxytocin mediates the antidepressant effects of mating behavior in male mice. Neurosci Res 68: 151–153.

Matsushita H, Matsuzaki M, Han XJ et al. (2012). Antidepressant-like effect of sildenafil through oxytocin-dependent cyclic AMP response element-binding protein phosphorylation. Neuroscience 200: 13–18.

Meyer-Lindenberg A, Kolachana B, Gold B et al. (2009). Genetic variants in AVPR1A linked to autism predict amygdala activation and personality traits in healthy humans. Mol Psychiatry 14: 968–975.

Meyer-Lindenberg A, Domes G, Kirsch P et al. (2011). Oxytocin and vasopressin in the human brain: social neuropeptides for translational medicine. Nat Rev Neurosci 12: 524–538.

Meynen G, Unmehopa UA, van Heerikhuize JJ et al. (2006). Increased arginine vasopressin mRNA expression in the human hypothalamus in depression: a preliminary report. Biol Psychiatry 60: 892–895.

Mizuguchi M, Otsuka N, Sato M et al. (1995). Neuronal localization of CD38 antigen in the human brain. Brain Res 697: 235–240.

Mizumoto Y, Kimura T, Ivell R (1997). A genomic element within the third intron of the human oxytocin receptor gene may be involved in transcriptional suppression. Mol Cell Endocrinol 135: 129–138.

Modahl C, Fein D, Waterhouse L (1992). Does oxytocin deficiency mediate social deficits in autism? J Autism Dev Disord 22: 449–451.

Modahl C, Green L, Fein D et al. (1998). Plasma oxytocin levels in autistic children. Biol Psychiatry 43: 270–277.

Montag C, Brockmann EM, Bayerl M et al. (2013). Oxytocin and oxytocin receptor gene polymorphisms and risk for schizophrenia: a case-control study. World J Biol Psychiatry 14: 500–508.

Morimoto I, Yamamoto S, Kai K et al. (2000). Centrally administered murine-leptin stimulates the hypothalamus–pituitary–adrenal axis through arginine-vasopressin. Neuroendocrinology 71: 366–374.

Munesue T, Yokoyama S, Nakamura K et al. (2010). Two genetic variants of CD38 in subjects with autism spectrum disorder and controls. Neurosci Res 67: 181–191.

Muthukumaraswamy SD, Johnson BW (2004). Changes in rolandic mu rhythm during observation of a precision grip. Psychophysiology 41: 152–156.

Muthukumaraswamy SD, Johnson BW, McNair NA (2004). Mu rhythm modulation during observation of an object-directed grasp. Brain Res Cogn Brain Res 19: 195–201.

Neumann ID (2008). Brain oxytocin: a key regulator of emotional and social behaviours in both females and males. J Neuroendocrinol 20: 858–865.

Neumann ID, Landgraf R (2012). Balance of brain oxytocin and vasopressin: implications for anxiety depression and social behaviors. Trends Neurosci 35: 649–659.

Newman-Norlund RD, van Schie HT, van Zuijlen AM et al. (2007). The mirror neuron system is more active during complementary compared with imitative action. Nat Neurosci 10: 817–818.

Oberman LM, Pineda JA, Ramachandran VS (2007). The human mirror neuron system: a link between action observation and social skills. Soc Cogn Affect Neurosci 2: 62–66.

Oberman LM, Ramachandran VS, Pineda JA (2008). Modulation of mu suppression in children with autism spectrum disorders in response to familiar or unfamiliar stimuli: the mirror neuron hypothesis. Neuropsychologia 46: 1558–1565.

Panksepp J (1993). Commentary on the possible role of oxytocin in autism. J Autism Dev Disord 23: 567–569.

Parker KJ, Kenna HA, Zeitzer JM et al. (2010). Preliminary evidence that plasma oxytocin levels are elevated in major depression. Psychiatry Res 178: 359–362.

Perry A, Bentin S, Shalev I et al. (2010a). Intranasal oxytocin modulates EEG mu/alpha and beta rhythms during perception of biological motion. Psychoneuroendocrinology 35: 1446–1453.

Perry A, Troje NF, Bentin S (2010b). Exploring motor system contributions to the perception of social information: evidence from EEG activity in the mu/alpha frequency range. Soc Neurosci 5: 272–284.

Pietrowsky R, Struben C, Molle M et al. (1996). Brain potential changes after intranasal vs intravenous administration of vasopressin: evidence for a direct nose-brain pathway for peptide effects in humans. Biol Psychiatry 39: 332–340.

Pineda JA (2005). The functional significance of mu rhythms: translating seeing and hearing into doing. Brain Res Brain Res Rev 50: 57–68.

Pineda JA, Hecht E (2009). Mirroring and mu rhythm involvement in social cognition: are there dissociable subcomponents of theory of mind? Biol Psychol 80: 306–314.

Preston SD, de Waal FB (2002). Empathy: its ultimate and proximate bases. Behav Brain Sci 25: 1–20, discussion 20–71.

Purba JS, Hoogendijk WJ, Hofman MA et al. (1996). Increased number of vasopressin- and oxytocin-expressing neurons in the paraventricular nucleus of the hypothalamus in depression. Arch Gen Psychiatry 53: 137–143.

Riebold M, Mankuta D, Lerer E et al. (2011). All-trans retinoic acid upregulates reduced CD38 transcription in lymphoblastoid cell lines from autism spectrum disorder. Mol Med 17: 799–806.

Rizzolatti G, Fadiga L, Gallese V et al. (1996). Premotor cortex and the recognition of motor actions. Brain Res Cogn Brain Res 3: 131–141.

Rodrigues SM, Saslow LR, Garcia N (2009). Oxytocin receptor genetic variation relates to empathy and stress reactivity in humans. Proc Natl Acad Sci U S A 106: 21437–21441.

Rubin LH, Carter CS, Bishop JR (2013). Peripheral vasopressin but not oxytocin relates to severity of acute psychosis in women with acutely-ill untreated first-episode psychosis. Schizophr Res 146: 138–143.

Scantamburlo G, Hansenne M, Fuchs S et al. (2007). Plasma oxytocin levels and anxiety in patients with major depression. Psychoneuroendocrinology 32: 407–410.

Schneiderman I, Zagoory-Sharon O, Leckman JF et al. (2012). Oxytocin during the initial stages of romantic attachment: relations to couples' interactive reciprocity. Psychoneuroendocrinology 37: 1277–1285.

Shalev I, Lerer E, Israel S et al. (2009). BDNF Val66Met polymorphism is associated with HPA axis reactivity to psychological stress characterized by genotype and gender interactions. Psychoneuroendocrinology 34: 382–388.

Shalev I, Israel F, Uzefovsky F et al. (2011). Vasopressin needs an audience: neuropeptide elicited stress responses are contingent upon perceived social evaluative threats. Horm Behav 60: 121–127.

Sigmund K, Hauert C (2002). Primer: altruism. Curr Biol 12: R270–R272.

Souza RP, Ismail P, Meltzer HY et al. (2010). Variants in the oxytocin gene and risk for schizophrenia. Schizophr Res 121: 279–280.

Swaab DF, Purba JS, Hofman MA (1995). Alterations in the hypothalamic paraventricular nucleus and its oxytocin neurons (putative satiety cells) in Prader–Willi syndrome: a study of five cases. J Clin Endocrinol Metab 80: 573–579.

Szeto A, McCabe PM, Nation DA et al. (2011). Evaluation of enzyme immunoassay and radioimmunoassay methods for the measurement of plasma oxytocin. Psychosom Med 73: 393–400.

Taylor SE, Saphire-Bernstein S, Seeman TE (2010). Are plasma oxytocin in women and plasma vasopressin in men biomarkers of distressed pair-bond relationships? Psychol Sci 21: 3–7.

Teltsh O, Kanyas-Sarner K, Rigbi A et al. (2011). Oxytocin and vasopressin genes are significantly associated with schizophrenia in a large Arab-Israeli pedigree. Int J Neuropsychopharmacol 1: 1–11.

Theodoridou A, Rowe AC, Penton-Voak IS et al. (2009). Oxytocin and social perception: oxytocin increases perceived facial trustworthiness and attractiveness. Horm Behav 56: 128–132.

Thibonnier M, Auzan C, Madhun Z et al. (1994). Molecular cloning, sequencing, and functional expression of a cDNA encoding the human V1a vasopressin receptor. J Biol Chem 269: 3304–3310.

Thibonnier M (2004). Genetics of vasopressin receptors. Curr Hypertens Rep 6: 21–26.

Thompson R, Gupta S, Miller K et al. (2004). The effects of vasopressin on human facial responses related to social communication. Psychoneuroendocrinology 29: 35–48.

Thompson RR, George K, Walton JC et al. (2006). Sex-specific influences of vasopressin on human social communication. Proc Natl Acad Sci U S A 103: 7889–7894.

Thompson RJ, Parker KJ, Hallmayer JF et al. (2011). Oxytocin receptor gene polymorphism (rs2254298) interacts with familial risk for psychopathology to predict symptoms of depression and anxiety in adolescent girls. Psychoneuroendocrinology 36: 144–147.

Tsao DY, Freiwald WA, Knutsen TA et al. (2003). Faces and objects in macaque cerebral cortex. Nat Neurosci 6: 989–995.

Uhart M, Chong RY, Oswald L et al. (2006). Gender differences in hypothalamic–pituitary–adrenal (HPA) axis reactivity. Psychoneuroendocrinology 31: 642–652.

Uzefovsky F, Shalev I, Israel S et al. (2012). Vasopressin selectively impairs emotion recognition in men. Psychoneuroendocrinology 37: 576–580.

Van Lange PAM (1999). The pursuit of joint outcomes and equality in outcomes: an integrative model of social value orientation. J Pers Soc Psychol 77: 337–349.

Verbalis JG (1999). The brain oxytocin receptor(s)? Front Neuroendocrinol 20: 146–156.

Walter NT, Montag C, Markett S et al. (2012). Ignorance is no excuse: moral judgments are influenced by a genetic variation on the oxytocin receptor gene. Brain Cogn 78: 268–273.

Walum H, Westberg L, Henningsson S et al. (2008). Genetic variation in the vasopressin receptor 1a gene (AVPR1A) associates with pair-bonding behavior in humans. Proc Natl Acad Sci U S A 105: 14153–14156.

Walum H, Lichtenstein P, Neiderhiser JM et al. (2012). Variation in the oxytocin receptor gene is associated with pair-bonding and social behavior. Biol Psychiatry 71: 419–426.

Wang Z, Smith W, Major DE (1994). Sex and species differences in the effects of cohabitation on vasopressin messenger RNA expression in the bed nucleus of the stria terminalis in prairie voles (*Microtus ochrogaster*) and meadow voles (*Microtus pennsylvanicus*). Brain Res 650: 212–218.

Wassink TH, Piven J, Vieland VJ et al. (2004). Examination of AVPR1a as an autism susceptibility gene. Mol Psychiatry 9: 968–972.

Waterhouse L, Fein D, Modahl C (1996). Neurofunctional mechanisms in autism. Psychol Rev 103: 457–489.

Weisman O, Zagoory-Sharon O, Feldman R (2012). Oxytocin administration to parent enhances infant physiological and behavioral readiness for social engagement. Biol Psychiatry 72: 982–989.

Weisman O, Zagoory-Sharon O, Schneiderman I et al. (2013). Plasma oxytocin distributions in a large cohort of women and men and their gender-specific associations with anxiety. Psychoneuroendocrinology 38: 694–701.

Welch GN, Upchurch Jr GR, Loscalzo J (1997). Homocysteine oxidative stress and vascular disease. Hosp Pract (Off Ed) 32: 81–82, 85, 88–92.

Winslow JT, Insel TR (1993). Effects of central vasopressin administration to infant rats. Eur J Pharmacol 233: 101–107.

Winslow JT, Hastings N, Carter CS et al. (1993). A role for central vasopressin in pair bonding in monogamous prairie voles. Nature 365: 545–548.

Wu S, Jia M, Ruan Y et al. (2005). Positive association of the oxytocin receptor gene (OXTR) with autism in the Chinese Han population. Biol Psychiatry 58: 74–77.

Wu N, Li Z, Su Y (2012). The association between oxytocin receptor gene polymorphism (OXTR) and trait empathy. J Affect Disord 138: 468–472.

Yirmiya N, Rosenberg C, Levi S et al. (2006). Association between the arginine vasopressin 1a receptor (AVPR1a) gene and autism in a family-based study: mediation by socialization skills. Mol Psychiatry 11: 488–494.

Young LJ (2007). Regulating the social brain: a new role for CD38. Neuron 54: 353–356.

Zak PJ (2005). Trust: a temporary human attachment facilitated by oxytocin. Behav Brain Sci 28: 368–369.

Zak PJ, Kurzban R, Matzner WT (2005). Oxytocin is associated with human trustworthiness. Horm Behav 48: 522–527.

Zak PJ, Stanton AA, Ahmadi S (2007). Oxytocin increases generosity in humans. PLoS One 2: e1128.

Zhong S, Monakhov M, Mok HP et al. (2012). U-shaped relation between plasma oxytocin levels and behavior in the trust game. PLoS One 7: e51095.

Zink CF, Kempf L, Hakimi S et al. (2011). Vasopressin modulates social recognition-related activity in the left temporoparietal junction in humans. Transl Psychiatry 1: e3.

Handbook of Clinical Neurology, Vol. 124 (3rd series)
Clinical Neuroendocrinology
E. Fliers, M. Korbonits, and J.A. Romijn, Editors

Chapter 5

Corticotropin-releasing hormone and the hypothalamic–pituitary–adrenal axis in psychiatric disease

MARIE NAUGHTON, TIMOTHY G. DINAN*, AND LUCINDA V. SCOTT
Department of Psychiatry, University College Cork, Cork, Ireland

INTRODUCTION

The hypothalamic–pituitary–adrenal (HPA) axis is the core endocrine stress system in humans (Rubin et al., 2001). It not only regulates the body's peripheral functions relating to metabolism and immunity but also has profound effects on the brain through the regulation of neuronal survival and neurogenesis in structures such as the hippocampus, where it plays a role in memory (for a comprehensive review on the impact on memory of HPA axis alterations in mental disorders see Wolf and Wingenfeld, 2011). While acute stress activates the sympathoadrenal medullary system, resulting in the components of the "fight or flight" response with a release of catecholamines, chronic stress results in alterations of the HPA axis which invoke a number of adaptive behavioral and physiologic changes to include an increase in the release of cortisol. For a detailed review of the function of the stress axis and the central role of the brain the reader is referred to Lupien et al. (2009) and McEwen and Gianaros (2010). Briefly, upon stress exposure, corticotropin-releasing hormone (CRH) is released from the hypothalamus and is transported to the anterior pituitary, where it stimulates the secretion of adrenocorticotropin (ACTH), which in turn stimulates the synthesis and release of glucocorticoids (GCs) from the adrenal cortex. The neuroendocrine stress response is counter-regulated by circulating GCs via a negative feedback mechanism targeting the pituitary, hypothalamus, and hippocampus. This negative feedback loop is essential for the regulation of the HPA axis and, therefore, for the regulation of the stress response (Carrasco and Van de Kar, 2003). Figure 5.1 illustrates the HPA axis and components of the axis will be elaborated on throughout the chapter.

CORTICOTROPIN-RELEASING HORMONE AND BASAL HYPOTHALAMIC–PITUITARY–ADRENAL AXIS ACTIVITY

CRH is a 41-amino acid peptide originally discovered and sequenced by Vale et al. (1981). Under basal conditions, CRH is mainly produced within the medial paraventricular nucleus (PVN) of the hypothalamus and is the dominant regulator of the axis (Pariante and Lightman, 2008; Binder and Nemeroff, 2010), mediating the endocrine response to stress. CRH release is triggered following any threat to homeostasis. Using immunohistochemical and radioimmunoassay techniques CRH was found not only to be confined to the median PVN region of the hypothalamus but also to be heterogeneously distributed throughout the central nervous system (CNS) (Boorse and Denver, 2006; for review see Binder and Nemeroff, 2010). CRH-containing interneurons are widely distributed in the neocortex and are believed to be important in several behavioral actions of the peptide, including effects on cognitive processing. Another brain region with a high density of CRH cell bodies is the bed nucleus of the stria terminalis (BNST), which projects to brainstem areas that are involved in autonomic functioning such as the parabrachial nuclei and dorsal vagal complex. CRH perikarya in the central nucleus of the amygdala send terminals to the parabrachial nucleus of the brainstem as well as to the BNST and the medial preoptic area, both of which, in turn, send terminals to the parvocellular region of the PVN and thus may influence both neuroendocrine and autonomic function (Gray and Bingaman, 1996). The presence of CRH immunoreactivity in the raphe nuclei and locus coeruleus (LC), the origin of the major serotonergic and noradrenergic

*Correspondence to: Professor Ted Dinan, Chairman, Department of Psychiatry, Cork University Hospital, Wilton, Cork, Ireland. Tel: +353-21-490-1224, E-mail: t.dinan@ucc.ie

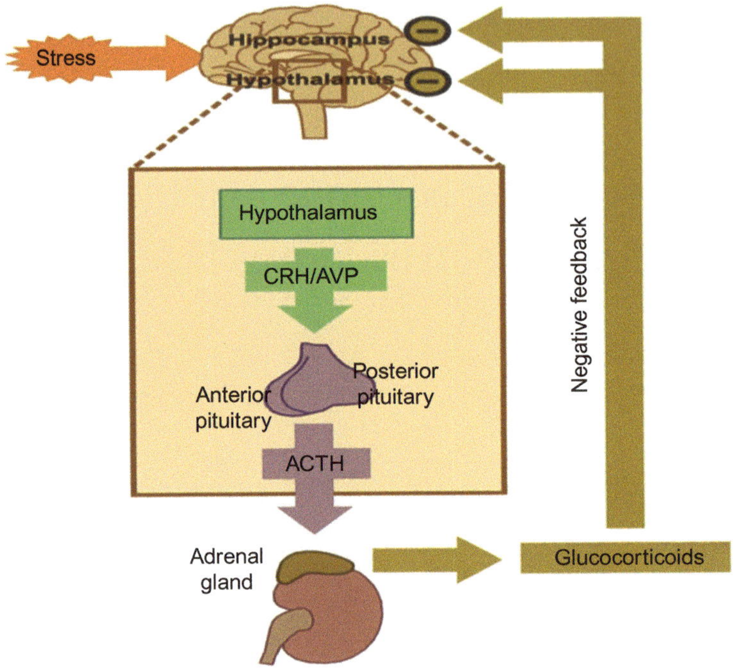

Fig. 5.1. The hypothalmic–pituitary–adrenal axis. (Designed by Dr Marcela Julio.)

pathways in brain, points to a role for CRH in modulating these monoaminergic systems which have long been implicated in the pathophysiology of depression and anxiety disorders (Arborelius et al., 1999).

Two different CRH receptors have been described, CRH1 and CRH2, both of which are positively coupled to adenylate cyclase (Hauger et al., 2003). CRH1 receptors are expressed in high density in the cerebral cortex, cerebellum, hippocampus, amygdale, and pituitary; the peripheral expression is less robust and concentrated in the skin, the ovaries or testes, and the adrenal gland. The CRH2 receptors are also expressed in the CNS but largely restricted to subcortical areas including the hypothalamus, amygdale, bed nucleus of the stria terminalis and raphe nucleus, and in peripheral tissues, such as the pituitary, heart, lungs ovaries, testes, and adrenal gland (Potter et al., 1994; Chalmers et al., 1995; Lovenberg et al., 1995; Sanchez et al., 1999; Hiroi et al., 2001). The CRH2 receptor is currently known to exist in two different isoforms in both rat and human (Chalmers et al., 1995; Grigoriadis et al., 1996).

In situations of chronic stress many parvicellular neurons coexpress vasopressin (AVP), which plays an important role in sustaining HPA activation through a synergistic action with CRH (Dinan and Scott, 2005). Vasopressin has ACTH-releasing properties when administered alone in humans, a response which may be dependent on the ambient endogenous CRH level. CRH and AVP act on the anterior pituitary corticotropes to stimulate the release of ACTH (Aguilera and Rabadan-Diehl, 2000). Following the combination of AVP and CRH, a much greater ACTH response is seen and both peptides are required for maximal pituitary–adrenal stimulation. The precise nature of this synergism is incompletely understood with most information derived from animal studies. It has been demonstrated that CRH, through cAMP, increases pro-opiomelanocortin (POMC) gene transcription and peptide synthesis and storage (Hammer et al., 1990). There may also be distinct corticotrope populations in the anterior pituitary, some of which require both AVP and CRH for ACTH release.

While CRH and AVP are the major secretagogue peptides of the HPA/stress system, glucocorticoids play a pivotal feedback role in the onset and termination of the stress response. The principal mechanism of action of cortisol throughout the body is through activation of intracellular receptors which subsequently translocate to the nucleus and bind to specific DNA sequences, thus modulating gene transcription. There are two receptors which bind cortisol (McEwen and Plapinger, 1970; see McEwen et al., 2012, for a comprehensive review). The type 1 receptor (MR), which is indistinguishable from the peripheral mineralocorticoid receptor, is distributed principally in the septohippocampal region and mediates tonic influences of cortisol or corticosterone; the type 2 or glucocorticoid receptor (GR) has a wider distribution and mediates stress-related changes in cortisol level. By binding to the GR and the MR, endogenous glucocorticoids serve as potent negative regulators of HPA axis activity (Holsboer, 2001). The MR has a 10-fold higher affinity for cortisol than GR does and governs basal diurnal fluctuation in cortisol, while GR becomes progressively occupied only at peaks of cortisol secretion and after a stressful stimulus (Reul and De Kloet, 1985; De Kloet et al., 2007). These receptor systems also provide negative feedback loops at a limbic, hypothalamic, and pituitary level.

The sensitivity of CRH and AVP transcription to glucocorticoid feedback is markedly different. GCs, while inhibiting the secretion of ACTH at the corticotrophs, promote AVP-mediated ACTH secretion via upregulation of the pituitary V1b receptor (Aguilera and Rabadan-Diehl, 2000). These effects may underpin the refractoriness of AVP-stimulated ACTH secretion to glucocorticoid feedback, suggesting that vasopressinergic regulation of the HPA axis is critical for sustaining corticotrope responsiveness in the presence of high circulating glucocorticoid levels during chronic stress.

A wide variety of neurotransmitters also influence the hypothalamic regulation of the HPA. These include serotonin, noradrenaline, acetylcholine, and opioids. In addition, the proinflammatory cytokines tumor necrosis factor (TNF), interleukin 1 (IL-1), and interleukin 6 (IL-6) are primary HPA stimulating cytokines that act directly on hypothalamic PVNs to stimulate CRH production (Chrousos, 1995; Dinan, 1996; Dentino et al., 1999).

HYPOTHALAMIC–PITUITARY–ADRENAL AXIS FUNCTIONING IN MAJOR DEPRESSION

Major depression, characterized by excessive sadness, loss of pleasure, and reduced energy, sustained over a period of 2 weeks, with a constellation of other neurovegetative and cognitive features, is the most common mood disorder, the 12 month prevalence rate of which is 10% (Kessler et al., 2005). Depressed individuals can have insomnia, anorexia, and motor agitation; or hypersomnia, hyperphagia, and "leaden paralysis". These two symptom clusters characterize the "melancholic specifier" and the "atypical specifier", respectively, of major depressive disorder (MDD) in the DSM-IV-TR (American Psychiatric Association, 2000). Most individuals have mixed features of melancholic and atypical depression with only 25–30% presenting with pure melancholic features (Gold and Chrousos, 2002). Many studies of melancholic depression support the hypothesis that relative HPA axis hyperactivity occurs in melancholic depression compared to nondepressed states, and that this is more likely to occur in the more severe form of melancholic depression (Lightman, 2008). Several well-established findings document dysregulation of the hypothalamic–pituitary–adrenal (HPA) system in a subgroup of patients with depressive illness. These include hypercortisolemia (Linkowski et al., 1985), escape of plasma cortisol from dexamethasone suppression (Kathol et al., 1989), blunted ACTH response to CRH (Gold and Chrousos, 1985), enlargement of the adrenal cortex (Nemeroff et al., 1992), and exaggerated cortisol response to ACTH (Jaeckle et al., 1987). Initial proposals for synthesizing these findings generated a model including: central CRH overdrive; downregulation of pituitary CRH receptors due to overexposure to CRH; hypertrophy, hyperplasia, and enhanced responsiveness of the adrenal cortex to ACTH, all due to increased stimulation by ACTH; and downregulation of cortisol receptors on negative feedback neurons in the hippocampus and elsewhere due to increased exposure to cortisol. These findings will be discussed in more detail in the following section.

The corticotropin-releasing hormone system and dexamethasone/corticotropin-releasing hormone studies in depression

Repeated findings from preclinical and clinical studies support a preeminent function for the CRH system in mediating the physiologic response to external stressors and in the pathophysiology of depression. In addition to its well-documented function as a hypothalamic hypophysiotropic factor that stimulates pituitary ACTH synthesis and secretion and thereby controls the activity of the HPA axis (Vale et al., 1981), CRH neurons also innervate the locus coeruleus, thus activating the other major stress response axis, the CNS noradrenergic and sympathetic nervous systems (Valintino et al., 1983). Furthermore, effects of CRH in limbic brain regions have been associated with increased fear, alertness, and decreased appetite and libido, all functions relevant in the flight or fight response and dysregulated in depression (Nemeroff, 1996). These effects seem to be mediated mainly by the CRH1 receptor (Heinricks et al., 1997; Arborelius et al., 1999; Reul and Holsboer, 2002). The function of the CRH2 receptor remains more obscure and likely to be dependant on context and brain region. Overall, it seems that the CRH1 receptor is the principal receptor mediating the stress response, whereas the CRH2 receptor modulates the effects of CRH1 signal transduction (Bale et al., 2000; Reul and Holsboer, 2002; Nemeroff and Vale, 2005).

Laboratory animal studies in which brain intercerebroventricular or brain region-specific microinjections of CRH have been used have revealed that in human beings CRH produces behavioral responses reminiscent of major depression, including increased anxiety, reduced slow wave sleep, psychomotor alterations, anhedonia, decreased appetite and libido (Dunn and Berridge, 1990; Keck, 2006). These studies are complemented by results obtained in transgenic animals either lacking or overexpressing CRH-system ligands or receptors, as well as from studies using selective CRH antagonists. Conditional CRH1 receptor knockout mice and CRH overexpressing mice restricted to forebrain areas have further shown that these anxiety- and depression-related phenotypes are specific to activation of the CRH1 receptor in limbic forebrain regions and independent of actions on HPA axis activity (Muller et al., 2003; Lu et al., 2008), although the latter endocrine effects of CRH may contribute to the depressive symptoms. Although a negative effect of GR activation on CRH expression has been described for the hypothalamus, glucocorticoids were shown to increase CRH expression in limbic areas, including the amygdala and the lateral septum (Schulkin et al., 1998; Kageyama and Suda, 2009). In support of the critical function on limbic CRH transmission, increased CRH expression in the amygdala induced by use of a lentiviral vector was shown to produce most of the behavioral effects that comprise the depressive syndrome, as well as HPA-axis hyperactivity (Keen-Rhinehart et al., 2009).

In humans, after intravenous administration of CRH, depressed patients exhibit a blunted ACTH but normal cortisol response in comparison to healthy controls (Gold et al., 1986; Holsboer et al., 1986; Krishnan et al., 1993). Moreover, a correlation between dexamethasone nonsuppression of cortisol (see below) and a blunted ACTH response to CRH challenge in patients with major depression has been reported (Krishnan et al., 1993). After clinical recovery, normalization of the blunted ACTH response to CRH is also observed (Amsterdam et al., 1988).

One plausible mechanism to explain the blunted ACTH response to CRH challenge observed in depressed patients is downregulation of pituitary CRH receptors, presumably secondary to increased hypothalamic CRH release. Support for hypersecretion of hypothalamic CRH in depression comes from a series of findings in depressed patients and suicide victims. Significantly elevated concentrations of CRH in the cerebrospinal fluid (CSF) of drug-free patients with major depression and of suicide victims compared with concentrations in patients with other psychiatric disorders and healthy controls has been repeatedly observed (Nemeroff et al., 1984; Arato et al., 1986, 1989; Banki et al., 1987, 1992a; France et al., 1988; Widerlov et al., 1988). However, other studies have been unable to replicate these observations (Kling et al., 1991, 1993; Molchan et al., 1993; Pitts et al., 1995). Gold and collaborators did not find any difference between CSF CRH concentrations in depressed patients and healthy controls, although depressed patients who were dexamethasone suppression test (DST) nonsuppressors had significantly higher CSF CRH concentrations as compared with depressed DST suppressors (Roy et al., 1987). Indeed, decreased CSF CRH concentrations have been observed in a group of depressed patients with normal plasma cortisol levels compared with healthy subjects (Geracioti et al., 1997). These discrepant findings are almost certainly due to the inclusion in these studies of patients with atypical depression or with only mild to moderate depression. Further support for the postulate that depression is associated with CRH hypersecretion may be derived from postmortem studies which revealed an increase in CRH concentrations and in CRH mRNA expression in the PVN of patients with depression (Raadsheer et al., 1994, 1995). Also, persistent elevations of CSF CRH concentrations in symptomatically improved depressed patients are associated with early relapse of depression (Banki et al., 1992c). The role of CRH in depression is comprehensively reviewed by Arborelius and colleagues (1999) and insights into the CRH system in depression from human genetic studies are elaborated on by Binder and Nemeroff (2010).

Dexamethasone (DEX), a potent synthetic glucocorticoid, binds primarily to glucocorticoid receptors on anterior pituitary corticotropes and, by feedback inhibition, suppresses ACTH and cortisol secretion (Cole et al., 2000). In the DEX/CRH test, when healthy subjects are treated with dexamethasone prior to CRH infusion, the release of ACTH is blunted and the extent of blunting is proportional to the dose of DEX (von Bardeleben and Holsboer, 1989). Paradoxically, when depressed patients are pretreated with DEX they show an enhanced ACTH response to CRH. This combined test appeared to be a very sensitive diagnostic measure for depression, especially when the patients were clustered into different age groups. Also, healthy nondepressed subjects at high familial risk for affective disorders exhibit disturbed HPA axis activity as induced by the combined DST/CRH test, suggesting that the potential for abnormalities in HPA axis function in depressed patients may be genetically transmitted (Holsboer et al., 1995).

There is evidence that, like measures of HPA axis activity, CSF CRH concentrations normalize when patients recover from depression. Thus, the elevated CSF CRH concentrations of drug-free depressed patients are significantly decreased 24 hours after a successful series of electroconvulsive therapy (ECT) treatments (Nemeroff et al., 1991). In a preliminary report, Kling et al. (1994a) observed a reduction in diurnal CSF CRH concentrations in depressed patients after successful ECT. In addition, normalization of elevated

CRH concentrations in CSF has also been reported after successful treatment of depression with fluoxetine (De Bellis et al., 1993). In another study, a significant reduction of elevated CSF CRH concentrations was found in 15 depressed women who remained depression-free for at least 6 months after antidepressant drug treatment (Banki et al., 1992b). In contrast, there was a tendency for increased CSF CRH concentrations in the nine patients who relapsed within 6 months. Although CSF CRH concentrations are not correlated with depression severity, these findings suggest that lack of normalization of CRH levels in CSF after antidepressant treatment may predict early relapse. Taken together the above studies indicate that elevated CRH concentrations in CSF appear to be a state, rather than a trait, marker in depression.

Neuropeptides such as CRH appear to be secreted directly into CSF from brain tissue, and neuropeptides found in CSF are not derived from the systemic circulation (Post et al., 1982). Studies using nonhuman primates suggest that CSF levels of CRH primarily reflect function of extrahypothalamic rather than hypothalamic CRH systems (Kalin, 1990). Thus, manipulations that enhance pituitary ACTH release, i.e., physostigmine administration or stress, are not accompanied by an increase in CSF CRH levels. A dissociation between the diurnal variation of CSF CRH and cortisol concentrations has also been described in both humans and primates (Kalin, 1990; Kling et al., 1994b).

Using magnetic resonance imaging (MRI) and computed tomography (CT), enlargement of both the pituitary and the adrenal gland has been observed in depressed patients (Krishnan et al., 1991; Axelson et al., 1992; Nemeroff et al., 1992; Rubin et al., 1995). In laboratory animals both hyperplasia and hypertrophy of the anterior pituitary, as well as adrenal gland hypertrophy, have been observed after enhanced stimulation of the pituitary–adrenal axis (Gertz et al., 1987; Sapolsky and Plotsky, 1990). Thus, these imaging findings lend further support to the hypothesis of increased hypothalamic CRH secretion in depression. Finally, Nemeroff and colleagues (1988) have found a marked decrease in CRH receptor-binding sites in the prefrontal cortex of depressed suicide victims, which they hypothesized develops as compensatory consequence of increased release of CRH in this brain region.

Adrenocorticotropin and cortisol in depression

Both the peak and the nadir in circulating cortisol concentrations are elevated in depression but overall there is little reduction in amplitude of the circadian rhythm, nor is its timing significantly shifted (Fig. 5.2). Linkowski et al. (1987) found increased 24 hour mean plasma cortisol, shorter nocturnal quiescence of cortisol secretion, decreased relative amplitude of the 24 hour cortisol rhythm, and advance of the cortisol nadir in patients with major depression. In 40 research diagnostic criteria (RDC) definite endogenous depressives compared to 40 matched controls, Rubin and colleagues (1987) reported hypercortisolism throughout the 24 hours in 15 of the patients, with no significant advance of the cortisol nadir. Overall the data indicate that HPA axis hyperactivity in depression occurs throughout both diurnal and nocturnal periods.

Wedekind and colleagues (2007) have found that basal HPA activity, as measured by aggregated nocturnal urinary cortisol levels, remains elevated even after remission of symptoms in patients with psychotic depression. This observation supports the concept that a dysfunctional regulation of the HPA system is possibly a trait-related, rather than a state-related, feature. This relationship has been further examined using the cortisol awakening response (CAR), which has allowed normal cortisol secretory dynamics to be examined in large populations. This increase in cortisol after awakening appears to be a distinct feature of the HPA axis, superimposing the circadian rhythmicity of cortisol secretion. The CAR can be measured in saliva, an easily accessible biologic fluid, and this has led to a wealth of information on factors that can influence cortisol secretion in

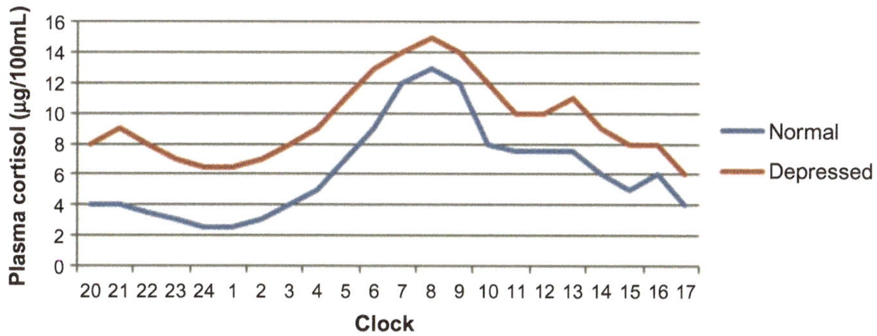

Fig. 5.2. Sample cortisol response in a depressed patient versus a nondepressed control.

a variety of contexts, as reviewed by Fries et al. (2009). Vreeburg et al. (2009), in an impressively large study, examined CARs in a currently depressed sample (n = 701), a remitted depressed sample (n = 579), and a healthy control sample (n = 307). Elevated CARs were found in the currently depressed and remitted group compared to the control group, supporting the concept that a dysfunctional regulation of the HPA system in such patients is a trait- rather than a state-related marker.

Dexamethasone suppression test

Dexamethasone (DEX), a potent synthetic glucocorticoid, binds primarily to glucocorticoid receptors on anterior pituitary corticotropes and, by feedback inhibition, suppresses ACTH and cortisol secretion. The degree and duration of suppression depends on a balance between the amount of DEX administered, its pharmacokinetics in a given subject, and the degree of suprapituitary drive. While low-dose and high-dose dexamethasone suppression tests (DSTs) have been used for the differential diagnosis of Cushing's disease, a low-dose DST has been used as a marker of HPA axis hyperactivity in mood disorders (Carroll et al., 1981). Its usefulness in a clinical setting is limited due to low specificity and sensitivity.

In normal individuals, following the administration of 1 mg dexamethasone at 11 p.m., cortisol remains suppressed to very low levels for the full 24 hours. In contrast, up to 70% of patients with major depression, particularly those with melancholic features, show cortisol nonsuppression or early escape from suppression during the 24 hours following DEX administration. Figure 5.3 gives a diagrammatic representation of the DST in depressed and control subjects. Studies in milder forms of depression indicate low levels of nonsuppression similar to those seen in many other psychiatric disorders. Of note is the fact that high degrees of nonsuppression are also found in mania.

The DST has been used also to follow the course of treatment in patients who demonstrated initial nonsuppression when depressed. With successful treatment, the ability to suppress cortisol production gradually normalizes, and such patients tend to remain in remission longer than patients who show clinical improvement but continue to have abnormal results in the DST. Interestingly, the normalization of the DST following effective treatment in depression contrasts with the cortisol and CAR responses discussed earlier, which are seen as trait-related markers. Overall, the DST has remained a vital tool in the exploration of HPA axis dysfunction in depression; however, it appears to have limited utility in the diagnostic process in psychiatry (Berger et al., 1984).

Adrenocorticotropin stimulation test

Exogenous adrenocorticotropin (ACTH) administration has been used as a direct test of adrenal cortical responsiveness in depression. Two strategies have been employed using either pharmacologic or physiologic doses of ACTH. In general, exaggerated ACTH release has been reported with the standard supramaximal stimulation dose of 250 μg ACTH, thus testing maximal adrenal secretory capacity. Thakore et al. (1997) examined the effects of antidepressant treatment on ACTH-induced cortisol release in a cohort of melancholically depressed subjects. After an intravenous bolus dose (250 μg) of tetracosactide, a potent stimulator of ACTH secretion, plasma levels of cortisol were measured at times 0, +30, +60, +90, +120 and +180 min. Patients were then randomized to receive either 50 mg of sertraline or 20 mg of paroxetine (both of which are selective serotonin reuptake inhibitors) until their depressive episode went into remission. They were retested at least 4 weeks after discontinuing the medication; overall, the mean time to retesting was 9.1 months. A reduction in ACTH-mediated cortisol release was seen post-treatment, supporting the view that higher ACTH-induced cortisol responses in depression are a state-dependent phenomenon in depression which normalizes with effective treatment. Several studies using much

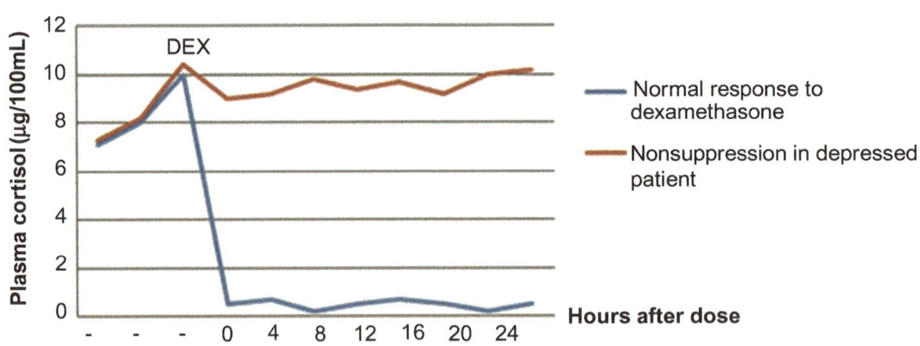

Fig. 5.3. Dexamethasone supression test in a sample depressed patient versus a healthy control.

lower, more physiologic doses of ACTH (Rubin et al., 1995) have also been conducted; these have failed to find differences between depressives and healthy controls. However, a characteristic feature of melancholic depression is an overactive HPA axis; therefore, the validity of using a low-dose ACTH test in this condition is questionable. Furthermore, Rubin et al. (1995) showed that depressed patients, while depressed, had blunted CRH/ACTH responses which normalized with recovery; however, their cortisol responses pre- and post-treatment were not significantly different from each other. The latter indicates that while depressed, patients secrete a greater amount of cortisol for a given dose of ACTH, suggesting a hyperresponsive adrenal cortex. This would be in keeping with their observation that adrenal gland volume is approximately 70% greater during the melancholic phase.

Vasopressin in depression

A potential role for AVP in affective illness was forwarded in 1978 by Gold and Goodwin. They described symptom complexes in affective illness that AVP is known to influence, notably memory processes, pain sensitivity, synchronization of biologic rhythms, and the timing and quality of REM sleep. A role for AVP was supported not only by the above spectrum of symptoms but also by dynamic tests of HPA activity, and in particular, the DEX/CRH test. The enhanced response to DEX/CRH seen in depression is thought to be due to enhanced AVP drive.

There are relatively few data on plasma AVP levels in depression. An early report found no change in plasma AVP levels in depression (Gjerris et al., 1985). In contrast, van Londen et al. (1997) reported basal plasma levels of AVP to be elevated. This study compared 52 patients with major depression and 37 healthy controls; AVP concentrations were found to be higher in the depressed cohort, with greater elevation in inpatient compared to outpatient depressives and in those with melancholic features. A number of studies have shown a significant positive correlation between peripheral plasma levels of AVP and hypercortisolemia in patients with unipolar depression. Apart from increased CRH production, an AVP-mediated overdrive in depression has been mooted as it is known that AVP, also a cortictroph secretagogue, is insensitive to GC feedback (Dinan et al., 2004; Dinan and Scott, 2005). This is compatible with findings of impaired feedback to hydrocortisone and prednisolone in severe depression (Young et al., 1991; Juruena et al., 2009) when another test of HPA axis functioning was used, namely the prednisolone suppression test. This test purports to assess both glucocorticoid receptors and mineralocorticoid receptors, in contrast to the DST, which probes the function of glucocorticoid receptors only. Since endogenous HPA axis feedback involves both glucocorticoid and mineralocorticoid receptors, and since there is some evidence that mineralocorticoid receptors can compensate for altered glucocorticoid receptor function, prednisolone should provide a more valid test of the HPA axis in depression.

A postmortem study of depressed subjects reported an increased number of vasopressin-expressing neurons in paraventricular hypothalamic neurons (Purba et al., 1996; Meynen et al., 2006). Dinan et al. (1999) examined a cohort of depressed subjects on two separate occasions, with CRH alone and with the combination of CRH and desmopressin (dDAVP), a vasopressin analog. A significant blunting of ACTH output to CRH alone was noted. Following the combination of CRH and dDAVP, the release of ACTH in depressives and healthy volunteers was indistinguishable. It was concluded that whilst the CRH1 receptor is downregulated in depression, a concomitant upregulation of the V_3 receptor takes place. The finding by Dinan and Scott (2005) that coadministration of desmopressin (dDAVP) with CRH normalizes the blunted ACTH response to CRH alone in melancholic depression supports a model whereby the stress response in depression is largely driven by AVP rather than CRH. This is consistent with the animal models of chronic stress, in which a switching from CRH to AVP regulation is observed. It is interesting that in CRH1 receptor-deficient mice, basal plasma AVP levels are significantly elevated, AVP mRNA is increased in the PVN, and there is increased AVP-like immunoreactivity in the median eminence (Muller et al., 2000).

In a further study, Dinan et al. (2004) provided evidence for the upregulation of the anterior pituitary V_3 receptor. Fourteen patients with major depression and 14 age- and sex-matched healthy comparison subjects were recruited. Desmopressin (dDAVP) 10 μg was given intravenously and ACTH and cortisol release was monitored for 120 min. There was an enhanced ACTH and cortisol release following dDAVP in the depressed subjects, indicating enhanced V_3 responsivity. Studies of AVP mediated ACTH and cortisol release are not consistent, however, with increased, decreased, or no effect on ACTH release being reported (Meller et al., 1987; Newport et al., 2003).

Melancholic depression with cardinal features of reduced sleep and appetite is the diametric opposite of atypical depression in which hyperphagia and hypersomnia are seen. In an individual these symptom clusters and features can change over time and respond selectively to different treatments. In more chronic clinical pictures atypical features are more commonly found, and O'Keane et al. (2012) have hypothesized that different pathophysiologic processes may underpin this difference: they suggest that there is a "switch" in the regulation of the HPA from CRH to AVP control as stress becomes more sustained or repeated.

Early life stress, depression, and the hypothalamic–pituitary–adrenal axis

Both animal and human studies give ample evidence that early life trauma can contribute to adult depression and it is postulated that the dysregulation of the HPA axis and CRH system in depression might reflect a susceptibility that can be programmed through early life events. Both rodents and nonhuman primates exposed to adverse experiences in early life exhibit evidence of hyperactivity of the CRH system as adults. Bonnet macaques reared under stressful conditions exhibit higher CSF CRH concentrations as adults than monkeys under nonstressful conditions (Coplan et al., 1996). In rodents, early life stress is associated with persistent alterations in the CRH system, including increased CRH concentrations, increased CRH mRNA expression, altered CRH receptor expression and binding in the hypothalamus and limbic brain regions (Plotsky and Meaney, 1993, Plotsky et al., 2005; Ladd et al., 2000). This overactivity of the CNS CRH system is paralleled by a hyperreactive HPA-axis response to stress in these animals, as evidenced by neonatal maternal separation (MS) models in rodents which elicits HPA axis changes that persist into adulthood (Lehmann et al., 2002; Lippmann et al., 2002; Neumann et al., 2005). MS results in higher levels of corticosterone. The sensitivity of the glucocorticoid feedback is decreased due to downregulation of glucocorticoid and mineralocorticoid receptor gene expression in the CNS, particularly the hippocampal region CA1 and the paraventricular nucleus (Aisa et al., 2008). An interesting study by O'Mahony et al. (2009), looking at the link between MS, the HPA axis, and the brain–gut axis, demonstrated that MS, as a model of early life stress in rats, led to an increased number of fecal boli in response to novel stress and elevated plasma corticosterone. Elevated proinflammatory cytokines TNF-α and IFN-γ and a trend toward an increase in IL-6 were also found following MS. Proinflammatory cytokines, especially IL-6, are potent activators of the HPA axis (Loizzo et al., 2002). This shows that there is an altered stress system in adulthood after early life stress and these changes may contribute to the susceptibility to depression.

The findings in laboratory animals are consistent with human studies which have shown that women who are sexually or physically abused in childhood exhibit a markedly enhanced activation of the HPA axis (Heim et al., 2008); if currently depressed, they exhibit the largest increase in ACTH secretion and heart rate, as well as a very large increase in cortisol secretion.

Early life trauma is a strong predictor of CSF CRH concentrations in adults (Carpenter et al., 2004; Lee et al., 2005, 2006) and is associated with an enhanced stress response to standardized psychosocial stressors, such as the Trier Social Stress Test. In the combined DEX/CRH stimulation test patients with early life trauma exhibit evidence of marked HPA-axis hyperactivity (Heim et al., 2000, 2001, 2008; Tyrka et al., 2008).

In a recent negative study controlling for childhood adversity, Carpenter et al. (2009) assessed 34 patients with major depressive disorder (MDD) and 34 age- and sex-matched control subjects who had no current or lifetime DSM-IV depressive disorder. Effect of diagnosis on cortisol response to the DEX/CRH test was examined in a repeated measure general linear model. The matched groups were equivalent with regard to childhood adversity. Cortisol response to the DEX/CRH test among subjects with current MDD was not significantly different from that seen in matched healthy controls. Independent of diagnosis, an exploratory analysis showed a trend-level association between maltreatment history and diminished cortisol response; no interactive effects with depression diagnosis were detected.

Overactivity of the CRH/CRH1 receptor system has been shown to be one of the long-term neurobiologic sequelae of early life trauma, a major risk factor for the development of affective disorders (Edwards et al., 2003; Nemeroff, 2004).Variation in the CRH1 receptor (CRHR1) gene has been shown to interact with early life stress to predict adult depression (Bradley et al., 2008; Tyrka et al., 2009). CRHR1 polymorphisms interact with childhood maltreatment to predict HPA axis reactivity linking depression and early life stress. DEX/CRH test and CRHR1 polymorphisms showed a significant interaction with maltreatment, with CRHR1 moderating the effect of childhood maltreatment on cortisol responses in the DEX/CRH test. This excessive HPA-axis activation could represent a mechanism of interactions of risk genes with stress in the development of mood and anxiety disorders.

Monoamines, the hypothalamic–pituitary–adrenal axis, and the effects of antidepressants

Serotonin (5-HT) input to the hypothalamus is an important stimulus to CRH release. Of the many 5-HT receptors, the 5-HT$_{1A}$ receptor appears dominant in this regard (Thakore et al., 1997). Stimulation of these receptors in humans activates the HPA axis and induces hypothermia. Lesch et al. (1990a) used ipsapirone, an azaspirone that acts as a partial agonist at the 5-HT$_{1A}$

receptor, as a challenge in 12 patients with unipolar depression and 12 matched healthy controls. Ipsapirone (0.3 mg/kg) or placebo was given in random order. High basal cortisol levels were present in the patients and their ACTH/cortisol and hypothermic responses to ipsapirone were attenuated compared to controls. The impaired HPA response in the depressed patients may have been due to a glucocorticoid-induced subsensitivity of postsynaptic 5-HT_{1A} receptors or defective postreceptor signaling pathways.

Lesch et al. (1990b) also examined the effect of amitriptyline treatment on 5-HT_{1A}-induced hypothermia. Patients with major depression were chronically treated with amitriptyline, and their temperature responses to ipsapirone challenge were monitored. Amitriptyline caused further blunting in 5-HT_{1A}-mediated hypothermia, supporting the view that effective antidepressant treatment downregulates 5-HT_{1A} receptors.

It was first demonstrated almost three decades ago that stimulation of noradrenergic α_2 receptors brought about the release of growth hormone (GH) and that such a response is blunted in major depression. This was interpreted as indicative of a subsensitivity of the α_2 receptor in depression. In terms of the serotonergic system, it is well established that 5-HT receptors stimulate prolactin release and this has been used to study the sensitivity of such receptors. Studies using a variety of probe drugs of the serotonergic system, such as d-fenfluramine, a 5-HT-releasing agent and reuptake inhibitor, demonstrate a 5-HT receptor subsensitivity in unipolar major depression (Cleare et al., 1996; Newman et al., 1998).

Antidepressants that block 5-HT and noradrenergic (NE) uptake increase HPA axis activity following acute administration (Laakmann et al., 1990). Chronic administration of antidepressants to depressed patients normalizes HPA activity when there is a remission of the depressive episode. Increased concentrations of hippocampal MR and GR occur transiently between 2 and 6 weeks following the start of antidepressant treatment and parallel the time course of clinical improvement of depressive symptoms. This suggests that increased corticosteroids may contribute to the depressive syndrome (Dinan, 1994).

The hypothalamic–pituitary–adrenal axis as a target for antidepressant treatment: CRH1 receptor antagonists and cortisol synthesis inhibitors

Corticotropin-releasing hormone receptor antagonists

The relevance of overactive limbic CRH1 receptor system in depression is underlined by the fact that selective CRH1 receptor antagonists exert antidepressant effects at doses that do not influence baseline or stimulated HPA axis activation (Zobel, 2000; Kunzel et al., 2003). Zobel and colleagues (2000), in an open labeled study, administered R121919, a CRH receptor antagonist, to 20 subjects with major depression which resulted in significant reductions in depression and anxiety scores using both patient and clinician ratings. These findings, along with the observed worsening of affective symptomatology after drug discontinuation, suggests that the pharmacologic principle of CRH1 receptor antagonism has considerable therapeutic potential in the treatment and prevention of diseases where exaggerated central CRH activity is present at baseline or following stress exposure. It is important to note that, although not all studies with CRH1 receptor antagonists are positive in major depression, lack of a CRH1 receptor positron emission tomography ligand renders choice of dose problematic (Binnerman et al., 2008).

Cortisol synthesis inhibitors

Results from open label and double-blind studies have indicated that cortisol synthesis inhibitors (CSIs) may be efficacious or of adjunctive value in patients with depression, including those refractory to standard treatments. This strategy was initially proposed by Murphy (1997). The main drugs used to date include ketoconazole, metyrapone, and aminoglutethimide (Thakore and Dinan, 1995; Wolkowitz et al., 1999). However, these studies are characterized by small sample sizes and there is a need for larger controlled studies.

In a review of this subject, Wolkowitz and Reus (1999) suggest that CSIs, when used alone, have a moderate antidepressant effect, but not necessarily sufficient to produce acceptable clinical remission. It also appears that hypercortisolemic patients, rather than normocortisolemic patients, are more likely to respond. On the other hand, a glucocorticoid synthesis inhibitor may be a useful adjunctive therapy in depressed patients with HPA axis hyperactivity who do not fully respond to conventional antidepressant treatment. Kling et al. (2009) argue that the mechanism of action of these agents may not be solely a function of inhibition of adrenal cortisol production.

HYPOTHALAMIC–PITUITARY–ADRENAL AXIS FUNCTIONING IN BIPOLAR DISORDER

Bipolar disorder, or bipolar affective disorder (BPAD), characterized by profound mood swings with episodes of elation/mania alternating with episodes of depression, has a lifetime prevalence of 1–2%. The diagnosis of BPAD is made on the basis of clinical history. Key features of mania include elevated, expansive or irritable mood accompanied by hyperactivity, pressure of speech, flight of ideas, grandiosity, hyposomnia and distractibility. In recent years, the concept of BPAD has been broadened to include subtypes with similar clinical courses, phenomenology, family histories, and treatment responses. These subtypes are thought to form a continuum of disorders that, while differing in severity, are related. Readers are referred to the Diagnostic and Statistical Manual of Mental Disorders of the American Psychiatric Association (DSM-IV-TR) for details of this further classification.

While the DST has been examined primarily in depression, studies in manic patients have also been undertaken. Rates of nonsuppression of the HPA axis in mania are in the 40–50% range, similar to those reported for depression. However, as in the case of depression, the abnormality usually normalizes with effective treatment.

Dexamethasone/corticotropin-releasing hormone test in bipolar disorder

Rybakowski and Twardowska (1999) examined 40 patients with depression, 16 bipolar patients, and 24 unipolar depression patients, both during an acute episode of depression and in remission. They found that during depressive episodes bipolar patients had a greater cortisol response to DEX/CRH than unipolar patients and there was a significant correlation between the endocrine response and the severity of depression. Furthermore, this anomaly persisted in the bipolar patients even in remission. Schmider et al. (1995) tested acute and remitted manic patients, those in acute depression and healthy controls. ACTH and cortisol release were significantly increased in both manic and depressed patients in comparison to healthy subjects. In remission manic patients had a decreased response but nonetheless a response significantly higher than seen in controls. The data indicate that in either an acute manic episode or remission, bipolar patients have abnormalities in HPA function.

A more comprehensive study of the DEX/CRH test in bipolar patients in remission was carried out by Watson and colleagues (2004). They examined 53 bipolar patients, of whom 27 were in remission, and 28 healthy controls. The bipolar patients had an enhanced cortisol response to the DEX/CRH relative to the controls. They concluded that the DEX/CRH test is abnormal in both remitted and nonremitted bipolar patients.

Patients with a rapid cycling form of bipolar I disorder were investigated by Watson et al. (2005). Five patients were sequentially tested with DEX/CRH. The results were stable over time, suggesting that the test yields results which are independent of the mood state in such patients.

Vasopressin in bipolar disorder

The release of ACTH from the anterior pituitary is under the joint control of CRH acting through CRH1 receptors and vasopressin acting on V_3 (sometimes called V_{1B}) receptors. The latter become more important during chronic stress. Dinan et al. (1999) demonstrated increased responsiveness in these receptors in patients with major depression. They used desmopressin to stimulate ACTH release and found that patients with major depression released more ACTH than did healthy subjects. Watson et al. (2006) examined DEX/negative feedback on AVP release in 64 patients with mood disorder and 21 controls. Forty-one patients were bipolar and the remainder had chronic depressive disorder. Twenty-one of the bipolar patients were in remission, 10 were depressed and 10 were rapid cycling. All subjects were administered DEX 1.5 mg at 23.00 h. On the following afternoon at 15.00 h blood was collected for AVP measurement. Post-DEX levels were significantly higher in patients with bipolar illness and chronic depression than matched healthy controls. It is not clear from the study whether or not AVP levels were elevated at baseline or whether the AVP emanates from the paraventricular nucleus or the magnocellular nucleus of the hypothalamus. However, it is tempting to speculate that the HPA overactivity seen in patients with bipolar illness may be driven by excess AVP.

Monoamines in bipolar disorder

Far fewer studies have been conducted in bipolar disorder than in depression. When healthy subjects are administered d-fenfluramine a significant increase in prolactin levels is observed. Unipolar depressives show a blunted response to this challenge. In a study of manic patients Thakore et al. (1996) found blunted responses similar to those found in depression.

HYPOTHALAMIC–PITUITARY–ADRENAL AXIS FUNCTIONING IN SCHIZOPHRENIA

Schizophrenia and psychosis are a group of illnesses that occur in approximately 1% of the adult population in most countries in which surveys have been conducted. This disorder usually begins during adolescence or young adulthood and is characterized by a spectrum of symptoms that typically include disordered thought, social withdrawal, hallucinations (aural and less commonly visual), delusion of persecution (paranoia), and bizarre behavior. These symptoms are sometimes categorized as "positive" (e.g., hallucinations) and "negative" (e.g., social withdrawal and apathy). So far there is no known cure and the disease is chronic and generally progressive. Nevertheless, the introduction of the phenothiazine neuroleptic chlorpromazine in 1952 initiated the era of pharmacotherapy in psychiatric medicine and currently there is a range of antipsychotic drugs used to treat psychotic disorders and schizophrenia. Investigators then began to define the mechanism of action of these groups of drugs hoping that the elucidation of their mechanism of action would give some insight into the psychopathology of schizophrenia. Thus the dopamine hypothesis of schizophrenia came into being, and it is only in recent years that other neurotransmitters and neuroendocrine pathways such as the HPA axis have been investigated in the illness.

HPA functioning in people with schizophrenia has been systematically reviewed by Bradley and Dinan (2010). They noted that the two most commonly studied areas of HPA axis functioning, i.e., the direct measurement of basal cortisol and the DST, provided highly heterogeneous results, making interpretation complex and firm conclusion on the state of HPA axis function in schizophrenia difficult. Other areas that have been investigated in the literature, such as the HPA axis response to psychological stressors, provide more consistent findings but the results are still open to interpretation.

Basal cortisol in schizophrenia

On review of the literature there is evidence of elevated basal cortisol in some, but not all, patients with schizophrenia. In a systematic review by Bradley and Dinan (2010), mean basal cortisol was statistically significantly elevated in schizophrenic patients compared to controls (area under the curve (AUC) or at a single time point) in 33 of 77 (44.2%) studies. Basal cortisol was not statistically different between schizophrenic patients and controls in 44 of 77 (57.1%) studies and basal cortisol was significantly lower in schizophrenic patients than in controls in 4 of 77 (5.2%) studies. Elevated basal cortisol was reported in studies assessing acutely psychotic patients, in patients described as stable, and in patients with prominent negative symptoms (Bradley and Dinan, 2010). These studies demonstrate that elevated cortisol can be present in patients with schizophrenia at different phases of the illness and with different levels and types of symptoms.

Measurement of HPA axis functioning in schizophrenia overall, however, is confounded by several factors which may account for the heterogeneous results described above. Most importantly, psychological stress influences cortisol secretion and, for this reason, a fair comparison of basal cortisol secretion between patients and controls would require both groups to be under similar levels of psychological stress. In practical terms this is difficult to achieve for a number of reasons. Schizophrenia patients are exposed to a broad range of psychological stressors, some true stressors experienced by the population at large but others which can be termed "pseudostressors" (Dinan, 2004). These emanate from subjective space and are only experienced by people with schizophrenia and other psychotic illness. They are represented by the core symptoms of schizophrenia such as delusions and hallucinations and can have profound emotional intensity. They are likely quite different to the stresses experienced by the rest of the population. It is likely that the type of psychotic symptoms and the perceptions of these symptoms as stressful or not by the individual will influence the cortisol level. For example, acute persecutory delusions are more likely to be perceived as stressful by the individual than delusions of grandeur and therefore may influence cortisol secretion differently. The type of stress and whether it is social-evaluative and uncontrollable has been demonstrated to influence cortisol level in healthy subjects (Dickerson and Kemeny, 2004) and may also be applicable to those with schizophrenia (Jones and Fernyhough, 2007). Acutely psychotic patients are frequently hospitalized and this can bring about different stressors to the control populations in the studies reviewed. Because of these differences it is difficult to interpret whether any difference in cortisol level is due to different levels of stress experienced by patients compared to control or to a difference in basal function of the HPA axis.

Although a study by Strous and colleagues (2004) found no difference in basal cortisol between first episode schizophrenia patients and controls, it did demonstrate significant associations of cortisol with negative ($r = 0.35$, $p = 0.036$), general ($r = 0.35$, $p = 0.034$), and total positive and negative syndrome scale (PANSS) scores ($r = 0.33$, $p = 0.045$). In this study, levels of the adrenal steroid dehydroepiandrosterone (DHEA) and its sulphated form DHEA-S were significantly elevated in schizophrenia patients. These steroids have antiglucocorticoid activity and have been shown to reduce cortisol levels in healthy individuals (Wolf et al., 1997). They could, therefore, explain the nonelevation of cortisol in this group.

Antipsychotic medications may influence cortisol levels. Evidence in healthy subjects suggests this effect is minimal with first generation drugs but significant with second generation antipsychotics (Cohrs et al., 2006). In order to examine basal cortisol in patients without the influence of antipsychotic medication, ideal subjects are first episode, drug-naive schizophrenic patients. In studies of this type that have assessed drug-naive schizophrenic patients, basal cortisol was generally found to be significantly elevated compared to controls. Collectively, these positive studies included 226 of the

312 (72.4%) first episode drug-naive patients studied in the literature, indicating that elevated cortisol secretion is a common finding in drug-naive schizophrenic patients (Bradley and Dinan, 2010).

The dexamethasone suppression test in schizophrenia

A systematic review of the literature (Bradley and Dinan, 2010) identified 85 studies including 2722 schizophrenia patients that employed the dexamethasone suppression test in schizophrenia. Nonsuppression of cortisol following DEX ranged from 0 to 81%. From the total group of schizophrenia patients studied, 731/2722 (26.9%) were classified as nonsuppressors. Since antipsychotic medication may affect DST results (Tandon et al., 1991), medicated and drug-free patients are looked at separately. For nonmedicated schizophrenia patients (no neuroleptic medication for ≥ 2 weeks prior to the DST), 227/773 (29.4%) were classified as nonsuppressors. Of the 1949 medicated schizophrenia patients, 504 (25.9%) were classified as nonsuppressors, suggesting that antipsychotic medications themselves had little effect on DST results.

Several studies correlated symptoms such as depressive, negative, and positive symptoms with DST outcomes; however, no clear picture of correlation emerged. Associations between negative symptoms and nonsuppression of post-DEX cortisol levels were found in just over half of the studies evaluated (Bradley and Dinan, 2010). In the same review seven studies found a correlation of nonsuppression with depression, whist 15 did not. Goldman et al. (1996) found no association of nonsuppression with a family history of depression. Seven studies found no correlation between positive symptoms and nonsuppression of cortisol, with one study finding a positive association (Jones et al., 1994), one a trend towards a positive correlation (Keshavan et al., 1989), and one a negative association (Newcomer et al., 1991). Two studies have found an association of post-DEX cortisol with suicide attempt (Jones et al., 1994; Plocka-Lewandowska et al., 2001); however, a study by Lewis et al. (1996) found no such association.

Studies have looked at the DST performed on newly admitted psychotic patients which was then repeated following varying lengths of antipsychotic treatment (Bradley and Dinan, 2010). These studies all demonstrated high nonsuppression rates on admission (range 30–81%) and in each case there was a large reduction in the percentage of nonsuppressors following antipsychotic treatment (range 0–46%), during which time there was an assumed clinical improvement. In the longest study over time, Plocka-Lewandowska et al. (2001) found nonsuppression in 41% of the acutely hospitalized sample; 25% were nonsuppressors 9 years later.

Basal measures of corticotropin-releasing hormone and adrenocorticotropin in schizophrenia

Twenty-one studies are reported in the literature, measuring basal CRH and/or ACTH in 539 schizophrenia patients (Bradley and Dinan, 2010). Basal CRH in CSF was measured in five studies; four found basal CRH to be similar in schizophrenia and controls and in only one, Banki et al. (1987), was there any evidence for increased basal CRH in schizophrenia. However, in this study, 6/23 (26%) schizophrenia patients had a CRH value higher than the greatest value in any control subject. Thus, the mean CRH for schizophrenia patients was statistically significantly higher than that for control subjects ($p < 0.001$) but the authors suggest the mean value was skewed by the data from the three schizophrenia patients with extremely high CRH values.

Seventeen studies measured basal ACTH (Bradley and Dinan, 2010). Eight of these studies found no difference between basal ACTH in schizophrenia patients and controls. These studies also all reported no difference in cortisol in these patients, indicating a normal effect of ACTH on cortisol secretion.

Corticotropin-releasing hormone test in schizophrenia

The CRH test has only been reported once in drug-free and medicated patients with schizophrenia and results did not differ from controls, suggesting an intact HPA axis function (Roy et al., 1986). However, this study was small (n = 9) and patients showed various levels of psychosis. There was a trend towards a negative ACTH response with psychosis rating ($r = -0.42$, $p < 0.1$) suggesting ACTH response may be blunted in a more uniformly psychotic cohort. More studies using the CRH test are needed before any firm conclusions can be drawn.

Dexamethasone/corticotropin-releasing hormone test in schizophrenia

The DEX/CRH test is thought to be more sensitive to subtle HPA system changes than the DST and was utilized by Lammers and colleagues (1995), who demonstrated that following pretreatment with DEX, schizophrenia patients released more cortisol in response to CRH than controls. This was particularly so in drug-free patients who had higher BPRS scores than in medicated, less severely ill patients. This suggests that illness phase may influence cortisol secretion. However, it is not possible to determine if the reduction in cortisol secretion was related to lower levels of symptoms or the use of medication, as many of the medicated patients were taking clozapine, a drug known to have a direct effect on cortisol secretion. This study suggests the mechanism of cortisol hypersecretion in these patients may be due, in part, to impaired negative feedback mechanisms since this occurred following DEX. Of note, the cortisol response was increased in patients despite no significant difference in ACTH AUC between patients and controls over the course of the test, suggesting an explanation other than increased ACTH for the increase in cortisol, which may agree with the findings of Ryan et al. (2004) and Breier et al. (1988a, b).

Effects of psychological stress on the hypothalamic–pituitary–adrenal axis in schizophrenia

Six studies that measured the effects of psychological stress on HPA function were identified in the literature (Bradley and Dinan, 2010). These included 89 patients with schizophrenia and 144 controls. The study by Albus et al. (1982) used a combination of psychological and physical stressors whilst that by Goldman et al. (2007) used the cold pressor test (immersion of a limb in iced water), which is reported to invoke an HPA response via both psychological and physical components (Bullinger et al., 1984). The study reported by Breier et al. (1988a) measured response to the psychological stress associated with lumbar puncture. The remaining studies used a public speaking task as the psychological stressor. The results from five of these studies were consistent, indicating that people with schizophrenia have a blunted cortisol response to psychological stress compared to controls. In one study (Brenner et al., 2009), there was a trend towards a lower cortisol response to psychological stress in schizophrenia patients. Two of the studies (Breier et al., 1988a; Goldman et al., 2007) also measured ACTH and both found this too had a blunted response to stress.

Since psychological stress may be important in the development and course of schizophrenia, studies that measure the reaction to psychological stress could be important in understanding how environmental factors contribute to pathogenesis of the disease. Studies measuring HPA axis response to psychological stress in schizophrenia patients produced consistent results demonstrating a blunted ACTH and cortisol response to stress. The results from the studies in the literature (Dinan and Bradley, 2010) were the same in drug-free, medicated, acutely psychotic, and more chronic patients, suggesting these factors did not affect the outcomes of the studies. Where heart rate and mean arterial pressure were measured, increases in schizophrenia patients were similar to those in controls, indicating that patients found the tests stressful and that the physical response to stress was intact. A variety of explanations have been suggested for these results. Breier et al. (1988a) suggested that the blunted ACTH and cortisol stress response may partly be due to the significant negative correlation between the levels of psychosis and stress-induced increases in ACTH, i.e.. the greater the psychosis, the less the increase in cortisol. They suggest it is possible that cognitive and/or neurobiologic processes associated with severe psychosis have a disruptive effect on mechanisms involved in mounting a neuroendocrine stress response. In the study reported by Jansen et al. (2000), the cortisol response in schizophrenia patients to a physical stress, exercise, was not different from controls and they suggest the blunted cortisol response to a psychological stress could be due to the different mechanisms of HPA axis activation to different types of stress. They cite evidence that physical stressors invoke a response via CRH and, as described in this review, CRH function appears relatively normal in schizophrenic patients, hence the normal stress response to exercise.

Psychological stress stimulates the HPA axis via AVP (Romero and Sapolsky, 1996), which may be diminished in schizophrenic patients (Marx and Lieberman, 1998), resulting in a blunted cortisol response to psychological stress. The hypothesis proposed by Ryan et al. (2004), that AVP may be responsible for hypersecretion of cortisol in first episode patients via its ability to stimulate cortisol release directly from the adrenal gland, is seemingly at odds with this. They too found a reduction of AVP compared to controls, although the effects of AVP would have been augmented by the increased level of ACTH seen in their patients, which together could have accounted for the increased cortisol level. Jansen et al. (2000) also suggest that the impaired stress response may be due to differences in coping strategies used by people with schizophrenia compared to controls. They found that schizophrenia patients used more passive and avoidance strategies compared to controls, which may invoke a different biologic stress response. In the study reported by Brenner et al. (2009), although there was a trend towards lower cortisol secretion during psychological stress, it was suggested that cortisol secretion may just be delayed. This is based on the observation that cortisol levels were significantly lower at one time point during the study but were subsequently similar, suggesting the cortisol response was initially delayed but ultimately caught up. This group proposes that this could be due to impairment in executive functions where patients were less able to think ahead before the task and therefore the stress-associated HPA axis response was delayed. This view may be supported by evidence from Gaab et al. (2005), who found that anticipatory cognitive appraisal of acute stressors explained up to one-third of the variance observed in cortisol response to stress. This cognitive appraisal could, of course, be impaired in schizophrenia patients. Brenner et al. (2009) also suggest that physiological differences in the HPA axis such as a hypoactivity could explain the delay in cortisol response. Hypofunction of the HPA axis during stress tests was also demonstrated by MacMillan et al. (2009) and Elzinga et al. (2008) in studies that found a blunted cortisol stress response to psychological stress in patients previously exposed to significant life stress. It is possible that in the schizophrenia cohorts studied, the rate of exposure to significant life stressors may have been greater in patients than in controls.

Evidence is accumulating of other factors that may significantly affect the integrity and function of the HPA axis and therefore significantly influence outcomes in these studies. Some of these factors are particularly applicable to people with schizophrenia. For example, HPA axis function may be influenced by exposure to maternal stressors such as fetal malnutrition (Phillips and Jones, 2006) and psychological stress experienced by the mother during pregnancy (Glover et al., 2010) in a process termed fetal programming. These stressors may be expressed as a low birth weight or minor physical anomalies in the infant. Studies have demonstrated that low birth weight is associated with enhanced HPA axis and autonomic response to experimentally-induced psychological stress (Phillips and Jones, 2006).

Schizophrenia is an illness associated with low birth weight (Cannon et al., 2002) and an increased incidence of minor physical anomalies (Compton and Walker, 2009), which may indicate increased exposure to a variety of maternal stressors. As these factors are more common in people with schizophrenia, they may have a greater influence on HPA function in schizophrenia patients and may contribute to the differences in HPA function seen between patients with schizophrenia and controls, as well as accounting for some of the variation in HPA function within the schizophrenia population. This potential is illustrated by a study by Mittal et al. (2007) which demonstrated that schizotypal adolescents express significantly more minor physical anomalies than normal controls and that these predict cortisol elevation.

Another factor is childhood trauma, which is associated with sensitization of the stress response, glucocorticoid resistance, increased CRH activity, immune activation, and reduced hippocampal volume (Heim et al., 2008). Several studies have directly illustrated the effect of childhood trauma on HPA axis function. In a study of young females maltreated during childhood (e.g., physical, sexual, and emotional abuse), a blunted cortisol response to the Trier Social Stress Test was found compared to controls (MacMillan et al., 2009). Elzinga et al. (2008) also measured the cortisol response to the Trier Social Stress Test in people exposed to various levels of adverse life events such as emotional, physical, or sexual abuse but with similar baseline cortisol levels. A significant blunted cortisol response was found in individuals with a history of adverse life events compared to individuals with no such events. This finding was primarily driven by the result in men.

Childhood trauma has been shown to be predictive, and probably causal, in the development of psychosis (Read et al., 2005; Shevlin et al., 2008) and it is therefore reasonable to assume that differences in the frequency and severity of childhood traumas in the schizophrenia populations studied could also explain some of the variance in the results. The potential for such an effect was seen in a recent study. Mondelli et al. (2010) found a significant negative correlation between the number of stressful life events and cortisol level ($r = -0.36$, $p < 0.014$) in first episode, psychotic patients including schizophrenia and schizophreniform disorder. The control group, as expected, had a positive correlation between the number of stressful life events and cortisol level ($r = 0.42$, $p < 0.013$). The authors hypothesized that the unexpected negative correlation in psychotic patients may be explained by the excessive load of stressful life events experienced in this patient group. Stressful life events were experienced on average at around three times the rate in the psychotic cohort compared to the controls and 85.7% of the psychotic patients had experienced a childhood trauma compared to 38.7% of controls. It is conceivable, given the evidence from Mondelli et al. (2010), that some patients with schizophrenia who have experienced significant levels of life stress or perhaps childhood trauma may actually hyposecrete cortisol. In many studies where the schizophrenia cohort would have included such patients, this would effectively reduce the mean cortisol level and therefore may explain why many studies found no significant differences in basal cortisol level between schizophrenia patients and controls.

Despite the heterogeneity of outcomes from HPA studies, there is evidence that people with schizophrenia experience periods of heightened cortisol secretion. This consistently occurs at the first episode but also occurs in some chronic patients with more stable clinical features. The variation in results may be explained by varying symptoms of illness as well as by medication use and exposure to other environmental factors known to influence HPA axis function. Some patients with schizophrenia may also experience low cortisol levels under certain conditions. This is particularly evidenced by a blunted cortisol response to a psychological stress test but may also be the case in acutely ill patients as many studies found no evidence of elevated cortisol in psychotic patients admitted to hospital where a stress response would reasonably be expected.

HYPOTHALAMIC–PITUITARY–ADRENAL AXIS FUNCTIONING IN ANXIETY DISORDERS

The HPA axis has been studied in a number of anxiety disorders, most notably post-traumatic stress disorder (PTSD). Anxiety is a normal human emotion which is triggered when we perceive threats of harm. We need a fear response in order to warn us of threats and enable us to prepare for them. This response is called the flight or fight response (and is related to the release of adrenaline with sympathetic actions throughout the body). When anxiety is abnormal it causes distress, harm, and interferes with our lives, often not protecting us and decreasing our performance levels. It has emotional, behavioral, cognitive, and physical components. Anxiety disorders are a constellation of disorders that have a variety of anxiety symptoms to the core of their psychopathology.

The CRH system has been implicated in the pathophysiology of anxiety disorders (Reul and Holsboer, 2002; Risbrough and Stein, 2006; Mathew et al., 2008). Increased CSF CRH concentrations have been reported in post-traumatic stress disorder (Bremner et al., 1997; Baker et al., 1999; Sautter et al., 2003), but studies in adults with panic disorder and generalized anxiety disorder have failed to show such abnormalities (Banki et al., 1992a; Jolkkonen et al., 1993; Fossey et al., 1996). Interestingly, there is evidence from a series of animal studies that the CRH system is also highly relevant to alcohol dependence (Heilig and Koop, 2007), particularly during alcohol withdrawal (Valdez and Koop, 2004). It is also important to note that benzodiazepine anxiolytics reduce CRH-ergic activity (Skelton et al., 2000). The HPA axis in panic disorder and post-traumatic stress disorder will be elaborated on below.

Panic disorder

Cortisol secretion has not been associated with panic attacks and dexamethasone nonsuppression occurs in only 17% of patients with panic disorder (Holsboer et al., 1987; Heninger, 1990; Abelson and Cameron, 1994; Graeff et al., 2005). CRH challenge studies have been inconsistent, with some demonstrating a decreased ACTH response when compared with healthy controls and others finding a normal response. Using the DEX/CRH test, patients with panic disorder show some dissociation between ACTH and cortisol response, with indications that the duration of psychopathology is a risk factor for increased reactivity of the HPA axis (Petrowski et al., 2012). Other than in PTSD, studies in anxiety states overall would suggest a normal or modestly enhanced HPA function in these disorders.

Post-traumatic stress disorder is characterized by the re-experiencing of an extremely traumatic event accompanied by symptoms of increased arousal and by avoidance of stimuli associated with the trauma. Neuroendocrine studies have shown profound alterations in HPA axis regulation in PTSD. The majority of studies demonstrate reduced baseline cortisol levels in addition to enhanced cortisol suppression to low-dose dexamethasone administration (Yehuda, 2006). Patients with PTSD show enhanced cortisol feedback inhibition of ACTH secretion at the level of the pituitary in PTSD (Yehuda et al., 2006) and a blunted ACTH response to CRH. Strohle and colleagues (2008), using the combined DEX/CRH test in eight drug-free patients with PTSD and matched healthy subjects, found a reduced ACTH response in the former group although de Kloet et al. (2008) failed to find such a difference. However, in the latter study, PTSD patients with comorbid MDD showed an attenuated ACTH response to the DEX/CRH test, unlike the PTSD without comorbid MDD group. A recent study (Vythilingam et al., 2010) demonstrated that low early morning plasma cortisol in PTSD is associated with comorbid depression but not with enhanced glucocorticoid feedback inhibition as demonstrated by the DST, supporting a central abnormality in glucocorticoid regulation.

A high coincidence of childhood abuse, major depressive disorder, and PTSD has been reported in patients with borderline personality disorder (BPD), characterized by affective instability, chronic suicidality, and self-harm. Animals exposed to early trauma show increased stress-induced HPA axis activity due to enhanced CRH drive and glucocorticoid feedback resistance. In humans, as indicated above, PTSD and MDD are associated with decreased and increased resistance to glucocorticoid feedback respectively. Childhood trauma in adults with personality disorder is associated with blunted cortisol and ACTH secretion following DEX/CRH challenge. These effects were independent of depression and PTSD (Rinne et al., 2002). Comorbid PTSD significantly attenuated the ACTH response. Hyperresponsiveness of the HPA axis in chronically abused subjects might be due to enhanced central drive to pituitary ACTH release (Lee et al., 2012). This study and a further study by Carvalho-Fernando et al. (2012) using the DST suggest that sustained childhood abuse rather than BPD, MDD, or PTSD pathology accounts for this effect.

CONCLUSION

An increasing volume of data provides support for the view that neuroendocrine and immune alterations occur in psychiatric illnesses and in recent years an enormous amount has been learned regarding the role of the HPA in such illnesses. Depression has received the greatest amount of attention to date. Disturbance in HPA function has long been recognized as a feature of major depression, especially in patients with melancholic features. It was hoped that the axis might yield diagnostic markers and clinically useful diagnostic tests for depression but this has yet to happen. The dexamethasone suppression test showed initial promise in the 1970s and 1980s to diagnose endogenous depression with accuracy and predict drug response and clinical relapse. Yet, after thousands of patients were tested, an American Psychiatric Association task force (1987) concluded that the test had a rather low sensitivity (40–50% for depression, 60–70% for endogenous forms), modest specificity (often 70%) and limited clinical utility. Thus its use has been mainly to further develop our knowledge of the HPA axis rather than to diagnosis depression in a real world setting. One valid argument for its low sensitivity and specificity is that such a test, which is essentially a peripheral HPA-axis test, may not give direct evidence on the state of the central projecting CRH and vasopressin neurons. It was also thought that the HPA axis might be an appropriate target for novel therapies in depression; sadly, however, CRH, V_3, and GR antagonists have been studied without resounding success. Interesting data to emerge from this field of research, however, indicates that early childhood trauma can permanently dysregulate the axis and lead to depression vulnerability.

The research to date in bipolar affective disorder indicates that in either an acute manic episode or in remission, bipolar patients have abnormalities in HPA function when compared to control subjects. Other than in PTSD, studies in anxiety disorders would suggest a normal or modestly enhanced HPA function in these disorders. Neuroendocrine studies have shown profound alterations in HPA axis regulation in PTSD, the majority of these demonstrating reduced baseline cortisol levels in addition to enhanced cortisol suppression to low-dose dexamethasone administration. Despite the heterogeneity in outcomes from studies of the HPA axis in schizophrenia, there is evidence of elevated cortisol secretion compared to controls. This occurs more often at the first episode but can also occur in some chronic patients with more stable clinical features. The variation in results in studies of schizophrenia may be explained by varying symptoms of illness as well as by medication use and exposure to other environmental factors known to influence HPA axis function.

Overall, findings conclude that a vast array of neuroendocrine abnormalities occur in the HPA axis in many psychiatric illnesses. What this means from a practical perspective in terms of novel therapeutic targets for effective future treatments to achieve symptom relief and remission remains to be seen. One way forward could be to assess the HPA axis dysregulation in different symptom clusters in depression rather than in using DSM-IV criteria as there may be different subgroups with different abnormalities of the axis. This has been suggested by others in psychiatry as a limiting factor to clear pathways forward for investigations of novel therapeutic targets (Kapur et al., 2012). The current diagnostic system was not designed to facilitate biologic differentiation and the biologic studies to date have not been able to propose a clinically viable alternative system. The lack of a gold standard has thus hampered progress in this field. Another real opportunity for psychiatry is to use the emerging advances in genetics, molecular biology, imaging, and cognitive science to supplement, rather than replace, the symptom-driven diagnosis and this is where research may be directed to in the future.

ACKNOWLEDGMENTS

The authors would like to thank Dr Marcela Julio for preparing the HPA axis image used in this chapter (Fig. 5.1) (http://www.facebook.com/imagenesciencia).

REFERENCES

Abelson JL, Cameron OG (1994). Adrenergic dysfunction in anxiety disorders. In: OG Cameron (Ed.), Adrenergic dysfunction and psychobiology. AP Press, Washington DC.

Aguilera G, Rabadan-Diehl C (2000). Vasopressinergic regulation of the hypothalamic–pituitary–adrenal axis: implications for stress adaptation. Regul Pept 96: 23–29.

Aisa B, Tordera R, Lasheras B et al. (2008). Effects of maternal separation on hypothalamic–pituitary–adrenal responses, cognition and vulnerability to stress in adult female rats. Neuroscience 154: 1218–1226.

Albus M, Ackenheil M, Engel RR et al. (1982). Situational reactivity of autonomic functions in schizophrenic patients. Psychiatr Res 6: 361–370.

American Psychiatric Association (2000). Diagnostic and Statistical Manual of Mental Disorders. 4th edn. Text Revision (DSM-IV-TR), American Psychiatric Association, Washington.

American Psychiatric Association Task Force on Laboratory Tests in Psychiatry (1987). The dexamethasone suppression test: an overview of its current status in psychiatry. Am J Psychiatry 144: 1253–1262.

Amsterdam JD, Maislin G, Winokur A et al. (1988). The CRH test before and after clinical recovery from depression. J Affect Disord 14: 213–222.

Arato M, Banki CM, Nemeroff CB et al. (1986). Hypothalamic–pituitary–adrenal axis and suicide. Ann N Y Acad Sci 487: 263–270.

Arato M, Banki CM, Bissette G et al. (1989). Elevated CSF CRH in suicide victims. Biol Psychiatry 24: 355–359.

Arborelius L, Owens MJ, Plotsky PM et al. (1999). The role of corticotrophin-releasing factor in depression and anxiety disorders. J Endocrinol 160: 1–12.

Axelson DA, Doraiswamy PM, Boyko OB et al. (1992). In vivo assessment of pituitary volume with magnetic resonance imaging and systematic stereology: relationship to dexamethasone suppression test results in patients. Psychiatry Res 46: 63–70.

Baker DG, West SA, Nicholson WE et al. (1999). Serial CSF corticotrophin-releasing hormone levels and adrenocortical activity in combat veterans with posttraumatic stress disorder. Am J Psychiatry 156: 585–588.

Bale TL, Contarino A, Smith GW et al. (2000). Mice deficient for corticotrophin-releasing hormone receptor-2 display anxiety-like behavior and are hypersensitive to stress. Nat Genet 24: 410–414.

Banki CM, Bissette G, Arato M et al. (1987). CSF corticotrophin-releasing factor-like immunoreactivity in depression and schizophrenia. Am J Psychiatry 144: 873–877.

Banki CM, Karmacsi L, Bissette G et al. (1992a). Cerebrospinal fluid neuropeptides in mood disorders and dementia. J Affect Disord 25: 39–46.

Banki CM, Karmacsi L, Bissette G et al. (1992b). Cerebrospinal-fluid neuropeptides: a biochemical subgrouping approach. Neuropsychobiology 26: 37–42.

Banki CM, Karmacsi L, Bissette G et al. (1992c). CSF corticotrophin-releasing hormone and somatostatin in major depression: response to antidepressant treatment and relapse. Eur Neuropsychopharmacol 2: 107–113.

Berger M, Pirke KM, Doerr P et al. (1984). The limited utility of the dexamethasone suppression test for the diagnostic process in psychiatry. Br J Psychiatry 145: 372–382.

Binder EB, Nemeroff CB (2010). The CRF system, stress, depression and anxiety – insights from human genetic studies. Mol Psychiatry 15: 574–588.

Binnerman B, Feltner D, Kolluri S et al. (2008). A 6-week randomised, placebo-controlled trial of CP-316,311 (a selective CRH1 antagonist) in the treatment of major depression. Am J Psychiatry 165: 617–620.

Boorse GC, Denver RJ (2006). Widespread tissue distribution and diverse functions of corticotrophin-releasing factor and related peptides. Gen Comp Endocrinol 146: 9–18.

Bradley AJ, Dinan TJ (2010). A systematic review of hypothalamic–pituitary–adrenal axis function in schizophrenia: implications for mortality. J Psychopharmacol 24 (Suppl): 91–118.

Bradley RG, Binder EB, Epstein MP et al. (2008). Influence of child abuse and adult depression: moderation by the corticotrophin releasing hormone receptor gene. Arch Gen Psychiatry 65: 190–200.

Breier A, Wolkowitz OM, Doran AR et al. (1988a). Neurobiological effects of lumbar puncture stress in psychiatric patients and healthy volunteers. Psychiatry Res 25: 187–194.

Breier A, Wolkowitz OM, Rapaport M et al. (1988b). Metabolic stress effects in normal volunteers and schizophrenic patients. Psychopharmacol Bull 24: 431–433.

Bremner JD, Licinio J, Darnell A et al. (1997). Elevated CSF corticotrophin-releasing factor concentrations in posttraumatic stress disorder. Am J Psychiatry 154: 624–629.

Brenner K, Liu A, Laplante DP et al. (2009). Cortisol response to a psychosocial stressor in schizophrenia: blunted, delayed, or normal? Psychoneuroendocrinology 34: 859–886.

Bullinger M, Naber D, Pickar D et al. (1984). Endocrine effects of the cold pressor test: relationships to subjective pain appraisal and coping. Psychiatry Res 12: 227–233.

Cannon M, Jones PB, Murray RM (2002). Obstetric complications and schizophrenia: historical and meta-analytic review. Am J Psychiatry 159: 1080–1092.

Carpenter LL, Tyrka AR, McDougle CJ et al. (2004). Cerebrospinal fluid corticotrophin-releasing factor and perceived early-life stress in depressed patients and healthy control subjects. Neuropsychopharmacology 29: 777–784.

Carpenter LL, Ross NS, Tyrka AR et al. (2009). Dex/CRH test cortisol response in outpatients with major depression and matched healthy controls. Psychoneuroendocrinology 34: 1208–1213.

Carrasco GA, Van de Kar LD (2003). Neuroendocrine pharmacology of stress. Eur J Pharmacol 463: 235–272.

Carroll BJ, Feinberg M, Greden JF et al. (1981). A specific laboratory test for the diagnosis of melancholia: standardization, validation, and clinical utility. Arch Gen Psychiatry 38: 15–22.

Carvalho-Fernando S, Beblo T, Schlosser N et al. (2012). Associations of childhood trauma with hypothalamic–pituitary–adrenal function in borderline personality disorder and major depression. Psychoneuroendocrinology 37: 1659–1668.

Chalmers DT, Lovenberg TW, De Souza EB (1995). Localisation of novel corticotrophin-releasing factor receptor (CRF2) mRNA expression to specific subsortical nuclei in rat brain: comparison with CRF-1 receptor mRNA. J Neurosci 15: 6340–6350.

Chrousos GP (1995). The hypothalamic–pituitary–adrenal axis and immune mediated inflammation. N Engl J Med 332: 1351–1362.

Cleare AJ, Murray RM, O'Keane V (1996). Reduced prolactin and cortisol responses to d-fenfluramine in depressed compared to healthy matched control subjects. Neuropsychopharmacology 14: 349–354.

Cohrs S, Röher C, Jordan W et al. (2006). The atypical antipsychotics olanzapine and quetiapine, but not haloperidol, reduce ACTH and cortisol secretion in healthy subjects. Psychopharmacology (Berl) 185: 11–18.

Cole MA, Kim PJ, Kalman BA et al. (2000). Dexamethasone suppression of cortciosteroid secretion: evaluation of the site of action by receptor measures and functional studies. Psychoneuroendocrinology 25: 151–167.

Compton MT, Walker EF (2009). Physical manifestations of neurodevelopmental disruption: are minor physical anomalies part of the syndrome of schizophrenia? Schizophr Bull 35: 425–436.

Coplan JD, Andrews MW, Rosenblum LA et al. (1996). Persistent elevations of cerebrospinal fluid concentrations of corticotrophin-releasing factor in adult nonhuman primates exposed to early-life stressors: implications for the pathophysiology of mood and anxiety disorders. Proc Natl Acad Sci U S A 93: 1619–1923.

De Bellis MD, Gold PW, Geracioti TD et al. (1993). Fluoxetine significantly reduces CSF CRH and AVP concentrations in patients with major depression. Am J Psychiatry 150: 656–657.

de Kloet ER, De Rijk RH, Meijer OC (2007). Therapy insight: is there an imbalanced response of mineralocorticoid and glucocorticoid receptors in depression? Nat Clin Pract Endocrinol Metabol 3: 168–179.

de Kloet C, Vermetten E, Lentjes E et al. (2008). Differences in response to the combined DEX-CRH test between PTSD patients with and without co-morbid depressive disorder. Psychoneuroendocrinology 33: 313–320.

Dentino AN, Pieper CF, Rao KMK et al. (1999). Association of interleukin-6 and other biological variables with depression in older people living in the community. J Am Geriatr Soc 47: 6–8.

Dickerson SS, Kemeny ME (2004). Acute stressors and cortisol responses: a theoretical integration and synthesis of laboratory research. Psychol Bull 150: 355–391.

Dinan TG (1994). Glucocorticoids in the genesis of depression: a psychobiological model. Br J Psychiatry 164: 365–371.

Dinan TG (1996). Serotonin and the regulation of hypothalamic–pituitary–adrenal function: a mini-review. Life Sci 58: 1683–1694.

Dinan TG (2004). Stress and the genesis of diabetes mellitus in schizophrenia. Br J Psychiatry 184 (Suppl 47): s72–s75.

Dinan TG, Scott LV (2005). Anatomy of melancholia: focus on the hypothalamic–pituitary–adrenal axis overactivation and the role of vasopressin. J Anatomy 207: 259–264.

Dinan TG, Lavelle E, Scott LV et al. (1999). Desmopressin normalises the blunted ACTH response to corticotropin-releasing hormone in melancholic depression: evidence of enhanced vasopressinergic responsivity. J Clin Endocrinol Metab 84: 2238–2246.

Dinan TG, O Brien S, Lavelle E et al. (2004). Further neuroendocrine evidence for enhanced vasopressin V3 receptor responses in melancholic depression. Psychol Med 34: 169–172.

Dunn AJ, Berridge CW (1990). Physiological and behavioural responses to corticotrophin-releasing factor administration: is CRH a mediator of anxiety or stress response? Brain Res Rev 15: 71–100.

Edwards VJ, Holden GW, Felitti VJ et al. (2003). Relationship between multiple forms of childhood maltreatment and adult mental health in community respondents: results from the adverse childhood experiences study. Am J Psychiatry 160: 1453–1460.

Elzinga BM, Roelofs K, Tollenaar MS et al. (2008). Diminished cortisol responses to psychosocial stress associated with lifetime adverse events: a study among healthy young subjects. Psychoneuroendocrinology 33: 227–237.

Fossey MD, Lydiard RB, Ballenger JC et al. (1996). Cerebrospinal fluid corticotrophin-releasing factor concentrations in patients with anxiety disorders and normal comparison subjects. Biol Psychiatry 39: 703–707.

France RD, Urban B, Krishnan KRR et al. (1988). CSF corticotropin-releasing factor-like immunoreactivity in chronic pain patients with and without major depression. Biol Psychiatry 23: 86–88.

Fries E, Dettenborn L, Kirschbaum C (2009). The cortisol awakening response (CAR): facts and future directions. Int J Psychophysiol 72: 67–73.

Gaab J, Rohleder N, Nater UM et al. (2005). Psychological determinants of the cortisol stress response: the role of anticipatory cognitive appraisal. Psychoneuroendocrinology 30: 599–610.

Geracioti TD, Loosen PT, Orth DN (1997). Low cerebrospinal fluid corticotropin-releasing hormone concentrations in eucortisolemic depression. Biol Psychiatry 42: 166–174.

Gertz BJ, Cantreras LN, McComb DJ et al. (1987). Chronic administration of corticotropin-releasing factor increases pituitary corticotroph number. Endocrinology 120: 381–388.

Gjerris A, Hammer M, Vendsborger P et al. (1985). Cerebrospinal fluid vasopressin: changes in depression. Br J Psychiatry 147: 696–701.

Glover V, O'Connor TG, O'Donnell K (2010). Prenatal stress and the programming of the HPA axis. Neurosci Biobehav Rev 35: 17–22.

Gold PW, Chrousos GP (1985). Clinical studies with corticotrophin releasing factor: implications for the diagnosis and pathophysiology of depression, Cushing's disease, and adrenal insufficiency. Psychoneuroendocrinology 10: 401–419.

Gold PW, Chrousos GP (2002). Organisation of the stress system and its dysregulation in melancholic and atypical depression: high vs low CRH/NE levels. Mol Psychiatry 7: 254–275.

Gold PW, Goodwin FK (1978). Vasopressin in affective illness. Lancet 10: 1233–1236.

Gold PW, Loriaux DL, Roy A et al. (1986). Responses to corticotropin-releasing hormone in the hypercortisolism of depression and Cushing's disease: pathophysiologic and diagnostic implications. N Engl J Med 314: 1329–1335.

Goldman M, Tandon R, Taylor SF et al. (1996). Dexamethasone non suppression and short rapid eye movement latency in schizophrenia: markers of an affective diathesis? Biol Psychiatry 40: 927–992.

Goldman MB, Gnerlich J, Hussain M (2007). Neuroendocrine responses to a cold pressor stimulus in polydipsic hyponatremic and in matched schizophrenic patients. Neuropsychopharmacology 32: 1611–1621.

Graeff FG, Garcia-Leal C, Del-Ben CM et al. (2005). Does the panic attack activate the hypothalamic–pituitary–adrenal axis? An Acad Bras Cienc 77: 477–491.

Gray TS, Bingaman EW (1996). The amygdala: corticotropin-releasing factor, steroids, and stress. Crit Rev Neurobiol 10: 155–168.

Grigoriadis DE, Liu X-J, Vaughan J et al. (1996). [125I]-Tyro-sauvagine: a novel high-affinity radioligand for the pharmacological and biochemical study of human corticotropin-releasing factor2á (CRH2á) receptors. Mol Pharmacol 50: 679–686.

Hammer GD, Fairchild-Huntress V, Low MJ (1990). Pituitary-specific and hormonally regulated gene expression directed by the rat proopiomelanocortin promoter in transgenic mice. Mol Endocrinol 4: 1689–1697.

Hauger RL, Grigoriadis DE, Dallman MF et al. (2003). International Union of Pharmacology XXXVI. Current status of the nomenclature for receptors for corticotropin-releasing factor and their ligands. Pharmacol Rev 55: 21–26.

Heilig M, Koop GF (2007). A key role for corticotropin-releasing factor in alcohol dependance. Trends Neurosci 30: 399–406.

Heim C, Newport DJ, Heit S et al. (2000). Pituitary–adrenal and autonomic responses to stress in women after sexual and physical abuse in childhood. JAMA 284: 592–597.

Heim C, Newport DJ, Bonsall R et al. (2001). Altered pituitary–adrenal axis responses to provocative challenge test in adult survivors of childhood abuse. Am J Psychiatry 158: 575–581.

Heim C, Newport DJ, Mletzko T et al. (2008). The link between childhood trauma and depression: insights from HPA axis studies in humans. Psychoneuroendocrinology 33: 693–710.

Heinricks SC, Lapsansky J, Lovenberg TW et al. (1997). Corticotropin-releasing factor CRH1 but not CRH2 receptors mediate anxiogenic-like behaviours. Regul Pept 71: 15–21.

Heninger GR (1990). A biological perspective on the co-morbidity of major depressive disorder and panic disorder. In: JD Maser, CR Cloninger (Eds.), Co-morbidity of mood and anxiety disorders. AP Press, Washington DC.

Hiroi N, Wong ML, Licinio J (2001). Expression of corticotrophin releasing hormone receptors type 1 and type 11 mRNA in suicide victims and controls. Mol Psychiatry 6: 540–546.

Holsboer F (2001). Stress, hypercortisolism and corticosteroid receptors in depression: implications for therapy. J Affect Disord 62: 77–91.

Holsboer F, Gerken A, von Bardeleben U et al. (1986). Human corticotropin-releasing hormone in depression. Biol Psychiatry 21: 601–611.

Holsboer F, von Bardeleben U, Buller et al. (1987). Stimulation response to corticotrophin-releasing hormone (CRH) in patients with depression, alcoholism and panic disorder. Horm Metab Res 16 (Suppl): 80–88.

Holsboer F, Lauer CJ, Schreiber W et al. (1995). Altered hypothalamic pituitary–adrenocortical regulation in healthy subjects at high familial risk for affective disorders. Neuroendocrinology 62: 340–347.

Jaeckle RS, Kathol RG, Lopez JF (1987). Enhanced adrenal sensitivity to exogenous ACTH1-24 stimulation in major depression: relationship to dexamethasone suppression test result. Arch Gen Psychiatry 43: 233–240.

Jansen LMC, Gispen-de Wied CC, Khan RS (2000). Selective impairments in the stress response in schizophrenia patients. Psychopharmacology (Berl) 149: 319–325.

Jolkkonen J, Lepola U, Bissette G et al. (1993). CSF corticotrophin-releasing factor is not affected in panic disorder. Biol Psychiatry 33: 136–138.

Jones SR, Fernyhough C (2007). A new look at the neural diathesis–stress model of schizophrenia: the primacy of social-evaluative and uncontrollable situations. Schizophr Bull 33: 1171–1177.

Jones JS, Stein DJ, Stanley B et al. (1994). Negative and depressive symptoms in suicidal schizophrenics. Acta Psychiatr Scand 89: 81–87.

Juruena MF, Pariante CP, Papadopolous AS et al. (2009). Prednisolone depression test in depression: prospective study of the role of HPA axis dysfunction in treatment resistance. Br J Psychiatry 194: 342–349.

Kageyama K, Suda T (2009). Regulatory mechanisms underlying corticotrophin-releasing factor gene expression in the hypothalamus. Endocr J 56: 335–344.

Kalin NH (1990). Behavioural and endocrine studies of corticotropin releasing hormone in primates. In: EB De Souza, CB Nemeroff (Eds.), Corticotropin-Releasing Factor: Basic and Clinical Studies of a Neuropeptide. CRC Press, Boca Raton, pp. 275–289.

Kapur S, Philips AG, Insel TR (2012). Why has it taken so long for biological psychiatry to develop clinical tests and what to do about it? Mol Psychiatry 17: 1174–1179.

Kathol RG, Jaeckle RS, Lopez JF et al. (1989). Pathophysiology of HPA axis abnormalities in patients with major depression: an update. Am J Psychiatry 146: 311–317.

Keck ME (2006). Corticotropin-releasing factor, vasopressin and receptor systems in depression and anxiety. Amino Acids 31: 241–250.

Keen-Rhinehart E, Michopoulos V, Toufexis DJ et al. (2009). Continuous expression of corticotrophin-releasing factor in the central nucleus of the amygdale emulates the dysregulation of the stress and reproductive axis. Mol Psychiatry 14: 37–50.

Keshavan MS, Brar J, Ganguli R et al. (1989). DST and schizophrenic symptomatology. Biol Psychiatry 26: 856–858.

Kessler RC, Chiu WT, Demier O et al. (2005). Prevalence, severity and comorbidity of 12-month DSM-IV disorders in the National Comorbidity Survey Replication. Arch Gen Psychiatry 62: 617–627.

Kling MA, Roy A, Doran AR et al. (1991). Cerebrospinal fluid immunoreactive corticotropin-releasing hormone and adrenocorticotropin secretion in Cushing's disease and major depression: potential clinical implications. J Clin Endocrinol Metab 72: 260–271.

Kling MA, Rubinow DR, Doran AR et al. (1993). Cerebrospinal fluid immunoreactive somatostatin concentrations in patients with Cushing's disease and major depression: relationship to indices of corticotropin-releasing hormone and cortisol secretion. Neuroendocrinology 57: 79–88.

Kling MA, De Bellis MD, O'Rourke DK et al. (1994a). Diurnal variation of cerebrospinal fluid immunoreactive corticotropin-releasing hormone levels in healthy volunteers. J Clin Endocrinol Metab 79: 233–239.

Kling MA, Geracioti TD, Licinio J et al. (1994b). Effects of electroconvulsive therapy on the CRH–ACTH–cortisol system in melancholic depression: preliminary findings. Psychopharmacol Bull 30: 489–494.

Kling MA, Coleman VH, Schulkin J (2009). Glucocorticoid inhibition in the treatment of depression; can we think outside the endocrine hypothalamus? Depress Anxiety 26: 641–649.

Krishnan KRR, Doraiswamy PM, Lurie SN et al. (1991). Pituitary size in depression. J Clin Endocrinol Metab 72: 256–259.

Krishnan KRR, Rayasam K, Reed DR et al. (1993). The corticotropin-releasing factor stimulation test in patients with major depression: relationship to dexamethasone suppression test results. Depression 1: 133–136.

Kunzel HE, Zobel AW, Nickel T et al. (2003). Treatment of depression with CRH-1 receptor antagonist R121919: endocrine changes and side effects. J Psychiatr Res 37: 525–533.

Laakmann G, Hinz A, Voderholzer U et al. (1990). The influence of psychotropic drugs and releasing hormones on anterior pituitary hormone secretion in healthy subjects and depressed patients. Pharmacopsychiatry 23: 18–26.

Ladd CO, Huot RL, Thrivikraman KV et al. (2000). Long-term behavioural and neuroendocrine adaptations to adverse early experience. Prog Brain Res 122: 81–103.

Lammers CH, Garcia-Borreguero D, Schmider J et al. (1995). Combined dexamethasone/corticotropin-releasing hormone test in patients with schizophrenia and in normal controls: II. Biol Psychiatry 38: 803–807.

Lee R, Geracioti Jr TD, Kasckow JW et al. (2005). Childhood trauma and personality disorder: positive correlations with adult CSF corticotrophin-releasing factor concentrations. Am J Psychiatry 162: 995–997.

Lee RJ, Gollan J, Kasckow J et al. (2006). CSF corticotrophin-releasing factor in personality disorder: relationship with self-reported parental care. Neuropsychopharmacology 31: 2289–2295.

Lee RJ, Hempel J, Tenharmsel A et al. (2012). The neuroendocrinology of childhood trauma in personality disorder. Psychoneuroendocrinology 44: 433–438.

Lehmann J, Russig H, Feldon J et al. (2002). Effect of a single maternal separation at different pup ages on the corticosterone stress response in adult and aged rats. Pharmacol Biochem Behav 73: 141–145.

Lesch KP, Mayer S, Disselkamp-Tietze D et al. (1990a). 5-HT1A receptor responsivity in unipolar depression: evaluation of ipsapirone-induced ACTH and cortisol secretion in patients and controls. Biol Psychiatry 28: 620–628.

Lesch KP, Disselkamp-Tietze J, Schmidtke A (1990b). 5-HT1A receptor function in depression: effect of chronic amitriptyline treatment. J Neural Transm 80: 157–161.

Lewis CF, Tandon R, Shipley JE et al. (1996). Biological predictors of suicidality in schizophrenia. Acta Psychiatr Scand 94: 416–420.

Lightman SL (2008). The neuroendocrinology of stress: a never ending story. J Neuroendocrinol 20: 880–884.

Linkowski P, Medlewicz J, Leclercq R et al. (1985). The 24-hour profile of adrenocorticotropin and cortisol in major depressive illness. J Clin Endocrinol Metab 61: 429–438.

Linkowski P, Mendlewicz J, Kerkhofs M et al. (1987). 24-hour profiles of adrenocorticotropin, cortisol, and growth hormone in major depressive illness: effect of antidepressant treatment. J Clin Endocrinol Metab 65: 141–152.

Loizzo A, Loizzo S, Lopez L et al. (2002). Naloxone prevents cell mediated immune alterations in adult mice following repeated mild stress in the neonatal period. Br J Pharmacol 135: 1219–1226.

Lovenberg TW, Liaw CW, Grigoriadis DE et al. (1995). Cloning and characterisation of a functionally distinct corticotrophin-releasing factor receptor subtype from rat brain. Proc Natl Acad Sci U S A 92: 836–840.

Lu A, Steiner MA, Whittle N et al. (2008). Conditional mouse mutants highlight mechanisms of corticotropin-releasing hormone effects on stress-coping behaviour. Mol Psychiatry 13: 1028–1042.

Lupien SJ, McEwen BS, Gunnar MR et al. (2009). Effects of stress throughout the lifespan on the brain, behaviour and cognition. Nat Rev Neurosci 10: 434–445.

MacMillan HL, Georgiades K, Duku EK et al. (2009). Cortisol response to stress in female youths exposed to childhood maltreatment: results of the youth mood project. Biol Psychiatry 66: 62–68.

Marx CE, Lieberman JA (1998). Psychoneuroendocrinology of schizophrenia. Psychiatr Clin North Am 21: 413–434.

Mathew SJ, Price RB, Charney DS (2008). Recent advances in the neurobiology of anxiety disorders: implications for novel therapeutics. Am J Med Genet C Semin Med Genet 148: 89–98.

McEwen BS, Gianaros PJ (2010). Central role of the brain in stress and adaptation: links to socioeconomic status, health, and disease. Ann N Y Acad Sci 1186 (1): 190–222.

McEwen BS, Plapinger L (1970). Association of corticosterone-1,2 3H with macromolecules extracted from brain cell nuclei. Nature 220: 911–912.

McEwen BS, Eiland L, Hunter RG et al. (2012). Stress and anxiety: structural plasticity and epigenetic regulation as a consequence of stress. Neuropharmacology 62: 3–12.

Meller WH, Kathol RL, Jaeckle RS et al. (1987). Stimulation of the pituitary–adrenal axis with arginine vasopressin in patients with depression. J Psychiatr Res 21: 269–277.

Meynen G, Unmehopa UA, van Heerikhuize JJ et al. (2006). Increased arginine vasopressin mRNA expression in the human hypothalamus in depression: a preliminary report. Biol Psychiatry 60: 892–895.

Mittal VA, Dhruv S, Tessner KD et al. (2007). The relations among putative biorisk markers in schizotypal adolescents: minor physical anomalies, movement abnormalities, and salivary cortisol. Biol Psychiatry 61: 1179–1186.

Molchan SE, Hill JL, Martinez RA et al. (1993). CSF somatostatin in Alzheimer's disease and major depression: relationship to hypothalamic–pituitary–adrenal axis and clinical measures. Psychoneuroendocrinology 19: 509–519.

Mondelli V, Dazzen P, Hepgul N et al. (2010). Abnormal cortisol levels during the day and cortisol awaking response in first episode psychosis: the role of stress and antipsychotic treatment. Schizophr Res 116: 234–242.

Muller MB, Landgraf R, Preil J et al. (2000). Selective activation of the hypothalamic vasopressinergic system in mice deficient for the corticotropin-releasing hormone receptor 1 is dependent on glucocorticoids. Endocrinology 141: 4262–4269.

Muller MB, Zimmerman S, Sillaber I et al. (2003). Limbic corticotrophin-releasing hormone receptor 1 mediates anxiety-related behaviour and hormonal adaptation to stress. Nat Neurosci 6: 1100–1107.

Murphy BE (1997). Antiglucocorticoid therapies in major depression. A review. Psychoneuroendorinology 22 (Suppl): S125–S132.

Nemeroff CB (1996). The corticotrophin-releasing factor (CRH) hypothesis of depression: new findings and new directions. Mol Psychiatry 1: 336–342.

Nemeroff CB (2004). Neurobiological consequences of childhood trauma. J Clin Psychiatry 64 (Suppl 1): 18–28.

Nemeroff CB, Vale WW (2005). The neurobiology of depression: inroads to treatment and new drug discovery. J Clin Psychiatry 66 (Suppl 7): 5–13.

Nemeroff CB, Widerlov E, Bissette G et al. (1984). Elevated concentrations of CSF corticotropin-releasing factor-like immunoreactivity in depressed patients. Science 226: 1342–1344.

Nemeroff CB, Owens MJ, Bissette G et al. (1988). Reduced corticotropin-releasing factor binding sites in the frontal cortex of suicide victims. Arch Gen Psychiatry 45: 577–579.

Nemeroff CB, Bissette G, Akil H et al. (1991). Neuropeptide concentrations in the cerebrospinal fluid of depressed patients treated with electroconvulsive therapy: corticotropin-releasing factor, â-endorphin and somatostatin. Br J Psychiatry 158: 59–63.

Nemeroff CB, Krishnan KRR, Reed D et al. (1992). Adrenal gland enlargement in major depression: a computed tomographic study. Arch Gen Psychiatry 49: 384–387.

Neumann ID, Wigger A, Kromer S et al. (2005). Differential effects of periodic maternal separation on adult stress coping in a rat model of extremes in trait anxiety. Neuroscience 132: 867–877.

Newcomer JW, Faustman WO, Whiteford HA et al. (1991). Symptomatology and cognitive impairment associate independently with post-dexamethasone cortisol concentrations in unmedicated schizophrenic patients. Biol Psychiatry 29: 855–864.

Newman ME, Shapira B, Lerer B (1998). Evaluation of central serotonergic function in affective and related disorders by the fenfluramine challenge test: a critical review. Int J Neuropsychopharmacol 1: 49–69.

Newport DJ, Heim C, Owens MJ et al. (2003). Cerebrospinal fluid corticotropin-releasing factor (CRH) concentration predicts pituitary response in the CRH stimulation test: a multipler regression analysis. Neuropsychopharmacology 28: 569–576.

O'Keane V, Frodl T, Dinan T (2012). A review of atypical depression in relation to the course of depression and changes in the HPA axis organization. Psychoneuroendocrinology 37: 1589–1599.

O'Mahony SM, Marchesi JR, Scully P et al. (2009). Early life stress alters behaviour, immunity and microbiota in rats: implications for irritable bowel syndrome and psychiatric illnesses. Biol Psychiatry 65: 263–267.

Pariante CM, Lightman SL (2008). The HPA axis in major depression: classical and new developments. Trends Neurosci 31: 464–468.

Petrowski K, Wintermann GB, Kirschbaum C et al. (2012). Dissocation between ACTH and cortisol response in DEX-CRH test in patients with panic disorder. Psychoneuroendocrinology 37: 1199–1208.

Phillips DIW, Jones A (2006). Fetal programming of autonomic and HPA function: do people who were small babies have enhanced stress responses? J Physiol 521: 45–50.

Pitts AF, Samuelson SD, Meller WH et al. (1995). Cerebrospinal fluid corticotropin-releasing hormone, vasopressin, and oxytocin concentrations in treated patients with major depression and controls. Biol Psychiatry 38: 330–335.

Plocka-Lewandowska M, Araszkiewicz A, Rybakowski JK (2001). Dexamethasone suppression test and suicide attempts in schizophrenic patients. Eur Psychiatry 16: 428–431.

Plotsky PM, Meaney MJ (1993). Early, postnatal experience alters hypothalamic corticotrophin-releasing factor (CRH) mRNA, median eminence CRH content and stress-induced release in adult rats. Mol Brain Res 18: 195–200.

Plotsky PM, Thrivikraman KV, Nemeroff CB et al. (2005). Long-term consequences of neonatal rearing on central corticotrophin-releasing factor systems in adult male rat offspring. Neuropsychopharmacology 30: 2192–2204.

Post RM, Gold P, Rubinow DR (1982). Peptides in cerebrospinal fluid of neuropsychiatric patients: an approach to central nervous system peptide function. Life Sci 31: 1–15.

Potter E, Sutton S, Donaldson C et al. (1994). Distribution of corticotrophin-releasing factor receptor mRNA expression in the rat brain and pituitary. Proc Natl Acad Sci U S A 91: 8777–8781.

Purba JS, Hoogendijk WJ, Hofman MA et al. (1996). Increased number of vasopressin- and oxytocin-expressing neurons in the paraventricular nucleus of the hypothalamus in depression. Arch Gen Psychiatry 53: 137–143.

Raadsheer FC, Hoogendijk WJ, Stam FC et al. (1994). Increased number of corticotropin-releasing hormone expressing neurons in the hypothalamic paraventricular nucleus of depressed patients. Clin Neuroendocrinol 60: 436–444.

Raadsheer FC, Van Heerikhuize JJ, Lucassen PJ et al. (1995). Corticotropin-releasing hormone mRNA levels in the paraventricular nucleus of patients with Alzheimer's disease and depression. Am J Psychiatry 152: 1372–1376.

Read J, van Os J, Morrison AP et al. (2005). Childhood trauma, psychosis and schizophrenia: a literature review with theoretical and clinical implications. Acta Psychiatr Scand 112: 330–350.

Reul JM, De Kloet ER (1985). Two receptor systems for corticosterone in the rat brain: microdistribution and differential occupation. Endocrinology 117: 2505–2511.

Reul JM, Holsboer F (2002). Corticotropin-releasing factor receptors 1 and 2 in anxiety and depression. Curr Opin Pharmacol 2: 23–33.

Rinne T, deKloet ER, Wouters L et al. (2002). Hyperresponsiveness of hypothalamic–pituitary–adrenal axis to combined dexamethasone/corticotrophin releasing hormone challenge in female borderline personality disorder subjects with a history of sustained childhood abuse. Biol Psychiatry 52: 1102–1112.

Risbrough VB, Stein MB (2006). Role of corticotrophin releasing factor in anxiety disorders: a translation research perspective. Horm Behav 50: 550–561.

Romero LM, Sapolsky RM (1996). Patterns of ACTH secretagog secretion in response to psychological stimuli. J Neuroendocrinol 8: 243–258.

Roy A, Pickar D, Doran A et al. (1986). The corticotropin-releasing hormone stimulation test in chronic schizophrenia. Am J Psychiatry 143: 1393–1397.

Roy A, Pickar D, Paul S et al. (1987). CSF corticotropin-releasing hormone in depressed patients and normal subjects. Am J Psychiatry 144: 641–645.

Rubin R, Poland RE, Lesser IM et al. (1987). Neuroendocrine aspects of primary endogenous depression. III. Cortisol secretion in relation to diagnosis and symptom patterns. Psychol Med 17: 609–619.

Rubin RT, Phillips JJ, Sadow TF et al. (1995). Adrenal gland volume in major depression: increase during the depressive episode and decrease with successful treatment. Arch Gen Psychiatry 52: 213–218.

Rubin R, Dinan TG, Scott LV (2001). The neuroendocrinology of affective disorders. In: D Pfaff, AP Arnold, AM Etgen (Eds.), Hormones, Brain and Behaviour. Academic Press, New York, pp. 467–514.

Ryan MCM, Sharifi N, Condren R et al. (2004). Evidence of basal pituitary–adrenal over activity in first episode, drug naive patients with schizophrenia. Psychoneuroendocrinology 29: 1065–1070.

Rybakowski JK, Twardowska K (1999). The dexamethasone/corticotrophin releasing hormone test in bipolar and unipolar depressive illness. J Psychiatr Res 33: 363–370.

Sanchez MM, Young LJ, Plotsky PM et al. (1999). Autoradiographic and in situ hybridization localization of corticotrophin-releasing factor 1 and 2 receptor in non-human primate brain. J Comp Neurol 408: 365–377.

Sapolsky RM, Plotsky PM (1990). Hypercortisolism and its possible neural basis. Biol Psychiatry 27: 937–952.

Sautter FJ, Bissette G, Willey J et al. (2003). Corticotropin-releasing factor in posttraumatic stress disorder (PTSD) with secondary psychotic symptoms, non-psychotic PTSD, and healthy control. Biol Psychiatry 54: 1382–1388.

Schmider J, Lammers CH, Gotthardt U et al. (1995). Combined dexamethasone/corticotropin-releasing hormone test in acute and remitted manic patients, and in normal controls: I. Biol Psychiatry 15: 797–802.

Schulkin J, Gold PW, McEwens BS (1998). Induction of corticotropin-releasing hormone gene expression by glucocorticoids: implications for understanding the states of fear and anxiety and allostatic load. Psychoneuroendocrinology 23: 219–243.

Shevlin M, Houston JE, Dorahy MJ et al. (2008). Cumulative traumas and psychosis: an analysis of the national comorbidity survey and the British Psychiatric Morbidity Survey. Schizophr Bull 34: 193–199.

Skelton KH, Nemeroff CB, Knight DL et al. (2000). Chronic administration of the triazolobenzodiazepine alprazolam produces opposite effects on corticotrophin-releasing factor and urocortin neuronal systems. J Neurosci 20: 1240–1248.

Strohle A, Scheel M, Modell S et al. (2008). Blunted ACTH response to dexamethasone suppression-CRH stimulation in post-traumatic stress disorder. J Psychiatr Res 42: 1185–1188.

Strous RD, Maayan R, Lapidus R et al. (2004). Increased circulatory dehydroepiandrosterone and dehydroepiandrosterone-sulphate in first episode schizophrenia: relationship to gender, aggression and symptomatology. Schizophr Res 71: 427–434.

Tandon R, Mazzara C, DeQuardo J et al. (1991). Dexamethasone suppression test in schizophrenia: relationship to symptomatology, ventricular enlargement and outcome. Biol Psychiatry 29: 953–964.

Thakore JH, Dinan TG (1995). Cortisol synthesis inhibition; a new treatment strategy for the clinical and endocrine manifestations of depression. Biol Psychiatry 37: 364–368.

Thakore JH, O'Keane V, Dinan TG (1996). D-Fenfluramine-induced prolactin responses in mania: evidence for serotonergic subsensitivity. Am J Psychiatry 154: 1460–1463.

Thakore JH, Barnes C, Joyce J et al. (1997). Effects of antidepressant treatment on corticotropin-induced cortisol responses in patients with melancholic depression. Psychiatry Res 73: 27–32.

Tyrka AR, Price LH, Gelernter J et al. (2009). Interaction of childhood maltreatment with the corticotropin-releasing hormone receptor gene: effects on hypothalamic-pituitary–adrenal axis reactivity. Biol Psychiatry 66: 681–685.

Valdez GR, Koop GF (2004). Allostatic and dysregulation of corticotrophin-releasing factor and neuropeptide Y systems: implications for the development of alcoholism. Pharmacol Biochem Behav 79: 671–689.

Vale W, Spiess J, Rivier C et al. (1981). Characterisation of a 41 residue ovine hypothalamic peptide that stimulates secretion of the corticotropin and betaendorphin. Science 213: 1394–1399.

Valintino RJ, Foote SL, Aston-Jones G (1983). Corticotropin-releasing factor activates noradrenergic neurons of the locus coeruleus. Brain Res 270: 363–367.

Van Londen L, Goekoop J, van Kempen G et al. (1997). Plasma levels of arginine vasopressin elevated in depression. Neuropsychopharmacology 17: 284–292.

von Bardeleben U, Holsboer F (1989). Cortisol response to a combined dexamethasone-human corticotropin-releasing hormone challenge in patients with depression. J Neuroendocrinol 1: 485–488.

Vreeburg SA, Hoogendijk WJ, van Pelt J et al. (2009). Major depressive disorder and hypothalamic–pituitary–adrenal axis activity: results from a large cohort study. Arch Gen Psychiatry 66: 617–626.

Vythilingam M, Gill JM, Luckenbaugh DA et al. (2010). Low early morning plasma cortisol in post-traumatic stress disorder is associated with co-morbid depression but not with enhanced glucocorticoid feedback inhibition. Psychoneuroendocrinology 35: 442–450.

Watson S, Gallagher P, Ritchie JC et al. (2004). Hypothalamic–pituitary–adrenal axis function in patients with bipolar disorder. Br J Psychiatry 184: 496–502.

Watson S, Thompson JM, Malik N et al. (2005). Temporal stability of the dex/CRH tests in rapid-cycling bipolar I disorder: a pilot study. Aust N Z J Psychiatry 39: 244–248.

Watson S, Gallagher P, Ferrier N et al. (2006). Post-dexamethasone arginine vasopressin levels in patients with severe mood disorders. J Psychiatr Res 40: 353–359.

Wedekind D, Preiss B, Cohrs S et al. (2007). Relationship between nocturnal urinary cortisol excretion and symptom severity in subgroups of patients with depressive episodes. Neuropsychobiology 56: 119–122.

Widerlov E, Bissette G, Nemeroff CB (1988). Monoamine metabolites, corticotropin releasing factor and somatostatin as CSF markers in depressed patients. J Affect Disord 14: 99–107.

Wolf OT, Wingenfeld K (2011). HPA axis alterations in mental disorders: impact on memory and its relevance for therapeutic interventions. CNS Neurosci Ther 17: 714–722.

Wolf OT, Köster B, Kirschbaum C et al. (1997). A single administration of dehydroepiandrosterone does not enhance memory performance in young healthy adults, but immediately reduces cortisol levels. Biol Psychiatry 42: 845–848.

Wolkowitz O, Reus V (1999). Treatment of depression with antiglucocorticoid drugs. Psychosom Med 61: 698–711.

Wolkowitz O, Reus V, Chan T et al. (1999). Antiglucocorticoid treatment of depression: double blind ketoconazole. Biol Psychiatry 451: 1070–1074.

Yehuda R (2006). Advances in understanding neuroendocrine alterations in PTSD and therapeutic implications. Ann N Y Acad Sci 1071: 137–166.

Yehuda R, Lang RK, Buchsbaum MS et al. (2006). Alterations in cortisol negative feedback inhibition using the ACTH response to cortisol administration in PTSD. Psychoneuroendocrinology 31: 447–451.

Young EA, Haskett RF, Murphy-Weinberg V et al. (1991). Loss of glucocorticoid fast feedback in depression. Arch Gen Psychiatry 48: 693–699.

Zobel AW, Nickel T, Kunzel HE et al. (2000). Effects of the high affinity corticotropin-releasing hormone receptor 1 antagonist R121919 in major depression: the first 20 patients treated. J Psychiatr Res 34: 171–181.

Handbook of Clinical Neurology, Vol. 124 (3rd series)
Clinical Neuroendocrinology
E. Fliers, M. Korbonits, and J.A. Romijn, Editors

Chapter 6

Genetic aspects of human obesity

RACHEL LARDER, CHUNG THONG LIM, AND ANTHONY P. COLL*
University of Cambridge Metabolic Research Laboratories, MRC Metabolic Diseases Unit, Wellcome Trust–MRC Institute of Metabolic Science, Addenbrooke's Hospital, Cambridge, UK

INTRODUCTION

Obesity: a heritable disorder

Obesity is defined by the World Health Organization as "a condition of abnormal or excessive fat accumulation in adipose tissue, to the extent that health may be impaired". In terms of body composition it is also defined as a body mass index (BMI, weight in kilograms divided by height in meters squared) of over 30 kg/m^2. Obesity is associated with a variety of social, economic, and psychological consequences (Gortmaker et al., 1993), and can lead both directly and indirectly to a number of medical problems including type 2 diabetes, hypertension, coronary heart disease, cancer, stroke, osteoarthritis, and reduced fertility (Kopelman, 2007). The economic implications of this metabolic condition are substantial, with an estimated US$190 billion spent on treatments for obesity and obesity-related conditions in the US in 2005 (Cawley and Meyerhoefer, 2012). As rates of obesity rise over the next decade, this figure too is expected to rise.

The recent increase in worldwide obesity rates has undoubtedly been driven by dramatic environmental changes that have promoted both an increase in caloric intake and a decrease in energy expenditure, the so-called "obesogenic environment". In the last 50 years, there has been a significant increase in both the availability and consumption of energy-dense food (Drewnowski, 2004), while physical activity in multiple aspects of daily life has decreased (Church et al., 2011).

However, whilst environment is clearly a major factor in determining whether an individual will gain weight, compelling data from human studies have revealed that genetic factors also play an important role in an individual's predisposition to gain weight. In the 1980s, Stunkard and colleagues showed that monozygotic twins have a much higher correlation of BMI than dizygotic twins, suggesting that BMI is heritable (Stunkard et al., 1986a). A similar study looking at families with adopted children supported this as the weight of the adoptees correlated strongly with the BMI of the biological parents, rather than that of the adoptive parents (Stunkard et al., 1986b). Further proof that BMI was highly heritable came from data showing that the BMI of monozygotic twins raised together correlated as highly as the BMI of identical twins raised in separate households, highlighting the importance of genetics as a powerful determination of body composition (Stunkard et al., 1990).

The critical role of the hypothalamus

The hypothalamus is a key region of the vertebrate brain involved in homeostatic regulation. It coordinates the control of multiple crucial functions such as regulation of body temperature, fertility, circadian rhythms, thirst, and food intake.

The role of the hypothalamus in appetite regulation is well documented. Initial clinical descriptions of hypothalamic–pituitary injury leading to obesity in the late 19th century attributed the dramatic weight gain following brain injury to a pituitary hormone imbalance, a condition that became known as Frohlich's syndrome. This theory was challenged in the early 1900s when it was observed that obesity was often seen in patients with tumors at the base of the brain that were near, but did not extend into, the pituitary. Further evidence from patients who had undergone hypophysectomy suggested that it was the basomedial hypothalamus, rather than the pituitary, that controlled weight gain as obesity only

*Correspondence to: Dr Anthony P. Coll, University of Cambridge Metabolic Research Laboratories, MRC Metabolic Diseases Unit, Level 4, Wellcome Trust–MRC Institute of Metabolic Science, Box 289, Addenbrooke's Hospital, Cambridge CB2 0QQ, UK. Tel: +44-1223-769041, E-mail: apc36@cam.ac.uk

occurred in those patients where there was damage to hypothalamus during surgery to remove the pituitary. In the 1920s, direct experimental evidence that the hypothalamus controlled weight gain was provided by work from animal models. While studying diabetes in dogs, Bailey and Bremer showed that electrolytic damage of the basomedial hypothalamus resulted in hyperphagia and obesity (Bailey and Bremer, 1921).

This was followed by a series of systematic experiments in rats in which various discrete regions of the hypothalamus were ablated and the effects on body weight analyzed (Hetherington and Ranson, 1940, 1942). These data identified that a specific region of the hypothalamus, called the ventromedial nucleus (VMN), was critical for maintaining normal body weight. Lesions anterior or dorsal to the VMN did not result in obesity but anything that targeted the VMN specifically, or the area just posterior of the VMN, resulted in weight gain (Graff and Stellar, 1962; Hennessy and Grossman, 1976).

Rats with VMN lesions were described as "ravenous" following the surgery (Brobeck et al., 1943). They were observed eating two or three times more than controls and would begin to seek food as soon as they came round from the anesthesia and would eat continuously for the first few hours postsurgery (Kennedy, 1950). Furthermore, when a meal pattern was established it was characterized by both increased meal frequency and meal size (Balagura and Devenport, 1970). This hyperphagia usually resulted in a doubling of body weight in the first month following surgery but would then usually (but not always) level out as daily food intake decreased to presurgery levels and body weight plateaued (King and Gaston, 1977), suggesting that the lesion had altered a set point for body weight regulation in the rats.

These data led to the proposal that the VMN served as a "satiety center" within the brain. Kennedy proposed that this region of the hypothalamus was sensitive to metabolites in the blood which signaled how much stored fat was present and consequently coordinated food intake and energy expenditure to ensure correct body weight regulation (Kennedy, 1953). If the satiety center was damaged then it could not respond properly to these signals and obesity, due to hyperphagia, was the consequence. This theory was supported by parabiosis experiments in rats in which the circulation systems of two rats were joined together and then the VMN of one of the rats was surgically damaged. This rat became hyperphagic and obese whereas the unlesioned animal became anorexic and lost weight (Hervey, 1959). Refusal of food would continue until death and could only be prevented by VMN lesion of the second rat so that it also became hyperphagic and obese. While this "lipostatic model" was hypothesized in the 1950s, it would be another 35 years before the identity of leptin as one of these "metabolic signals" would be determined.

Some of the initial studies by Hetherington and Ranson had noted that hypothalamic lesions in regions other than the VMH led to a reduction in food intake (Hetherington and Ranson, 1940), suggesting that different regions of the hypothalamus were involved in bidirectional modulation of feeding behavior. Building on this, work done by Anand and Brobeck in rats and cats identified the location of a "feeding center" within the lateral hypothalamus (LH) and established the "dual feeding center" hypothesis. Damage to the LH resulted in animals developing anorexia and losing substantial amounts of weight (Anand and Brobeck, 1951). Furthermore, animals rendered hyperphagic by VMH lesions could have their overeating habits reversed if the LH was subsequently damaged.

Integration and coordination of peripheral signals

The 1990s brought forth a number of seminal discoveries that underpin our current understanding of hypothalamic obesity. Highlights include the cloning of a family of genes that encode the melanocortin receptors (Mountjoy et al., 1992), the discovery that *ob/ob* mice harbor a loss-of-function mutation in the gene encoding the secreted peptide leptin (Zhang et al., 1994), and the report that mice genetically engineered to lack Mc4r develop hyperphagia and severe obesity (Huszar et al., 1997). An intense and exciting period of synergy between human genetic and model organism studies blossomed with the leptin-melanocortin system becoming the best characterized central pathway known to control energy balance (Cone, 2006; Friedman, 2009) and a canonical model emerged, centered around two neuronal populations. Within the hypothalamus leptin inhibits orexigenic AgrP/NPY neurons and excites anorexigenic POMC neurons, with both sets of neurons projecting to second-order neuronal populations throughout the brain. Building on this basic model, the last decade has seen exponential growth in the neuronal and molecular architecture underpinning the neuronal circuitry at play within the hypothalamus. In particular, huge technological leaps in the ability to express, remove, and stimulate single species of receptors in discrete neuronal populations have meant we have a very detailed understanding of the intricacies of the neuronal connectivity within the rodent hypothalamus (Yeo and Heisler, 2012). The challenge is to continue to be able to harness these data generated in model organisms into knowledge that is transferrable and relevant to the increasing disease burden of human obesity.

HUMAN MONOGENIC OBESITY

Monogenic obesity is defined as obesity resulting from mutation within a single gene. Several of the best characterized monogenic obesity syndromes are characterized by severe, early onset obesity as well as other endocrine disorders.

Congenital leptin deficiency

Leptin is a hormone produced by the white adipose tissue and is a key regulator of energy homeostasis. The leptin receptor is a class I cytokine receptor with multiple isoforms. The main signaling form (ObRb) is found in the hypothalamus and immune cells. Leptin regulates energy intake and expenditure through actions at multiple sites, including several hypothalamic regions, the mesolimbic system, and the brainstem.

Congenital leptin deficiency was first reported in two severely obese cousins (an 8-year-old girl weighing 86 kg and a 2-year-old boy weighing 29 kg) from a highly consanguineous family of Pakistani origin in 1997 (Montague et al., 1997). Both children had normal weight at birth but developed severe, intractable obesity with marked hyperphagia from an early age. Direct sequencing of genomic DNA identified a homozygous frameshift mutation (ΔG133) in the leptin gene, leading to a truncated protein that was not secreted, and the children had undetectable levels of serum leptin. The children had no additional features suggestive of pleiotropic genetic syndrome associated with obesity, and both karyotyping and computed tomography (CT) of the brain were normal. Detailed phenotyping also revealed normal basal temperature, no defect in basal metabolic rate (BMR) or free-living energy expenditure, increased fat mass, normal bone density and mineralization, hyperinsulinemia consistent with obesity, normal linear growth, hypogonadotropic hypogonadism, and impaired T cell-mediated immunity. Interestingly, and in contrast to leptin-deficient *ob/ob* mice which have elevated levels of corticosterone, serum cortisol levels were normal in the children.

The dramatic and life-saving effect of treating congenital leptin deficiency with subcutaneous injections of recombinant human leptin has been well documented (Farooqi et al., 2002; Paz-Filho et al., 2010). Leptin therapy significantly reduces the body weight (decreases fat mass and increases lean mass), normalizes hyperphagia, permitting normal pubertal development, and improves hyperinsulinemia, hyperlipidemia, and T cell-mediated immunity (Farooqi et al., 2002).

Leptin-receptor deficiency

Leptin-receptor deficiency was first identified shortly after the discovery of congenital leptin deficiency in three affected female subjects in a consanguineous family of Algerian origin with a strong prevalence of morbid obesity occurring early in life (Clement et al., 1998). These index cases were found to be homozygous for a mutation that truncated the receptor before the transmembrane domain. The three subjects had normal birth weight but rapidly developed severe obesity within the first months of life. Biochemically, the three sisters had serum leptin concentrations which were much higher than the measured normal range.

As with leptin-deficient subjects, a strong hyperphagic phenotype and a failure to progress through normal pubertal development were evident. The index cases had no mammary glands development, sparse pubic and no axillary hair, and remained amenorrheic at the age of 13.5 years (one of the three sisters) and 19 (two of the three sisters). Biochemical parameters were in keeping with hypogonadotropic hypogonadism.

Further analysis of these and other leptin receptor-deficient patients (Clement et al., 1998; Farooqi et al., 2007b) has revealed that patients with leptin receptor deficiency have some neuroendocrine features not seen with leptin deficiency; these include frank hypothalamic hypothyroidism, mild growth retardation with impaired basal and stimulated growth hormone secretion, and decreased IGF-1 levels (Farooqi and O'Rahilly, 2005).

Pro-opiomelanocortin (POMC) deficiency

Pro-opiomelanocortin is a large precursor protein expressed within the hypothalamus, pituitary and brainstem that undergoes extensive post-translational processing and modification to generate a range of smaller biologically active peptides. Sequential cleavage involving, among others, prohormone convertase enzymes (PC1 and PC3) generates a family of melanocortins, adrenocorticotropic hormone (ACTH), α-, β- and γ-melanocyte-stimulating hormones (MSH) and the opioid-receptor ligand β-endorphin.

Two obese patients with POMC mutations were first identified in 1998 (Krude et al., 1998). Both patients had normal birth weight but developed severe, early onset obesity associated with increased appetite. The first patient, a girl, had pale skin and red hair pigmentation and developed cholestasis at 3 weeks. Circulating levels of both pituitary-derived ACTH and cortisol were undetectable in this patient, even after stimulation. Other anterior pituitary hormones were normal, as was MRI of the pituitary. Sequencing of the *POMC* gene showed two compound heterozygous mutations with exon 3 which interfered with appropriate synthesis of ACTH and α-MSH.

The second patient, a boy, suffered perinatally from transient hypoglycemia. An episode, initially considered to be a febrile seizure, at 12 months, brought him to medical attention, again with more detailed assessment revealing hypoglycemia and hyponatremia secondary to adrenal insufficiency. Sequencing revealed a homozygous mutation in exon 2 which abolished POMC translation. Again, after appropriate steroid replacement his subsequent development was uneventful apart from abnormal eating behavior and significant weight gain.

Subsequent studies have reported three additional unrelated European children with congenital POMC deficiency who were either homozygous or compound heterozygous for *POMC* mutation (Krude et al., 2003). All these children presented in neonatal life with adrenal crisis and hypocortisolism due to ACTH deficiency and lack of stimulation at the adrenal MC2 receptor. They also had pale skin and red hair due to the lack of MSH action at melanocortin-1 receptors in the skin and hair follicles, thus leading to alteration in pigmentation. The sixth patient identified, a 2-year-old child of Turkish origin, harbored a homozygous frameshift mutation in *POMC* leading to congenital POMC deficiency (Farooqi et al., 2006). He presented with similar clinical phenotypes to those observed in the first five cases. Interestingly, he did not have red hair although closer examination of the scalp revealed brown hair with dark red roots. This is an important finding as previously red hair was considered a key clinical feature and useful diagnostic clue to identify this rare monogenic obesity syndrome.

The availability of a large number of family members in this Turkish pedigree provided the opportunity to address whether heterozygous loss of one copy of the *POMC* gene was sufficient to alter obesity risk (Farooqi et al., 2006). Parametric linkage analysis of the trait "overweight" provided statistically significant evidence of linkage with the POMC locus, thus concluding that loss of one copy of the *POMC* gene predisposes to obesity in humans.

Both α- and β-MSH bind to the MC4R with high affinity and similar IC_{50} values (Abbott et al., 2000). Studies in rodents have elucidated the central role of α-MSH in the regulation of food intake by activation of brain MC4R (Abbott et al., 2000). β-MSH is absent in rodents because of the lack of a proximal dibasic site that is necessary for the proteolytic cleavage event that produces β-MSH in humans. However, in humans, five unrelated probands presenting with severe early onset obesity were identified as heterozygous for a rare missense variant in the region encoding β-MSH, Tyr221Cys (Lee et al., 2006). The subjects were hyperphagic, showed increased linear growth, and had normal basal metabolic rate compared to that predicted using age- and gender-specific equations. The variant β-MSH peptide was impaired in its ability to bind to and activate signaling from the MC4R. This finding supports a role for β-MSH in the control of human energy homeostasis.

Prohormone converatse 1/3 (PC1/3) deficiency

Prohormone convertase 1 (PC1/3), also known as proprotein convertase 1/3, is the major endopeptidase or processing enzyme of precursor proteins in the regulated secretory pathway. It is expressed in the brain, enteroendocrine cells, and neuroendocrine system (Creemers et al., 1998; Creemers and Khatib, 2008) and is encoded by the *proprotein convertase subtilisin/kexin type 1* (*PCSK1*) gene. PC1/3 is itself synthesized as an inactive precursor (proPC1/3), then undergoes several autocatalytic maturation events, first within the endoplasmic reticulum (ER) and then within the trans-Golgi network and secretory vesicles of the regulated secretory pathway to generate fully active enzyme that is stored in mature secretory granules (Muller and Lindberg, 1999). PC1/3 is largely responsible for the first step of biosynthesis of insulin as it is involved in performing proteolytic cleavage of proinsulins to their intermediate forms. It is also involved in cleaving POMC, prorenin, proenkephalin, prosomatostatin, and progastrin to their intermediate or active forms.

The association between endopeptidase and obesity was first identified when mutation of carboxypeptidase E (CPE) was shown to cause obesity in *fat/fat* mice (Cool et al., 1997). CPE is required for further proteolytic cleavage of certain proteins that have been initially processed by PC1/3 to be active. Heterozygote *PCSK1* mice are not growth retarded but tend to be mildly obese, hyperphagic, and have multiple endocrinologic defects such as abnormal proinsulin processing (Lloyd et al., 2006). Interestingly, *PCSK1-null* mice are not obese but display abnormal growth retardation and neuroendocrine abnormalities (Zhu et al., 2002).

Mutations causing partial PCSK1 deficiency are present in 0.83% of extreme obesity phenotypes and there is an increased risk of obesity by 8.7 fold in heterozygous carriers of mutations in the *PCSK1* gene (Creemers et al., 2012). The first human case of congenital PC1/3 deficiency was identified in a woman with severe early onset obesity, hypogonadotropic hypogonadism, postprandial hyperglycemia, hypocortisolemia and elevated plasma proinsulin, and POMC concentrations with low levels of insulin (Jackson et al., 1997). She was found to have a compound heterozygous mutation of the *PC1/3* that causes failure of propeptide cleavage of PC1/3 and its retention in the ER as well as impaired catalytic activity.

A second case of congenital PC1/3 deficiency was identified in a child who was a compound heterozygote for two loss-of-function mutations, leading to a truncated protein with impaired function (Jackson et al., 2003). In addition to the phenotypes observed in the first case, this child also suffered from severe small-intestinal absorptive dysfunction and diarrhea. Malabsorption and diarrhea were likely due to the failure of enteroendocrine cells to produce hormones that need to be processed by PC1/3 for their function (Martin et al., 2013). These include cholecystokinin (CCK), gastrin, glucagon-like peptides (GLP), ghrelin, peptide YY (PYY), glucose-dependent insulinotropic polypeptide (GIP), secretin, somatostatin and chromogranins. Similar phenotypes were also observed in a third case of congenital PC1/3 deficiency, a 6-year-old boy from a consanguineous union of parents of North African origin (Farooqi et al., 2007a). This patient had a homozygous missense mutation of the *PC1*, leading to impaired catalytic function but with a preserved intracellular trafficking.

Since then, more rare variants of *PCSK1* mutations, both heterozygous and homozygous, have been identified (Creemers et al., 2012; Martin et al., 2013). PC1/3-deficient patients were identified to have a general pattern of endocrinopathies that develop in an age-dependent manner (Martin et al., 2013). Neonates had severe malabsorptive diarrhea and failure to thrive, required prolonged parenteral nutrition support, and had high mortality. As the disease progressed, additional endocrine abnormalities developed, including diabetes insipidus, growth hormone deficiency, primary hypogonadism, adrenal insufficiency and hypothyroidism.

Melanocortin-4-receptor (MC4R) deficiency

MC4R deficiency is the most common monogenic obesity disorder that has been identified so far. The prevalence of MC4R mutations has varied from 0.5–1% of obese adults to 6% in subjects with severe obesity starting in childhood (Farooqi et al., 2003; Lubrano-Berthelier et al., 2003). In a general population, the estimated prevalence of such mutations is of at least 1 in 2000 (Farooqi and O'Rahilly, 2005).

Patients with MC4R deficiency are obese and tall. They exhibit hyperphagia that often begins in the first year of life, but it is less severe than that seen in leptin deficiency. They also exhibit both increased fat and lean mass, the latter a feature that is not seen in leptin-deficient patients. Resting metabolic rate is similar to that predicted on the basis of age- and sex-specific equations after correction for lean body mass. Bone mineral density is markedly increased. The accelerated linear growth remains to be fully explained and may be a consequence of the disproportionate early hyperinsulinemia. Secondary sexual characteristics are appropriate for age in affected children and infertility or reproductive dysfunctions in adults are not observed in general.

Biochemically, patients with MC4R deficiency are normoglycemic but have significantly elevated plasma insulin concentrations compared to those in obese subjects matched for age. Children with MC4R deficiency have higher plasma insulin concentrations compared to adults with similar deficiency, indicating an age-dependent effect. Serum lipid concentrations and urinary 24 hour free cortisol excretion are within normal ranges and serum leptin concentrations are appropriate for the fat mass. Levels of free thyroxine are normal although some patients with MC4R deficiency may have a slight elevation of thyrotropin concentrations. Gonadotropin secretion and levels of sex steroids are normal.

Both homozygous and heterozygous forms of mutation have been identified. Some heterozygotes or "obligate carriers" show incomplete penetrance and are not obese, suggesting that both genetic and environmental factors may have important effects in some pedigrees (Farooqi et al., 2002).

Therapies for melanocortin pathway disorders

MC4R is a G-protein-coupled receptor with a known ligand, making it a promising and potential target for the development of drugs to treat MC4R deficient patients. At present, there is no specific treatment available for POMC-deficient patients, although selective small molecule MC4R agonists are being developed (Farooqi and O'Rahilly, 2005). Trials with intranasal ACTH4-10 treatment, a melanocortin fragment for which an anorectic effect has been previously described, did not have any effect on the body weight or metabolic rate of the POMC-deficient patients (Krude et al., 2003). A more recent trial using daily nasal administration of ACTH/MSH 4-10 also failed to have an anorectic effect in obese males, despite demonstrating statistically significant decrease in body fat in normal-weight males (Fehm et al., 2001; Hallschmid et al., 2006), raising concerns that the obese subjects might somehow be resistant to the effects of the agonist compared to normal weight cohorts. Further problems come from the fact that ACTH/MSH 4-10 has been shown to lack potent activity at the MC4 receptor *in vitro* and that it undergoes rapid degradation under physiologic conditions (Emmerson et al., 2007). In addition, the administration of peptides by the nasal route is generally characterized by low bioavailability, short duration of action, and a great degree of intra- and intersubject variability in its actions (Emmerson et al., 2007).

More recently, a study using highly selective MC4R agonists in *MC4R* deficient patients has also highlighted an ongoing challenge with such a strategy. This treatment resulted in increased blood pressure through changes in sympathetic neural activity, as well as the predictable yawning, stretching, and penile erection also recognized to occur with activation of central MC4R (Greenfield et al., 2009). These side-effects happen because MC4R is widely expressed in both the central and peripheral nervous system, particularly in the spinal cord and autonomic nervous system, where it is known to mediate various physiologic responses including nociception (Vergoni et al., 1998; Gautron et al., 2012).

An effective MC4R agonist will be one that is able to cross the blood–brain barrier with good bioavailability, able to be selective at receptors expressed at very discrete regions of hypothalamus that regulate energy homeostasis, has a convenient mode of administration, and has an acceptable side-effect profile. This has yet to be developed despite extensive research on both small molecule- and peptide-based strategies.

Brain-derived neurotrophic factor (BDNF) and obesity

Brain-derived neurotrophic factor (BDNF) is a member of the neurotrophin family of growth factors that regulates the development, survival, and differentiation of neurons through its high-affinity tyrosine kinase receptor, tropomyosin-related kinase B (TrkB) (Tapia-Arancibia et al., 2004). BDNF has been implicated in memory and various neurologic disorders such as epilepsy, Alzheimer's disease, psychiatric problems such as depression, schizophrenia, and bipolar disorder, as well as Rett's syndrome and Huntington's disease (Zuccato and Cattaneo, 2009). BDNF has also been implicated in the regulation of body weight as its expression is reduced by fasting (Xu et al., 2003), and BDNF administration causes weight loss in wild-type mice through a reduction in food intake (Pelleymounter et al., 1995).

In humans, a *de novo* heterozygous missense mutation (Y722C) in the coding region of *NTRK2*, the gene coding for TrkB receptor, was identified in a child with severe obesity, hyperphagia, developmental delay, impaired learning, short-term memory and nociception (Yeo et al., 2004). This mutation resulted in impaired receptor phosphorylation and signaling through MAPK, thus leading to impaired function of TrkB (Gray et al., 2007). This was followed by the identification of a functional loss of one copy of the *BDNF* gene in an 8-year-old girl with severe obesity, hyperphagia, hyperactive behavior, impaired cognitive function, and poor memory performance (Gray et al., 2006). She also had normal basal metabolic rate compared to the predicted values by age- and sex-specific equations. The patient harbored a *de novo* chromosomal inversion, 46,XX,inv(11)(p13p15.3), a region which encompasses the *BDNF* locus. Fluorescence *in situ* hybridization (FISH) technology further revealed a proximal inversion breakpoint that lies 850 kb telomeric of the 5′ end of the *BDNF* gene. The patient's genomic DNA was heterozygous for the BDNF polymorphism but monoallelic expression was seen in peripheral lymphocytes. Biochemically, the patient had fasting hyperinsulinemia with normoglycemia. Full serum lipid profile, thyroid function tests, and IGF-1 concentrations were all in the normal range. As expected, serum concentration of BDNF proteins was reduced compared with age- and BMI-matched subjects.

Src homology 2 B adapter protein 1 (SH2B1) and obesity

SH2B1 is a cytoplasmic adapter protein that modulates signaling by a variety of ligands that bind to tyrosine kinase receptors or JAK-associated cytokine receptors, including leptin, insulin, growth hormone, and nerve growth factor. Targeted deletion of *SH2B1* in mice results in impaired leptin signaling and severe obesity (Ren et al., 2005). *Sh2b1*-null mice are also insulin resistant and exhibit impaired insulin signaling (Morris et al., 2009).

In humans, heterozygous mutations in *SH2B1* were identified in five probands of mixed European descent from the Genetics of Obesity Study (GOOS) cohort (Doche et al., 2012). These include a frameshift mutation (F344LfsX20) which leads to a truncated protein product and three missense mutations (P90H, T175N, and P322S), all of which were absent from 500 control subjects. The probands were apparently unrelated over three generations based on medical history and all the mutations were inherited from overweight or obese parents. The carriers were hyperphagic and had reduced final height as adults. They were also reported to have delayed speech and language development, social isolation, and aggressive behavior by the healthcare professionals and family members. Biochemically, the mutation carriers had high levels of fasting insulin with normoglycemia. Liver function tests and lipid profiles were all within normal range. Recombinant *Sh2b1β* treatment has been shown to reverse obesity observed in *Sh2b1* knockout mice via neuron-specific restoration (Ren et al., 2007). However, its role in humans is yet to be elucidated.

Single-minded 1 (SIM1) and obesity

SIM1 is a basic helix-loop-helix transcription factor involved in the development and function of the paraventricular nucleus (PVN) of the hypothalamus. It is expressed in the PVN, basomedial amygdala, anterior hypothalamus, and lateral hypothalamic area (Holder et al., 2004). *Sim1* homozygous null mice ($Sim1^{-/-}$) die in the perinatal period and their hypothalami lack paraventricular, anterior periventricular, and supraoptic nuclei secondary to failure of terminal migration and differentiation of Sim1-expressing neurons (Michaud et al., 2001). *Sim1* heterozygous mice (*Sim1*$^{+/-}$) survive but develop severe obesity associated with increased food intake without any measureable deficit in energy expenditure. Consistently, *Sim1* overexpression in mice leads to a decrease in food intake. Conditional postnatal deletion of *Sim1* in postmitotic neurons recapitulates the obesity phenotype, indicating that *Sim1* has a physiologic role in energy homeostasis that is distinct from its role in hypothalamic development (Tolson et al., 2010). The obesity phenotype of *Sim1*$^{+/-}$ mice closely resembles that of mice lacking the MC4R, with increased linear growth and increased food intake (although unlike *MC4R*-null mice, *Sim1*$^{+/-}$ mice do not have reduced energy expenditure), suggesting that SIM1 may influence energy homeostasis by interacting with pathways involved in central melanocortin signaling. In addition, SIM1 is expressed in regions that also have high concentrations of *MC4R*-expressing neurons, and *Sim1*$^{+/-}$ mice have significantly reduced levels of *MC4R* mRNA in the PVN (Tolson et al., 2010). Administration of melanotan II, a potent melanocortin-receptor agonist, results in a reduction in food intake in WT mice but not in *Sim1*$^{+/-}$ mice, suggesting that SIM1 lies downstream of MC4R (Kublaoui et al., 2006a). Increased SIM1 expression is also able to compensate for impaired MC4R signaling, as evident in transgenic overexpression of human SIM1 via a BAC transgene in *Agouti* mice. These mice overexpress the melanocortin-receptor antagonist Agouti and share many of the features of *MC4R*-null mice. Overexpression of SIM1 ameliorates their phenotype by normalizing their food intake (Kublaoui et al., 2006b).

In humans, several chromosomal deletions involving 6q14-q21, a region that includes *SIM1*, have been identified in obese patients presenting with developmental delay and a Prader–Willi-like syndrome (Faivre et al., 2002), although the specific role of *SIM1* haploinsufficiency in the development of Prader–Willi syndrome has not definitively been established. A *de novo* balanced translocation between the short arm of chromosome 1 and the long arm of chromosome 6 (46, XX, t(1;6) (p22.1;q16.2)), leading to disruption of *SIM1*, was identified in a female patient with severe early onset obesity (Holder et al., 2000). The patient was referred to pediatric geneticist at the age of 18 months due to excessive growth. Birth weight was normal but rapidly became excessive and accelerated growth was noted at age 3 months. The patient had no features suggestive of Prader–Willi or other pleiotropic obesity syndromes. Abdominal computed tomography and pituitary magnetic resonance imaging were normal. Further follow-up on this patient revealed mild hyperphagia beginning at the age of 4 years and no evidence of developmental delay, preschool difficulties or precocious puberty. Biochemically, the patient had slightly elevated serum insulin with normoglycemia. Serum leptin concentrations were also within the acceptable levels for her weight.

A recent study identified 13 different heterozygous variants in *SIM1* in 28 unrelated severely obese patients (Ramachandrappa et al., 2013). Nine of these heterozygous variants lead to reduced ability of SIM1 to activate a SIM1-responsive reporter gene. The phenotypic similarities between these patients and MC4R-deficient patients are suggestive that some of the effects of SIM1 deficiency on energy homeostasis are mediated by altered melanocortin signaling. All SIM1 probands in this study had increased body weight with a normal linear growth. The mean percentage of body fat of these patients was high and comparable to that seen in patients with heterozygous MC4R variants. Hyperphagia was observed in childhood and continued into later life in the subset of adult subjects of the SIM1 variant carriers. Eleven of the 13 probands who agreed to further studies had a degree of cognitive deficit as reported by physicians and family members. These include impaired concentration, memory deficit, emotional lability, and in some cases, autistic spectrum behavior. Basal metabolic rate was identical to that predicted based on the basis of age, sex, and body composition in children and adults. Mean systolic blood pressure was also lower in SIM1 variant carriers than in obese controls, thereby suggesting autonomic dysfunction in these patients. Biochemically, all except one of the SIM1 deficient patients were euglycemic. In contrast to MC4R deficiency, they were all hyperinsulinemic but to a degree that was consistent with the obesity. TSH and free thyroxine levels were normal.

Another recent study also identified four rare variants in four severely obese children with Prader–Willi-like syndrome features and four other rare variants in seven morbidly obese adults (Bonnefond et al., 2013). Three of these mutations showed strong loss-of-function effects and assessment of the carriers' relatives revealed a significant contribution of these *SIM1* rare variants to intrafamily risk of obesity. The variants with mild or no effects on SIM1 activity were not associated with obesity within families. These genetic and functional studies demonstrate a firm link between SIM1 loss of function and severe obesity, associated with or, independent of, Prader–Willi-like features.

Melanocortin-2-receptor accessory protein 2 (MRAP2) and obesity

It is increasingly recognized that accessory proteins can modulate GPCR trafficking and ligand binding and signaling (Hay et al., 2006). MC2R accessory protein (MRAP) is an accessory protein for MC2R that is required for the trafficking of MC2R to the surface of adrenal cells and for signaling in response to ACTH. Loss of either MC2R or MRAP in humans causes severe resistance to ACTH with resulting glucocorticoid deficiency (Metherell et al., 2005). MRAP2 is expressed in the brain, most prominently in the pons and cerebellum but also in regions involved in energy homeostasis, such as the hypothalamus and brainstem (Asai et al., 2013). Within the hypothalamic PVN, MRAP2 and MC4R mRNAs are coexpressed in many cells.

Mrap-null (*Mrap*$^{-/-}$) mice appeared normal at birth, with normal weight gain and postweaning food intake during early life and gradually became extremely obese on a regular chow *ad libitum*. Heterozygous mice were also significantly heavier than the wild-type animals on standard chow. *Mrap*$^{-/-}$ mice also had increased length and fat mass with decreased lean mass. Thyroid function tests were normal and the hypothalamic pituitary adrenal axis was also intact.

MRAP2 has been shown to interact directly with MC4R, enhancing the MC4R-mediated generation of the second messenger cyclic adenosine monophosphate (Asai et al., 2013). In addition, MRAP2 proteins also allow for developmental control of MC4R activity (Sebag et al., 2013). In zebrafish, MRAP2a blocks the function of MC4R and stimulates growth during larval development. At later stage of development, MRAP2b enhances responsiveness to α-MSH once the zebrafish begins feeding, thus increasing the capacity for regulated feeding and growth. These findings suggest that alterations in MC4R signaling may be one mechanism underlying the association between MRAP2 disruption and obesity. In humans, four rare heterozygous variants were found in unrelated, nonsyndromic, severely obese individuals from the GOOS cohort and Swedish obese children's cohort (Asai et al., 2013). Only one of the variants is clearly disruptive and pathogenic. These findings suggest that *MRAP2* gene may also contribute to energy homeostasis in humans and disruption of MRAP2, although rare, may lead to obesity.

HUMAN PLEIOTROPIC AND "SYNDROMIC" OBESITY

Pleiotropic and "syndromic" obesity are Mendelian disorders in which patients are obese but have additional characteristics such as dysmorphic features and organ-specific developmental abnormalities. There are at least 30 of these rare syndromes, caused by discrete genetic defects or chromosomal abnormalities, both autosomal and X-linked (O'Rahilly et al., 2003). Table 6.1 summarizes some of these disorders.

Bardet–Biedl syndrome

Bardet–Biedl syndrome (BBS) is a rare, typically autosomal recessive, syndrome with a prevalence of less than one in every 100 000 of the population (Farooqi and O'Rahilly, 2005). Some families exhibit triallelic inheritance. The detailed biochemical mechanism that leads to BBS is still unknown. Mutations in *BBS* genes have been found to lead to ciliopathic disorder (Ansley et al., 2003). The gene products of *BBS* genes, called BBS proteins, are located on the basal bodies and centrosomes of the cell. They interact with pericentriolar material 1 protein (PCM1), which is necessary for ciliogenesis. Other developmental abnormalities associated with *BBS* mutation include loss of microtubule anchoring at the centrosome of the cell and defects in cytokinesis and apoptosis (Kim et al., 2004).

BBS is characterized by six main features: central obesity, learning disabilities, hypogonadism in males, renal abnormalities, polydactyly and rod-cone dystrophy, which is the most frequent phenotype. BBS patients have early onset obesity, and the disorder usually arises within the first few years of life. Interestingly, only 52% of post-pubertal BBS patients are found to be clinically obese, thereby suggesting that this syndrome can present with a heterogeneous phenotype (Mutch and Clement, 2006).

Although BBS proteins are involved in microtubule-based protein or vesicle trafficking, the precise molecular function of each BBS protein has not been characterized and the pathophysiologic mechanisms leading to each component of the BBS phenotype are unclear. Using a BBS knockout mice model, it was shown that leptin-receptor signaling was attenuated in the hypothalamus, rendering the animals leptin resistant (Seo et al., 2009). The knockout mice also had significantly lower expression of the anorexigenic *Pomc* gene in the hypothalamus compared to wild-type controls. After fasting, the orexigenic AgRP and NPY mRNA levels increased and POMC mRNA levels decreased in wild-type mice. In the BBS knockout mice, while the changes in AgRP and NPY mRNA levels in response to fasting were similar, the POMC mRNA levels were somehow significantly less than that of wild-type controls. Further molecular studies showed that BBS proteins physically interact with the leptin receptors (Seo et al., 2009). Loss of BBS proteins affects the trafficking of leptin receptors and thus perturbs the signaling. These findings suggest that attenuated leptin-receptor signaling might account for the pathophysiology of obesity phenotype observed in BBS patients.

Prader–Willi syndrome

Prader–Willi syndrome (PWS) is a complex genetic disorder caused by a loss of one or more paternal genes in

Table 6.1

Table summarizing a number of pleiotropic and "syndromic" disorders associated with obesity

Pleiotropic and "syndromic" disorders	Genetics background	Associated phenotypes
WAGR syndrome (Wilms' tumor, Aniridia, Genitourinary anomalies, mental Retardation)	Heterozygous contiguous gene deletions (*WT1* and *PAX6*) in the 11p13 region	Low-normal birth weight but subsequently develops marked obesity, Wilms' tumor, aniridia, genitourinary anomalies, mental retardation
Pseudohypoparathyroidism type 1A (PHP1A)	Maternally transmitted mutation in GNAS1	Obesity, hyperphagia
Albright's hereditary osteodystrophy	Autosomal dominant disorder, germline mutations in GNAS1	Obesity, short adult stature, brachydactyly, ectopic ossifications
Fragile X syndrome	X-linked, single gene disorder of the *FMR1* gene	Obesity, moderate to severe mental retardation, macro-orchidism, large ears, macrocephaly, mandibular prognathism, high pitched jocular speech
Börjeson–Forssman–Lehmann syndrome	Mutations in a novel, widely expressed zinc-finger gene plant homodomain (PHD)-like finger (PHF6)	Obesity, moderate to severe mental retardation, epilepsy, hypogonadism, marked gynecomastia
Alström syndrome	Homogeneous autosomal recessive disorder	Childhood obesity, hyperinsulinemia, chronic hyperglycemia, neurosensory deficits, male hypogonadism, mild to moderate developmental delay, short stature, hypothyroidism
Cohen syndrome	Autosomal recessive disorder, mutation in the *COH1* gene	Mental retardation, microcephaly, characteristic facial features, progressive retinochoroidal dystrophy
Ulnar–mammary syndrome	Autosomal dominant disorder, mutations in Tbx3 (member of the T-box family of transcription factors)	Defects in limb, tooth, hair, apocrine gland and genital development

(Source: O'Rahilly et al., 2003; Farooqi and O'Rahilly, 2005.)

the region 15q11-15q13 (Nicholls and Knepper, 2001). It is the most common syndromal cause of human obesity, with an estimated prevalence of about 1 in 25000 births. PWS patients are characterized by diminished fetal activity, hypogonadotropic hypogonadism, mild mental disability, short stature, muscular hypotonia and obesity secondary to hyperphagia. Children with PWS present with rapid weight gain in childhood along with a marked increase in appetite. Interestingly, PWS patients have high ghrelin levels despite their high BMI (Cummings et al., 2002; Goldstone et al., 2005). The high ghrelin levels might in part contribute to their hyperphagia and will normalize following recovery to ideal body weight. PWS adults were shown to have an impaired postprandial suppression of plasma ghrelin which was associated with a blunted postprandial PYY response. In contrast, PWS children were shown to have a postprandial decrease in ghrelin levels which was associated with increase in PYY levels, thereby implying that the appetite regulation of these peptides is operative during childhood and progressively deteriorates in adulthood when hyperphagia and obesity worsen (Bizzarri et al., 2010). The negative correlation between ghrelin and adiposity is generally preserved among PWS patients, suggesting that there is no complete ghrelin dysregulation in PWS. There is no cure for PWS at present and a multidisciplinary approach remains the key management strategy. Studies have shown that administration of octreotide, a somatostatin agonist, in PWS subjects leads to suppression of ghrelin levels, both acyl and des-acyl ghrelin (Haqq et al., 2003; De Waele et al., 2008). However, these effects did not impact on the weight and appetite behavior of PWS patients.

NEW TECHNOLOGIES TO IDENTIFY GENETIC COMPONENTS OF OBESITY

Before the human genome project was completed, the majority of our understanding about the genetic determinants of body weight came from analysis of human pedigrees and restriction fragment length polymorphisms (RFLPs), or studies of rodent models. These techniques lend themselves to the understanding of monogenic causes of obesity but are not suited to identify the genetic factors contributing to "common polygenic obesity". Some traits, such as BMI, may be influenced by the simultaneous presence of genetic variation at multiple loci, so-called polygenic variants. Individually, each single nucleotide polymorphism (SNP) has a small impact but together the effect can be cumulative such that if an individual harbors several polygenic variants that separately confer an increased BMI then obesity is more likely to occur in that person. The completion of both the Human Genome and HapMap projects (McPherson et al., 2001; International HapMap, 2003), in conjunction with the development of experimental and computational techniques, has helped identify several of these polygenic variants/loci using techniques such as genome-wide association studies (GWAS).

Genome-wide association studies

GWAS is a high throughput approach that uses statistical methods to scan the entire human genome of a large number of individuals (>10 000) and identify SNPs associated with a trait such as increased BMI. Because the technique is hypothesis-free it allows the identification of molecules and pathways that would not have been implicated as important for energy balance by other conventional methods. The association of the previously unknown gene FTO (fat mass and obesity associated gene) with increased BMI is perhaps the most "well known" obesity GWAS gene (Frayling et al., 2007). A link between the FTO locus and obesity was first reported in 2007 following a GWAS study of ~500 000 SNPs in a large cohort of adults with increased BMI and type 2 diabetes (T2D). The report identified a SNP in the first intron of FTO that showed a strong association with both increased BMI and T2D. Individuals homozygous for this risk allele (A) were ~3 kg heavier than those homozygous for the nonrisk allele (T). Various other SNPs close to the FTO locus were subsequently identified and in addition to being associated with increased BMI, have also been linked to increased energy intake with patients reporting increased food intake and decreased satiety if they had one FTO risk allele (Speakman et al., 2008; Haupt et al., 2009) and an increased preference for energy-dense food (Cecil et al., 2008; Timpson et al., 2008). While the associations were repeated multiple times in various different ethnic groups, it was not until the first report of an Fto knockout mouse that biological evidence was provided to link perturbation of Fto expression with energy homeostasis (Fischer et al., 2009). Following on from *FTO*, GWAS has been successful in identifying over 50 loci associated with obesity (Loos, 2012). However, while some of these loci are close to genes known to be involved in regulating energy homeostasis (e.g., *MC4R*, *BDNF*, *LEP*) there is still much analysis required to determine if the robust statistical associations with novel loci such as *MTCH2* and *TMEM18* can lead to meaningful biological insights.

Copy number variants

While most of the GWAS results have been highly reproducible, they only explain approximately 2% of the expected heritability of BMI (Manolio et al., 2009; Loos, 2012). The discovery of common copy number variants (CNVs) in the general population has led to speculation that they may explain some of the "missing heritability" left by obesity GWAS SNPs. A CNV is defined as "a DNA segment of one kilobase or larger that is present at a variable copy number in comparison with a reference genome" and they have been related to several complex traits such as schizophrenia and autism (Redon et al., 2006). In obesity, the presence of common CNVs at both GWAS-confirmed obesity loci (*NEGR1* (Speliotes et al., 2010; Wheeler et al., 2013), *SH2B1* (Bochukova et al., 2010), *GPRC5B* (Willer et al., 2009), *PPYR1* (Sha et al., 2009)), and novel chromosomal positions (olfactory receptor genes *OR40*, *OR4S2*, and *OR4C6* on Chr 11q11 (Jarick et al., 2011)) have established this tool as a useful identifier of genetic loci that may contribute to common polygenic obesity.

Whole exome sequencing

In recent years advances in sequencing technologies have transformed the study of human genetics by dramatically decreasing both the cost and time to get data (Metzker, 2010). The advent of whole exome sequencing (WES) has allowed researchers to focus on the functional coding regions of the human genome and identify rare variants that may contribute to complex disease such as obesity (Kaiser, 2010). The advantage of WES over GWAS analysis is that it is not biased to SNPs that are common but will identify all coding region polymorphisms regardless of how frequently they occur – a "common disease, rare variant" approach rather than the "common disease, common variant" hypothesis seen with GWAS. Of course, once variants are identified there lies the further challenge of determining whether it is possible to ascribe functional consequence of these mutations. Although the development of this technology is very recent, there are already several reports detailing the use of WES to identify novel mutations in genes such as *LEPR*, *CYB5A*, *RNF10*, and *KSR2* in obese patients (Pearce et al., 2013; Gill et al., 2014; Huang et al., 2014) as well as identifying novel mutations in patients with pleiotropic obesity syndromes such as Bardet–Biedl syndrome (Scheidecker et al., 2014). No doubt more novel genes that contribute to polygenic obesity will be identified in the future as data become available from projects such as UK10K (www.uk10k.org) which will utilize new sequencing technologies to compare the exomes of 10 000 obese patients and controls.

CONCLUSIONS

Our understanding of the molecular and pathophysiologic mechanisms of human obesity continues to evolve and grow. Unquestionably, highly detailed dissection of these pathways in model organisms have formed the vanguard in this progress but equally rapid computational and technological advances mean we are entering an era where new genetic vistas will become apparent. We suggest that these developments will allow a more nuanced and rational classification of causes of obesity. Embracing this heterogeneity is an essential step in developing targeted, effective intervention for a clinical problem where there remains major unmet need.

References

Abbott CR, Rossi M, Kim M et al. (2000). Investigation of the melanocyte stimulating hormones on food intake. Lack of evidence to support a role for the melanocortin-3-receptor. Brain Res 869: 203–210.

Anand BK, Brobeck JR (1951). Hypothalamic control of food intake in rats and cats. Yale J Biol Med 24: 123–140.

Ansley SJ, Badano JL, Blacque OE et al. (2003). Basal body dysfunction is a likely cause of pleiotropic Bardet–Biedl syndrome. Nature 425: 628–633.

Asai M, Ramachandrappa S, Joachim M et al. (2013). Loss of function of the melanocortin 2 receptor accessory protein 2 is associated with mammalian obesity. Science 341: 275–278.

Bailey P, Bremer F (1921). Experimental diabetes insipidus. Arch Int Med 28: 773–803.

Balagura S, Devenport LD (1970). Feeding patterns of normal and ventromedial hypothalamic lesioned male and female rats. J Comp Physiol Psychol 71: 357–364.

Bizzarri C, Rigamonti AE, Luce A et al. (2010). Children with Prader–Willi syndrome exhibit more evident meal-induced responses in plasma ghrelin and peptide YY levels than obese and lean children. Eur J Endocrinol 162: 499–505.

Bochukova EG, Huang N, Keogh J et al. (2010). Large, rare chromosomal deletions associated with severe early-onset obesity. Nature 463: 666–670.

Bonnefond A, Raimondo A, Stutzmann F et al. (2013). Loss-of-function mutations in SIM1 contribute to obesity and Prader–Willi-like features. J Clin Invest 123: 3037–3041.

Brobeck JR, Tepperman J, Long CN (1943). Experimental hypothalamic hyperphagia in the albino rat. Yale J Biol Med 15: 831–853.

Cawley J, Meyerhoefer C (2012). The medical care costs of obesity: an instrumental variables approach. J Health Econ 31: 219–230.

Cecil JE, Tavendale R, Watt P et al. (2008). An obesity-associated FTO gene variant and increased energy intake in children. N Engl J Med 359: 2558–2566.

Church TS, Thomas DM, Tudor-Locke C et al. (2011). Trends over 5 decades in U.S. occupation-related physical activity and their associations with obesity. PLoS One 6: e19657.

Clement K, Vaisse C, Lahlou N et al. (1998). A mutation in the human leptin receptor gene causes obesity and pituitary dysfunction. Nature 392: 398–401.

Cone RD (2006). Studies on the physiological functions of the melanocortin system. Endocr Rev 27: 736–749.

Cool DR, Normant E, Shen F et al. (1997). Carboxypeptidase E is a regulated secretory pathway sorting receptor: genetic obliteration leads to endocrine disorders in Cpe(fat) mice. Cell 88: 73–83.

Creemers JW, Khatib AM (2008). Knock-out mouse models of proprotein convertases: unique functions or redundancy? Front Biosci 13: 4960–4971.

Creemers JW, Jackson RS, Hutton JC (1998). Molecular and cellular regulation of prohormone processing. Semin Cell Dev Biol 9: 3–10.

Creemers JW, Choquet H, Stijnen P et al. (2012). Heterozygous mutations causing partial prohormone convertase 1 deficiency contribute to human obesity. Diabetes 61: 383–390.

Cummings DE, Clement K, Purnell JQ et al. (2002). Elevated plasma ghrelin levels in Prader Willi syndrome. Nat Med 8: 643–644.

De Waele K, Ishkanian SL, Bogarin R et al. (2008). Long-acting octreotide treatment causes a sustained decrease in ghrelin concentrations but does not affect weight, behaviour and appetite in subjects with Prader–Willi syndrome. Eur J Endocrinol 159: 381–388.

Doche ME, Bochukova EG, Su HW et al. (2012). Human SH2B1 mutations are associated with maladaptive behaviors and obesity. J Clin Invest 122: 4732–4736.

Drewnowski A (2004). Obesity and the food environment: dietary energy density and diet costs. Am J Prev Med 27: 154–162.

Emmerson PJ, Fisher MJ, Yan LZ et al. (2007). Melanocortin-4 receptor agonists for the treatment of obesity. Curr Top Med Chem 7: 1121–1130.

Faivre L, Cormier-Daire V, Lapierre JM et al. (2002). Deletion of the SIM1 gene (6q16.2) in a patient with a Prader–Willi-like phenotype. J Med Genet 39: 594–596.

Farooqi IS, O'Rahilly S (2005). Monogenic obesity in humans. Annu Rev Med 56: 443–458.

Farooqi IS, Matarese G, Lord GM et al. (2002). Beneficial effects of leptin on obesity, T cell hyporesponsiveness, and neuroendocrine/metabolic dysfunction of human congenital leptin deficiency. J Clin Invest 110: 1093–1103.

Farooqi IS, Keogh JM, Yeo GS et al. (2003). Clinical spectrum of obesity and mutations in the melanocortin 4 receptor gene. N Engl J Med 348: 1085–1095.

Farooqi IS, Drop S, Clements A et al. (2006). Heterozygosity for a POMC-null mutation and increased obesity risk in humans. Diabetes 55: 2549–2553.

Farooqi IS, Volders K, Stanhope R et al. (2007a). Hyperphagia and early-onset obesity due to a novel homozygous missense mutation in prohormone convertase 1/3. J Clin Endocrinol Metab 92: 3369–3373.

Farooqi IS, Wangensteen T, Collins S et al. (2007b). Clinical and molecular genetic spectrum of congenital deficiency of the leptin receptor. N Engl J Med 356: 237–247.

Fehm HL, Smolnik R, Kern W et al. (2001). The melanocortin melanocyte-stimulating hormone/adrenocorticotropin (4-10) decreases body fat in humans. J Clin Endocrinol Metab 86: 1144–1148.

Fischer J, Koch L, Emmerling C et al. (2009). Inactivation of the Fto gene protects from obesity. Nature 458: 894–898.

Frayling TM, Timpson NJ, Weedon MN et al. (2007). A common variant in the FTO gene is associated with body mass index and predisposes to childhood and adult obesity. Science 316: 889–894.

Friedman JM (2009). Leptin at 14 y of age: an ongoing story. Am J Clin Nutr 89: 973S–979S.

Gautron L, Lee CE, Lee S et al. (2012). Melanocortin-4 receptor expression in different classes of spinal and vagal primary afferent neurons in the mouse. J Comp Neurol 520: 3933–3948.

Gill R, Him Cheung Y, Shen Y et al. (2014). Whole-exome sequencing identifies novel LEPR mutations in individuals with severe early onset obesity. Obesity (Silver Spring) 22: 576–584.

Goldstone AP, Patterson M, Kalingag N et al. (2005). Fasting and postprandial hyperghrelinemia in Prader–Willi syndrome is partially explained by hypoinsulinemia, and is not due to peptide YY3-36 deficiency or seen in hypothalamic obesity due to craniopharyngioma. J Clin Endocrinol Metab 90: 2681–2690.

Gortmaker SL, Must A, Perrin JM et al. (1993). Social and economic consequences of overweight in adolescence and young adulthood. N Engl J Med 329: 1008–1012.

Graff H, Stellar E (1962). Hyperphagia, obesity, and finickiness. J Comp Physiol Psychol 55: 418–424.

Gray J, Yeo GS, Cox JJ et al. (2006). Hyperphagia, severe obesity, impaired cognitive function, and hyperactivity associated with functional loss of one copy of the brain-derived neurotrophic factor (BDNF) gene. Diabetes 55: 3366–3371.

Gray J, Yeo G, Hung C et al. (2007). Functional characterization of human NTRK2 mutations identified in patients with severe early-onset obesity. Int J Obes (Lond) 31: 359–364.

Greenfield JR, Miller JW, Keogh JM et al. (2009). Modulation of blood pressure by central melanocortinergic pathways. N Engl J Med 360: 44–52.

Hallschmid M, Smolnik R, McGregor G et al. (2006). Overweight humans are resistant to the weight-reducing effects of melanocortin4-10. J Clin Endocrinol Metab 91: 522–525.

Haqq AM, Stadler DD, Rosenfeld RG et al. (2003). Circulating ghrelin levels are suppressed by meals and octreotide therapy in children with Prader–Willi syndrome. J Clin Endocrinol Metab 88: 3573–3576.

Haupt A, Thamer C, Staiger H et al. (2009). Variation in the FTO gene influences food intake but not energy expenditure. Exp Clin Endocrinol Diabetes 117: 194–197.

Hay DL, Poyner DR, Sexton PM (2006). GPCR modulation by RAMPs. Pharmacol Ther 109: 173–197.

Hennessy JW, Grossman SP (1976). Overeating and obesity produced by interruption of the caudal connections of the hypothalamus: evidence of hormonal and metabolic disruption. Physiol Behav 17: 103–109.

Hervey GR (1959). The effects of lesions in the hypothalamus in parabiotic rats. J Physiol 145: 336–352.

Hetherington AW, Ranson SW (1940). Hypothalamic lesions and adiposity in the rat. Anat Rec 78: 149–172.

Hetherington AW, Ranson SW (1942). The relation of various hypothalmic lesions to adiposity in the rat. J Comp Neurol 76: 475–499.

Holder Jr JL, Butte NF, Zinn AR (2000). Profound obesity associated with a balanced translocation that disrupts the SIM1 gene. Hum Mol Genet 9: 101–108.

Holder Jr JL, Zhang L, Kublaoui BM et al. (2004). Sim1 gene dosage modulates the homeostatic feeding response to increased dietary fat in mice. Am J Physiol Endocrinol Metab 287: E105–E113.

Huang K, Nair AK, Muller YL et al. (2014). Whole exome sequencing identifies variation in CYB5A and RNF10 associated with adiposity and type 2 diabetes. Obesity (Silver Spring) 22: 984–988.

Huszar D, Lynch CA, Fairchild-Huntress V et al. (1997). Targeted disruption of the melanocortin-4 receptor results in obesity in mice. Cell 88: 131–141.

International HapMap Consortium (2003). The International HapMap Project. Nature 426: 789–796.

Jackson RS, Creemers JW, Ohagi S et al. (1997). Obesity and impaired prohormone processing associated with mutations in the human prohormone convertase 1 gene. Nat Genet 16: 303–306.

Jackson RS, Creemers JW, Farooqi IS et al. (2003). Small-intestinal dysfunction accompanies the complex endocrinopathy of human proprotein convertase 1 deficiency. J Clin Invest 112: 1550–1560.

Jarick I, Vogel CI, Scherag S et al. (2011). Novel common copy number variation for early onset extreme obesity on chromosome 11q11 identified by a genome-wide analysis. Hum Mol Genet 20: 840–852.

Kaiser J (2010). Human genetics. Affordable "exomes" fill gaps in a catalog of rare diseases. Science 330: 903.

Kennedy GC (1950). The hypothalamic control of food intake in rats. Proc R Soc Lond B Biol Sci 137: 535–549.

Kennedy GC (1953). The role of depot fat in the hypothalamic control of food intake in the rat. Proc R Soc Lond B Biol Sci 140: 578–596.

Kim JC, Badano JL, Sibold S et al. (2004). The Bardet–Biedl protein BBS4 targets cargo to the pericentriolar region and is required for microtubule anchoring and cell cycle progression. Nat Genet 36: 462–470.

King BM, Gaston MG (1977). Reappearance of dynamic hyperphagia during the static phase in medial hypothalamic lesioned rats. Physiol Behav 18: 945–950.

Kopelman P (2007). Health risks associated with overweight and obesity. Obes Rev 8 (Suppl 1): 13–17.

Krude H, Biebermann H, Luck W et al. (1998). Severe early-onset obesity, adrenal insufficiency and red hair pigmentation caused by POMC mutations in humans. Nat Genet 19: 155–157.

Krude H, Biebermann H, Gruters A (2003). Mutations in the human proopiomelanocortin gene. Ann N Y Acad Sci 994: 233–239.

Kublaoui BM, Holder Jr JL, Gemelli T et al. (2006a). Sim1 haploinsufficiency impairs melanocortin-mediated anorexia and activation of paraventricular nucleus neurons. Mol Endocrinol 20: 2483–2492.

Kublaoui BM, Holder Jr JL, Tolson KP et al. (2006b). SIM1 overexpression partially rescues agouti yellow and diet-induced obesity by normalizing food intake. Endocrinology 147: 4542–4549.

Lee YS, Challis BG, Thompson DA et al. (2006). A POMC variant implicates beta-melanocyte-stimulating hormone in the control of human energy balance. Cell Metab 3: 135–140.

Lloyd DJ, Bohan S, Gekakis N (2006). Obesity, hyperphagia and increased metabolic efficiency in Pc1 mutant mice. Hum Mol Genet 15: 1884–1893.

Loos RJ (2012). Genetic determinants of common obesity and their value in prediction. Best Pract Res Clin Endocrinol Metab 26: 211–226.

Lubrano-Berthelier C, Cavazos M, Dubern B et al. (2003). Molecular genetics of human obesity-associated MC4R mutations. Ann N Y Acad Sci 994: 49–57.

Manolio TA, Collins FS, Cox NJ et al. (2009). Finding the missing heritability of complex diseases. Nature 461: 747–753.

Martin MG, Lindberg I, Solorzano-Vargas RS et al. (2013). Congenital proprotein convertase 1/3 deficiency causes malabsorptive diarrhea and other endocrinopathies in a pediatric cohort. Gastroenterology 145: 138–148.

McPherson JD, Marra M, Hillier L et al. (2001). A physical map of the human genome. Nature 409: 934–941.

Metherell LA, Chapple JP, Cooray S et al. (2005). Mutations in MRAP, encoding a new interacting partner of the ACTH receptor, cause familial glucocorticoid deficiency type 2. Nat Genet 37: 166–170.

Metzker ML (2010). Sequencing technologies – the next generation. Nat Rev Genet 11: 31–46.

Michaud JL, Boucher F, Melnyk A et al. (2001). Sim1 haploinsufficiency causes hyperphagia, obesity and reduction of the paraventricular nucleus of the hypothalamus. Hum Mol Genet 10: 1465–1473.

Montague CT, Farooqi IS, Whitehead JP et al. (1997). Congenital leptin deficiency is associated with severe early-onset obesity in humans. Nature 387: 903–908.

Morris DL, Cho KW, Zhou Y et al. (2009). SH2B1 enhances insulin sensitivity by both stimulating the insulin receptor and inhibiting tyrosine dephosphorylation of insulin receptor substrate proteins. Diabetes 58: 2039–2047.

Mountjoy KG, Robbins LS, Mortrud MT et al. (1992). The cloning of a family of genes that encode the melanocortin receptors. Science 257: 1248–1251.

Muller L, Lindberg I (1999). The cell biology of the prohormone convertases PC1 and PC2. Prog Nucleic Acid Res Mol Biol 63: 69–108.

Mutch DM, Clement K (2006). Unraveling the genetics of human obesity. PLoS Genet 2: e188.

Nicholls RD, Knepper JL (2001). Genome organization, function, and imprinting in Prader–Willi and Angelman syndromes. Annu Rev Genomics Hum Genet 2: 153–175.

O'Rahilly S, Farooqi IS, Yeo GS et al. (2003). Minireview: human obesity-lessons from monogenic disorders. Endocrinology 144: 3757–3764.

Paz-Filho G, Mastronardi C, Delibasi T et al. (2010). Congenital leptin deficiency: diagnosis and effects of leptin replacement therapy. Arq Bras Endocrinol Metabol 54: 690–697.

Pearce LR, Atanassova N, Banton MC et al. (2013). KSR2 mutations are associated with obesity, insulin resistance, and impaired cellular fuel oxidation. Cell 155: 765–777.

Pelleymounter MA, Cullen MJ, Wellman CL (1995). Characteristics of BDNF-induced weight loss. Exp Neurol 131: 229–238.

Ramachandrappa S, Raimondo A, Cali AM et al. (2013). Rare variants in single-minded 1 (SIM1) are associated with severe obesity. J Clin Invest 123: 3042–3050.

Redon R, Ishikawa S, Fitch KR et al. (2006). Global variation in copy number in the human genome. Nature 444: 444–454.

Ren D, Li M, Duan C et al. (2005). Identification of SH2-B as a key regulator of leptin sensitivity, energy balance, and body weight in mice. Cell Metab 2: 95–104.

Ren D, Zhou Y, Morris D et al. (2007). Neuronal SH2B1 is essential for controlling energy and glucose homeostasis. J Clin Invest 117: 397–406.

Scheidecker S, Etard C, Pierce NW et al. (2014). Exome sequencing of Bardet–Biedl syndrome patient identifies a null mutation in the BBSome subunit BBIP1 (BBS18). J Med Genet 51: 132–136.

Sebag JA, Zhang C, Hinkle PM et al. (2013). Developmental control of the melanocortin-4 receptor by MRAP2 proteins in zebrafish. Science 341: 278–281.

Seo S, Guo DF, Bugge K et al. (2009). Requirement of Bardet–Biedl syndrome proteins for leptin receptor signaling. Hum Mol Genet 18: 1323–1331.

Sha BY, Yang TL, Zhao LJ et al. (2009). Genome-wide association study suggested copy number variation may be associated with body mass index in the Chinese population. J Hum Genet 54: 199–202.

Speakman JR, Rance KA, Johnstone AM (2008). Polymorphisms of the FTO gene are associated with variation in energy intake, but not energy expenditure. Obesity (Silver Spring) 16: 1961–1965.

Speliotes EK, Willer CJ, Berndt SI et al. (2010). Association analyses of 249,796 individuals reveal 18 new loci associated with body mass index. Nat Genet 42: 937–948.

Stunkard AJ, Foch TT, Hrubec Z (1986a). A twin study of human obesity. JAMA 256: 51–54.

Stunkard AJ, Sorensen TI, Hanis C et al. (1986b). An adoption study of human obesity. N Engl J Med 314: 193–198.

Stunkard AJ, Harris JR, Pedersen NL et al. (1990). The body-mass index of twins who have been reared apart. N Engl J Med 322: 1483–1487.

Tapia-Arancibia L, Rage F, Givalois L et al. (2004). Physiology of BDNF: focus on hypothalamic function. Front Neuroendocrinol 25: 77–107.

Timpson NJ, Emmett PM, Frayling TM et al. (2008). The fat mass- and obesity-associated locus and dietary intake in children. Am J Clin Nutr 88: 971–978.

Tolson KP, Gemelli T, Gautron L et al. (2010). Postnatal Sim1 deficiency causes hyperphagic obesity and reduced Mc4r and oxytocin expression. J Neurosci 30: 3803–3812.

Vergoni AV, Bertolini A, Mutulis F et al. (1998). Differential influence of a selective melanocortin MC4 receptor antagonist (HS014) on melanocortin-induced behavioral effects in rats. Eur J Pharmacol 362: 95–101.

Wheeler E, Huang N, Bochukova EG et al. (2013). Genome-wide SNP and CNV analysis identifies common and low-frequency variants associated with severe early-onset obesity. Nat Genet 45: 513–517.

Willer CJ, Speliotes EK, Loos RJ et al. (2009). Six new loci associated with body mass index highlight a neuronal influence on body weight regulation. Nat Genet 41: 25–34.

Xu B, Goulding EH, Zang K et al. (2003). Brain-derived neurotrophic factor regulates energy balance downstream of melanocortin-4 receptor. Nat Neurosci 6: 736–742.

Yeo GS, Heisler LK (2012). Unraveling the brain regulation of appetite: lessons from genetics. Nat Neurosci 15: 1343–1349.

Yeo GS, Connie Hung CC, Rochford J et al. (2004). A de novo mutation affecting human TrkB associated with severe obesity and developmental delay. Nat Neurosci 7: 1187–1189.

Zhang Y, Proenca R, Maffei M et al. (1994). Positional cloning of the mouse obese gene and its human homologue. Nature 372: 425–432.

Zhu X, Zhou A, Dey A et al. (2002). Disruption of PC1/3 expression in mice causes dwarfism and multiple neuroendocrine peptide processing defects. Proc Natl Acad Sci U S A 99: 10293–10298.

Zuccato C, Cattaneo E (2009). Brain-derived neurotrophic factor in neurodegenerative diseases. Nat Rev Neurol 5: 311–322.

Handbook of Clinical Neurology, Vol. 124 (3rd series)
Clinical Neuroendocrinology
E. Fliers, M. Korbonits, and J.A. Romijn, Editors

Chapter 7

Sleep characteristics and insulin sensitivity in humans

ESTHER DONGA[1] AND JOHANNES A. ROMIJN[2]*
[1]*Department of Endocrinology, Leiden University Medical Center, Leiden, The Netherlands*
[2]*Department of Medicine, Academic Medical Center, University of Amsterdam, Amsterdam, The Netherlands*

INTRODUCTION

The diurnal variation of the geophysical position of the earth in relation to the sun has imposed considerable evolutionary pressure. The diurnal variation in light exposure has caused adaptations to light and dark in virtually all biological systems, including mammals. Light is sensed by the eyes and transmitted to different areas in the brain via the optic nerves. The specialized retinal cells that detect light are called photoreceptors. In addition to the well-known rods and cones, a third photoreceptor cell was identified in the 1990s, namely the melanopsin-containing ganglion cells. Mice genetically lacking these cells still have normal vision but they are unable to synchronize their clocks to light (Güler et al., 2008). All connections go from these few melanopsin cells via the retinohypothalamic tract to the hypothalamic pacemaker neurons in the suprachiasmatic nucleus (SCN), above the optic chiasm, which serves as the central biological clock (Hattar et al., 2002). This circadian clock allows light-sensitive organisms to synchronize their daily molecular oscillations, behavioral and physiologic rhythms with the daily rotations of the earth (Bass, 2012). This retinohypothalamic pathway is also present in humans (Sadun et al., 1984). The central clock has a complex output through hypothalamic nuclei and, subsequently, through the autonomic nervous system throughout the body. In addition to this central biological clock there are molecular systems in peripheral organs which also serve diurnal clock functions in interaction with the central biological clock (Bass, 2012).

Sleep is a profound alteration from the normal conscious state. Before the introduction of electric light, sleep largely coincided with the geophysical night. In the Western lifestyle, the introduction of electric light has enabled people to shorten this geophysically imposed night, and this has been reinforced by the introduction of television. As a consequence, perceived duration of nights and the duration of sleep have been shortened considerably in recent decades. These changes in sleep characteristics have largely coincided with epidemic increases in insulin resistance, hypertension, type 2 diabetes mellitus, and obesity. In recent years, much attention has been focused on the effects of sleep duration and sleep quality on glucose metabolism and insulin sensitivity, which is the focus of this review.

SLEEP PHYSIOLOGY AND GLUCOSE HOMEOSTASIS

Maintenance of constant blood glucose levels is essential for daily functioning, particularly for the brain, since the brain can neither synthesize nor store glucose. Therefore, the consolidation of human sleep to a single, relatively long period without food ingestion imposed adaptations to the effects of fasting during the night. An important example of this adaptation is the diurnal variation of glucose tolerance. Changes in glucose tolerance do not occur at random, but show a clear 24 hour rhythm, reaching a minimum around midnight. These diurnal variations in glucose tolerance are directly controlled by the endogenous biological clock, located in the SCN of the hypothalamus. Importantly, the SCN regulates diurnal rhythms both in sleep/wake behavior and in glucose homeostasis (Kalsbeek et al., 2010).

Sleep architecture, the structure and pattern of sleep, can be recorded by polysomnography. Throughout a normal night of sleep, several cycles occur consisting of nonrapid eye movement (NREM) sleep, stage 1 to 3, and rapid eye movement (REM) sleep. Stage 3 NREM sleep, also called slow-wave sleep, is thought to be the most restorative part of sleep, in which several (neuro)

*Correspondence to: Johannes A. Romijn, Department of Medicine, Academic Medical Center, University of Amsterdam, PO Box 22660, 1100 DD Amsterdam, The Netherlands. Tel: +31-20-5662171, Fax: +31-20-6919658, E-mail: j.a.romijn@amc.uva.nl

endocrine and metabolic changes occur. For example, brain glucose utilization decreases and secretion of counter-regulatory hormones is stimulated at the onset of slow wave sleep, leading to decreased glucose tolerance (Scheen and Van Cauter, 1998).

How are sleep/wake behavioral states and glucose homeostasis linked? The hypothalamic neuropeptide orexin provides a critical link between central regulation of sleep/wakefulness and peripheral energy homeostasis (reviewed in Sakurai et al., 2010). Orexin-expressing neurons are located in the lateral hypothalamic area and have projections to the entire neuroaxis, predominantly to the arcuate nucleus of the hypothalamus, locus coeruleus, tuberomammillary nucleus, and paraventricular nucleus of the thalamus. During sleep, orexin neurons are inactive, whereas during waking hours, they are active. These neurons are thought to act as a stabilizer of our sleep/wakefulness state. For example, patients with narcolepsy are deficient in orexin and are characterized by disorganization of sleep and wakefulness (Peyron et al., 2000). Narcoleptic patients have excessive daytime sleepiness, sleep attacks, and often disturbed nocturnal sleep. Interestingly, the metabolic syndrome is more frequently present in these patients, despite decreased caloric intake. Therefore, orexin neurons have been suggested to play a role in energy homeostasis. Indeed, electrophysiologic studies have shown that the activity of orexin neurons is influenced by peripheral metabolic factors. Orexin activity was inhibited by glucose and leptin and stimulated by ghrelin, which provides a direct link between energy homeostasis and regulation of behavioral states (Sakurai, 2005).

SLEEP DEPRIVATION AND INSULIN RESISTANCE

In the last five decades, mean self-reported sleep duration in the US has decreased by 1.5–2 hours per night, which is primarily due to voluntary sleep restriction (National Sleep Foundation, 2005). Although sleep loss has been studied scientifically for over a 100 years, the impact of insufficient sleep on general health, mood, and performance has only begun to be widely appreciated in the past two decades. Accumulating evidence from both epidemiologic and experimental studies has pointed to a strong association between sleep duration and insulin resistance.

Experimental studies on the effects of sleep deprivation on glucose metabolism

In early experimental studies using a model of prolonged total sleep deprivation in healthy males, sleep deprivation decreased glucose tolerance (VanHelder et al., 1993). However, little attention was paid to these studies, since total sleep deprivation was considered to be an unlikely and nonphysiologic condition. The first detailed laboratory study that assessed the effects of partial sleep deprivation was reported in 1999 (Spiegel et al., 1999). In 11 healthy young men, sleep was restricted to 4 hours per night during 6 consecutive nights. In these participants, partial sleep deprivation reduced glucose tolerance by ~40%. Several other controlled experimental studies have reported similar results for impaired glucose tolerance after 5–14 nights of 4–5.5 hours of sleep restriction per day (Nedeltcheva et al., 2009; Buxton et al., 2010; Reynolds et al., 2012). Moreover, even a single night of partial sleep restriction induces insulin resistance in healthy subjects. During one of our studies, sleep was restricted to 4 hours on a single night, followed by assessment of insulin sensitivity the next morning using hyperinsulinemic euglycemic clamp studies (Fig. 7.1). Partial sleep restriction reduced insulin sensitivity of both hepatic and peripheral glucose metabolism by 20–25%. In addition, the effect of insulin on suppression of lipolysis was reduced by 19%, reflected in plasma levels of free fatty acids (Donga et al., 2010a). The observations from these studies are consistent in indicating that sleep restriction during one or more consecutive nights reduces insulin sensitivity of multiple metabolic pathways in otherwise healthy subjects.

Epidemiologic studies on the association between sleep duration and glucose metabolism

In accordance with these experimental studies in healthy subjects, epidemiologic studies have documented a correlation between sleep duration and impaired glucose metabolism. In the Sleep Heart Healthy Study, a community-based prospective study, 722 men and 764 women, aged between 53 and 93 years, were included. Mean total sleep duration in the study population was 7 hours (h) per night, with 27% of the subjects sleeping less than 6 h per night. Both short (<6 h/night) and long (>9 h/night) sleep duration were associated with an increased risk for diabetes mellitus and impaired glucose tolerance. Adjusted odds ratios (OR) for diabetes and impaired glucose tolerance were 1.66 (95% CI (confidence interval) 1.15–2.39) and 1.58 (95% CI 1.15–2.18), respectively, for short sleepers and 1.79 (95% CI 1.08–2.96) and 1.88 (95% CI 1.21–2.91) for long sleepers. (Gottlieb et al., 2005). A recent meta-analysis of 10 prospective population-based studies showed a consistent pattern of an increased risk of developing type 2 diabetes in short and long sleepers. The relative risk for type 2 diabetes was 1.28 (95% CI 1.03–1.60) for short sleep

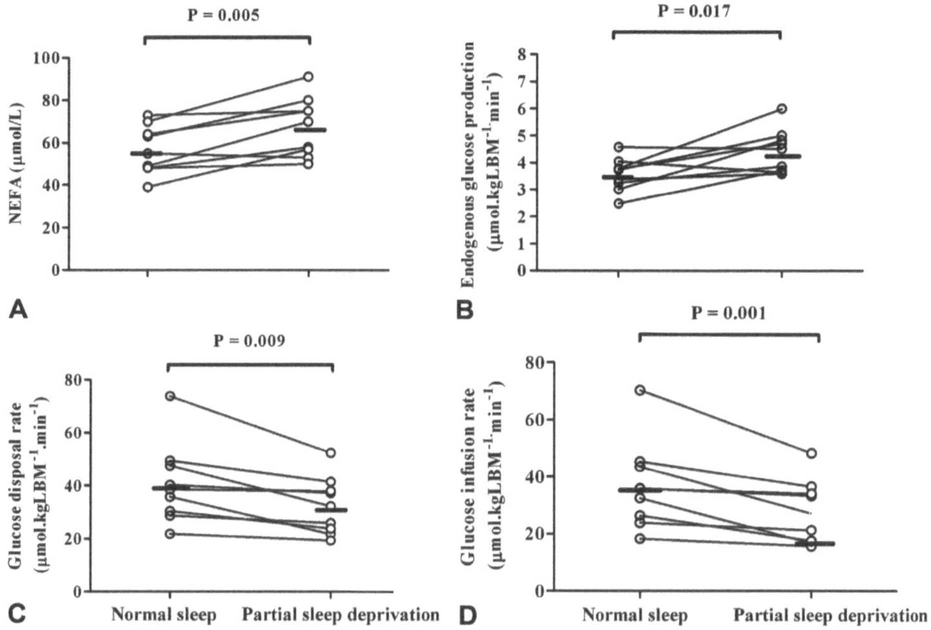

Fig. 7.1. Metabolic consequences of sleep restriction to 4 hours during a single night. (**A**) Individual values obtained during steady state of the hyperinsulinemic euglycemic clamp studies of nonesterified fatty acids. (**B**) Endogenous glucose production. (**C**) Glucose disposal rate. (**D**). The glucose infusion rate after a night of normal sleep duration versus after a night of partial sleep deprivation in healthy subjects (n = 9). Black horizontal lines represent the mean of the values of nine subjects. There were significant differences ($p < 0.05$) for all these parameters after partial sleep deprivation. (Reproduced from Donga et al., 2010a. © Copyright 2010 Endocrine Society.)

duration (5–6 h/night) and 1.48 (95% CI 1.13–1.96) for long sleep duration (>8 h/night) (Cappuccio et al., 2010).

An important limitation of most epidemiologic studies was that sleep duration was self-reported using standardized questionnaires, but not objectively measured by polysomnography or actigraphy. One of the few studies that used objective sleep parameters was the ancillary study to the Coronary Artery Risk Development In young Adults (CARDIA) study. In total, 571 participants, 40 of whom had type 2 diabetes, were included for sleep analysis using both wrist actigraphy and sleep questionnaires. Habitual sleep duration was assessed during 6 days of wrist actigraphy. Insulin resistance was estimated by the homeostatic model assessment (HOMA) index. Results were stratified for diabetes status and showed that habitual sleep duration was not associated with markers of glucose metabolism in adults, irrespective of the presence of diabetes (Knutson et al., 2011). The results from this CARDIA study were in contrast with previously mentioned prospective studies. However, different methods were used to quantify insulin sensitivity, which could contribute to these conflicting data. Therefore, additional research is needed to better define the relation between objectively measured sleep duration in epidemiologic studies and insulin resistance.

SLEEP DISORDERS AND INSULIN RESISTANCE

Population-based studies on sleep disorders and insulin resistance

Sleep disorders are highly prevalent and lead to decreased sleep duration and/or sleep quality. Subjective sleep disorders, such as difficulty in falling asleep or maintaining sleep, have been independently associated with an increased risk for development of type 2 diabetes mellitus (Nilsson et al., 2004; Meisinger et al., 2005).

Sleep-disordered breathing, also called obstructive sleep apnea, is one of the most common and treatable sleep disorders and is highly prevalent among obese individuals. Sleep apnea is characterized by a reduction or cessation of airflow, which results in oxygen desaturation and frequent microarousals. Intermittent hypoxemia is thought to play a role in the development of insulin resistance by promoting the release of inflammatory cytokines, such as interleukin 6 and tumor necrosis factor α. Furthermore, microarousals are thought to increase sympathetic activity, thereby contributing to a decrease in insulin sensitivity (Pamidi and Tasali, 2012). Population-based studies have shown that sleep-disordered breathing is an independent risk factor for development of impaired glucose metabolism. Among individuals with moderate to severe sleep apnea, a significantly higher prevalence of 15–30% of type 2 diabetes was reported, compared to individuals without sleep apnea. Moreover, severe sleep apnea was associated with worsening of insulin resistance (Botros et al., 2009; Pamidi and Tasali, 2012). Most of these studies included middle aged individuals, most with at least mild obesity. However, a recent study in healthy, young lean men with obstructive sleep apnea documented that insulin sensitivity after an oral glucose load was 27% lower in participants with confirmed sleep apnea compared to matched controls without sleep apnea (Pamidi et al., 2012). These results show that sleep-disordered breathing impairs glucose metabolism even at a young age, in the absence of obesity and other known risk factors for type 2 diabetes mellitus.

Intervention studies assessing sleep quality and insulin sensitivity

There are differences between objective and subjective sleep qualities. Examples of decreased objective sleep quality are a decrease in total amount of slow wave sleep and fragmentation of sleep due to frequent (micro) arousals. Both are present in patients with sleep-disordered breathing. Interestingly, the duration of slow-wave sleep also decreases with aging, which in turn is associated with increased incidence of diabetes mellitus.

As previously mentioned, several changes occur at the initiation of slow-wave sleep in hormonal secretion and metabolism, which could affect glucose metabolism. Tasali et al. have examined the effects of selective slow wave sleep suppression on glucose tolerance in nine healthy volunteers (Tasali et al., 2008). During 3 consecutive nights, slow-wave sleep was suppressed by acoustic stimuli, without any change in total sleep time. Insulin sensitivity was decreased by 25%, assessed by an intravenous glucose tolerance test after 3 nights. The degree of insulin resistance was strongly correlated to the amount of slow-wave sleep reduction.

Sleep fragmentation is characterized by repetitive interruptions of sleep. Stamatakis and Punjabi examined the effect of sleep fragmentation on insulin sensitivity in 11 healthy volunteers (Stamatakis and Punjabi, 2010). Sleep fragmentation was achieved by acoustic and mechanical stimuli across all sleep stages. After 2 nights of sleep fragmentation, insulin sensitivity and glucose effectiveness were significantly decreased without any change in total sleep duration. Sleep fragmentation increased morning cortisol plasma levels and sympathetic activity, assessed from heart rate variability indices, which were both thought to contribute to impaired glucose metabolism. In conclusion, in addition to reduced sleep duration, impaired sleep quality decreases insulin sensitivity and may therefore contribute to diabetes risk.

DIABETES MELLITUS, METABOLIC DYSREGULATION, AND SLEEP DISORDERS

What is the impact of sleep disorders on glucose metabolism in patients already known to have diabetes mellitus? The findings mentioned above raise the possibility that impaired sleep characteristics could also adversely affect glycemic control in patients with established diabetes. Conversely, type 1 and type 2 diabetes mellitus are both associated with an increased prevalence of sleep disturbances (Skomro et al., 2001; Van Dijk et al., 2011), which in turn could aggravate glucose regulation.

In type 2 diabetes mellitus, a cross-sectional study in 161 patients classified 71% of participants as having poor sleep quality, assessed by validated questionnaires. Furthermore, perceived depth of sleep in these patients was associated with decreased glycemic control, assessed by HbA1c values (Knutson et al., 2006).

In patients with type 2 diabetes, sleep-disordered breathing is a very common and often unrecognized condition. Overall prevalence of obstructive sleep apnea is even estimated to be present in ~71% of patients with type 2 diabetes (Pamidi and Tasali, 2012). Increasing severity of untreated sleep apnea was associated with worsening of glycemic control in patients with type 2 diabetes mellitus, independently of body mass index (BMI) and other confounders (Aronsohn et al., 2010).

Nonetheless, there is controversy over whether treatment of sleep apnea improves glucose metabolism. Some studies have documented improvement in insulin sensitivity after treatment with continuous positive airway pressure (CPAP) (Harsch et al., 2004). In contrast, a double-blind, placebo-controlled trial by West et al. showed no significant improvement in insulin sensitivity or HbA1c after 3 months' treatment with CPAP (West et al., 2007). Differences in study population, duration of CPAP treatment, and lack of data of compliance to CPAP might explain these conflicting results.

In type 1 diabetes mellitus, only a few studies are available on sleep characteristics. Van Dijk et al. reported, in a cross-sectional study, that 35% of patients with type 1 diabetes mellitus have poor sleep quality, defined by validated sleep questionnaires. Interestingly, an increased risk of sleep apnea was also observed in patients with type 1 diabetes mellitus compared to healthy, matched controls; it was not related to BMI. There was no relation between poor sleep quality or increased risk for sleep apnea and HbA1c values in these patients (Van Dijk et al., 2011). However, the effects of decreased sleep quality may not simply be reflected in

HbA1c values, as adaptations in intensive insulin treatment regimens could potentially counteract the effects of impaired sleep characteristics on glycemic control.

We have assessed the effect of sleep restriction on insulin sensitivity in patients with type 1 diabetes. In seven patients with type 1 diabetes mellitus without known diabetic complications, sleep duration was restricted to 4 hours during a single night and insulin sensitivity was assessed by hyperinsulinemic euglycemic clamp studies the following morning. A single night of partial sleep restriction decreased peripheral insulin sensitivity in these patients by 14–21% (Donga et al., 2010b). This experimental study proves that sleep restriction, even during a single night, reduces insulin sensitivity in patients with type 1 diabetes, similar to earlier findings in healthy subjects.

In conclusion, sleep duration and/or sleep quality are determinants of insulin sensitivity in patients with type 1 and 2 diabetes mellitus. In addition, various aspects of diabetes could be linked to increased prevalence of sleep disturbances. This might be due to the disease itself as well as diabetic complications such as autonomic dysfunction and neuropathic pain. Impaired sleep and diabetes mellitus might potentiate each other in some patients, thereby creating a negative vicious circle. Optimizing sleep duration and sleep quality could therefore be considered as a potential therapeutic target to improve glucoregulation in patients with diabetes mellitus.

POTENTIAL MECHANISMS LINKING IMPAIRED SLEEP AND INSULIN RESISTANCE

The mechanisms underlying the impact of decreased sleep duration and/or sleep quality on glucose metabolism are not yet fully understood. Much of the evidence linking impaired sleep to metabolic dysfunction is based on epidemiologic studies, which do not provide insight into underlying causal pathophysiologic mechanisms. Furthermore, most experimental intervention studies were neither designed nor powered to address these underlying mechanisms.

Altered secretion patterns of counter-regulatory hormones have been postulated to contribute to induction of insulin resistance by reduced sleep duration and/or sleep quality. In humans, diurnal variations in cortisol levels strongly correlate with variations in plasma glucose and insulin levels (Van Cauter et al., 1991). However, data in experimental studies have shown conflicting results regarding cortisol levels after sleep intervention. Some studies in humans showed an increase in afternoon and evening cortisol levels after sleep restriction (Spiegel et al., 1999; Buxton et al., 2010), whereas others found no significant differences in 24 hour cortisol levels (Nedeltcheva et al., 2009; Schmid et al., 2011). Growth hormone levels were extendedly elevated at nighttime after 6 nights of sleep restriction in one study (Spiegel et al., 2000), but showed no significant increase in another study after 14 nights of bedtime restriction (Nedeltcheva et al., 2009). In the study by Nedeltcheva et al. (2009), the most consistent effect of sleep restriction on counter-regulatory hormones was an increase in plasma catecholamine concentrations within the normal range, and this might be a reflection of increased sympathetic activity.

Changes in autonomic nervous system activity could play an important role in inducing insulin resistance after impaired sleep. The ratio between sympathetic and parasympathetic activity of the autonomic nervous system is called sympathovagal balance, and can be indirectly derived from heart rate variability analysis. Experimental studies have shown that both sleep restriction and sleep fragmentation induce a shift in sympathovagal balance, towards increased sympathetic nervous system activity (Spiegel et al., 1999; Stamatakis and Punjabi, 2010). It is currently unclear to what extent these heart rate variability analyses are predictive of changes in sympathetic activity at other peripheral tissues involved in glucose metabolism and insulin sensitivity. This is a major problem at present. Although the neuroanatomy of the autonomic nervous system has largely been delineated, major questions remain with respect to the tissue-specific contribution of diurnal variations in autonomic nervous activity in humans. This is related to current limitations in experimental design in both human and experimental models regarding assessment of tissue-specific activity of the autonomic nervous system. Nonetheless, we speculate that autonomic nervous activity may play a very important role in explaining the induction of insulin resistance in multiple metabolic pathways after sleep restriction and/or impaired sleep quality. The effects of the central nervous system on peripheral insulin sensitivity in multiple organs are very powerful in experimental models (Kreier et al., 2002; Coomans et al., 2011). Moreover, in mice malfunction of the SCN induces the development of insulin resistance, which underscores a direct role of the central pacemaker in the ontogeny of metabolic disorders (Coomans et al., 2013).

Another potential pathway involved in the interaction between disturbed sleep and metabolic dysregulation could be dysregulation of hormones controlling appetite. Sleep restriction for 2 nights in healthy young men increased the orexigenic hormone ghrelin by 28% and decreased the anorexigenic hormone leptin by 18%. Feelings of hunger and appetite, particularly for carbohydrate-rich foods, were increased after sleep restriction (Spiegel et al., 2004). These changes could

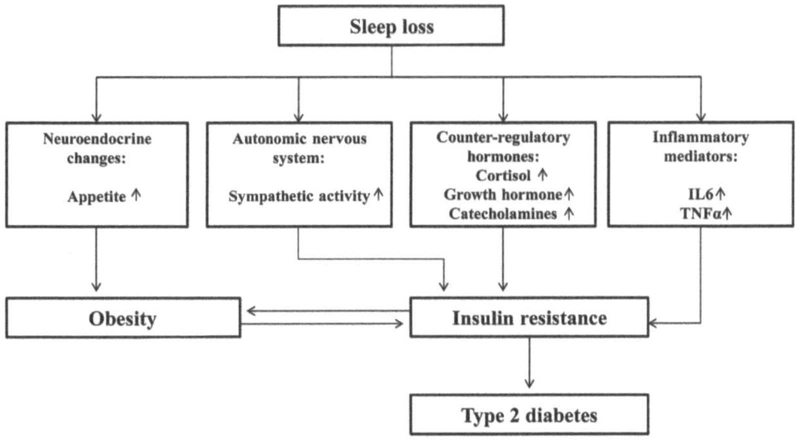

Fig. 7.2. Potential pathways for insulin resistance after sleep deprivation.

contribute to altered eating behavior and energy balance, leading to weight gain and, in turn, decreased insulin sensitivity.

Sleep deprivation and sleep-disordered breathing also increase plasma levels of the proinflammatory cytokines tumor necrosis factor α and interleukin-6 (Vgontzas et al., 2004; Pamidi and Tasali, 2012). Impaired sleep duration and/or quality could therefore promote a proinflammatory state which is associated with insulin resistance and type 2 diabetes.

In summary, endocrine changes do not provide a simple explanation for reduced insulin sensitivity after sleep restriction or decreased sleep quality. The mechanisms underlying impaired glucose metabolism are likely to be multifactorial (Fig. 7.2). We hypothesize that alterations in the activity of the autonomic nervous system might prove to be an important factor.

FUTURE RESEARCH SUGGESTIONS

Current experimental evidence in humans indicates that impaired sleep duration and/or sleep characteristics induce insulin resistance. This is supported by the association in epidemiologic studies between impaired sleep and insulin resistance and type 2 diabetes. An important difference between intervention studies in humans and the epidemiologic studies is that intervention studies can only assess the acute effects of impaired sleep characteristics on glucose metabolism. On the other hand, most epidemiologic studies lack objective assessments of sleep parameters. Future research should therefore focus on the long-term consequences and possible adaptations to impaired sleep, using both objective and subjective sleep parameters. Furthermore, study populations should be better phenotyped, since large interindividual differences exist in normal sleep patterns. For instance, short sleep in naturally short sleepers might have less impact on glucose homeostasis compared to chronically sleep-deprived individuals.

Further insights into the mechanisms inducing insulin resistance after sleep interventions require experimental studies in animal models, with impaired function of the suprachiasmatic nuclei and other disruptions in diurnal control mechanisms of peripheral glucose metabolism. Another option might be to investigate the morphologic integrity of the SCN in diabetes mellitus patients. For instance, earlier studies in human SCN showed cell loss in senescence (Swaab et al., 1985). Since then, this has been explained by a loss of immunocytochemical SCN markers (indicative for diminished neuronal activity) rather that a loss of cells. In addition, it remains a major challenge to design new methods to assess the interaction between the central nervous system and individual organs and to assess the effects of disturbed sleep on these interactions in more detail.

Finally, intervention studies in patients with diabetes mellitus are needed to define and assess the effect of improvement of sleep characteristics on glycemic control and diabetic complications.

References

Aronsohn RS, Whitmore H, Van Cauter E et al. (2010). Impact of untreated obstructive sleep apnea on glucose control in type 2 diabetes. Am J Respir Crit Care Med 181: 507–513.

Bass J (2012). Circadian topology of metabolism. Nature 491: 348–356.

Botros N, Concato J, Mohsenin V et al. (2009). Obstructive sleep apnea as a risk factor for type 2 diabetes. Am J Med 122: 1122–1127.

Buxton OM, Pavlova M, Reid EW et al. (2010). Sleep restriction for 1 week reduces insulin sensitivity in healthy men. Diabetes 59: 2126–2133.

Cappuccio FP, D'Elia L, Strazzullo P et al. (2010). Quantity and quality of sleep and incidence of type 2 diabetes:

a systematic review and meta-analysis. Diabetes Care 33: 414–420.

Coomans CP, Biermasz NR, Geerling JJ et al. (2011). Stimulatory effect of insulin on glucose uptake by muscle involves the central nervous system in insulin-sensitive mice. Diabetes 60: 3132–3140.

Coomans CP, van den Berg SA, Lucassen EA et al. (2013). The suprachiasmatic nucleus controls circadian energy metabolism and hepatic insulin sensitivity. Diabetes 62: 1102–1108.

Donga E, van Dijk M, van Dijk JG et al. (2010a). A single night of partial sleep deprivation induces insulin resistance in multiple metabolic pathways in healthy subjects. J Clin Endocrinol Metab 95: 2963–2968.

Donga E, van Dijk M, van Dijk JG et al. (2010b). Partial sleep restriction decreases insulin sensitivity in type 1 diabetes. Diabetes Care 33: 1573–1577.

Gottlieb DJ, Punjabi NM, Newman AB et al. (2005). Association of sleep time with diabetes mellitus and impaired glucose tolerance. Arch Intern Med 165: 863–867.

Güler AD, Ecker JL, Lall GS et al. (2008). Melanopsin cells are the principal conduits for rod–cone input to non-image-forming vision. Nature 453: 102–105.

Harsch IA, Schahin SP, Radespiel-Troger M et al. (2004). Continuous positive airway pressure treatment rapidly improves insulin sensitivity in patients with obstructive sleep apnea syndrome. Am J Respir Crit Care Med 169: 156–162.

Hattar S, Liao HW, Takao M et al. (2002). Melanopsin-containing retinal ganglion cells: architecture, projections, and intrinsic photosensitivity. Science 295: 1065–1070.

Kalsbeek A, Yi CX, La Fleur SE et al. (2010). The hypothalamic clock and its control of glucose homeostasis. Trends Endocrinol Metab 21: 402–410.

Knutson KL, Ryden AM, Mander BA et al. (2006). Role of sleep duration and quality in the risk and severity of type 2 diabetes mellitus. Arch Intern Med 166: 1768–1774.

Knutson KL, Van Cauter E, Zee P et al. (2011). Cross-sectional associations between measures of sleep and markers of glucose metabolism among subjects with and without diabetes: the Coronary Artery Risk Development in Young Adults (CARDIA) sleep study. Diabetes Care 34: 1171–1176.

Kreier F, Fliers E, Voshol PJ et al. (2002). Selective parasympathetic innervation of subcutaneous and intra-abdominal fat – functional implications. J Clin Invest 110: 1243–1250.

Meisinger C, Heier M, Loewel H (2005). Sleep disturbance as a predictor of type 2 diabetes mellitus in men and women from the general population. Diabetologia 48: 235–241.

National Sleep Foundation (2005). Sleep in America Poll. National Sleep Foundation, Washington DC.

Nedeltcheva AV, Kessler L, Imperial J et al. (2009). Exposure to recurrent sleep restriction in the setting of high caloric intake and physical inactivity results in increased insulin resistance and reduced glucose tolerance. J Clin Endocrinol Metab 94: 3242–3250.

Nilsson PM, Roost M, Engstrom G et al. (2004). Incidence of diabetes in middle-aged men is related to sleep disturbances. Diabetes Care 27: 2464–2469.

Pamidi S, Tasali E (2012). Obstructive sleep apnea and type 2 diabetes: is there a link? Front Neurol 3: 126.

Pamidi S, Wroblewski K, Broussard J et al. (2012). Obstructive sleep apnea in young lean men: impact on insulin sensitivity and secretion. Diabetes Care 35: 2384–2389.

Peyron C, Faraco J, Rogers W et al. (2000). A mutation in a case of early onset narcolepsy and a generalized absence of hypocretin peptides in human narcoleptic brains. Nat Med 9: 991–997.

Reynolds AC, Dorrian J, Liu PY et al. (2012). Impact of five nights of sleeprestriction on glucose metabolism, leptin and testosterone in young adult men. PLoS One 7: e41218.

Sadun AA, Schaechter JD, Smith LE (1984). A retinohypothalamic pathway in man: light mediation of circadian rhythms. Brain Res 302: 371–377.

Sakurai T (2005). Roles of orexin/hypocretin in regulation of sleep/wakefulness and energy homeostasis. Sleep Med Rev 9: 231–241.

Sakurai T, Mieda M, Tsujino N (2010). The orexin system: roles in sleep/wake regulation. Ann N Y Acad Sci 1200: 149–161.

Scheen AJ, Van Cauter E (1998). The roles of time of day and sleep quality in modulating glucose regulation: clinical implications. Horm Res 49: 191–201.

Schmid SM, Hallschmid M, Jauch-Chara K et al. (2011). Disturbed glucoregulatory response to food intake after moderate sleep restriction. Sleep 34: 371–377.

Skomro RP, Ludwig S, Salamon E et al. (2001). Sleep complaints and restless legs syndrome in adult type 2 diabetics. Sleep Med 2: 417–422.

Spiegel K, Leproult R, Van Cauter E (1999). Impact of sleep debt on metabolic and endocrine function. Lancet 354: 1435–1439.

Spiegel K, Leproult R, Colecchia EF et al. (2000). Adaptation of the 24-h growth hormone profile to a state of sleep debt. Am J Physiol Regul Integr Comp Physiol 279: R874–R883.

Spiegel K, Tasali E, Penev P et al. (2004). Brief communication: sleep curtailment in healthy young men is associated with decreased leptin levels, elevated ghrelin levels, and increased hunger and appetite. Ann Intern Med 141: 846–850.

Stamatakis KA, Punjabi NM (2010). Effects of sleep fragmentation on glucose metabolism in normal subjects. Chest 137: 95–101.

Swaab DF, Fliers E, Partiman TS (1985). The suprachiasmatic nucleus of the human brain in relation to sex, age and senile dementia. Brain Res 342: 37–44.

Tasali E, Leproult R, Ehrmann DA et al. (2008). Slow-wave sleep and the risk of type 2 diabetes in humans. Proc Natl Acad Sci U S A 105: 1044–1049.

Van Cauter E, Blackman JD, Roland D et al. (1991). Modulation of glucose regulation and insulin secretion by circadian rhythmicity and sleep. J Clin Invest 88: 934–942.

van Dijk M, Donga E, van Dijk JG et al. (2011). Disturbed subjective sleep characteristics in adult patients with long-standing type 1 diabetes mellitus. Diabetologia 54: 1967–1976.

VanHelder T, Symons JD, Radomski MW (1993). Effects of sleep deprivation and exercise on glucose tolerance. Aviat Space Environ Med 64: 487–492.

Vgontzas AN, Zoumakis E, Bixler EO et al. (2004). Adverse effects of modest sleep restriction on sleepiness, performance, and inflammatory cytokines. J Clin Endocrinol Metab 89: 2119–2126.

West SD, Nicoll DJ, Wallace TM et al. (2007). Effect of CPAP on insulin resistance and HbA1c in men with obstructive sleep apnoea and type 2 diabetes. Thorax 62: 969–974.

Handbook of Clinical Neurology, Vol. 124 (3rd series)
Clinical Neuroendocrinology
E. Fliers, M. Korbonits, and J.A. Romijn, Editors

Chapter 8

Hypothalamic–pituitary hormones during critical illness: a dynamic neuroendocrine response

LIES LANGOUCHE* AND GREET VAN DEN BERGHE
Laboratory and Department of Intensive Care Medicine, University of Leuven, Leuven, Belgium

INTRODUCTION

Critical illness is the medical condition in which a patient, because of major surgery or severe illness, requires immediate intensive medical support of vital organ functions in order to survive. Independent of the underlying condition, critical illness is characterized by a uniform dysregulation of the hypothalamic–pituitary–peripheral axes. In the majority of these axes a clear biphasic pattern can be distinguished (Fig. 8.1). The early phase of illness is characterized by an actively secreting pituitary in the presence of low peripheral target hormones. The acute endocrine alterations can be considered beneficial, as they appear to delay costly anabolism and facilitate the release of substrates as fuel to vital tissues in order to improve survival. In the prolonged phase of critical illness, when recovery does not quickly ensue, a uniform hypothalamic–pituitary suppression occurs, further contributing to the low levels of peripheral target hormones. The ongoing hypercatabolism, despite the administration of artificial nutrition, leads to substantial loss of lean body mass. Ultimately, this may compromise recovery of vital functions and delay rehabilitation. The severity of the neuroendocrine alterations is associated with a high risk of morbidity and mortality in the intensive care unit (ICU). However, before therapeutic intervention can be considered, a thorough understanding of the underlying pathophysiology is essential.

This chapter will focus on each hypothalamic–pituitary axis separately. For each axis the impact of acute and more prolonged critical illness will be described as well as the current therapeutic knowledge.

THE THYROID AXIS

At the level of the hypothalamus, thyrotropin-releasing hormone (TRH) is released and stimulates the thyrotropic cells in the pituitary to synthesize and secrete thyroid-stimulating hormone (TSH). TSH, secreted in a pulsatile and diurnal pattern, in turn drives the thyroid gland to synthesize and secrete thyroid hormones (Yen, 2001). The thyroid gland predominantly secretes the inactive thyroid hormone T_4, which is converted to the active thyroid hormone T_3 in the peripheral target organs (Goldsmith et al., 1951). Different types of deiodinases (D1–D3) are responsible for the peripheral activation of T_4 to T_3 or to the biologically inactive reverse T_3 (rT_3) (Bianco et al., 2002; Friesema et al., 2005). Thyroid hormones are essential for the regulation of energy metabolism and have profound effects on differentiation and growth (Yen, 2001). Thyroid hormones exert an inhibitory feedback control on both TRH and TSH secretion (Weintraub et al., 1989; Bodenner et al., 1991).

The thyroid axis in acute critical illness

During acute illness or severe physical stress, circulating T_3 levels rapidly decline, whereas reverse T_3 levels are upregulated. These changes are predominantly caused by an altered peripheral conversion of T_4 due to a decreased D1 activity and an increased D3 activity (Peeters et al., 2003; Rodriguez-Perez et al., 2008). Circulating T_4 levels have been shown to rise only transiently, although T_4 levels may also decrease in the more severely ill (Van den Berghe, 2000).

*Correspondence to: Lies Langouche, PhD, Senior Researcher, Laboratory of Intensive Care Medicine, K.U.Leuven,OandN1 Herestraat 49 bus 503,B-3000 Leuven, Belgium. Tel: +32-16-330524, Fax: +32-16-330932, E-mail: lies.langouche@med.kuleuven.be

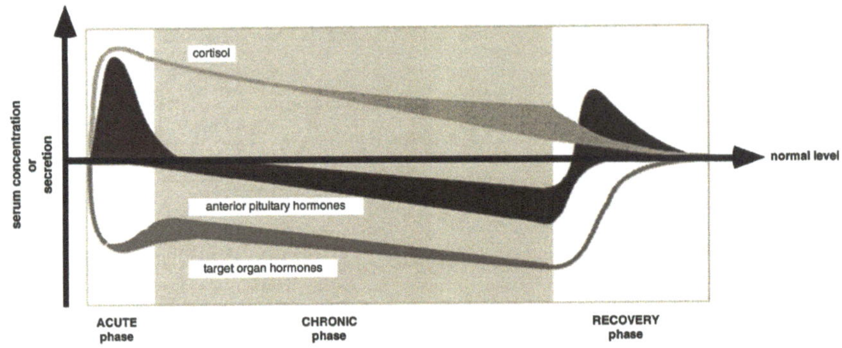

Fig. 8.1. Simplified scheme of the neuroendocrine changes during the acute, chronic, and recovery phase of critical illness. In the acute phase of illness (first hours to a few days after onset), the secretory activity of the anterior pituitary is essentially maintained or amplified, whereas anabolic target organ hormones are inactivated. In the chronic phase of protracted critical illness (intensive care-dependent for weeks), the secretory activity of the anterior pituitary appears uniformly suppressed in relation to reduced circulating levels of target organ hormones. Impaired anterior pituitary hormone secretion allows the respective target organ hormones to decrease proportionately over time, with cortisol being a noteworthy exception, the circulating levels of which remain elevated. The onset of recovery is characterized by restored levels of target hormones and pituitary hormones. Shaded areas represent the range within which the hormonal changes occur. (Reproduced from Van den Berghe et al., 1998b. © Copyright 1998, Endocrine Society.)

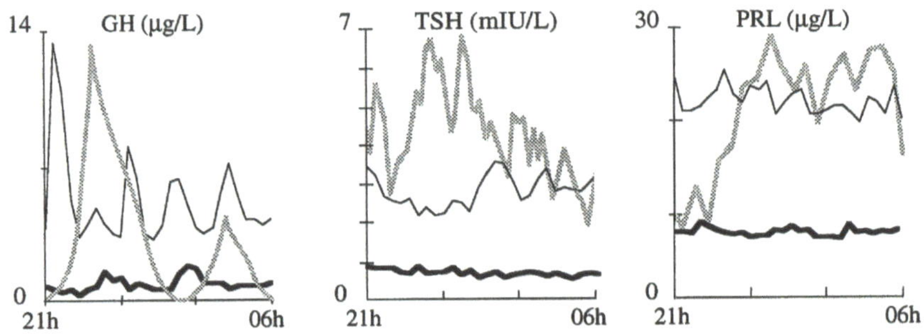

Fig. 8.2. Representative nocturnal serum concentration profiles of GH, TSH, and PRL. The upper panels illustrate the differences between the acute phase (thin black line) and the chronic phase (thick black line) of critical illness within an intensive care setting. The gray lines illustrate normal patterns. (Reproduced from Van den Berghe et al., 1998b. © Copyright 1998, Endocrine Society.)

Regardless of normal single-sample TSH levels, the TSH profile is already affected in this acute phase of illness, as the normal nocturnal TSH surge is absent (Fig. 8.2) (Romijn and Wiersinga, 1990; Persani et al., 1995). In the presence of the low T_3 levels, TSH levels thus remain low-normal, a constellation referred to as "the low T_3 syndrome", "euthyroid sick syndrome", or "nonthyroidal illness".

The decrease in circulating T_3 and T_4 during the first 24 hours after the insult reflects the severity of illness and correlates with mortality (Schlienger et al., 1991; Rothwell et al., 1993; Rothwell and Lawler, 1995; Ray et al., 2002). Whether or not these changes reflect a beneficial and adaptive response to the severity of illness or rather contribute to adverse outcome remains currently unclear and the mechanisms behind the observed changes are still not fully understood. Cytokines have been demonstrated to affect deiodinase activity and are able to mimic the acute stress response of the thyroid axis (Van der Poll et al., 1990; Boelen et al., 1995; Michalaki et al., 2001). However, cytokine antagonists failed to restore normal thyroid function after endotoxemic challenge (Van der Poll et al., 1995, 1999). Other potential factors of the low T_3 syndrome include low concentrations of thyroid hormone-binding proteins, reduced thyroid hormone uptake and metabolism by elevated levels of free fatty acids and bilirubin (Lim et al., 1993; Mebis and Van den Berghe, 2011). Another major contributor to the low T_3 syndrome might be the poor nutritional status and reduced food intake of severely ill patients. Indeed, starvation in healthy individuals induces changes in the thyroid axis that are very similar to the ones observed in acutely ill patients (Gardner et al., 1979; Boelen et al., 2008). Furthermore, circulating T_3 levels and tissue deiodinase activity could be normalized by administration of artificial feeding in a rabbit model of critical illness (Mebis et al., 2012).

The thyroid axis during prolonged critical illness

Patients who need prolonged intensive care and enter a more chronic phase of illness display a dramatically reduced pulsatile TSH secretion in addition to the absent nocturnal TSH surge (Fig. 8.2) (Van den Berghe et al., 1997b). Furthermore, circulating T_3 and in addition T_4 levels are low, with the decline in T_3 in particular correlating positively with the diminished pulsatile TSH release (Van den Berghe et al., 1997b). Despite low circulating thyroid hormone levels, and thus reduced negative feedback, pulsatile TSH and hypothalamic TRH expression are low, pointing to a central suppression of the thyroid axis (Fliers et al., 1997; Mebis et al., 2009a).

The prognostic value of the altered thyroid axis with regard to mortality in the prolonged phase of illness is illustrated by lower TSH, T_4 and T_3, and higher rT_3 levels in patients who ultimately die as compared with those surviving prolonged critical illness (Peeters et al., 2005). The fall in T_4 in particular has been shown to relate to the severity of illness (Wartofsky and Burman, 1982; Maldonado et al., 1992).

Several factors can be listed as being possibly involved in the low TSH pulsatility. These are an impaired capacity of the pituitary to synthesize or release TSH, inadequate TRH-induced stimulation of TSH, an altered setpoint for feedback inhibition, or an elevated somatostatin tone. Endogenous dopamine and/or hypercortisolism could also be involved since they are known to provoke or severely aggravate hypothyroidism in critical illness (Faglia et al., 1973; Van den Berghe et al., 1994a). Fliers et al. described reduced hypothalamic TRH gene expression in patients who died after prolonged illness, a phenomenon also observed in rabbits with prolonged illness (Fliers et al., 1997; Mebis et al., 2009a). As bolus injections of TRH induced a normal to high TSH response in patients with prolonged critical illness, an impaired pituitary capacity to synthesize TSH seems less likely (Van den Berghe et al., 1998a; Weekers et al., 2002). Furthermore, a continuous infusion of TRH together with the synthetic GH secretagogue GHRP-2 to patients with prolonged illness was able to restore the pulsatile TSH secretion and to normalize circulating T_3 and T_4 without affecting rT_3 (Van den Berghe et al., 1996a, b, 1999). The suppression of the central part of the thyroid axis, which is most pronounced in the prolonged or more severely ill patients, seems thus primarily of hypothalamic nature. Observations in an animal model of prolonged critical illness argue against an altered setpoint of the thyroid feedback mechanism or a local hypothalamic thyrotoxicosis, as low hypothalamic TRH gene expression coincided with low T_4 and low-normal T_3 tissue levels, and upregulated D2 and thyroid hormone transporters within the hypothalamus (Mebis et al., 2009a). Different mechanisms are more likely involved in the TRH suppression during prolonged critical illness.

In addition to the central suppression, the peripheral levels of thyroid hormones remains altered in the prolonged phase of critical illness (Peeters et al., 2003). Remarkably, type 2 deiodinase (D2) activity increases in the prolonged phase of critical illness and does not appear to play a role in the pathogenesis of the low T_3 syndrome in patients with prolonged critical illness (Mebis et al., 2007). Regulation of thyroid hormone action at the level of the thyroid hormone transporters as well as thyroid hormone receptors also appears to be altered by critical illness in some tissues, possibly causing an upregulated thyroid hormone sensitivity in response to low T_3 levels (Thijssen-Timmer et al., 2007; Mebis et al., 2009b).

Therapeutic potential

The peripheral changes in thyroid hormone metabolism occur already very early in the process of critical illness and are present in all forms of acute stress and illness. Furthermore, such changes are very similar to the changes evoked by short-term fasting and likely partly brought about by the lack of normal nutritional intake during acute illness. The immediate fall in circulating T_3 during starvation has been interpreted as an attempt of the body to reduce its energy expenditure, and prevent protein wasting (Gardner et al., 1979). Hence, the rapid changes during the acute phase of illness could be looked upon as a beneficial and adaptive response that does not warrant intervention (De Groot, 1999).

During the prolonged phase of illness a hypothalamic suppression also occurs which could benefit from treatment. From the current literature, however, it remains controversial whether administration of thyroid hormone to critically ill patients is beneficial or harmful. A few randomized controlled studies assessed the impact of either T_3 or T_4 administration in critically ill patients (Brent and Hershman, 1986; Acker et al., 2000; Bettendorf et al., 2000). Administration of T_3 substitution doses to pediatric cardiac surgery patients has been associated with improved postoperative cardiac function, but these patients were treated with dopamine which induces iatrogenic hypothyroidism (Bettendorf et al., 2000). Treatment with T_4 failed to demonstrate a clinical benefit, although this could be partly due to impaired conversion of T_4 to T_3 (Brent and Hershman, 1986). Furthermore, administration of pharmacologic doses of T_4 consequently resulted in a progressive and sustained suppression of TSH levels (Brent and Hershman, 1986; Acker et al., 2000). As an alternative for T_4 and T_3 treatment, the administration of the hypothalamic releasing factor TRH might be a better choice as this allows the body to use its feedback mechanisms, preventing overtreatment. Indeed, the continuous infusion of TRH in combination with a GH secretagogue restored T_3 and T_4 hormone levels to low-normal physiologic levels without an additional rise in rT_3 (Van den Berghe et al., 1998a). Animal work further demonstrated that this combination therapy of hypothalamic releasing factors was associated with a normalization of D1 and D3 activity levels (Debaveye et al., 2005). However, sufficiently powered randomized controlled trials in patients are currently lacking.

THE SOMATOTROPIC AXIS

Growth hormone (GH) is produced in the somatotropic cells of the pituitary. Basal GH gene expression and release is stimulated by the hypothalamic GH releasing hormone (GHRH), while hepatic insulin-like growth factor 1 (IGF-1) attenuates GH gene expression and secretion via direct feedback inhibition at the level of the somatotropes (Ohlsson et al., 2009). In addition, ghrelin, a hormone secreted by the gut which is also involved in appetite regulation, stimulates GH release (not synthesis) in synergy with GHRH (Popovic et al., 2003). Ghrelin acts via a G-protein-coupled receptor which is located in the hypothalamus and the pituitary (Howard et al., 1996; Kojima et al., 1999). The pulsatility of GH release is regulated by an interplay between GHRH, responsible for the basal GH release, and somatostatin, determining the interpulse troughs (Tannenbaum and Ling, 1984; Tannenbaum et al., 2003). The regulation of the physiologic pulsatile release of GH, with peaks followed by undetectable levels, is important for its metabolic effects (Van den Berghe et al., 2000).

The main anabolic functions of GH are largely mediated through stimulation of IGF-1 production, of which the bioavailability and activity in turn are controlled by several IGF-binding proteins (IGFBPs). In health, around 80% of IGF-1 is bound to IGFBP-3 and acid labile subunit (ALS) in a large ternary complex, extending the half-life of IGF-1 (Baxter and Martin, 1989). A small fraction of IGF-1 is bound to the small inhibitory IGFBPs 1, 2, 4 and 6 in binary complexes (Baxter, 2001). In addition, GH has several direct actions, among them the stimulation of lipolysis and hepatic protein synthesis.

The somatotropic axis in acute critical illness

During the early phase of critical illness, GH secretion is elevated with an increased pulse amplitude, elevated peak levels and high interpulse concentrations (Fig. 8.2) (Ross et al., 1991; Van den Berghe et al., 1998b). Concurrently, a state of peripheral GH resistance develops, with low levels of IGF-1, GH-dependent IGFBP-3 and acid-labile subunit (Ross et al., 1991; Baxter et al., 1998; Baricevic et al., 2007). The serum concentration of the small binding proteins IGFBP-2, -4 and -6, but especially IGBP-1, on the other hand, are elevated (Timmins et al., 1996; Baxter et al., 1998; Mesotten et al., 2004; Baricevic et al., 2007). Admission GH levels have been demonstrated to be an independent predictor of ICU mortality (Schuetz et al., 2009).

The primary event in these changes is thought to be reduced GH-receptor expression levels triggered by circulating cytokines such as TNF-α (Ross et al., 1991; Hermansson et al., 1997; Defalque et al., 1999). The consequently lower circulating IGF-1 levels and thus suppressed negative feedback at the level of the pituitary might then drive the abundant release of GH in the acute phase of illness. This constellation, with high GH levels and low IGF-1 levels, has been considered as adaptive to the illness. The inhibition of the IGF-1-mediated anabolic effects and stimulation of the direct GH effects (lipolysis and insulin antagonism) will theoretically result in postponing of costly anabolism, and an increased availability of energy substrates (Arnold et al., 1993).

The somatotropic axis during prolonged critical illness

In prolonged critical illness, when patients do not recover within a few days, the pulsatile GH secretion is dramatically reduced during critical illness, while nonpulsatile, basal GH release is elevated (Fig. 8.2) (Van den Berghe et al., 2002; Weekers et al., 2002). Remarkably, a rather modest GH response was observed after the injection of a GHRH bolus in prolonged critically ill patients (Van den Berghe et al., 1997b). This suggests that sufficient GHRH is still available to stimulate GH gene transcription and release. The diminished GH pulse amplitude and elevated interpulse levels may correspond with low somatostatin levels during critical illness, but also lack of active ghrelin might play a role. The injection of a GHRP-2 bolus, a synthetic small peptide ghrelin-receptor ligand, induced a GH secretion response which was several-fold higher than normal (Van den Berghe et al., 1997a, 1998a; Weekers et al., 2002). Such hyperresponse can be reconciled with chronically reduced circulating ghrelin levels or ghrelin activity (Popovic et al., 2005). Low circulating ghrelin levels have indeed been reported in prolonged critically ill patients and animals (Wu et al., 2004; Nematy et al., 2006).

The suppressed pulsatile fraction of GH release correlates with the continuously low IGF-1, IGFBP-3, and ALS levels, suggesting that the loss of pulsatile GH release now contributes to the low peripheral target hormones levels (Van den Berghe et al., 1997a, 1998a, 1999).

The administration of GH secretagogues increased IGF-1 and IGFBP levels, which indicates that GH responsiveness at least partially recovers in the chronic phase of critical illness (Van den Berghe et al., 1997a, 1998a).

The low circulating IGF-1 and ternary complex-binding protein levels are tightly related to the biochemical markers of impaired anabolism, such as low serum osteocalcin and leptin concentrations (Van den Berghe et al., 1999). Thus the chronic GH deficiency with reduced anabolism and ongoing catabolism contributes to the pathogenesis of the "wasting syndrome" that characterizes prolonged critical illness. Clinical recovery is indeed also preceded by a rapid normalization of the somatotropic changes.

Therapeutic potential

To convert the negative effects of peripheral GH deficiency and counteract the ongoing hypercatabolism, GH administration was considered as a potential treatment therapy in the ICU. Although several small studies showed promising results (Herndon et al., 1990; Ziegler et al., 1990; Voerman et al., 1995), a multicenter randomized controlled trial with pharmacologic doses of GH had to be stopped after an interim safety control, when an elevated mortality rate in GH-treated patients was found (Takala et al., 1999). The exact mechanism by which GH had caused this mortality increase could not be identified, but it was speculated that the administration of such high doses (up to 20-fold substitution dose) may have worsened hypoglutaminemia and hyperglycemia, both associated with worse outcome (Oudemans-van Straaten et al., 2001; Van den Berghe et al., 2001b; Rodas et al., 2012). The combined administration of GH and IGF-1, additive in their anabolic actions, might have been a better option, because they neutralize each other's side-effects (Kupfer et al., 1993). A more recent trial, where GH was administered in intravenous pulses to mimic the endogenous secretion pattern, in combination with glutamine substitution and tight glycemic control, demonstrated attenuated catabolism but aggravated insulin resistance in trauma patients (Duska et al., 2008). Treatment with hypothalamic-releasing factors to reactivate the pituitary may therefore be more effective and safer than administration of pituitary or peripheral hormones. Indeed, infusions of GH secretagogues not only restored the pulsatile GH secretion, but evoked an increase of IGF-1, IGFBP-3, and ALS, indicative of a restored peripheral responsiveness (Van den Berghe et al., 1997a, 1998a). Furthermore, ghrelin – the recently identified endogenous GH secretagogue – has demonstrated beneficial effects on catabolic protein loss in several animal models of critical illness (Hill et al., 2012). Further studies to prove the therapeutic potential of ghrelin to counteract the wasting syndrome in critically ill humans are, however, still lacking.

THE GONADAL AXIS

The hypothalamic gonadotropin-releasing hormone (GnRH) controls the reproductive axis. GnRH stimulates synthesis and secretion of luteinizing hormone (LH) and follicle-stimulating hormone (FSH) from the gonadotropic cells in the pituitary. Whereas one single GnRH pulse stimulates LH and FSH secretion, the pulsatile chronic GnRH stimulation will stimulate production of both pituitary hormones. In females, LH mediates ovarian androgen production, whereas FSH stimulates the aromatization of androgens to estrogens in the ovary. In men, LH stimulates androgen production by the testicular Leydig cells, whereas the combined action of FSH and testosterone on the Sertoli cells supports spermatogenesis. Sex steroids exert a negative feedback on GnRH and gonadotropin secretion.

The gonadal axis in acute critical illness

The reproductive hormone levels change dramatically with the onset of severe illness. Acute physical stress causes an immediate fall in the serum levels of testosterone, even though LH levels are normal to elevated (Wang et al., 1978a, b; Christeff et al., 1988; Dong et al., 1992; Spratt et al., 1993a, b). In contrast, estrogen levels rise (Plymate et al., 1987; Christeff et al., 1988; Spratt et al., 1993b; Fourrier et al., 1994). The severity of the changes was associated with the severity of the illness (Christeff et al., 1988; Spratt et al., 1993b).

The reduction in the anabolic hormone testosterone may be viewed as an adaptive attempt to reduce energy consumption and conserve substrates for more vital functions, whereas elevation of the anti-inflammatory and vasoprotective estrogen could be protective during acute illness (Gilliver, 2010). The mechanism behind these acute changes is not clear. An immediate suppression of androgen production in Leydig cells by cytokines (Guo et al., 1990) or an increased aromatization of androgens to estrogens (Spratt et al., 2006) is potentially involved.

The gonadal axis during prolonged critical illness

When critical illness is prolonged circulating levels of testosterone become extremely low or even undetectable, now in the presence of suppressed mean LH concentrations and pulsatile LH release (Vogel et al., 1985; Woolf et al., 1985; Semple et al., 1987; Spratt et al., 1993a; Van den Berghe et al., 1994b, 2001a). In contrast with the acute phase of illness, estrogen levels reach low-normal levels (Spratt et al., 1993a; Fourrier et al., 1994; Van den Berghe et al., 2001a).

The reduced pulsatility in LH release in prolonged illness points to an impaired compensatory GnRH response under reduced testosterone negative feedback (Van den Berghe et al., 1994b). However, exogenous GnRH is only partially and transiently effective in correcting these abnormalities, which indicates that the changes in LH and testosterone result from combined central and peripheral defects within the male gonadal axis (Van den Berghe et al., 2001a). Cytokine reduced androgen production and alterations in the aromatization of androgens might still be involved (Guo et al., 1990; Spratt et al., 2006).

Therapeutic potential

No conclusive clinical benefit has been demonstrated for androgen treatment in prolonged critical illness (Angele et al., 1998; Ferrando et al., 2001). Admission estradiol is a marker of injury severity and a predictor of death in the critically ill patient, but whether this is cause or consequence is unclear (Dossett et al., 2008). Animal models have demonstrated that estradiol improved survival after trauma/hemorrhage and sepsis, which raised clinical interest for intravenous estrogen therapy in critically ill patients (Wigginton et al., 2010). However, results from clinical trials in critically ill humans are lacking. A safer approach might be the administration of the hypothalamic-releasing factor GnRH. Indeed, exogenous pulsatile GnRH administration partially overcomes the hypogonadotropic hypogonadism in prolonged critically ill men (Van den Berghe et al., 2001a). Clearly, further studies are needed to clarify the therapeutic potential of these hormones.

THE LACTOTROPIC AXIS

Unlike the secretion of the other pituitary hormones, prolactin (PRL) secretion is primarily under a tonic inhibitory control by the hypothalamus (Freeman et al., 2000). PRL has a high basal expression and secretion which is predominantly inhibited by dopamine, whereas estrogen, TRH, but also IGF-1 can stimulate PRL gene expression and secretion (Lamberts and Macleod, 1990). The diurnal pulsatile pattern of PRL secretion is mainly under the control of estrogens (Juneja et al., 1991). The main function of prolactin is to stimulate lactation, but it is also known as stress hormone with immune-enhancing properties (Russell, 1989).

The lactotropic axis in acute critical illness

In response to acute stress, prolactin levels rise (Fig. 8.2) (Noel et al., 1972; Van den Berghe et al., 1998b). The PRL surge following acute stress is thought to result from a general increase in the adrenergic activity of the hypothalamus, which inhibits dopaminergic neurons and leads to the secretion of PRL-releasing factors such as vasoactive intestinal peptide and oxytocin (Reis et al., 1998).

The lactotropic axis during prolonged critical illness

The basal and pulsatile PRL release becomes suppressed in the prolonged phase of critical illness (Fig. 8.2) (Van den Berghe et al., 1997b, 1998b; Weekers et al., 2002; Felmet et al., 2005). The increased dopaminergic tone which is present during critical illness might inhibit PRL production and secretion (Benedict and Grahame-Smith, 1978; Freeman et al., 2000; Ben-Jonathan and Hnasko, 2001). The low estrogen, somatostatin, TRH, or IGF-I levels present during prolonged critical illness might also play a role (Vale et al., 1973; Van den Berghe et al., 1994b, 2001a; Stefaneanu et al., 1999; Fruchtman et al., 2000; Mesotten and Van den Berghe, 2006).

Whether the blunted prolactin secretion contributes to the immune suppression or increased susceptibility to infection associated with prolonged critical illness remains unknown. Exogenous dopamine, a frequently used inotropic drug, not only further suppressed prolactin secretion, but concomitantly aggravated T lymphocyte dysfunction and disturbed neutrophil chemotaxis (Devins et al., 1992; Van den Berghe et al., 1994a). Prolonged hypoprolactinemia was more common in critically ill children with multiorgan failure and was associated independently with prolonged lymphopenia and lymphoid depletion (Felmet et al., 2005).

Therapeutic potential

The high PRL levels observed during the acute face of critical illness could be considered beneficial as these could potentially boost the immune system. In contrast, the low PRL levels observed during the prolonged phase of critical illness could be linked to the ongoing immune suppression and increased susceptibility to infection, and therefore might warrant intervention. However, although PRL displayed beneficial immunomodulatory effects in animal models of severe illness, no clinical trials have examined PRL treatment during human critical illness (Zellweger et al., 1996, 1998). Treatment with the hypothalamic-releasing factor GHRH and with the synthetic GH secretagogue GHRP-2 induced a small but significant increase in basal PRL release without affecting pulsatile secretion (Van den Berghe et al., 1997b).

THE ADRENAL AXIS

The hypothalamic corticotropin-releasing hormone (CRH) controls synthesis and secretion of adrenocorticotropic hormone (ACTH) from the corticotropes in the pituitary. CRH regulates conversion of the prohormone pro-opiomelanocortin (POMC) to ACTH (Kageyama and Suda, 2009). ACTH stimulates the adrenal cortex to produce cortisol (Cooper and Stewart, 2003). Cortisol itself exerts a negative feedback control on both hormones. In a stress-free healthy human, cortisol is released in a circadian rhythm. Only free cortisol is biologically active, but more than 90% of circulating cortisol is bound to binding proteins, predominantly corticosteroid-binding globulin (CBG) but also albumin (Burchard, 2001).

The adrenal axis in acute critical illness

In the acute phase of critical illness, cortisol levels rise in response to an increased release of CRH and ACTH (Vermes and Beishuizen, 2001). However, in a recent study, plasma ACTH concentrations were found to be suppressed already from ICU admission in a heterogeneous critically ill patient population (Boonen et al., 2013). The diurnal variation in cortisol secretion is lost (Cooper and Stewart, 2003). Moreover, CBG concentrations and CBG-binding affinity for cortisol are substantially decreased in the critically ill, in part due to elastase-induced cleavage, resulting in proportionally much higher levels of free cortisol (Pemberton et al., 1988; Hammond et al., 1990; Beishuizen et al., 2001; Hamrahian et al., 2004; Vanhorebeek et al., 2006). An appropriate activation of the hypothalamic–pituitary–adrenal axis is essential for survival. The elevation of cortisol will acutely improve the hemodynamic status of the patient by induction of fluid retention and sensitization of the vasopressor response to catecholamines, cortisol will promote the acute provision of energy by mobilizing carbohydrate, fat, and protein stores to release glucose, free fatty acids, and amino acids, and it can protect against excessive inflammation by suppression of the inflammatory response (Van den Berghe et al., 1998b; Marik and Zaloga, 2002).

The adrenal axis during prolonged critical illness

During the chronic phase of critical illness, high cortisol levels are sustained while ACTH levels remain low (Bornstein and Chrousos, 1999; Vermes and Beishuizen, 2001; Boonen et al., 2013). CBG levels recover in the chronic phase of illness (Beishuizen et al., 2001; Vanhorebeek et al., 2006).

This dissociation of ACTH and cortisol release suggests the presence of ACTH-independent pathways (Bornstein et al., 2008). A changed adrenal sensitivity for ACTH might be involved, or ACTH-independent activators for cortisol production such as cytokines, neuropeptides, or vasoactive factors might stimulate high cortisol levels (Bornstein et al., 2008). Alternatively, a reduced cortisol removal might theoretically also lead to the observed cortisol-ACTH dissociation. A recent study using state-of-the-art cortisol tracer technique showed that daytime cortisol production was only mildly elevated in critically ill patients (Boonen et al., 2013). Cortisol breakdown, on the other hand, was substantially reduced, and was attributable to suppressed expression and activity of A-ring reductases in the liver and by suppressed activity of 11β-hydroxysteroid dehydrogenase type 2 in the kidney (Boonen et al., 2013).

Whether the continued elevation in cortisol is beneficial in prolonged critical illness remains uncertain. It could theoretically be involved in the increased susceptibility to infectious complications associated with prolonged critical illness. On the other hand, sustaining elevated cortisol levels by reducing its breakdown can be interpreted as an energy-sparing mechanism of the organism in times of sustained stress. Furthermore, this peripheral regulation of cortisol levels might also explain why ACTH levels remain low, through a continuous negative feedback inhibition at the hypothalamic–pituitary level. However, sustained low ACTH levels might induce adrenal atrophy, which predisposes to adverse outcome (Barquist and Kirton, 1997).

Therapeutic potential

Both very high and low cortisol levels have been associated with poor outcome (Finlay and McKee, 1982; Rothwell et al., 1991; Span et al., 1992; Annane et al., 2000, 2002; Sam et al., 2004). High cortisol levels point to more severe stress, whereas low levels at baseline and/or upon ACTH stimulation have been interpreted as indicating an inability to sufficiently respond to stress. This effect is termed "relative adrenal insufficiency". Although the concept of relative adrenal insufficiency has been related to increased morbidity and mortality in several studies, its actual existence remains controversial and no consensus has been reached on its definition.

Initial trials studying administration of high doses of glucocorticoids have clearly shown that this strategy is ineffective, and perhaps even harmful (Minneci et al., 2004; Roberts et al., 2004). In contrast, studies using lower-dose glucocorticoid therapy – although still supraphysiologic – for presumed relative adrenal insufficiency suggested beneficial effects, at least in patients with septic shock (Annane et al., 2002; Minneci et al.,

2004). More recently, a randomized controlled trial on "low-dose" glucocorticoid replacement therapy in patients with septic shock could not demonstrate a beneficial effect on outcome (Sprung et al., 2008). Importantly, administration of hydrocortisone in the so-called "replacement dose" resulted in several-fold higher total and free cortisol levels, indicating the need of a re-evaluation of the doses used (Vanhorebeek et al., 2006).

CONCLUSIONS

In conclusion, the hypothalamic–pituitary–peripheral axes are substantially altered by critical illness. During the acute phase of illness, the pituitary is actively secreting, whereas many peripheral effector organs become resistant and concentrations of anabolic target hormones are low. In contrast, prolonged, intensive care-dependent critical illness is characterized by a uniform suppression of the neuroendocrine axes, predominantly of hypothalamic origin, which can contribute to low serum levels of the respective target-organ hormones. The onset of recovery is characterized by restored sensitivity of the anterior pituitary to reduced feedback control.

The acute – predominantly peripheral – adaptations are probably beneficial in the struggle for short-term survival whereas the chronic neuroendocrine suppression may no longer be beneficial, as they could mediate the general wasting syndrome of prolonged critical illness and hereby contribute to adverse outcome. Attempts to reverse the neuroendocrine alterations with hormonal therapies have demonstrated that the timing, the choice of hormone and corresponding dosage are of crucial importance. As most hypothalamic–pituitary axes show a decreased activity during prolonged critical illness, treatment with hypothalamic-releasing factors may be more effective and safer than administration of pituitary or peripheral hormones, because this allows functioning of the respective negative feedback mechanisms and could thereby avoid toxic side-effects of peripheral hormone activity.

Infusions of GH secretagogues, TRH or GnRH have been shown to reactivate the corresponding pituitary axes, resulting in elevated levels of the peripheral effector hormones. Concomitant infusion of GHRP-2 and TRH reactivated both the somatotropic and thyrotropic axes, but avoided the rise of inactive rT_3 levels seen with TRH alone (Van den Berghe et al., 1998a). This combined intervention was associated with a reduced hypercatabolism and stimulated anabolism (Van den Berghe et al., 1999). Additional coactivation of the gonadal axis by administering GnRH together with GHRP-2 and TRH in prolonged critically ill men at least partially restored the three pituitary axes and appeared to more effectively induce an anabolic response (Van den Berghe et al., 2002). These results emphasize the interactions between the different endocrine axes and the potential of jointly correcting all hypothalamic–pituitary defects instead of a single hormone treatment (Van den Berghe et al., 1998a, 1999, 2002). However, no outcome studies adequately powered for clinically relevant endpoints have yet been performed.

References

Acker CG, Singh AR, Flick RP et al. (2000). A trial of thyroxine in acute renal failure. Kidney Int 57: 293–298.

Angele MK, Ayala A, Cioffi WG et al. (1998). Testosterone: the culprit for producing splenocyte immune depression after trauma hemorrhage. Am J Physiol 274: C1530–C1536.

Annane D, Sebille V, Troche G et al. (2000). A 3-level prognostic classification in septic shock based on cortisol levels and cortisol response to corticotropin. JAMA 283: 1038–1045.

Annane D, Sebille V, Charpentier C et al. (2002). Effect of treatment with low doses of hydrocortisone and fludrocortisone on mortality in patients with septic shock. JAMA 288: 862–871.

Arnold J, Campbell IT, Samuels TA et al. (1993). Increased whole body protein breakdown predominates over increased whole body protein synthesis in multiple organ failure. Clin Sci (Lond) 84: 655–661.

Baricevic I, Jones DR, Dordevic B et al. (2007). Differential influence of open surgery and sepsis on the circulating insulin-like growth factors and their binding proteins as representative metabolic markers. Clin Biochem 40: 1122–1128.

Barquist E, Kirton O (1997). Adrenal insufficiency in the surgical intensive care unit patient. J Trauma 42: 27–31.

Baxter RC (2001). Changes in the IGF-IGFBP axis in critical illness. Best Pract Res Clin Endocrinol Metab 15: 421–434.

Baxter RC, Martin JL (1989). Structure of the Mr 140,000 growth hormone-dependent insulin-like growth factor binding protein complex: determination by reconstitution and affinity-labeling. Proc Natl Acad Sci U S A 86: 6898–6902.

Baxter RC, Hawker FH, To C et al. (1998). Thirty-day monitoring of insulin-like growth factors and their binding proteins in intensive care unit patients. Growth Horm IGF Res 8: 455–463.

Beishuizen A, Thijs LG, Vermes I (2001). Patterns of corticosteroid-binding globulin and the free cortisol index during septic shock and multitrauma. Intensive Care Med 27: 1584–1591.

Benedict CR, Grahame-Smith DG (1978). Plasma noradrenaline and adrenaline concentrations and dopamine-beta-hydroxylase activity in patients with shock due to septicaemia, trauma and haemorrhage. Q J Med 47: 1–20.

Ben-Jonathan N, Hnasko R (2001). Dopamine as a prolactin (PRL) inhibitor. Endocr Rev 22: 724–763.

Bettendorf M, Schmidt KG, Grulich-Henn J et al. (2000). Triiodothyronine treatment in children after cardiac surgery: a double-blind, randomised, placebo-controlled study. Lancet 356: 529–534.

Bianco AC, Salvatore D, Gereben B et al. (2002). Biochemistry, cellular and molecular biology, and physiological roles of the iodothyronine selenodeiodinases. Endocr Rev 23: 38–89.

Bodenner DL, Mroczynski MA, Weintraub BD et al. (1991). A detailed functional and structural analysis of a major thyroid hormone inhibitory element in the human thyrotropin beta-subunit gene. J Biol Chem 266: 21666–21673.

Boelen A, Platvoet-ter Schiphorst MC, Bakker O et al. (1995). The role of cytokines in the lipopolysaccharide-induced sick euthyroid syndrome in mice. J Endocrinol 146: 475–483.

Boelen A, Wiersinga WM, Fliers E (2008). Fasting-induced changes in the hypothalamus–pituitary–thyroid axis. Thyroid 18: 123–129.

Boonen E, Vervenne H, Meersseman P et al. (2013). Reduced cortisol metabolism during critical illness. N Engl J Med 368: 1477–1488.

Bornstein SR, Chrousos GP (1999). Clinical review 104: adrenocorticotropin (ACTH)- and non-ACTH-mediated regulation of the adrenal cortex: neural and immune inputs. J Clin Endocrinol Metab 84: 1729–1736.

Bornstein SR, Engeland WC, Ehrhart-Bornstein M et al. (2008). Dissociation of ACTH and glucocorticoids. Trends Endocrinol Metab 19: 175–180.

Brent GA, Hershman JM (1986). Thyroxine therapy in patients with severe nonthyroidal illnesses and low serum thyroxine concentration. J Clin Endocrinol Metab 63: 1–8.

Burchard K (2001). A review of the adrenal cortex and severe inflammation: quest of the "eucorticoid" state. J Trauma 51: 800–814.

Christeff N, Benassayag C, Carli-Vielle C et al. (1988). Elevated oestrogen and reduced testosterone levels in the serum of male septic shock patients. J Steroid Biochem 29: 435–440.

Cooper MS, Stewart PM (2003). Corticosteroid insufficiency in acutely ill patients. N Engl J Med 348: 727–734.

De Groot LJ (1999). Dangerous dogmas in medicine: the nonthyroidal illness syndrome. J Clin Endocrinol Metab 84: 151–164.

Debaveye Y, Ellger B, Mebis L et al. (2005). Tissue deiodinase activity during prolonged critical illness: effects of exogenous thyrotropin-releasing hormone and its combination with growth hormone-releasing peptide-2. Endocrinology 146: 5604–5611.

Defalque D, Brandt N, Ketelslegers JM et al. (1999). GH insensitivity induced by endotoxin injection is associated with decreased liver GH receptors. Am J Physiol 276: E565–E572.

Devins SS, Miller A, Herndon BL et al. (1992). Effects of dopamine on T-lymphocyte proliferative responses and serum prolactin concentrations in critically ill patients. Crit Care Med 20: 1644–1649.

Dong Q, Hawker F, McWilliam D et al. (1992). Circulating immunoreactive inhibin and testosterone levels in men with critical illness. Clin Endocrinol (Oxf) 36: 399–404.

Dossett LA, Swenson BR, Heffernan D et al. (2008). High levels of endogenous estrogens are associated with death in the critically injured adult. J Trauma 64: 580–585.

Duska F, Fric M, Waldauf P et al. (2008). Frequent intravenous pulses of growth hormone together with glutamine supplementation in prolonged critical illness after multiple trauma: effects on nitrogen balance, insulin resistance, and substrate oxidation. Crit Care Med 36: 1707–1713.

Faglia G, Ferrari C, Beck-Peccoz P et al. (1973). Reduced plasma thyrotropin response to thyrotropin releasing hormone after dexamethasone administration in normal subjects. Horm Metab Res 5: 289–292.

Felmet KA, Hall MW, Clark RS et al. (2005). Prolonged lymphopenia, lymphoid depletion, and hypoprolactinemia in children with nosocomial sepsis and multiple organ failure. J Immunol 174: 3765–3772.

Ferrando AA, Sheffield-Moore M, Wolf SE et al. (2001). Testosterone administration in severe burns ameliorates muscle catabolism. Crit Care Med 29: 1936–1942.

Finlay WE, McKee JI (1982). Serum cortisol levels in severely stressed patients. Lancet 1: 1414–1415.

Fliers E, Guldenaar SE, Wiersinga WM et al. (1997). Decreased hypothalamic thyrotropin-releasing hormone gene expression in patients with nonthyroidal illness. J Clin Endocrinol Metab 82: 4032–4036.

Fourrier F, Jallot A, Leclerc L et al. (1994). Sex steroid hormones in circulatory shock, sepsis syndrome, and septic shock. Circ Shock 43: 171–178.

Freeman ME, Kanyicska B, Lerant A et al. (2000). Prolactin: structure, function, and regulation of secretion. Physiol Rev 80: 1523–1631.

Friesema EC, Jansen J, Visser TJ (2005). Thyroid hormone transporters. Biochem Soc Trans 33: 228–232.

Fruchtman S, Jackson L, Borski R (2000). Insulin-like growth factor I disparately regulates prolactin and growth hormone synthesis and secretion: studies using the teleost pituitary model. Endocrinology 141: 2886–2894.

Gardner DF, Kaplan MM, Stanley CA et al. (1979). Effect of tri-iodothyronine replacement on the metabolic and pituitary responses to starvation. N Engl J Med 300: 579–584.

Gilliver SC (2010). Sex steroids as inflammatory regulators. J Steroid Biochem Mol Biol 120: 105–115.

Goldsmith RE, Stanbury JB, Brownell GL (1951). The effect of thyrotropin on the release of hormone from the human thyroid. J Clin Endocrinol Metab 11: 1079–1094.

Guo H, Calkins JH, Sigel MM et al. (1990). Interleukin-2 is a potent inhibitor of Leydig cell steroidogenesis. Endocrinology 127: 1234–1239.

Hammond GL, Smith CL, Paterson NA et al. (1990). A role for corticosteroid-binding globulin in delivery of cortisol to activated neutrophils. J Clin Endocrinol Metab 71: 34–39.

Hamrahian AH, Oseni TS, Arafah BM (2004). Measurements of serum free cortisol in critically ill patients. N Engl J Med 350: 1629–1638.

Hermansson M, Wickelgren RB, Hammarqvist F et al. (1997). Measurement of human growth hormone receptor messenger ribonucleic acid by a quantitative polymerase chain reaction-based assay: demonstration of reduced expression after elective surgery. J Clin Endocrinol Metab 82: 421–428.

Herndon DN, Barrow RE, Kunkel KR et al. (1990). Effects of recombinant human growth hormone on donor-site healing in severely burned children. Ann Surg 212: 424–429.

Hill NE, Murphy KG, Singer M (2012). Ghrelin, appetite and critical illness. Curr Opin Crit Care 18: 199–205.

Howard AD, Feighner SD, Cully DF et al. (1996). A receptor in pituitary and hypothalamus that functions in growth hormone release. Science 273: 974–977.

Juneja HS, Karanth S, Dutt A et al. (1991). Diurnal variations and temporal coupling of bioactive and immunoactive luteinizing hormone, prolactin, testosterone and 17-beta-estradiol in adult men. Horm Res 35: 89–94.

Kageyama K, Suda T (2009). Role and action in the pituitary corticotroph of corticotropin-releasing factor (CRF) in the hypothalamus. Peptides 30: 810–816.

Kojima M, Hosoda H, Date Y et al. (1999). Ghrelin is a growth-hormone-releasing acylated peptide from stomach. Nature 402: 656–660.

Kupfer SR, Underwood LE, Baxter RC et al. (1993). Enhancement of the anabolic effects of growth hormone and insulin-like growth factor I by use of both agents simultaneously. J Clin Invest 91: 391–396.

Lamberts SW, Macleod RM (1990). Regulation of prolactin secretion at the level of the lactotroph. Physiol Rev 70: 279–318.

Lim CF, Docter R, Visser TJ et al. (1993). Inhibition of thyroxine transport into cultured rat hepatocytes by serum of nonuremic critically ill patients: effects of bilirubin and nonesterified fatty acids. J Clin Endocrinol Metab 76: 1165–1172.

Maldonado LS, Murata GH, Hershman JM et al. (1992). Do thyroid function tests independently predict survival in the critically ill? Thyroid 2: 119–123.

Marik PE, Zaloga GP (2002). Adrenal insufficiency in the critically ill: a new look at an old problem. Chest 122: 1784–1796.

Mebis L, Van den Berghe G (2011). Thyroid axis function and dysfunction in critical illness. Best Pract Res Clin Endocrinol Metab 25: 745–757.

Mebis L, Langouche L, Visser TJ et al. (2007). The type II iodothyronine deiodinase is up-regulated in skeletal muscle during prolonged critical illness. J Clin Endocrinol Metab 92: 3330–3333.

Mebis L, Debaveye Y, Ellger B et al. (2009a). Changes in the central component of the hypothalamus–pituitary–thyroid axis in a rabbit model of prolonged critical illness. Crit Care 13: R147.

Mebis L, Paletta D, Debaveye Y et al. (2009b). Expression of thyroid hormone transporters during critcal illness. Eur J Endocrinol 161: 243–250.

Mebis L, Eerdekens A, Guiza F et al. (2012). Contribution of nutritional deficit to the pathogenesis of the nonthyroidal illness syndrome in critical illness: a rabbit model study. Endocrinology 153: 973–984.

Mesotten D, Van den Berghe G (2006). Changes within the GH/IGF-I/IGFBP axis in critical illness. Crit Care Clin 22: 17–28, v.

Mesotten D, Wouters PJ, Peeters RP et al. (2004). Regulation of the somatotropic axis by intensive insulin therapy during protracted critical illness. J Clin Endocrinol Metab 89: 3105–3113.

Michalaki M, Vagenakis AG, Makri M et al. (2001). Dissociation of the early decline in serum T(3) concentration and serum IL-6 rise and TNFalpha in nonthyroidal illness syndrome induced by abdominal surgery. J Clin Endocrinol Metab 86: 4198–4205.

Minneci PC, Deans KJ, Banks SM et al. (2004). Meta-analysis: the effect of steroids on survival and shock during sepsis depends on the dose. Ann Intern Med 141: 47–56.

Nematy M, O'Flynn JE, Wandrag L et al. (2006). Changes in appetite related gut hormones in intensive care unit patients: a pilot cohort study. Crit Care 10: R10.

Noel GL, Suh HK, Stone JG et al. (1972). Human prolactin and growth hormone release during surgery and other conditions of stress. J Clin Endocrinol Metab 35: 840–851.

Ohlsson C, Mohan S, Sjogren K et al. (2009). The role of liver-derived insulin-like growth factor-I. Endocr Rev 30: 494–535.

Oudemans-van Straaten HM, Bosman RJ, Treskes M et al. (2001). Plasma glutamine depletion and patient outcome in acute ICU admissions. Intensive Care Med 27: 84–90.

Peeters RP, Wouters PJ, Kaptein E et al. (2003). Reduced activation and increased inactivation of thyroid hormone in tissues of critically ill patients. J Clin Endocrinol Metab 88: 3202–3211.

Peeters RP, Wouters PJ, van Toor H et al. (2005). Serum rT3 and T3/rT3 are prognostic markers in critically ill patients and are associated with post-mortem tissue deiodinase activities. J Clin Endocrinol Metab 90: 4559–4565.

Pemberton PA, Stein PE, Pepys MB et al. (1988). Hormone binding globulins undergo serpin conformational change in inflammation. Nature 336: 257–258.

Persani L, Terzolo M, Asteria C et al. (1995). Circadian variations of thyrotropin bioactivity in normal subjects and patients with primary hypothyroidism. J Clin Endocrinol Metab 80: 2722–2728.

Plymate SR, Vaughan GM, Mason AD et al. (1987). Central hypogonadism in burned men. Horm Res 27: 152–158.

Popovic V, Miljic D, Micic D et al. (2003). Ghrelin main action on the regulation of growth hormone release is exerted at hypothalamic level. J Clin Endocrinol Metab 88: 3450–3453.

Popovic V, Miljic D, Pekic S et al. (2005). Low plasma ghrelin level in gastrectomized patients is accompanied by enhanced sensitivity to the ghrelin-induced growth hormone release. J Clin Endocrinol Metab 90: 2187–2191.

Ray DC, Macduff A, Drummond GB et al. (2002). Endocrine measurements in survivors and non-survivors from critical illness. Intensive Care Med 28: 1301–1308.

Reis FM, Ribeiro-de-Oliveira JA, Machado LJ et al. (1998). Plasma prolactin and glucose alterations induced by surgical stress: a single or dual response? Exp Physiol 83: 1–10.

Roberts I, Yates D, Sandercock P et al. (2004). Effect of intravenous corticosteroids on death within 14 days in 10008 adults with clinically significant head injury (MRC CRASH trial): randomised placebo-controlled trial. Lancet 364: 1321–1328.

Rodas PC, Rooyackers O, Hebert C et al. (2012). Glutamine and glutathione at ICU admission in relation to outcome. Clin Sci (Lond) 122: 591–597.

Rodriguez-Perez A, Palos-Paz F, Kaptein E et al. (2008). Identification of molecular mechanisms related to nonthyroidal illness syndrome in skeletal muscle and adipose tissue from patients with septic shock. Clin Endocrinol (Oxf) 68: 821–827.

Romijn JA, Wiersinga WM (1990). Decreased nocturnal surge of thyrotropin in nonthyroidal illness. J Clin Endocrinol Metab 70: 35–42.

Ross R, Miell J, Freeman E et al. (1991). Critically ill patients have high basal growth hormone levels with attenuated oscillatory activity associated with low levels of insulin-like growth factor-I. Clin Endocrinol (Oxf) 35: 47–54.

Rothwell PM, Lawler PG (1995). Prediction of outcome in intensive care patients using endocrine parameters. Crit Care Med 23: 78–83.

Rothwell PM, Udwadia ZF, Lawler PG (1991). Cortisol response to corticotropin and survival in septic shock. Lancet 337: 582–583.

Rothwell PM, Udwadia ZF, Lawler PG (1993). Thyrotropin concentration predicts outcome in critical illness. Anaesthesia 48: 373–376.

Russell DH (1989). New aspects of prolactin and immunity: a lymphocyte-derived prolactin-like product and nuclear protein kinase C activation. Trends Pharmacol Sci 10: 40–44.

Sam S, Corbridge TC, Mokhlesi B et al. (2004). Cortisol levels and mortality in severe sepsis. Clin Endocrinol (Oxf) 60: 29–35.

Schlienger JL, Sapin R, Capgras T et al. (1991). Evaluation of thyroid function after myocardial infarction. Ann Endocrinol (Paris) 52: 283–288.

Schuetz P, Muller B, Nusbaumer C et al. (2009). Circulating levels of GH predict mortality and complement prognostic scores in critically ill medical patients. Eur J Endocrinol 160: 157–163.

Semple CG, Robertson WR, Mitchell R et al. (1987). Mechanisms leading to hypogonadism in men with burns injuries. Br Med J (Clin Res Ed) 295: 403–407.

Span LF, Hermus AR, Bartelink AK et al. (1992). Adrenocortical function: an indicator of severity of disease and survival in chronic critically ill patients. Intensive Care Med 18: 93–96.

Spratt DI, Cox P, Orav J et al. (1993a). Reproductive axis suppression in acute illness is related to disease severity. J Clin Endocrinol Metab 76: 1548–1554.

Spratt DI, Longcope C, Cox PM et al. (1993b). Differential changes in serum concentrations of androgens and estrogens (in relation with cortisol) in postmenopausal women with acute illness. J Clin Endocrinol Metab 76: 1542–1547.

Spratt DI, Morton JR, Kramer RS et al. (2006). Increases in serum estrogen levels during major illness are caused by increased peripheral aromatization. Am J Physiol Endocrinol Metab 291: E631–E638.

Sprung CL, Annane D, Keh D et al. (2008). Hydrocortisone therapy for patients with septic shock. N Engl J Med 358: 111–124.

Stefaneanu L, Powell-Braxton L, Won W et al. (1999). Somatotroph and lactotroph changes in the adenohypophyses of mice with disrupted insulin-like growth factor I gene. Endocrinology 140: 3881–3889.

Takala J, Ruokonen E, Webster NR et al. (1999). Increased mortality associated with growth hormone treatment in critically ill adults. N Engl J Med 341: 785–792.

Tannenbaum GS, Ling N (1984). The interrelationship of growth hormone (GH)-releasing factor and somatostatin in generation of the ultradian rhythm of GH secretion. Endocrinology 115: 1952–1957.

Tannenbaum GS, Epelbaum J, Bowers CY (2003). Interrelationship between the novel peptide ghrelin and somatostatin/growth hormone-releasing hormone in regulation of pulsatile growth hormone secretion. Endocrinology 144: 967–974.

Thijssen-Timmer DC, Peeters RP, Wouters P et al. (2007). Thyroid hormone receptor isoform expression in livers of critically ill patients. Thyroid 17: 105–112.

Timmins AC, Cotterill AM, Hughes SC et al. (1996). Critical illness is associated with low circulating concentrations of insulin-like growth factors-I and -II, alterations in insulin-like growth factor binding proteins, and induction of an insulin-like growth factor binding protein 3 protease. Crit Care Med 24: 1460–1466.

Vale W, Blackwell R, Grant G et al. (1973). TRF and thyroid hormones on prolactin secretion by rat anterior pituitary cells in vitro. Endocrinology 93: 26–33.

Van den Berghe G (2000). Novel insights into the neuroendocrinology of critical illness. Eur J Endocrinol 143: 1–13.

Van den Berghe G, de Zegher F, Lauwers P (1994a). Dopamine and the sick euthyroid syndrome in critical illness. Clin Endocrinol (Oxf) 41: 731–737.

Van den Berghe G, de Zegher F, Lauwers P et al. (1994b). Luteinizing hormone secretion and hypoandrogenaemia in critically ill men: effect of dopamine. Clin Endocrinol (Oxf) 41: 563–569.

Van den Berghe G, de Zegher F, Vlasselaers D et al. (1996a). Thyrotropin-releasing hormone in critical illness: from a dopamine-dependent test to a strategy for increasing low serum triiodothyronine, prolactin, and growth hormone concentrations. Crit Care Med 24: 590–595.

Van den Berghe G, de Zegher F, Bowers CY et al. (1996b). Pituitary responsiveness to GH-releasing hormone, GH-releasing peptide-2 and thyrotrophin-releasing hormone in critical illness. Clin Endocrinol (Oxf) 45: 341–351.

Van den Berghe G, de Zegher F, Veldhuis JD et al. (1997a). The somatotropic axis in critical illness: effect of continuous growth hormone (GH)-releasing hormone and GH-releasing peptide-2 infusion. J Clin Endocrinol Metab 82: 590–599.

Van den Berghe G, de Zegher F, Veldhuis JD et al. (1997b). Thyrotrophin and prolactin release in prolonged critical illness: dynamics of spontaneous secretion and effects of growth hormone-secretagogues. Clin Endocrinol (Oxf) 47: 599–612.

Van den Berghe G, de Zegher F, Baxter RC et al. (1998a). Neuroendocrinology of prolonged critical illness: effects of exogenous thyrotropin-releasing hormone and its combination with growth hormone secretagogues. J Clin Endocrinol Metab 83: 309–319.

Van den Berghe G, de Zegher F, Bouillon R (1998b). Clinical review 95: acute and prolonged critical illness as different neuroendocrine paradigms. J Clin Endocrinol Metab 83: 1827–1834.

Van den Berghe G, Wouters P, Weekers F et al. (1999). Reactivation of pituitary hormone release and metabolic improvement by infusion of growth hormone-releasing peptide and thyrotropin-releasing hormone in patients with protracted critical illness. J Clin Endocrinol Metab 84: 1311–1323.

Van den Berghe G, Baxter RC, Weekers F et al. (2000). A paradoxical gender dissociation within the growth hormone/insulin-like growth factor I axis during protracted critical illness. J Clin Endocrinol Metab 85: 183–192.

Van den Berghe G, Weekers F, Baxter RC et al. (2001a). Five-day pulsatile gonadotropin-releasing hormone administration unveils combined hypothalamic–pituitary–gonadal defects underlying profound hypoandrogenism in men with prolonged critical illness. J Clin Endocrinol Metab 86: 3217–3226.

Van den Berghe G, Wouters P, Weekers F et al. (2001b). Intensive insulin therapy in critically ill patients. N Engl J Med 345: 1359–1367.

Van den Berghe G, Baxter RC, Weekers F et al. (2002). The combined administration of GH-releasing peptide-2 (GHRP-2), TRH and GnRH to men with prolonged critical illness evokes superior endocrine and metabolic effects compared to treatment with GHRP-2 alone. Clin Endocrinol (Oxf) 56: 655–669.

Van der Poll T, Romijn JA, Wiersinga WM et al. (1990). Tumor necrosis factor: a putative mediator of the sick euthyroid syndrome in man. J Clin Endocrinol Metab 71: 1567–1572.

Van der Poll T, Van Zee KJ, Endert E et al. (1995). Interleukin-1 receptor blockade does not affect endotoxin-induced changes in plasma thyroid hormone and thyrotropin concentrations in man. J Clin Endocrinol Metab 80: 1341–1346.

Van der Poll T, Endert E, Coyle SM et al. (1999). Neutralization of TNF does not influence endotoxininduced changes in thyroid hormone metabolism in humans. Am J Physiol 276: R357–R362.

Vanhorebeek I, Peeters RP, Vander Perre S et al. (2006). Cortisol response to critical illness: effect of intensive insulin therapy. J Clin Endocrinol Metab 91: 3803–3813.

Vermes I, Beishuizen A (2001). The hypothalamic–pituitary–adrenal response to critical illness. Best Pract Res Clin Endocrinol Metab 15: 495–511.

Voerman BJ, Strack van Schijndel RJ, Groeneveld AB et al. (1995). Effects of human growth hormone in critically ill nonseptic patients: results from a prospective, randomized, placebo-controlled trial. Crit Care Med 23: 665–673.

Vogel AV, Peake GT, Rada RT (1985). Pituitary-testicular axis dysfunction in burned men. J Clin Endocrinol Metab 60: 658–665.

Wang C, Chan V, Tse TF et al. (1978a). Effect of acute myocardial infarction on pituitary-testicular function. Clin Endocrinol (Oxf) 9: 249–253.

Wang C, Chan V, Yeung RT (1978b). Effect of surgical stress on pituitary-testicular function. Clin Endocrinol (Oxf) 9: 255–266.

Wartofsky L, Burman KD (1982). Alterations in thyroid function in patients with systemic illness: the "euthyroid sick syndrome". Endocr Rev 3: 164–217.

Weekers F, Van Herck E, Coopmans W et al. (2002). A novel in vivo rabbit model of hypercatabolic critical illness reveals a biphasic neuroendocrine stress response. Endocrinology 143: 764–774.

Weintraub BD, Wondisford FE, Farr EA et al. (1989). Pre-translational and post-translational regulation of TSH synthesis in normal and neoplastic thyrotrophs. Horm Res 32: 22–24.

Wigginton JG, Pepe PE, Idris AH (2010). Rationale for routine and immediate administration of intravenous estrogen for all critically ill and injured patients. Crit Care Med 38: S620–S629.

Woolf PD, Hamill RW, McDonald JV et al. (1985). Transient hypogonadotropic hypogonadism caused by critical illness. J Clin Endocrinol Metab 60: 444–450.

Wu R, Zhou M, Cui X et al. (2004). Upregulation of cardiovascular ghrelin receptor occurs in the hyperdynamic phase of sepsis. Am J Physiol Heart Circ Physiol 287: H1296–H1302.

Yen PM (2001). Physiological and molecular basis of thyroid hormone action. Physiol Rev 81: 1097–1142.

Zellweger R, Wichmann MW, Ayala A et al. (1996). Prolactin: a novel and safe immunomodulating hormone for the treatment of immunodepression following severe hemorrhage. J Surg Res 63: 53–58.

Zellweger R, Wichmann MW, Ayala A et al. (1998). Metoclopramide: a novel and safe immunomodulating agent for restoring the depressed macrophage immune function after hemorrhage. J Trauma 44: 70–77.

Ziegler TR, Young LS, Ferrari-Baliviera E et al. (1990). Use of human growth hormone combined with nutritional support in a critical care unit. JPEN J Parenter Enteral Nutr 14: 574–581.

Handbook of Clinical Neurology, Vol. 124 (3rd series)
Clinical Neuroendocrinology
E. Fliers, M. Korbonits, and J.A. Romijn, Editors

Chapter 9

Central regulation of the hypothalamo–pituitary–thyroid (HPT) axis: focus on clinical aspects

E. FLIERS[1]*, A. BOELEN[1], AND A.S.P. VAN TROTSENBURG[2]

[1]*Department of Endocrinology and Metabolism, Academic Medical Center, University of Amsterdam, Amsterdam, The Netherlands*

[2]*Department of Paediatric Endocrinology, Academic Medical Center, University of Amsterdam, Amsterdam, The Netherlands*

INTRODUCTION AND OUTLINE

The tripeptide thyrotropin-releasing hormone (TRH) was first isolated from the hypothalamus in the late 1960s, and its neuronal expression in various hypothalamic nuclei was demonstrated when immunocytochemistry became available for neuroanatomic studies in the 1970s. These studies helped establish the pivotal role for TRH neurons in the hypothalamic paraventricular nucleus (PVN) in the neuroendocrine regulation of the hypothalamo–pituitary–thyroid (HPT) axis. The demonstration of an inverse relationship between plasma thyroid hormone concentrations and TRH mRNA expression in the PVN during experimentally induced hyper- and hypothyroidism (Segerson et al., 1987) confirmed the central role of TRH neurons in the HPT axis as a classic neuroendocrine feedback loop. The neuroanatomic distribution of TRH neurons in the human hypothalamus was reported only in the 1990s (for review see Fliers et al., 2006). Meanwhile, studies in rodents had revealed a novel role for hypophysiotropic TRH neurons in the integration of energy metabolism, modulating plasma thyroid hormone concentrations according to the availability or absence of food. Additional metabolic effects of central TRH were uncovered, including effects on core body temperature and food intake (for review see Lechan and Fekete, 2006).

Recent experiments in rodents have shown profound intrahypothalamic effects of thyroid hormone on energy balance and hepatic glucose metabolism via the autonomic nervous system (ANS) (Klieverik et al., 2009; Lopez et al., 2010). Thus, both neuroendocrine and neural pathways are involved in communicating the intrahypothalamic effects of thyroid hormone within the organism.

Insufficient TRH release from the hypothalamus and/or insufficient thyroid-stimulating hormone (TSH) release from the anterior pituitary gland underlie the clinical picture of central hypothyroidism (CeH). In the Netherlands, CeH is detected within the framework of the Dutch neonatal screening program. Over the years, this strategy has yielded novel insights into the presentation, differential diagnosis, and management of congenital CeH. In adults, CeH occurs in the context of pituitary insufficiency resulting from the mass effects of pituitary macroadenomas, or it may develop as an adverse effect of transsphenoidal surgery or external radiotherapy. Central hyperthyroidism is a rare entity resulting from a TSH-secreting adenoma, or thyrotropinoma. Clinical aspects, including treatment, of these pituitary adenomas will be discussed in the final paragraph of this chapter.

HYPOTHALAMUS AND PITUITARY

The hypothalamic thyrotropin-releasing hormone neuron

Thyrotropin-releasing hormone (TRH) is a pivotal hypothalamic neuropeptide in the regulation of energy metabolism, both through its role in the regulation of TSH-producing cells in the anterior pituitary and through central effects on feeding behavior, thermogenesis and autonomic regulation. Thyroid hormone, principally T_3, inhibits TRH gene expression in hypophysiotropic neurons within the parvocellular subdivision of the hypothalamic PVN. Conversely, TRH gene expression increases in response to low serum T_4 and T_3 levels.

*Correspondence to: Eric Fliers, MD PhD, Academic Medical Center, University of Amsterdam, Meibergdreef 9, 1105AZ Amsterdam, The Netherlands. Tel: +31-20-5666071, Fax: +31-20-6917682, E-mail: e.fliers@amc.uva.nl

In this way, hypophysiotropic TRH neurons help to keep serum concentrations of thyroid hormone in a narrow range. Intrahypothalamic T_3 is produced locally, at least in part, from T_4 by the action of the deiodinating enzyme type 2 deiodinase (D2), which is expressed in tanycytes and other glial cells. This role for D2 is supported by observations in mice lacking D2 (D2 knockout mice), showing maintained TSH secretion in the presence of high serum T_4 (Schneider et al., 2001). T_3 regulates TRH gene expression via its nuclear receptor, the thyroid hormone receptor (TR), and the TRβ2 isoform is required for negative regulation of TRH gene expression (Abel et al., 2001).

Hypophysiotropic TRH neurons in the PVN receive monosynaptic input from leptin-responsive neurons in the arcuate nucleus, where the blood–brain barrier is absent. These neurons express either α-melanocyte-stimulating hormone (α-MSH) and cocaine- and amphetamine-regulated transcript (CART), inhibiting food intake and promoting energy expenditure, or neuropeptide Y (NPY) and agouti-related protein (AGRP), stimulating food intake and decreasing energy expenditure (for review see Lechan and Fekete, 2006). Various studies in rodents and humans have shown a major downregulation of the HPT axis at the central and peripheral level during fasting, presumably as a homeostatic mechanism to reduce catabolism. Both direct and indirect hypothalamic effects of decreased serum leptin levels during fasting, including increased NPY expression and increased local T_3 concentrations in the arcuate nucleus, contribute to downregulation of hypophysiotropic TRH neurons in the PVN (for review see Boelen et al., 2008).

Indirect evidence points to similar hypothalamic mechanisms in humans. A 60 hour fast decreases TSH serum concentrations in association with decreased TSH pulse amplitude (Romijn et al., 1990), while a more recent study showed that a 72 hour fast in healthy men induces a major decrease in serum TSH (and T_3), reflected by a decrease in integrated TSH area under the curve by over 70% and a loss of pulsatility characteristics. In the latter study, the administration of leptin blunted the fall in TSH secretion (but not in serum T_3), showing a role for leptin in the regulation of starvation-induced alterations in TSH levels and pulsatility in humans (Chan et al., 2003). A recent study using functional magnetic resonance imaging (fMRI) in young men showed a decrease in the hypothalamic blood oxygen level-dependent (BOLD) signal in response to an oral glucose load after a 60 hour fast (Snel et al., 2012). However, direct evidence for the involvement of hypophysiotropic TRH neurons in the human hypothalamus in the resetting of the HPT axis during fasting is not available.

In addition to starvation, the TRH neuron is inhibited during a variety of nonthyroidal illnesses including infection and inflammation. In these circumstances the resetting of the HPT axis may reflect a homeostatic mechanism to reduce metabolic rate and protein catabolism. For example, chronic inflammation in mice reduces hypothalamic TRHmRNA expression independently of diminished food intake (Boelen et al., 2006). Likewise, hypothalamic TRHmRNA expression as visualized by *in situ* hybridization is decreased in a rabbit model of prolonged critical illness (Mebis et al., 2009). In acute inflammation, cytokine-mediated induction of type 2 deiodinase in hypothalamic tanycytes has been proposed to represent a key factor inhibiting hypophysiotropic TRH neurons during inflammation and/or infection (Fekete et al., 2004).

Only a few studies have addressed TRH neurons in the human PVN. The first report of the neuroanatomic distribution of TRH neurons in the human hypothalamus was provided in the 1990s, and later studies using double immunocytochemistry showed these neurons to be innervated by NPY, AGRP and α-MSH. In addition to TRH neurons, dense TRH immunopositive fiber networks were observed in the human hypothalamus, e.g., in the perifornical area and the ventromedial nucleus, where TRH may be involved in central functions including temperature regulation (Fliers et al., 1994; Mihály et al., 2000). After the development of TRHmRNA *in situ* hybridization in human hypothalamus (Guldenaar et al., 1996), a quantitative *in situ* hybridization study reported decreased TRH mRNA expression in the PVN of patients with prolonged critical illness in close correlation with serum TSH and T_3 (Fliers et al., 1997). In the same period, clinical studies in critically ill patients showed that the administration of TRH partially restores serum concentrations of TSH, T_4 and T_3 (Van den Berghe et al., 1998; Fliers et al., 2001), supporting a role for hypothalamic TRH neurons in the decreased TSH release during critical illness.

Studies in rodents have shown that the central administration of TRH reduces food intake both in ad libitum feeding and in fasting animals. Although the neuroanatomic pathway for this anorectic effect of TRH is unknown, a role for nonhypophysiotropic TRH neurons in the anterior parvocellular subdivision of the PVN has been proposed (Lechan and Fekete, 2006). Additional central effects of TRH include the stimulation of thermogenesis. The intracerebroventricular administration of TRH increases rectal temperature in Syrian hamsters, and this thermogenic effect of central TRH administration can be blocked both by sympathetic denervation of brown adipose tissue and by the administration of β-adrenergic blockers, pointing to a role for the ANS (Chi and Lin, 1983; Shintani et al., 2005). Additional effects of central TRH via the autonomic nervous system include increased gastric motility and emptying, as well as stimulation of hepatic blood flow and of exocrine and endocrine pancreatic secretion. Although TRH neurons in the brainstem are known to be involved in these effects, a role for TRH neurons in the PVN has been proposed, thereby integrating and coordinating the effects of TRH on energy metabolism via the HPT axis and the gastrointestinal system through the autonomic nervous system (Lechan and Fekete, 2006).

Pituitary

Thyrotropin (thyroid-stimulating hormone (TSH)) is produced and secreted by the thyrotropic cells of the anterior pituitary and is the classic ligand for the TSH receptor (TSHR) in the thyroid. TSH is a heterodimeric glycoprotein consisting of an α subunit and a β subunit. The α subunit is shared with other glycoprotein hormones (i.e. follicle-stimulating hormone (FSH), luteinizing hormone (LH), and chorionic gonadotropin (CG)), whereas the TSHβ subunit is unique, determining the specificity of TSH. The human α and β subunit amino acid chains of TSH form three loops, the so-called cysteine knot. The gene for the human α subunit is located on chromosome 6, while the gene for the β subunit is located on chromosome 1. Mutations in both subunits have been reported and may affect TSH activity to a certain extent depending of the amino acid substitution (Szkudlinski et al., 2002). TSH plays a critical role in the regulation of the thyroid gland, as it activates the TSHR on the thyroid epithelial cell surface which stimulates the synthesis and secretion of thyroid hormones from the follicular thyrocytes.

Thyrotropic cells in the anterior pituitary constitute less than 10% of the total number of adenohypophysial cells and are preferentially located in the anteromedial and anterolateral portions of the pituitary. Binding of TRH to the TRH membrane receptor (TRHR) on the thyrotrophs stimulates synthesis and release of TSH. This response is dual, as the rapid release of presynthesized, stored TSH is followed by increased TSH gene expression and hormone production. The TRHR is a member of the seven-transmembrane-spanning, GTP-binding, G protein-coupled receptor family. TRH binds to the transmembrane helix 3 which results in activation of a G protein and subsequently in activation of IP_3 (inositol 1,4,5-triphosphate). IP_3 stimulates the release of intracellular Ca^{2+}, which is involved in rapid TSH release. Simultaneously, protein kinase C is activated which is involved in the increase in TSH production. TRH also stimulates the glycosylation of TSH, necessary for full biological activity of the hormone (Chiamolera and Wondisford, 2009).

The hypothalamic neuropeptide somatostatin inhibits TSH secretion. In rodents, infusion of somatostatin impairs the TSH surge in the morning and suppresses high serum TSH levels in hypothyroidism. Other inhibiting factors which may decrease serum TSH include dopamine, which inhibits TSH secretion, and glucocorticoids, which decrease the sensitivity of the pituitary to TRH. In contrast, estrogens increase the sensitivity of the pituitary to TRH. Experimental studies suggest an additional modulating role for pituitary peptides in the regulation of TSH secretion. For example, neuromedin B, a bombesin-related peptide which is highly concentrated in the pituitary and expressed in the thyrotrophic cell (Steel et al., 1988), inhibits TSH secretion (Ortiga-Carvalho et al., 1997). Finally, pituitary proteins including Pit-1 and RXRγ are expressed in thyrotropic cells and modulate TSHβ gene expression (Shupnik, 2000).

In the context of the HPT axis, TSH secretion is inhibited by circulating thyroid hormones. This so-called negative feedback regulation occurs via binding of T_3 to the TRβ2 expressed in the pituitary and – as in the hypothalamus – involves deiodination of the prohormone T_4 to the biologically active T_3. A number of changes in physiologic TSH regulation have been observed during illness and starvation. In prolonged critical illness, low circulating thyroid hormone levels appear even positively correlated with reduced pulsatile TSH secretion (Van den Berghe et al., 1997). In these conditions, the pituitary is unresponsive to low serum thyroid hormone concentrations, implying that the physiologic negative feedback mechanism of the HPT axis is overruled. Several mechanisms underlying this phenomenon have been proposed. First, increased pituitary type 2 deiodinase (D2) may be involved, as it may increase local T_3 concentrations, finally resulting in decreased pituitary TSHβ mRNA expression as has been observed in rodents (for review see Boelen et al., 2011). A role for T_3 is supported by the fact that lacking the TRβ gene prevents the illness-induced decrease of pituitary TSHβ mRNA in rodents (Boelen et al., 2009). The administration of exogenous TSH to mice during acute illness diminishes the illness-induced decrease in serum T_4 (Rocchi et al., 2007), supporting the concept that central downregulation of the HPT axis at the pituitary level is one of the determinants of decreased serum thyroid hormone levels in this setting.

Pulsatility and diurnal rhythm

A circadian pattern of TSH secretion in humans was first reported in the early 1990s, with lower plasma TSH levels during the daytime, an increase in the early evening and peak values ("nocturnal TSH surge") around the beginning of the sleep period (Brabant et al., 1990; Allan and Czeisler, 1994). Superimposed upon this diurnal rhythm is a clear ultradian rhythm as first reported by Parker et al. (1976). Later studies showed that the ultradian TSH release follows a high-frequency (approximately 10 pulses per hour) and low-amplitude (0.4 mU/L) pulsatile pattern superimposed on the low-frequency high-amplitude (1.0 mU/L) pattern of the circadian TSH rhythm (for a review see Kalsbeek and Fliers, 2013).

Searching for the neural origin of the diurnal TSH rhythm, experimental studies in rodents have focused on the suprachiasmatic nucleus (SCN), which is the biological clock of the brain. Immunocytochemical studies revealed that SCN fibers contact TRH neurons in the PVN, which may represent the anatomic basis for the daily rhythms in hypothalamic TRH mRNA and plasma TSH (for a review see Kalsbeek and Fliers, 2013). Additional studies using the retrograde transneuronal tracer pseudorabies virus (PRV) showed multisynaptic connections between the hypothalamic SCN and the thyroid gland via sympathetic and parasympathetic outflow. Interestingly, pre-autonomic neurons in the PVN, including TRH immunoreactive neurons, were labeled after injection of the retrograde tracer in the thyroid gland. Thus, the PVN is connected with the thyroid gland not only via the classic neuroendocrine pathway involving TRH and TSH, but also via a neural pathway involving the sympathetic and parasympathetic branch of the autonomic nervous system (Kalsbeek et al., 2000). Frequent blood sampling in rats revealed diurnal rhythms of TSH and thyroid hormones with peak levels during the light period and trough levels in the early dark period in these nocturnal animals. Thermic ablation of the SCN eliminated the diurnal peak in plasma TSH and thyroid hormones, confirming that the SCN drives the diurnal variation (Kalsbeek et al., 2000). Ablation of the SCN also abolished the diurnal rhythm of the activity of D2, which deiodinates the prohormone T_4 into the biologically active T_3, in a number of brain areas including the pineal gland, hypothalamus, and neocortex (Kalsbeek et al., 2005). These results indicate that the bioavailability of T_3 in various brain areas may show a diurnal rhythm that is driven by the SCN.

In humans, the physiologic meaning of the TSH rhythm is still elusive, and the mechanisms responsible for rhythmic TSH release are incompletely understood. In spite of the clear diurnal variation in plasma TSH levels, a diurnal rhythm in plasma T_4 and T_3 concentrations in humans is less obvious (Greenspan et al., 1986), which may be a consequence, at least in part, of the relatively long half-lives of these hormones in humans. Although variation in TSH sensitivity of the thyroid gland over the clock has been suggested as an explanation based on animal experimental studies (Kalsbeek et al., 2000), there is no evidence to support this concept in humans.

There are several physiologic (Behrends et al., 1998) and pathologic conditions that alter the TSH rhythm (for a review see Roelfsema and Veldhuis, 2013). For example, the nocturnal TSH surge is absent in major depression (Bartalena et al., 1990), suggestive of a role for hypothalamic TRH in HPT axis changes in depression. This notion was supported by the observation of decreased TRHmRNA expression in the PVN of patients with major depression (Alkemade et al., 2003). A decreased or even absent nocturnal TSH surge was also reported in a variety of nonthyroidal illnesses (NTI), occurring independently of pituitary responsiveness to TRH (Romijn and Wiersinga, 1990), which also pointed to a role for hypothalamic TRH. Finally, patients with critical illness show markedly decreased TSH pulsatility with an absent nocturnal TSH surge and decreased TSH pulse amplitude (Van den Berghe et al., 1997). Clinical studies in the intensive care setting showed that the continuous intravenous administration of TRH to patients with prolonged critical illness partially restored the serum concentrations of TSH as well as T_4 and T_3 (Van den Berghe et al., 1998; Fliers et al., 2001). Together with decreased TRHmRNA expression in the PVN of patients with critical illness (Fliers et al., 1997), these observations support a major role for hypothalamic TRH in the decreased TSH release during critical illness.

In addition to critical illness, the nocturnal TSH surge is also diminished in various endocrine pathologies. Although the diurnal TSH rhythm persists in primary hypothyroidism, the acrophase occurs earlier in most patients and both basal and pulsatile TSH secretion rates are increased resulting from increased burst mass with unaltered burst frequency (Roelfsema et al., 2010). A different pattern is seen in CeH, with a lower absolute and relative nocturnal rise in TSH (Adriaanse et al., 1992). Finally, physiologic conditions may affect the TSH rhythm. Examples of physiologic modulation are the decreased nocturnal TSH surge during fasting (Romijn et al., 1990) and the increased TSH surge during the first night of sleep deprivation (Goichot et al., 1998). The clear sex difference in the diurnal TSH and thyroid hormone rhythm reported in rodents, however, is absent in humans (Roelfsema et al., 2009).

Neural connections of hypothalamic nuclei with adipose tissue and liver

In the 1970s, the hypothalamus was recognized as the origin of neuroendocrine regulation of the pituitary gland via so-called releasing and inhibiting factors reaching the anterior pituitary gland via the portal circulation in the median eminence, thereby modulating a vast array of physiologic processes throughout the body via hormones. More recently, neural processes emanating from the hypothalamus have been increasingly recognized as additional modulators of peripheral organ function. In particular, intrahypothalamic effects of classic hormones appeared to include stimulation or inhibition of (pre-)autonomic pathways that reach metabolic organs such as the liver or adipose tissue via the sympathetic and parasympathetic branch of the ANS. The hypothalamic arcuate nucleus is localized at the base of the hypothalamus, where the blood–brain barrier is largely absent. Neurons in the arcuate nucleus express a variety of hormone receptors, including insulin, estrogen, and thyroid hormone receptors (see, for example, Alkemade et al., 2005; for reviews see Fliers et al., 2010, and Kalsbeek et al., 2010). Via the arcuate nucleus, endocrine information from the periphery is conveyed to several upstream hypothalamic nuclei including the PVN. In addition to the classic neuroendocrine route via the median eminence to the pituitary, the PVN contains pre-autonomic neurons that project to motor neurons of both branches of the ANS. In turn, these ANS branches reach peripheral organs, including the liver, and brown and white adipose tissue, and profound metabolic effects of hypothalamic hormones via these neuronal pathways have been reported. A well-known example is the major reduction of endogenous glucose production in the liver following central administration of insulin without concomitant changes in the plasma concentrations of insulin. The mechanism for this central effect of insulin, first reported in 2002 (Obici et al., 2002), has been largely unraveled and includes insulin receptors and potassium-dependent ATP channels in the arcuate nucleus, neuropeptidergic projections from the arcuate nucleus to the PVN, and autonomic output from the hypothalamus to the liver (for a review see Yi et al., 2010).

Similar pathways have been reported for the metabolic effects of thyroid hormone in the liver and adipose tissue. Thyrotoxicosis in humans is associated with increased glucose production and decreased hepatic insulin sensitivity (Cavallo-Perin et al., 1988; Jenkins et al., 2000). In view of the abundant expression of thyroid hormone receptors in the arcuate nucleus and PVN (Fliers et al., 2006; Lechan and Fekete, 2006), a role for hepatic autonomic innervation in the metabolic effects of thyrotoxicosis seemed plausible. Indeed, experiments in rats first showed that administration of T_3 in the PVN increases endogenous glucose production, while decreasing hepatic insulin sensitivity, via a sympathetic input to the liver. These effects occurred independently of plasma T_3 concentrations (Klieverik et al., 2009)

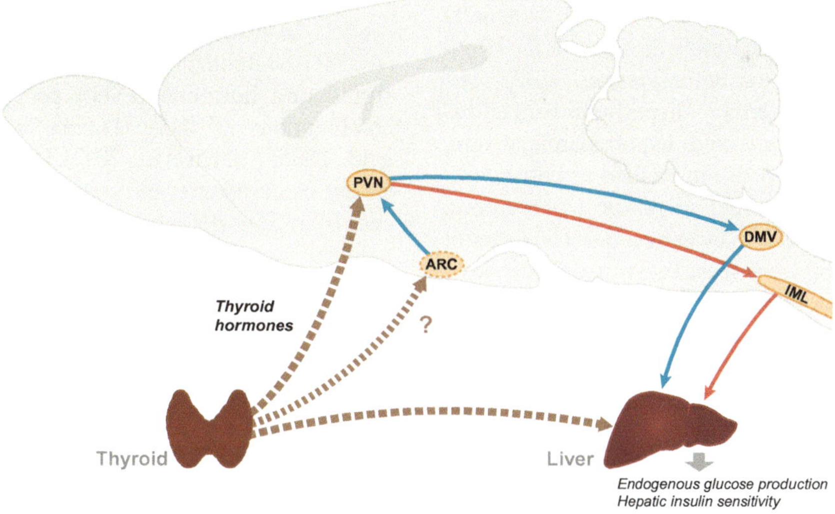

Fig. 9.1. Schematic representation of humoral and neural pathways for thyroid hormone to affect hepatic glucose metabolism. Sympathetic outflow from the hypothalamic paraventricular nucleus (PVN) via the IML to the liver is represented in red, whereas parasympathetic outflow via the dorsal motor nucleus of the vagus (DMV) to the liver is indicated in blue. Thyrotoxicosis results in increased endogenous glucose production and reduced hepatic insulin sensitivity by thyroid hormone's effects on the liver. Administration of T_3 within the hypothalamic PVN increases hepatic glucose production, without affecting plasma concentrations of glucoregulatory hormones, and this effect can be blocked by a selective hepatic sympathectomy. After a selective hepatic parasympathectomy during systemic thyrotoxicosis, plasma insulin increases without affecting glucose production, indicating hepatic insulin resistance. Both sympathetic and parasympathetic outflow from the hypothalamic PVN to the liver modulates hepatic glucose metabolism. Although the blood–brain barrier is largely absent in the arcuate nucleus (ARC) and the ARC expresses TRs, there are no data at present on the hepatic effects of T_3 administration to the ARC selectively. (Reproduced, with permission, from Fliers et al., 2010.)

(Fig. 9.1). Later studies reported that administration of T_3 in the ventromedial hypothalamus increases sympathetic input to brown adipose tissue by modulating intrahypothalamic AMP-activated protein kinase (AMPK) and fatty acid metabolism, thereby increasing thermogenesis in brown adipose tissue (Lopez et al., 2010). Although it is currently unknown if similar pathways are functional in humans, the induction of brown adipose tissue by thyroid hormone in a patient with severe insulin resistance might support this concept (Skarulis et al., 2010). In conclusion, hypothalamic bioavailability of thyroid hormone can now be regarded as a determinant of both neuroendocrine and neural effects of thyroid hormone in peripheral organs (Fliers et al., 2010).

CENTRAL HYPOTHYROIDISM

Central hypothyroidism (CeH) is a rare condition. Yet, the prevalence is probably higher than the previously suggested 1 in 50,000 to 1 in 20,000. Analysis of the Dutch neonatal congenital hypothyroidism screening program revealed the prevalence of permanent congenital CeH to be approximately 1 in 18 000 (Lanting et al., 2005; van Tijn et al., 2005; Kempers et al., 2006). If the prevalence of acquired CeH equals this number, the combined prevalence of permanent congenital and acquired CeH may be as high as 1 in 8000–10 000.

CeH can be defined as reduced thyroid hormone secretion resulting from deficient stimulation of an intrinsically normal thyroid gland by TSH. It is usually caused by a functional or anatomic disorder of the hypothalamus, the pituitary, or both (Persani, 2012). The terms "secondary" and "tertiary" hypothyroidism to distinguish between pituitary and hypothalamic origin, respectively, are no longer recommended as both conditions involve insufficient TSH secretion. Furthermore, stimulation of TSH secretion by exogenous TRH cannot reliably distinguish between these two causes because of overlap in the TSH response pattern. TSH secretion may be insufficient not only in quantitative terms, but also in qualitative terms, resulting from decreased biological activity (Persani, 2012).

Central hypothyroidism in neonates and children

Worldwide, most national neonatal screening programs for congenital hypothyroidism are TSH-based and effectively detect neonates with congenital hypothyroidism of thyroidal origin (CH-T). However, only screening programs that combine measurement of thyroxine (T_4) or free T_4 (FT_4) with the measurement of TSH are able to detect congenital CeH.

Based on an analysis of the first 10 years of the total T_4 + TSH-based Northwest Regional Screening Program (covering the US states of Oregon, Montana, Alaska, and Idaho between 1975 and 1985), the prevalence of permanent congenital CeH was calculated to be approximately 1 in 29 000 (Hanna et al., 1986). Analysis of the first years of the Dutch primarily T_4-based neonatal screening program (with TSH measurement in the 20% lowest T_4 concentrations) between 1981 and 1989 revealed a comparable prevalence (1 in 26 000) (Vulsma et al., 1990). With the aim of decreasing the number of false-positives due to low total T_4 concentrations because of low thyroxine-binding globulin (TBG) concentrations, measurement of TBG in the lowest 5% T_4 concentrations was added as a diagnostic step to the Dutch screening program in 1995. This modification turned the Dutch screening into a cost-effective program, detecting CH-T as well as congenital CeH (Lanting et al., 2005). The prevalence of permanent congenital CeH, calculated from three separate analyses from the Netherlands covering the periods 1994–1996, 1995–2000, and 2002–2004, is approximately 1 in 18 000 (Lanting et al., 2005; van Tijn et al., 2005; Kempers et al., 2006). This is higher than the aforementioned prevalences and also higher than the recently reported 1 in 30 833 prevalence in Japan, which currently employs a FT_4 + TSH-based neonatal screening program (Adachi et al., 2012).

Approximately 75% of neonates with permanent congenital CeH have hypothyroidism within the framework of multiple pituitary hormone deficiency (MPHD), i.e., growth hormone (GH), adrenocorticotropic hormone (ACTH), luteinizing hormone (LH), and follicle-stimulating hormone (FSH), prolactin deficiency, or combinations of these (Hanna et al., 1986; van Tijn et al., 2005; Adachi et al., 2012). In this setting, vasopressin deficiency is rare, and seems to occur only in the presence of midline defects such as septo-optic dysplasia and holoprosencephaly (Mehta et al., 2009a). Seventy-five percent of neonates with MPHD have morphologic abnormalities of the hypothalamo–pituitary region, for example, an ectopic posterior pituitary gland (located in the floor of the third ventricle), a thin or absent pituitary stalk, or a small anterior pituitary gland (van Tijn et al., 2005). Although mutations in genes involved in hypothalamo–pituitary development, including POU1F2, PROP1, HESX1, LHX3, and LHX4, explain some cases of congenital hypothyroidism within the framework of MPHD, the cause in the majority of cases (>90%) remains elusive (Mehta and Dattani, 2008).

The remaining 25% of neonates with permanent congenital CeH have an isolated form. Among the established causes of isolated congenital CeH are mutations of the TRHR and TSH β subunit genes. In both cases homozygous or compound heterozygous mutations can result in severe hypothyroidism (Medeiros-Neto et al., 1996; Collu et al., 1997; Karges et al., 2004). Recently, a novel X-linked syndrome of apparently isolated congenital CeH was discovered, caused by loss-of-function mutations in the immunoglobulin superfamily member 1 (IGSF1) gene. Sun et al. (2012) described 11 families, from the Netherlands (8), the United Kingdom (1) and Italy (2), in which 26 male cases were diagnosed with CeH either in the neonatal period after an abnormal neonatal screening result, or later in life. In addition to CeH, all adult patients had macroorchidism, and 16 of the 26 males had prolactin deficiency. IGSF1 is a membrane glycoprotein highly expressed in the anterior pituitary gland and the identified mutations impair its trafficking to the cell surface in heterologous HEK293 cells. Studies in mice suggest that the mechanism underlying the CeH may be decreased pituitary TRHR expression. Although this may also explain the prolactin deficiency, the cause of the macroorchidism remains unexplained. In a follow-up study, a subset of female mutation carriers also exhibited CeH (Joustra et al., 2013).

The signs and symptoms of CeH in the neonatal period do not differ from the classic signs and symptoms of CH-T, and include feeding problems, temperature instability, prolonged jaundice, enlarged tongue, umbilical hernia, and muscular hypotonia. However, these signs and symptoms can be subtle and easily missed. If congenital CeH is part of MPHD there may be signs and symptoms of cortisol, growth hormone, and testosterone deficiency as well, such as hypoglycemia, sepsis-like illness, conjugated hyperbilirubinemia, and micropenis. In addition, there may be midline defects.

When the diagnosis of congenital CeH is suspected in the neonatal period based on an abnormal neonatal congenital hypothyroidism screening result or on clinical grounds, the diagnosis should be confirmed by measurement of plasma or serum FT_4 (or total T_4+TBG) and TSH concentrations. A FT_4 concentration clearly below the age-specific reference interval in the presence of a TSH concentration within, below, or just above the reference interval supports the diagnosis. It should be realized, however, that neonates with a normal HPT axis may have low (free) triiodothyronine and $(F)T_4$ concentrations because of nonthyroidal illness (NTI). So, if a low FT4 concentration is found and the medical history reveals severe birth asphyxia, (proven) bacterial sepsis, or major surgery in the first days of life, NTI is a more likely diagnosis than CeH. In this situation FT_4 measurement should be repeated once or twice weekly until a normal concentration is found. If not, the diagnosis congenital CeH should be reconsidered. When the medical history does not point to NTI, a second alternative diagnosis to consider is transient congenital CeH related to maternal Graves' disease (Matsuura and Konishi, 1990; Kempers et al., 2003). The mother's medical history followed by an immediate TSH measurement in the mother will provide the diagnosis in most cases. Rare causes of a low plasma or serum FT_4 concentration in the presence of a normal or mildly elevated TSH concentration are a defective monocarboxylate transporter 8 (MCT8) and resistance to thyroid hormone caused by a mutation in the thyroid hormone receptor α (TRα) gene (Friesema et al., 2004; Bochukova et al., 2012). Finding a high serum or plasma T_3 concentration is suggestive of a MCT8 defect, while clinical signs suggestive of severe congenital hypothyroidism in the presence of only slightly lowered $(F)T_4$ concentrations may be a clue to a TRα mutation.

When the serum or plasma FT_4 concentration is clearly below the age-specific reference interval, and NTI and transient CeH related to maternal Graves' disease are ruled out or unlikely, the further diagnostic workup should focus on the question whether the neonate has MPHD or isolated congenital CeH. Considering the high mortality (probably more than 10%) and morbidity, including brain damage resulting from hypoglycemia and thyroid hormone deficiency, of congenital MPHD (Nebesio et al., 2010), this workup should be immediate and carried out by an experienced pediatric endocrinologist. First, neonates suspected of MPHD should be admitted for repeated prefeeding blood glucose measurements and assessment of the water balance. The next diagnostic step consists of testing the integrity of the hypothalamo–pituitary–adrenal axis. Although random cortisol measurements have been proposed (Mehta et al., 2009b), dynamic testing, i.e., the plasma or serum cortisol response to intravenous CRH or (low dose) ACTH, probably provides a more accurate answer (van Tijn et al., 2008a). If the cortisol response is abnormal, treatment with cortisol should be started immediately and parents should receive proper instructions with respect to increasing the cortisol dose during stressful situations. Within 6–8 hours after the start of cortisol treatment, thyroxine treatment should be started. If the cortisol response is normal, thyroxine treatment can be started immediately. If there is prefeeding hypoglycemia due to central adrenal insufficiency, the effect of cortisol treatment should be awaited. If glucose concentrations do not improve, the integrity of the hypothalamo–pituitary/GH-IGF-1 axis should be assessed by GH stimulation tests, and GH treatment may have to be started. However, if glucose metabolism normalizes, assessment of the hypothalamo–pituitary/GH-IGF-1 axis can be postponed until, or after, the age of 3 months. The same applies for the evaluation of the hypothalamo–pituitary–gonadal axis. In 2008, van Tijn et al. showed that the serum or plasma response to intravenous TRH test predicts the presence or absence of MPHD or isolated CeH (van Tijn et al., 2008b). However, some neonates and children with MPHD have a normal TSH response (Mehta et al., 2003), casting doubt on the clinical usefulness of this test. On the other hand, magnetic resonance imaging (MRI) of the brain and especially the hypothalamo–pituitary region provides indispensable information. Neuroradiologic features are predictive not only of the presence, but also of the type, of hypopituitarism (Mehta et al., 2009b).

Serum or plasma FT_4 concentrations just below or fluctuating around the lower border of the age-specific reference interval in a neonate represent a diagnostic challenge. Although by definition 2.5% of a group of individuals has a FT_4 concentration below the reference interval, the possible serious sequelae of a missed diagnosis of MPHD in a neonate warrant a very low threshold for the abovementioned diagnostic workup and, if necessary, treatment.

The prevalence of CeH in older children is not clear, but theoretically should be the sum of (un)diagnosed cases of congenital CeH and acquired CeH. The Dutch neonatal screening program does not detect all neonates with CeH and may be false-negative in prematurely born children with this condition (Lanting et al., 2005; Kempers et al., 2006). Among the causes of acquired CeH in children are tumor, trauma, infection, and inflammation, as well as surgery and irradiation of the hypothalamus and pituitary gland. As in neonates, a diagnosis of CeH should be confirmed by measuring and interpreting plasma or serum FT_4 and TSH concentrations. CeH is characterized by a FT_4 concentration below the reference interval, and a lowered, normal, or even slightly elevated TSH concentration. When plasma TSH and FT_4 are both within the age-specific reference interval, a diagnosis of hypothyroidism is highly unlikely. Rare exceptions may be children and adolescents who underwent cranial irradiation, in whom a plasma FT_4 concentration in the lower third of the reference interval may point to mild CeH, especially when the preirradiation FT_4 was substantially higher. In these cases it may be worthwhile to evaluate the nocturnal TSH surge (Rose, 2001). Newly diagnosed CeH is a clear indication for MRI of the brain and hypothalamo–pituitary region. In addition, there should always be appropriate attention paid to the integrity of the other hypothalamic/pituitary functions.

CeH is an indisputable indication for thyroxine treatment. Both neonates and children should be treated with thyroxine tablets. Thyroxine is effectively converted into triiodothyronine, the active hormone, by the type 1 and 2 deiodinases in the target tissues. During childhood and adolescence, the thyroxine replacement dose gradually decreases from approximately 10–12 μg/kg/day shortly after birth, to 5 μg/kg/day between ages 1 and 5 years, and to 2–3 μg/kg/day for adolescents (Van Vliet, 2005). However, the correct dose should always be sought by means of dose adjustments guided by measurement of the plasma or serum FT_4 concentrations, initially at 4–6 week intervals and later at 6 month intervals. In neonates most centers use intervals of 2–3 weeks initially and later 3 months. It is important to realize that in CeH the TSH concentration cannot be used to guide thyroxine dose adjustments. In this condition the FT_4 concentration should be kept around or just above the age-specific mean. In CeH adrenal insufficiency of central origin should be excluded before starting thyroxine treatment, as thyroxine treatment can provoke an adrenal crisis.

Central hypothyroidism in adults

The most frequent cause (>50%) of CeH in adults is pituitary macroadenoma, either secreting or nonsecreting. In these cases, insufficient TSH secretion may result from compression of the anterior pituitary, the pituitary stalk or the hypothalamus. Less frequently, CeH may result from pituitary apoplexy or from extrasellar tumors including meningiomas, gliomas, and metastases. Another fairly common iatrogenic cause of CeH is external radiotherapy for tumors of the head and neck. CeH may be present in up to 65% of patients irradiated for brain tumors (Constine et al., 1993), while a more recent study reported a prevalence of 72% in adults treated with radiotherapy for a brain tumor in childhood (van Santen et al., 2003). Another iatrogenic cause is surgery for pituitary tumors. Excision of large macroadenomas may cause CeH in up to 10% of cases, but this is unusual after selective removal of a microadenoma.

The clinical features of CeH are similar to those of primary hypothyroidism, but generally the clinical picture is less severe (Persani, 2012). Symptoms and signs arising from the pituitary or hypothalamic lesions may be present, while additional pituitary insufficiency may obscure or aggravate the hypothyroidism. When CeH is suspected, laboratory tests should be performed. The diagnosis is based on the demonstration of low serum thyroid hormone concentrations in the presence of inappropriately low serum TSH. The prerequisite is low serum T_4. The reduction of serum T_4 is usually less pronounced than in primary hypothyroidism. Measurement of serum T_3 is less useful as it may be (still) within the reference range. Serum TSH may be undetectable, but it is typically within the reference range, or even slightly elevated in some patients. The measurement of serum TSH does not allow the distinction between hypothalamic or pituitary hypothyroidism. Once the diagnosis is established, therapy should be started and directed toward restoring and maintaining euthyroidism.

Thyroid hormone replacement is associated with the suppression of residual TSH secretion, and clinicians cannot rely on the determination of serum TSH to monitor substitution therapy (Persani, 2012). For challenges related to the treatment of patients with CeH the reader is referred to the chapter in this volume by Dr Persani (Ch. 27).

CENTRAL HYPERTHYROIDISM

Thyrotropin (TSH)-secreting pituitary adenomas, or thyrotropinomas, account for 1–3% of all pituitary adenomas and most commonly present as central hyperthyroidism in a patient with a pituitary macroadenoma (Beck-Peccoz and Persani, 2008). It is unclear at present whether the incidence of TSH-secreting adenomas is increasing, or whether these adenomas are more readily diagnosed as a result of improved hormone assays and imaging techniques. Although some TSH-secreting pituitary adenomas may co-secrete growth hormone or prolactin, the majority do not show any associated hypersecretion. In patients with TSH-secreting pituitary adenomas, acromegaly and hyperprolactinemia were present in 19% and 21%, respectively. Basal plasma TSH was elevated in 42% of these patients with a subnormal response to TRH in 81%, and basal α subunit (αSU) was elevated in 30% with an over two-fold TRH-stimulated αSU response in 44% (Socin et al., 2003). Detailed analysis of the TSH secretion pattern in these patients revealed increased TSH pulse frequency with a preserved diurnal rhythm, albeit at a higher mean plasma concentration (Roelfsema et al., 2008).

Patients most commonly present with symptoms resulting from tumor growth, such as headache or visual field abnormalities, or from thyroid hormone overproduction, such as palpitations, weight loss, or nervousness. A small goiter is common, reflecting the growth-promoting role of TSH in the thyroid gland. The pituitary adenoma is usually rather large, and the majority are locally invasive macroadenomas at presentation (Brucker-Davis et al., 1999). In most, but not all, patients plasma concentrations of TSH and αSU are slightly elevated and the combination of elevated plasma FT_4 with nonsuppressed TSH points to central hyperthyroidism. A similar combination of plasma FT_4 and TSH may be present in patients with resistance to thyroid hormone (RTH) due to a mutation in the thyroid hormone receptor, but in RTH patients αSU is typically within the reference range (Socin et al., 2003).

The diagnostic evaluation should include plasma T_4, T_3, TSH, and αSU. The combination of elevated T_4 and T_3 (and αSU), in combination with high or inappropriately normal plasma TSH and the presence of a pituitary tumor confirms the diagnosis of a TSH-secreting pituitary adenoma. The TSH response to TRH is blunted in the majority of patients (Socin et al., 2003), while the molar ratio of αSU to TSH is mostly > 1. IGF-1 and prolactin should be measured to exclude acromegaly and hyperprolactinemia. In patients with large pituitary tumors, endocrine evaluation should be performed to exclude pituitary insufficiency.

The most appropriate treatment for a patient with a thyrotropinoma is transsphenoidal resection of the tumor, although additional therapeutic modalities are needed as surgical cures occur in fewer than 40% (Melmed and Kleinberg, 2011). However, large prospective trials are lacking due to the rarity of this disorder. In one study, >50% of the patients showed evidence of residual tumor when assessed by MRI at 6 months after surgery (Losa et al., 1996). There are no large series addressing the efficacy of radiotherapy alone, and radiotherapy has mostly been used as adjunctive therapy to noncurative surgery. The most effective medical therapy is with long-acting somatostatin analogs, such as octreotide LAR or lanreotide SR, which reduce TSH and αSU secretion in most cases, while inducing pituitary tumor shrinkage and vision improvement in 42% and 66%, respectively (Beck-Peccoz and Persani, 2008). Whether or not somatostatin analog treatment should have a role as primary therapy for TSH-secreting pituitary adenomas remains to be established, but in most centers it is used to restore euthyroidism before surgery and to treat patients who are not cured by transsphenoidal surgery. A case of cure of a TSH-secreting pituitary adenoma by medical therapy without any preceding surgical procedure was recently reported (Fliers et al., 2012), which may facilitate future research on the efficacy and safety of discontinuation of somatostatin analogs in patients with TSH-secreting pituitary adenomas.

CONCLUSION

TRH neurons in the hypothalamic PVN play a key role in setpoint regulation of the HPT axis, adapting thyroid hormone concentrations to environmental factors such as caloric deprivation or infection. Intrahypothalamic thyroid hormone has a negative feedback action on hypophysiotropic TRH neurons. For this role, intrahypothalamic deiodination of T_4 into T_3 and the TRβ2 are essential. Recent studies have shown additional roles for intrahypothalamic thyroid hormone, including modulation of metabolism in adipose tissue and liver via the ANS. Thus, both neuroendocrine and neural pathways communicate the effects of intrahypothalamic bioavailability of thyroid hormone. Congenital or acquired dysfunction of the hypothalamus or pituitary gland may result in CeH, with an estimated prevalence in the

Netherlands of 1 in 18000. CeH is detected within the framework of the Dutch neonatal screening program. In most neonates congenital CeH is accompanied by additional anterior pituitary hormone deficiencies. A novel mutation was recently identified as a cause of X-linked, isolated CeH. In adults, the most frequent cause of acquired CeH is a pituitary adenoma. Central hyperthyroidism is mostly caused by a TSH-secreting pituitary tumor. Although transsphenoidal pituitary surgery is the treatment of choice, medical therapy with somatostatin analogs is effective in many patients. Future studies will indicate if somatostatin analog treatment should have a role as primary therapy.

References

Abel ED, Ahima RS, Boers ME et al. (2001). Critical role for thyroid hormone receptor beta 2 in the regulation of paraventricular thyrotropin-releasing hormone neurons. J Clin Invest 107: 1017–1023.

Adachi M, Soneda A, Asakura Y et al. (2012). Mass screening of newborns for congenital hypothyroidism of central origin by free thyroxine measurement of blood samples on filter paper. Eur J Endocrinol 166: 829–838.

Adriaanse R, Romijn JA, Endert E et al. (1992). The nocturnal thyroid-stimulating hormone surge is absent in overt, present in mild primary and equivocal in central hypothyroidism. Acta Endocrinol 126: 206–212.

Alkemade A, Unmehopa U, Brouwer J et al. (2003). Decreased thyrotropin-releasing hormone gene expression in the hypothalamic paraventricular nucleus of patients with major depression. Mol Psychiatry 8: 838.

Alkemade A, Vuijst CL, Unmehopa UA et al. (2005). Thyroid hormone receptor expression in the human hypothalamus and anterior pituitary. J Clin Endocrinol Metab 90: 904–912.

Allan JS, Czeisler CA (1994). Persistence of the circadian thyrotropin rhythm under constant conditions and after light-induced shifts of circadian phase. J Clin Endocrinol Metab 79: 508–512.

Bartalena L, Placidi GF, Martino E et al. (1990). Nocturnal serum thyrotropin (TSH) surge and the TSH response to TSH-releasing hormone: dissociated behavior in untreated depressives. J Clin Endocrinol Metab 71: 650–655.

Beck-Peccoz P, Persani L (2008). Thyrotropinomas. Endocrinol Metab Clin North Am 37: 123–134.

Behrends J, Prank K, Dogu E et al. (1998). Central nervous system control of thyrotropin secretion during sleep and wakefulness. Horm Res 49: 173–177.

Bochukova E, Schoenmakers N, Agostini M et al. (2012). A mutation in the thyroid hormone receptor alpha gene. N Engl J Med 366: 243–249.

Boelen A, Kwakkel J, Wiersinga WM et al. (2006). Chronic local inflammation in mice results in decreased TRH and type 3 deiodinase mRNA expression in the hypothalamic paraventricular nucleus independently of diminished food intake. J Endocrinol 191: 707–714.

Boelen A, Wiersinga WM, Fliers E (2008). Fasting-induced changes in the hypothalamus-pituitary-thyroid axis. Thyroid 18: 123.

Boelen A, Kwakkel J, Chassande O et al. (2009). Thyroid hormone receptor β mediates acute illness-induced alterations in central thyroid hormone metabolism. J Neuroendocrinol 21: 465–472.

Boelen A, Kwakkel J, Fliers E (2011). Beyond low plasma T3: local thyroid hormone metabolism during inflammation and infection. Endocr Rev 32: 670–693.

Brabant G, Prank K, Ranft U et al. (1990). Physiological regulation of circadian and pulsatile thyrotropin secretion in normal man and woman. J Clin Endocrinol Metab 70: 403–409.

Brucker-Davis F, Oldfield EH, Skarulis MC et al. (1999). Thyrotropin-secreting pituitary tumors: diagnostic criteria, thyroid hormone sensitivity, and treatment outcome in 25 patients followed at the National Institutes of Health. J Clin Endocrinol Metab 84: 476–486.

Cavallo-Perin P, Bruno A, Boine L et al. (1988). Insulin resistance in Graves' disease: a quantitative in-vivo evaluation. Eur J Clin Invest 18: 607–613.

Chan JL, Heist K, DePaoli AM et al. (2003). The role of falling leptin levels in the neuroendocrine and metabolic adaptation to short-term starvation in healthy men. J Clin Invest 111: 1409–1421.

Chi M, Lin M (1983). Involvement of adrenergic receptor mechanisms within hypothalamus in the fever induced by amphetamine and thyrotropin-releasing hormone in the rat. J Neural Transm 58: 213–222.

Chiamolera MI, Wondisford FE (2009). Thyrotropin-releasing hormone and the thyroid hormone feedback mechanism. Endocrinology 150: 1091–1096.

Collu R, Tang J, Castagné J et al. (1997). A novel mechanism for isolated central hypothyroidism: inactivating mutations in the thyrotropin-releasing hormone receptor gene. J Clin Endocrinol Metab 82: 1561–1565.

Constine LS, Woolf PD, Cann D et al. (1993). Hypothalamic-pituitary dysfunction after radiation for brain tumors. N Engl J Med 328: 87–94.

Fekete C, Gereben B, Doleschall M et al. (2004). Lipopolysaccharide induces type 2 iodothyronine deiodinase in the mediobasal hypothalamus: implications for the nonthyroidal illness. Syndrome Endocrinol 145: 1649–1655.

Fliers E, Noppen NW, Wiersinga WM et al. (1994). Distribution of thyrotropin-releasing hormone (TRH)-containing cells and fibers in the human hypothalamus. J Comp Neurol 350: 311–323.

Fliers E, Guldenaar SEF, Wiersinga WM et al. (1997). Decreased hypothalamic thyrotropin-releasing hormone gene expression in patients with nonthyroidal illness. J Clin Endocrinol Metab 82: 4032–4036.

Fliers E, Alkemade A, Wiersinga WM (2001). The hypothalamic–pituitary–thyroid axis in critical illness. Best Pract Res Clin Endocrinol Metab 15: 453–464.

Fliers E, Unmehopa UA, Alkemade A (2006). Functional neuroanatomy of thyroid hormone feedback in the human hypothalamus and pituitary gland. Mol Cell Endocrinol 251: 1–8.

Fliers E, Klieverik LP, Kalsbeek A (2010). Novel neural pathways for metabolic effects of thyroid hormone. Trends Endocrinol Metab 21: 230–236.

Fliers E, van Furth WR, Bisschop PH (2012). Cure of a thyrotrophin (TSH)-secreting pituitary adenoma by medical therapy. Clin Endocrinol (Oxf) 77: 788–790.

Friesema EC, Grueters A, Biebermann H et al. (2004). Association between mutations in a thyroid hormone transporter and severe X-linked psychomotor retardation. Lancet 364: 1435–1437.

Goichot B, Weibel L, Chapotot F et al. (1998). Effect of the shift of the sleep-wake cycle on three robust endocrine markers of the circadian clock. Am J Physiol Endocrinol Metab 275: E243–E248.

Greenspan SL, Klibansk A, Schoenfeld DA et al. (1986). Pulsatile secretion of thyrotropin in man. J Clin Endocrinol Metab 63: 661–668.

Guldenaar SE, Veldkamp B, Bakker O et al. (1996). Thyrotropin-releasing hormone gene expression in the human hypothalamus. Brain Res 743: 93–101.

Hanna CE, Krainz PL, Skeels MR et al. (1986). Detection of congenital hypopituitary hypothyroidism: ten-year experience in the Northwest Regional Screening Program. J Pediatr 109: 959–964.

Jenkins RC, Valcavi R, Zini M et al. (2000). Association of elevated insulin-like growth factor binding protein-1 with insulin resistance in hyperthyroidism. Clin Endocrinol (Oxf) 52: 187–195.

Joustra SD, Schoenmakers N, Persani L et al. (2013). The IGSF1 deficiency syndrome: characteristics of male and female patients. J Clin Endocrinol Metab 98: 4942–4952.

Kalsbeek A, Fliers E, Franke AN et al. (2000). Functional connections between the suprachiasmatic nucleus and the thyroid gland as revealed by lesioning and viral tracing techniques in the rat. Endocrinology 141: 3832–3841.

Kalsbeek A, Buijs RM, van Schaik R et al. (2005). Daily variations in type ii iodothyronine deiodinase activity in the rat brain as controlled by the biological clock. Endocrinology 146: 1418–1427.

Kalsbeek A, Bruinstroop E, Yi CX et al. (2010). Hypothalamic control of energy metabolism via the autonomic nervous system. Ann N Y Acad Sci 1212: 114–129.

Kalsbeek A, Fliers E (2013). Daily regulation of hormone profiles. In: A Kramer, M Merrow (Eds.), Circadian Clocks, Handbook of Experimental Pharmacology 217. Springer-Verlag, Berlin Heidelberg, pp. 185–226.

Karges B, LeHeup B, Schoenle E et al. (2004). Compound heterozygous and homozygous mutations of the TSHbeta gene as a cause of congenital central hypothyroidism in Europe. Horm Res 62: 149–155.

Kempers MJ, van Tijn DA, van Trotsenburg AS et al. (2003). Central congenital hypothyroidism due to gestational hyperthyroidism: detection where prevention failed. J Clin Endocrinol Metab 88: 5851–5857.

Kempers MJ, Lanting CI, van Heijst AF et al. (2006). Neonatal screening for congenital hypothyroidism based on thyroxine thyrotropin and thyroxine-binding globulin measurement: potentials and pitfalls. J Clin Endocrinol Metab 91: 3370–3376.

Klieverik LP, Janssen SF, van Riel A et al. (2009). Thyroid hormone modulates glucose production via a sympathetic pathway from the hypothalamic paraventricular nucleus to the liver. Proc Natl Acad Sci U S A 106: 5966–5971.

Lanting CI, van Tijn DA, Loeber JG et al. (2005). Clinical effectiveness and cost-effectiveness of the use of the thyroxine/thyroxine-binding globulin ratio to detect congenital hypothyroidism of thyroidal and central origin in a neonatal screening program. Pediatrics 116: 168–173.

Lechan RM, Fekete C (2006). The TRH neuron: a hypothalamic integrator of energy metabolism. Prog Brain Res 153: 209–235.

Lopez M, Varela L, Vazquez MJ et al. (2010). Hypothalamic AMPK and fatty acid metabolism mediate thyroid regulation of energy balance. Nat Med 16: 1001–1008.

Losa M, Giovanelli M, Persani L et al. (1996). Criteria of cure and follow-up of central hyperthyroidism due to thyrotropin-secreting pituitary adenomas. J Clin Endocrinol Metab 81: 3084–3090.

Matsuura N, Konishi J (1990). Transient hypothyroidism in infants born to mothers with chronic thyroiditis – a nationwide study of twenty-three cases. Endocrinol Jpn 37: 369–379.

Mebis L, Debaveye Y, Ellger B et al. (2009). Changes in the central component of the hypothalamus–pituitary–thyroid axis in a rabbit model of prolonged critical illness. Crit Care 13: R147.

Medeiros-Neto G, Herodotou DT, Rajan S et al. (1996). A circulating biologically inactive thyrotropin caused by a mutation in the beta subunit gene. J Clin Invest 97: 1250–1256.

Mehta A, Dattani MT (2008). Developmental disorders of the hypothalamus and pituitary gland associated with congenital hypopituitarism. Best Pract Res Clin Endocrinol Metab 22: 191–206.

Mehta A, Hindmarsh PC, Stanhope RG et al. (2003). Is the thyrotropin-releasing hormone test necessary in the diagnosis of central hypothyroidism in children. J Clin Endocrinol Metab 88: 5696–5703.

Mehta A, Gevers EF, Dattani MT (2009a). Congenital disorders of the hypothalamo–pituitary–somatotrope axis. In: C Brook, P Clayton, R Brown (Eds.), Brook's clinical pediatric endocrinology. 6th edn. John Wiley & Sons Ltd, Chichester, pp. 60–105.

Mehta A, Hindmarsh PC, Mehta H et al. (2009b). Congenital hypopituitarism: clinical molecular and neuroradiological correlates. Clin Endocrinol 71: 376–382.

Mihály E, Fekete C, Tatro JB et al. (2000). Hypophysiotropic thyrotropin-releasing hormone-synthesizing neurons in the human hypothalamus are innervated by neuropeptide Y agouti-related protein and alpha-melanocyte-stimulating hormone. J Clin Endocrinol Metab 85: 2596–2603.

Melmed S, Kleinberg D (2011). Pituitary masses and tumors. In: S Melmed et al. (Eds.), William's Textbook of Endocrinology. 11th edn. Elsevier Saunders, Philadelphia, pp. 229–290.

Nebesio TD, McKenna MP, Nabhan ZM et al. (2010). Newborn screening results in children with central hypothyroidism. J Pediatr 156: 990–993.

Obici S, Zhang BB, Karkanias G et al. (2002). Hypothalamic insulin signaling is required for inhibition of glucose production. Nat Med 8: 1376.

Ortiga-Carvalho TM, Curty FH, Nascimento-Saba CC et al. (1997). Pituitary neuromedin B content in experimental fasting and diabetes mellitus and correlation with thyrotropin secretion. Metabolism 46: 149–153.

Parker DC, Pekary AE, Hershman JM (1976). Effect of normal and reversed sleep-wake cycles upon nyctohemeral rhythmicity of plasma thyrotropin: evidence suggestive of an inhibitory influence in sleep. J Clin Endocrinol Metab 43: 318–329.

Persani L (2012). Central hypothyroidism: pathogenic diagnostic and therapeutic challenges. J Clin Endocrinol Metab 97: 3068–3078.

Rocchi R, Kimura H, Tzou S et al. (2007). Toll-like receptor-MyD88 and Fc receptor pathways of mast cells mediate the thyroid dysfunctions observed during nonthyroidal illness. Proc Natl Acad Sci U S A 104: 6019–6024.

Roelfsema F, Veldhuis JD (2013). Thyrotropin secretion patterns in health and disease. Endocr Rev 34: 619–657.

Roelfsema F, Pereira AM, Keenan DM et al. (2008). Thyrotropin secretion by thyrotropinomas is characterized by increased pulse frequency delayed diurnal rhythm enhanced basal secretion spikiness and disorderliness. J Clin Endocrinol Metab 93: 4052–4057.

Roelfsema F, Pereira AM, Veldhuis JD et al. (2009). Thyrotropin secretion profiles are not different in men and women. J Clin Endocrinol Metab 94: 3964–3967.

Roelfsema F, Pereira AM, Adriaanse R et al. (2010). Thyrotropin secretion in mild and severe primary hypothyroidism is distinguished by amplified burst mass and basal secretion with increased spikiness and approximate entropy. J Clin Endocrinol Metab 95: 928–934.

Romijn JA, Wiersinga WM (1990). Decreased nocturnal surge of thyrotropin in nonthyroidal illness. J Clin Endocrinol Metab 70: 35–42.

Romijn JA, Adriaanse R, Brabant G et al. (1990). Pulsatile secretion of thyrotropin during fasting: a decrease of thyrotropin pulse amplitude. J Clin Endocrinol Metab 70: 1631–1636.

Rose SR (2001). Cranial irradiation and central hypothyroidism. Trends Endocrinol Metab 12: 97–104.

Schneider MJ, Fiering SN, Pallud SE et al. (2001). Targeted Disruption of the type 2 selenodeiodinase gene (DIO2). Results in a phenotype of pituitary resistance to T4. Mol Endocrinol 15: 2137–2148.

Segerson TP, Kauer J, Wolfe HC et al. (1987). Thyroid hormone regulates TRH biosynthesis in the paraventricular nucleus of the rat hypothalamus. Science 238: 78–80.

Shintani M, Tamura Y, Monden M et al. (2005). Thyrotropin-releasing hormone induced thermogenesis in Syrian hamsters: site of action and receptor subtype. Brain Res 1039: 22–29.

Shupnik MA (2000). Thyroid hormone suppression of pituitary hormone gene expression. Rev Endocr Metab Disord 1: 35–42.

Skarulis MC, Celi FS, Mueller E et al. (2010). Thyroid hormone induced brown adipose tissue and amelioration of diabetes in a patient with extreme insulin resistance. J Clin Endocrinol Metab 95: 256–262.

Snel M, Wijngaarden MA, Bizino MB et al. (2012). Food cues do not modulate the neuroendocrine response to a prolonged fast in healthy men. Neuroendocrinol 96: 285–293.

Socin HV, Chanson P, Delemer B et al. (2003). The changing spectrum of TSH-secreting pituitary adenomas: diagnosis and management in 43 patients. Eur J Endocrinol 148: 433–442.

Steel JH, Noorden SV, Ballesta JO et al. (1988). Localization of 7B2 neuromedin B and neuromedin U in specific cell types of rat mouse and human pituitary in rat hypothalamus and in 30 human pituitary and extrapituitary tumors. Endocrinology 122: 270–282.

Sun Y, Bak B, Schoenmakers N et al. (2012). Loss-of-function mutations in IGSF1 cause an X-linked syndrome of central hypothyroidism and testicular enlargement. Nat Genet 44: 1375–1381.

Szkudlinski MW, Fremont V, Ronin C et al. (2002). Thyroid-stimulating hormone and thyroid-stimulating hormone receptor structure-function relationships. Physiol Rev 82: 473–502.

Van den Berghe G, De Zegher F, Veldhuis JD et al. (1997). Thyrotrophin and prolactin release in prolonged critical illness: dynamics of spontaneous secretion and effects of growth hormone-secretagogues. Clin Endocrinol 47: 599–612.

Van den Berghe G, De Zegher F, Baxter RC et al. (1998). Neuroendocrinology of prolonged critical illness: effects of exogenous thyrotropin-releasing hormone and its combination with growth hormone secretagogues. J Clin Endocrinol Metab 83: 309–319.

van Santen HM, Vulsma T, Dijkgraaf MG et al. (2003). No damaging effect of chemotherapy in addition to radiotherapy on the thyroid axis in young adult survivors of childhood cancer. J Clin Endocrinol Metab 88: 3657–3663.

van Tijn DA, de Vijlder JJ, Verbeeten B et al. (2005). Neonatal detection of congenital hypothyroidism of central origin. J Clin Endocrinol Metab 90: 3350–3359.

van Tijn DA, de Vijlder JJ, Vulsma T (2008a). Role of corticotropin-releasing hormone testing in assessment of hypothalamic-pituitary-adrenal axis function in infants with congenital central hypothyroidism. J Clin Endocrinol Metab 93: 3794–3803.

van Tijn DA, de Vijlder JJ, Vulsma T (2008b). Role of the thyrotropin-releasing hormone stimulation test in diagnosis of congenital central hypothyroidism in infants. J Clin Endocrinol Metab 93: 410–419.

Van Vliet G (2005). Hypothyroidism in infants and children. In: LE Braverman, RD Utiger (Eds.), The Thyroid. A Fundamental and Clinical Text. 9th edn. Lippincott Williams & Wilkins, Philadelphia, pp. 1029–1047.

Vulsma T, Delemarre HA, De Muinck Keizer SMPF et al. (1990). Detection and classification of congenital thyrotropin deficiency in the Netherlands. In: HA Drexhage, JJM de Vijlder, WM Wiersinga (Eds.), The thyroid gland, environment and autoimmunity. Elsevier Science Publishers BV, Amsterdam, pp. 343–346.

Yi CX, la Fleur SE, Fliers E et al. (2010). The role of the autonomic nervous liver innervation in the control of energy metabolism. Biochim Biophys Acta 1802: 416–431.

Section 2

Disorders of the pituitary gland

Handbook of Clinical Neurology, Vol. 124 (3rd series)
Clinical Neuroendocrinology
E. Fliers, M. Korbonits, and J.A. Romijn, Editors

Chapter 10

Evaluation of pituitary function

KELLY CHEER AND PETER J. TRAINER*
Department of Endocrinology, Christie Hospital NHS Foundation Trust, Manchester, UK

INTRODUCTION

Investigation of a patient with a potential pituitary problem can be complex, and requires specialist oversight and interpretation. Clinicians should, at a minimum, have reason to suspect a pituitary disorder before proceeding to biochemical testing, and as such, investigations are used to confirm or refute a suspected diagnosis. Testing for growth hormone (GH) deficiency in patients complaining of lethargy, for example, is inappropriate in the absence of reason to suspect pituitary disease. This chapter aims to give a rational, reliable and strategic approach to pituitary investigation with understanding of the underlying physiology, thereby increasing confidence when seeing patients with pituitary dysfunction or reading about dynamic pituitary function tests in clinical letters.

REASONS FOR UNDERTAKING PITUITARY INVESTIGATIONS

The initial presentation of pituitary disease can be very varied, and the indications for assessment of pituitary function include:

1. symptoms and signs of pituitary hormone excess
2. symptoms and signs of pituitary hormone deficiency or insufficiency
3. visual field defect, classically a bitemporal hemianopia due to compression of the optic chiasm
4. acute presentation: severe headache in pituitary apoplexy, cranial nerve palsies, or rarely with cerebrospinal fluid rhinorrhea
5. incidental finding of pituitary mass on magnetic resonance imaging (MRI) brain scan performed for another indication
6. as part of long-term monitoring program for endocrine "late effects" of cancer therapies
7. following traumatic brain injury
8. following postpartum hemorrhage (Sheehan's syndrome)
9. in the presence of congenital pituitary abnormalities.

Disorders of hypersecretion of pituitary hormones are dealt with separately in this volume, and their investigation will not be covered here, but all can be associated with varying degrees of hypopituitarism.

Pituitary hormone insufficiency or hypopituitarism can arise in a number of ways. Firstly, it can present with symptoms or signs due to deficiency of the hormones from the target organ, such as hypogonadism causing erectile dysfunction or amenorrhea, or hypoadrenalism leading to tiredness and postural hypotension. In cases with a pituitary origin, deficiency of the target organ hormones are found, without the expected compensatory rise in pituitary hormone levels, e.g., low follicle-stimulating hormone (FSH) and luteinizing hormone (LH) in the context of hypogonadism. Secondly, patients can present with mechanical signs of an underlying pituitary macroadenoma, such as headache, visual field defect, or rarely, with cranial nerve palsies; it is important in these circumstances to establish whether the pituitary adenoma is secretory or nonfunctioning.

In addition to a wide range of clinical presentations, with improved resolution of MRI scans, it is increasingly common to investigate patients with an incidental finding of a pituitary adenoma found on brain scans; pituitary function testing is often required. Even with a nonfunctioning pituitary adenoma, hypopituitarism may occur due to compression of the residual healthy pituitary gland and pituitary stalk.

A further group of patients is those in whom development of pituitary hormone deficiency may be anticipated. External craniospinal radiotherapy is the best studied cause of insidious pituitary failure; this can

*Correspondence to: Peter J. Trainer, Professor of Endocrinology, Christie Hospital NHS Foundation Trust, Wilmslow Road, Manchester, M20 4BX, UK. Tel: +44-161-446-3666, E-mail: peter.trainer@manchester.ac.uk

occur in association with radiotherapy for a primary brain or pituitary tumor or as scatter therapy from head and neck carcinoma treatment. As the long-term prognosis for those with central nervous system (CNS) tumors improves, so cancer survivors will increasingly experience endocrine "late-effects" and require screening for pituitary dysfunction (Cohen, 2005). It is important to ensure such patients are screened and followed up annually, so that potentially undiagnosed pituitary insufficiencies are detected in a timely manner.

In virtually all causes of hypopituitarism, deficiencies develop in a predictable order, with GH deficiency before gonadotropin deficiency, and then later thyroid-stimulating hormone (TSH) and adrenocorticotropic hormone (ACTH) deficiencies. In addition to the effect of radiotherapy, hypopituitarism may be the presenting feature of tumors themselves, such as craniopharyngiomas, or rarely, of pituitary metastases from another primary site such as breast or prostate carcinoma. It is important to remember that neither a pituitary adenoma nor pituitary radiotherapy will cause the development of diabetes insipidus (DI) (although this can occur following pituitary surgery), so if DI is present, then the hypopituitarism has not occurred due to radiotherapy or the presence of a pituitary adenoma.

Traumatic brain injury is a further cause of pituitary failure, which can be subtle in its presentation and, if untreated, have an effect on the progress and outcome (Mesquita et al., 2010). The diagnosis is made more complex as the order of development of pituitary hormone deficiencies may vary from that seen in other conditions. Pituitary involvement can also occur in a number of inflammatory conditions with progressive hypothalamic and pituitary destruction, such as sarcoidosis, Langerhans cell histiocytosis (formerly called histiocytosis X) or Wegener's granulomatosis (Lury, 2005). These conditions can be associated with the development of DI. Lymphocytic hypophysitis is a rare cause of hypopituitarism, but important to recognize as ACTH deficiency tends to occur early in its natural history.

APPROACH TO THE PATIENT IN PITUITARY CLINIC

There are clearly a number of reasons why a patient may be referred for assessment of pituitary function, with symptoms which can be very nonspecific, such as lethargy. As such, the most crucial first step in evaluating a patient is a thorough history of the symptoms and their medications, particularly exogenous glucocorticoids including inhaled glucocorticoid therapy. Family history of pituitary problems should be documented, due to the association of genetic forms of pituitary dysfunction, such as multiple endocrine neoplasia type 1 and inherited forms of acromegaly.

PRINCIPLES OF PITUITARY ASSESSMENT

The evaluation of pituitary function is often based on clinical assessment, amalgamated with relevant laboratory and radiologic investigations. Testing should assess the target organ hormone, along with the relevant pituitary hormone. For example, for a patient with hypogonadism, LH and FSH levels paired with the testosterone or estradiol level will indicate whether the cause is primary gonadal failure (elevated gonadotropins) or secondary to pituitary dysfunction (low-normal or suppressed gonadotropins). Dynamic pituitary testing may be needed in addition to basal pituitary function testing. MRI scanning is required to look for potential causes of pituitary dysfunction, and is described in a separate chapter (Chapter 11). Visual field assessment should be carried out and ophthalmology input may be required, particularly if any detected pituitary lesion is in close proximity to the optic chiasm. If needed, further investigations should be arranged depending on the cause of pituitary dysfunction, such as chest radiography and serum angiotensin-converting enzyme (ACE) levels for sarcoidosis, or assessment for other systemic involvement in Wegener's granulomatosis or Langerhans cell histiocytosis.

BASAL PITUITARY BLOOD TESTS

Basal pituitary blood tests are usually the first-line investigation. They should be taken with the patient resting, and because of the circadian variations, these investigations should ideally be performed between 7 a.m. and 9 a.m. The circadian decline in cortisol and testosterone make the interpretation of samples drawn later in the day very challenging. The baseline panel of blood tests should include:

1. TSH and free T_4 levels
2. LH, FSH, and testosterone for men, along with sex hormone-binding globulin (SHBG)
3. LH, FSH, and estradiol for women (but not for women taking estrogen therapy as either a combined oral contraceptive pill or hormone replacement therapy)
4. prolactin
5. cortisol (ACTH levels if primary cortisol deficiency is suspected)
6. insulin-like growth factor 1 (IGF-1)
7. if diabetes insipidus is suspected, paired serum and urine osmolalities along with urea and electrolyte levels should be performed.

If appropriate, estrogen therapy should be discontinued at least 6 weeks before measuring LH, FSH, estradiol, and cortisol (see below).

In many circumstances, dynamic testing is necessary to diagnose (or equally, to exclude) underlying pituitary hypofunction. As a general rule, if the diagnostic suspicion is of underlying hormone insufficiency, then the tests will aim to stimulate production of the particular hormone, and for hormone excesses, to inhibit their secretion.

EVALUATION OF THE PITUITARY–ADRENAL AXIS

Whilst ACTH deficiency may present late on in the development of hypopituitarism, the consequences of cortisol deficiency can be fatal, and therefore timely recognition of ACTH deficiency is important. There is, however, some controversy related to the best way to test this.

There are a number of means of testing the hypothalamic–pituitary–adrenal (HPA) axis. The aim of testing is to judge the ability to mount an adequate cortisol response to physiologic stress, and therefore judge the need for the patient to have emergency steroid cover at times of intercurrent illness, and possibly even initiate lifelong hydrocortisone replacement therapy.

Cortisol production in health

Cortisol production from the adrenal glands occurs in response to ACTH secretion from the pituitary, which in turn is secreted after stimulation by corticotropin-releasing hormone (CRH) and also by vasopressin (AVP) from the hypothalamus.

There is a clinical spectrum associated with ACTH deficiency. With severe deficiency, patients present with features of marked hypotension, oliguria, and deranged electrolytes. In other circumstances, patients may have adequate HPA axis function for day-to-day health, but lack the reserve to increase cortisol levels in times of stress, such as surgery, trauma, or sepsis. Whilst this latter group of patients may not need long-term replacement of hydrocortisone, they must be aware of the need to introduce "sick-day" hydrocortisone in the case of intercurrent illness or physiologic stress.

It is important to note that there can be considerable variation in serum cortisol measurements between laboratories, according to the assays utilized and their methodology (Clarke et al., 1998). Comparison of cortisol responses between different laboratories can therefore be difficult. When measuring a serum cortisol level, this refers to the total cortisol level, the majority of which is bound to cortisol-binding globulin (CBG) in the serum and therefore biologically inactive. Only 5–10% of cortisol is in the free and biologically active form. Serum cortisol measurements are therefore affected by changes in CBG level, and this is increased in pregnancy, or with oral estrogen therapy. Oral estrogen should therefore be discontinued where possible, at least 6 weeks prior to evaluation of the HPA axis, as the induction of CBG will artificially increase the total cortisol level. As well as CBG, around 15% of cortisol is bound to albumin and as such, testing can be more complex in patients with disorders affecting their albumin, such as nephrotic syndrome or hepatic cirrhosis. Salivary cortisol is a surrogate marker for serum free cortisol and is a potential means of minimizing the effect of estrogen therapy on CBG (Laudat et al., 1988).

Measurement of serum cortisol

An early morning cortisol level is an essential part of a screen of pituitary function, and is taken at the time of maximal activity of the endogenous HPA axis. An early morning serum cortisol of < 100 nmol/L strongly suggests ACTH deficiency in the context of suspected pituitary pathology, and treatment with glucocorticoid therapy should almost always commence immediately. Jones et al. (1994) suggest that an early morning cortisol level of 400 nmol/L or higher is likely to be representative of normal HPA axis function, and further dynamic testing of the HPA axis is not required.

If a patient is suspected to have hypoadrenalism in any situation where they are acutely unwell, then samples should be sent immediately for cortisol and ACTH, regardless of the time of day. The stress of acute illness will override the usual circadian rhythm and HPA axis activity will be maximal, and as such, a random serum cortisol measurement would generally be greater than 500 nmol/L. If adrenal insufficiency is suspected (either primary or secondary) in the context of acute illness, treatment with glucocorticoids should be commenced without delay; such treatment can be potentially life-saving. If the random cortisol, sent prior to administration of hydrocortisone, is > 500 nmol/L, i.e., shows an appropriate response to the intercurrent illness, then glucocorticoid therapy can be stopped. If not, then steroids are continued until the patient recovers from their stressful event, at which point dynamic testing of the HPA axis can be performed safely. The serum ACTH level can be useful to determine whether hypoadrenalism is primary or secondary in etiology, but does not affect management in the acute situation.

Insulin tolerance test

The insulin tolerance test (ITT) is widely held as the "gold standard" investigation for assessing ACTH and cortisol reserve. An ITT aims to simulate a physiologic stress response in a controlled environment, by means of inducing a hypoglycemic episode with intravenous short-acting insulin. With an intact HPA axis, hypoglycemia causes ACTH and GH release from the pituitary, and a subsequent cortisol release from the adrenal glands. An ITT therefore tests the functioning of the entire HPA axis: hypothalamus, pituitary, and adrenal glands need to function appropriately in order for a cortisol response to be generated. It is important to note that the results cannot be interpreted unless adequate hypoglycemia, with serum blood glucose < 2.2 mmol/L, is achieved; this may be difficult in patients with Cushing's disease or acromegaly, in whom insulin resistance may be high. Serum cortisol should rise to > 500 nmol/L following the episode of hypoglycemia. Those with a suboptimal response may require steroid cover for major illnesses or intercurrent stresses, but may not require daily hydrocortisone administration. The nature of hypoglycemia means that the test should not be undertaken in patients with a history of heart disease, epilepsy or unexplained blackouts; caution must also be shown in older patients.

The indications, contraindications, preparation and protocol for the ITT are shown in Table 10.1 (note that the ITT is also used as a test for growth hormone deficiency).

Short synacthen test

The most commonly used test of cortisol reserve is the short synacthen test (SST), which is fundamentally a test of adrenal reserve, and was introduced as a test for primary adrenal failure (Speckart et al., 1971). Synacthen is a synthetic peptide which acts as a mimetic of ACTH, and therefore tests the adrenal glands directly, rather than via stimulation of pituitary ACTH secretion. In the context of longstanding ACTH deficiency, the adrenal glands will atrophy, and as such, the zona fasciculata will no longer be able to produce cortisol in response to stimulation by ACTH. If ACTH deficiency has developed acutely and recently, for example, in the context of recent pituitary surgery, or pituitary apoplexy, then a synacthen test is not the correct investigation to choose to measure the HPA axis, as the adrenals may not have atrophied and thus will respond normally to synthetic ACTH. In this situation, an ITT should be utilized.

The major advantages of a SST are that it is relatively simple to perform and requires fewer blood samples than an ITT, but with the downside that it does not test growth hormone reserve. It is a safe procedure, but there have been rare reports of anaphylaxis following injection of synacthen, and caution should be taken in patients

Table 10.1

Protocol for insulin tolerance test

Indication	Assessment of cortisol and GH reserve
Contraindications	Ischemic heart disease – normal ECG required prior to test
	Epilepsy or history of seizures
	Untreated hypothyroidism (impairs GH and cortisol response): this should be corrected before the test is performed
	Untreated severe hypoadrenalism with baseline serum cortisol < 100 nmol/L
	Glycogen storage diseases
Preparation	Fast from midnight (water only permitted)
	Stop estrogen therapy 6 weeks prior if possible
	Treatments for hypoglycemia freely available on ward area
Test protocol	Supervise at all times, with one-to-one nursing
	Confirm baseline ECG normal
	Insert intravenous cannula: samples taken for basal cortisol, GH, and glucose
	Administer 0.15 units/kg of short-acting insulin as intravenous bolus dose (consider higher starting dose if insulin resistance suspected – 0.3 units/kg)
	Repeat initial dose after 45 minutes if not hypoglycemic
	Assess bedside blood glucose reading every 5 minutes
	Serum samples for glucose, cortisol and GH at 30, 45, 60, 90, and 120 minutes
	If symptomatic of hypoglycemia and capillary blood glucose reading < 2.2 mmol/L, draw glucose sample and label as hypoglycemic sample
	Reverse hypoglycemia with oral or intravenous glucose according to patient
Interpretation	Documented hypoglycemia with plasma glucose < 2.2 mmol/L necessary
	Adequate cortisol response – serum cortisol peaks above 500 nmol/L
	Peak GH of < 3 μg/L indicates severe GH deficiency

GH, growth hormone; ECG, electrocardiogram.

Table 10.2

Protocol for short synacthen test

Indication	Diagnosis of hypoadrenalism Alternative to insulin tolerance test in diagnosis of secondary hypoadrenalism due to pituitary hypofunction (not in early postoperative period) Ascertain function of adrenals after withdrawal of glucocorticoid therapy
Contraindications	Previous anaphylaxis to synacthen Cautions: allergy, asthma, atopy
Preparation	Omit evening dose of glucocorticoids where applicable on day before test Discontinue estrogen therapy for 6 weeks where possible
Test protocol	Patients can eat and drink as desired through the test Baseline blood sample for cortisol Administer synacthen 250 μg by intramuscular injection Draw samples for cortisol at 30 and 60 minutes
Interpretation	Cortisol should rise to 500 nmol/L by 30 minutes

with atopic tendencies. This may become relevant when assessing patients with asthma who may have been taking large doses of glucocorticoid therapy. The greatest concern with the SST is the risk of false reassurance, i.e., a normal response with cortisol > 500 nmol/L in a patient with ACTH deficiency, as the onset of ACTH deficiency is too recent for the adrenal glands to have atrophied. Table 10.2 shows important information about performing a SST.

Low-dose short synacthen test

The use of a low-dose short synacthen test (LDSST) has been reported in the literature, typically involving administration of a dose of 1 μg of synacthen, compared to 250 μg used in the standard short synacthen test. The basis for the test is that the 250 μg dose of synacthen is supraphysiologic and far exceeds the plasma levels of ACTH achieved during stress. Chronically understimulated adrenal glands may respond normally to the higher dose of ACTH administered in the standard synacthen test, but only normal adrenal glands will respond to the lower dose. Abdu et al. (1998) studied cortisol responses in patients with suspected or proven pituitary disease, using standard and low-dose ACTH tests, and insulin tolerance tests. They showed that the LDDST had a sensitivity of 100% when an adequate response was defined as 30 min cortisol of > 600 nmol/L; all these patients had a peak serum cortisol of > 500 nmol/L on insulin tolerance tests.

The adoption of the LDDST is hindered by the lack of a commercially available 1 μg preparation of synacthen. Current use requires the 250-fold dilution of the 1 mL vial of standard synacthen, and concerns therefore rise about the accuracy of dilution.

Glucagon stimulation test

Glucagon administration induces ACTH and growth hormone release, but is a less potent stimulus of cortisol secretion, and results should be interpreted with caution. The test is safe and therefore an alternative when the ITT is contraindicated and has the advantage of assessing growth hormone as well as cortisol. The technical aspects of the procedure are discussed in the section on growth hormone deficiency.

Metyrapone test

Metyrapone inhibits 11β-hydroxylase, the final enzyme in the cortisol synthesis pathway, and is used in the treatment of Cushing's syndrome. Administration of metyrapone therefore inhibits conversion of 11-deoxycortisol to cortisol, and the reduction in cortisol drives an increase in ACTH production from an intact pituitary. This then leads to an increase in 11-deoxycortisol, which does not have glucocorticoid activity, and therefore will not cause suppression of ACTH. In patients with hypopituitarism, there is no increase in ACTH, and therefore no increase in the level of 11-deoxycortisol. Clinical use of the test has been limited due to cross-reactivity of 11-deoxycortisol and cortisol in many cortisol assays, a problem that is overcome by measurement using mass spectroscopy (Monaghan et al., 2013).

It is important to remember that a suboptimal peak cortisol response on a dynamic pituitary test does not equate to inadequate cortisol secretion, and should not result in automatic lifelong replacement with glucocorticoids (Paisley et al., 2009).

EVALUATION OF THE PITUITARY–THYROID AXIS

The diagnosis of TSH deficiency relies on the basal measurement of TSH and free T_4. In secondary hypothyroidism, TSH is typically in the lower part of the reference range, with a low or low-normal free T_4. To wait until free T_4 is subnormal before instigating therapy is inappropriate, but there is a lack of consensus on the best way to earlier recognize TSH deficiency and instigate replacement therapy. Dynamic tests are of no value in the diagnosis of TSH deficiency; TRH testing can be misleading as patients with TSH deficiency can have false-positive responses. In patients with suspected pituitary disease thyroid function tests should be interpreted in conjunction with other tests of the other pituitary axes and, when possible, with trend of serial thyroid function tests.

There has been little progress in improving the diagnosis of TSH deficiency for several decades, but Jostel et al. (2009) have described a novel approach to the diagnosis. The TSH index uses a mathematical algorithm to correct the TSH concentration for the thyroxine level, analogous to body mass index (BMI) being weight corrected for height. Further work is required to validate the TSH index.

Once diagnosed, TSH deficiency is treated with levothyroxine to place the free T_4 and T_3 in the upper part of the reference range. TSH cannot be used to monitor therapy; doses of levothyroxine should be adjusted according to the serum free T_4 and possibly free T_3.

EVALUATION OF THE PITUITARY–GONADAL AXIS

Secondary (hypogonadal) hypogonadism is deficiency in estrogen or testosterone as a result of low FSH and LH levels from the pituitary. In women, regular menstruation implies normal gonadotroph function; in this situation measurement of estradiol and gonadotropins adds little to the clinical assessment. Premenopausal women with secondary hypogonadism present with amenorrhea, but may also report reduced libido and vaginal dryness. Gonadotropin deficiency is asymptomatic in postmenopausal women, but the absence of elevated LH and FSH levels serves as a marker of hypopituitarism and necessitates detailed investigation of pituitary function.

In male patients, secondary hypogonadism may present with erectile dysfunction, loss of libido, loss of general well-being and a change in body hair distribution and growth. Diagnosis is made from a low 9 a.m. testosterone with low or inappropriately normal LH and FSH levels. This should be distinguished from primary hypogonadism, where testicular failure results in a low testosterone with elevated gonadotropin levels.

Dynamic tests, such as the gonadotropin-releasing hormone (GnRH) test, have no part to play in the diagnosis of hypogonadism in adult patients; the diagnosis relies on the ratio of basal levels of sex steroids and gonadotropins. The GnRH test is, however, used in the pediatric setting to study isolated hypogonadotropic hypogonadism (IHH) and delayed puberty, although the evidence of its value is uncertain. It has no role in the diagnosis of hypogonadism that results from pituitary damage. Similarly, basal serum inhibin B level is emerging as a test which may have a role in differentiating IHH from delayed puberty (Harrington and Palmert, 2012).

Estradiol and testosterone bind to sex hormone-binding globulin (SHBG), which can complicate the interpretation of serum sex steroid levels. Approximately 99% of testosterone is bound to SHBG with only the free hormone being biologically active. Measurement of SHBG is therefore required in patients with low serum testosterone as it may reveal serum free testosterone to be better than suggested by the total serum testosterone. Obesity, insulin resistance and hypothyroidism are all causes of a low serum SHBG.

PROLACTIN

Prolactin deficiency in itself is not associated with a clinical syndrome, but measurement of prolactin levels can be helpful to distinguish the cause of hypogonadism. In contrast with other pituitary hormones, prolactin is under inhibitory control from the hypothalamus in the form of dopamine. Prolactin can therefore be elevated in the presence of a nonfunctioning pituitary tumor or other pituitary masses which causes compression on the pituitary stalk and a loss of inhibitory input from the hypothalamus.

Testing of prolactin levels is not without its potential problems. As mentioned in the introductory section, it is important to take a careful drug history from patients with pituitary problems, as a number of medications will interact with prolactin levels (Melmed et al., 2011); these are listed in Table 10.3. Other medical problems can also be associated with hyperprolactinemia; these include chronic renal failure, hepatic failure, primary hypothyroidism. Prolactin increase may be physiologic, such as in pregnancy, as a response to high levels of estrogen from the placenta, and in addition, during lactation. Finally, prolactin can be raised in response to stress, or may be idiopathic in nature.

Table 10.3

Drug causes of hyperprolactinemia

Antipsychotics: phenothiazines, flupentixol, risperidone
Antidepressants: tricyclic antidepressants, SSRIs, MAOIs
GI medications: metoclopramide, domperidone, omeprazole, H_2 antagonists
Opiates
Estrogens
Antihypertensives: verapamil, methyldopa, reserpine
Protease inhibitors
Cocaine

SSRIs, selective serotonin reuptake inhibitors; MAOIs, monoamine oxidase inhibitors; GI, gastrointestinal.

In addition to considering other medical conditions and their treatment, there are two potential laboratory pitfalls associated with measurement of prolactin. Hyperprolactinemia may be due to the presence in serum of macroprolactin which is detected to varying degrees by assays, but lacks bioactivity, and therefore is not clinically significant. Polyethylene glycol precipitation is the most commonly used method in laboratories to confirm macroprolatin, and will usually be performed as standard on samples with elevated prolactin results (and no symptoms or signs of hyperprolactinemia).

The second potential pitfall is the "hook" effect. This occurs when the prolactin level is extremely high, as may be the case with large prolactin-secreting macroadenomas. This causes saturation of the assay antibody and thus prolactin is reported spuriously as normal or even low. The hook effect can be overcome by repeating the analysis on a 1:100 diluted serum.

GROWTH HORMONE DEFICIENCY IN ADULTS

Normal growth hormone secretion is pulsatile and therefore random measurement of growth hormone levels is generally not helpful for diagnostic purposes. Actions of growth hormone are mediated via IGF-1, and IGF-1 levels below the reference range for the patient may be indicative of growth hormone deficiency. A "normal" IGF-1 level does not exclude the diagnosis of growth hormone deficiency as 30% of patients with severe growth hormone deficiency have IGF-1 levels within the normal reference range (Hoffman et al., 1994); dynamic testing is therefore needed to confirm the diagnosis. The gold standard investigation to evaluate growth hormone is the insulin tolerance test, the protocol for which is given in Table 10.1. Severe growth hormone deficiency is defined as a peak growth hormone response of 3 μg/L or less, and therapy with growth hormone may be considered (Ho, 2007).

Glucagon stimulation test

A glucagon stimulation test can be used to evaluate the growth hormone and cortisol axes when an insulin tolerance test is contraindicated, or a second diagnostic test is needed. The test is contraindicated in those patients with pheochromocytoma or insulinoma, or patients with glycogen storage disorders or prolonged starvation, in whom glycogen stores cannot be mobilized in response to the test. Table 10.4 shows guidelines for performing a glucagon stimulation test.

Table 10.4

Protocol for glucagon stimulation test

Indication	Assessment of GH and cortisol when hypoglycemia contraindicated Second stimulation test for growth hormone
Contraindications	Pheochromocytoma or insulinoma Starvation > 48 hours or glycogen storage disorders Untreated hypocortisolemia with 9 a.m. cortisol < 100 nmol/L Untreated hypothyroidism
Preparation	Fast from midnight until end of test
Test protocol	Cannulation: baseline bloods for glucose, GH and cortisol Intramuscular injection of glucagon 1 mg (1.5 mg if weight > 90 kg) Further blood for glucose, GH, and cortisol at 30, 60, 90, 120, 150, and 180 minutes Supervised meal following test Warn patients they may experience nausea during the test
Interpretation	Adequate cortisol response defined as peak > 500 nmol/L Severe growth hormone deficiency if peak GH < 3 μg/L

GH, growth hormone.

Arginine stimulation test

Arginine is administered as an intravenous infusion after fasting from midnight, at a dose of 0.5 g/kg to a maximum dose of 30 g. Potential side-effects of arginine include transient flushing, nausea, hypotension, and a metallic taste in the mouth. Growth hormone deficiency is again defined as a peak growth hormone level of < 3 μg/L after arginine stimulation. The arginine stimulation test can be used in patients with a history of seizures, but caution should be taken in interpreting the results in patients who have had cranial irradiation, in whom this can be a less sensitive assessment.

All these tests can produce false-negative results (Biller et al., 2002) and as such, two tests indicative of growth hormone deficiency are generally required to make the diagnosis prior to consideration of treatment, unless the patient in question has other pituitary deficiencies, in which case, only one test is needed. The decision to treat relies on more than biochemical evidence of growth hormone deficiency; for example, in the UK, evidence of impairment of quality of life is required which is best assessed with the Adult Growth Hormone Deficiency Assessment (AGHDA) (McKenna et al., 1999).

ANTIDIURETIC HORMONE DEFICIENCY: DIABETES INSIPIDUS

Diabetes insipidus (DI) can be classified as cranial or nephrogenic. In cranial DI, there is deficiency of antidiuretic hormone (ADH) from the posterior pituitary gland itself. In nephrogenic DI, the pituitary produces adequate ADH, but the kidneys are resistant to its action.

A water deprivation test is used for assessment of patients with suspected DI, but must be done according to a strict protocol to ensure patient safety and accurate results. The test is contraindicated when patients are dehydrated or have untreated anterior pituitary hormone deficiencies. All fluids are withheld from 8 a.m. on the day of the test, and food is restricted to dry food such as toast. A baseline weight is taken with an empty bladder, and if the patient's weight falls below 97% of this baseline weight, the investigation is terminated. Urine output is measured on an hourly basis for volume and urine osmolality, along with interval serum samples for serum osmolality and sodium.

In a normal response, the volume of urine output for each subsequent hour will reduce, and urine osmolality will increase as urine becomes more concentrated in response to ADH. A urine osmolality of > 600 mOsm/kg with serum osmolality of > 295 mOsm/kg is considered to be a normal response to water deprivation, and the test can be terminated.

Diabetes insipidus is suggested if, after the period of dehydration, the plasma osmolality is > 295 mOsm/kg but urine osmolality remains < 600 mOsm/kg, i.e., the urine is still inappropriately dilute in the context of increasing serum osmolality. In this case, synthetic ADH is administered as DDAVP 2 μg intramuscular injection. Once DDAVP has been given, the patient can eat and drink, although fluid intake should be limited to 1 L in the first hour. In the case of cranial DI, the kidneys will still be responsive to ADH, the urine will concentrate, and urine osmolality will increase to > 750 mOsm/kg. In nephrogenic DI, the kidneys do not respond to the DDAVP and urine osmolality remains < 300 mOsm/kg.

A summary of the various outcomes can be seen in Table 10.5. When performed correctly, the water deprivation test has a sensitivity and specificity of 95% for diagnosis of DI, and for differentiating between cranial and nephrogenic causes. Interpretation can be difficult when patients have a partial response to water deprivation, but do not quite reach the criteria for diagnosis, and consultation with an endocrinologist is needed. Hypertonic saline infusion is an alternative test but relies on access to a vasopressin assay (Fenske and Allolio, 2012).

Table 10.5

Summary of the outcomes for a water deprivation test

Postdehydration osmolality (mOsm/kg)		Post-DDAVP osmolality (mOsm/kg)	Diagnosis
Serum	Urine	Urine	
283–295	>750	>750	Normal
>295	<300	<300	Nephrogenic DI
>295	<300	>750	Cranial DI
<295	300–750	<750	Primary polydipsia or partial cranial DI

DDAVP, desmopressin (synthetic ADH); DI, diabetes inspidus.

CONCLUSION

Pituitary function testing should only be undertaken in patients in whom there is reason to suspect pituitary pathology, be it secondary to a mass, radiotherapy, or trauma. Testing should not be undertaken for nonspecific symptoms such as lethargy. Investigation should be performed and interpreted in the context of history and symptoms. With a logical approach to investigation, and well-performed biochemical evaluation, deficiencies can be correctly identified and treatment commence appropriately.

References

Abdu TAM, Elhadd TA, Neary R et al. (1998). Comparison of low dose short synacthen test (1 mcg), conventional dose short synacthen test (250 mcg) and insulin tolerance test for the assessment of the hypothalamo–pitutiary–adrenal axis in patients with pituitary disease. J Clin Endocrinol Metab 84: 838–843.

Biller B, Samuels M, Zagar A et al. (2002). Sensitivity and specificity of six tests for the diagnosis of adult growth hormone deficiency. J Clin Endocrinol Metab 87: 2067–2079.

Clarke PM, Neylon I, Rattatt PR et al. (1998). Defining the normal cortisol response to the short synacthen test: implications for the investigation of hypothalamic–pituitary disorders. Clin Endocrinol 49: 287–292.

Cohen LE (2005). Endocrine late effects of cancer treatment. Endocrinol Metab Clin North Am 34: 769–789.

Fenske W, Allolio B (2012). Current state and future perspectives in the diagnosis of diabetes insipidus: a clinical review. J Clin Endocrinol Metab 97: 3426–3437.

Harrington J, Palmert MR (2012). Distinguising constitutional delay of growth and puberty from isolated hypogonadotrohic hypogonadism: critical appraisal of available diagnostic tests. J Clin Endocrinol Metab 97: 3056–3067.

Ho KKY (2007). Consensus guidelines for the diagnosis and treatment of adults with GH deficiency II: a statement of the GH Research Society in association with the European Society for Pediatric Endocrinology, Lawson Wilkins Society, European Society of Endocrinology, Japan Endocrine Society and Endocrine Society of Australia. Eur J Endocrinol 157: 695–700.

Hoffman DM, O'Sullivan AJ, Baxter RC et al. (1994). Diagnosis of growth hormone deficiency in adults. Lancet 343: 1064–1068.

Jones SL, Trainer PJ, Perry L et al. (1994). An audit of the insulin tolerance test in adult subjects in an acute investigation unit over one year. Clin Endocrinol 42: 123–128.

Jostel A, Ryder WDJ, Shalet SM (2009). The use of thyroid function tests in the diagnosis of hypopituitarism: definition and evaluation of the TSH index. Clin Endocrinol 71: 529–534.

Laudat MH, Cerdas S, Fournier C et al. (1988). Salivary cortisol measurement: a practical approach to assess pituitary-adrenal function. J Clin Endocrinol Metab 66: 343–348.

Lury KM (2005). Inflammatory and infectious processes involving the pituitary gland. Top Magn Reson Imaging 16: 301–306.

McKenna SP, Doward LC, Alonso J et al. (1999). The QoL-AGHDA: an instrument for the assessment of quality of life in adults with growth hormone deficiency. Qual Life Res 8: 373–383.

Melmed S, Casanueva F, Hoffman A et al. (2011). Diagnosis and treatment of hypoprolactinaemia: an Endocrine Society clinical practice guideline. J Clin Endocrinol Metab 96: 273–288.

Mesquita J, Varela A, Medina JL (2010). Trauma and the endocrine system. Endocrinol Nutr 57: 492–499.

Monaghan PJ, Keevil BG, Trainer PG (2013). Mass spectrometry for the endocrine clinic – much to digest. Clin Endocrinol 78: 344–346.

Paisley AN, Rowles SV, Brandon D et al. (2009). A subnormal peak cortisol response to stimulation testing does not predict a subnormal cortisol production rate. J Clin Endocrinol Metab 95: 1757–1760.

Speckart PF, Nocolff JT, Bethune JE (1971). Screening for adrenocortical insufficiency with cosyntropin (synthetic ACTH). Arch Intern Med 128: 761–763.

Further reading

Ham GN, Ginsberg JP, Hendell CD et al. (2005). Growth hormone releasing hormone plus arginine stimulation testing in young adults treated in childhood with cranio-spinal radiation therapy. Clin Endocrinol 62: 628–632.

Kanis JA, McCloskey EV, Johansson H et al. (2010). Development and use of FRAX in osteoporosis. Osteoporos Int 21 (Suppl 2): S407–S413.

Levy A (2004). Pituitary disease: presentation, diagnosis and management. J Neurol Neurosurg Psychiatry 75: iii47–iii52.

Molitch M, Clemmons D, Malozowski S et al. (2011). Evaluation and treatment of adult growth hormone deficiency: an Endocrine Society clinical practice guideline. J Clin Endocrinol Metab 96: 1587–1609.

National Institute for Clinical Excellence (2003). Human growth hormone in adults with growth hormone deficiency. Available online at: http://www.nice.org.uk/nicemedia/live/11504/32665/32665.pdf (accessed April 28, 2014).

Handbook of Clinical Neurology, Vol. 124 (3rd series)
Clinical Neuroendocrinology
E. Fliers, M. Korbonits, and J.A. Romijn, Editors

Chapter 11

Imaging of pituitary pathology

MICHAEL BUCHFELDER* AND SVEN SCHLAFFER
Department of Neurosurgery, University of Erlangen-Nürnberg, Erlangen, Germany

INTRODUCTION

With its wide availability, magnetic resonance imaging (MRI) is currently (and will be for the immediate future) the standard imaging procedure for suspected or proven sellar lesions. Before computed tomography (CT) and MRI were available, pituitary mass lesions could not be directly depicted. Conventional X-ray-studies, pneumoencephalography, and catheter angiography could only reveal indirect signs of a space-occupying intra-, para-, or suprasellar lesion. It was only when CT and, later, low- and high-field MRI were introduced into clinical practice that the radiological diagnosis of pituitary adenomas was refined. A much better morphologic contribution to differential diagnosis of a lesion was obtained and the effects of therapies such as medical treatment, irradiation, and surgery could be monitored. To date, magnetic resonance tomography is generally considered the imaging method of choice for its premium resolution of lesions and surrounding anatomic structures. This is mainly due to the improved soft tissue contrast this method offers. Other advantages are the possibility of direct multiplanar imaging in sagittal, coronal, and axial orientations and the avoidance of ionizing radiation. The goal of all imaging studies in this region is to indicate precisely the location, extent, and size of a sella region tumor. These features, along with signal characteristics, frequently not only allow us to determine the presence of a lesion but also indicate the nature of the lesion. For this purpose, its extension in relation to the various surrounding structures, its structure, and its enhancement pattern must be recognized in order to help in differential diagnosis and in treatment planning (Zimmerman, 1990; Siegel and Hudgins, 1996; Naidich and Russell, 1999; Bonneville et al., 2005; Buchfelder and Schlaffer, 2010). The use of CT has so far been restricted to situations in which MRI is unavailable or when contraindications to MRI exist (e.g., cardiac pacemakers, cochlear implants, ferromagnetic aneurismal clips).

MAGNETIC RESONANCE IMAGING

Magnetic resonance imaging (MRI) techniques for detection and depiction of pituitary adenomas have witnessed rapid evolution, from the first noncontrast MRI studies in late 1980s to the introduction of thin-section contrast MRI scans in early 1990s and, more recently, the use of high magnetic field strength technology. Since it is possible to evaluate the sella and perisellar lesions with high soft tissue contrast and excellent anatomic resolution, T1-weighted spin echo images constitute the basis of any imaging of pathology in and around the sella (Siegel and Hudgins, 1996; Naidich and Russell, 1999; Bonneville et al., 2005; Buchfelder and Schlaffer, 2010). In these pulse sequences air appears black, cerebrospinal fluid (CSF) and fluid-filled cysts are dark, while fat is bright. Large arteries such as the carotid artery appear black for their flow void signal. Gadolinium is used as a contrast medium. Usually, the contrast medium is injected intravenously as a bolus and the postcontrast MR scans are thus obtained with some delay. This is needed to differentiate pituitary gland, blood vessels, and other tissues.

Standard MR imaging of the sella and perisellar region should consist of 2–3 mm thin-section T1-weighted turbo spin echo (TSE) images with and without contrast enhancing medium (i.e., gadolinium) in sagittal and coronal planes on a high (512) matrix. T2-weighted TSE or fast spin echo (FSE) images can provide additional information about the composition of the lesion, such as cystic components and regressive changes. Hyperintense regions in T2-weighted images

*Correspondence to: Prof. Dr. Michael Buchfelder, Professor and Chairman, Department of Neurosurgery, University of Erlangen-Nürnberg, Schwabachanlage 6, 91054 Erlangen, Germany. Tel: +49-9131-85-34566, Fax: +49-9131-85-34476, E-mail: michael.buchfelder@uk-erlangen.de

denote hemorrhages or protein-rich fluids. Bonneville et al. (2002, 2005), in particular, have demonstrated the value of axial T1- and T2-weighted images in which structure, symmetry, or asymmetry of the posterior pituitary can be appreciated.

In addition, MR angiography can provide detailed information about the arterial blood vessels surrounding the sella region, such as potential displacements and encasements. More recently, experience with 3 tesla (3 T) imaging of the normal and pathologic pituitary region has been reported (Wolfsberger et al., 2004; Linn et al., 2011). The higher magnetic field allows a more detailed imaging of the anatomic structures. The cranial nerves embedded in the cavernous sinus can be directly depicted. The acquisition time is shorter and the higher resolution in modified protocols allows for depiction of minute lesions which escape detection in standard 1.5 T images (Portocarrero-Ortiz et al., 2010). Unfortunately, the signal-to-noise ratio also increases resulting in significant artifacts being produced. For this reason, increasing the field strength even more, to 7 T, does not offer advantages for the diagnostic imaging of sellar lesions. The purpose of ultra-high magnetic fields is thus functional imaging (fMRI) rather than depiction of anatomic and pathologic structures (Beisteiner et al., 2011). Experience with sellar lesions and 7 T imaging is in any case very limited. Furthermore, there are several potential artifacts that need to be considered, such as pixel-to-pixel inhomogeneity, partial volume effects, chemical shifting, and others. The reader of images should be familiar with the typical appearances and major variants of the tumors, hyperplasias, malformations, vascular lesions, and inflammations that occur in the sellar and perisellar region (Naidich and Russell, 1999; Bonneville et al., 2005).

In the perisellar region, MR imaging can provide information about the size, location, and extent of a lesion, be it para- (lateral), supra- (superior), retro- (posterior to the dorsum sellae), or infrasellar (sphenoidal). Moreover, there may be indications of potential invasion of a pituitary adenoma into either the sphenoid cavity or cavernous sinus.

The normal pituitary

The correct interpretation of MR images requires a profound knowledge of the normal anatomy of the sellar region and of the normal anatomic variants, such as shape of the bony sella, structure of the sellar diaphragm, pneumatization of the sphenoid sinus, and venous drainage at the sellar floor (Zimmerman, 1990; Siegel and Hudgins, 1996; Naidich and Russell, 1999). The normal adult anterior pituitary usually has an isointense signal compared with the white matter of the temporal lobe on T1-weighted images, whereas the posterior lobe has a hyperintense signal ("bright spot"). The volume distribution is symmetric. The signal is homogenous throughout the gland in both T1- and T2-weighted images (Fig. 11.1). While in young children the gland has an upwards convex shape, in the adult the gland mostly has a plain surface. The volume of the sellar content does not exceed the level of the sella entrance. The maximum

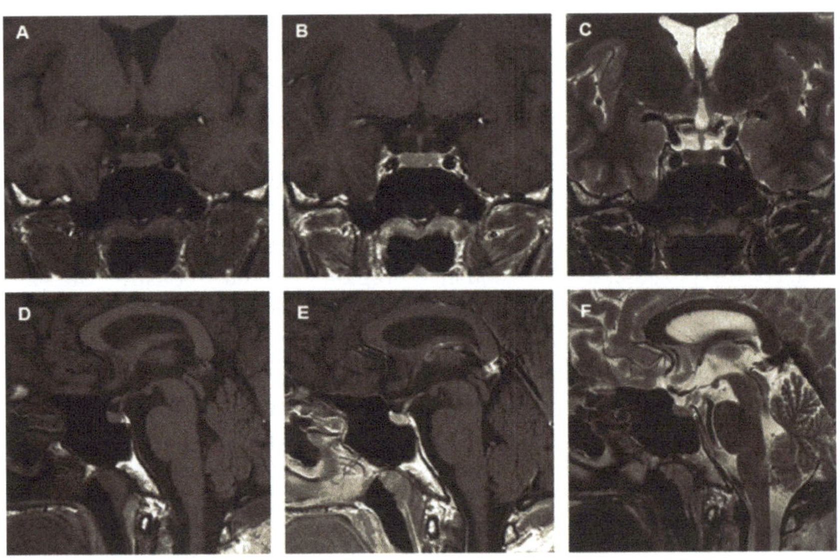

Fig. 11.1. MRI of the normal pituitary gland (1.5 T). In the upper panel (**A–C**) coronal MR images of the normal pituitary gland are shown in nonenhanced T1-weighted (**A**), T1-weighted with contrast medium (**B**), and T2-weighted (**C**) images. In the lower panel (**D–F**) the respective midsagittal plane of the same patient is displayed using T1-weighted images without (**D**) and with contrast medium (**E**). (**F**) represents the respective T_2-weighted images.

height of the gland in males appears to be 8 mm. The infundibulum is slender and descends from the tuber cinereum exactly in the midline. Its diameter is less than 2 mm at its pituitary insertion, but may measure slightly more than 3 mm at the optic chiasm (Simmons et al., 1992). In infants, the gland mostly has a convex surface and a higher signal intensity. A physiologic size increase occurs during pregnancy. Elster et al. (1991) estimated the size increase to be 0.08 mm per week during pregnancy. The surface of the gland becomes convex and often reaches into the opticochiasmatic cistern. The maximal craniocaudal dimension, however, rarely exceeds 10 mm and is reached immediately postpartum. Within 6 months after delivery, the gland regains the size and shape of an inconspicuous adult pituitary (Elster et al., 1991; Dinç et al., 1998; Wolpert et al., 1984). Some bulging of the arachnoid of the opticochiasmatic cistern into the sella is frequently observed as a normal variant; this is known as "cisternal herniation" or partial "empty sella". It results from a deficient sellar diaphragm which allows the CSF pulsations to exert some pressure onto the upper surface of the pituitary gland. The herniation can be expressed to a variable degree, with an empty sella being the extreme variant (Siegel and Hudgins, 1996). Sellar size increase may result. In an empty sella, which exhibits the signal intensity of CSF on MRI, the infundibulum travels through the intrasellar space to reach the flattened remnants of the pituitary which lie compressed against the sella floor at the bottom. The pituitary stalk is visible within the herniation. This "infundibulum sign", initially described on CT (Haughton et al., 1980), is also pathognomonic in MRI.

IMAGING OF PITUITARY ADENOMAS

Standard imaging techniques and protocols as described above are usually used to depict lesions in and around the sella turcica.

Pituitary microadenomas

Pituitary microadenomas are defined as lesions with a diameter below 10 mm and are usually localized within the sella (Hardy, 1969). They can be reliably visualized directly once the tumor diameter exceeds 4 mm, mostly as distinct rounded, oval, triangular, or flattened regions, hypointense to the anterior gland, but isointense to the gray matter of the temporal cortex on T1-weighted images (Fig. 11.2). If the microadenoma is isointense to the gland on T1-weighted images, other sequences may be required for direct depiction. Indirect signs, such as deviation of the infundibulum, upward bulging of the sellar content into the opticochiasmatic cistern, or a localized depression of the sellar floor may indicate a small space-occupying sellar lesion, even

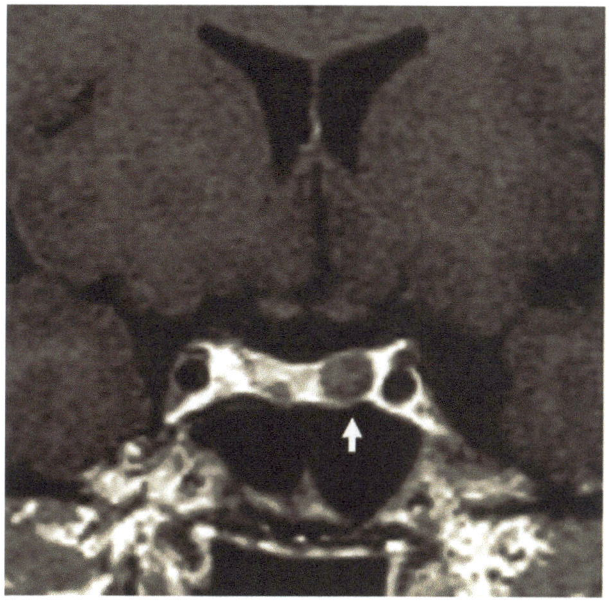

Fig. 11.2. Depiction of a pituitary microadenoma (1.5 T). In this coronal T1-weighted image with contrast enhancement a pituitary microadenoma appears as a distinct zone of lower signal intensity (arrow).

if this is not directly depicted. Hemorrhagic or cystic changes within an adenoma may alter its signal intensity so that a minority is hyperintense in the T1-weighted images. The appearance of microadenomas on T2-weighted FSE or TSE images is more variable (Bonneville et al., 2005). If no adenoma is visible on nonenhanced scans, the use of paramagnetic contrast medium can be helpful. Microadenomas are usually depicted as areas of relative hypoenhancement. The dose of gadolinium has to be selected very carefully. If clinical features suggest the presence of a functioning pituitary adenoma various suggestions have been made for techniques aimed at increasing the sensitivity of direct depiction. This is particularly a problem with ACTH-secreting microadenomas in Cushing's disease. At a rough estimate, only 50% of these tiny tumors are clearly visible on standard T1-weighted TSE images before and after contrast enhancement (Buchfelder et al., 1993). Dynamic scanning and delayed postcontrast images can depict some microadenomas that are not visible on images obtained with the routine parameters. Dynamic scans are taken in 20–30 second intervals after intravenous injection of gadolinium contrast agents. The pituitary stalk and posterior lobe enhance approximately 20 seconds after injection and the anterior pituitary after 80 seconds (Naidich and Russell, 1999; Bonneville et al., 2005). However, one needs to be particularly careful with this technique since artifacts also increase dramatically and the resolution of dynamic images is relatively low. Other proposals to increase the detection rate of small tumors include variations of the signal acquisition protocols. Batista et al. (2005) and Chowdhury et al. (2010) describe an increase in the detection rate achieved by applying variations of T1-weighted sequences. With postcontrast spoiled gradient-recalled acquisitions in the steady state (SPGR) sequences, investigators from the US National Institutes of Health (NIH) could depict several tiny microadenomas which had not been visible on standard T1-weighted TSE sequences. Likewise, the use of a high-field (3 T) magnet and a modified protocol which included the application of corticotropin-releasing hormone (Erickson et al., 2010) or the use of half-dose gadolinium led to an increase in the sensitivity of direct depiction of microadenomas in patients with suspected Cushing's disease (Portocarrero-Ortiz et al., 2010). In some patients in whom standard 1.5 T MRI only reveals some inhomogeneity of the intrasellar space, microadenomas can be clearly depicted by a modified 3 T protocol (Fig. 11.3).

However, there is a huge potential for misinterpretation, since such apparent tiny regions of altered signal intensity within the pituitary gland might also represent other lesions, e.g., cysts of the pars intermedia, or may be due to artifacts, such as partial volume effects. Thus, the findings have to be appreciated with considerable care (Naidich and Russell, 1999; Bonneville et al., 2005). Another problem that can be encountered is a rather diffuse enlargement of the entire pituitary gland, which is suggestive of a small pituitary adenoma of some 8–12 mm, although the tumor, despite all efforts, cannot be directly delineated. A potential explanation for this is a hyperplasia of the pituitary resulting from primary thyroid, adrenal, or gonadal failure; these can be ruled out by appropriate laboratory tests (Freda et al., 2011).

Pituitary macroadenomas

Pituitary macroadenomas are defined as sellar mass lesions with a diameter greater than 10 mm (Hardy, 1969). They can be reliably visualized with routine MR imaging and thus, the task is rather to describe precisely the extent, location, structure, and relation to the surrounding anatomic structures; this helps with the differential diagnosis. It is characteristic for pituitary adenomas that the bulk of the lesion is located within the sella. The tumors may be inhomogenous, since regressive changes may be present (Fig. 11.4). Mostly the extension is superior, into the opticochiasmatic cistern, and may reach up to the foramen of Monro, producing hydrocephalus. The amount of compression and deformation of the optic chiasm is best appreciated on the coronal sections. With increasing suprasellar tumor size the chiasm is increasingly displaced, becomes more and more flattened, and, in some huge adenomas with grotesque suprasellar extension, the visual pathways may escape direct depiction. There may be tumor extension downwards, into the sphenoid sinus, with expansion of the sella or even perforation of the sellar floor which is indicative of invasion. Lateral tumor expansion is also frequent and often the question arises if the tumor is still encased or exhibits an invasive growth pattern (Ahmadi et al., 1986). Only total encasement of the carotid artery, with tumor clearly visible lateral of the vessel, proves invasion beyond doubt. Even grotesque lateral extension sometimes turns out not to have resulted in invasion. A line drawn between the sectioned portions of the intracavernous carotid artery frequently helps to make the distinction (Knosp et al., 1993). In several instances, imaging with very high field strength (3 T) offers the possibility of direct depiction of the invasion site, which is usually not possible with 1.5 T images

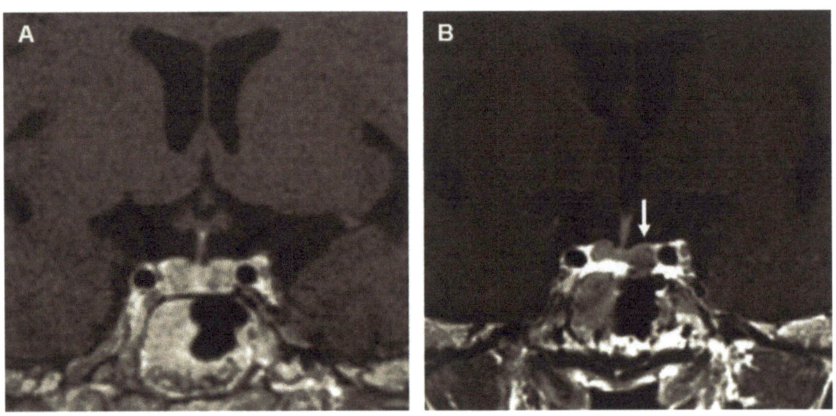

Fig. 11.3. Comparison of 1.5 T and 3 T images used for the depiction of a small pituitary microadenoma. In some cases a higher field strength offers advantages in the direct depiction of small intrasellar lesions, such as in ACTH-secreting microadenomas which are suspected on the basis of dynamic endocrine tests. While standard T1-weighted 1.5 T images may reveal only some intrasellar inhomogeneity (**A**), in the high-field investigation the zone of lesser signal intensity (arrow) becomes much clearer (**B**).

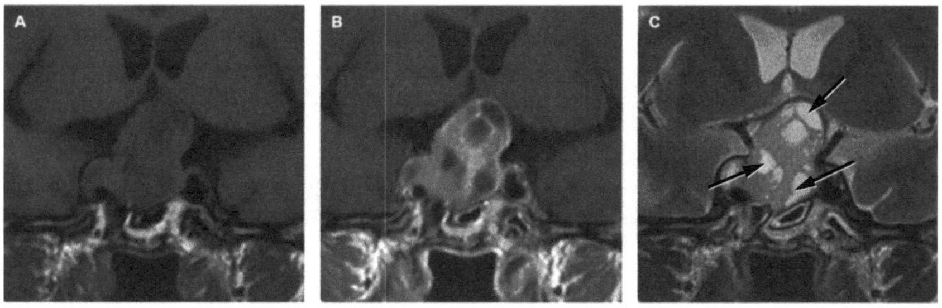

Fig. 11.4. Depiction of a pituitary macroadenoma (1.5 T). In the native coronal T1-weighted image (**A**) the macroadenoma appears as a space-occupying intra- and suprasellar tumor. After administration of contrast media (**B**) it shows a mixed signal intensity indicating cystic portions. In the T2-weighted turbo spin echo (TSE) image (**C**) these cystic portions of the macroadenoma show a bright signal (arrows), indicating protein-rich content.

(Wolfsberger et al., 2004; Yoneoka et al., 2008; Linn et al., 2011). Normally, a compression, flattening and distortion of the normal pituitary gland can be seen, preferably to the posterior or lateral periphery of the tumor. This is particularly well shown on contrast-enhanced images, on which the pituitary gland enhances more quickly and appears much brighter than the adenoma. An easier distinction is made if there is only unilateral parasellar extension of the adenoma. Pituitary adenomas are virtually never invasive into the very cavernous sinus that is covered by a thin layer of compressed and displaced pituitary, and consistently have their major parasellar extension on the contralateral side of the location of the deformed residual normal gland (Buchfelder and Schlaffer, 2010). Pituitary surgeons regularly appreciate this growth pattern to identify and protect normal gland tissue. Recently, the signal intensity of growth hormone-secreting pituitary adenomas in T2-weighted images was used to predict the response of the tumors and their secretion to somatostatin analogs. Heck et al. (2012) demonstrated much better treatment effects in hypointense tumors in contrast to hyperintense ones. They relate the different response to the ultrastructure of the tumors which is revealed by their presentation on MRI. All sparsely granulated adenomas exhibited a hyperintense signal in T2-weighted images. However, there were also isointense tumors which did not allow reliable prediction.

Postoperative magnetic resonance imaging and monitoring of effects of other treatment

Any monitoring of a treatment effect requires the comparison of images before and after the intervention. A baseline postoperative imaging investigation should be performed some 2–3 months after surgery, preferably with standard sequences, comparable to the initial, pretreatment MR scan, unless the clinical situation requires earlier postoperative imaging. After total transsphenoidal resection of a large pituitary adenoma, there is no more tumor visible. The space previously occupied by tumor is filled with CSF. The decompressed optic chiasm and the stretched infundibulum become more clearly visible. The typical postoperative appearance after such a total resection is a more or less an expressed cisternal herniation or secondary "empty sella" (Fig. 11.5). Tissue changes within the sphenoid sinus are suggestive of a transsphenoidal intervention. A virgin sellar floor and sphenoid sinus is usually observed after transcranial procedures. In patients with asymmetric, large tumors, particularly those who harbor invasive tumors, comparison of pre- and postoperative images usually reveals the residual tumor that could not be resected (Fig. 11.6). Residual tumor harbors the same signal intensity as the tumor preoperatively (Bonneville et al., 2005). Extreme caution has to be applied with early postoperative imaging: The postoperative tumor cavity is often filled with blood a few hours and days after surgery and the resorption of implanted packing material used to obtain hemostasis and blood takes several weeks (Dina et al., 1993). Thus, an early postoperative scan might still show a considerable space-occupying lesion which appears quite threatening. Moreover, a variety of other postoperative artifacts and structural changes appear even in the delayed postoperative images recommended here; these need to be differentiated from residual pituitary tumor. Drilling of the skull base, which is frequently required during such operations, may leave metal debris, which produces particularly nasty artifacts. Autologous or heterologous implants, such as fascia lata, fat, or other muscle may have been used to reconstruct the sellar floor or even been introduced into the tumor cavity to prevent a CSF leak (Figs 11.5 and 11.6). Fluid retentions and mucosal regenerates within the paranasal cavities are another type of structure that should not be confused with residual or recurrent tumor (Hald et al., 1992; Mikhael and Ciric, 1988; Steiner et al., 1992, 1994; Chakrabortty et al., 1993). A postoperative asymmetry of the cavernous sinus or even its expansion can be the result of a

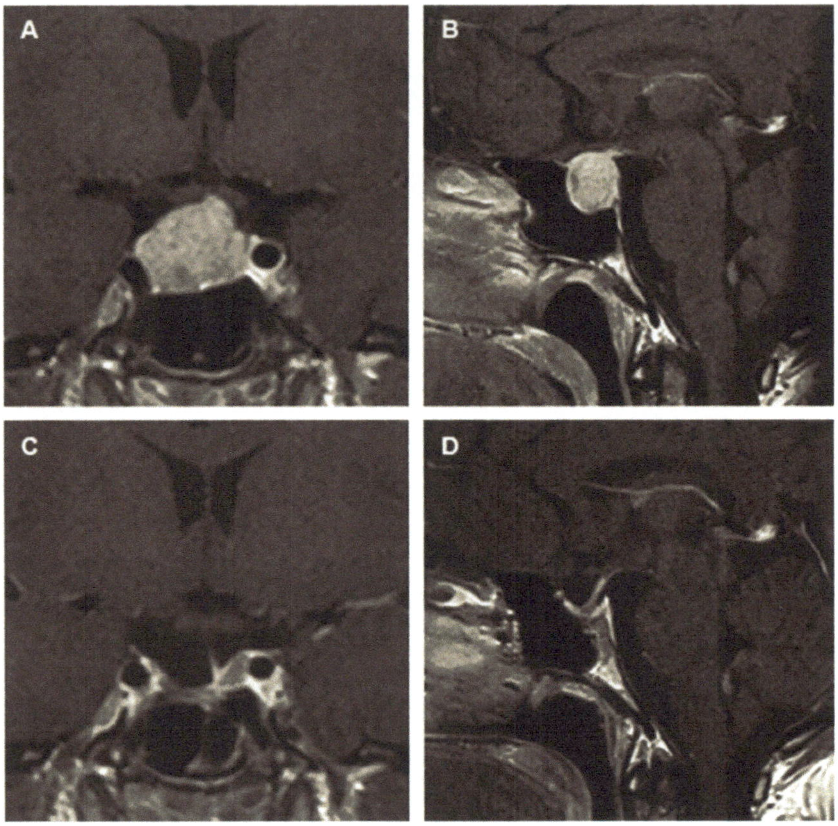

Fig. 11.5. Pre- and postoperative images of a pituitary macroadenoma (1.5 T). In the coronal (**A**) and sagittal (**B**) T1-weighted images the macroadenoma appears as a homogenous space-occupying intra- and suprasellar lesion. Three months following transsphenoidal surgery, delayed postoperative imaging reveals complete tumor resection (**C** and **D**) images with the optic chiasm properly displayed and the pituitary stalk reaching the flattened remnants of the normal gland at the bottom of the sella. The normal postoperative situation is also called "empty sella" or cisternal herniation.

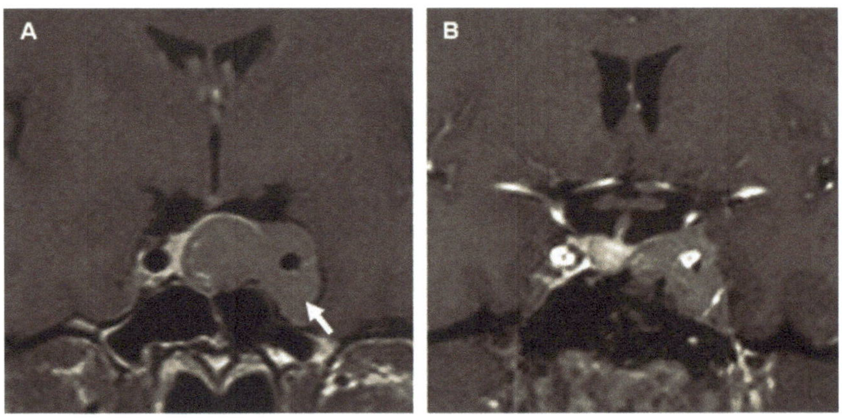

Fig. 11.6. Partial resection of an invasive pituitary adenoma (1.5 T). A large, homogenous, intra-, supra-, and parasellar pituitary adenoma is preoperatively visible in the T1-weighted coronal MR sections (**A**) with tumor also laterally from the carotid artery (arrows), indicating invasion. Three months postoperatively (**B**), the intra- and suprasellar tumor portions are resected, resulting in cisternal herniation. However, there is residual tumor within the cavernous sinus. The hyperintense, deformed normal pituitary is localized on the right side and still visible.

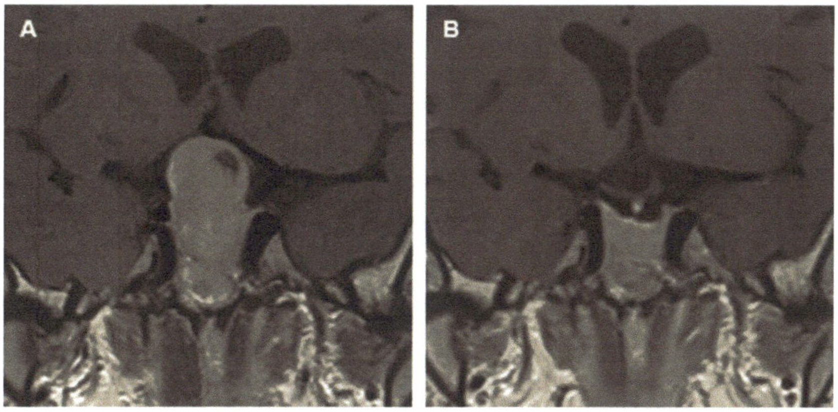

Fig. 11.7. Monitoring the effects of dopamine agonists. Tumor volume response to medical treatments and irradiation may also be monitored by repeat MRI investigations. This is a macroprolactinoma presenting as a large intra- and suprasellar tumor (**A**) treated by the dopamine agonist cabergoline. Rapid and consistent shrinkage occurred after 3 months (**B**) of medical therapy. Prolactin levels decreased from 5510 ng/mL to 171 ng/mL between examinations of the patient. Only an adherence to identical acquisition parameters and imaging protocols at each investigation provides reliable data.

successful decompression of a previously distorted venous compartment (Buchfelder and Schlaffer, 2010). In further follow-up it is essential that the whole series of postoperative scans is obtained using identical scanning parameters. The entire set needs to be evaluated rather than only the current and previous examination, since most pituitary adenomas are slowly growing and changes become more apparent as time advances. The baseline delayed postoperative image, obtained 2–3 months after the surgery, has a pivotal role to play here. Sometimes it is initially impossible to differentiate implanted materials and scarring from residual tumor. Only the progression of a tumor over time, while implants regress, resolves the diagnostic dilemma. The same principles apply to the follow-up of pituitary tumors after irradiation or during medical treatments (Fig. 11.7). Only keeping the same acquisition parameters and imaging protocols enables a proper evaluation (Tofts and Collins, 2011). The orientation of sections used to cut through the pituitary gland and pathology, respectively, play an important role. There should be strict adherence to proposals such as the one made by Bonneville et al. (2005), who suggested cutting perpendicular to a line introduced just below the corpus callosum (Fig. 11.8). Gantry variations and inconsistencies are probably the most frequent reasons for misinterpretations and errors in volume determinations based on 2D analysis. Considering follow-up intervals, a safe practice would be to obtain a repeat MR scan every year following the initial delayed postoperative baseline investigation. In patients with residual tumor this practice should be continued, while in patients in whom no tumor is visible after 5 years, the follow-up intervals can be extended (Soto-Ares et al., 2002; Greenman et al., 2003). While in functioning pituitary adenomas hormonal laboratory parameters can also be used to monitor tumor growth (Zirkzee et al., 2004), in nonfunctioning adenomas imaging surveillance is required for life. Reddy et al. (2011) found that 25% of both tumor size expansions and recurrences occurred more than

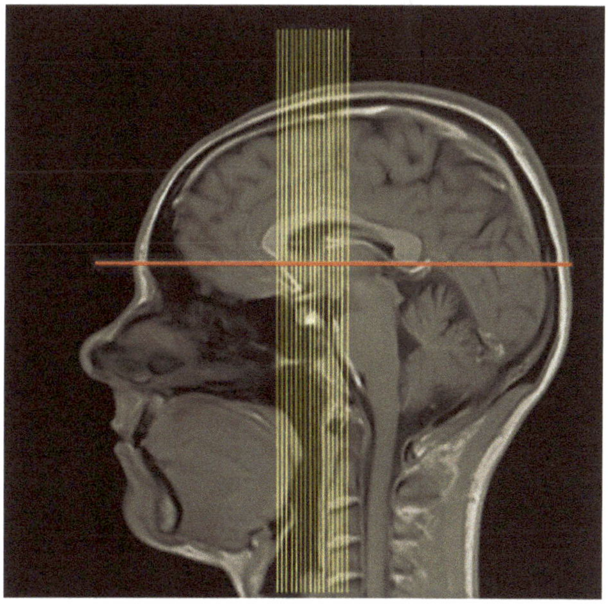

Fig. 11.8. Orientation of sections involving the pituitary gland and lesions. A practical proposal to enable comparisons of images obtained at different times is to cut through any pituitary pathology perpendicular to a line which is introduced just below the corpus callosum. Reliably reproducible coronal sections result. Since gantry variations represent the main reason for interpretation errors in the evaluation of treatment effects and follow-up examinations, keeping an identical acquisition standard is highly recommended.

10 years after the operations on the tumors. Volumetry is the most precise evaluation of the tumor dimension over observation time (Benesch et al., 1995). However, it requires specific software and is extremely time, staff, and labor intensive, so that its use to date has been reserved for research projects and clinical studies. It is not routinely used in clinical practice.

Intraoperative magnetic resonance imaging

While standardized pre- and postoperative imaging is widely available, intraoperative MR imaging is still restricted to limited surgical centers. However, in recent years, their number has increased dramatically. The major asset to the surgeon is that in pituitary surgery intraoperative imaging can provide information about residual tumor that is hidden initially but can be dealt with while the patient is still in the operating room. Therefore, this technique helps to increase the rate of complete tumor resection. In contrast to delayed, standard postoperative images, intraoperatively, blood and residual tumor need to be differentiated. Thus, in contrast to routine delayed postoperative MR scanning, sagittal and coronal T2-weighted TSE sequences are the ideal sequences to fulfill this task since they have the best potential for differentiation (Fahlbusch et al., 2005; Buchfelder and Schlaffer, 2012). The surgeon needs to compare the pre- and intraoperative situation. Especially in large asymmetric adenomas of giant size, where intraoperative estimation of the resection extent is difficult, tumor residuals can be localized and more aggressively resected in a targeted fashion. An overview of the results of the literature suggests that a total resection can be achieved in approximately an additional 30% of patients operated upon, but this obviously depends on the patients selected to undergo surgery when equipment is available (Buchfelder and Schlaffer, 2012). Moreover, the systems allow navigated surgery for which the MRI dataset serves as a basis. In any anatomic variants, such as kinks of the carotid artery into the sella and poor pneumatization of the sphenoid sinus and in reoperations, in which the first intervention has caused tissue defects and scars, image-derived information about anatomic structures to be expected is appreciated.

DIFFERENTIAL DIAGNOSIS

Magnetic resonance imaging in particular may help with the differential diagnosis of lesions in and around the sella turcica, and a huge body of literature covers this issue (Zimmerman, 1990; Donovan and Nesbit, 1996; Siegel and Hudgins, 1996; Naidich and Russell, 1999; Ruscalleda, 2005). There is no single radiologic sign which could predict the nature of a lesion with 100% reliability. Thus, only the consideration of clinical features and laboratory findings together with the imaging results provides enough information to establish a reliable differential diagnosis (Zimmerman, 1990; Naidich and Russell, 1999; Ruscalleda, 2005). After all, the radiologic differential diagnosis is restricted to the nonhormone-secreting lesions, since pituitary adenomas with hormonal oversecretion can be identified by their typical symptoms and laboratory investigations. Although a large variety of different lesions may occur in and around the sella turcica (Table 11.1), data about the relative frequency is useful in clinical practice. Up to 90% of pituitary pathology verified histologically represents

Table 11.1

Differential diagnosis of sellar and perisellar tumors

Benign solid parasellar tumors
- Meningiomas
- Hamartomas
- Paragangliomas
- Lipomas
- Neurinomas/ schwannomas
- Granular cell tumors
- Ganglion cell tumors

Malignant solid parasellar tumors
- Gliomas
- Germ cell tumors
- Primary lymphomas
- Metastases
- Supratentorial primitive neuroectodermal tumors (PNET)
- Ependymoblastomas

Low-grade solid malignant tumors
- Craniopharyngiomas
- Chordomas/chondrosarcomas
- Hemangiopericytomas
- Langerhans cell histiocytosis (LCH)

Non-neoplastic pathologies of the parasellar region
- Granulomatous, infectious and inflammatory lesions
- Pituitary abscess, bacterial and fungal
- Tuberculosis
- Sarcoidosis
- Wegener's granulomatosis
- Mucocele (sphenoid sinus)
- Hypophysitis (lymphocytic, granulomatous, xanthomatous)
- Tolosa–Hunt-syndrome

Vascular lesions
- Aneurysms
- Cavernous sinus thrombosis

Entirely cystic lesions
- Rathke's cleft cyst
- Arachnoid cyst
- Colloid cyst
- Epidermoid cyst

Miscellaneous
- Empty sella syndrome
- Arachnoiditis opticochiasmatica

pituitary adenomas, with craniopharyngiomas and Rathke's cleft cysts together accounting for less than 10(–20)%. Meningiomas, chordomas, hypophysitis, and all other entities are extremely rare (Freda et al., 1996; Buchfelder and Schlaffer, 2010; Famini et al., 2011). Zimmerman (1990) has grouped pathologic conditions with clinical presentations, age preponderance, and typical presenting MRI features. Craniopharyngiomas, chiasmal gliomas and hypothalamic astrocytomas are more frequent lesions in the pediatric population. Almost exclusively, craniopharyngiomas and aneurysms have a ring-like calcification, whereas scattered calcification may be found in meningiomas, craniopharyngiomas, pituitary adenomas, teratomas, dermoids, and chordomas.

A thickening of the infundibulum is typically seen in germinomas, infundibulohypophysitis, craniopharyngiomas, and in some metastases. Craniopharyngiomas, pituitary adenomas, meningiomas, optic-hypothalamic gliomas and epidermoids may be partially cystic. A really low density in CT and a high signal intensity in T2-weighted MRI is characteristic for lipomas, dermoids, and teratomas. However, even collision or duplicate lesions have been described in a few patients who turned out to harbor at least two different pathologies within the sellar region (Koutourousiou et al., 2010).

Craniopharyngiomas and Rathke's cleft cyst

Craniopharyngiomas and Rathke's cleft cysts both arise from Rathke's pouch remnants and thus can extend from the bottom of the sella throughout the third ventricle. Craniopharyngiomas may have an intrasellar component, but are usually predominantly suprasellarly localized (Fig. 11.9). They frequently consist of cystic, solid, and calcified portions. Calcification is a typical feature of craniopharyngioma but may also (though much less frequently) be seen in aneurysms or pituitary macroadenomas. Microcalcifications are visible in some 75% of craniopharyngiomas, but these will escape MRI detection. CT is very sensitive in picking up

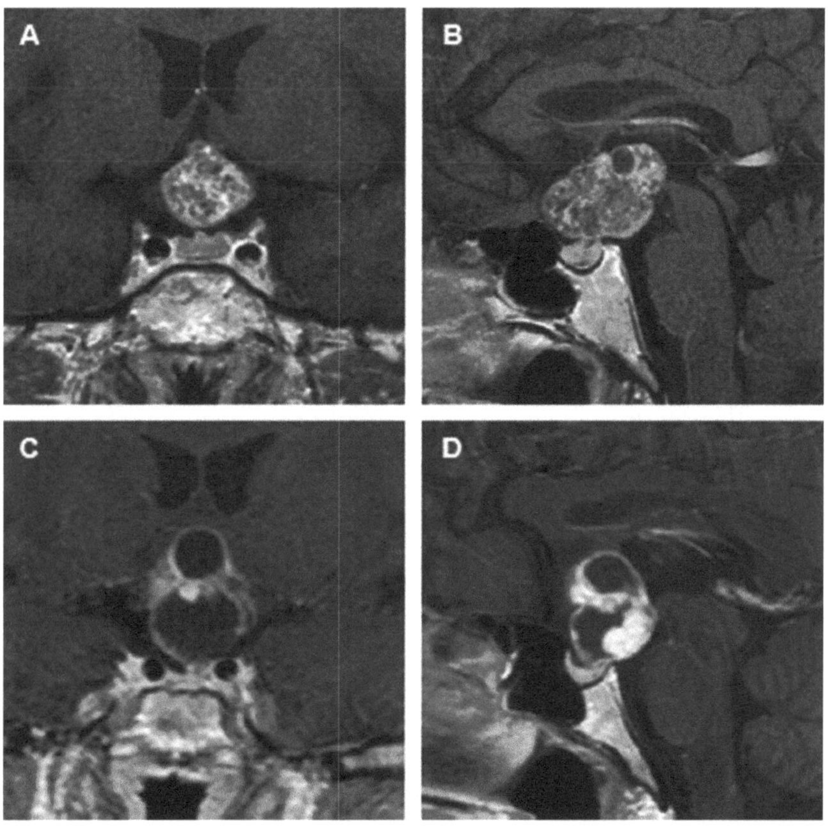

Fig. 11.9. Depiction of a craniopharyngioma (1.5 T). Since craniopharyngiomas originate from residuals of the cranopharyngeal duct, which are found from the base of the sella up to the hypothalamus, the radiologic appearance can vary considerably in structure, size, localization, and extension. However, the combination of cystic portions, solids tumor, and calcification (which is represented as a lack of signal on MRI), as seen in this patient in coronal (**A**) and sagittal (**B**) projections, is virtually pathognomonic for craniopharyngioma. Radiologically the more solid adamantinomatous subtype (**A** and **B**) may be differentiated from the papillary (**C** and **D**), predominantly cystic subtype.

calcifications and may aid in the differential diagnosis. Both the solid components and the cyst walls enhance with contrast. The variable histologic appearances of craniopharyngiomas are also reflected in their MRI characteristics. Cysts, although mostly hypointense on T1-weighted images, may also be isointense, hyperintense, or of mixed intensity. The shape of the lesion may be well confined and oval, while other craniopharyngiomas are quite extensive, lobulated, and thus harbor an irregular surface. Only a small minority of craniopharyngiomas have only intrasellar extension. On the other hand, they rarely invade the cavernous sinus (Donovan and Nesbit, 1996; Siegel and Hudgins, 1996). Sartoretti-Schaefer et al. (1997) have established criteria to differentiate the two histologic subtypes of craniopharyngiomas, the adamantinomatous and the squamous-papillary, on the basis of MRI features. Statistically significant parameters for differentiation were the encasement of blood vessels, the lobulated or round shape, and the presence of hypointense cysts or predominantly solid appearance.

In contrast, Rathke's cleft cysts are entirely cystic lesions, well-defined midline lesions and completely lack solid and calcified portions. The signal characteristics of the cyst fluid are variable but mostly homogenous throughout the cyst. The cyst wall may be thick or thin, with or without contrast enhancement. The midline anterior displacement of the infundibulum is a typical sign (Fig. 11.10). The normal gland is located at the bottom of the sella. The cyst is located in the upper portion of the sella, above the pituitary gland or within the suprasellar cistern. Although the entity can be suggested on the basis of imaging, the final differentiation from cystic craniopharyngioma requires histologic investigation of the epithelial membrane. Many Rathke's cleft cysts are small and asymptomatic. They are usually discovered as incidental findings.

Suprasellar and parasellar meningiomas

Suprasellar meningiomas have a wide attachment to the dura of the planum sphenoidale from which they arise ("dural tail"). In the past, skull X-rays demonstrated a hyperostosis of the tuberculum sellae. While in pituitary adenomas the normal pituitary is dislocated into the superolateral periphery of the tumor, in suprasellar meningiomas the normal gland is usually compressed dorsally of the lesion into the basis of the sella (Fig. 11.11). Moreover, meningiomas usually have a bright signal enhancement after application of contrast medium. Most of them are homogenous in structure, but cysts may be present in a few. Parasellar meningiomas originate from the walls of the cavernous sinus. They invade and distend the cavernous sinus and might narrow the carotid artery, which they encase, in its intracavernous portion up to total obstruction (Zimmerman, 1990; Donovan and Nesbit, 1996; Siegel and Hudgins, 1996). The mass lesion is predominantly lateral to the sella. The involvement of the arteries can be used for the differential diagnosis, since pituitary adenomas very rarely cause vascular arterial compromise (Molitch et al., 2012). In adults, meningiomas are the third most frequent parasellar lesion, after pituitary adenomas and craniopharyngiomas. Petroclival meningiomas may have cavernous sinus extension.

Chordomas and chondrosarcomas of the clivus

Chordomas and chondrosarcomas of the clivus distend or may even destroy the clivus and the dorsum sellae (Fig. 11.12). They usually involve the sphenoid sinus bone more than the sella. Their signal intensity, enhancement, and density characteristics are nonspecific and variable. The most reliable feature is bone destruction of the

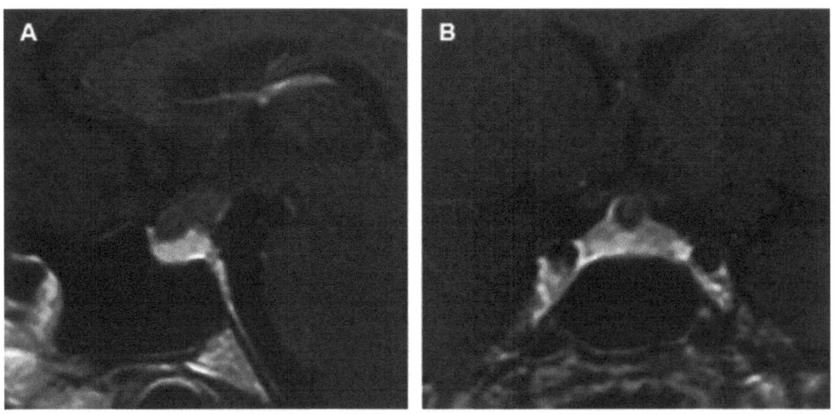

Fig. 11.10. Depiction of Rathke's cleft cyst (1.5 T). In contrast to craniopharyngioma, Rathke's cleft cysts lack solid tumor portions and calcifications. The normal pituitary is located at the bottom of the sella, which is particularly visible on sagittal sections (**A**). This lesion exerts compression on the visual pathways. The chiasmal compromise is recognizable on the coronal projections (**B**).

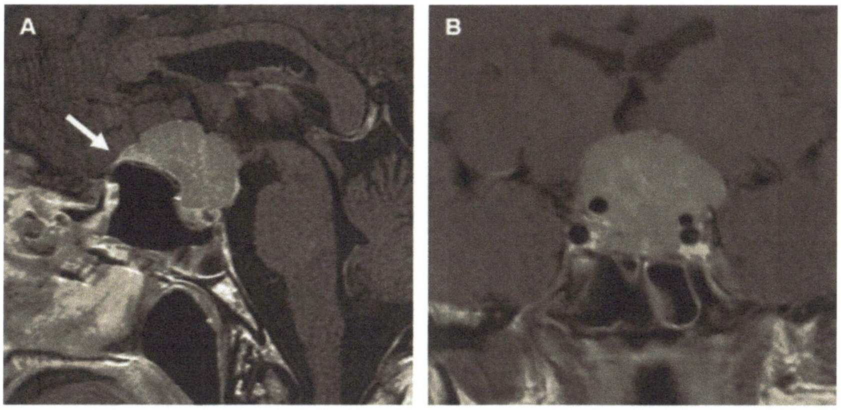

Fig. 11.11. Depiction of a suprasellar meningioma (1.5 T). Suprasellar meningiomas arise from the planum sphenoidale of the anterior skull base and extend into the intra- and suprasellar space. They are characterized by a wide dural attachment ("dural tail", arrow) in the sagittal sections (**A**), while they might resemble pituitary adenomas in coronal projections (**B**). The normal gland lies in the posterior inferior zone of the lesion and the expansion of the sella is mostly much less than in pituitary adenomas of similar size.

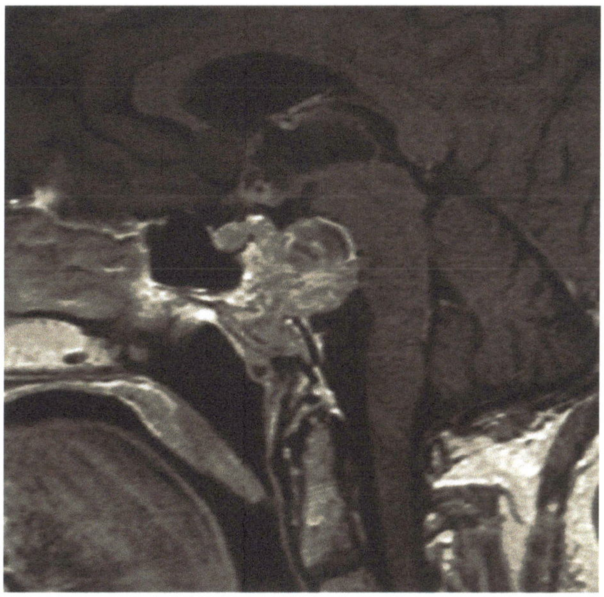

Fig. 11.12. Depiction of a chordoma of the clivus (1.5 T). Chordomas of the clivus are skull base tumors which arise from stem cells of the chorda dorsalis in front of the basilar artery. In the T1- weighted sagittal images they represent distensions of the clivus with secondary displacement of the pituitary gland.

clivus. Sellar erosion occurs, but is less frequent (Donovan and Nesbit, 1996). The pituitary gland is usually pushed upwards. Sometimes it sits like a cap at the top of the tumor. Virtually all chordomas and chondrosarcomas also have parasellar components. Imaging does not allow differentiation of chordomas and chondrosarcomas. The distinction remains a matter of microscopic tissue investigation. Differentiation from nasopharyngeal malignant neoplasms and invasive pituitary adenomas may be impossible.

Hypophysitis

In patients with a suspected lymphocytic infundibulohypophysitis the infundibulum is usually enlarged and enhances brightly with contrast medium. There is usually also some space-occupying effect on the normal pituitary which appears expanded (Fig. 11.13). A differentiation between normal and pathologic tissues is radiologically impossible. It rather appears that the entire content of the sella is expanded. The "inflammatory triangle" has been denoted a pathognomonic sign. A size reduction and reduction of contrast enhancement has been observed after high-dose corticosteroid application over several weeks (Zimmerman, 1990; Siegel and Hudgins, 1996). However, there is no single radiologic sign that can differentiate (lymphocytic) hypophysitis from nonfunctioning pituitary adenomas with sufficient reliability. Therefore, Gutenberg et al. (2009) have developed a radiologic score with several features, encompassing posterior pituitary bright spot, pituitary mass volume and symmetry, stalk size, and mucosal swelling within the sphenoid sinus which might help to differentiate the respective lesions. Lumbar puncture is sometimes useful in verifying a suspected lymphocytic hypophysitis when it reveals monolymphocitic cells in the CSF.

Aneurysms

"Sellar" aneurysms can originate from the carotid or anterior cerebral arteries. While contrast-enhanced CT does not allow a differentiation from tumor, MRI is usually diagnostic because of its ability to detect flowing blood. The hypointense "flow void" signal is caused by rapid blood flow within the lumen of the aneurysm. Usually, when the diagnosis is suspected, digital subtraction angiography is performed to better delineate the size, location, and shape of the aneurysm. A partially

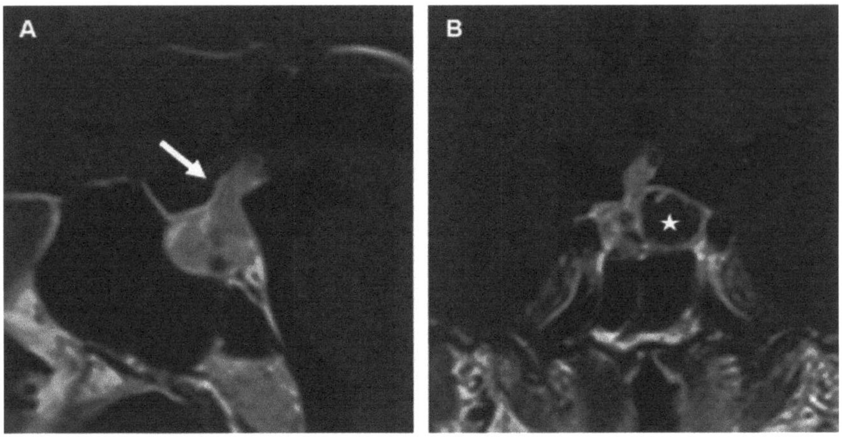

Fig. 11.13. Depiction of lymphycytic hypophysitis (1.5 T). Lymphocytic hypophysitis has a variety of shapes and presentations. An enlargement of the pituitary and infundibulum (**A**) with strong contrast enhancement (**B**) is best seen in the sagittal sections. The infundibulum and contrast enhancement may have a "tongue-like" extension into the basal hypothalamus (arrow). Sometimes, the posterior lobe is no more visible as a "bright spot" or necrotic areas (*) are present.

thrombosed aneurysm may cause confusion, since the portions of it which contain the thrombus are not perfused. Moreover, the aneurysm wall may contain a curvilinear ring of calcification in individual cases, and thus the differential diagnosis from a craniopharyngioma might be problematical. Angiography does not depict the thrombosed portions and thus a completely thrombosed aneurysm may be virtually undetectable (Zimmerman, 1990; Siegel and Hudgins, 1996).

Other intrasellar lesions

Metastases to the pituitary gland occur in many disseminated cancers but are rarely revealed in surgical series (He et al., 2014). Colloid cysts do not appear in the pathologic literature since they lack membranes with which they could be histologically proven. They represent large pars intermedia cysts and are consequently located in between the anterior and posterior pituitary lobes. They are often shaped like an American football (Buchfelder and Schlaffer, 2010). Microinfarctions may mimic small cysts or pituitary microadenomas

Other suprasellar lesions

Langerhans cell histiocytosis may cause a thickened, brightly enhancing infundibulum and hypothalamic granulomas. Other granulomatous diseases and tuberculosis may also affect the suprasellar region and exhibit enhancing thickening of the pituitary stalk and enhancing, thickened meninges of the opticochiasmatic cistern. Hypothalamic hamartomas cause precocious puberty or gelastic seizures. They resemble brain in their signal characteristics, do not enhance, and sometimes present as exophytic lumps with some connection to the tuber cinereum. Intrinsic hypothalamic hamartomas may only be recognized by an asymmetry of the third ventricle or opticochiasmatic cistern. Suprasellar germinomas occur predominantly in children and adolescents. They are mostly asymmetrically shaped and have a high contrast affinity. They grow in the midline along the infundibulum and distend it. Usually, they are ill defined against the hypothalamus (Fig. 11.14). Some may grow throughout the third ventricle and reach the pineal region. Unfortunately, they may metastasize via the CSF circulation. Thus, MRI of the spine is usually also performed to rule out seeding once the diagnosis of germinoma has been made. Opticohypothalamic gliomas are intrinsic tumors of the central nervous system. They expand the structures that they invade. Hypothalamic gliomas may cause obstructive hydrocephalus. They are never well delineated and may present with inhomogenous consistency, signal intensity, and enhancement pattern. They are more frequent in the pediatric age group and in adolescents. The predominantly suprasellar location is characteristic (Donovan and Nesbit, 1996; Siegel and Hudgins, 1996).

Incidentalomas

With the virtually ubiquitous availability of MRI and its frequent use for a variety of complaints (e.g., headaches), there are many small incidental findings picked up by sophisticated imaging that lack clinical correlates. The US Endocrine Society has suggested reinvestigation of incidental lesions less than 10 mm in diameter 1 year after the initial recognition and of larger ones 6 months after their detection (Freda et al., 2011). The physician is thus presented with ever more difficult decision-making challenges concerning when to intervene.

COMPUTED TOMOGRAPHY

While a normal axial CT scan of the brain, used, for instance, to rule out a space-occupying intracerebral

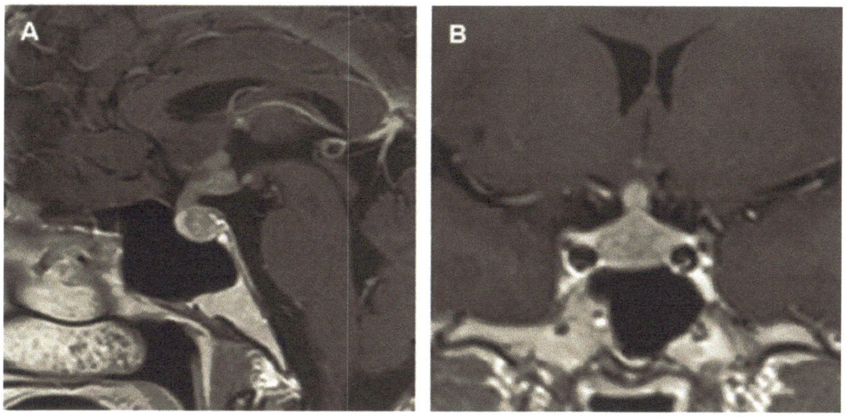

Fig. 11.14. Depiction of a suprasellar germinoma (1.5 T). Suprasellar germinomas often also have an intrasellar component. Thus, both the pituitary gland and infundibulum are expanded. The ill-defined margins against the hypothalamus are revealed in sagittal (**A**) and coronal (**B**) projections.

hematoma after a trauma and employing fairly thick slices, is unable to detect small pituitary tumors or microadenomas, it can reveal large pituitary macroadenoms and other perisellar tumors such as meningiomas or craniopharyngiomas. Mainly this happens as an incidental finding when a CT is obtained for some other reason, e.g., after a head injury. However, a proper thin-collimation CT in the coronal sections can clearly and precisely demonstrate microadenomas, especially when contrast medium is applied.

With the availability of the multislice CT scanner (64 or 128 slices), scanning became very fast and image quality improved considerably. Especially when contraindications for MRI exist (e.g., claustrophobia or presence of a pacemaker), contrast-enhanced CT can provide reliable information about sellar region tumors (Zimmerman, 1990) when a three-dimensional dataset is available and sagittal and coronal reconstructions are possible. In contrast to MR studies, the strength of a CT scan is the precise information about osseous structures and paranasal cavities (e.g., septation of the sphenoid sinus, calcification of craniopharyngiomas, destruction of skull base by malignomas), which adds valuable information for surgical approaches. Furthermore a preoperative three-dimensional dataset can be used for neuronavigation during surgery in particular cases (e.g., McCune–Albright syndrome).

Unfortunately CT is associated with radiation exposure, so that as a standard imaging procedure MR is clearly preferable (Naidich and Russell, 1999).

SPECT/PET

While this chapter focuses on the current use of MR imaging, molecular imaging needs to be mentioned in the context of pituitary tumor imaging. The presence of some dopamine receptors can be detected in almost all patients with hormonally active, and also in nonfunctioning pituitary adenomas. However, high numbers of high-affinity D2 receptors can be detected only in prolactin-secreting and some GH-secreting adenomas. Different single positron emission computed tomography (SPECT) methods are used: 123I-S-(K)-N-((1-ethyl-2-pyrrolidinyl)methyl)-2-hydroxy-3-iodo-6-methoxybenzamide (123I-IBZM) and 123I-epidepride. 123I-epidepride is generally superior to 123I-IBZM for the visualization of D2 receptors on pituitary macroadenomas (de Herder et al., 2007). However, the mere presence of receptors unfortunately does not prove their functionality or provide information about post-translational processing. Thus, 123I-IBZM and 123I-epidepride scintigraphy are not suitable investigations to predict the response to dopaminergic treatment in pituitary tumor patients. A more practical application of the techniques is the discrimination of nonfunctioning pituitary macroadenomas from other nontumor pathologies in the sellar region. Dopamine D2 receptors on pituitary tumors can also be studied using positron emission tomography (PET) with 11C-N-raclopride and 11C-N-methylspiperone (Khan et al., 2008). Because of the high cost and their poor special resolution, these investigations are not so far used in daily clinical practice, but they are undoubtedly valuable tools for research purposes. Earlier hope that they might be helpful in aiding nonsurgical treatment options in pituitary tumors have hitherto been rather disappointing (Ikeda et al., 2010). More recently, Ikeda et al. (2010) have shown some value for composite methionine-positron emission tomography and high-field MRI to detect the extent of early-stage ACTH-secreting pituitary adenomas in patients with Cushing's disease. The diagnostic accuracy was 73% when FDG-PET was fused with 3 T MRI and up to 100% when methionine PET was used. Experience is still limited. Taguchi et al. (2010) claim some diagnostic value of 18 F-dihydroxyphenylalanine PET for growth hormone-producing pituitary adenomas. The method is costly but might help in the differential diagnosis or as a research tool. Van den Bergh et al. (2011) have determined the protein synthesis rate in pituitary adenomas by tyrosine PET.

SUMMARY

For the investigation of pituitary adenomas and all sellar and perisellar pathologies the imaging method of choice consists of high-field magnetic resonance (MR) scanning with T1-weighted images and a field strength of 1.5 tesla in sagittal and coronal planes, together with a T2-weighted sequence in the axial plane. Pre- and post-contrast images are recommended. With this protocol, differential diagnosis and appropriate follow-up investigations can be performed. Whether the use of dynamic sequences increases the recognition of tiny lesions unrecognized by standard imaging remains a matter of controversy. However, images obtained with a 3 T scanner can sometimes detect small lesions and depict invasion sites of tumors into the cavernous sinus. For follow-up investigations, it is important to compare the whole set of investigations, from pretreatment up to the most recent follow-up images, to determine whether a tumor is prestable, decreased, or increasing in size. The strength of computed tomography (CT) is the depiction of calcification and structural bony abnormalities. It is only used in selected cases but can still be very valuable for surgical navigation planning. CT is still the second choice of method to directly depict a tumor and is used if, for example, the presence of a cardiac pacemaker prohibits the acquisition of MR data. Although some rules can help in the differential diagnosis, there is no single sign that is 100% specific and would allow error-free classification on the basis of imaging alone. All other modern imaging techniques, such as positron emission tomography (PET) and single photon emission computed tomography (SPECT), are still experimental and to date have rarely been used for clinical decision making. Earlier hopes that they might be helpful in aiding nonsurgical treatment options in pituitary tumors have so far been rather disappointing.

References

Ahmadi J, North CM, Segall HD et al. (1986). Cavernous sinus invasion by pituitary adenomas. AJR Am J Roentgenol 148: 257–262.

Batista D, Courkoutsakis NA, Oldfield EH et al. (2005). Detection of adrenocorticotropin-secreting pituitary adenomas by magnetic resonance imaging in children and adolescents with Cushing disease. J Clin Endocrinol Metab 90: 5134–5140.

Beisteiner R, Robinson S, Wurnig M et al. (2011). Clinical fMRI: evidence for a 7 T benefit over 3 T. Neuroimage 57: 1015–1021.

Benesch H, Felber SR, Finkenstedt G et al. (1995). MR volumetry for monitoring intramuscular bromocriptine treatment in macroprolactinomas. J Comput Assist Tomogr 19: 866–870.

Bonneville F, Narboux Y, Cattin F et al. (2002). Preoperative location of the pituitary bright spot in patients with pituitary macroadenomas. AJNR Am J Neuroradiol 23: 528–532.

Bonneville JF, Bonneville F, Cattin F (2005). Magnetic resonance imaging of pituitary adenomas. Eur Radiol 15: 543–548.

Buchfelder M, Schlaffer S (2010). Modern imaging of pituitary adenomas. Front Horm Res 38: 109–120.

Buchfelder M, Schlaffer S (2012). Intraoperative magnetic resonance imaging during surgery for pituitary adenomas: pros and cons. Endocrine 42: 483–495.

Buchfelder M, Nistor R, Fahlbusch R et al. (1993). The accuracy of CT and MR evaluation of the sella turcica for detection of adrenocorticotropic hormone secreting adenomas in Cushing's disease. AJNR Am J Neuroradiol 14: 943–948.

Chakrabortty S, Oi S, Yamaguchi M et al. (1993). Growth-hormone producing pituitary adenomas: MR characteristics and pre- and postoperative evaluation. Neurol Med Chir (Tokyo) 33: 81–85.

Chowdhury IN, Sinaii N, Oldfield EH et al. (2010). A change in pituitary magnetic resonance imaging protocol detects ACTH-secreting tumours in patients with previously negative results. Clin Endocrinol 72: 502–506.

de Herder WW, Reijs AEM, Feelders RA et al. (2007). Diagnostic imaging of dopamine receptors in pituitary adenomas. Eur J Endocrinol 156: S53–S56.

Dina TS, Feaster SH, Laws ER et al. (1993). MR of the pituitary gland postsurgery: serial MR studies following transsphenoidal resection. AJNR Am J Neuroradiol 14: 763–769.

Dinç H, Esen F, Demirci A et al. (1998). Pituitary dimensions and volume measurements in pregnancy and post partum. Acta Radiol 39: 64–69.

Donovan JL, Nesbit GM (1996). Distinction of masses involving the sella and suprasellar space: specificity of imaging feature. Am J Roentgenol 167: 597–603.

Elster AD, Sanders TG, Vines FS et al. (1991). Size and shape of the pituitary gland during pregnancy and post partum: measurement with MR imaging. Radiology 181: 531–535.

Erickson D, Erickson B, Watson R et al. (2010). 3 Tesla magnetic resonance imaging with and without corticotrophin releasing hormone stimulation for the detection of microadenomas in Cushing's syndrome. Clin Endocrinol 72: 793–799.

Fahlbusch R, Keller von B, Ganslandt O et al. (2005). Transsphenoidal surgery in acromegaly investigated by intraoperative high-field magnetic resonance imaging. Eur J Endocrinol 153: 239–248.

Famini P, Maya MM, Melmed S (2011). Pituitary magnetic resonance imaging for sellar and parasellar masses: ten-year experience in 2598 patients. J Clin Endocrinol Metab 96: 1633–1641.

Freda P, Wardlaw SL, Post KD (1996). Unusual causes of sellar/parasellar masses in a large transsphenoidal surgical series. J Clin Endocrinol Metab 81: 3455–3459.

Freda P, Beckers AM, Katznelson L et al. (2011). Endocrine Society. Pituitary incidentalomas: an Endocrine Society clinical practice guideline. J Clin Endocrinol Metab 96: 894–904.

Greenman Y, Quaknine G, Veshchev I et al. (2003). Postoperative surveillance of clinically non-functioning pituitary macroadenomas: markers of tumour quiescence and regrowth. Clin Endocrinol 58: 763–769.

Gutenberg A, Larsen J, Lupi I et al. (2009). A radiologic score to distinguish autoimmune hypophysitis from nonsecreting pituitary adenoma preoperatively. AJNR Am J Neuroradiol 30: 1766–1772.

Hald JK, Nahstadt PH, Kollovold T et al. (1992). MR imaging of pituitary macroadenomas before and after transsphenoidal surgery. Acta Radiol 33: 369–399.

Hardy J (1969). Transsphenoidal surgery of the normal and pathological pituitary. Clin Neurosurg 16: 185–217.

Haughton VM, Rosenbaum AE, Williams AL et al. (1980). Recognizing the empty sella by CT: the infundibulum sign. AJNR Am J Neuroradiol 1: 527–529.

He W, Chen F, Dalm B et al. (2014). Metastatic involvement of the pituitary gland: a systematic review with pooled individual patient data analysis. Pituitary [E-pub ahead of print].

Heck A, Ringstadt G, Fougner SL et al. (2012). Intensity of pituitary adenoma on T2-weighted magnetic resonance imaging predicts the response to octreotide treatment in newly diagnosed acromegaly. Clin Endocrinol 77: 72–78.

Ikeda H, Abe T, Watanabe K (2010). Usefulness of composite methionine-positron emission tomography/3.0-tesla-manetic resonance imagingto detect the localization and extent of early-stage Cushing adenoma. J Neurosurg 112: 750–755.

Khan S, Llloyd C, Szysko T et al. (2008). PET imaging in endocrine tumours. Minerva Endocrinol 33: 41–42.

Knosp E, Steiner E, Kitz K et al. (1993). Pituitary adenomas with invasion of the cavernous sinus space: a magnetic resonance imaging classification compared with surgical findings. Neurosurgery 33: 610–617.

Koutourousiou M, Kontogeorgos G, Wesseling P et al. (2010). Collision sellar lesions: experience with eight cases and review of the literature. Pituitary 13: 8–17.

Linn J, Peters F, Lummel N et al. (2011). Detailed imaging of the normal anatomy and pathologic conditions of the cavernous sinus region at 3 Tesla using a contrast-enhanced MR-angiography. Neuroradiology 53: 947–954.

Mikhael MA, Ciric IS (1988). MR imaging of pituitary macroadenomas before and after surgical and/or medical treatment. J Comput Assist Tomogr 12: 441–445.

Molitch ME, Cowen L, Stadiem R et al. (2012). Tumors invading the cavernous sinus that cause internal carotid artery compression are rarely pituitary adenomas. Pituitary 15: 598–600.

Naidich MJ, Russell EJ (1999). Current approaches to imaging of the sellar region and pituitary. Endocrinol Metab Clin North Am 28: 45–79.

Portocarrero-Ortiz L, Bonfacio-Delgadillo D, Sotomajor-González A et al. (2010). A modified protocol using half-dose gadolinium in dynamic 3-Tesla magnetic resonance imaging for detection of ACTH-secreting pituitary tumours. Pituitary 13: 230–235.

Reddy R, Cudlip S, Byrne JV et al. (2011). Can we ever stop imaging in surgically treated and radiotherapy-naïve patients with non-functioning pituitary adenomas? Eur J Endocrinol 165: 739–744.

Ruscalleda J (2005). Imaging of parasellar lesions. Eur Radiol 15: 549–559.

Sartoretti-Schaefer S, Wichmann W, Aguzzi A et al. (1997). MR differentiation of adamantinomatous and squamous-papillary craniopharyngiomas. AJNR Am J Neuroradiol 18: 77–87.

Siegel JL, Hudgins PA (1996). Imaging of the sellar region. In: G Tindall, P Cooper, DL Barrow (Eds.), The Practice of Neurosurgery. Williams and Wilkins, Baltimore, pp. 1071–1090.

Simmons GE, Suchnicki JE, Rak KM et al. (1992). MR imaging of the pituitary stalk: size, shape, and enhancement pattern. AJR Am J Roentgenol 159: 375–377.

Steiner E, Knosp E, Herold CJ et al. (1992). Pituitary adenomas: findings of postoperative MR images. Radiology 185: 521–527.

Steiner E, Math G, Knosp E et al. (1994). MR-appearance of the pituitary gland before and after resection of pituitary macroadenomas. Clin Radiol 49: 524–530.

Soto-Ares G, Cortet-Rudelli C, Assaker R et al. (2002). MRI protocol technique in the optimal therapeutic strategy of non-functioning pituitary adenomas. Eur J Endocrinol 146: 179–186.

Taguchi T, Takao T, Iwasaki Y et al. (2010). Diagnostic value of 18 F-dihydroxyphenylalanine positron emission tomography for growth hormone-producing pituitary adenoma. Pituitary 13: 78–79.

Tofts PS, Collins DL (2011). Multicentre imaging measurements for oncology and in the brain. Br J Radiol 84: S213–S226.

Van den Bergh ACM, Pruim J, Links TP et al. (2011). Tyrosine positron emission tomography and protein synthesis rate in pituitary adenoma: different effects of surgery and radiation therapy. Radiother Oncol 98: 213–216.

Wolfsberger S, Ba-Ssalamah A, Pinker K et al. (2004). Application of three-tesla magnetic resonance imaging for diagnosis and surgery of sellar lesions. J Neurosurg 100: 278–286.

Wolpert SM, Molitch ME, Goldman JA et al. (1984). Size, shape, and appearance of the normal female pituitary gland. Am J Roentgenol 143: 377–381.

Yoneoka Y, Watanabe N, Matsuzawa H et al. (2008). Preoperative depiction of cavernous sinus invasion by pituitary macroadenoma using three-dimensional anisotropy contrast periodically rotated overlapping parallel lines with reconstruction imaging on a 3-tesla system. J Neurosurg 108: 37–41.

Zimmerman RA (1990). Imaging of intrasellar, suprasellar, and parasellar tumours. Semin Roentgenol 25: 174–197.

Zirkzee EJM, Corssmit EPM, Biermasz NR et al. (2004). Pituitary magnetic resonance imaging is not required in the postoperative follow-up of acromegalic patients with long-term biochemical cure after transsphenoidal surgery. J Clin Endocrinol Metab 89: 4320–4324.

Handbook of Clinical Neurology, Vol. 124 (3rd series)
Clinical Neuroendocrinology
E. Fliers, M. Korbonits, and J.A. Romijn, Editors

Chapter 12

Nonfunctioning pituitary tumors

MARK E. MOLITCH*

Division of Endocrinology, Metabolism and Molecular Medicine, Northwestern University Feinberg School of Medicine, Chicago, IL, USA

INTRODUCTION

Clinically nonfunctioning adenomas (CNFAs) range from being completely asymptomatic, and therefore detected either at autopsy or as incidental findings on head magnetic resonance imaging (MRI) or computed tomography (CT) scans performed for other reasons, to causing significant hypothalamic/pituitary dysfunction and visual symptoms due to their large size. About three-quarters of CNFAs are actually gonadotroph adenomas, based upon the finding that patients with such tumors produce intact gonadotropins or their glycoprotein subunits (α and/or β) *in vivo* or *in vitro* as shown by measurement of these products in the blood, staining on immunohistochemistry, or measurement of the mRNA for these products in the tumors (Young et al., 1996). Rarely, clinically nonfunctioning tumors will be found to stain positively for adrenocorticotropic hormone (ACTH), growth hormone (GH), prolactin (PRL), or thyrotropin (TSH) but do not secrete these hormones in sufficient quantities to cause clinical syndromes; such tumors are referred to as "silent" corticotroph, somatotroph, lactotroph, or thyrotroph adenomas (Yamada et al., 1991). In this discussion, both asymptomatic and symptomatic tumors will be reviewed.

A number of other lesions may be found in the sellar area that may mimic a pituitary adenoma, including aneurysms of the internal carotid artery, craniopharyngiomas, Rathke's cleft cysts, meningiomas of the tuberculum sellae, gliomas of the hypothalamus and optic nerves, dysgerminomas, cysts, hamartomas, metastases, sarcoidosis, eosinophilic granulomas, lymphocytic hypophysitis, sphenoid sinus mucoceles, and focal areas of infarction (Freda et al., 1996; Naidich and Russell, 1999; Valassi et al., 2010; Famini et al., 2011). Some normal individuals statistically must have pituitaries that exceed the normal size boundary of 9 mm plus three standard deviations in healthy subjects (Doraiswamy et al., 1992; Tsunoda et al., 1997; Elster et al., 1999), and Chanson et al. (2001) have reported several such patients. These individuals with "normal pituitary hypertrophy" had pituitaries that on MRI had homogeneous isointense signals and enhanced homogeneously with contrast (Chanson et al., 2001). Furthermore, surgical specimens showed normal pituitary tissue in two cases (Chanson et al., 2001). Artifacts mimicking pituitary lesions include beam-hardening effects with CT and susceptibility distortions with MRI (Naidich and Russell, 1999). (For more details regarding the radiologic characteristics of adenomas versus other lesions, see Ch. 11.)

Recent series have shown that symptomatic pituitary tumors are more common than previously thought (Table 12.1). Daly et al. (2006) found an overall prevalence of 94 cases per 100 000 population in Belgium, with CNFAs comprising 14.7% of these case, for a prevalence of 13.8 cases per 100 000 population of CNFAs. Subsequent similar analyses in Switzerland, the UK, and Finland showed overall prevalences of CNFAs in the same range of 23.8, 13.4, and 25.2 per 100 000 population, respectively (Fontana and Gaillard, 2009; Fernandez et al., 2010; Raappana et al., 2010).

THE ASYMPTOMATIC, INCIDENTAL, CLINICALLY NONFUNCTIONING ADENOMA (PITUITARY INCIDENTALOMA)

Autopsy findings

Pituitary adenomas have been found at autopsy in 1.5–27% of subjects not suspected of having pituitary

*Correspondence to: Mark E. Molitch, Division of Endocrinology, Metabolism and Molecular Medicine, Northwestern University Feinberg School of Medicine, 645 N. Michigan Ave., Suite 530, Chicago, IL 60611, USA. Tel: +1-312-503-4130, Fax: +1-312-926-8693, E-mail: molitch@northwestern.edu

Table 12.1

Prevalence of clinical nonfunctioning pituitary adenomas

Series	Country	Overall pituitary adenoma (prevalence per 100 000 population)	Percentage with clinically nonfunctioning adenomas	Clinically nonfunctioning pituitary adenoma (prevalence per 100 000 population)
Daly et al., 2006	Belgium	94	14.7%	13.8
Fontana and Gaillard, 2009	Switzerland	80	29.5%	23.8
Fernandez et al., 2010	UK	78	17.2%	13.4
Raappana et al., 2010	Finland	68	37.0%	25.2

disease while alive (see Table 12.2). The average frequency of finding an adenoma for these studies, which examined a total of 21604 pituitaries, was 10%. The tumors were distributed equally throughout the age groups (range 16–86 years) and between the sexes. In the studies in which PRL immunohistochemistry was performed, 22–66% stained positively for PRL (Kovacs et al., 1980; Burrow et al., 1981; Mosca et al., 1981; Coulon et al., 1983; DeStephano et al., 1984; El-Hamid et al., 1988; Gorczyca and Hardy, 1988; Scheithauer et al., 1989; Marin et al., 1992; Sano et al., 1993; Teramoto et al., 1994; Uie et al., 1994; Camaris et al., 1995; Kurosaki et al., 2001; Buurman and Saeger, 2006; Kim et al., 2007). Buurman and Saeger (2006) provide a detailed breakdown of the 334 pituitary adenomas found in 316 pituitaries of the 3048 autopsy cases they examined, finding 39.5% staining for PRL, 13.8% staining for ACTH, 7.2% staining for gonadotropins or α subunits, 1.8% staining for GH, 0.6% staining for TSH, and 3.0% staining for multiple hormones.

In these postmortem studies all but seven of the tumors were less than 10 mm in diameter. The relative lack of macroadenomas in these studies suggests that growth from micro- to macroadenomas must be an exceedingly uncommon event and/or that virtually all macroadenomas come to clinical attention and therefore are not included in autopsy findings. There is a separate report of an additional three macroadenomas being found at autopsy (Auer et al., 1996).

CT and MRI scans in normal individuals

Three series have evaluated CT scans of the sellar area in normal subjects who were having such scans for reasons unrelated to possible pituitary disease. Chambers et al. (1982) found discrete areas of low density > 3 mm in diameter in 10 of 50 such subjects. In our study of 107 normal women, we found 7 who had focal hypodense areas and 5 who had focal high density regions > 3 mm in diameter (Wolpert et al., 1984). In a third study, Peyster et al. (1986) found focal hypodense areas > 3 mm in diameter in only 8 of 216 subjects.

Two similar studies have been carried out using MRI. Chong et al. (1994) found focal pituitary gland hypodensities 2–5 mm (mean 3.9 mm) in 20 of 52 normal subjects with nonenhanced images using a 1.5 T scanner and 3 mm thick sections. With similar scans but with gadolinium-DTPA enhancement, Hall et al. (1994) found that in 100 normal volunteers focal areas of decreased intensity ≥ 3 mm in diameter compatible with the diagnosis of adenoma were present in 34, 10, and 2 volunteers, depending upon whether there was agreement on the diagnosis between 1, 2, or 3 independent reviewing neuroradiologists.

Sellar lesions > 10 mm in diameter have not been found in these studies of consecutive normal individuals, similar to the very limited number found at autopsy. However, Nammour et al. (1997) found that of 3550 consecutive CT scans done in men with a mean age of 57 years for the symptoms of change in mental status, headache, or possible metastases, 7 (0.2%) were discovered to have pituitary macroadenomas ranging from 1.0 to 2.5 cm in size; all were thought to be CNFAs after hormonal evaluation. Similarly, when nonenhanced MRI scans were performed without specific views of the sellar area in asymptomatic normal subjects, macroadenomas were found in 0.16% of 3672 subjects in a study by Yue et al. (1997) and in 0.3% of 2000 subjects in a study by Vernooij et al. (2007). Furthermore, macroadenomas have been reported as incidental findings (Chacko and Chandy, 1992).

In the nine series of patients reported with pituitary incidentalomas (Reincke et al., 1990; Donovan and Corenblum, 1995; Nishizawa et al., 1998; Feldkamp et al., 1999; Igarashi et al., 1999; Sanno et al., 2003; Fainstein Day et al., 2004; Arita et al., 2006; Aghakhani et al., 2011), 304 of the 454 patients (67%) had macroadenomas (Table 12.3). Several of these patients had tumors 2 cm or more in maximum diameter. This proportion of patients with macroadenomas found clinically, therefore, is much greater than would be expected based on the autopsy findings, suggesting that the mass effects of such tumors may have caused some of the symptomatology causing the patients to have the scans in the first place.

Table 12.2

Frequency of pituitary adenomas found at autopsy

Series†	Number of pituitaries examined	Number of adenomas found	Frequency (%)	Number of macroadenomas found	Stain for prolactin (%)‡
Susman (1933)	260	23	8.8	–	–
Close (1934)	250	23	9.2	–	–
Costello (1936)	1000	225	22.5	0	–
Sommers (1959)	400	26	6.5	0	–
Hardy (1968)	1000	27	2.7%	0	–
McCormick (1971)	1600	140	8.8	0	–
Haugen (1973)	170	33	19.4	–	–
Kovacs (1980)	152	20	13.2	2	53
Landolt (1980)	100	13	13.0	0	–
Mosca (1980)	100	24	24.0	0	23
Burrow (1981)	120	32	26.7	0	41
Parent (1981)	500	42	8.4	1	–
Muhr (1981)	205	3	1.5	0	–
Max (1981)	500	9	1.8	–	–
Schwezinger (1982)	5100	485	9.5	0	–
Chambers (1982)	100	14	14.0	0	–
Coulon (1983)	100	10	10.0	0	60
DeStephano (1984)	100	14	14.0	0	23
Siqueira (1984)	450	39	8.7	0	–
Char (1986)	350	35	10.0	0	–
Gorczyca (1988)	100	27	27.0	0	30
El-Hamid (1988)	486	97	20.0	0	48
Scheithauer (1989)	251	41	16.3	0	66
Kontogeorgos et al., (1991)	470	49	10.4	0	–
Marin (1992)	210	35	16.7	0	32
Sano (1993)	166	15	9.0	0	47
Teramoto (1994)	1000	51	5.1	0	30
Uie (1994)	1117	36	3.2	0	33
Camararis (1995)	423	14	3.2	0	44
Tomita (1999)	100	24	24.0	–	50
Kurosaki (2001)	692	79	11.4	1	24
Buurman (2006)	3048	334	11.0	3	40
Rittierodt (2007)	228	7	3.0	0	–
Kim (2007)	120	7	5.8	0	29
Furgal-Borzych et al., (2007)	151	47	31.1	0	–
Aghakhani (2011)	485	61	12.6	–	–
Total	21 604	2161	10.0%	7	

†Each series is identified by the first author and date.
‡Prolactin + indicates the percentage of tumors that had positive immunostaining for prolactin, indicating that they were prolactinomas.

Endocrinologic evaluation of the asymptomatic incidental mass

As the most common lesion in the sella is a pituitary adenoma, it is reasonable to evaluate patients for hormone oversecretion, regardless of the size of the lesion seen. Many of the changes occurring with hormone oversecretion syndromes may be quite subtle and only slowly progressive; therefore, screening for hormonal oversecretion is warranted even in patients with no clinical evidence of hormone oversecretion. "Silent" somatotroph and corticotroph adenomas have been reported many times but it is not clear whether such patients with minimal clinical evidence of hormone oversecretion are free from the increased risk for the more subtle cardiovascular, bone, oncological, and possibly other adverse effects we usually associate with such tumors. Indeed, there is emerging evidence that subclinical Cushing's

Table 12.3

Changes in pituitary incidentaloma size

	Microadenomas				Macroadenomas				
Series	Total	Increased	Decreased	No change	Total	Enlarged	Decreased	No change	Years followed
Reincke et al. (1990)	7	1	1	5	7	2	0	5	8
Donovan and Corenblum (1995)	15	0	0	15	16	4*	0	12	6–7
Nishizawa et al. (1998)	–	–	–	–	28	2*	0	26	5.6
Feldkamp et al. (1999)	31	1	1	29	19	5	1	13	2.7
Igarashi et al. (1999)	1	0	0	1	22	6	10	6	5.1
Sanno et al. (2003)	74	10	7	57	165	20*	22	123	2.3
Fainstein Day et al. (2004)	11	1	0	10	7	1	0	6	3.2
Arita et al. (2006)	5	2	0	3	37	19*	0	18	5.2
Karavitaki (2007)	16	2	1	13	24	12	4	8	3.6
Dekkers et al. (2007)	–	–	–	–	28	14	8	6	7.1
Anagnostis et al. (2011)	6	0	1	5	3	1	0	2	4.0
Total	166	17 (10%)	11 (7%)	138 (83%)	356	86 (24%)	45 (13%)	225 (63%)	

*Seven cases in these series had tumor enlargement due to apoplexy.

syndrome due to adrenal incidentalomas is associated with significantly increased prevalences of diabetes, hypertension, obesity, osteoporosis, and cardiovascular risk (Angeli and Terzolo, 2002). Whether there is a similar increased risk for these comorbidities with "silent" corticotroph adenomas is unknown. Furthermore, some have reported that silent corticotroph adenomas have a worse prognosis than noncorticotroph CNFAs with respect to aggressiveness following initial surgery (Scheithauer et al., 2000; Bradley et al., 2003), but this has not been found in other series (Soto-Ares et al., 2002; Dekkers et al., 2006a). Progression to overt Cushing's disease over time was reported in 4 of 22 (18%) cases in one series (Baldeweg et al., 2005). It is not clear how many of these patients have nonsuppressible serum cortisol levels or elevated urinary free cortisol levels, but Lopez et al. (2004) have found suppressed ACTH secretion and hypocortisolism in 2 of 12 patients following resection of silent ACTH-secreting adenomas.

Screening for hormone oversecretion in such patients has been questioned as to its cost-effectiveness (Chanson and Young, 2003; Krikorian and Aron, 2006; Randall et al., 2010). However, evidence from the series of Fainstein Day et al. (2004) suggests such screening is worthwhile, as 7 of their 46 patients turned out to have prolactinomas, and of the 13 who ended up going to surgery and having immunohistochemistry performed, 2 (15%) were GH positive, 3 (23%) were gonadotropin positive, and 4 (31%) were plurihormonal adenomas.

A serum PRL should be obtained but it may be difficult to distinguish between prolactin production by a tumor versus hyperprolactinemia from stalk dysfunction in the case of macroadenomas, especially those with suprasellar extension. For such tumors, PRL levels are usually over 200 ng/mL with prolactinomas, and lower numbers suggest stalk dysfunction (Arafah et al., 1995; Karavitaki et al., 2006; Hong et al., 2010). For very large tumors, the sample should be diluted 1:100 to avoid the "hook effect" (St-Jean et al., 1996; Barkan and Chandler, 1998). An IGF-1 is probably sufficient to screen for acromegaly but if this cannot be performed, it may be necessary to demonstrate nonsuppression of GH levels during an oral glucose tolerance test (Growth Hormone Research Society and Pituitary Society Consensus Conference, 2004). The best screening tests for Cushing's syndrome have traditionally been the overnight dexamethasone suppression test and the 24 hour urinary free cortisol and, more recently, the assessment of a midnight salivary cortisol (Findling and Raff, 2006; Carroll et al., 2008; Nieman et al., 2008). An abnormal midnight salivary cortisol has been found to have greater than 93% specificity and sensitivity for diagnosing Cushing's syndrome (Carroll et al., 2008). Because the patients may have little in the way of symptoms, it is likely that the 24 hour urinary free cortisol will be normal and an overnight 1 mg dexamethasone suppression test or a midnight salivary cortisol might be better tests to diagnose early Cushing's disease. Because of the high variability of ACTH levels, even in patients with overt Cushing's disease, ACTH levels are probably not worth measuring. However, some participants on the Endocrine Society Taskforce that wrote the pituitary incidentaloma clinical practice guideline of the society felt that measurement of ACTH levels might be useful (Freda et al., 2011). Any abnormality found on such screening would then need to be pursued with more definitive testing (see Ch. 10). Most clinically nonfunctioning adenomas are gonadotroph adenomas, as shown by immunohistochemistry (Young et al., 1996). However, as gonadotropin oversecretion rarely causes clinical symptoms and the finding of such hormone oversecretion would not influence therapy, there is no reason to screen for this.

Microadenomas have generally been thought to not cause disruption of normal pituitary function. Of the 22 patients with suspected microadenomas evaluated in the series of Reincke et al. (1990) and Donovan and Corenblum (1995), all had normal pituitary function. However, Wichers-Rother et al. (2004) found high frequencies of hormone loss in 25 patients with nonfunctioning microadenomas, with deficiencies of GH being present in 80%, of gonadotropins in 24%, of ACTH in 28%, and of TSH in 12%. Yuen et al. (2008) subsequently found deficiencies of one or more pituitary hormones in 50% of 38 patients with clinically nonfunctioning microadenomas. Larger lesions are much more likely to cause varying degrees of hypopituitarism because of compression of the hypothalamus, the hypothalamic–pituitary stalk, or the pituitary itself. Of the various series reported, up to 41% of patients with macroadenomas were found to have hypopituitarism (Reincke et al., 1990; Donovan and Corenblum, 1995; Fainstein Day et al., 2004; Arita et al., 2006). Thus, all patients with macroadenomas should be screened for hypopituitarism but whether all patients with microadenomas should be similarly screened is controversial (Freda et al., 2011). Specifics of diagnostic testing for hypopituitarism can be found elsewhere in this volume (see Ch. 10).

Natural history and follow-up of incidental clinically nonfunctioning adenomas

Clinically, patients with incidental macroadenomas are commonly seen in everyday practice. In 11 series totaling 522 patients that have been reported with pituitary CNFAs that were not treated either surgically or medically, thereby giving an indication of their natural history, 356 (70%) had macroadenomas and 166 (30%) had microadenomas (Table 12.3) (Reincke et al., 1990; Donovan and Corenblum, 1995; Nishizawa et al., 1998; Feldkamp et al., 1999; Igarashi et al., 1999; Sanno et al., 2003; Fainstein Day et al., 2004; Arita et al., 2006; Karavitaki et al., 2007; Dekkers et al., 2007; Aghakhani et al., 2011). However, these were not all true incidentalomas. Many had tumors 2 cm or more in diameter and many were symptomatic with hypopituitarism or visual field defects but for a variety of reasons surgery was not carried out. For example, in the series reported by Karavitaki et al. (2007), only one-half of the 24 macroadenomas were incidental findings and 11 had varying degrees of hypopituitarism. In this series, 5 of the patients had major visual field defects but did not have surgery either because of major comorbidities (3 patients) or because the patient did not wish surgery (2 patients) (Karavitaki et al., 2007). Similarly, in the series of 28 patients with macroadenomas reported by Dekkers et al. (2007) (62), only 6 (21%) were truly incidentalomas, with 44% having hypopituitarism and 46% having visual field defects.

Of the 166 patients with microadenomas reported in these series, 17 (10%) experienced tumor growth, 11 (7%) showed evidence of a decrease in tumor size, and 138 (83%) remained unchanged in size in follow-up MRI scans over periods of up to 8 years (Reincke et al., 1990; Donovan and Corenblum, 1995; Feldkamp et al., 1999; Igarashi et al., 1999; Sanno et al., 2003; Fainstein Day et al., 2004; Arita et al., 2006; Karavitaki et al., 2007; Aghakhani et al., 2011). Of the 356 patients with macroadenomas, 86 (24%) showed evidence of tumor enlargement, 45 (13%) showed evidence of a decrease in tumor size, and 225 (63%) remained unchanged in size on follow-up MRI scans over periods of 8 years (Reincke et al., 1990; Donovan and Corenblum, 1995; Nishizawa et al., 1998; Feldkamp et al., 1999; Igarashi et al., 1999; Sanno et al., 2003; Fainstein Day et al., 2004; Arita et al., 2006; Karavitaki et al., 2007; Dekkers et al., 2007; Aghakhani et al., 2011). In their review of the data from these series, Fernández-Balsells et al. (2011) expressed the incidence of tumor enlargement per 100 patient-years as 12.53 for macroadenomas and 3.32 for microadenomas. The duration of follow-up of these patients was variable and in their analysis, Dekkers et al. (2007) suggested that with longer follow-up up to 50% of patients with macroadenomas will have an increase in tumor size. It should be mentioned that of the 86 macroadenomas with tumor size increase, in 7 this was due to a hemorrhage into the tumor. The incidence of pituitary apoplexy per 100 patient-years was 1.1 for macroadenomas and 0.4 for microadenomas (Fernández-Balsells et al., 2011).

Attempts have been made to look at the growth rates of those tumors that do grow. Dekkers et al. estimated a growth rate of 0.6 mm per year or 236 mm^3 per year (Dekkers et al., 2007). Of their 14 patients who experienced tumor growth, 2 showed evidence of growth by 2 years, 3 more by 3 years, and then 1 each at 4 years, 5 years, 6 years, 7 years, 12 years, 15 years, 17 years, 20 years, and 22 years. In contrast, Karavitaki et al. (2007) found that all but 4 of their 12 patients who experienced macroadenoma regrowth did so by 5 years, although patients also had evidence of tumor growth at 6 and 8 years. Honegger et al. (2008) found tumor volume doubling times ranging from 0.8 to 27.2 years, emphasizing the tremendous variability in tumor size increases; there was no correlation between initial tumor size and the rate of tumor volume doubling. These data suggest that at least for patients with macroadenomas, surveillance MRI scans should be carried out for at least 22 years, although the frequency of scanning can certainly be reduced after the first few years if there is no evidence of tumor growth.

Management of incidental clinically nonfunctioning adenomas

Therapy is indicated for tumors that are hypersecreting. Therefore, prolactinomas would generally be treated with dopamine agonists and those producing GH or ACTH would be treated with surgery (Biller et al., 2008; Melmed et al., 2009, 2011). For tumors not oversecreting these hormones, the indications for surgery are based primarily on size, size change, and mass effects of the tumors.

For patients with microadenomas, the data presented above suggest that significant tumor enlargement will occur in only 10%. Therefore, surgical resection is generally not indicated and yearly repeat scanning for 2–3 years is indicated to detect tumor enlargement; subsequently this can be done at less frequent intervals (Fig. 12.1). Surgery is done only for significant tumor enlargement. However, the rate of growth is generally quite slow so that the decision and timing of any surgery would depend on the rate and amount of growth as well as any clinical consequences, such as the development of visual field defects.

Tumors greater than 1 cm in diameter have already indicated a propensity for growth. A careful evaluation of the mass effects of these tumors is indicated, including evaluation of pituitary function and visual field examination if the tumor abuts the chiasm. If there are visual field defects, surgery is indicated (Freda et al., 2011). Because hypopituitarism is potentially correctable with tumor resection (see below), this is also an indication for surgery. In my opinion, tumors larger than 2 cm should also be considered for surgery simply because of their already demonstrated propensity for growth. Similarly, if a tumor is found to be abutting the optic chiasm, even though testing shows normal visual fields, consideration should be given to surgery (Freda et al., 2011). If surgery is not done, then visual fields should be tested at 6–12 monthly intervals thereafter. If a completely asymptomatic lesion is thought

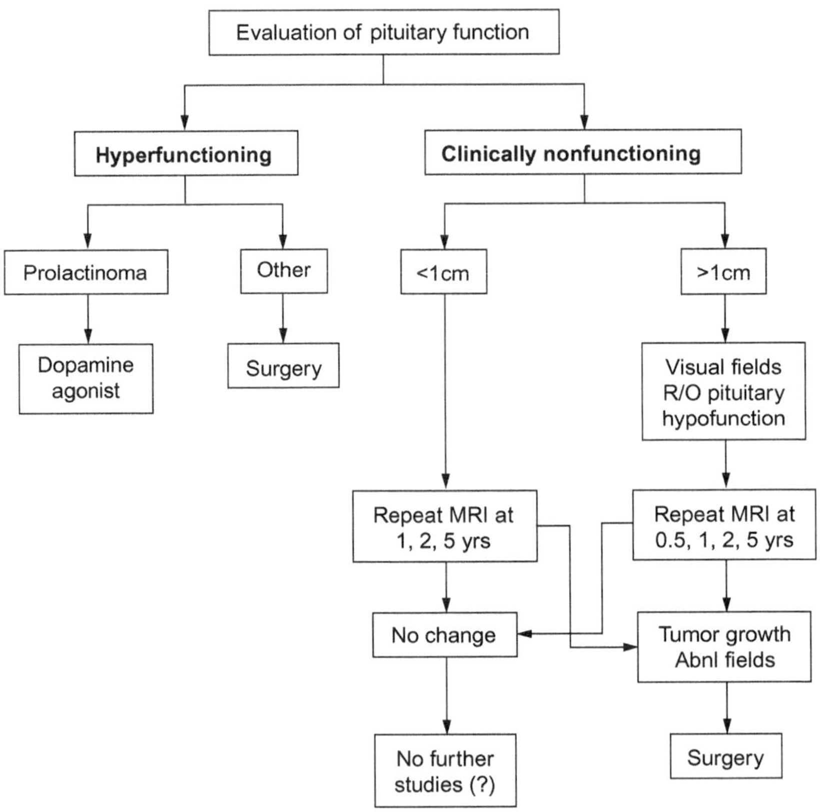

Fig. 12.1. Flow diagram indicating the approach to the patient found to have a pituitary incidentaloma. The first step is to evaluate patients for pituitary hyperfunction and then treat those found to be hyperfunctioning. Patients with tumors that are clinically nonfunctioning who have macroadenomas are evaluated further for evidence of chiasmal compression and hypopituitarism. Scans are then repeated at progressively longer intervals to assess for enlargement of the tumors. (Reproduced from Molitch, 2008.)

to be a pituitary adenoma on the basis of radiologic and clinical findings, then a decision could be made to simply repeat scans on a yearly basis, surgery being deferred until there is evidence of tumor growth. Some would obtain the initial follow-up scan at 6 months, to detect potential rapid growers (Freda et al., 2011). As indicated above, significant tumor growth can be expected in approximately 20% of such patients. Hemorrhage into such tumors is uncommon but anticoagulation may predispose to this complication; surgery would prevent such a complication. When there is no evidence of visual field defects or hypopituitarism and the patient is asymptomatic, an attempt at medical therapy with a dopamine agonist or octreotide is reasonable, realizing that only about 10–20% of such patients will respond with a decrease in tumor size (see below). Surgery is indicated if surveillance scans show evidence of tumor enlargement. As with microadenomas, the decision to proceed with surgery is affected by the rate and extent of growth and any clinical consequences, such as compression of the optic chiasm or the development of pituitary hormone deficiencies, as well as the patient's comorbidities and risks for surgery (Fig. 12.1).

SYMPTOMATIC CLINICALLY NONFUNCTIONING ADENOMAS

Presenting symptoms

CNFAs usually present because of symptoms due to the mass effects of the tumor. Data from 12 representative series are shown in Table 12.4 (this table is not meant to be comprehensive but to show the range of effects of the tumors). Clearly, the most common symptoms and signs are visual field disturbance, headache, and hypopituitarism, although the last was often only found with detailed testing. Testing of pituitary function in 13 series (Arafah, 1986; Ebersold et al., 1986; Comtois et al., 1991; Tominaga et al., 1995; Colao et al., 1998; Greenman et al., 2003; Nomikos et al., 2004; Wichers-Rother et al., 2004; Dekkers et al., 2006a; Losa et al., 2008; Anagnostis et al., 2010; Brochier et al., 2010; Chen et al., 2011) showed that the loss of hormones was on the order of GH > LH/FSH > ACTH > TSH (Table 12.5). In many older series, testing for GH deficiency in adults was not carried out because they thought that the finding of GH deficiency would not change therapy. However, the multitude of studies carried out in recent years showing potential benefit of GH therapy in GH-deficient adults suggest that GH testing should be done in all such patients if such therapy would be appropriate for the given patient (Molitch et al., 2011). When it is done, the majority of patients will be found to be GH deficient (Table 12.5).

A number of studies have recently shown that patients with CNFAs have decreased quality of life and some psychological issues, including increased anxiety and harm avoidance (Sievers et al., 2009), increased fatigue, decreased social functioning (Dekkers et al., 2006b), poor sleep quality (Biermasz et al., 2011) and memory (Brummelman et al., 2011). Because these were generally mixed populations of patients with respect to hypopituitarism, hormone replacement, prior surgery, and prior radiotherapy, it is difficult to pinpoint an exact cause for these issues and all of these factors may be important in individual patients.

Diagnostic evaluation

MRI with gadolinium is preferred to CT, as it can reveal far more anatomic detail of the lesion itself and its relationship to surrounding structures as well as providing characteristics distinguishing an adenoma from other mass lesions, such as craniopharyngiomas or Rathke's cleft cysts (Elster, 1993; Naidich and Russell, 1999). MRI can demonstrate the decreased signal of flowing blood and therefore can better determine the presence of aneurysms. Aneurysms and adenomas may coexist, however, and rarely magnetic resonance arteriography may be necessary. Somatostatin receptor imaging shows that most CNFAs have somatostatin receptors (DeHerder et al., 2006) and dopamine receptor imaging shows that many CNFAs have dopamine receptors (DeHerder et al., 2007), but the imaging with these compounds is considerably less sensitive than MRI or CT for showing anatomic detail and as yet does not accurately predict those tumors that might respond to somatostatin analogs or dopamine agonists with size reduction (DeHerder et al., 2006, 2007).

The endocrinologic evaluation for hormonal over- and undersecretion is discussed above.

Treatment

Treatment options for CNFAs include no treatment with careful observation, surgery with or without postoperative radiotherapy, radiotherapy alone, and medical therapy. Transsphenoidal resection is usually recommended for patients with enlarging tumors or tumors with mass effects. Rarely, craniotomy with subfrontal resection is required for very large tumors but this technique is associated with higher morbidity. The effectiveness of each treatment modality can be assessed clinically, with assessment of resolution or amelioration of symptoms (e.g., headache) and signs (e.g., visual field defects), biochemically, with assessment of reversal of pituitary hormonal deficits, and structurally, with assessment of tumor size by MRI. Postoperative MRIs to assess completeness of resection should ideally be done at least 3–4 months after

Table 12.4

Signs and symptoms in patients with clinically nonfunctioning adenomas

Series†	Toronto Erlichman (1979)	Rochester Ebersold (1986)	Montreal Comtois (1991)	Cardiff Shone (1991)	Naples Colao (1998)	Tel Aviv Greenman (2003)	Erlangen Nomikos (2004)	Bonn Wichers-Rother (2004)	Milan Losa (2008)	Paris Brochier (2010)	Beijing Chen (2011)	Thessalonki Anagnostis (2011)
No. of patients	153	100	126	35	84	122	721	155	491	142	385	114
Symptoms/signs (%)												
Visual field defects	66	68	78	71	39	18	31	51	59	54	61	55
Hypopituitarism	58	61	75	89	74	34	48	85	71	82	61	50
Headaches	44	36	56	17	75	29	19	55	–	25	62	53
Visual acuity reduction	–	–	54	26	32	–	–	–	0	–	–	–
Ophthalmoplegia	–	5	12	11	16	14	–	–	4	–	15	–
Apoplexy	–	5	8	–	–	4	10	–	10	7	10	3

†Each series is identified by the city of the first named hospital as well as the last name of the first author and the year of publication. In the series from Rochester (Ebersold et al., 1986), although all were clinically nonfunctioning adenomas, on immunohistochemistry 82 were null-cell adenomas, 9 were prolactinomas, 2 were silent corticotroph adenomas, 4 were gonadotroph adenomas, 1 was a plurihormonal adenoma, and 2 were not assessed. Hypopituitarism is counted if one or more pituitary hormones were noted to be deficient.

Table 12.5

Frequency of endocrine deficiencies and hyperprolactinemia at presentation in patients with clinically nonfunctioning adenomas

Series†	Rochester Ebersold (1986)	Cleveland Arafah (1986)	Montreal Comtois (1991)	Hiroshima Tominaga (1995)	Naples Colao (1998)	Tel Aviv Greenman (2003)	Erlangen Nomikos (2004)	Bonn Wichers-Rother (2004)	Leiden Dekkers (2006a)	Milan Losa (2008)	Paris Brochier (2010)	Beijing Chen (2011)	Thessaloniki Anagnostis (2011)
Number of patients	100	26	126	33	84	122	721	155	109	491	142	385	114
Hormone lost													
LH/FSH	36	96	75	52	56	69	78	54	77	71	75	41	26
ACTH	17	62	36	48	23	27	32	32	40	24	53	33	16
TSH	32	81	18	19	8	29	20	30	47	25	43	35	15
GH	NT	100	NT	97	93	NT	0	85	48	NT	77	61	55
Hyperprolactinemia	NT	46	65	42	42	43	28	–	NT	43	54	NT	19

†Each series is identified by the city of the first named hospital as well as the last name of the first author and year of publication.
NT, not tested; LH, luteinizing hormone; FSH, follicle-stimulating hormone; ACTH, adrenocorticotropic hormone; TSH, thyroid-stimulating hormone; GH, growth hormone.

surgery (Steiner et al., 1992; Dina et al., 1993) since the appearance can be deceiving in the immediate postoperative period because of surgical packing, edema, and debris. Once this postoperative baseline has been established, MRIs should be repeated yearly for 3–5 years to detect regrowth and then less frequently if stable. Tumors should be followed by MRI in a similar fashion following radiotherapy or with medical treatment.

Surgery

Outcome data from series in which transsphenoidal surgery was performed are often difficult to analyze, as many series mix results from patients who may or may not have received postoperative radiotherapy, often without an early postoperative scan to determine the presence of residual tumor. Whether patients were cured or not, resection of the bulk of the tumor mass often resulted in improvement of pituitary function with resolution of hyperprolactinemia that was due to stalk dysfunction (Table 12.5). In general, surgical success is highly dependent upon the skill and experience of the neurosurgeon as well as specific tumor characteristics, such as size, invasiveness, and parasellar extension (Ciric et al., 1997; Barker et al., 2003; Dekkers et al., 2008; Roelfsema et al., 2012; Swearingen, 2012). The best way to assess completeness of resection is to obtain an MRI 3–4 months postoperatively and then follow the tumor with serial MRIs, as noted above.

Resolution of one or more hormonal deficits by surgery can be expected in 15–50% of patients, with resolution of hyperprolactinemia in more than two-thirds of patients (Table 12.6). Conversely, in 2–15% of patients surgery may induce an additional loss of one or more pituitary hormones when they were normal preoperatively (Losa et al., 2008; Comtois et al., 1991; Gsponer et al., 1999; Nomikos et al., 2004; Brochier et al., 2010; Chen et al., 2011). Transient diabetes insipidus (DI) can be expected in about 20–30% of patients (Nomikos et al., 2004; Chen et al., 2011) but permanent DI occurs in 0.5–5% (Comtois et al., 1991; Gsponer et al., 1999; Nomikos et al., 2004; Nemergut et al., 2005; van den Bergh et al., 2007; Losa et al., 2008; Anagnostis et al., 2010; Chen et al., 2011). Other complications of surgery occurring in less than 1% of patients each include postoperative CSF leak, meningitis, seizures, rebleeding, cranial nerve injury, and visual function deterioration (Ciric et al., 1997; Barker et al., 2003; Losa et al., 2008; Swearingen, 2012). Mortality remains about 0.3–1.3% and is seen more often in patients with very large tumors who require craniotomy (Ciric et al., 1997; Barker et al., 2003; Losa et al., 2008; Chen et al., 2011). Patients with giant tumors, defined as being > 4 cm in diameter, do particularly poorly, with very low cure rates and relatively high morbidity and mortality rates, and combined transsphenoidal/transcranial approaches are often used (Mohr et al., 1990; Alleyne et al., 2002; Garibi et al., 2002; Mortini et al., 2007; Nishioka et al., 2012).

Radiotherapy

Radiotherapy as primary therapy is rarely done and mainly restricted to those patients who would not be able to tolerate surgery. More commonly, radiotherapy is given as adjunctive therapy postoperatively. Prior to the institution of policies of routine surveillance with MRI scans, many series reported postoperative recurrence rates in patients who did or did not have routine postoperative irradiation. In many of those series the criteria for choosing which patient received postoperative irradiation were not specified. Reports from nine series of patients, showed that the recurrence rates for tumors following surgery were 8.0% of 552 patients who received routine postoperative radiotherapy and 16.3% of 428 patients who did not receive such routine radiotherapy (Ebersold et al., 1986; Comtois et al., 1991; Shone et al., 1991; Brada et al., 1993; Bradley et al., 1994; Breen et al., 1998; Colao et al., 1998; Gittoes et al., 1998; Park et al., 2004) (Table 12.6). More recent series have reported similar data stratified by the

Table 12.6

Results of transsphenoidal surgery for clinically nonfunctioning pituitary adenomas on pituitary function

Series†	Cleveland Arafah (1986)	Montreal Comtois (1991)	Madrid Marzuela et al., (1994)	Hiroshima Tominaga (1995)	Naples Colao (1998)	Erlangen Nomikos (2004)	Milan Losa (2008)
Hormone axis normalized							
LH/FSH	32	11	29	13	34	15	33
ACTH	38	41	57	73	10	35	42
TSH	57	14	13	67	57	33	36
Hyperprolactinemia	58	NT	NT	85	68	95	91

†Each series is identified by the city of the first named hospital as well as last name of the first author and year of publication.
LH, luteinizing hormone; FSH, follicle-stimulating hormone; ACTH, adrenocorticotropic hormone; TSH, thyroid-stimulating hormone.

presence or absence of tumor on the initial postoperative MRI (Lillehei et al., 1998; Gsponer et al., 1999; Turner et al., 1999; Woollons et al., 2000; Soto-Ares et al., 2002; Greenman et al., 2003; Alameda et al., 2005; Dekkers et al., 2006a; van den Bergh et al., 2007; Losa et al., 2008; Brochier et al., 2010). For those patients with no visible tumor on postoperative MRI, 1 of 14 (7.1%) who received routine radiotherapy had a recurrence and only 14% of 615 patients who did not receive routine radiotherapy experienced recurrences (Table 12.7). For those patients who had visible tumor on postoperative MRI, 11.2% of 339 patients who received routine radiotherapy had growth of tumor remnants whereas 50.1% of 487 patients who did not receive routine radiotherapy experienced growth of tumor remnants (Table 12.7). Thus, the risk of tumor recurrence/growth is very low when no tumor is visible postoperatively, but such a risk is considerably higher when a tumor remnant is visible and this risk can be substantially reduced with postoperative irradiation. Besides postoperative irradiation, other factors such as the age of the patient, gender, tumor size, invasiveness, and histology were not shown to have consistent prognostic significance with respect to tumor regrowth (Roelfsema et al., 2012).

Conventional radiotherapy has substantial complications, including a risk of hypopituitarism in over 50% of patients, a two-fold increase in stroke, and a three- to four-fold increase in the development of brain tumors (Hansen and Molitch, 1998; Gittoes, 2003). However, with respect to this increase in second tumors, there is a clear ascertainment bias due to the routine surveillance MRIs done in patients who have had pituitary irradiation compared to none in the general population (Gittoes, 2003). Cognitive dysfunction is rare when conventional radiotherapy doses are used (Hansen and Molitch, 1998; Gittoes, 2003; Brummelman et al., 2011).

Table 12.7

Regrowth of pituitary adenomas following transsphenoidal surgery with and without postoperative radiotherapy and with and without residual tumor visible postoperatively

Series†	Postop MRI status unknown				No visible tumor on postop MRI				Visible tumor on postop MRI			
	RT		No RT		RT		No RT		RT		No RT	
	Total	Regrowth	Total	Regrowth	Total	Regrowth	Total	Regrowth	Total	Regrowth	Total	Regrowth
Ebersold (1986)	50	9	42	5								
Comtois (1991)			71	15								
Shone (1991)	10	0	22	5								
Bradley (1994)			73	8								
Colao (1998)	57	9	25	4								
Gittoes (1998)	63	4	63	27								
Park (2004)	44	1	132	26								
Brada (1993)	208	6										
Breen (1998)	120	15										
Lillehei (1998)							32	2				
Turner (1999)							57	13			8	8
Gsponer (1999)							21	2	8	0	21	9
Woollons (2000)							11	2	41	13	11	8
Soto-Ares (2002)							17	0			34	13
Greenman (2003)							30	6	14	5	78	41
Alameda (2005)					11	0	11	1	22	1	5	3
Dekkers (2006a)							27	0	6	0	64	9
Ferrante (2006)							73	14	76	14	77	45
van den Bergh (2007)					3	1	15	0	76	3	28	16
Losa (2008)							279	36	81	1	76	46
Brochier (2010)							42	10	15	1	85	46
TOTAL	552	44 (8%)	428	90 (16.3%)	14	1 (7%)	615	86 (14%)	339	38 (11.2%)	487	244 (50.1%)

†Each series is identified by the last name of the first author and the year of publication. The numbers shown are numbers of cases. For each category is shown the total number of cases evaluated and the number in whom there was recurrence/regrowth of the tumors.
Postop, postoperative; RT, radiotherapy; No RT, no radiotherapy; MRI, magnetic resonance imaging.

Over the past decade, the mode of radiotherapy has changed in most institutions around the world from conventional to stereotactic (Sheehan et al., 2005; Kanner et al., 2009; Jagannathan et al., 2009). With conventional radiotherapy, radiation is given through two or three ports, 5 days per week for 5 weeks, thereby giving a total dose of approximately 45 Gy over 25 sessions. With stereotactic radiotherapy (often referred to as "radiosurgery"), after tumor anatomy is defined by MRI, radiation is given in a single fraction or fractionated either via linear accelerator (LINAC) beams that are shaped or through multiple cobalt sources (gamma knife) in higher doses to the tumor with lower doses to the surrounding brain tissue or with protons using proton beam therapy (for review of these techniques see Kanner et al., 2009). In a recent review summarizing 28 series in which 904 patients with CNFAs were treated with these newer stereotactic techniques, tumor growth was controlled in 88–97% but this was only over a 4 year period of time (Kanner et al., 2009). Because of the short period of follow-up, the true long-term tumor size control rate is as yet unknown.

The complication rates of stereotactic radiotherapy are also only beginning to be appreciated after this short follow-up period. In one detailed report it was found that after a median of 80.5 months following gamma knife radiotherapy, new pituitary hormone deficiencies developed in 39% of patients (Gopalan et al., 2011). Overall, the reported incidences of new hormonal losses varied from 0% in the first few years to up to 66% after 17 years following stereotactic radiotherapy (Darzy and Shalet, 2009). Thus, the ultimate rate for hypopituitarism is likely to be quite similar to that for conventional radiotherapy. Although cognitive and memory problems have been issues with radiotherapy in the past (Hansen and Molitch, 1998), doses used in recent years with fractionated conventional radiotherapy (Brummelman et al., 2011) and gamma knife stereotactic radiotherapy (Tooze et al., 2012) do not seem to cause problems. The optic nerves and optic chiasm are radiosensitive and so stereotactic radiotherapy is not given to tumors < 5 mm from the optic chiasm (Tishler et al., 1993; Leber et al., 1998). However, the cranial nerves coursing through the cavernous sinus (III, IV, V_1, V_2, VI) are much more radioresistant (Tishler et al., 1993; Leber et al., 1998); therefore, stereotactic radiotherapy is quite good for treating residual tumor in the cavernous sinus. Acute symptoms of headache, fatigue, nausea and vomiting, developing usually within 2 but up to 10 days following irradiation, have been reported in up to 50% of patients (St. George et al., 2002). An MRI should be done to exclude an apoplexy under such circumstances. Dexamethasone in doses of about 8 mg per day has been used for these acute complications but results have been variable.

Surgery for tumor regrowth has had variable success. In a series of 42 such patients, Benveniste et al. (2005) reported that visual loss improved in only 57%, residual tumor was still present in 75%, and late relapse following this second surgery occurred in 15%. Stereotactic radiotherapy has also been given to a limited number of patients who have had prior surgery and prior conventional radiotherapy who have continued tumor growth or hypersecretion with some success and no increase in adverse events (Swords et al., 2003).

Medical therapy

CNFAs have been shown to have high affinity dopaminergic binding sites using [^{3}H]spiperone as a radioligand with affinities similar to that seen for normal pituitary tissue and prolactinomas; however, the number of binding sites was only 18.8% of that seen in prolactinomas but similar to that seen in normal pituitaries (Bevan and Burke, 1986). In a more recent study, Pivonello et al. (2004) showed that the dopamine D_2 receptor was expressed in 67% of 18 cases. Clinically, however, bromocriptine has been shown to be effective in shrinking CNFAs in less than 20% of cases (Grossman et al., 1985; Bevan and Burke, 1986; Van Schaardenburg et al., 1989). On the other hand, Greenman et al. (2005) reported giving bromocriptine to patients with CNFAs who had residual tumor on MRI following surgery, finding that the tumor mass decreased or remained stable in 18 of 20 patients. In this same study, in 13 subjects bromocriptine was started when tumor remnant growth became evident during the course of routine follow-up and this growth stabilized or decreased in 8 (62%) (Greenman et al., 2005). In contrast, tumor size increased in 29 of 47 (62%) subjects who had neither bromocriptine nor radiotherapy (Greenman et al., 2005). Cabergoline, a long-acting dopamine agonist with greater activity on prolactinomas than bromocriptine has also been found to have somewhat better *in vivo* activity on CNFAs. Lohmann et al. (2001) found that a $>10\%$ (range 10–18%) tumor shrinkage in 7 of 13 patients treated with cabergoline 1 mg/week for 1 year, and Pivonello et al. (2004) found a more significant tumor size reduction (range 28.6–62.4%) in 5 of 9 patients with cabergoline 3 mg/week for 1 year. In a review of the literature, Colao et al. (2008) summarized 13 studies of patients receiving bromocriptine, finding that 8 of 128 (6.2%) patients experienced further tumor growth, 36 of 128 (28.1%) experienced tumor size reduction, and the remainder showed no change in tumor size; similarly, in three studies of patients receiving cabergoline, 3 of 23 (13%) patients experienced further tumor growth, 12 of 23 (52.2%) experienced tumor size reduction, and the remainder showed no change in tumor size.

Caution is needed when using high doses of cabergoline, however, in that patients using very high doses (>3 mg/day) for Parkinson's disease have been found to be at an increased risk for fibrotic cardiac valvular lesions (Schade et al., 2007; Zanettini et al., 2007).

Somatostatin receptors are also present in most CNFAs (De Bruin et al., 1992; Duet et al., 1994). Octreotide, a somatostatin analog, has been shown to produce a modest reduction in tumor size with improvement in visual field defects. In their review of 11 studies, Colao et al. (2008) found that visual fields improved in 27 of 84 patients (32.1%) but the changes in tumor volume were much less impressive, with tumor reduction occurring in only 5 of 100 patients, tumor increase occurring in 12 of 100 patients and no size change in the remainder. Interestingly, Andersen et al. (2001) found a 60% response rate with the combination of octreotide 200 µg three times daily subcutaneously and cabergoline 0.5 mg daily, the six responders being those who had elevated blood levels of gonadotropin subunits and the four nonresponders having no elevation of gonadotropin subunits.

Management of the symptomatic patient

Transsphenoidal resection is usually recommended for patients with enlarging tumors, tumors that cause hypopituitarism, or tumors that press on the optic chiasm. Because of its generally limited success, medical therapy would be an option only in those patients with comorbidities so severe as to preclude surgery. Postoperative MRIs should be done 3–4 months after surgery to assess completeness of resection and then repeated yearly for 3–5 years and subsequently less frequently if there is no evidence of growth. If tumor resection is shown to be complete by postoperative MRI scans, routine radiotherapy may well not be needed as the tumor recurrence rate for such patients is only 14% and patients can be followed with periodic MRI scans as stated above. Radiotherapy or repeat surgery, or even medical therapy, can be given at the time tumor size increase is documented. If tumor resection is incomplete, postoperative radiotherapy may be indicated, as radiotherapy can reduce the recurrence rate from 50% to about 11%. However, the recent demonstration by Greenman et al. (2005) regarding the control of postoperative growth suggests that perhaps a trial of a dopamine agonist is reasonable in the patient with a small residual tumor; irradiation (stereotactic) would then only be given to those not responding to the dopamine agonist.

References

Aghakhani K, Kadivar M, Kazemi-Esfeh S et al. (2011). Prevalence of pituitary incidentaloma in the Iranian cadavers. Indian J Pathol Microbiol 54: 692–694.

Alameda C, Lucas T, Peneda E et al. (2005). Experience in the management of 51 non-functioning pituitary adenomas: indications for post-operative radiotherapy. J Endocrinol Invest 23: 18–22.

Alleyne Jr CH, Barrow DL, Oyesiku NM (2002). Combined transsphenoidal and pterional craniotomy approach to giant pituitary tumors. Surg Neurol 57: 380–390.

Anagnostis P, Adamidou F, Polyzos SA et al. (2010). Non-functioning pituitary adenomas: a single center experience. Exp Clin Endocrinol Diabetes 118: 1–6.

Anagnostis P, Adamidou F, Polyzos SA et al. (2011). Pituitary incidentalomas: a single-centre experience. Int J Clin Pract 65: 172–177.

Andersen M, Bjerre P, Schrøder HD et al. (2001). In vivo secretory potential and the effect of combination therapy with octreotide and cabergoline in patients with clinically non-functioning pituitary adenomas. Clin Endocrinol 56: 23–30.

Angeli A, Terzolo M (2002). Adrenal incidentaloma – a modern disease with old complications. J Clin Endocrinol Metab 87: 4869–4871.

Arafah AM (1986). Reversible hypopituitarism in patients with large nonfunctioning pituitary adenomas. J Clin Endocrinol Metab 62: 1173–1179.

Arafah MB, Neki KE, Gold RS et al. (1995). Dynamics of prolactin secretion in patients with hypopituitarism and pituitary macroadenomas. J Clin Endocrinol Metab 80: 3507–3512.

Arita K, Tominaga A, Sugiyama K et al. (2006). Natural course of incidentally found nonfunctioning pituitary adenoma, with special reference to pituitary apoplexy during follow-up examination. J Neurosurg 104: 884–891.

Auer RN, Alakija P, Sutherland GR (1996). Asymptomatic large pituitary adenomas discovered at autopsy. Surg Neurol 46: 28–31.

Baldeweg SE, Pollock JR, Powell M et al. (2005). A spectrum of behaviour in silent corticotroph pituitary adenomas. Br J Neurosurg 19: 38–42.

Barkan AL, Chandler WF (1998). Giant pituitary prolactinoma with falsely low serum prolactin: the pitfall of the "high-dose hook effect": case report. Neurosurgery 42: 913–915.

Barker II FG, Klibanski A, Swearingen B (2003). Transsphenoidal surgery for pituitary tumors in the United States, 1996–2000; mortality, morbidity, and the effects of hospital and surgeon volume. J Clin Endocrinol Metab 88: 4709–4719.

Benveniste RJ, King WA, Walsh J et al. (2005). Repeated transsphenoidal surgery to treat recurrent or residual pituitary adenoma. J Neurosurg 102: 1004–1012.

Bevan JS, Burke CW (1986). Non-functioning pituitary adenomas do not regress during bromocriptine therapy but possess membrane-bound dopamine receptors which bind bromocriptine. Clin Endocrinol 25: 561–572.

Biermasz NR, Joustra SD, Donga E et al. (2011). Patients previously treated for nonfunctioning pituitary macroadenomas have disturbed sleep characteristics, circadian movement rhythm, and subjective sleep quality. J Clin Endocrinol Metab 96: 1524–1531.

Biller BMK, Grossman AB, Stewart PM et al. (2008). Treatment of adrenocorticotropin-dependent Cushing's syndrome: a consensus statement. J Clin Endocrinol Metab 93: 2454–2462.

Brada M, Rajan B, Traish D et al. (1993). The long-term efficacy of conservative surgery and radiotherapy in the control of pituitary adenomas. Clin Endocrinol 38: 571–578.

Bradley KM, Adams CBT, Potter CPS et al. (1994). An audit of selected patients with non-functioning pituitary adenoma treated by transsphenoidal surgery without irradiation. Clin Endocrinol 41: 655–659.

Bradley KJ, Wass JAH, Turner HE (2003). Non-functioning pituitary adenomas with positive immunoreactivity for ACTH behave more aggressively than ACTH immunonegative tumours but do not recur more frequently. Clin Endocrinol 58: 59–64.

Breen P, Flickinger JC, Kondziolka D et al. (1998). Radiotherapy for nonfunctional pituitary adenoma: analysis of long-term tumor control. J Neurosurg 89: 933–938.

Brochier S, Galland F, Kujas M et al. (2010). Factors predicting relapse of nonfunctioning pituitary macroadenomas after neurosurgery: a study of 142 patients. Eur J Endocrinol 163: 193–200.

Brummelman P, Elderson MF, Dullaart RPF et al. (2011). Cognitive functioning in patients treated for nonfunctioning pituitary macroadenoma and the effects of pituitary radiotherapy. Clin Endocrinol 74: 481–487.

Burrow GN, Wortzman G, Rewcastle NB et al. (1981). Microadenomas of the pituitary and abnormal sellar tomograms in an unselected autopsy series. N Engl J Med 304: 156–158.

Buurman H, Saeger W (2006). Subclinical adenomas in postmortem pituitaries: classification and correlations to clinical data. Eur J Endocrinol 154: 753–758.

Camaris C, Balleine R, Little D (1995). Microadenomas of the human pituitary. Pathology 27: 8–11.

Carroll T, Raff H, Findling JW (2008). Late-night salivary cortisol measurement in the diagnosis of Cushing's syndrome. Nat Clin Pract Endocrinol Metab 4: 344–350.

Chacko AG, Chandy MJ (1992). Incidental pituitary macroadenomas. Br J Neurosurg 6: 233–236.

Chambers EF, Turski PA, LaMasters D et al. (1982). Regions of low density in the contrast-enhanced pituitary gland: normal and pathologic processes. Radiology 144: 109–113.

Chanson P, Young J (2003). Pituitary incidentalomas. Endocrinologist 13: 124–135.

Chanson P, Daujat F, Young J et al. (2001). Normal pituitary hypertrophy as a frequent cause of pituitary incidentalomas: a follow-up study. J Clin Endocrinol Metab 86: 3009–3015.

Char G, Persaud V (1986). Asymptomatic microadenomas of the pituitary gland in an unselected autopsy series. W Indian Med J 35: 275–279.

Chen L, White WL, Spetzler RF et al. (2011). A prospective study of nonfunctioning pituitary adenomas: presentation, management and clinical outcome. J Neurooncol 102: 129–138.

Chong BW, Kucharczyk AW, Singer W et al. (1994). Pituitary gland MR: a comparative study of healthy volunteers and patients with microadenomas. AJNR Am J Neuroradiol 15: 675–679.

Ciric I, Ragin A, Baumgartner C et al. (1997). Complications of transsphenoidal surgery: results of a national survey, review of the literature, and personal experience. Neurosurgery 40: 225–236.

Close HG (1934). The incidence of adenoma of the pituitary body in some types of new growth. Lancet 1: 732–734.

Colao A, Cerbone G, Cappabianca P et al. (1998). Effect of surgery and radiotherapy on visual and endocrine function in nonfunctioning pituitary adenomas. J Endocrinol Invest 21: 284–290.

Colao A, DiSomma C, Pivonello R et al. (2008). Medical therapy for clinically non-functioning pituitary adenomas. Endocr Relat Cancer 15: 905–915.

Comtois R, Beauregard H, Somma M et al. (1991). The clinical and endocrine outcome to transsphenoidal microsurgery of nonsecreting pituitary adenomas. Cancer 68: 860–866.

Costello RT (1936). Subclinical adenoma of the pituitary gland. Am J Pathol 12: 205–215.

Coulon G, Fellmann D, Arbez-Gindre F et al. (1983). Les adenome hypophysaires latents. Etude autopsique. Sem Hop Paris 59: 2747–2750.

Daly AF, Rixhon M, Adam C et al. (2006). High prevalence of pituitary adenomas: a cross-sectional study in the Province of Liège, Belgium. J Clin Endocrinol Metab 91: 4769–4775.

Darzy KH, Shalet SM (2009). Hypopituitarism following radiotherapy. Pituitary 12: 40–50.

De Bruin TWA, Kwekkeboom DJ, Van't Verlaat JW et al. (1992). Clinically nonfunctioning pituitary adenoma and octreotide response to long term high dose treatment, and studies in vitro. J Clin Endocrinol Metab 75: 1310–1317.

De Herder WW, Kwekkeboom DJ, Feelders RA et al. (2006). Somatostatin receptor imaging for neuroendocrine tumors. Pituitary 9: 243–248.

De Herder WW, Reijs AE, Feelders RA et al. (2007). Diagnostic imaging of dopamine receptors in pituitary adenomas. Eur J Endocrinol 156 (Suppl 1): S53–S56.

Dekkers OM, Pereira AM, Roelfsema F et al. (2006a). Observation alone after transsphenoidal surgery for nonfunctioning pituitary macroadenoma. J Clin Endocrinol Metab 91: 1796–1801.

Dekkers OM, van der Klaauw AA, Pereira AM et al. (2006b). Quality of life is decreased after treatment for nonfunctioning pituitary macroadenoma. J Clin Endocrinol Metab 91: 3364–3369.

Dekkers OM, Hammer S, de Keizer RJW et al. (2007). The natural course of non-functioning pituitary macroadenomas. Eur J Endocrinol 156: 217–224.

Dekkers OM, Pereira AM, Romijn JA (2008). Treatment and follow-up of clinically nonfunctioning pituitary macroadenomas. J Clin Endocrinol Metab 93: 3717–3726.

DeStephano DB, Lloyd RV, Pike AM et al. (1984). Pituitary adenomas. An immunohistochemical study of hormone production and chromogranin localization. Am J Pathol 116: 464–472.

Dina TS, Feaster SH, Laws ER (1993). MR of the pituitary gland postsurgery: serial MR studies following transsphenoidal resection. AJNR Am J Neuroradiol 14: 763–769.

Donovan LE, Corenblum B (1995). The natural history of the pituitary incidentaloma. Arch Intern Med 153: 181–183.

Doraiswamy PM, Potts JM, Axelson DA et al. (1992). MR assessment of pituitary gland morphology in healthy volunteers: age- and gender-related differences. AJNR Am J Neuroradiol 13: 1295–1299.

Duet M, Mundler O, Ajzenberg C et al. (1994). Somatostatin receptor imaging in non-functioning pituitary adenomas: value of an uptake index. Eur J Nucl Med 21: 647–650.

Ebersold MJ, Quast LM, Laws ER et al. (1986). Long-term results in transsphenoidal removal of nonfunctioning pituitary adenomas. J Neurosurg 64: 713–719.

El-Hamid MWA, Joplin GF, Lewis PD (1988). Incidentally found small pituitary adenomas may have no effect on fertility. Acta Endocrinol 117: 361–364.

Elster AD (1993). Modern imaging of the pituitary. Radiology 187: 1–14.

Elster AD, Chen MY, Williams DWD et al. (1999). Pituitary gland: MR imaging of physiologic hypertrophy in adolescence. Radiology 174: 681–685.

Erlichman C, Meakin JW, Simpson WJ (1979). Review of 154 patients with non-functioning pituitary tumors. Int J Radiat Oncol Biol Phys 5: 1981–1986.

Fainstein Day P, Guitelman M, Artese R et al. (2004). Retrospective multicentric study of pituitary incidentalomas. Pituitary 7: 145–148.

Famini P, Maya MM, Melmed S (2011). Pituitary magnetic resonance imaging for sellar and parasellar masses: ten-year experience in 2598 patients. J Clin Endocrinol Metab 96: 1633–1641.

Feldkamp J, Santen R, Harms E et al. (1999). Incidentally discovered pituitary lesions: high frequency of macroadenomas and hormone-secreting adenomas – results of a prospective study. Clin Endocrinol 51: 109–113.

Fernandez A, Karavitaki N, Wass JAH (2010). Prevalence of pituitary adenomas: a community-based cross-sectional study in Abanbury (Oxfordshire, UK). Clin Endocrinol 72: 877–882.

Fernández-Balsells MM, Murad MH, Barwise A et al. (2011). Natural history of nonfunctioning pituitary adenomas and incidentalomas: a systematic review and metaanalysis. J Clin Endocrinol Metab 96: 905–912.

Ferrante E, Ferraroni M, Castrignanò T et al. (2006). Non-functioning pituitary adenoma database: a useful resource to improve the clinical management of pituitary tumors. Eur J Endocrinol 155: 823–829.

Findling JW, Raff H (2006). Cushing's syndrome: important issues in diagnosis and management. J Clin Endocrinol Metab 91: 3746–3753.

Fontana E, Gaillard R (2009). Epidemiology of pituitary adenoma: results of the first Swiss study. Rev Med Suisse 5: 2172–2174.

Freda PU, Wardlaw SL, Post KD (1996). Unusual causes of sellar/parasellar masses in a large transsphenoidal surgical series. J Clin Endocrinol Metab 81: 3455–3459.

Freda PU, Beckers M, Katznelson L et al. (2011). Pituitary incidentaloma: an Endocrine Society clinical practice guideline. J Clin Endocrinol Metab 96: 894–904.

Furgal-Borzych A, Lis GJ, Litwin JA et al. (2007). Increased incidence of pituitary microadenomas in suicide victims. Neuropsychobiology 55: 163–166.

Garibi J, Pomposos I, Villar G et al. (2002). Giant pituitary adenomas: clinical characteristics and surgical results. Br J Neurosurg 16: 133–139.

Gittoes NJL (2003). Radiotherapy for non-functioning pituitary tumors – when and under what circumstances. Pituitary 6: 103–108.

Gittoes NJL, Bates AS, Tse W et al. (1998). Radiotherapy for non-functioning pituitary tumours. Clin Endocrinol 48: 331–337.

Gopalan R, Schlesinger D, Vance ML et al. (2011). Long-term outcomes after gamma knife radiosurgery for patients with a nonfunctioning pituitary adenoma. Neurosurgery 69: 284–293.

Gorczyca W, Hardy J (1988). Microadenomas of the human pituitary and their vacularization. Neurosurgery 22: 1–6.

Greenman Y, Ouaknine G, Veshchev I et al. (2003). Postoperative surveillance of clinically nonfunctioning pituitary macroadenomas: markers of tumour quiescence and regrowth. Clin Endocrinol 58: 763–769.

Greenman Y, Tordjman K, Osher E et al. (2005). Postoperative treatment of clinically nonfunctioning pituitary adenomas with dopamine agonists decreases tumor remnant growth. Clin Endocrinol 63: 39–44.

Grossman A, Ross R, Charlesworth M et al. (1985). The effect of dopamine agonist therapy on large functionless pituitary tumours. Clin Endocrinol 22: 679–686.

Growth Hormone Research Society and Pituitary Society Consensus Conference (2004). Biochemical assessment and long-term monitoring in patients with acromegaly. J Clin Endocrinol Metab 89: 3099–3102.

Gsponer J, de Tribolet N, Déruaz J-P et al. (1999). Diagnosis, treatment and outcome of pituitary tumors and other abnormal intrasellar masses. Medicine 78: 236–269.

Hall WA, Luciano MG, Doppman JL et al. (1994). Pituitary magnetic resonance imaging in normal human volunteers: occult adenomas in the general population. Ann Intern Med 120: 817–820.

Hansen LK, Molitch ME (1998). Postoperative radiotherapy for clinically nonfunctioning pituitary adenomas. Endocrinologist 8: 71–78.

Hardy J (1968). Transsphenoidal microsurgery of the normal and pathological pituitary. Clin Neurosurg 16: 185–217.

Haugen OA (1973). Pituitary adenomas and the histology of the prostate in elderly men. Acta Pathol Microbiol Scand Pathol 81: 425–434.

Honegger J, Zimmermann S, Psaras T et al. (2008). Growth modeling of non-functioning pituitary adenomas in patients referred for surgery. Eur J Endocrinol 158: 287–294.

Hong JW, Lee MK, Kim SH et al. (2010). Discrimination of prolactinoma from hyperprolactinemic non-functioning adenoma. Endocrine 37: 140–147.

Igarashi T, Saeki N, Yamaura A (1999). Long-term magnetic resonance imaging follow-up of asymptomatic sellar tumors. Their natural history and surgical indications. Neurol Med Chir (Tokyo) 39: 592–599.

Jagannathan J, Yen C-P, Pouratian N et al. (2009). Stereotactic radiosurgery for pituitary adenomas: a comprehensive review of indications, techniques and long-term results using the gamma knife. J Neurooncol 92: 345–356.

Kanner AA, Corn BW, Greenman Y (2009). Radiotherapy of nonfunctioning and gonadotroph adenomas. Pituitary 12: 15–22.

Karavitaki N, Thanabalasingham G, Shore HC et al. (2006). Do the limits of serum prolactin in disconnection hyperprolactinaemia need re-definition? A study of 226 patients with histologically verified non-functioning pituitary macroadenoma. Clin Endocrinol 65: 524–529.

Karavitaki N, Collison K, Halliday J et al. (2007). What is the natural history of nonoperated, nonfunctioning pituitary adenomas? Clin Endocrinol 67: 938–943.

Kim J-H, Seo J-S, Lee B-W et al. (2007). The characteristics of incidental pituitary microadenomas in 120 Korean forensic autopsy cases. J Korean Med Sci 22 (Suppl): S61–S65.

Kontogeorgos G, Kovacs K, Horvath E et al. (1991). Multiple adenomas of the human pituitary. A retrospective autopsy study with clinical implications. J Neurosurg 74: 243–247.

Kovacs K, Ryan N, Horvath E (1980). Pituitary adenomas in old age. J Gerontol 35: 16–22.

Krikorian A, Aron D (2006). Evaluation and management of pituitary incidentalomas – revisiting an acquaintance. Nat Clin Pract Endocrinol Metab 2: 138–145.

Kurosaki M, Saeger W, Lüdecke DK (2001). Pituitary tumors in the elderly. Pathol Res Pract 197: 493–497.

Landolt AM (1980). Biology of pituitary microadenomas. In: G Faglia, MA Giovanelli, RM MacLeod (Eds.), Pituitary Microadenomas. Academic Press, New York, pp. 107–122.

Leber KA, Berglöff J, Pendl G (1998). Dose-response tolerance of the visual pathways and cranial nerves of the cavernous sinus to stereotactic radiosurgery. J Neurosurg 88: 43–50.

Lillehei KO, Kirschman DL, Kleinschmidt-DeMasters BK et al. (1998). Reassessment of the role of radiation therapy in the treatment of endocrine-inactive pituitary macroadenomas. Neurosurgery 43: 432–439.

Lohmann T, Trantakis C, Biesold M et al. (2001). Minor tumour shrinkage in nonfunctioning pituitary adenomas by long-term treatment with the dopamine agonist cabergoline. Pituitary 4: 173–178.

Lopez JA, Kleinschmidt-DeMasters BK, Sze C-I et al. (2004). Silent corticotroph adenomas: further clinical and pathological observations. Hum Pathol 35: 1137–1147.

Losa M, Mortini P, Barzaghi R et al. (2008). Early results of surgery in patients with nonfunctioning pituitary adenoma and analysis of the risk of tumor recurrence. J Neurosurg 108: 525–532.

Marin F, Kovacs KT, Scheithauer BW et al. (1992). The pituitary gland in patients with breast carcinoma: a histologic and immunocytochemical study of 125 cases. Mayo Clin Proc 67: 949–956.

Marzuela M, Astigarraga B, Vicente A et al. (1994). Recovery of visual and endocrine function following transsphenoidal surgery of large nonfunctioning pituitary adenomas. J Endocrinol Invest 17: 703–707.

Max MB, Deck MDF, Rottenberg DA (1981). Pituitary metastasis: incidence in cancer patients and clinical differentiation from pituitary adenoma. Neurology 31: 998–1002.

McCormick WF, Halmi NS (1971). Absence of chromophobe adenomas from a large series of pituitary tumors. Arch Pathol 92: 231–238.

Melmed S, Colao A, Barkan A et al. (2009). A guideline for acromegaly management: an update. J Clin Endocrinol Metab 94: 1509–1517.

Melmed S, Casanueva FF, Hoffman AR et al. (2011). Diagnosis and treatment of hyperprolactinemia: an Endocrine Society clinical practice guideline. J Clin Endocrinol Metab 96: 273–288.

Mohr G, Hardy J, Comtois R et al. (1990). Surgical management of giant pituitary adenomas. Can J Neurol Sci 17: 62–66.

Molitch ME (2008). Nonsecreting tumors and pituitary incidentalomas. Endocrinol Metab Clin N Am 37: 151–171.

Molitch ME, Clemmons DR, Malozowski S et al. (2011). Evaluation and treatment of adult growth hormone deficiency: an Endocrine Society clinical practice guideline. J Clin Endocrinol Metab 96: 1587–1609.

Mortini P, Barzaghi R, Losa M et al. (2007). Surgical treatment of giant pituitary adenomas: strategies and results in a series of 95 consecutive patients. Neurosurgery 60: 993–1002.

Mosca L, Solcia E, Capella C et al. (1980). Pituitary adenomas: surgical versus post-mortem findings today. In: G Faglia, MA Giovanelli, RM MacLeod (Eds.), Pituitary Microadenomas. Academic Press, New York, pp. 137–142.

Muhr C, Bergstrom K, Grimelius L et al. (1981). A parallel study of the roentgen anatomy of the sella turcica and the histopathology of the pituitary gland in 205 autopsy specimens. Neuroradiology 21: 55–65.

Naidich MJ, Russell EJ (1999). Current approaches to imaging of the sellar region and pituitary. Endocrinol Metab Clin N Amer 28: 45–79.

Nammour GM, Ybarra J, Naheedy MH et al. (1997). Incidental pituitary macroadenoma: a population-based study. Am J Med Sci 314: 287–291.

Nemergut EC, Zuo Z, Jane JA et al. (2005). Predictors of diabetes insipidus after transsphenoidal surgery: a review of 881 patients. J Neurosurg 103: 448–454.

Nieman LK, Biller BM, FIndling JW et al. (2008). The diagnosis of Cushing's syndrome: an Endocrine Society clinical practice guideline. J Clin Endocrinol Metab 93: 1526–1540.

Nishioka H, Hara T, Usui M et al. (2012). Simultaneous combined supra-infrasellar approach for giant/large multilobulated pituitary adenomas. World Neurosurg 77: 533–539.

Nishizawa S, Ohta S, Yokoyama T et al. (1998). Therapeutic strategy for incidentally found pituitary tumors ("pituitary incidentalomas"). Neurosurgery 43: 1344–1348.

Nomikos P, Ladar C, Fahlbusch R et al. (2004). Impact of primary surgery on pituitary function in patients with nonfunctioning pituitary adenomas – a study on 721 patients. Acta Neurochir (Wien) 146: 27–35.

Parent AD, Bebin J, Smith RR (1981). Incidental pituitary adenomas. J Neurosurg 54: 228–231.

Park P, Chandler WF, Barkan AL et al. (2004). The role of radiation therapy after surgical resection of nonfunctional pituitary macroadenomas. Neurosurgery 55: 100–107.

Peyster RG, Adler LP, Viscarello RR et al. (1986). CT of the normal pituitary gland. Neuroradiology 28: 161–165.

Pivonello R, Matrone C, Filippella M et al. (2004). Dopamine receptor expression and function in clinically nonfunctioning pituitary tumors: comparison with the effectiveness of cabergoline treatment. J Clin Endocrinol Metab 89: 1674–1683.

Raappana A, Koivukangas J, Ebeling T et al. (2010). Incidence of pituitary adenomas in northern Finland in 1992–2007. J Clin Endocrinol Metab 95: 4268–4275.

Randall BR, Kraus KL, Simard MF et al. (2010). Cost of evaluation of patients with pituitary incidentalomas. Pituitary 13: 383–384.

Reincke M, Allolio B, Saeger W et al. (1990). The "incidentaloma" of the pituitary gland. Is neurosurgery required? JAMA 263: 2772–2776.

Rittierodt M, Hori A (2007). Pre-morbid morphological conditions of the human pituitary. Neuropathology 27: 43–48.

Roelfsema F, Biermasz NR, Pereira AM (2012). Clinical factors involved in the recurrence of pituitary adenomas after surgical remission: a structure review and meta-analysis. Pituitary 15: 71–83.

Sanno N, Oyama K, Tahara S et al. (2003). A survey of pituitary incidentaloma in Japan. Eur J Endocrinol 149: 123–127.

Sano T, Kovacs KT, Scheithauer BW et al. (1993). Aging and the human pituitary gland. Mayo Clin Proc 68: 971–977.

Schade R, Andersohn F, Suissa S et al. (2007). Dopamine agonists and the risk of cardiac-valve regurgitation. N Engl J Med 356: 29–38.

Scheithauer BW, Kovacs KT, Randall RV et al. (1989). Effects of estrogen on the human pituitary: a clinicopathologic study. Mayo Clin Proc 64: 1077–1084.

Scheithauer BW, Jaap AJ, Horvath E et al. (2000). Clinically silent corticotroph tumors of the pituitary gland. Neurosurgery 47: 723–729.

Schwezinger G, Warzok R (1982). Hyperplasien und adenome der hypophyse im unselektierten sektionsgut. Zentralbi Allg Pathol 126: 495–498.

Sheehan JP, Niranjan A, Sheehan JM et al. (2005). Stereotactic radiosurgery for pituitary adenomas: an intermediate review of its safety, efficacy, and role in the neurosurgical treatment armamentarium. J Neurosurg 102: 678–691.

Shone GR, Richards SH, Hourihan MD et al. (1991). Nonsecretory adenomas of the pituitary treated by transethmoidal sellotomy. J R Soc Med 84: 140–143.

Sievers C, Ising M, Pfister H et al. (2009). Personality in patients with pituitary adenomas is characterized by increased anxiety-related traits: comparison of 70 acromegalic patients with patients with non-functioning pituitary adenomas and age- and gender-matched controls. Eur J Endocrinol 160: 367–373.

Siqueira MG, Guembarovski AL (1984). Subclinical pituitary microadenomas. Surg Neurol 22: 134–140.

Sommers SC (1959). Pituitary cell relations to body states. Lab Invest 8: 588–621.

Soto-Ares G, Cortet-Rudelli C, Assaker R et al. (2002). MRI protocol technique in the optimal therapeutic strategy of non-functioning pituitary adenomas. Eur J Endocrinol 146: 179–186.

St. George EJ, Kudhail J, Perks J et al. (2002). Acute symptoms after gamma knife radiosurgery. J Neurosurg 97 (Suppl 5): 631–634.

Steiner E, Knosp E, Herold CJ et al. (1992). Pituitary adenomas: findings of postoperative MR imaging. Radiology 185: 521–527.

St-Jean E, Blain F, Comtois R (1996). High prolactin levels may be missed by immunoradiometric assay in patients with macroprolactinomas. Clin Endocrinol 44: 305–309.

Susman W (1933). Pituitary adenoma. Br Med J 2: 1215.

Swearingen B (2012). Update on pituitary surgery. J Clin Endocrinol Metab 97: 1073–1081.

Swords FM, Allan CA, Plowman PN et al. (2003). Stereotactic radiosurgery SVI: a treatment for previously irradiated pituitary adenomas. J Clin Endocrinol Metab 88: 5334–5340.

Teramoto A, Hirakawa K, Sanno N et al. (1994). Incidental pituitary lesions in 1,000 unselected autopsy specimens. Radiology 193: 161–164.

Tishler RB, Loeffler JS, Lunsford LD et al. (1993). Tolerance of cranial nerves of the cavernous sinus to radiosurgery. Int J Radiat Oncol Biol Phys 27: 215–221.

Tominaga A, Uozumi T, Arita K et al. (1995). Anterior pituitary function in patients with nonfunctioning pituitary adenoma: results of longitudinal follow-up. Endocr J 42: 421–427.

Tomita T, Gates E (1999). Pituitary adenomas and granular cell tumors. Incidence, cell type, and location of tumor in 100 pituitary glands at autopsy. Am J Clin Pathol 111: 817–825.

Tooze A, Hiles CL, Sheehan JP (2012). Neurocognitive changes in pituitary adenoma patients after gamma knife radiosurgery: a preliminary study. World Neurosurg 78: 122–128.

Tsunoda A, Okuda O, Sato K (1997). MR height of the pituitary gland as a function of age and sex: especially physiological hypertrophy in adolescence and in climacterium. AJNR Am J Neuroradiol 18: 551–554.

Turner HE, Stratton IM, Byrne JV et al. (1999). Audit of selected patients with nonfunctioning pituitary adenomas treated without irradiation – a follow-up study. Clin Endocrinol 51: 281–284.

Uie Y, Kanzaki M, Yabana T (1994). Incidental adenomas of the human pituitary gland. Endocr Pathol 5: 90–99.

Valassi E, Biller BMK, Klibanski A et al. (2010). Clinical features of nonpituitary sellar lesions in a large surgical series. Clin Endocrinol 73: 798–807.

Van den Bergh ACM, van den Berg G, Schoorl MA et al. (2007). Immediate postoperative radiotherapy in residual nonfunctioning pituitary adenoma: beneficial effect on local control without additional negative impact on pituitary function and life expectancy. Int J Radiat Oncol Biol Phys 67: 863–869.

Van Schaardenburg D, Roelfsema F, Van Seters AP et al. (1989). Bromocriptine therapy for non-functioning pituitary adenoma. Clin Endocrinol 30: 475–484.

Vernooij MW, Ikram A, Tanghe HL et al. (2007). Incidental findings on brain MRI in the general population. N Engl J Med 357: 1821–1828.

Wichers-Rother M, Hoven S, Kristof RA et al. (2004). Non-functioning pituitary adenomas: endocrinological and clinical outcome after transsphenoidal and transcranial surgery. Exp Clin Endocrinol Diabetes 112: 323–327.

Wolpert SM, Molitch ME, Goldman JA et al. (1984). Size, shape and appearance of the normal female pituitary gland. AJNR Am J Neuroradiol 5: 263–267.

Woollons AC, Hunn MK, Rajapakse YR et al. (2000). Non-functioning pituitary adenomas: indications for postoperative radiotherapy. Clin Endocrinol 53: 713–717.

Yamada S, Kovacs K, Horvath E et al. (1991). Morphological study of clinically nonsecreting pituitary adenomas in patients under 40 years of age. J Neurosurg 75: 902–905.

Young WF, Scheithauer BW, Kovacs KT et al. (1996). Gonadotroph adenoma of the pituitary gland: a clinicopathologic analysis of 100 cases. Mayo Clin Proc 71: 649–656.

Yue NC, Longsteth Jr WT, Elster AD et al. (1997). Clinically serious abnormalities found incidentally at MR imaging of the brain: data from the Cardiovascular Health Study. Radiology 202: 41–46.

Yuen KCJ, Cook DM, Sahasranam P et al. (2008). Prevalence of GH and other anterior pituitary hormone deficiencies in adults with nonsecreting pituitary microadenomas and normal serum IGF-1 levels. Clin Endocrinol 69: 292–298.

Zanettini R, Antonini A, Gatto G et al. (2007). Valvular heart disease and the use of dopamine agonists for Parkinson's disease. N Engl J Med 356: 39–46.

Handbook of Clinical Neurology, Vol. 124 (3rd series)
Clinical Neuroendocrinology
E. Fliers, M. Korbonits, and J.A. Romijn, Editors

Chapter 13

Hyperprolactinemia and prolactinoma

JOHANNES A. ROMIJN*
Department of Medicine, Academic Medical Center, University of Amsterdam, Amsterdam, The Netherlands

INTRODUCTION

Prolactin is produced by lactotroph cells in the anterior lobe of the pituitary. These cells constitute 10–30% of adenohypophyseal glandular cells (Horvath et al., 1999). Production of prolactin by these pituitary cells is presumed to be the main source of circulating prolactin, and in conditions of hyperprolactinemia.

In humans, the best known functions of prolactin are related to lactation and reproduction. During pregnancy, prolactin levels increase and the combined actions of progesterone and prolactin generate alveoli in the mammary glands, which secrete milk during lactation. In animals, a multiplicity of prolactin actions have been documented; this contrasts with the lack of strong arguments suggesting a major involvement in human pathophysiology other than the well-known effects on lactation and reproduction (Goffin et al., 2002).

In general, doctors perceive hyperprolactinemia and prolactinoma as relatively simple entities with a simple treatment (dopamine agonists). However, there are many tips and tricks that should be considered in diagnostic and therapeutic approaches to these patients.

CAUSES OF HYPERPROLACTINEMIA

The physiological regulation of prolactin secretion enables us to understand the causes of hyperprolactinemia and the fundamentals of the treatment of hyperprolactinemia. Uniquely for pituitary hormones, the secretion of prolactin is tonically inhibited by dopamine from hypothalamic nuclei, which reaches the pituitary through the pituitary stalk (Freeman et al., 2000). Prolactin secretion is stimulated by estrogen (Yen et al., 1974). Thyrotropin-releasing hormone (TRH) is another stimulator of prolactin secretion, which seems relevant from a clinical perspective only in hyperprolactinemia in some cases of primary hypothyroidism (Raber et al., 2003). In animal experiments other factors have been studied in the regulation of prolactin secretion, but their relevance with respect to human pathophysiology is unclear.

There are many causes for hyperprolactinemia (Serri et al., 2003). Irrespective of the cause of hyperprolactinemia, the signs and symptoms of patients presenting with true hyperprolactinemia are similar.

- *Physiological causes* include pregnancy, nipple stimulation, and stress. During pregnancy high estrogen levels induce lactotroph hyperplasia and hyperprolactinemia. However, in general, estrogen substitution and oral contraceptives have no, or only minimal, effect on prolactin levels (Goh et al., 1989; Molitch, 2005a).
- *Pathological causes* of hyperprolactinemia include diseases of the pituitary and hypothalamus. In prolactinomas the primary cause of hyperprolactinemia is excessive and autonomic production of prolactin by lactotroph cells. In other conditions, hyperprolactinemia is secondary to circumstances that stimulate secretion of prolactin by intrinsically normal lactotroph cells, or that are the result of decreased clearance of prolactin.
- *Pharmacological causes* are drugs with central effects including dopamine antagonist activity, resulting in lack of inhibition of prolactin secretion by intrinsically normal lactotroph cells. There are many classes of drugs that induce hyperprolactinemia, including antipsychotics, antidepressants, opiates and cocaine, antihypertensive drugs, drugs acting on gastrointestinal motility, and estrogens (Molitch, 2005a).
- *Decreased clearance of prolactin*. Hyperprolactinemia may be present in renal or hepatic insufficiency, associated with decreased plasma clearance of prolactin.

*Correspondence to: Johannes A. Romijn, Department of Medicine, Academic Medical Center, University of Amsterdam, PO Box 22660, 1100 DD Amsterdam, The Netherlands. Tel: +31-20-5662171, Fax: +31-20-6919658, E-mail: j.a.romijn@amc.uva.nl

CLINICAL FEATURES OF HYPERPROLACTINEMIA

In men, hyperprolactinemia causes loss of libido, erectile dysfunction with hypogonadism, manifested by oligo- and azoospermia. This is associated with lowered testosterone levels (de Rosa et al., 2003), which may also be associated with osteoporosis. Unlike in women, galactorrhea is uncommon in men with hyperprolactinemia.

In women, the presenting symptoms of hyperprolactinemia are galactorrhea, menstrual disorders including primary and secondary amenorrhea, delayed menarche, and infertility. Some women may complain of decreased libido and dyspareunia and may suffer from osteoporosis due to estrogen deficiency.

These hypogonadal effects of hyperprolactinemia are explained by central and possibly peripheral effects of prolactin. Increased prolactin levels decrease gonadotropin pulsatile secretion through inhibition of hypothalamic GnRH release (Milenković et al., 1994). In humans, hyperprolactinemia is associated with alterations in luteinizing hormone (LH) pulsatility. Upon return of prolactin levels to physiological values, LH pulses recover their physiological distribution associated with normalization of menstrual cycle and fertility (Sartorio et al., 2000). In addition, there may be direct effects of hyperprolactinemia on testes and ovaries.

Some women present with nonpuerperal galactorrhea in the presence of regular menstrual cycles. This discrepancy is clinically relevant, since prolactin levels and prolactin secretion patterns will be normal in most of these women (Kleinberg et al., 1977; Kapcala et al., 1984). This so-called "idiopathic galactorrhea" is estimated to be present in 40–50% of all women with nonpuerperal galactorrhea (Sakiyama and Quan, 1983).

The analysis of hyperprolactinemia in women with normal menstrual cycles and no galactorrhea may be an important pitfall in the interpretation of hyperprolactinemia. The pretest likelihood that this combination of hyperprolactinemia with absence of signs and symptoms is associated with clinically relevant disease is very low. In most of these instances, the measurement of prolactin levels was not indicated in retrospect, and may have yielded the laboratory equivalent of an incidentaloma. In such circumstances, the presence of macroprolactinemia should be considered (Leslie et al., 2001; see Laboratory assessment).

In patients with macroprolactinomas (>1 cm), additional features may be present due to mass effects and/or invasivity of the pituitary adenomas in surrounding tissues. These features include visual field defects, decreased visual acuity, headaches, and varying degrees of hypopituitarism. In some cases, the effects of acute apoplexy with bleeding within the macroprolactinoma may cause the presenting symptoms.

In general, hyperprolactinemia is the consequence of physiologic, pathologic, and pharmacologic causes. Therefore, the assessment of patients with hyperprolactinemia should include a search for clues for primary and secondary causes of hyperprolactinemia. It is imperative to obtain a detailed medical history and physical examination in each patient with hyperprolactinemia, with a detailed analysis of all drugs.

An important clinical issue for planning appropriate therapy is the differentiation between a nonfunctioning macroadenoma with compression of the pituitary stalk and a macroprolactinoma. This "stalk effect" of nonfunctioning macroadenomas is thought to result in loss of the inhibitory effect of dopamine on prolactin secretion, and, as a result, in hyperprolactinemia. Based on a large series of histologically confirmed cases (n = 226) with nonfunctioning pituitary macroadenomas, serum prolactin > 2000 mU/L (> 94 μg/L) is almost never (<2%) encountered in patients with nonfunctioning pituitary macroadenomas (Karavitaki et al., 2006). After exclusion of acromegaly, in patients with prolactin levels above these values the most likely diagnosis is prolactinoma.

LABORATORY ASSESSMENT OF HYPERPROLACTINEMIA

A single venipuncture obtained in minimized stress conditions suffices for blood sampling for measurement of prolactin (Melmed et al., 2011). Alternatively, in cases of doubt, two samples can be taken 15 minutes apart using a cannula to be sure that the second sample is obtained in a pain-free manner.

In most laboratories, automated assays are currently used to measure serum prolactin levels, mostly two-site immunoassay sandwich methods. Nonetheless, in clinical practice there is some variation in the results between different laboratories. Therefore, each laboratory should establish its own reference values. This is especially relevant for patients with only mildly elevated prolactin levels. In general, hyperprolactinemia is present, if the measured value exceeds the upper value of normal. In addition, there are two major analytic issues that need to be considered in the interpretation of the results of prolactin assays: macroprolactinemia and the "high-dose hook" effect.

Macroprolactinemia

There are three variants of circulating prolactin. Monomeric prolactin is the most abundant circulating form in normal and hyperprolactinemic subjects. The second variant, "big" prolactin, is considered to be a covalently bound dimer of prolactin. The third variant, macroprolactinemia, is an antigen–antibody complex of monomeric prolactin and IgG with only limited bioavailability, but with an extended half-life compared to monomeric prolactin, resulting in increased prolactin levels measured with varying degrees by different assays (Hattori and Inagaki, 1997). The classic signs and symptoms of hyperprolactinemia are caused by excess of monomeric prolactin. If macroprolactinemia is not differentiated from true hyperprolactinemia it may result in unnecessary investigation and inappropriate treatment. The problem is even more complicated, since macroprolactinemia may sometimes coincide with true hyperprolactinemia (Leslie et al., 2001; Suliman et al., 2003). In addition, women with macroprolactinemia may suffer from menstrual irregularities. However, long term follow-up of patients with macroprolactinemia and normal concentrations of monomeric prolactin did not reveal symptomatic progression (Wallace et al., 2010). Precipitation of macroprolactin with polyethylene glycol (PEG) is currently used to screen for the presence of circulating macroprolactin (Leslie et al., 2001; Melmed et al., 2011).

High-dose hook effect

In patients with large macroprolactinomas with exceedingly high prolactin levels two-site immunometric assays may sometimes yield spuriously low values of prolactin levels (Hobbs et al., 1995). This "high-dose hook effect" is explained by a very high concentration of antigen, in this case prolactin, in the fluid phase of the assay which saturates the antibodies and consequently prevents the binding to the antibody in the solid phase, and thus inhibits the formation of the "sandwich complexes" (Fig. 13.1). This leads to a loss of the labeled antibody in the fluid phase and results in falsely low values in the ultimate measurement. These hook effects differ between the different assay systems used in clinical practice. Dilution of the sample 1:100 is the test of choice to unmask this hook effect (Melmed et al., 2011). Since it is very unpractical to do this in all samples offered for prolactin analysis, the clinician should consider this possibility in macroadenomas. The failure to recognize this hook effect may have serious consequences, since the treatment of choice of large prolactinomas is treatment with a dopamine agonist, whereas in other macroadenomas with serious mass effects neursurgical treatment will be considered.

Dynamic tests of prolactin secretion

There are no indications for testing patients with hyperprolactinemia with TRH or dopamine agonists (Melmed et al., 2011).

In case a macroprolactinoma or another pathologic cause of hyperprolactinemia is considered, other pituitary functions should be assessed appropriately, since mass effects may result in varying degrees of hypopituitarism. In patients with microprolactinomas, no additional testing of pituitary functions is necessary, because of the absence of mass effects by the microadenoma. If secondary causes of hyperprolactinemia are considered, assessment of renal and liver functions should be included in the analyses. It should be realized that primary hypothyroidism and pregnancy may also be causes of hyperprolactinemia.

RADIOLOGIC DIAGNOSIS OF PROLACTINOMAS

There are no radiologic clues that differentiate prolactinomas from other pituitary adenomas. Therefore, the diagnosis of prolactinoma is based on the combination

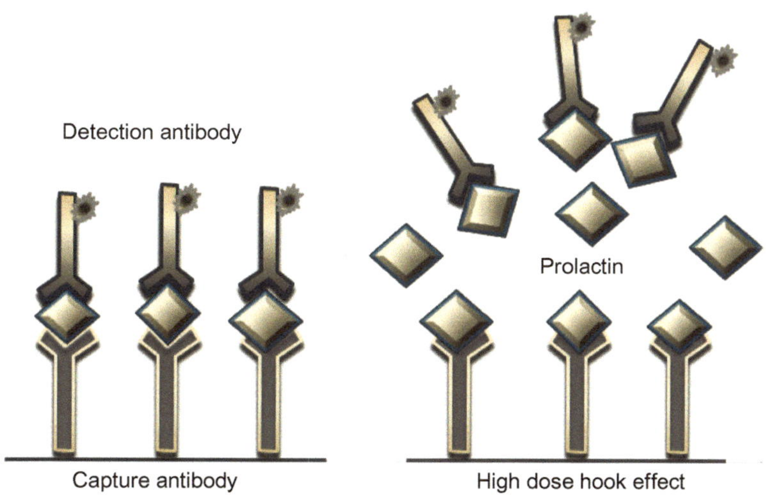

Fig. 13.1. High-dose hook effect in two-site immunometric assays: a very high concentration of prolactin saturates the antibodies in the fluid phase of the assay. Consequently, the binding of these antibodies to the antibodies in the solid phase and the formation of the "sandwich complexes" is prevented. As a result, spuriously low values of prolactin will be measured.

of data derived from a proper medical history and physical examination, the presence of hyperprolactinemia, the exclusion of nonprolactinoma-related causes of hyperprolactenemia, and radiologic indications of the presence of a pituitary adenoma. Magnetic resonance imaging (MRI) assessment of the pituitary and hypothalamic areas is also crucial to exclude other pathologic lesions causing hyperprolactinemia. Prolactinomas are classified according to their diameter on MRI into microprolactinomas (< 10 mm in diameter), macroprolactinomas (≥ 10 mm in diameter), and giant prolactinomas (> 40 mm in diameter).

Idiopathic hyperprolactinemia is interpreted to be present in the absence of pituitary and central lesions on MRI and the absence of secondary causes of hyperprolactinemia. Although the explanation of idiopathic hyperprolactinemia has not been fully elucidated, subliminal microprolactinomas, escaping detection by MRI, can be involved, at least in some patients. This bears a resemblance to the absence of measurable pituitary lesions on MRI in a considerable portion of patients with Cushing's disease. Idiopathic hyperprolactinemia is a very benign condition, since < 10% of the patients ultimately have microprolactinomas (Sluijmer and Lappöhn, 1992).

PREVALENCE RATES OF PROLACTINOMAS

Assessment of the incidence and prevalence rates of prolactinomas is not straightforward. In a meta-analysis, an overall estimated prevalence of pituitary adenomas of 14.4% in autopsy studies and 22.5% in radiologic studies was reported (Ezzat et al., 2004). Prolactinomas are the most frequent pituitary adenomas, being present in ~60% of all clinically evident pituitary tumors (Daly et al., 2006). In community surveys, prolactinomas were present in 44–62 of 100 000 inhabitants (Daly et al., 2006; Fernandez et al., 2010).

In general, patients with hyperprolactinemia are treated by gynecologists, internists, general practitioners, and neurologists. Consequently, selection bias may easily occur when patients are selected from these different populations. Alternatively, linkage of datasets from different sources (general hospitals, referral centers, and general practitioner practices) will be cumbersome. Frequently, the source population required to calculate the denominator is not clearly defined and may vary according to the different treatment settings. For these reasons, we assessed the prevalence rates by a pharmacoepidemiologic approach (Kars et al., 2009). The incidence and prevalence rates of dopamine agonist-treated hyperprolactinemia, estimated using this pharmacoepidemiologic approach, showed an overall preponderance in women, with a strong peak in women aged 25–34 years, whereas in men there was no relationship between age and prevalence (Kars et al., 2009). The estimated incidence rate of dopamine agonist-treated hyperprolactinemia for women was 8.7/100 000 person-years and for men, 1.4/100 000 person-years. The highest incidence rate was found in women 25–34 years of age: 23.9/100 000 person-years. The mean prevalence of treated female patients was almost five times higher (93.9/100 000) compared with male patients (19.6/100 000). Remarkably, there were no higher incidence rates in women after menopause compared with men of similar age (Kars et al., 2009).

Within the experience of clinical endocrinology, the prevalence of macroprolactinomas is similar in men and women, whereas microprolactinomas and nontumoral hyperprolactinemia occur much more frequently in women (Colao et al., 2003a). The high incidence of prolactinomas in women during their reproductive period, but not after menopause, is remarkable and unique for pituitary adenomas. This is probably related to the observation that estrogen receptors are expressed mainly in prolactin-producing pituitary adenomas, but rarely in other adenomas (Burdman et al., 2008).

TREATMENT OF HYPERPROLACTINEMIA AND PROLACTINOMA

The goals of treatment are reduction/normalization of prolactin levels to restore signs and symptoms, especially of gonadal dysfunction and galactorrhea, and to reduce the size of the prolactinoma, and, consequently, to reduce mass effects, if present.

Normalization of prolactin levels

Dopamine agonists are the first-line treatment for the majority of patients with prolactinomas and idiopathic hyperprolactinemia. Bromocriptine was the first dopamine agonist to be introduced into clinical practice. It is a semisynthetic ergot derivative ergoline, a D2 receptor agonist with antagonist properties at D1 receptors. Bromocriptine has a short elimination half-life and therefore has to be prescribed two to three times daily in daily doses of between 2.5 and 15 mg (Webster et al., 1994). Subsequently, cabergoline was introduced; this has a high affinity and selectivity for D2 receptors. In contrast to bromocriptine, carbergoline has an extremely long plasma half-life of ~65 hours, which allows once or twice weekly administration (Webster et al., 1994). Most efficacy studies have assessed bromocriptine and cabergoline, and fewer studies have evaluated other dopamine agonists such as pergolide and quinagolide. The general conclusion is that cabergoline is more effective in reducing prolactin levels and prolactinoma size than bromocriptine and has fewer side-effects (Webster et al., 1994; Verhelst et al., 1999; Gillam et al., 2006; Melmed et al., 2011; Wang et al., 2012). Cabergoline is therefore the preferred drug of treatment. In Europe, quinagolide is also available, but the general opinion is that cabergoline is more efficacious with fewer side-effects (Colao and Savastano, 2011). Many prospective and some retrospective studies in more than 1700 patients with hyperprolactinemic disorders, predominantly micro- and macroprolactinomas, have documented that cabergoline normalized prolactin levels in 85% (range 61–96) of the patients (reviewed by Gillam et al., 2006). The effect of cabergoline is more or less similar in macroprolactinomas. For instance, normoprolactinemia was achieved within 6 months in 81% of patients receiving 0.25–2 mg cabergoline weekly in a prospective study of 26 treatment-naive patients with macroprolactinomas (Colao et al., 2000).

Reduction of prolactinoma size

Dopamine agonists are very effective in reducing the size of prolactinomas. Initially, studies documented that bromocriptine in daily dosages ranging between 5 and 12.5 mg reduced the size of macroprolactinomas in almost all patients (Molitch et al., 1985). Tumor size was reduced by more than 50% in 46% of patients and to about 50% of the original size in an additional 18%. In the remaining 36% of the patients, tumor size reduction was less; in no patient did the tumor enlarge during therapy (Molitch et al., 1985). Cabergoline is more effective in reducing prolactinoma size than bromocriptine (Di Sarno et al., 2001). In a prospective study in 26 patients with macroprolactinomas, cabergoline (0.25–2 mg weekly) induced normoprolactinemia within 6 months in 81% of patients, 92% of whom exhibited significant tumor shrinkage (Colao et al., 2000). In general, cabergoline reduces the size of prolactinomas by 55–100% (Gillam et al., 2006). The reduction in size of prolactinomas can be evident within days to weeks (Shimon et al., 2007). However, in other patients this effect may require a longer duration of treatment, up to 6–12 months (Molitch et al., 1985).

Recently, it has become apparent that much higher doses of cabergoline can be used than described in earlier publications. This has changed the perspective of the efficacy of cabergoline. By using the individualized, dose-escalating regimen of cabergoline described below, Ono et al. documented that all prolactinomas contracted, and 52% of microadenomas and 38% of macroadenomas disappeared. Moreover, all macroadenomas (maximal diameter 16 mm prior to treatment) contracted to values < 10 mm (Ono et al., 2010).

Several studies have assessed the effects of cabergoline in patients with giant prolactinomas. Since these are very rare patients, the patient series were small (4–12 patients) (Minniti et al., 2002; Corsello et al., 2003; Shimon et al., 2007). The mean dose of cabergoline ranged between 1.6 and 3.4 mg/week, although in incidental patients 7 mg/week was used. These studies document that cabergoline is effective and safe in normalizing prolactin levels in 18 of 26 patients, with a significant reduction in prolactinoma size in 19 of 25 patients.

In some cases, macroprolactinomas may invade the sellar floor and the sphenoid sinus. In cases where the arachnoid has also been violated, there is the potential risk for developing a fistula of cerebrospinal fluid (CSF). The plugging effect of the macroadenoma may prevent leakage of CSF. However, in those circumstances, any significant reduction in the size of the macroprolactinoma can cause CSF rhinorrhea. This has been described in anecdotal reports after rapid reduction in prolactinoma size upon treatment with bromocriptine or cabergoline. The management of dopamine agonist-induced rhinorrhea is surgical repair, which was ultimately required in nearly 90% of patients reviewed in the literature (Lam et al., 2012).

It should be realized that the responses to dopamine agonists can be discrepant. Sometimes, there may be normalization of prolactin levels, but no response of prolactinoma size or, conversely, a considerable reduction in prolactinoma size but no response of prolactin levels.

Dopamine agonist-resistant prolactinomas

Resistance to dopamine agonists has been arbitrarily defined as a failure to normalize prolactin levels and to decrease macroprolactinoma size by $\geq 50\%$ (Molitch, 2005b). Of note, it may occasionally be difficult to discriminate between noncompliance and insufficient therapeutic efficacy. Bromocriptine fails to normalize prolactin levels in about one-quarter of patients and pergolide and cabergoline in 10–15% of patients. Bromocriptine fails to decrease prolactinoma size by at least 50% in about one-third of the patients and pergolide and cabergoline in 10–15% of the patients (Di Sarno et al., 2001; Molitch 2005b).

Recently, however, Ono et al. have challenged this concept of resistance by proposing an individualized, escalating, high-dose regimen of cabergoline in patients with a poor response to cabergoline (Ono et al., 2008, 2010). In their studies, cabergoline was started at a standard weekly dosage and schedule, i.e., 0.25–0.5 mg twice a week. Bromocriptine that had been given to resistant patients was switched to cabergoline after a 2–4 week washout period. Serum prolactin level was measured every 2 weeks for the first month and once a month thereafter until the end of the study. Cabergoline was incrementally dose adjusted on the basis of individual post-treatment prolactin values until amelioration of hyperprolactinemia. This adjustment was carried out on the day of, or within a few days after, blood withdrawal for prolactin measurement. For the initial 3 months, cabergoline was increased by 0.5 mg/week at 2–4 week intervals at a maximum incremental rate of 3 mg/week per 3 months. Cabergoline was maintained at the dose at which prolactin was first normalized. If prolactin was not normalized at 3, 6, or 9 months, cabergoline was further increased at a maximum of 3 mg/week in the next 3 months. In retrospect, earlier studies have used lower-dose regimens, which were not able to normalize prolactin levels in all patients. Using this individualized, dose-escalating regimen of cabergoline, prolactin levels normalized in > 99% of 150 patients, irrespective of previous treatment, micro- or macroprolactinomas, intolerance or resistance to other dopamine agonists (Ono et al., 2008). Moreover, cabergoline normalized menstrual cycles in all patients with micro- (n = 56) and macroprolactinomas (n = 29) (Ono et al., 2010). Using this individualized, stepwise approach of dose titration of cabergoline, they documented that 25 of 26 patients considered to have dopamine agonist resistance achieved normalization of prolactin levels within 12 months, with a mean dose of cabergoline of 5.2 ± 0.6 mg per week (range 3–12 mg/week), which is higher than the usual dose of cabergoline to treat nonresistant prolactinomas. The rate of prolactin normalization gradually increased to 35, 73, and 89% at 3, 6, and 9 mg/week, respectively, finally reaching 96% at the highest dose of 12 mg/week.

In a subsequent study, Ono et al. documented that cabergoline can induce and promote successful pregnancy in a large majority of infertile women with prolactinoma irrespective of prolactinoma size or bromocriptine resistance and intolerance (Ono et al., 2010). Therefore, it may be necessary to revise the criteria of dopamine agonist resistance. In case of an individualized, stepwise approach with higher ultimate dosages of cabergoline than previously used, the prevalence of dopamine resistance, at least in Japanese patients, is much lower than until recently appreciated.

From these observations, the following approaches can be considered in patients who are considered to be resistant to dopamine agonist treatment. When bromocriptine is used, a switch to treatment with cabergoline should be considered, since cabergoline is more effective in reducing prolactin levels and prolactinoma size (Colao and Savastano, 2011). When cabergoline is used, a stepwise increase in the dose should be considered. It should be realized that using this stepwise approach it may take 9–12 months to normalize prolactin levels in "resistant" patients. In the individualized, stepwise approach, resistance to cabergoline appears to be rare. In patients with microprolactinomas and resistance to dopamine agonist treatment who do not desire pregnancy, estrogen (in premenopausal women) or testosterone replacement (in men) can be considered (Melmed et al., 2011). In incidental, symptomatic patients in whom hyperprolactinemia and/or macroprolactinoma cannot be controlled by maximal tolerable doses of cabergoline, neurosurgical treatment may be considered.

Symptomatic patients with idiopathic hyperprolactinemia

One study assessed the natural history of idiopathic hyperprolactinemia. Although pituitary structure was only assessed by CT scanning, the data may serve to appreciate the benign nature of this syndrome. Serum prolactin levels remained the same, decreased, or returned to normal in 34 of 41 patients with idiopathic hyperprolactinemia during prolonged follow-up to 11 years (mean, 5.5 years) (Martin et al., 1985). In 30% hyperprolactinemia normalized spontaneously during follow-up (Schlechte et al., 1989). Only 17% of all patients had serum prolactin levels that were significantly higher (greater than 50% of their original value). No patient reported worsening of signs or symptoms, and in many patients improvement (n = 16) or complete resolution (n = 8) of the amenorrhea and/or galactorrhea occurred. In only one patient was there development of a pituitary lesion. There were 27 spontaneous or bromocriptine-induced uncomplicated pregnancies and deliveries without development of a pituitary tumor. Therefore, these data question the benefit of chronic medical therapy for all patients with this condition.

Pregnancy

There are three issues to consider in the relationship between pregnancy and prolactinomas: (1) the induction of pregnancy, (2) the effects of dopamine agonists on fetal development and pregnancy outcome, and (3) the effects of pregnancy on prolactinoma size.

Induction of pregnancy

In most instances (>90% of cases), fertility is restored if (near) normal prolactin levels are achieved by bromocriptine or cabergoline (Melmed et al., 2011). Although the experience with bromocriptine extends over several decades, which have demonstrated the safety of bromocriptine for both the pregnancy and the fetus (Gillam et al., 2006), in recent years it has become evident that cabergoline is more effective even in patients with bromocriptine inefficacy or intolerance. Cabergoline can induce and promote successful pregnancy in a large majority of infertile women with prolactinomas, irrespective of tumor size or bromocriptine resistance and intolerance (Ono et al., 2010). Therefore, in cases where treatment with bromocriptine is not effective in restoring fertility, carbergoline can be chosen as an alternative. Because patients with prolactinoma do not respond uniformly to the same dose of dopamine agonists, the dose of cabergoline should be adjusted appropriately to induce pregnancy in individual patients. The dosages of cabergoline used to induce pregnancy were 0.125–4.0 mg/week (Robert et al., 1996), 0.25–7.0 mg/week (Ricci et al., 2002), 0.25–5.0 mg/week (Colao et al., 2008) and 0.25–9 mg/week (Ono et al., 2010).

Effects of dopamine agonists on fetal development and pregnancy outcome

In general, the advice is to stop dopamine agonist treatment as soon a menstrual period is missed and pregnancy is diagnosed to limit fetal exposure. This approach has been used in over 6000 pregnancies with bromocriptine and in over 600 pregnancies with cabergoline, without evidence of increases in abortions or congenital malformations (Melmed et al., 2011). Because there has been more extensive experience with bromocriptine, there is a slight preference for the use of bromocriptine for pregnancy induction in women with prolactinomas. Quinagolide has a high incidence of adverse outcomes and is not recommended for pregnancy induction.

Effects of pregnancy on prolactinoma size

During normal pregnancy, there is hypertrophy of the pituitary and a gradual increase in prolactin levels, which is explained by the effects of prolonged exposure to high estrogen levels. In healthy women the pituitary gland enlarges throughout pregnancy, but should probably not exceed 10 mm during most of this period. Immediately postpartum, a diameter of up to 12 mm may be observed in healthy women (Elster et al., 1991). Histologically, there is progressive lactotroph cell hyperplasia involving the majority of acini throughout the pituitary. Around term, even 60–70% of adenohypophysial cells may be immunopositive for prolactin (Horvath et al., 1999). These changes are reversible after cessation of pregnancy.

In pregnant patients with prolactinomas, the question is whether to stop or to continue treatment with dopamine agonists, once dopamine agonist treatment has induced pregnancy. For microprolactinomas there is only a minimal change of symptomatic enlargement during pregnancy (<2%) (Casanueva et al., 2006; Molitch, 2006). Therefore, in patients with microprolactinomas it is safe to stop treatment with dopamine agonists as soon as pregnancy is confirmed. For macroprolactinomas pregnancy may induce further enlargement caused by the stimulatory effect of high estrogen levels and the discontinuation of the dopamine agonist. In patients with macroprolactinomas previously treated by surgery or radiotherapy, the risk of symptomatic growth of macroprolactinomas during pregnancy is only small, at ~5% (Gillam et al., 2006). However, in patients with macroprolactinomas without previous surgery or radiotherapy the risk of symptomatic enlargement during pregnancy is higher (~30%) (Gillam et al., 2006). For these patients, there is the balanced choice between careful watching without dopamine agonist treatment versus continued treatment during pregnancy with a dopamine agonist. Careful observation is needed with monthly evaluations and repeated visual field assessments. In case of sudden headache or visual field defects, MRI assessment is indicated (Melmed et al., 2011). If treatment with a dopamine agonist is continued during pregnancy, the preference is for bromocriptine rather than for cabergoline. The reason for this preference is that there is more experience in this circumstance with bromocriptine. The continued use of bromocriptine has been documented in ~100 pregnancies, and only two incidental cases of congenital abnormalities have been reported (Molitch, 2006; Melmed et al., 2011).

Drug-induced hyperprolactinemia

In the assessment of symptomatic patients with hyperprolactinemia, it is critical to assess the medication. Many drugs can induce hyperprolactinemia (see above). It may not be easy to differentiate this potential side-effect of such drugs from other causes of hyperprolactinemia. Moreover, incidentalomas of the pituitary are present in ~5% of the general population and may be falsely labeled as microprolactinomas in cases of drug-induced hyperprolactinemia. Sometimes, it may be possible to relate the onset of hyperprolactinemia with the start of a certain drug. To confirm the diagnosis of symptomatic medication-induced hyperprolactinemia, it is necessary to stop the medication for 3–5 days to determine whether prolactin levels normalize (Molitch, 2005a). If a drug cannot be stopped, consideration can be given to obtaining a MRI of the sellar region, to exclude pathologic lesions as a cause of hyperpolactinemia. If drug-induced hyperprolactinemia is confirmed, several strategies are possible. First, the medication might be switched to medication that does not cause hyperprolactinemia. Second, if the medication cannot be stopped because there is an imperative indication, the problem resulting from hyperprolactinemia can be symptomatically treated by substitution of estrogen/

progesterone or testosterone. Finally, rarely, adding a dopamine agonist very cautiously can be considered, but this may antagonize the effects of the drug that induced hyperprolactinemia (Molitch, 2005a).

ADVERSE EFFECTS OF DOPAMINE AGONISTS

Dopamine agonists are associated with gastrointestinal, neurologic, and cardiovascular side-effects. The most common side-effects are nausea and incidentally vomiting, headache, dizziness, anxiety, and depression. These side-effects tend to occur at the start of therapy and upon increases of the dose. To minimize them it is recommended to start with a low dose and to take the drug at bedtime. Bromocriptine is associated with more side-effects than cabergoline. Side-effects cause cessation of treatment in 12% of patients who use bromocriptine versus only 4% of patients who use cabergoline (Webster et al., 1994). In recent years, higher dosages of cabergoline have been proposed to control prolactinomas which are resistant to lower doses of cabergoline. In a prospective study with a stepwise, individualized, dose-escalating regimen, with ultimately a maximum dose of cabergoline ranging between 0.5 and 12 mg/week, ~10% of the patients experienced side-effects. However, these side-effects were minimal and transient, disappearing within 2–4 weeks. Moreover, none of the patients dropped out during treatment (Ono et al., 2010).

In Parkinson's disease there is an increased risk of valvular regurgitation in patients who use pergolide or cabergoline (Schade et al., 2007; Zanettini et al., 2007). Cabergoline has a high affinity for 5-hydroxytryptamine (serotonin) receptor 2B (HTR2B) located on heart valves (Roth, 2007). Activation of these receptors might lead to mitogenesis and fibroblast proliferation. Histopathologic investigations of cardiac valves obtained from patients after treatment with cabergoline for Parkinson's disease resemble the histologic abnormalities observed in the valves from patients with carcinoid disease (Simula et al., 2002; Pinero et al., 2005). These changes induce inadequate closure of the valves and clinically relevant regurgitation that is asymptomatic in most cases (Colao and Savastano, 2011).

Many studies have attempted to answer the question as to how far these observations in patients with Parkinson's disease are relevant for patients with prolactinomas treated with dopamine agonists. It is clear that there are major differences in clinical characteristics between the patients with Parkinson's disease and patients with prolactinomas and that these affect cardiovascular outcome parameters. Patients with Parkinson's disease are much older, are slightly more likely to be male, suffer from comorbidities, and are treated with a much higher cumulative dose of cabergoline. In patients with Parkinson's disease the cumulative dose of cabergoline ranged between 2600 and 6700 mg, whereas in patients treated for prolactinoma the cumulative dose is much lower, at 200–500 mg (Kars et al., 2008). Colao recently reviewed all 11 studies published on the relationship between the use of dopamine agonists, especially cabergoline, in patients with prolactinomas and valve regurgitation. There are a number of methodological issues relating to these studies. In some studies the assessment of the echocardiography was not done blinded. The number of patients included in each study was relatively limited, and the combined total number of subjects for all the studies was only 795 (Colao and Savastano, 2011). This may be insufficient to detect side-effects, which occur only in a limited percentage of the patients or in incidental patients. Most studies were cross-sectional and there are only limited prospective observations. A recent prospective study documented that follow-up of 2 years of patients on dopamine agonist treatment was associated with an increased prevalence of valvular calcification. Nonetheless, there was no increased prevalence of clinically relevant valvular dysfunction or progression of existing mild valvular dysfunction in those patients (Delgado et al., 2012). The interim conclusion may be that at present there is no compelling evidence that cabergoline results in clinically relevant valvular disease in patients with prolactinomas. However, there are suggestions that long-term use of cabergoline may result in subclinical valvular changes on echocardiography. Considering the higher dosages of cabergoline that are used by patients who cannot be controlled with lower dosages, it might be prudent to include echocardiography during follow-up of patients on high-dose cabergoline. There is uncertainty with respect to the dose and duration of cabergoline treatment, which requires echocardiographic follow-up, but according to Endocrine Society guidelines patients receiving typical doses of cabergoline (1–2 mg/week) likely will not require regular echocardiographic screening (Melmed et al., 2011).

WITHDRAWAL OF DOPAMINE AGONIST TREATMENT

Once prolactin secretion is adequately controlled, the dose of cabergoline can often be significantly decreased (Verhelst et al., 1999). Moreover, in many patients prolactin levels normalize during dopamine agonist treatment and on MRI the signs of pituitary adenoma may disappear, even with macroprolactinomas (Ono et al., 2010). This has prompted several studies to assess the effects of withdrawal of dopamine agonists (Di Sarno et al., 2000; Colao et al., 2003b; Biswas et al., 2005).

The prevalence of prolactinomas in women is related to pre-versus postmenopausal status. During the reproductive phase of women, the prevalence rates of prolactinomas are much higher than in men, whereas after menopause there are no differences between men and women (Kars et al., 2009). In accordance, in one retrospective study, 45% of women with microprolactinomas who passed through the menopause showed normalization of prolactin levels (Karunakaran et al., 2001). The effects of menopause should therefore be included when considering whether to stop dopamine agonist treatment. In postmenopausal women with microprolactinomas, stopping dopamine agonist treatment can be considered even if prolactin levels are increased after stopping treatment. These patients are usually asymptomatic. Moreover, untreated microprolactinomas only very rarely grow (Schlechte et al., 1989).

There are no randomized controlled studies comparing different withdrawal strategies after successful treatment of hyperprolactinemia. In 2003, Colao et al. demonstrated that dopamine agonist treatment can be successfully withdrawn in a considerable proportion of patients, provided they fulfill certain clinical criteria, including significant tumor reduction on radiologic imaging and a prolonged period of normoprolactinemia during treatment (Colao et al., 2003b). This study involved selected patients, since only 58% of newly diagnosed patients were included; this may have resulted in an overestimation of the proportion of patients with a favorable outcome after stopping dopamine agonist treatment. Recently, we performed a systematic review and meta-analysis to estimate the pooled proportion of patients with persistent normoprolactinemia after withdrawal from dopamine agonists. Our study showed that withdrawal was associated with persisting normoprolactinemia in only 21% of all patients. Success rates were higher in patients treated for idiopathic hyperprolactinemia, after treatment with cabergoline, and in patients with treatment duration of more than 2 years (Dekkers et al., 2010). Therefore, the success rate of stopping dopamine agonists is much lower than initially appreciated. Remarkably, the effect of menopausal status on success rates of withdrawal of dopamine agonists has not been assessed in detail.

CONCLUSION

At present the majority of patients with prolactinomas can be adequately treated with cabergoline, and in selected cases with bromocriptine. Research is needed to specify the answers to several uncertainties. These include uncertainties in the prediction of successful withdrawal of dopamine agonist treatment in patients with prolactinomas. In addition, there is uncertainty with respect to the dose and duration of cabergoline treatment, which requires echocardiographic follow-up.

References

Biswas M, Smith J, Jadon D et al. (2005). Long-term remission following withdrawal of dopamine agonist therapy in subjects with microprolactinomas. Clin Endocrinol (Oxf) 63: 26–31.

Burdman JA, Pauni M, Heredia Sereno GM et al. (2008). Estrogen receptors in human pituitary tumors. Horm Metab Res 40: 524–527.

Casanueva FF, Molitch ME, Schlechte JA et al. (2006). Guidelines of the Pituitary Society for the diagnosis and management of prolactinomas. Clin Endocrinol (Oxf) 65: 265–273.

Colao A, Savastano S (2011). Medical treatment of prolactinomas. Nat Rev Endocrinol 7: 267–278.

Colao A, Di Sarno A, Landi ML et al. (2000). Macroprolactinoma shrinkage during cabergoline treatment is greater in naive patients than in patients pretreated with other dopamine agonists: a prospective study in 110 patients. J Clin Endocrinol Metab 85: 2247–2252.

Colao A, Di Sarno A, Cappabianca P et al. (2003a). Gender differences in the prevalence, clinical features and response to cabergoline in hyperprolactinemia. Eur J Endocrinol 148: 325–331.

Colao A, Di Sarno A, Cappabianca P et al. (2003b). Withdrawal of long-term cabergoline therapy for tumoral and nontumoral hyperprolactinemia. N Engl J Med 349: 2023–2033.

Colao A, Abs R, Bárcena DG et al. (2008). Pregnancy outcomes following cabergoline treatment: extended results from a 12-year observational study. Clin Endocrinol (Oxf) 68: 66–71.

Corsello SM, Ubertini G, Altomare M et al. (2003). Giant prolactinomas in men: efficacy of cabergoline treatment. Clin Endocrinol (Oxf) 58: 662–670.

Daly AF, Rixhon M, Adam C et al. (2006). High prevalence of pituitary adenomas: a cross-sectional study in the province of Liege, Belgium. J Clin Endocrinol Metab 91: 4769–4775.

De Rosa M, Zarrilli S, Di Sarno A et al. (2003). Hyperprolactinemia in men: clinical and biochemical features and response to treatment. Endocrine 20: 75–82.

Dekkers OM, Lagro J, Burman P et al. (2010). Recurrence of hyperprolactinemia after withdrawal of dopamine agonists: systematic review and meta-analysis. J Clin Endocrinol Metab 95: 43–51.

Delgado V, Biermasz NR, van Thiel SW et al. (2012). Changes in heart valve structure and function in patients treated with dopamine agonists for prolactinomas, a 2-year follow-up study. Clin Endocrinol (Oxf) 77: 99–105.

Di Sarno A, Landi ML, Marzullo P et al. (2000). The effect of quinagolide and cabergoline, two selective dopamine receptor type 2 agonists, in the treatment of prolactinomas. Clin Endocrinol (Oxf) 531: 53–60.

Di Sarno A, Landi ML, Cappabianca P et al. (2001). Resistance to cabergoline as compared with bromocriptine in hyperprolactinemia: prevalence, clinical definition, and therapeutic strategy. J Clin Endocrinol Metab 86: 5256–5261.

Elster AD, Sanders TG, Vines FS et al. (1991). Size and shape of the pituitary gland during pregnancy and post partum: measurement with MR imaging. Radiology 181: 531–535.

Ezzat S, Asa SL, Couldwell WT et al. (2004). The prevalence of pituitary adenomas: a systematic review. Cancer 101: 613–619.

Fernandez A, Karavitaki N, Wass JA (2010). Prevalence of pituitary adenomas: a community-based, cross-sectional study in Banbury (Oxfordshire, UK). Clin Endocrinol (Oxf) 72: 377–382.

Freeman ME, Kanyicska B, Lerant A et al. (2000). Prolactin: structure, function and regulation of secretion. Physiol Rev 80: 1523–1631.

Gillam MP, Molitch ME, Lombardi G et al. (2006). Advances in the treatment of prolactinomas. Endocr Rev 27: 485–534.

Goffin V, Binart N, Touraine P et al. (2002). Prolactin: the new biology of an old hormone. Annu Rev Physiol 64: 47–67.

Goh HH, McCarthy TG, Dramusic V et al. (1989). Prolactin levels in menopausal women on long-term estrogen implants. Gynecol Endocrinol 3: 153–157.

Hattori N, Inagaki C (1997). Anti-prolactin (PRL) autoantibodies cause asymptomatic hyperprolactinemia: bioassay and clearance studies of PRL-immunoglobulin G complex. J Clin Endocrinol Metab 82: 3107–3110.

Hobbs GA, Watson M, Hess P (1995). A "hook effect" detected in a two-step immunoassay. J Clin Ligand Assay 18: 200–204.

Horvath E, Kovacs K, Scheithauer BW (1999). Pituitary hyperplasia. Pituitary 1: 169–179.

Kapcala LP, Molitch ME, Arno J et al. (1984). Twenty-four-hour prolactin secretory patterns in women with galactorrhea, normal menses, normal random prolacting levels and abnormal sellar tomograms. J Endocrinol Invest 7: 455–460.

Karavitaki N, Thanabalasingham G, Shore HC et al. (2006). Do the limits of serum prolactin in disconnection hyperprolactinaemia need re-definition? A study of 226 patients with histologically verified non-functioning pituitary macroadenoma. Clin Endocrinol (Oxf) 65: 524–529.

Kars M, Pereira AM, Bax JJ et al. (2008). Cabergoline and cardiac valve disease in prolactinoma patients: additional studies during long-term treatment are required. Eur J Endocrinol 159: 363–367.

Kars M, Souverein PC, Herings RM et al. (2009). Estimated age- and sex-specific incidence and prevalence of dopamine agonist-treated hyperprolactinemia. J Clin Endocrinol Metab 94: 2729–2734.

Karunakaran S, Page RC, Wass JA (2001). The effect of the menopause on prolactin levels in patients with hyperprolactinaemia. Clin Endocrinol (Oxf) 54: 295–300.

Kleinberg DL, Noel GL, Frantz AG (1977). Galactorrhea: a study of 235 cases, including 48 with pituitary tumors. N Engl J Med 296: 589–600.

Lam G, Mehta V, Zada G (2012). Spontaneous and medically induced cerebrospinal fluid leakage in the setting of pituitary adenomas: review of the literature. Neurosurg Focus 32: E2.

Leslie H, Courtney CH, Bell PM et al. (2001). Laboratory and clinical experience in 55 patients with macroprolactinemia identified by a simple polyethylene glycol precipitation method. J Clin Endocrinol Metab 86: 2743–2746.

Martin TL, Kim M, Malarkey WB (1985). The natural history of idiopathic hyperprolactinemia. J Clin Endocrinol Metab 60: 855–858.

Melmed S, Casanueva FF, Hoffman AR et al. (2011). Endocrine Society. Diagnosis and treatment of hyperprolactinemia: an Endocrine Society clinical practice guideline. J Clin Endocrinol Metab 96: 273–288.

Milenković L, D'Angelo G, Kelly PA et al. (1994). Inhibition of gonadotropin hormone-releasing hormone release by prolactin from GT1 neuronal cell lines through prolactin receptors. Proc Natl Acad Sci U S A 91: 1244–1247.

Minniti G, Jaffrain-Rea ML, Santoro A et al. (2002). Giant prolactinomas presenting as skull base tumors. Surg Neurol 57: 99–103.

Molitch ME (2005a). Review. Medication-induced hyperprolactinemia. Mayo Clin Proc 80: 1050–1057.

Molitch ME (2005b). Pharmacologic resistance in prolactinoma patients. Pituitary 8: 43–52.

Molitch ME (2006). Pituitary disorders during pregnancy. Endocrinol Metab Clin North Am 35: 99–116, vi.

Molitch ME, Elton RL, Blackwell RE et al. (1985). Bromocriptine as primary therapy for prolactin-secreting macroadenomas: results of a prospective multicenter study. J Clin Endocrinol Metab 60: 698–705.

Ono M, Miki N, Kawatama T et al. (2008). Prospective study of high dose cabergoline treatment of prolactinomas in 150 patients. J Clin Endocrinol Metab 93: 4721–4727.

Ono M, Miki N, Amano K et al. (2010). Individualized high-dose cabergoline therapy for hyperprolactinemic infertility in women with micro- and macroprolactinomas. J Clin Endocrinol Metab 95: 2672–2679.

Pinero A, Marcos-Alberca P, Fortes J (2005). Cabergoline-related severe restrictive mitral regurgitation. N Engl J Med 353: 1976–1977.

Raber W, Gessi A, Nowotny P et al. (2003). Hyperprolactinemia in hypothyroidism: clinical significance and impact of TSH normalization. Clin Endocrinol (Oxf) 58: 185–191.

Ricci E, Parazzini F, Motta T et al. (2002). Pregnancy outcome after treatment in early weeks of gestation. Reprod Toxicol 16: 791–793.

Robert E, Musatti L, Piscitelli G et al. (1996). Pregnancy outcome after treatment with the ergot derivative, cabergoline. Reprod Toxicol 10: 333–337.

Roth BL (2007). Drugs and valvular heart disease. N Engl J Med 356: 6–9.

Sakiyama R, Quan M (1983). Galactorrhea and hyperprolactinemia. Obstet Gynecol Surv 38: 689–700.

Sartorio A, Pizzocaro A, Liberati D et al. (2000). Abnormal LH pulsatility in women with hyperprolactinaemic amenorrhoea normalizes after bromocriptine treatment:

deconvolution-based assessment. Clin Endocrinol (Oxf) 52: 703–712.

Schade R, Andersohn F, Suissa S et al. (2007). Dopamine agonists and the risk of cardiac-valve regurgitation. N Engl J Med 356: 29–38.

Schlechte J, Dolan K, Sherman B et al. (1989). The natural history of untreated hyperprolactinemia: a prospective analysis. J Clin Endocrinol Metab 68: 412–418.

Serri O, Chik CL, Ur E et al. (2003). Diagnosis and management of hyperprolactinemia. Can Med Assoc J 169: 575–581.

Shimon I, Benbassat C, Hadani M (2007). Effectiveness of long-term cabergoline treatment for giant prolactinoma: study of 12 men. Eur J Endocrinol 156: 225–231.

Simula DV, Edwards WD, Tazelaar HD et al. (2002). Surgical pathology of carcinoid heart disease: a study of 139 valves from 75 patients spanning 20 years. Mayo Clin Proc 77: 139–147.

Sluijmer AV, Lappöhn RE (1992). Clinical history and outcome of 59 patients with idiopathic hyperprolactinemia. Fertil Steril 58: 72–77.

Suliman AM, Smith TP, Gibney J et al. (2003). Frequent misdiagnosis and mismanagement of hyperprolactinemic patients before the introduction of macroprolactin screening: application of a new strict laboratory definition of macroprolactinemia. Clin Chem 49: 1504–1509.

Verhelst J, Abs R, Lane MA et al. (1999). Cabergoline in the treatment of hyperprolactinemia: a study in 455 patients. J Clin Endocrinol Metab 84: 2518–2522.

Wallace IR, Satti N, Courtney CH et al. (2010). Ten-year clinical follow-up of a cohort of 51 patients with macroprolactinemia establishes it as a benign variant. J Clin Endocrinol Metab 95: 3268–3271.

Wang AT, Mullan RJ, Lane MA et al. (2012). Treatment of hyperprolactinemia: a systematic review and meta-analysis. Syst Rev 1: 33.

Webster J, Piscitelli G, Polli A et al. (1994). For the cabergoline comparative study group. A comparison of cabergoline and bromocriptine in the treatment of hyperprolactinemic amenorrhea. N Engl J Med 331: 904–909.

Yen SS, Ehara Y, Siler TM (1974). Augmentation of prolactin secretion by estrogen in hypogonadal women. J Clin Invest 53: 652–655.

Zanettini R, Antonini A, Gatto G et al. (2007). Valve heart disease and the use of dopamine agonists for Parkinson's disease. N Engl J Med 356: 39–46.

Handbook of Clinical Neurology, Vol. 124 (3rd series)
Clinical Neuroendocrinology
E. Fliers, M. Korbonits, and J.A. Romijn, Editors

Chapter 14

Acromegaly

PHILIPPE CHANSON[1,2,3,*], SYLVIE SALENAVE[1], AND PETER KAMENICKY[1,2,3]

[1]*Department of Endocrinology and Disorders of Reproduction, Hôpital Bicêtre and Reference Center for Rare Endocrine Disorders of Growth, Le Kremlin-Bicêtre, Paris, France*

[2]*Faculty of Medicine, Université Paris-Sud 11, Le Kremlin-Bicêtre, Paris, France*

[3]*INSERM U693, Le Kremlin-Bicêtre, Paris, France*

INTRODUCTION

Acromegaly is a rare disease characterized by progressive somatic disfigurement, mainly involving the face and extremities, together with systemic manifestations related to organ overgrowth. Acromegaly is associated with severe comorbidity and premature death if not adequately treated. It is due to excessive secretion of growth hormone (GH), originating from a pituitary adenoma in the vast majority of cases.

EPIDEMIOLOGY

The prevalence of acromegaly was long estimated to be about 40–70 cases per million inhabitants (Holdaway and Rajasoorya, 1999), but recent studies have suggested that it may be much higher: around 80–130 cases per million in Belgium or the UK, for example (Daly et al., 2006; Fernandez et al., 2010). A prevalence of around 1000 per million inhabitants was found in a German study based on screening with IGF-1 measurement in the general population (Schneider et al., 2008). In a recent Finnish study the adjusted annual incidence of acromegaly was estimated to be 3.4 cases per million (Raappana et al., 2010). These new data need to be confirmed. Owing to the insidious clinical onset, acromegaly is often diagnosed late. Older series, in the 1980s, suggested a mean diagnostic delay of 6–10 years after onset, at an average age of about 40 years (Nabarro, 1987; Ezzat et al., 1994; Melmed, 2002). More recent retrospective studies suggest that this interval has fallen to around 3–5 years on average (Nachtigall et al., 2008; Reid et al., 2010).

PATHOPHYSIOLOGY

Acromegaly related to a pituitary tumor

In more than 95% of cases, acromegaly is secondary to GH hypersecretion by a benign monoclonal pituitary adenoma which develops from somatotroph cells (Asa et al., 2002; Heaney and Melmed, 2004; Melmed, 2009, 2011; Dworakowska et al., 2011).

SOMATOTROPH PITUITARY ADENOMAS

Pure somatotroph pituitary adenomas (60%) contain either cells rich in secretory granules that show diffuse immunostaining or cells poor in secretory granules with scattered immunolabeling (Asa et al., 2002). Some of these pure somatotroph adenomas also express free α subunits shared by several glycoprotein hormones, including follicle-stimulating hormone (FSH), luteinizing hormone (LH), thyroid-stimulating hormone (TSH), and chorionic gonadotropin hormone (CG), co-localized in the same cells or even in the same granules as GH (Beck-Peccoz et al., 1985). Mixed GH- and prolactin (PRL)-secreting adenomas are frequent (25%). Some adenomas contain both cell types, while other develop from a mammosomatotropic stem cell and consist of more mature monomorphic cells that express both GH and PRL (Asa et al., 2002). Mixed GH- and TSH-secreting adenomas are rare and cause acromegaly associated with hyperthyroidism and inappropriate TSH secretion (Beck-Peccoz et al., 1996; Socin et al., 2003). Very rarely, corticotrophin (ACTH) hypersecretion may also be found.

*Correspondence to: Prof. Philippe Chanson, Assistance Publique-Hôpitaux de Paris, Hôpital BicêtreService d'Endocrinologie et des Maladies de la Reproduction, 78 rue du Général Leclerc, F-94275 Le Kremlin-Bicêtre, France. Tel: +33-1-45213708, Fax: +33-1-45212212, E-mail: philippe.chanson@bct.aphp.fr

The pituitary versus hypothalamic origin of pituitary somatotroph adenomas is controversial (Melmed, 2006, 2009). Evidence pointing to a hypothalamic origin suggests that the main actor is growth hormone-releasing hormone (GHRH), which can cause not only hyperplasia of somatotroph cells but also, as demonstrated in animal models, true adenomas (albeit such adenomas have never been observed in humans). In contrast, the monoclonal nature of these tumors and the absence of relapse after total adenoma removal point instead to a pituitary origin (Melmed, 2002). In fact, the initiation and/or progression of neoplastic transformation of normal somatotrophs could be due to a polyclonal hyperplastic response of these cells secondary to hypothalamic dysregulation. The prerequisite for an abnormal response to pathologic GHRH secretion may be the existence of a mutation in somatotroph cells. Most human somatotroph adenomas seem to be associated with clonal expansion (Herman et al., 1990) of cells bearing a somatic mutation. However, as in other types of pituitary adenoma, it has proved difficult to isolate a single causative factor in sporadic pituitary tumorigenesis (Lytras and Tolis, 2006; Horvath and Stratakis, 2008; Melmed, 2009). A mutated Gsα protein has been identified in up to 40% of somatotroph adenomas. Mutations at two critical sites (gsp mutations) inhibit GTPase activity and lead to constitutive adenyl cyclase activation (Vallar et al., 1987). Cell-cycle disruption also seems to play an important role (Melmed, 2009), as demonstrated in MEN1 or in patients with mutations of *CDKN1B* which encodes the cyclin-dependent kinase inhibitor $p27^{KIP1}$, an atypical tumor suppressor playing a key role in cell-cycle regulation, cell proliferation, and differentiation. Overexpression of HMGA2, a nuclear architectural protein (De Martino et al., 2009), and an activating gene (*PTTG*, pituitary tumor transforming gene), also play a role. This latter gene (a securin homolog) is overexpressed in functional pituitary tumors, and this could lead to aneuploidy (Vlotides et al., 2007); the degree of overexpression correlates with tumor size and invasiveness (Filippella et al., 2006). Premature senescence likely explains the persistence of a benign phenotype and the rarity of progression to carcinoma. Epigenetic mechanisms may also contribute to cell proliferation by silencing genes such as *CDKN2A*, which encodes p16, a CDK inhibitor that inhibits cell proliferation. Finally, even if it is clear that somatotroph cells are altered in somatotroph adenomas, the sequence of events leading to their clonal expansion seems to be multifactorial. An activated oncogene may be necessary to initiate tumorigenesis, while promotion of cell growth may require GHRH or other growth factors such as bFGF (basic fibroblast growth factor) (Asa and Ezzat, 2002; Melmed, 2003).

Growth hormone-secreting carcinomas

The large majority of these somatotroph adenomas are benign. Pituitary carcinomas are exceptional (fewer than 20 published cases) and the presence of distant metastases is generally required to support the diagnosis of malignancy (review in Chanson et al., 2009a).

Genetic syndromes associated with acromegaly

Syndromes associated with acromegaly have been recently reviewed (see Chanson et al., 2009a; Daly et al., 2009; Dworakowska et al., 2011; Gadelha et al., 2013).

McCune–Albright syndrome, which is associated with multiple fibrous bone dysplasia, precocious puberty and café-au-lait spots, can be accompanied by acromegaly. This syndrome is related to a somatic mutation that activates the α subunit of Gs protein (Chanson et al., 2007; Collins et al., 2012). More than 20% of these patients develop acromegaly (Chanson et al., 2007; Collins et al., 2012; Vortmeyer et al., 2012).

Acromegaly can also be associated with hyperparathyroidism, neuroendocrine tumors (e.g., gastrinoma, insulinoma, or a nonfunctional pancreatic tumor), adrenal and other endocrine or nonendocrine tumors in patients with multiple endocrine neoplasia (MEN) type 1, which is generally related to a germline mutation of the *MEN1* gene encoding menin (Verges et al., 2002; Thakker et al., 2012). Mutations in *CDKN1B* are responsible for a very rare MEN syndrome, MEN type 4, which combines hyperparathyroidism, pituitary adenomas (including acromegaly), and other endocrine or nonendocrine neoplasia (Pellegata et al., 2006; Georgitsi, 2010).

When acromegaly is associated with bilateral pigmented micronodular adrenal hyperplasia (causing ACTH-independent hypercorticism) and with cutaneous lesions or cardiac myxomas, the patient should be screened for the Carney complex, which is often related to a germline mutation of the regulatory 1-α subunit of protein kinase A (PRKAR1A) (Kirschner et al., 2000; Bertherat, 2006).

Recently, familial acromegaly related to germline mutations of the *AIP* (aryl hydrocarbon receptor interacting protein) gene was described (Vierimaa et al., 2006; Chahal et al., 2011; Beckers et al., 2013). These mutations can also, albeit rarely, be found in some apparently sporadic cases of acromegaly, particularly in young patients (Barlier et al., 2007; Cazabat et al., 2007, 2011, 2012; Daly et al., 2007).

Extrapituitary acromegaly

This topic has been recently reviewed by Chanson et al. (2009a).

Acromegaly can also be due to eutopic hypothalamic GHRH hypersecretion (gangliocytoma, hamartoma, choristoma, glioma, etc.) or, more often, to ectopic, peripheral GHRH hypersecretion (pancreatic or bronchial carcinoid tumor) that stimulates normal somatotrophs to become hyperplastic and to hypersecrete GH (Gola et al., 2006; Biermasz et al., 2007; Garby et al., 2012). The diagnosis is based on plasma GHRH assay (revealing excess secretion) and on identification of the GHRH-secreting endocrine tumor (Thorner et al., 1982). The prognosis is generally good (Garby et al., 2012).

GH can also be hypersecreted by an ectopic pituitary adenoma (sphenoidal sinus, petrous temporal bone, nasopharyngeal cavity) or, in exceptional cases, by a peripheral tumor (pancreatic islet tumor or lymphoma) (Melmed et al., 1985; Beuschlein et al., 2000).

SIGNS AND SYMPTOMS

Acromegaly is generally suspected on clinical signs and symptoms which are important to recognize (Molitch, 1992; Melmed, 2002; Giustina et al., 2003; Colao et al., 2004; Chanson and Salenave, 2008; Nachtigall et al., 2008; Melmed et al., 2009; Reid et al., 2010; Ribeiro-Oliveira and Barkan, 2012).

The dysmorphic syndrome

The extremities (hands and feet) are broadened, the fingers are widened, thickened and stubby, and the soft tissues are thickened. The patient may have had to enlarge his or her ring in recent years, or to change shoe size. The facial aspect is characteristic, and patients with established acromegaly are generally alike in this respect: the nose is widened and thickened, the cheekbones are prominent, the forehead bulges, the lips are thick and the facial lines are marked (Fig. 14.1). The forehead and overlying skin are thickened, sometimes leading to frontal bossing. There is a tendency towards mandibular overgrowth with prognathism, maxillary widening, teeth separation, and jaw malocclusion. Photographs show a slow, insidious transformation over several years (Fig. 14.1). As a result, the diagnosis is often raised by a doctor who has never seen the patient before. Deformations can also affect the rest of the skeleton, and dorsal kyphosis with distortion of rib cage may be observed in severe chronic forms, leading to the classic "Punchinello" aspect, especially when GH hypersecretion begins prior to closure of the epiphyses. Acral changes are the most common problem leading to diagnosis, which is made by a general practitioner in the

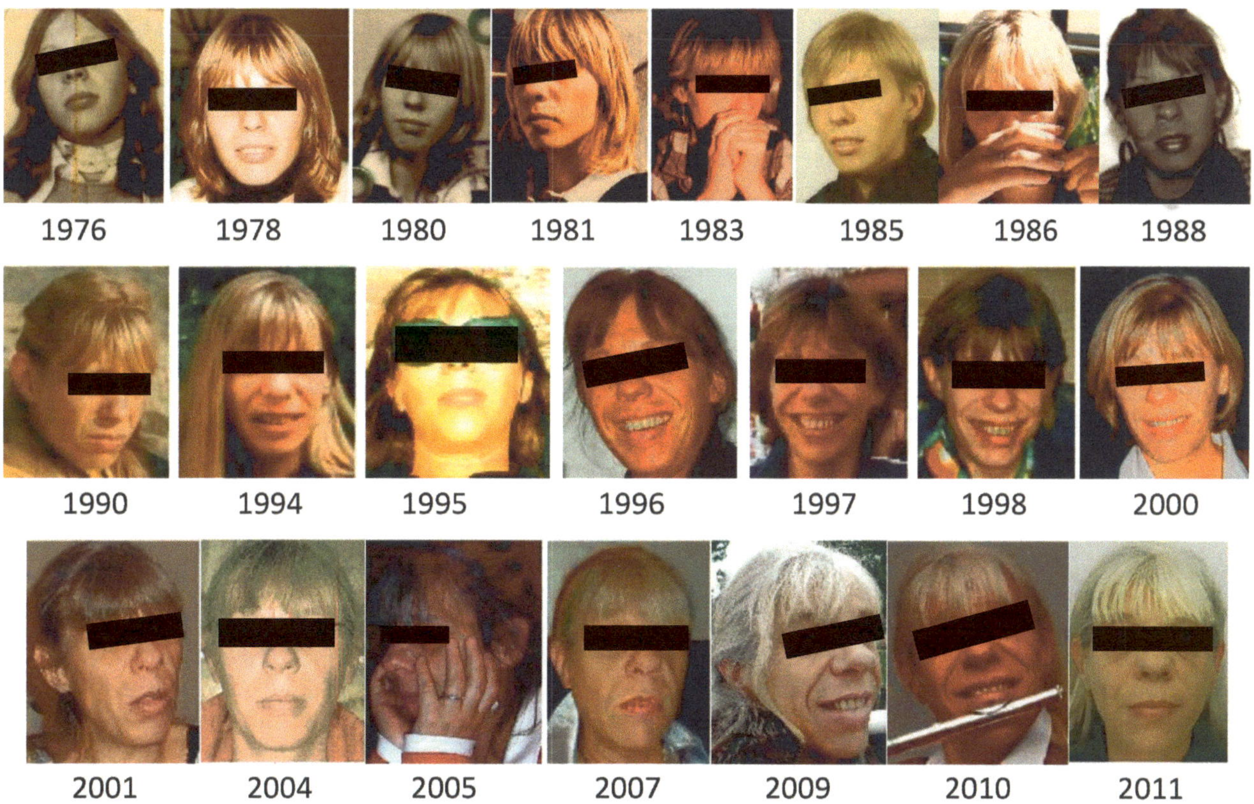

Fig. 14.1. Series of photographs of a patient with acromegaly showing the progressive changes in facial appearance. It is possible to assume that the disease began between 1988 and 1990, 12 years before the diagnosis, made in 2011. In 1990, the patient had to cut the ring that she bought in 1988 because it was too narrow.

majority of cases (Nachtigall et al., 2008; Reid et al., 2010). However, most patients complain of a delay in diagnosis, despite clear signs and symptoms: this is because many physicians are unaware of this disease, having never encountered such a patient during their training or practice (Danzig, 2007). In our experience it is not uncommon that a patient makes the diagnosis himself by searching the Internet.

Symptoms

Acromegaly can cause a variety of symptoms, such as headache (whether the pituitary adenoma is large or small): this is the second initial complaint leading to diagnosis of acromegaly) (Nachtigall et al., 2008); malodorous sweating (especially at night); acroparesthesia (carpal tunnel syndrome); and joint pain. A gradual deepening of the voice is also observed.

Skin changes

Nearly 70% of patients have sweaty, oily skin. Skin thickening is due to glycosaminoglycan deposition and to increased collagen production by connective tissue. Skin tags are frequent and may be a marker of colonic polyps. Raynaud's disease is present in one-third of cases.

Bone changes

Craniofacial

In response to both GH and IGF-1, new periosteal bone formation leads to an increase in skeletal growth, especially at the level of the mandible (prognathism). Jaw thickening, teeth separation, frontal bossing, malocclusion, and nasal bone hypertrophy are the standard facial bony deformities in acromegaly.

Radiography shows a thickening of the cranial vault and protuberances, frontal internal hyperostosis, and condensation of the walls of the sella turcica with clinoid hypertrophy. Hypertrophy of the sinuses, especially the frontal sinuses, is also clearly visible. This, along with laryngeal hypertrophy, explains why the voice tends to become deeper and acquires a sonorous resonance.

Extremities

These changes are not only due to soft tissue hypertrophy and excessive growth of bone and cartilage but also to bone deformation. Indeed, radiographic findings are abnormal in half these patients, showing distal tufting of the phalanges, widening of the base of the phalanges with osteophyte formation, enthesopathy (mineralization of ligamentous insertions), widening of the diaphysis in cortical bone, and widening of joint spaces due to cartilage hypertrophy.

Trunk

Bony deformation also affects the spine, with upper dorsal kyphosis and compensatory lumbar hyperlordosis. Vertebral enlargement, widened intervertebral spaces and osteophyte formation are also observed. The thorax is deformed by protuberance of the lower portion of the sternum, and by elongation and divergence of the ribs, due to overgrowth of the chondrocostal joints.

Limbs

Imaging studies show diaphyseal cortical thickening of the long bones and widened joint spaces, sometimes with osteophytes.

Bone mineral density

Bone remodeling is stimulated in acromegaly (Ueland et al., 2006; Giustina et al., 2008). Cortical bone thickens (as measured by the metacarpal index and histomorphometric parameters) and its porosity is diminished. Trabecular bone mass may be decreased, normal, or increased. Measurement of spinal bone mass can give contradictory results, probably because acromegaly is often associated with other endocrine disorders that interfere with bone mass. In general, bone mass is normal in the lumbar spine in patients with isolated acromegaly but is decreased in patients with associated hypogonadism, as is also generally the case in patients with hypogonadism, whatever its cause. In a recent study (Mazziotti et al., 2008), although bone mineral density (BMD) was not significantly different between acromegalic patients and control subjects, the prevalence of vertebral fractures was higher in acromegalic patients (57.5% versus 22.6%). Fractures were associated with higher serum IGF-1 values, a longer duration of active disease, and a longer history of untreated hypogonadism. This higher prevalence of vertebral fractures persists despite biochemical control of acromegaly (Claessen et al., 2013a).

Rheumatologic complications

Peripheral arthropathy

This topic has been reviewed in detail by Liote and Orcel (2000) and Kropf et al. (2013).

Peripheral joint symptoms are very frequent. Arthralgia and myalgia occur in 30–70% of patients. All the joints can be affected, especially the large joints (knees, shoulders, hands, wrists, and hips). Acromegalic arthropathy develops within an average of 10 years after diagnosis. The arthralgia is mainly mechanical, degenerative, and noninflammatory in origin, but features of osteoarthritis may develop in some patients. Joint mobility (especially of the shoulders) can be limited in the later stages of the disease. Joint effusion is rare and synovial aspiration shows a generally degenerative picture with no evidence of inflammation; it may also reveal calcium microcrystals (associated chrondrocalcinosis).

Physical examination of the joints often provides little information. The abnormalities are generally minor despite subjective functional discomfort. The shoulders and hips may show a loss of mobility and function. In contrast, some patients have joint hyperlaxity. There is no correlation between the presence (or severity) of arthropathy and the age of onset of acromegaly, or the mean GH or IGF-1 concentration at baseline or during follow-up. Arthropathy appears to be more frequent after age 45 years.

Radiologic studies show a widening of the joint spaces, reflecting hypertrophy of the hyaline cartilage, as well as the presence of osteophytes, bone proliferation at the attachment sites of tendons and ligaments, periarticular calcium deposition, and exostosis of the bone surface (Fig. 14.2). The joint space subsequently diminishes due to destructive arthropathy. Sonography shows a thickening of the cartilage in the shoulder, wrist and knee joints, which improves during treatment for acromegaly.

The arthropathy progresses inexorably in advanced stages and unpredictably in minor forms. It is not influenced by successful treatment of acromegaly, with the exception of diffuse articular symptoms and some sites of pain (Wassenaar et al., 2009). It considerably impairs patients' quality of life (Miller et al., 2008; Claessen et al., 2013b).

Spinal involvement

The estimated prevalence of spinal involvement is about 40–50% (Scarpa et al., 2004). Backache is more frequent at the level of the lumbar spine than the cervical or dorsal spine. The pain is mainly mechanical in nature, but inflammatory features can occur (16%). Spinal involvement may be accompanied by nerve compression. Occasionally, bilateral intermittent claudication reveals lumbar spinal stenosis. Pain may also be related to increased prevalence of vertebral fractures despite normal BMD (Mazziotti et al., 2008; Wassenaar et al., 2011).

Radiologic examination shows typical features, including ossification of the anterior and lateral surfaces of the vertebral bodies, contributing to enlargement of their anteroposterior diameter, as well as a biconcave vertebral aspect and scalloping of the vertebral bodies (exaggerated concavity of the posterior vertebral wall). The mechanism is poorly understood but may involve hypertrophy of the intraspinal soft tissues (ligamentous hypertrophy, epidural lipomatosis) or bone. In more severe cases, ossification of the anterior surface of the vertebral bodies can bridge the disc space and give an aspect of diffuse idiopathic skeletal hyperostosis.

Neuropathies

Symptomatic carpal tunnel syndrome is frequent. Nerve conduction studies have shown that the vast majority of acromegalic patients have subclinical abnormalities of nerve conduction. Magnetic resonance imaging (MRI) shows a higher amplitude and intensity of the median nerve signal in patients with symptomatic carpal tunnel syndrome compared to asymptomatic patients (Jenkins et al., 2000). The mechanism appears to involve median nerve edema more than extrinsic compression due to an excess of connective tissue, bone, or synovial hypertrophy, or an increase in extracellular fluid within the carpal tunnel itself with Schwann cell demyelination. The nerve edema, which can easily be evaluated with ultrasonography (Tagliafico et al., 2008a), improves when GH and IGF-1 levels fall, suggesting that hormonal control is a key prerequisite for improving these patients' neurologic status. Sometimes, however, the carpal tunnel syndrome persists.

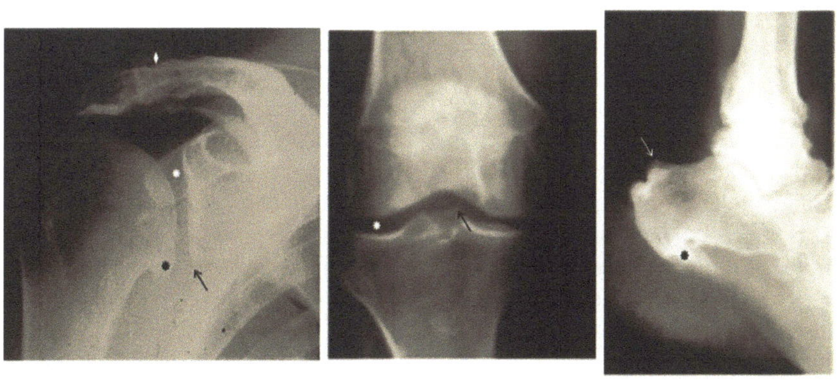

Fig. 14.2. Osteoarticular consequences of acromegaly at the level of shoulder, knee and foot. Radiographs show thickening of the interarticular spaces (white stars), osteophytes (black arrows), bone proliferation at the attachment sites of tendons and ligaments (black stars), periarticular calcium deposition (white diamond), and exostosis of the bone surface (white arrow).

Ulnar nerve neuropathy at the cubital tunnel is also frequent in patients with acromegaly (Tagliafico et al., 2008b) and improves with treatment of acromegaly.

Cardiovascular manifestations

Arterial hypertension

Hypertension occurs in 20–50% of patients. Its prevalence increases with time after the onset of acromegaly, as well as with the GH level and age. It is at least partly due to chronic hypervolemia (the plasma volume is 10–40% above normal, owing to increased renal sodium reabsorption at the distal tubule level) (Chanson et al., 1990). Body fluid expansion is related to enhanced epithelial sodium channel (ENaC) activity (Kamenicky et al., 2008, 2011, 2014). Hypertension can also result from endothelial dysfunction (Maison et al., 2000). Neither renin–angiotensin–aldosterone nor sympathetic systems seem to be involved in the pathogenesis of hypertension in this setting. Insulin resistance and diabetes may also play a role in the onset of hypertension (Colao et al., 2000; Jaffrain-Rea et al., 2001). Sleep apnea syndrome is likely to contribute to the pathogenesis of hypertension.

Specific cardiomyopathy

Cardiac disorders are a consistent feature. Many lines of evidence, especially from experimental studies, point to the existence of specific cardiac disorders in acromegaly, independently of coronary involvement (found in only a minority of patients nowadays) and valve disorders (Sacca et al., 1994; Clayton, 2003; Colao et al., 2004).

Initially the cardiac involvement is asymptomatic (at least at rest), and consists mainly of myocardial hypertrophy (of the interventricular septum and left ventricular posterior wall), as assessed by echocardiography or magnetic resonance imaging (MRI) (Fig. 14.3), but the dimensions of the left ventricle are normal (concentric hypertrophy). Myocardial hypertrophy can occur in the absence of hypertension and even in young patients (<30 years), reflecting the impact of the GH excess itself on the myocardium. Hypertension further aggravates cardiac hypertrophy. Echocardiography and isotope studies show altered diastolic function (abnormal left and right ventricle filling) related to abnormal relaxation: parietal stiffness is, at least in part, probably linked to edematous infiltration of the ventricular wall (Gouya et al., 2008) and perhaps also to a certain degree of fibrosis. Clinical symptoms such as dyspnea during exercise may be observed in patients who are asymptomatic at rest. Systolic function is normal at this stage (again, at least at rest), thanks to increased myocardial contractility. A hyperkinetic syndrome (increased cardiac index) is

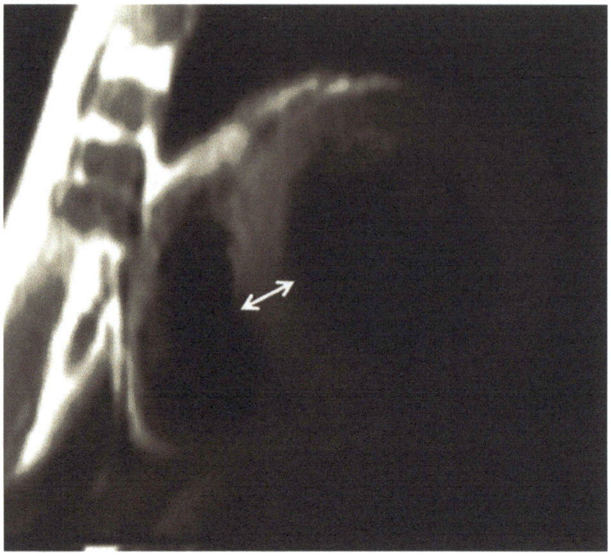

Fig. 14.3. Cardiac hypertrophy in a patient with acromegaly as assessed by cardiac MRI. Short-axis black-blood single-shot fast spin echo acquisition showing the two ventricular cavities with enlarged walls and hypertrophic interventricular septum (arrow).

always present. Systolic function is altered during exercise, however. Even at this early stage, arrhythmias and/or conduction disorders may occur. Their prevalence in acromegaly was long underestimated. In fact, ventricular premature complexes occur in about 40% of patients with acromegaly. In one study systematic 24 hour Holter electrocardiogram (ECG) recordings showed complex ventricular arrhythmias in 48% of patients compared to only 12% of controls (Kahaly et al., 1992). Most of these arrhythmias are subclinical and persist despite successful treatment of acromegaly. Myocardial remodeling, hypertrophy, and fibrosis are all likely to play a role in their onset.

Congestive heart failure can occur if the cardiac disorders progress (if GH hypersecretion persists and, probably, if other risk factors such as diabetes, hypertension, and sleep apnea are also present). Functional signs first appear on effort before becoming permanent. At this stage echocardiography shows variable degrees of cavity dilation. Fortunately, these severe forms are now far less frequent (prevalence 3%) (Bihan et al., 2004).

A number of cardiovascular parameters improve during effective treatment of acromegaly, even if some lesions may appear to be irreversible in certain patients. In general, younger patients and patients with a relatively short history of acromegaly show better "recovery" (from diastolic disorders, myocardial hypertrophy, or systolic dysfunction) (Colao et al., 2004). In contrast, when dilated congestive heart failure occurs, cardiac function (especially systolic function) may show a short-term improvement, allowing some patients to survive or to avoid heart transplantation, but the longer-term prognosis is worse than that of patients with heart failure due to other causes (5 year mortality rate 37%) (Bihan et al., 2004).

There is controversy surrounding the cardiovascular (ischemic) risk carried by patients with acromegaly. An increased prevalence of hypertension, a history of diabetes mellitus, and decreased levels of high-density lipoprotein, low-density lipoprotein, and total cholesterol were found in patients with acromegaly, leading to significantly higher Framingham risk scores than in controls (Berg et al., 2010). Biomarkers of cardiovascular disease were also found to be altered in another study (Boero et al., 2009). However, carotid atherosclerosis and carotid internal media thickening are not more extensive in acromegalic patients than in nonacromegalic subjects (Otsuki et al., 2001; Bogazzi et al., 2007; Paisley et al., 2011), and no increase in the prevalence of coronary artery disease is found in patients with newly diagnosed acromegaly (Cannavo et al., 2006; Bogazzi et al., 2007; Akutsu et al., 2010). The reasons for this apparent discordance are unclear, but it suggests that the known atherogenic effects of hypertension, insulin resistance and diabetes induced by GH excess are counterbalanced by cardioprotective factors such as decreased C-reactive protein (CRP) (Parkinson et al., 2002; Sesmilo et al., 2002; Verhelst et al., 2013).

Valve disease

The increased prevalence of valve disorders can also contribute to the onset or aggravation of heart disease in patients with acromegaly (Colao et al., 2003). The risk of valve disease increases with time since onset (Pereira et al., 2004), and these abnormalities often persist after effective treatment of acromegaly (Maione et al., 2012). They may be related to fibrotic changes.

Metabolic complications

Physiologically, GH increases blood glucose levels, exerts a lipolytic effect, and promotes triglyceride hydrolysis into free fatty acids and glycerol.

GH excess leads to insulin resistance at the level of the liver or in the periphery, leading to hyperinsulinemia. The prevalence of diabetes in acromegalic patients ranges from 20% to 56%, and that of glucose intolerance from 16% to 46%, depending on the series (Colao et al., 2004). As long as the compensatory increase in insulin secretion by pancreatic β cells counterbalances the reduction in insulin sensitivity, glucose tolerance remains normal. Impaired glucose tolerance occurs when insulin secretion is altered, and this is followed by the onset of diabetes (Alexopoulou et al., 2014).

Acromegaly is associated with a decrease in fat mass (both visceral and subcutaneous) but an increase in intermuscular fat mass (which may contribute to insulin resistance) and also in lean body mass (Freda et al., 2008; Katznelson, 2009).

Hypercalciuria is frequent in patients with acromegaly and may be associated with an increased incidence of nephrolithiasis. It is related to an IGF-1-mediated and PTH-independent increase in calcitriol synthesis which is responsible for both absorptive hypercalciuria and increased fasting plasma calcium linked to enhanced distal tubular calcium reabsorption (Kamenicky et al., 2012, 2014).

Respiratory complications

Sleep apnea affects 60–80% of all patients with acromegaly at diagnosis of acromegaly (more often men) (Attal and Chanson, 2010). Sleep apnea is more likely to be sought in patients who snore (reported by 78% of patients with acromegaly) and in those with daytime sleepiness (51%), or morning fatigue and morning headache (16%). Sleep apnea may be a contributory factor in hypertension and cardiovascular disease. In most cases the apnea is obstructive, but one-third of patients have central apnea. Obstructive apnea is linked to anatomic changes due to mandibular and maxillary growth, soft-tissue thickening (especially of the palate and uvula) and changes in the angles of the different bone segments, leading to hypercollapsibility of the posterior and lateral hypopharyngeal walls. Hypertrophy of the tongue also plays a role (Herrmann et al., 2004), as does hypertrophy of the submaxillary glands.

Changes in respiratory function are frequent but less well documented. Anatomic modifications of thoracic bones and cartilage (leading to profound changes in the geometry of the rib cage) and mechanical changes in thoracic elasticity and inspiratory muscles can lead to ventilatory disorders. Respiratory muscle strength is also abnormal. Altered mechanical and energetic properties of some upper airway dilator muscles have recently been demonstrated (Attal et al., 2009). The inspiratory time is shorter and the breathing frequency may increase.

Patients with acromegaly often have an increase in their total lung capacity (81% of men and 56% of women), owing to an increase in alveolar volume. An obstruction is found in 20–30% of patients (small airway or upper airway narrowing). Subclinical hypoxemia may be present. No ventilation-perfusion mismatching has been demonstrated.

The apnea-hypopnea index improves during effective treatment of acromegaly, along with the obstructive apnea index and oximetry values (Ip et al., 2001;

Herrmann et al., 2004; Attal and Chanson, 2010). However, while apnea can disappear in some patients whose acromegaly is cured, others require nocturnal positive end expiratory pressure for persistent sleep apnea.

Neoplasia and acromegaly

Conclusions concerning cancer-related mortality vary from one epidemiologic study to another, but it seems clear that there is an excess of cancer-related mortality in patients with uncontrolled disease. There is also some controversy regarding the incidence of each individual cancer type in acromegalic patients.

Gastrointestinal tumors

The importance of colon cancer risk in acromegaly is controversial (Melmed, 2001; Renehan and Brennan, 2008). The relative risk of colon cancer, as compared with the general population, has been widely overestimated at 10–20, whereas it is, in reality, probably only 2–3 (Jenkins and Besser, 2001; Renehan et al., 2003). There are various potential biological mechanisms which may explain the association of increased risk colon cancer in acromegaly: direct effects of GH and/or IGF-1 are the most evident but others, such as hyperinsulinemia, changes in IGFBP-3, increase in IGF-2 and IGFBP-2 levels, altered bile acid secretions and local immune response, increased large bowel length, may be important (Renehan and Brennan, 2008). Moreover, acromegaly, independently of hormone secretion, may be a disease that heralds genetic and/or epigenetic alterations predisposing patients to cancer development (Loeper and Ezzat, 2008). As colon cancer may be the consequence of colon polyp degeneration, many studies have examined the prevalence of colon polyps in patients with acromegaly. Prospective studies show that up to 45% of patients with acromegaly have colonic polyps, which are adenomatous in 24% of cases (Delhougne et al., 1995) and can arise in all parts of the colon. There is no clear correlation between GH or IGF-1 concentrations and the incidence of colonic polyps. Colonoscopy guidelines for acromegalic patients are controversial. The recommendations of the British Society of Gastroenterology (Cairns et al., 2010) are as follows:

1. Patients with acromegaly should be offered regular colonoscopic screening, starting at the age of 40 years. It is a recommendation grade B (evidence obtained from at least one well-designed controlled study without randomization – IIa – or from at least one other type of well-designed quasi-experimental study – IIb – or from a well-designed nonexperimental descriptive study, such as comparative studies, correlation studies and case studies – III).
2. The frequency of repeat colonoscopy should depend on the findings at the original screening and the activity of the underlying acromegaly (recommendation grade B):
   - patients with an adenoma at first screening or elevated serum IGF-1 level above the maximum of the age-corrected normal range should be offered 3 yearly screening
   - patients with a negative first colonoscopy or a hyperplastic polyp or normal growth hormone/IGF-1 levels should be offered screening every 5–10 years. Total colonoscopy is required rather than sigmoidoscopy, although the former is associated with technical difficulties.

However, these recommendations may be revised in the future due to the modest increase in colon cancer risk in acromegaly and the potential harms of the procedure related, in particular, to the increased length of the colon in patients with acromegaly (Renehan and Brennan, 2008).

Thyroid nodules

Goiter is found in 25–90% of patients with acromegaly. The risk of developing thyroid nodules increases with the time since onset of acromegaly. Multinodular goiter is autonomous in 10–20% of patients, sometimes causing patent thyrotoxicosis. Thyroid nodules are generally harmless, but the risk of thyroid cancer may be slightly higher than in the general population (Renehan and Brennan, 2008; dos Santos et al., 2013).

Other cancers (lung, breast, prostate, etc.) are not over-represented in patients with acromegaly (Melmed, 2001).

DIAGNOSIS OF ACROMEGALY

The diagnosis of acromegaly is based on clinical grounds with biochemical confirmation (Chanson and Salenave, 2008; Ribeiro-Oliveira and Barkan, 2012). Clinical diagnosis is suggested by the typical disfigurement due to progressive acral enlargement and modification of the facial appearance, as assessed on serial photographs (Fig. 14.1). Biochemical diagnosis is based on increased serum GH concentrations that are not suppressed by an oral glucose load (oral glucose tolerance test, OGTT). An increase (with reference to the age-adjusted normal range) in the serum concentration of IGF-1, the main GH-dependent growth factor, confirms the diagnosis.

Growth hormone assays

The first GH assays, which began to be used 35 years ago, were poorly sensitive polyclonal competitive radioimmunoassays (RIAs). Over the last 25 years, more sensitive noncompetitive two-site antibody radioimmunometric assays (IRMAs) have been introduced. Fifteen years ago, nonisotopic two-site antibody assays began to become available, with the major practical advantage that some of them could be automated. This last type of assay is the most widely used today, at least in Europe. Differences in analytical methods (RIA, IRMA, immunochemiluminescent assay (ICMA), and enzyme-linked immunoabsorbent assay (ELISA)) are one explanation for the variability of published GH results. GH circulates in plasma as a mixture of different molecular forms: 22 kDa GH, 20 kDa GH, a GH-binding protein (GH-BP)-linked form, as well as dimers and polymers. The use of polyclonal or monoclonal antibodies specific for the predominant 22 kDa isoform or for several of different isoforms also contributes to the observed variability.

Many standard GH preparations have been used to calibrate GH assays. In Europe, however, manufacturers were recently advised to calibrate their GH assay kits using International Standard (IS) 98/574, which was established with recombinant GH. According to a European consensus statement on the standardization of GH assays (Trainer et al., 2006), which was followed by an international consensus statement on the standardization and evaluation of GH and IGF assays (Clemmons, 2011), the use of the second International Standard (IS) for GH (WHO IS 98/574), a recombinant material consisting of 22 kDa GH with better than 95% purity, is recommended as a single calibrant for GH immunoassays, and it is also recommended to report GH concentrations in μg/L of IS 98/574 (1 μg corresponding to 3 IU of somatropin). Assays now need to achieve a lower limit of quantification of 0.05 μg/L with an interassay CV of < 20% (Giustina et al., 2010; Clemmons, 2011).

Which growth hormone cutoff to use for diagnosis?

The basal plasma GH level (in the morning, for example, or at randomly selected times) is elevated in acromegaly. However, high GH concentrations can also be found in healthy subjects, owing to the episodic nature of GH secretion, which can fluctuate between undetectable levels (most of the time) and peaks of up to 30 μg/L (90 mIU/L). According to a 2000 consensus statement (Giustina et al., 2000), basal GH and IGF-1 must be measured when acromegaly is suspected. A GH concentration below 0.4 μg/L plus a normal IGF-1 level rules out acromegaly. In contrast, if the GH is above 0.4 μg/L and/or if the IGF-1 is elevated (as compared with age-adjusted normal range), an oral 75 g OGTT must be performed. If the lowest GH value (nadir) during the OGTT is below 1 μg/L, acromegaly can be ruled out. If it remains above 1 μg/L, acromegaly is confirmed. As discussed below (see below, Difficult and borderline clinical situations), with the generalized use of very sensitive assays (Cazabat et al., 2008), this cutoff probably needs to be reduced to 0.3 μg/L or even lower (Ribeiro-Oliveira and Barkan, 2012). However, it is probably unrealistic to propose a universal cutoff under OGTT, as its determination is highly dependent on the assay that was used for its calculation (Arafat et al., 2008). Finally, it should be kept in mind that in the vast majority of cases, GH levels, both baseline and after OGTT, are clearly elevated and the biochemical diagnosis of acromegaly is straightforward and does not raise any question (Carmichael et al., 2009). It must also be remembered that, paradoxically, the OGTT stimulates GH secretion in about 10% of patients with acromegaly.

IGF-1 assays

The IS 02/254 WHO reference standard has recently become available. It is a ~97% pure recombinant standard which has been well characterized. Its use is now recommended for manufacturers of IGF-1 assays (Giustina et al., 2010; Clemmons, 2011). The IGF-1 level increases in parallel to the log of the GH concentration and must be determined by using age-adjusted norms, since levels fall with age. Normative data should be based on a random selection of individuals from the background population, and this population sample should include representatives of all age groups (Clemmons, 2011). Pregnancy, puberty, and the postpubertal period are accompanied by high IGF-1 concentrations. The concentration of IGFBP3, the main IGF carrier protein, is usually increased in patients with acromegaly, but this marker offers little further diagnostic information.

Stimulation tests

Up to 50% of patients have an increase in their GH concentration after thyrotropin-releasing hormone (TRH) and/or gonadotropin-releasing hormone (GnRH) stimulation. These tests have no diagnostic value, however, and neither does the response to GHRH.

Difficult and borderline clinical situations

A few patients with clear clinical signs of acromegaly and high IGF-1 levels have low GH output, including a nadir of < 1 μg/L during the OGTT (Dimaraki et al., 2002; Arafat et al., 2008; Ribeiro-Oliveira et al., 2011b; Ribeiro-Oliveira and Barkan, 2012). Thus, some authors have suggested that when chemiluminescence or fluorometric assays with very low detection limits are used, the OGTT positivity cutoff should not be 1 μg/L but rather 0.3 μg/L (Trainer, 2002). However, it must be kept in mind that when GH output is low (basal GH levels < 5 μg/L), the OGTT may sometimes be misleading, as GH levels may be suppressed to less than 0.3 μg/L in some patients with true acromegaly (Ribeiro-Oliveira et al., 2011a; Ribeiro-Oliveira and Barkan, 2012).

Other difficult problems include a typical clinical picture of acromegaly in a patient with normal IGF-1 and GH concentrations. This situation probably corresponds to cases in which acromegaly has resolved spontaneously, probably through necrosis of a pituitary adenoma.

GH and/or IGF-1 measurement cannot be used for diagnosis or to assess treatment efficacy in patients with certain conditions such as diabetes mellitus, chronic renal failure, or pregnancy, and also at the time of puberty.

Some somatotroph adenomas (silent somatotroph adenomas) are not accompanied by systemic GH hypersecretion or are associated with only very mild GH hypersecretion. In such cases, clinical signs of acromegaly are absent or very mild, but GH immunolabeling of the excised adenoma (generally resected because of local tumor symptoms) is positive (Kovacs et al., 1989). However, subtle abnormalities of GH/IGF-1 hypersecretion are frequent in such cases (Dimaraki et al., 2002; Sakharova et al., 2005).

Differential diagnosis

Lastly, some patients have acromegaloid features despite normal GH and IGF-1 levels. Some such cases are attributed to severe insulin resistance.

TUMORAL AND FUNCTIONAL PITUITARY ASSESSMENT

Once the diagnosis has been established, and before initiating treatment for acromegaly, patients must undergo a two-pronged workup focusing on tumor mass effects (headaches, changes in the visual field and acuity, MRI abnormalities) and pituitary function.

MRI can show whether the pituitary tumor is a micro- or macroadenoma, if it expands upwards, downwards or laterally, in the cavernous sinus, and if it is invasive (Fig. 14.4). Conversely, the lack of clear signs of an adenoma on MRI, or an aspect showing a bulging, hyperplastic pituitary, suggests that the acromegaly is secondary to ectopic GHRH secretion (Chanson et al., 2009a).

Further evaluation is necessary to determine whether other pituitary hormones are deficient or secreted in excess: PRL hypersecretion is present in 30% of cases and may be either functional, secondary to impairment of hypothalamic dopamine production or compression of the pituitary stalk by the tumor, thereby impairing dopamine transport to the pituitary, or due to a mixed adenoma (see above).

PROGNOSIS AND OUTCOME

Acromegaly is associated with increased mortality (Dekkers et al., 2008). According to series published in the 1980s and 1990s, about 60% of patients die from cardiovascular disease, 25% from respiratory complications, and 15% from cancer. Left untreated, patients with acromegaly would die about 10 years earlier than healthy subjects (Holdaway and Rajasoorya, 1999). Several studies have shown that cerebrovascular disorders are a frequent cause of death, especially among women, but these studies involved patients treated in various ways (craniotomy, radiotherapy) many years ago, and a deleterious effect of these treatments (especially radiotherapy) cannot be ruled out (Ayuk et al., 2004; Kauppinen-Makelin et al., 2005). In two recent meta-analyses (Dekkers et al., 2008; Holdaway et al.,

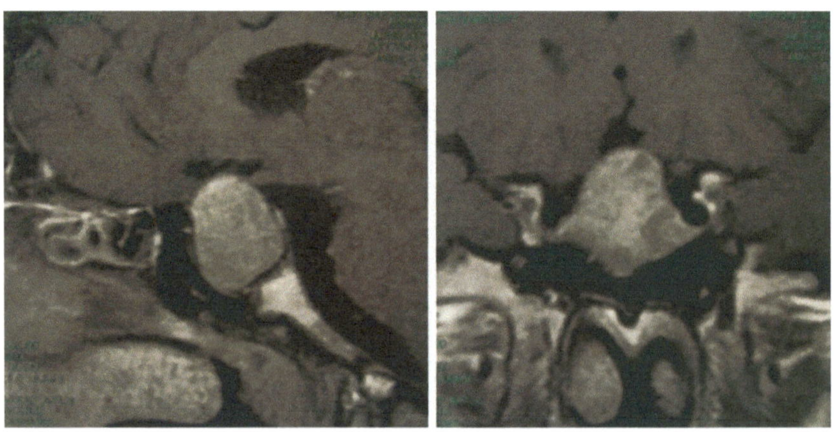

Fig. 14.4. Pituitary macroadenoma measuring 26×22×17 mm with suprasellar extension compressing the optic chiasm at time of diagnosis in a patient in whom acromegaly was diagnosed because of visual fields defects. Invasion of left cavernous sinus was confirmed at surgery.

2008), the standardized mortality ratio (the ratio of observed mortality in the acromegalic population to expected mortality in the general population) was 1.72 (CI 95% 1.62–1.83), but survival has been better in more recent studies. The post-treatment GH concentration is probably the best predictor of survival for all causes of death, independently of the types of complication. Thus, life expectancy outcomes can be stratified according to the post-treatment GH concentration: if GH secretion is controlled (<2 μg/L, or IGF-1 normalization), life expectancy merges with that of the matched general population (Ayuk et al., 2004; Holdaway et al., 2004). It is interesting to note that the recent meta-analysis by Holdaway et al., who investigated the relationship between mortality and GH and IGF-1 levels in patients with acromegaly, confirmed that the criteria proposed almost a decade ago, and still in use today, i.e., serum growth hormone (GH) levels < 2.5 μg/L and age-adjusted normal levels of insulin-like growth factor I (IGF-1), are adequate to predict vital outcome. Holdaway's meta-analysis also showed that the prognosis of acromegaly has improved in the last 20 years. This improvement may be related both to more aggressive treatment of the disease and to better management of comorbidities (Chanson and Maison, 2009). High GH/IGF-1 concentrations, arterial hypertension, and cardiomyopathy are factors of poor prognosis, while the duration of symptoms and other factors (diabetes, lipid disorders, and cancer) are less important (Melmed, 2001). Quality of life is also altered in acromegaly, and is improved by effective treatment (Biermasz et al., 2004). Finally, it must be stressed that, with the current therapeutic strategy, the vast majority of acromegalic patients achieve very good control of GH/IGF-1 secretion and have no problems relating to tumor growth. Adverse effects are infrequent and minor, even in the very long term. This picture is very different from the situation only 20 years ago, before the advent of somatostatin analogs. In addition, the use of more stringent criteria to define cure, together with aggressive treatment of comorbidities, has significantly improved the outlook of patients with acromegaly (Maison and Chanson, 2006; Arosio et al., 2012). However, even patients who are cured or well-controlled may have sequelae (joint pain, deformities, and altered quality of life) (Annamalai et al., 2013).

MANAGEMENT AND TREATMENT

There is now a general consensus on management of acromegaly (Melmed, 2009; Chanson et al., 2009b; Giustina et al., 2010; Sherlock et al., 2011; Chanson and Kamenicky, 2012; Ribeiro-Oliveira and Barkan 2012).

Treatment aims

The clinical aims are to relieve symptoms, to reduce the volume of the pituitary tumor, to avoid tumor relapse, and to improve long-term morbidity and mortality (Melmed et al., 2002). Recent epidemiologic studies have helped to refine the definitions of "cure" and "good disease control", which are now far more precise: the GH concentration (in a random sample) must return to less than 1 μg/L in the new sensitive assays which are now used worldwide (if the OGTT is used, the nadir needs to be less than 0.4 μg/L) and the IGF-1 level must return to normal (Giustina et al., 2010). A stepwise therapeutic strategy based on surgery and/or radiotherapy and/or medical treatment allows these goals to be achieved.

Surgery is generally the first-line treatment

Tumor excision, usually by the transsphenoidal route, is the most rapid way of reducing GH and IGF-1 concentrations in patients with acromegaly (see below, Fig. 14.5). Nevertheless, levels normalize in only 40–70% of cases (Fahlbusch et al., 1992; Swearingen et al., 1998; Biermasz et al., 2000; Nomikos et al., 2005; Jane et al., 2011), depending on the size of the tumor (microadenomas are more amenable to cure), the preoperative GH concentration (the success rate is higher when GH concentrations are low, i.e. < 10 μg/L or 30 mIU/L), and the surgeon's experience. Endoscopic techniques are now used in the majority of expert centers (Cappabianca et al., 2004). They have not improved surgical outcomes (Jane et al., 2011) but they may attenuate local adverse effects.

Surgical outcome is carefully assessed at 3 months. When surgery fails to achieve good disease control, or when surgery is impossible or contraindicated, patients are offered radiotherapy and/or pharmacologic treatments.

Radiotherapy

Radiotherapy techniques have evolved. Radiation can be delivered in a single sitting by stereotactic radiosurgery or in fractionated form. Radiosurgery is a term used to define radiation delivered at a high dose to a typically small target in a single or few sittings. Fractionated radiotherapy refers to radiation therapy delivered at smaller doses delivered over typically 5–6 weeks in 25–30 treatments. In order to minimize the dose to surrounding tissue, techniques for stereotactic localization are now used. Different forms of stereotactic radiosurgery (SRS) are available, with radiation delivered as photons (Gamma Knife, Linac, CyberKnife) or charged particles (protons). Stereotactic fractionated radiotherapy (SFR) is a hybrid form that has been developed which employs stereotactic localization techniques

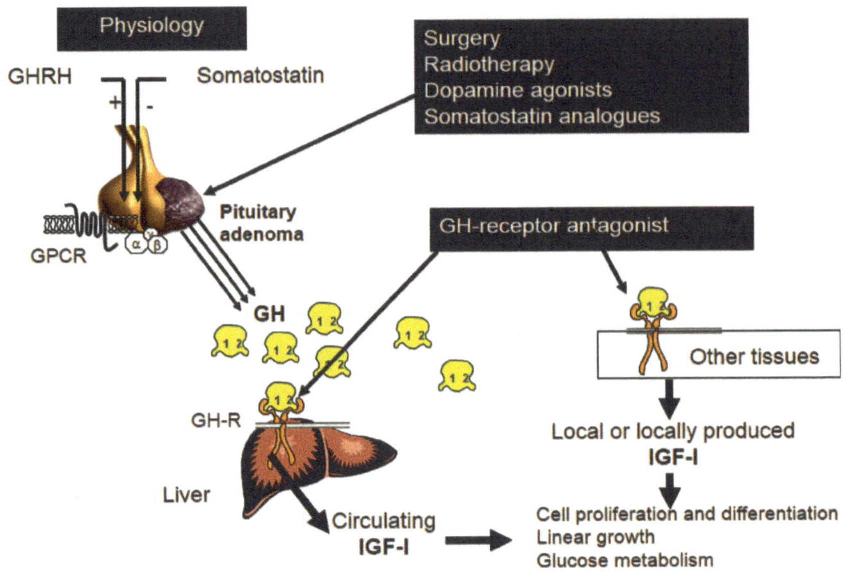

Fig. 14.5. Site of action of the different therapeutic tools in acromegaly. Surgery, radiotherapy, somatostatin analogs and dopamine agonists act at the level of the pituitary adenoma, while GH-receptor antagonists act in the periphery by blocking the GH receptor and thus impairing the effects of GH on the different tissues. (Redrawn from Chanson et al., 2009c, with permission.)

with fractionated therapy, administered by 3D-conformal radiation therapy, intensity modulated radiation therapy, or proton radiation therapy (Shih and Loeffler, 2008).

For acromegaly, SFR delivers an average total dose of 50 Gy in about 25 daily sessions. SRS delivers 20–25 Gy in one (or a few) session(s). The choice between the two methods, SFR and SRS, depends on the size of the tumor and its location relative to the optic apparatus (Loeffler and Shih, 2011). If the lesion measures less than 3 cm and is situated more than 3–5 mm from the optic chiasm, SFR is indicated, while SRS can be proposed in other cases. Fractioned irradiation resulted in GH concentrations below 2 μg/L and normal IGF-1 levels in 22% of patients at 2 years, 60% at 10 years, and 77% at 20 years in a UK study (Jenkins et al., 2006), and, respectively, 23% of patients at 5 years, 42% at 10 years, and 61% at 15 years in an Italian study (Minniti et al., 2005). Here again, the baseline GH concentration seemed to be predictive of treatment outcome (Barrande et al., 2000; Jenkins et al., 2006).

SRS provides more focused irradiation. In a French series of over 80 patients, the efficacy of Gamma Knife irradiation was close to that of fractionated radiotherapy (Castinetti et al., 2005). In a recent meta-analysis, good control of acromegaly (without complementary treatment) was achieved in 48–53% of cases after a mean follow-up of 4 years. This apparently better efficacy of SRS is probably explained by the fact that tumors treated with this approach are relatively small (Yang et al., 2010). The efficacy of the different SRS techniques seems to be similar, with excellent tumor control in > 95% of cases (Castinetti et al., 2005; Castinetti et al., 2009; Yang et al., 2010; Loeffler and Shih, 2011).

Whatever the technique, radiotherapy leads to variable degrees of anterior pituitary insufficiency in 50–80% to 100% of patients after 10–15 years. Complications such as radionecrosis and optic neuropathy are now very rare. In contrast, the risk of stroke may be increased, sometimes many years after irradiation (Brada et al., 1999): stroke may be 1.7 to 2.8 times more frequent (Loeffler and Shih, 2011; Minniti et al., 2011) and is considered to be an important factor of excess mortality in these patients (Sherlock et al., 2010), together with cortisol deficiency and inadequate hormonal substitution. Whether this increased cerebrovascular risk is related to older techniques of radiotherapy only and will not be associated with procedures used nowadays is presently unknown.

Medical treatment

Dopamine agonists

Bromocriptine moderately attenuates the symptoms of acromegaly and reduces GH concentrations, but it normalizes IGF-1 levels in only 10% of patients. Cabergoline appears to be more effective (Abs et al., 1998; Newman, 1999). In a meta-analysis of all published studies, IGF-1 normalization was achieved in 34% of cases (Sandret et al., 2011). Multivariate analysis showed that efficacy depended on the initial IGF-1 concentration, the treatment duration, and the basal concentration of PRL (and, to a lesser degree, the dose of cabergoline) (Sandret et al., 2011). The use of very high doses of dopamine agonists in patients with Parkinson's disease has been linked to an increase in the risk of cardiac valve disease (Schade et al., 2007; Zanettini et al., 2007), but this has not been confirmed in hyperprolactinemic patients (Molitch, 2008; Valassi et al., 2010) or in acromegalic patients treated with cabergoline (Maione et al., 2012).

Somatostatin analogs

Somatostatin analogs (SAs) suppress GH secretion by binding to somatostatin receptors, of which there are five subtypes (SST), on somatotroph adenoma cells. These drugs exert their antisecretory and antitumoral effects by acting on sst2 and 5 (Fig. 14.5).

Octreotide (Sandostatin®) can be injected subcutaneously (SC), generally by the patient him/herself, at a dose of 100–200 μg two or three times a day. This was the first such analog to be marketed, in the 1980s, and represented a real therapeutic advance (Lamberts et al., 1996). Sustained-release lanreotide (Somatuline® SR 30 mg) was the first slow-release preparation to be marketed. It was injected intramuscularly every 10–14 days (the frequency of injections depends on the impact on the GH concentration). Lanreotide is now available for deep SC injection every 28 days, at doses of 60, 90, and 120 mg (Somatuline® Autogel® 60, 90, or 120 mg). Octreotide LAR (Sandostatin® LAR 10–20 or 30 mg), the sustained-release version of octreotide, is administered intramuscularly, once a month. Treatment is usually started at the median dose and is then adjusted (decreased or increased) according to the GH concentration. Alternatively, it is possible to increase or decrease the frequency of injections. Intensity of pituitary adenoma on T2-weighted MRI could help to predict the response to SAs (Puig-Domingo et al., 2010; Heck et al., 2012).

The two SAs, which have largely similar efficacy (Murray and Melmed, 2008), drive GH concentrations below 2 μg/L (5 mIU/L) in 60–70% of patients and normalize IGF-1 levels in 50–80% of patients (Bevan, 2005; Freda et al., 2005; Chanson, 2008). In a recent meta-analysis, overall control rate achieved were 56% and 55%, respectively (Carmichael et al., 2014).

Several long-term studies have shown that the cure rate improves over time (Ayuk et al., 2002; Cozzi et al., 2003; Maiza et al., 2007).

SA therapy must be continued indefinitely because, theoretically, it only suspends GH hypersecretion. In fact, it seems possible, in a handful of very good responders, to lengthen the interval between injections, or even to halt the treatment permanently with no subsequent increase in GH/IGF-1 concentrations (Ronchi et al., 2008; Ramirez et al., 2012).

SAs also have effects on tumor volume, which falls in 20–70% of patients (Bevan, 2005; Melmed et al., 2005). A recent meta-analysis showed that this was achieved in 53% of patients treated with octreotide (Giustina et al., 2012), and in 66% of patients treated with octreotide LAR. The overall weighted mean percentage reduction in tumor size was 37.4%, rising to 50.6% with octreotide LAR (Giustina et al., 2012). The reduction in tumor volume is larger when a SA is the first-line treatment (Colao et al., 2001). Even when SA therapy does not lead to a reduction in tumor volume, it does at least control tumor volume in the vast majority of cases (Bevan, 2005).

SA therapy is indicated after surgical failure, but it can sometimes be used for first-line treatment, especially when severe comorbidities create a risk of perioperative complications. Thus, when heart failure or respiratory problems are associated with acromegaly (Colao et al., 2004; Attal and Chanson, 2010), it is preferable to prepare the patient for surgery by administering a SA for a few months first. In some cases, when the tumor is very large and extends outside the sella, and is not completely extractable by surgery (which would justify postoperative SA therapy), a SA can be administered first in the hope of controlling GH hypersecretion and tumor growth, thus avoiding the need for surgery, particularly as it has long been considered that the effectiveness of first-line therapy (*de novo* patients) is equivalent to that of second-line treatment after surgery or radiotherapy (Newman et al., 1998; Colao et al., 2001; Bevan et al., 2002; Attanasio et al., 2003; Sheppard, 2003). In fact, according to the meta-analysis published by Freda et al., IGF-1 is more likely to normalize after second-line treatment than after first-line drug therapy (Freda et al., 2005).

When SA therapy fails to achieve remission, some authors report that an increase in the SA dose or a shorter interval between injections can rescue some patients (Giustina et al., 2009). When SA therapy is inadequate, another possibility is debulking surgery aimed at lowering hormone levels, in which case retreatment with an SA often provides disease control (Petrossians et al., 2005; Karavitaki et al., 2008). It is also possible to combine cabergoline with an SA: according to a recent meta-analysis of studies in which cabergoline was added to a SA that had failed to achieve remission, IGF-1 levels normalized in half the patients (Sandret et al., 2011).

There is some controversy surrounding the ability of preoperative SA therapy to improve surgical outcome: some studies (Barkan et al., 1988; Stevenaert et al., 1992; Colao et al., 1997; Abe and Ludecke, 2001; Lucas et al., 2003; Carlsen et al., 2008; Mao et al., 2010) indicate that, in some patients with isolated somatotroph macroadenomas, surgery provides better control of acromegaly when patients are pretreated with a SA, while other studies showed no difference (Biermasz et al., 1999; Kristof et al., 1999; Losa et al., 2006). A recent meta-analysis of all controlled trials showed a borderline effect, with a pooled odds ratio (OR) of 1.62 (95% CI, 0.93–2.82) for biochemical cure with SA treatment; analysis of the three prospective controlled trials showed a statistically significant effect (OR 3.62, 95% CI, 1.88–6.96) (Pita-Gutierrez et al., 2013).

SAs have mild adverse effects, consisting mainly of transient gastrointestinal disorders (abdominal bloating, nausea, diarrhea), and also gallstones in 10–20% of cases (ursodeoxycholic acid has no proven effectiveness in this setting) (Attanasio et al., 2008). Gallstone complications are rare, and guidelines for their monitoring have been much relaxed in recent years (Chanson et al., 2009c). Changes in glucose metabolism are sometimes observed, including impaired glucose tolerance or even diabetes in patients who are overweight. In other cases, however, glucose tolerance improves following the reduction in insulin resistance due to lowering of GH concentrations. Overall, according to a recent meta-analysis, the consequences are very minor in terms of fasting glucose and HbA1c levels (Mazziotti et al., 2009). Regular metabolic monitoring is nevertheless justified.

SAs bind with high affinity to somatostatin receptor subtypes sst2 and 5. As a significant proportion of somatotroph adenomas appear partially or totally resistant to octreotide or lanreotide, possibly owing to variable expression or a reduced density of receptor subtypes (Hofland and Lamberts, 2003), it was logical to develop new, more potent somatostatin analogs able to bind with greater affinity to certain ssts. Pasireotide is a new somatostatin analog that binds with high affinity to sst1, 2, 3, and 5 (Bruns et al., 2002) and that potently inhibits GH and IGF1 in different animal models (Hofland et al., 2004). According to the only available study, pasireotide seems more effective than octreotide in terms of the GH/IGF-1 levels achieved (Petersenn et al., 2010); gastrointestinal adverse events occurred at a similar frequency, but glucose metabolism was more strongly altered on pasireotide.

GH-RECEPTOR ANTAGONISTS

Pegvisomant (Somavert®) has a different mechanism of action. It acts in the periphery, blocking the effects of GH on its target organs by binding to GH receptors and preventing their dimerization (Fig. 14.5); this blocks GH signal transduction and inhibits GH activity, including IGF-1 production (Kopchick et al., 2002). As pegvisomant inhibits the action of GH but not its secretion, GH concentrations cannot be used to evaluate treatment efficacy. IGF-1 is used as a surrogate marker, together with clinical parameters. Pegvisomant is administered subcutaneously at a daily dose of 10, 15, or 20 mg (sometimes more), the dose being adapted to the hormone response (IGF-1 normalization). Pegvisomant is highly effective, as IGF-1 levels normalize in more than 90% of patients (Trainer et al., 2000; van der Lely et al., 2001a). In fact, efficacy may be slightly lower in routine practice, as shown by observational studies (Moore et al., 2009), where normalization of IGF-1 was achieved in around 70% of cases (Colao et al., 2006; Schreiber et al., 2007; Marazuela et al., 2009; van der Lely et al., 2012). For the moment this treatment is reserved for patients in whom somatostatin analogs fail. A small increase in tumor volume is observed in a few patients. In a series of 304 patients in whom tumor volume was monitored for at least 3 years, an increase in tumor volume occurred in nine cases, within 8 months after commencing pegvisomant. This is likely related to rebound expansion after discontinuation of somatostatin analogs and/or to the natural history of aggressively growing pituitary tumors (Jimenez et al., 2008); this latter situation may justify combination with a somatostatin analog to reduce tumor volume (van der Lely et al., 2001b). Tumor volume must therefore be monitored (by MRI) during this treatment. Available clinical data on pegvisomant concern a relatively small number of patients and relatively short treatment periods. Adverse effects are limited to rare liver enzyme elevations: in 2.5% of patients, according a large surveillance study (van der Lely et al., 2012), they generally normalize either spontaneously or after treatment interruption. Exceptional cases of true hepatitis have been reported (Schreiber et al., 2007; van der Lely et al., 2012). Gilbert disease has been suggested as a risk factor for severe hepatitis (Bernabeu et al., 2009, 2010) but this was not confirmed by a recent Italian study (Filopanti et al., 2014).

SA-pegvisomant combination therapy has also been developed (Feenstra et al., 2005): a slow-release formulation of the SA is given once a month at the highest marketed dose (30 mg octreotide LAR or 120 mg lanreotide Autogel) and pegvisomant is injected once a week at escalating doses until the IGF-1 level normalizes. IGF-1 normalization was obtained in all the patients with a median weekly pegvisomant dose of 60 mg (Neggers et al., 2007). This decreased dose requirement during combined therapy might be partially explained by an increase of about 20% in serum levels of pegvisomant (Jorgensen et al., 2005). Biochemical hepatic anomalies were quite frequent (although always transient) with this combination, and appeared to be twice as common in acromegalic patients with diabetes (Neggers et al., 2009). By comparison with octreotide monotherapy, this combination appears to have a greater positive impact on quality of life for a given degree of IGF-1 normalization (Neggers et al., 2008), an effect that might be due to an extrahepatic action of pegvisomant (Neggers and van der Lely, 2009).

Cabergoline-pegvisomant combination therapy has also been proposed. In a multicenter, open-label, prospective clinical trial (Higham et al., 2012), the combination of cabergoline and low-dose pegvisomant (10 mg/day) was associated with a significant decrease in IGF-1 levels compared with cabergoline alone, and 68% of patients achieved normalization. Then, when cabergoline was withdrawn and pegvisomant continued as monotherapy, only 26% of patients maintained normal IGF-1 levels. The addition of cabergoline may be interesting when pegvisomant alone has achieved minimally increased IGF-1 levels (Bernabeu et al., 2013).

Current therapeutic strategy

The advantages, disadvantages and costs of the different treatment options must be taken into account.

A therapeutic strategy is proposed by the authors of this review in Figure 14.6. Currently, if surgical treatment fails to cure acromegaly, medical treatment with somatostatin analogs is preferred to radiotherapy. In some selected patients (those with moderately increased IGF-1), cabergoline may be tried first. If somatostatin analog therapy fails, it may be interesting to propose repeat surgery in order to remove a large tumor remnant, before trying a somatostatin analog again. Combination with cabergoline may also be interesting if an SA is only partially effective. Otherwise, pegvisomant can be tried before resorting to radiotherapy. If possible, surgical debulking may also be interesting to propose for reducing the dose of pegvisomant necessary to achieve normal IGF-1. The cost of these medical treatments, which may be required indefinitely, must be weighed up against the risks of radiotherapy. In any event, medical treatment will be necessary while waiting for the benefits of radiotherapy to emerge.

If surgery and radiotherapy are both contraindicated, first-line somatostatin therapy may be proposed.

All these treatments must be reassessed on a yearly basis. After radiotherapy, if a medical treatment is necessary while waiting for the effects of irradiation, its regular withdrawal is necessary for assessing the persistence of active disease.

CONCLUSION

Acromegaly is a rare disease usually due to growth hormone (GH) hypersecretion by a pituitary adenoma. In addition to this dysmorphic syndrome, acromegaly has cardiovascular, respiratory, rheumatologic, and metabolic consequences and is associated with a risk of neoplasia. A detailed workup of the various organs potentially involved in these complications is recommended. Treatment is aimed at correcting (or preventing) tumor compression of neighboring tissues by excising the culprit lesion, and at reducing GH and IGF-1 levels to normal

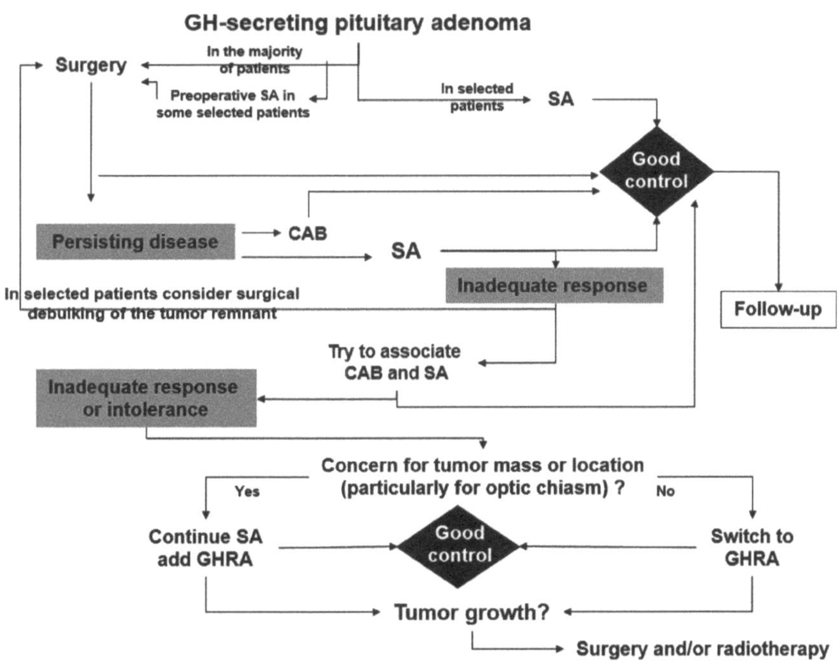

Fig. 14.6. Strategy proposed by the authors for the current management of acromegaly. SA, somatostatin analogs; CAB, cabergoline; GHRA, GH-receptor antagonist (pegvisomant).

values. Transsphenoidal surgery is often the first-line treatment. When surgery fails to correct GH/IGF-1 hypersecretion, medical treatment with somatostatin analogs and/or cabergoline can be used. The GH receptor antagonist pegvisomant is used to treat patients who are resistant to or intolerant of somatostatin analogs. Radiotherapy is nowadays proposed as third-line treatment. The prognosis of acromegaly has improved in recent years: adequate hormonal control is achieved in most cases, providing a life expectancy similar to that of the general population. It remains to be shown whether the criteria used to define control of disease in terms of mortality also apply to optimal management of comorbidities.

References

Abe T, Ludecke DK (2001). Effects of preoperative octreotide treatment on different subtypes of 90 GH-secreting pituitary adenomas and outcome in one surgical centre. Eur J Endocrinol 145: 137–145.

Abs R, Verhelst J, Maiter D et al. (1998). Cabergoline in the treatment of acromegaly: a study in 64 patients. J Clin Endocrinol Metab 83: 374–378.

Akutsu H, Kreutzer J, Wasmeier G et al. (2010). Acromegaly per se does not increase the risk for coronary artery disease. Eur J Endocrinol 162: 879–886.

Alexopoulou O, Bex M, Kamenicky P et al. (2014). Prevalence and risk factors of impaired glucose tolerance and diabetes mellitus at diagnosis of acromegaly: a study in 148 patients. Pituitary 17: 81–89.

Annamalai AK, Webb A, Kandasamy N et al. (2013). A comprehensive study of clinical, biochemical, radiological, vascular, cardiac, and sleep parameters in an unselected cohort of patients with acromegaly undergoing presurgical somatostatin receptor ligand therapy. J Clin Endocrinol Metab 98: 1040–1050.

Arafat AM, Mohlig M, Weickert MO et al. (2008). Growth hormone response during oral glucose tolerance test: the impact of assay method on the estimation of reference values in patients with acromegaly and in healthy controls, and the role of gender, age, and body mass index. J Clin Endocrinol Metab 93: 1254–1262.

Arosio M, Reimondo G, Malchiodi E et al. (2012). Predictors of morbidity and mortality in acromegaly: an Italian survey. Eur J Endocrinol 167: 189–198.

Asa SL, Ezzat S (2002). The pathogenesis of pituitary tumours. Nat Rev Cancer 2: 836–849.

Asa SL, Kovacs KT, Melmed S (2002). Functional anatomy of the hypothalamic pituitary axis. In: S Melmed (Ed.), The Pituitary. Blackwell Science Inc., Malden, MA, pp. 3–44.

Attal P, Chanson P (2010). Endocrine aspects of obstructive sleep apnea. J Clin Endocrinol Metab 95: 483–495.

Attal P, Claes V, Bobin S et al. (2009). Growth hormone excess and sternohyoid muscle mechanics in rats. Eur Respir J 34: 967–974.

Attanasio R, Baldelli R, Pivonello R et al. (2003). Lanreotide 60 mg, a new long-acting formulation: effectiveness in the chronic treatment of acromegaly. J Clin Endocrinol Metab 88: 5258–5265.

Attanasio R, Mainolfi A, Grimaldi F et al. (2008). Somatostatin analogs and gallstones: a retrospective survey on a large series of acromegalic patients. J Endocrinol Invest 31: 704–710.

Ayuk J, Stewart SE, Stewart PM et al. (2002). Long-term safety and efficacy of depot long-acting somatostatin analogs for the treatment of acromegaly. J Clin Endocrinol Metab 87: 4142–4146.

Ayuk J, Clayton RN, Holder G et al. (2004). Growth hormone and pituitary radiotherapy, but not serum insulin-like growth factor-I concentrations, predict excess mortality in patients with acromegaly. J Clin Endocrinol Metab 89: 1613–1617.

Barkan AL, Lloyd RV, Chandler WF et al. (1988). Preoperative treatment of acromegaly with long-acting somatostatin analog SMS 201–995: shrinkage of invasive pituitary macroadenomas and improved surgical remission rate. J Clin Endocrinol Metab 67: 1040–1048.

Barlier A, Vanbellinghen JF, Daly AF et al. (2007). Mutations in the aryl hydrocarbon receptor interacting protein gene are not highly prevalent among subjects with sporadic pituitary adenomas. J Clin Endocrinol Metab 92: 1952–1955.

Barrande G, Pittino-Lungo M, Coste J et al. (2000). Hormonal and metabolic effects of radiotherapy in acromegaly: long-term results in 128 patients followed in a single center [In Process Citation]. J Clin Endocrinol Metab 85: 3779–3785.

Beckers A, Aaltonen LA, Daly AF et al. (2013). Familial isolated pituitary adenomas (FIPA) and the pituitary adenoma predisposition due to mutations in the aryl hydrocarbon receptor interacting protein (AIP) gene. Endocr Rev 34: 239–277.

Beck-Peccoz P, Bassetti M, Spada A et al. (1985). Glycoprotein hormone alpha-subunit response to growth hormone (GH)-releasing hormone in patients with active acromegaly. Evidence for alpha-subunit and GH coexistence in the same tumoral cell. J Clin Endocrinol Metab 61: 541–546.

Beck-Peccoz P, Brucker-Davis F, Persani L et al. (1996). Thyrotropin-secreting pituitary adenomas. Endocr Rev 17: 610–638.

Berg C, Petersenn S, Lahner H et al. (2010). Cardiovascular risk factors in patients with uncontrolled and long-term acromegaly: comparison with matched data from the general population and the effect of disease control. J Clin Endocrinol Metab 95: 3648–3656.

Bernabeu I, Cameselle-Teijeiro J, Casanueva FF et al. (2009). Pegvisomant-induced cholestatic hepatitis with jaundice in a patient with Gilbert's syndrome. Eur J Endocrinol 160: 869–872.

Bernabeu I, Marazuela M, Lucas T et al. (2010). Pegvisomant-induced liver injury is related to the UGT1A1*28 polymorphism of Gilbert's syndrome. J Clin Endocrinol Metab 95: 2147–2154.

Bernabeu I, Alvarez-Escola C, Paniagua AE et al. (2013). Pegvisomant and cabergoline combination therapy in acromegaly. Pituitary 16: 101–108.

Bertherat J (2006). Carney complex (CNC). Orphanet J Rare Dis 1: 21.

Beuschlein F, Strasburger CJ, Siegerstetter V et al. (2000). Acromegaly caused by secretion of growth hormone by a non-Hodgkin's lymphoma. N Engl J Med 342: 1871–1876.

Bevan JS (2005). Clinical review. The antitumoral effects of somatostatin analog therapy in acromegaly. J Clin Endocrinol Metab 90: 1856–1863.

Bevan JS, Atkin SL, Atkinson AB et al. (2002). Primary medical therapy for acromegaly: an open, prospective, multicenter study of the effects of subcutaneous and intramuscular slow- release octreotide on growth hormone, insulin-like growth factor-I, and tumor size. J Clin Endocrinol Metab 87: 4554–4563.

Biermasz NR, van Dulken H, Roelfsema F (1999). Direct postoperative and follow-up results of transsphenoidal surgery in 19 acromegalic patients pretreated with octreotide compared to those in untreated matched controls. J Clin Endocrinol Metab 84: 3551–3555.

Biermasz NR, van Dulken H, Roelfsema F (2000). Ten-year follow-up results of transsphenoidal microsurgery in acromegaly. J Clin Endocrinol Metab 85: 4596–4602.

Biermasz NR, van Thiel SW, Pereira AM et al. (2004). Decreased quality of life in patients with acromegaly despite long-term cure of growth hormone excess. J Clin Endocrinol Metab 89: 5369–5376.

Biermasz NR, Smit JW, Pereira AM et al. (2007). Acromegaly caused by growth hormone-releasing hormone-producing tumors: long-term observational studies in three patients. Pituitary 10: 237–249.

Bihan H, Espinosa C, Valdes-Socin H et al. (2004). Long-term outcome of patients with acromegaly and congestive heart failure. J Clin Endocrinol Metab 89: 5308–5313.

Boero L, Manavela M, Gomez Rosso L et al. (2009). Alterations in biomarkers of cardiovascular disease (CVD) in active acromegaly. Clin Endocrinol (Oxf) 70: 88–95.

Bogazzi F, Battolla L, Spinelli C et al. (2007). Risk factors for development of coronary heart disease in patients with acromegaly: a five-year prospective study. J Clin Endocrinol Metab 92: 4271–4277.

Brada M, Burchell L, Ashley S et al. (1999). The incidence of cerebrovascular accidents in patients with pituitary adenoma. Int J Radiat Oncol Biol Phys 45: 693–698.

Bruns C, Lewis I, Briner U et al. (2002). SOM230: a novel somatostatin peptidomimetic with broad somatotropin release inhibiting factor (SRIF) receptor binding and a unique antisecretory profile. Eur J Endocrinol 146: 707–716.

Cairns SR, Scholefield JH, Steele RJ et al. (2010). Guidelines for colorectal cancer screening and surveillance in moderate and high risk groups (update from 2002). Gut 59: 666–689.

Cannavo S, Almoto B, Cavalli G et al. (2006). Acromegaly and coronary disease: an integrated evaluation of conventional coronary risk factors and coronary calcifications detected by computed tomography. J Clin Endocrinol Metab 91: 3766–3772.

Cappabianca P, Cavallo LM, de Divitiis E (2004). Endoscopic endonasal transsphenoidal surgery. Neurosurgery 55: 933–940, discussion 940–931.

Carlsen SM, Lund-Johansen M, Schreiner T et al. (2008). Preoperative octreotide treatment in newly diagnosed acromegalic patients with macroadenomas increases cure short-term postoperative rates: a prospective, randomized trial. J Clin Endocrinol Metab 93: 2984–2990.

Carmichael JD, Bonert VS, Mirocha JM et al. (2009). The utility of oral glucose tolerance testing for diagnosis and assessment of treatment outcomes in 166 patients with acromegaly. J Clin Endocrinol Metab 94: 523–527.

Carmichael JD, Bonert VS, Nuño M et al. (2014). Acromegaly clinical trialmethodology impact on reported biochemical efficacy rates of somatostatin receptor ligand treatments: a meta-analysis. J Clin Endocrinol Metab 99: 1825–1833.

Castinetti F, Taieb D, Kuhn JM et al. (2005). Outcome of gamma knife radiosurgery in 82 patients with acromegaly: correlation with initial hypersecretion. J Clin Endocrinol Metab 90: 4483–4488.

Castinetti F, Nagai M, Morange I et al. (2009). Long-term results of stereotactic radiosurgery in secretory pituitary adenomas. J Clin Endocrinol Metab 94: 3400–3407.

Cazabat L, Libe R, Perlemoine K et al. (2007). Germline inactivating mutations of the aryl hydrocarbon receptor-interacting protein gene in a large cohort of sporadic acromegaly: mutations are found in a subset of young patients with macroadenomas. Eur J Endocrinol 157: 1–8.

Cazabat L, Souberbielle JC, Chanson P (2008). Dynamic tests for the diagnosis and assesment of treatment efficacy in acromegaly. Pituitary 11: 129–139.

Cazabat L, Bouligand J, Chanson P (2011). AIP mutation in pituitary adenomas. N Engl J Med 364: 1973–1974, author reply 1974–1975.

Cazabat L, Bouligand J, Salenave S et al. (2012). Germline AIP mutations in apparently sporadic pituitary adenomas: prevalence in a prospective single-center cohort of 443 patients. J Clin Endocrinol Metab 97: E663–E670.

Chahal HS, Stals K, Unterlander M et al. (2011). AIP mutation in pituitary adenomas in the 18th century and today. N Engl J Med 364: 43–50.

Chanson P (2008). Emerging drugs for acromegaly. Expert Opin Emerg Drugs 13: 273–293.

Chanson P, Kamenicky P (2012). Treatment of acromegaly: a critical analysis of the last ten years. Ann Endocrinol (Paris) 73: 99–106.

Chanson P, Maison P (2009). Does attainment of target levels of growth hormone and insulin-like growth factor I improve acromegaly prognosis? Nat Clin Pract Endocrinol Metab 5: 70–71.

Chanson P, Salenave S (2008). Acromegaly. Orphanet J Rare Dis 3: 17.

Chanson P, Timsit J, Masquet C et al. (1990). Cardiovascular effects of the somatostatin analog octreotide in acromegaly. Ann Intern Med 113: 921–925.

Chanson P, Salenave S, Orcel P (2007). McCune–Albright syndrome in adulthood. Pediatr Endocr Rev 4 (Suppl4): 453–463.

Chanson P, Salenave S, Droumaguet C et al. (2009a). Rare causes of acromegaly. In: JA Wass (Ed.), Acromegaly: a handbook of history, current therapy and future prospects, Bioscientifica Ltd, Bristol, UK, pp. 70–98.

Chanson P, Salenave S, Kamenicky P et al. (2009b). Pituitary tumours: acromegaly. Best Pract Res Clin Endocrinol Metab 23: 555–574.

Chanson P, Bertherat J, Beckers A et al. (2009c). French consensus on the management of acromegaly. Ann Endocrinol (Paris) 70: 92–106.

Claessen KM, Kroon HM, Pereira AM et al. (2013a). Progression of vertebral fractures despite long-term biochemical control of acromegaly: a prospective follow-up study. J Clin Endocrinol Metab 98: 4808–4815.

Claessen KM, Ramautar SR, Pereira AM et al. (2013b). Increased clinical symptoms of acromegalic arthropathy in patients with long-term disease control: a prospective follow-up study. Pituitary 17: 44–52.

Clayton RN (2003). Cardiovascular function in acromegaly. Endocr Rev 24: 272–277.

Clemmons DR (2011). Consensus statement on the standardization and evaluation of growth hormone and insulin-like growth factor assays. Clin Chem 57: 555–559.

Colao A, Ferone D, Cappabianca P et al. (1997). Effect of octreotide pretreatment on surgical outcome in acromegaly. J Clin Endocrinol Metab 82: 3308–3314.

Colao A, Baldelli R, Marzullo P et al. (2000). Systemic hypertension and impaired glucose tolerance are independently correlated to the severity of the acromegalic cardiomyopathy. J Clin Endocrinol Metab 85: 193–199.

Colao A, Ferone D, Marzullo P et al. (2001). Long-term effects of depot long-acting somatostatin analog octreotide on hormone levels and tumor mass in acromegaly. J Clin Endocrinol Metab 86: 2779–2786.

Colao A, Spinelli L, Marzullo P et al. (2003). High prevalence of cardiac valve disease in acromegaly: an observational, analytical, case-control study. J Clin Endocrinol Metab 88: 3196–3201.

Colao A, Ferone D, Marzullo P et al. (2004). Systemic complications of acromegaly: epidemiology, pathogenesis, and management. Endocr Rev 25: 102–152.

Colao A, Pivonello R, Auriemma RS et al. (2006). Efficacy of 12-month treatment with the GH receptor antagonist pegvisomant in patients with acromegaly resistant to long-term, high-dose somatostatin analog treatment: effect on IGF-I levels, tumor mass, hypertension and glucose tolerance. Eur J Endocrinol 154: 467–477.

Collins MT, Singer FR, Eugster E (2012). McCune–Albright syndrome and the extraskeletal manifestations of fibrous dysplasia. Orphanet J Rare Dis 7 (Suppl 1): S4.

Cozzi R, Attanasio R, Montini M et al. (2003). Four-year treatment with octreotide-long-acting repeatable in 110 acromegalic patients: predictive value of short-term results? J Clin Endocrinol Metab 88: 3090–3098.

Daly AF, Rixhon M, Adam C et al. (2006). High prevalence of pituitary adenomas: a cross-sectional study in the province of Liège, Belgium. J Clin Endocrinol Metab 91: 4769–4775.

Daly AF, Vanbellinghen JF, Khoo SK et al. (2007). Aryl hydrocarbon receptor-interacting protein gene mutations in familial isolated pituitary adenomas: analysis in 73 families. J Clin Endocrinol Metab 92: 1891–1896.

Daly AF, Tichomirowa MA, Beckers A (2009). The epidemiology and genetics of pituitary adenomas. Best Pract Res Clin Endocrinol Metab 23: 543–554.

Danzig J (2007). Acromegaly. BMJ 335: 824–825.

De Martino I, Visone R, Wierinckx A et al. (2009). HMGA proteins up-regulate CCNB2 gene in mouse and human pituitary adenomas. Cancer Res 69: 1844–1850.

Dekkers OM, Biermasz NR, Pereira AM et al. (2008). Mortality in acromegaly: a metaanalysis. J Clin Endocrinol Metab 93: 61–67.

Delhougne B, Deneux C, Abs R et al. (1995). The prevalence of colonic polyps in acromegaly: a prospective colonoscopic and pathological study in 103 patients. J Clin Endocrinol Metab 80: 3223–3226.

Dimaraki EV, Jaffe CA, DeMott-Friberg R et al. (2002). Acromegaly with apparently normal GH secretion: implications for diagnosis and follow-up. J Clin Endocrinol Metab 87: 3537–3542.

dos Santos MC, Nascimento GC, Nascimento AG et al. (2013). Thyroid cancer in patients with acromegaly: a case-control study. Pituitary 16: 109–114.

Dworakowska D, Korbonits M, Aylwin S et al. (2011). The pathology of pituitary adenomas from a clinical perspective. Front Biosci (Schol Ed) 3: 105–116.

Ezzat S, Forster MJ, Berchtold P et al. (1994). Acromegaly. Clinical and biochemical features in 500 patients. Medicine (Baltimore) 73: 233–240.

Fahlbusch R, Honegger J, Buchfelder M (1992). Surgical management of acromegaly. Endocrinol Metab Clin North Am 21: 669–692.

Feenstra J, de Herder WW, ten Have SM et al. (2005). Combined therapy with somatostatin analogues and weekly pegvisomant in active acromegaly. Lancet 365: 1644–1646.

Fernandez A, Karavitaki N, Wass JA (2010). Prevalence of pituitary adenomas: a community-based, cross-sectional study in Banbury (Oxfordshire, UK). Clin Endocrinol (Oxf) 72: 377–382.

Filippella M, Galland F, Kujas M et al. (2006). Pituitary tumour transforming gene (PTTG) expression correlates with the proliferative activity and recurrence status of pituitary adenomas: a clinical and immunohistochemical study. Clin Endocrinol (Oxf) 65: 536–543.

Filopanti M, Barbieri AM, Mantovani G et al. (2014). Role of UGT1A1 and ADH gene polymorphisms in pegvisomant-induced liver toxicity in acromegalic patients. Eur J Endocrinol 170: 249–256.

Freda PU, Katznelson L, van der Lely AJ et al. (2005). Long-acting somatostatin analog therapy of acromegaly: a meta-analysis. J Clin Endocrinol Metab 90: 4465–4473.

Freda PU, Shen W, Heymsfield SB et al. (2008). Lower visceral and subcutaneous but higher intermuscular adipose tissue depots in patients with growth hormone and insulin-like growth factor I excess due to acromegaly. J Clin Endocrinol Metab 93: 2334–2343.

Gadelha MR, Trivellin G, Hernandez Ramirez LC et al. (2013). Genetics of pituitary adenomas. Front Horm Res 41: 111–140.

Garby L, Caron P, Claustrat F et al. (2012). Clinical characteristics and outcome of acromegaly induced by ectopic secretion of growth hormone-releasing hormone (GHRH): a French nationwide series of 21 cases. J Clin Endocrinol Metab 97: 2093–2104.

Georgitsi M (2010). MEN-4 and other multiple endocrine neoplasias due to cyclin-dependent kinase inhibitors (p27 (Kip1) and p18(INK4C)) mutations. Best Pract Res Clin Endocrinol Metab 24: 425–437.

Giustina A, Barkan A, Casanueva FF et al. (2000). Criteria for cure of acromegaly: a consensus statement. J Clin Endocrinol Metab 85: 526–529.

Giustina A, Casanueva FF, Cavagnini F et al. (2003). Diagnosis and treatment of acromegaly complications. J Endocrinol Invest 26: 1242–1247.

Giustina A, Mazziotti G, Canalis E (2008). Growth hormone, insulin-like growth factors, and the skeleton. Endocr Rev 29: 535–559.

Giustina A, Bonadonna S, Bugari G et al. (2009). High-dose intramuscular octreotide in patients with acromegaly inadequately controlled on conventional somatostatin analogue therapy: a randomised controlled trial. Eur J Endocrinol 161: 331–338.

Giustina A, Chanson P, Bronstein MD et al. (2010). A consensus on criteria for cure of acromegaly. J Clin Endocrinol Metab 95: 3141–3148.

Giustina A, Mazziotti G, Torri V et al. (2012). Meta-analysis on the effects of octreotide on tumor mass in acromegaly. PLoS One 7: e36411.

Gola M, Doga M, Bonadonna S et al. (2006). Neuroendocrine tumors secreting growth hormone-releasing hormone: pathophysiological and clinical aspects. Pituitary 9: 221–229.

Gouya H, Vignaux O, Le Roux P et al. (2008). Rapidly reversible myocardial edema in patients with acromegaly: assessment with ultrafast T2 mapping in a single-breath-hold MRI sequence. AJR Am J Roentgenol 190: 1576–1582.

Heaney AP, Melmed S (2004). Molecular targets in pituitary tumours. Nat Rev Cancer 4: 285–295.

Heck A, Ringstad G, Fougner SL et al. (2012). Intensity of pituitary adenoma on T2-weighted magnetic resonance imaging predicts the response to octreotide treatment in newly diagnosed acromegaly. Clin Endocrinol (Oxf) 77: 72–78.

Herman V, Fagin J, Gonski R et al. (1990). Clonal origin of pituitary adenomas. J Clin Endocrinol Metab 71: 1427–1430.

Herrmann BL, Wessendorf TE, Ajaj W et al. (2004). Effects of octreotide on sleep apnoea and tongue volume (magnetic resonance imaging) in patients with acromegaly. Eur J Endocrinol 151: 309–315.

Higham CE, Atkinson AB, Aylwin S et al. (2012). Effective combination treatment with cabergoline and low-dose pegvisomant in active acromegaly: a prospective clinical trial. J Clin Endocrinol Metab 97: 1187–1193.

Hofland LJ, Lamberts SW (2003). The pathophysiological consequences of somatostatin receptor internalization and resistance. Endocr Rev 24: 28–47.

Hofland LJ, van der Hoek J, van Koetsveld PM et al. (2004). The novel somatostatin analog SOM230 is a potent inhibitor of hormone release by growth hormone- and prolactin-secreting pituitary adenomas in vitro. J Clin Endocrinol Metab 89: 1577–1585.

Holdaway IM, Rajasoorya C (1999). Epidemiology of acromegaly. Pituitary 2: 29–41.

Holdaway IM, Rajasoorya RC, Gamble GD (2004). Factors influencing mortality in acromegaly. J Clin Endocrinol Metab 89: 667–674.

Holdaway IM, Bolland MJ, Gamble GD (2008). A meta-analysis of the effect of lowering serum levels of GH and IGF-I on mortality in acromegaly. Eur J Endocrinol 159: 89–95.

Horvath A, Stratakis CA (2008). Clinical and molecular genetics of acromegaly: MEN1, Carney complex, McCune–Albright syndrome, familial acromegaly and genetic defects in sporadic tumors. Rev Endocr Metab Disord 9: 1–11.

Ip MSM, Tan KCB, Peh WCG et al. (2001). Effect of Sandostatin® LAR® on sleep apneoa in acromegaly: correlation with computerized tomographic cephalometry and hormonal activity. Clin Endocrinol 55: 477–483.

Jaffrain-Rea ML, Moroni C, Baldelli R et al. (2001). Relationship between blood pressure and glucose tolerance in acromegaly. Clin Endocrinol (Oxf) 54: 189–195.

Jane Jr JA, Starke RM, Elzoghby MA et al. (2011). Endoscopic transsphenoidal surgery for acromegaly: remission using modern criteria, complications, and predictors of outcome. J Clin Endocrinol Metab 96: 2732–2740.

Jenkins PJ, Besser M (2001). Clinical perspective: acromegaly and cancer: a problem. J Clin Endocrinol Metab 86: 2935–2941.

Jenkins PJ, Sohaib SA, Akker S et al. (2000). The pathology of median neuropathy in acromegaly. Ann Intern Med 133: 197–201.

Jenkins PJ, Bates P, Carson MN et al. (2006). Conventional pituitary irradiation is effective in lowering serum growth hormone and insulin-like growth factor-I in patients with acromegaly. J Clin Endocrinol Metab 91: 1239–1245.

Jimenez C, Burman P, Abs R et al. (2008). Follow-up of pituitary tumor volume in patients with acromegaly treated with pegvisomant in clinical trials. Eur J Endocrinol 159: 517–523.

Jorgensen JO, Feldt-Rasmussen U, Frystyk J et al. (2005). Cotreatment of acromegaly with a somatostatin analog and a growth hormone receptor antagonist. J Clin Endocrinol Metab 90: 5627–5631.

Kahaly G, Olshausen KV, Mohr-Kahaly S et al. (1992). Arrhythmia profile in acromegaly. Eur Heart J 13: 51–56.

Kamenicky P, Viengchareun S, Blanchard A et al. (2008). Epithelial sodium channel is a key mediator of growth hormone-induced sodium retention in acromegaly. Endocrinology 149: 3294–3305.

Kamenicky P, Blanchard A, Frank M et al. (2011). Body fluid expansion in acromegaly is related to enhanced epithelial sodium channel (ENaC) activity. J Clin Endocrinol Metab 96: 2127–2135.

Kamenicky P, Blanchard A, Gauci C et al. (2012). Pathophysiology of renal calcium handling in acromegaly: what lies behind hypercalciuria? J Clin Endocrinol Metab 97: 2124–2133.

Kamenicky P, Mazziotti G, Lombès M et al. (2014). Growth hormone, insulin-like growth factor-1, and the kidney: pathophysiological and clinical implications. Endocr Rev 35: 234–281.

Karavitaki N, Turner HE, Adams CB et al. (2008). Surgical debulking of pituitary macroadenomas causing acromegaly improves control by lanreotide. Clin Endocrinol (Oxf) 68: 970–975.

Katznelson L (2009). Alterations in body composition in acromegaly. Pituitary 12: 136–142.

Kauppinen-Makelin R, Sane T, Reunanen A et al. (2005). A nationwide survey of mortality in acromegaly. J Clin Endocrinol Metab 90: 4081–4086.

Kirschner LS, Carney JA, Pack SD et al. (2000). Mutations of the gene encoding the protein kinase A type I-alpha regulatory subunit in patients with the Carney complex. Nat Genet 26: 89–92.

Kopchick JJ, Parkinson C, Stevens EC et al. (2002). Growth hormone receptor antagonists: discovery, development, and use in patients with acromegaly. Endocr Rev 23: 623–646.

Kovacs K, Lloyd R, Horvath E et al. (1989). Silent somatotroph adenomas of the human pituitary. A morphologic study of three cases including immunocytochemistry, electron microscopy, in vitro examination, and in situ hybridization. Am J Pathol 134: 345–353.

Kristof RA, Stoffet-Wagner B, Klingmüller D et al. (1999). Does octreotide treatment improve the surgical results of macroadenomas in acromegaly? A randomised study. Acta Neurochir (Wien) 141: 399–405.

Kropf LL, Madeira M, Vieira Neto L et al. (2013). Functional evaluation of the joints in acromegalic patients and associated factors. Clin Rheumatol 32: 991–998.

Lamberts SW, van der Lely AJ, de Herder WW et al. (1996). Octreotide. N Engl J Med 334: 246–254.

Liote F, Orcel P (2000). Osteoarticular disorders of endocrine origin. Baillieres Best Pract Res Clin Rheumatol 14: 251–276.

Loeffler JS, Shih HA (2011). Radiation therapy in the management of pituitary adenomas. J Clin Endocrinol Metab 96: 1992–2003.

Loeper S, Ezzat S (2008). Acromegaly: re-thinking the cancer risk. Rev Endocr Metab Disord 9: 41–58.

Losa M, Mortini P, Urbaz L et al. (2006). Presurgical treatment with somatostatin analogs in patients with acromegaly: effects on the remission and complication rates. J Neurosurg 104: 899–906.

Lucas T, Astorga R, Catala M (2003). Preoperative lanreotide treatment for GH-secreting pituitary adenomas: effect on tumour volume and predictive factors of significant tumour shrinkage. Clin Endocrinol (Oxf) 58: 471–481.

Lytras A, Tolis G (2006). Growth hormone-secreting tumors: genetic aspects and data from animal models. Neuroendocrinology 83: 166–178.

Maione L, Garcia C, Bouchachi A et al. (2012). No evidence of a detrimental effect of cabergoline therapy on cardiac valves in patients with acromegaly. J Clin Endocrinol Metab 97: E1714–E1719.

Maison P, Chanson P (2006). Less is more risky? Growth hormone and insulin-like growth factor 1 levels and cardiovascular risk. Nat Clin Pract Endocrinol Metab 2: 650–651.

Maison P, Demolis P, Young J et al. (2000). Vascular reactivity in acromegalic patients: preliminary evidence for regional endothelial dysfunction and increased sympathetic vasoconstriction. Clin Endocrinol (Oxf) 53: 445–451.

Maiza JC, Vezzosi D, Matta M et al. (2007). Long-term (up to 18 years) effects on GH/IGF-1 hypersecretion and tumour size of primary somatostatin analogue (SSTa) therapy in patients with GH-secreting pituitary adenoma responsive to SSTa. Clin Endocrinol (Oxf) 67: 282–289.

Mao ZG, Zhu YH, Tang HL et al. (2010). Preoperative lanreotide treatment in acromegalic patients with macroadenomas increases short-term postoperative cure rates: a prospective, randomised trial. Eur J Endocrinol 162: 661–666.

Marazuela M, Lucas T, Alvarez-Escola C et al. (2009). Long-term treatment of acromegalic patients resistant to somatostatin analogues with the GH receptor antagonist pegvisomant: its efficacy in relation to gender and previous radiotherapy. Eur J Endocrinol 160: 535–542.

Mazziotti G, Bianchi A, Bonadonna S et al. (2008). Prevalence of vertebral fractures in men with acromegaly. J Clin Endocrinol Metab 93: 4649–4655.

Mazziotti G, Floriani I, Bonadonna S et al. (2009). Effects of somatostatin analogs on glucose homeostasis: a metaanalysis of acromegaly studies. J Clin Endocrinol Metab 94: 1500–1508.

Melmed S (2001). Acromegaly and cancer: not a problem? J Clin Endocrinol Metab 86: 2929–2934.

Melmed S (2002). Acromegaly. In: S Melmed (Ed.), The Pituitary, Blackwell Science Inc., Malden, MA, pp. 419–454.

Melmed S (2003). Mechanisms for pituitary tumorigenesis: the plastic pituitary. J Clin Invest 112: 1603–1618.

Melmed S (2006). Medical progress: Acromegaly. N Engl J Med 355: 2558–2573.

Melmed S (2009). Acromegaly pathogenesis and treatment. J Clin Invest 119: 3189–3202.

Melmed S (2011). Pathogenesis of pituitary tumors. Nat Rev Endocrinol 7: 257–266.

Melmed S, Ezrin C, Kovacs K et al. (1985). Acromegaly due to secretion of growth hormone by an ectopic pancreatic islet-cell tumor. N Engl J Med 312: 9–17.

Melmed S, Casanueva FF, Cavagnini F et al. (2002). Guidelines for acromegaly management. J Clin Endocrinol Metab 87: 4054–4058.

Melmed S, Sternberg R, Cook D et al. (2005). A critical analysis of pituitary tumor shrinkage during primary medical therapy in acromegaly. J Clin Endocrinol Metab 90: 4405–4410.

Melmed S, Colao A, Barkan A et al. (2009). Guidelines for acromegaly management: an update. J Clin Endocrinol Metab 94: 1509–1517.

Miller A, Doll H, David J et al. (2008). Impact of musculoskeletal disease on quality of life in long-standing acromegaly. Eur J Endocrinol 158: 587–593.

Minniti G, Jaffrain-Rea ML, Osti M et al. (2005). The long-term efficacy of conventional radiotherapy in patients with GH-secreting pituitary adenomas. Clin Endocrinol (Oxf) 62: 210–216.

Minniti G, Scaringi C, Maurizi Enrici R (2011). Radiation techniques for acromegaly. Radiat Oncol 6: 167.

Molitch ME (1992). Clinical manifestations of acromegaly. Endocrinol Metab Clin North Am 21: 597–614.

Molitch ME (2008). The cabergoline-resistant prolactinoma patient: new challenges. J Clin Endocrinol Metab 93: 4643–4645.

Moore DJ, Adi Y, Connock MJ et al. (2009). Clinical effectiveness and cost-effectiveness of pegvisomant for the treatment of acromegaly: a systematic review and economic evaluation. BMC Endocr Disord 9: 20.

Murray RD, Melmed S (2008). A critical analysis of clinically available somatostatin analog formulations for therapy of acromegaly. J Clin Endocrinol Metab 93: 2957–2968.

Nabarro JD (1987). Acromegaly. Clin Endocrinol (Oxf) 26: 481–512.

Nachtigall L, Delgado A, Swearingen B et al. (2008). Changing patterns in diagnosis and therapy of acromegaly over two decades. J Clin Endocrinol Metab 93: 2035–2041.

Neggers SJ, van der Lely AJ (2009). Somatostatin analog and pegvisomant combination therapy for acromegaly. Nat Rev Endocrinol 5: 546–552.

Neggers SJ, van Aken MO, Janssen JA et al. (2007). Long-term efficacy and safety of combined treatment of somatostatin analogs and pegvisomant in acromegaly. J Clin Endocrinol Metab 92: 4598–4601.

Neggers SJ, van Aken MO, de Herder WW et al. (2008). Quality of life in acromegalic patients during long-term somatostatin analog treatment with and without pegvisomant. J Clin Endocrinol Metab 93: 3853–3859.

Neggers SJ, de Herder WW, Janssen JA et al. (2009). Combined treatment for acromegaly with long-acting somatostatin analogs and pegvisomant: long-term safety for up to 4.5 years (median 2.2 years) of follow-up in 86 patients. Eur J Endocrinol 160: 529–533.

Newman CB (1999). Medical therapy for acromegaly. Endocrinol Metab Clin North Am 28: 171–190.

Newman CB, Melmed S, George A et al. (1998). Octreotide as primary therapy for acromegaly. J Clin Endocrinol Metab 83: 3034–3040.

Nomikos P, Buchfelder M, Fahlbusch R (2005). The outcome of surgery in 668 patients with acromegaly using current criteria of biochemical "cure". Eur J Endocrinol 152: 379–387.

Otsuki M, Kasayama S, Yamamoto H et al. (2001). Characterization of premature atherosclerosis of carotid arteries in acromegalic patients. Clin Endocrinol (Oxf) 54: 791–796.

Paisley AN, Banerjee M, Rezai M et al. (2011). Changes in arterial stiffness but not carotid intimal thickness in acromegaly. J Clin Endocrinol Metab 96: 1486–1492.

Parkinson C, Drake WM, Wieringa G et al. (2002). Serum lipoprotein changes following IGF-I normalization using a growth hormone receptor antagonist in acromegaly. Clin Endocrinol (Oxf) 56: 303–311.

Pellegata NS, Quintanilla-Martinez L, Siggelkow H et al. (2006). Germ-line mutations in p27Kip1 cause a multiple endocrine neoplasia syndrome in rats and humans. Proc Natl Acad Sci U S A 103: 15558–15563.

Pereira AM, van Thiel SW, Lindner JR et al. (2004). Increased prevalence of regurgitant valvular heart disease in acromegaly. J Clin Endocrinol Metab 89: 71–75.

Petersenn S, Schopohl J, Barkan A et al. (2010). Pasireotide (SOM230) demonstrates efficacy and safety in patients with acromegaly: a randomized, multicenter, phase II trial. J Clin Endocrinol Metab 95: 2781–2789.

Petrossians P, Borges-Martins L, Espinoza C et al. (2005). Gross total resection or debulking of pituitary adenomas improves hormonal control of acromegaly by somatostatin analogs. Eur J Endocrinol 152: 61–66.

Pita-Gutierrez F, Pertega-Diaz S, Pita-Fernandez S et al. (2013). Place of preoperative treatment of acromegaly with somatostatin analog on surgical outcome: a systematic review and meta-analysis. PLoS One 8: e61523.

Puig-Domingo M, Resmini E, Gomez-Anson B et al. (2010). Magnetic resonance imaging as a predictor of response to somatostatin analogs in acromegaly after surgical failure. J Clin Endocrinol Metab 95: 4973–4978.

Raappana A, Koivukangas J, Ebeling T et al. (2010). Incidence of pituitary adenomas in Northern Finland in 1992–2007. J Clin Endocrinol Metab 95: 4268–4275.

Ramirez C, Vargas G, Gonzalez B et al. (2012). Discontinuation of octreotide LAR after long term, successful treatment of patients with acromegaly: is it worth trying? Eur J Endocrinol 166: 21–26.

Reid TJ, Post KD, Bruce JN et al. (2010). Features at diagnosis of 324 patients with acromegaly did not change from 1981 to 2006: acromegaly remains under-recognized and under-diagnosed. Clin Endocrinol (Oxf) 72: 203–208.

Renehan AG, Brennan BM (2008). Acromegaly, growth hormone and cancer risk. Best Pract Res Clin Endocrinol Metab 22: 639–657.

Renehan AG, O'Connell J, O'Halloran D et al. (2003). Acromegaly and colorectal cancer: a comprehensive review of epidemiology, biological mechanisms, and clinical implications. Horm Metab Res 35: 712–725.

Ribeiro-Oliveira Jr A, Barkan A (2012). The changing face of acromegaly – advances in diagnosis and treatment. Nat Rev Endocrinol 8: 605–611.

Ribeiro-Oliveira Jr A, Faje A, Barkan A (2011a). Postglucose growth hormone nadir and insulin-like growth factor-1 in naive-active acromegalic patients: do these parameters always correlate? Arq Bras Endocrinol Metabol 55: 494–497.

Ribeiro-Oliveira Jr A, Faje AT, Barkan AL (2011b). Limited utility of oral glucose tolerance test in biochemically active acromegaly. Eur J Endocrinol 164: 17–22.

Ronchi CL, Rizzo E, Lania AG et al. (2008). Preliminary data on biochemical remission of acromegaly after somatostatin analogs withdrawal. Eur J Endocrinol 158: 19–25.

Sacca L, Cittadini A, Fazio S (1994). Growth hormone and the heart. Endocr Rev 15: 555–573.

Sakharova AA, Dimaraki EV, Chandler WF et al. (2005). Clinically silent somatotropinomas may be biochemically active. J Clin Endocrinol Metab 90: 2117–2121.

Sandret L, Maison P, Chanson P (2011). Place of cabergoline in acromegaly: a meta-analysis. J Clin Endocrinol Metab 96: 1327–1335.

Scarpa R, De Brasi D, Pivonello R et al. (2004). Acromegalic axial arthropathy: a clinical case-control study. J Clin Endocrinol Metab 89: 598–603.

Schade R, Andersohn F, Suissa S et al. (2007). Dopamine agonists and the risk of cardiac-valve regurgitation. N Engl J Med 356: 29–38.

Schneider HJ, Sievers C, Saller B et al. (2008). High prevalence of biochemical acromegaly in primary care patients with elevated IGF-1 levels. Clin Endocrinol (Oxf) 69: 432–435.

Schreiber I, Buchfelder M, Droste M et al. (2007). Treatment of acromegaly with the GH receptor antagonist pegvisomant in clinical practice: safety and efficacy evaluation from the German Pegvisomant Observational Study. Eur J Endocrinol 156: 75–82.

Sesmilo G, Fairfield WP, Katznelson L et al. (2002). Cardiovascular risk factors in acromegaly before and after normalization of serum IGF-I levels with the GH antagonist pegvisomant. J Clin Endocrinol Metab 87: 1692–1699.

Sheppard MC (2003). Primary medical therapy for acromegaly. Clin Endocrinol (Oxf) 58: 387–399.

Sherlock M, Ayuk J, Tomlinson JW et al. (2010). Mortality in patients with pituitary disease. Endocr Rev 31: 301–342.

Sherlock M, Woods C, Sheppard MC (2011). Medical therapy in acromegaly. Nat Rev Endocrinol 7: 291–300.

Shih HA, Loeffler JS (2008). Radiation therapy in acromegaly. Rev Endocr Metab Disord 9: 59–65.

Socin HV, Chanson P, Delemer B et al. (2003). The changing spectrum of TSH-secreting pituitary adenomas: diagnosis and management in 43 patients. Eur J Endocrinol 148: 433–442.

Stevenaert A, Harris AG, Kovacs K et al. (1992). Presurgical octreotide treatment in acromegaly. Metabolism 41: 51–58.

Swearingen B, Barker 2nd FG, Katznelson L et al. (1998). Long-term mortality after transsphenoidal surgery and adjunctive therapy for acromegaly. J Clin Endocrinol Metab 83: 3419–3426.

Tagliafico A, Resmini E, Nizzo R et al. (2008a). Ultrasound measurement of median and ulnar nerve cross-sectional area in acromegaly. J Clin Endocrinol Metab 93: 905–909.

Tagliafico A, Resmini E, Nizzo R et al. (2008b). The pathology of the ulnar nerve in acromegaly. Eur J Endocrinol 159: 369–373.

Thakker RV, Newey PJ, Walls GV et al. (2012). Clinical practice guidelines for multiple endocrine neoplasia type 1 (MEN1). J Clin Endocrinol Metab 97: 2990–3011.

Thorner MO, Perryman RL, Cronin MJ et al. (1982). Somatotroph hyperplasia. Successful treatment of acromegaly by removal of a pancreatic islet tumor secreting a growth hormone-releasing factor. J Clin Invest 70: 965–977.

Trainer PJ (2002). Editorial: acromegaly – consensus, what consensus? J Clin Endocrinol Metab 87: 3534–3536.

Trainer PJ, Drake WM, Katznelson L et al. (2000). Treatment of acromegaly with the growth hormone-receptor antagonist pegvisomant. N Engl J Med 342: 1171–1177.

Trainer PJ, Barth J, Sturgeon C et al. (2006). Consensus statement on the standardisation of GH assays. Eur J Endocrinol 155: 1–2.

Ueland T, Fougner SL, Godang K et al. (2006). Serum GH and IGF-I are significant determinants of bone turnover but not bone mineral density in active acromegaly: a prospective study of more than 70 consecutive patients. Eur J Endocrinol 155: 709–715.

Valassi E, Klibanski A, Biller BM (2010). Clinical review. Potential cardiac valve effects of dopamine agonists in hyperprolactinemia. J Clin Endocrinol Metab 95: 1025–1033.

Vallar L, Spada A, Giannattasio G (1987). Altered Gs and adenylate cyclase activity in human GH-secreting pituitary adenomas. Nature 330: 566–568.

van der Lely AJ, Hutson RK, Trainer PJ et al. (2001a). Long-term treatment of acromegaly with pegvisomant, a growth hormone receptor antagonist. Lancet 358: 1754–1759.

van der Lely AJ, Muller A, Janssen JA et al. (2001b). Control of tumor size and disease activity during cotreatment with octreotide and the growth hormone receptor antagonist pegvisomant in an acromegalic patient. J Clin Endocrinol Metab 86: 478–481.

van der Lely AJ, Biller BM, Brue T et al. (2012). Long-term safety of pegvisomant in patients with acromegaly: comprehensive review of 1288 subjects in ACROSTUDY. J Clin Endocrinol Metab 97: 1589–1597.

Verges B, Boureille F, Goudet P et al. (2002). Pituitary disease in MEN type 1 (MEN1): data from the France-Belgium MEN1 multicenter study. J Clin Endocrinol Metab 87: 457–465.

Verhelst J, Velkeniers B, Maiter D et al. (2013). Active acromegaly is associated with decreased hs-CRP and NT-proBNP serum levels: insights from the Belgian registry of acromegaly. Eur J Endocrinol 168: 177–184.

Vierimaa O, Georgitsi M, Lehtonen R et al. (2006). Pituitary Adenoma predisposition caused by germline mutations in the AIP gene. Science 312: 1228–1230.

Vlotides G, Eigler T, Melmed S (2007). Pituitary tumor-transforming gene: physiology and implications for tumorigenesis. Endocr Rev 28: 165–186.

Vortmeyer AO, Glasker S, Mehta GU et al. (2012). Somatic GNAS mutation causes widespread and diffuse pituitary disease in acromegalic patients with McCune–Albright syndrome. J Clin Endocrinol Metab 97: 2404–2413.

Wassenaar MJ, Biermasz NR, van Duinen N et al. (2009). High prevalence of arthropathy, according to the definitions of radiological and clinical osteoarthritis, in patients with long-term cure of acromegaly: a case-control study. Eur J Endocrinol 160: 357–365.

Wassenaar MJ, Biermasz NR, Hamdy NA et al. (2011). High prevalence of vertebral fractures despite normal bone mineral density in patients with long-term controlled acromegaly. Eur J Endocrinol 164: 475–483.

Yang I, Kim W, De Salles A et al. (2010). A systematic analysis of disease control in acromegaly treated with radiosurgery. Neurosurg Focus 29: E13.

Zanettini R, Antonini A, Gatto G et al. (2007). Valvular heart disease and the use of dopamine agonists for Parkinson's disease. N Engl J Med 356: 39–46.

Handbook of Clinical Neurology, Vol. 124 (3rd series)
Clinical Neuroendocrinology
E. Fliers, M. Korbonits, and J.A. Romijn, Editors

Chapter 15

Cushing's disease

NICHOLAS A. TRITOS[1,2]* AND BEVERLY M.K. BILLER[1,2]
[1]*Neuroendocrine Unit, Massachusetts General Hospital, Boston, MA, USA*
[2]*Harvard Medical School, Boston, MA, USA*

INTRODUCTION

Cushing's disease (CD) is characterized by pathologic hypercortisolism as a result of excess corticotropin (ACTH), secreted by tumorous pituitary corticotrophs (pituitary adenoma or rarely carcinoma) (Cushing, 1932, 1969; Raff and Findling, 2003). Patients with CD represent approximately 70–75% of all patients with endogenous Cushing's syndrome (CS) (Nieman and Ilias, 2005; Nieman et al., 2008). Other causes of ACTH-dependent CS include the ectopic ACTH syndrome (about 10% of patients), characterized by ectopic ACTH secretion from benign or malignant tumors (bronchial or thymic carcinoids, small cell lung carcinomas, medullary thyroid carcinomas, islet cell tumors, pheochromocytomas, and others) and the very rare syndrome of ectopic corticotropin-releasing hormone (CRH) secretion from a tumor in the chest, abdomen, or elsewhere (Ilias et al., 2005; Biller et al., 2008). In approximately 20% of the patients with CS, there is autonomous cortisol secretion by one or both adrenal glands, caused by a benign adenoma (10%), carcinoma (5%), or a variety of pathologic conditions characterized by bilateral adrenal hyperplasia (macronodular adrenal hyperplasia, primary pigmented nodular adrenal disease, and the Carney complex, or the McCune–Albright syndrome). In the general population, exogenous glucocorticoid administration (iatrogenic or, rarely, surreptitious) is by far the most common cause of CS (Raff and Findling, 2003; Nieman and Ilias, 2005).

The incidence of endogenous CS has been estimated to be approximately 2.5 per million per year in the general population, with CD representing approximately 70% of the cases (Cavagnini and Pecori Giraldi, 2001; Lindholm et al., 2001). Recent data suggest that the prevalence of CS may be substantial (2–5%) in certain patient groups, including those with poorly controlled diabetes mellitus, osteoporosis, or hirsutism (Bals-Pratsch et al., 1996; Catargi et al., 2003; Nieman and Ilias, 2005). Of note, a larger proportion (40%) of patients with adrenal adenomas has been found to be the underlying cause of hypercortisolism in these studies.

Since the description of the index case by Harvey Cushing 100 years ago, there has been major progress in the diagnosis and management of the eponymous syndrome (Cushing 1932, 1969; Raff and Findling, 2003). However, there are still numerous outstanding issues remaining. The diagnosis of CD can be one of the most difficult challenges facing a clinical endocrinologist. Equally importantly, the management of CD is often complex and may require considerable expertise from many disciplines, including neurosurgery, radiation or medical oncology, and endocrinology (Arnaldi et al., 2003; Biller et al., 2008). The underlying etiology and pathogenesis of CD remain obscure. In the present review the authors will discuss the diagnosis and management of endogenous CD, highlighting recent developments and areas for future study.

CLINICAL FEATURES AND DIAGNOSIS

Clinical features

The clinical aspects of CD or CS can be very diverse and may include metabolic (central obesity, hyperglycemia), catabolic (osteoporosis, proximal myopathy, skin thinning, striae, excessive bruising, decreased linear growth in children), hyperandrogenic (hirsutism, acne), hypogonadal (including infertility), cardiovascular (hypertension, deep venous thrombosis and pulmonary

*Correspondence to: Nicholas A. Tritos, MD, DSc, FACP, Neuroendocrine Unit, Massachusetts General Hospital and Harvard Medical School, Zero Emerson Place, Suite 112, Boston, MA 02114, USA. Tel: +1 617 726 7948, Fax: +1 617 726 1241, E-mail: ntritos@partners.org

embolism, coronary heart disease, stroke), immunosuppressive (frequent, systemic or opportunistic infections), mineralocorticoid (edema, hypokalemia) and neuropsychiatric (decreased memory, insomnia, anxiety, irritability, depression, psychosis) features (Arnaldi et al., 2003; Nieman and Ilias, 2005). In addition, patients with large pituitary tumors may have evidence of mass effect, including headache or visual impairment as a result of impingement on the optic apparatus. The syndrome may be unmistakable in florid cases, including Harvey Cushing's famous index patient (Minnie G.) (Cushing, 1932). In particular, the presence of facial plethora and rounding, central adiposity, wide pigmented striae, spontaneous bruising and proximal myopathy are relatively specific features of CD/CS, but are present in a minority of patients (Nieman and Ilias, 2005; Nieman et al., 2008). Declining growth velocity is common in children. A high index of suspicion is needed in order to consider the diagnosis in patients with more subtle presentations, including those with intermittent hypercortisolism (Alexandraki et al., 2009).

The Endocrine Society guidelines recommend that evaluation for CS be considered in several patient groups, including those with features unusual for their age (such as early onset hypertension or osteoporosis), patients who accumulate multiple suggestive features over time, children with decreased growth velocity and weight gain, as well as patients with incidentally found adrenal lesions (particularly those having radiographic features suggestive of adenomas) (Nieman et al., 2008). The 1990 NIH criteria for the diagnosis of the polycystic ovary syndrome stipulate that other conditions associated with hyperandrogenic anovulation, including CS, be considered and excluded (Azziz, 2005). Sensitive case finding, rather than broad screening, is prudent in patients with poorly controlled diabetes mellitus, osteoporosis, or hirsutism.

Diagnosis

A stepwise approach is recommended in the evaluation of patients with suspected CS (Nieman et al., 2008). Pathologic hypercortisolism should first be established, before undertaking investigations to identify the underlying cause.

Laboratory tests used to evaluate the presence of hypercortisolism include 24 hour urine free cortisol (UFC), serving as an index of endogenous cortisol secretory output, late-night salivary cortisol (or midnight serum cortisol), used as a measure of nadir cortisol levels, and one of several dexamethasone suppression tests (DST), evaluating the presence of autonomous function of the hypothalamic–pituitary–adrenal (HPA) axis (Raff and Findling, 2003; Elamin et al., 2008; Nieman et al., 2008). Randomly measured serum cortisol levels are of no diagnostic value, as there is extensive overlap between CS and healthy individuals.

Repeat testing is often required, including patients with borderline abnormal or discordant test results, or those with suspected cyclic or intermittent hypercortisolism (Kidambi et al., 2007). The differential diagnosis of CS is extensive and includes several conditions with shared clinical and biochemical features (pregnancy, depression, alcohol excess, severe obesity, polycystic ovary syndrome, poorly controlled type 2 diabetes mellitus, familial glucocorticoid resistance) as well as conditions characterized by biochemical hypercortisolism in the absence of clinical features of CS (regular and strenuous exercise, acute illness or injury, surgery, hospitalization, anorexia nervosa) (Raff and Findling, 2003; Nieman et al., 2008). These conditions are often termed "pseudo-Cushing" states and can be quite difficult to distinguish from true CS. Exogenous glucocorticoid administration should be considered in every patient with suspected CS. Although this information may be generally apparent from patient's history, glucocorticoid administration may also be less evident to the patient (involving use of topical creams, parenteral therapeutic injections, over the counter supplements) or, rarely, surreptitious.

Cortisol can be measured in 24 hour urine specimens using liquid chromatography followed by tandem mass spectrometry (LC/MS-MS) or high pressure liquid chromatography (HPLC) (Raff and Findling, 2003; Nieman et al., 2008). Older methods use immunoassays to measure 24 hour UFC and report higher cortisol values. These are less specific for cortisol as a result of cross-reactivity with other steroids. Urine free cortisol represents unconjugated cortisol filtered through the renal glomeruli in the urine, and represents approximately 1% of daily cortisol secreted by the adrenals. The test is reliable only in the presence of adequate kidney function and should not be used in patients with renal insufficiency (estimated creatinine clearance less than 30 mL/min). Proper patient instruction is essential in order to assure the adequacy of collected specimens, which require refrigeration. Patients should also be counseled to avoid high fluid intake (> 5 L/day), which can increase 24 hour UFC (Raff and Findling, 2003; Nieman et al., 2008). To examine this possibility, it is also advisable to measure urine volume. In addition, measuring urine creatinine is helpful to assure adequacy of the collection. Two to three specimens, collected on separate days, are needed to optimize diagnostic accuracy of this test (95% sensitivity and 81% specificity) (Pecori Giraldi et al., 2007; Nieman et al., 2008).

Blunting of the diurnal rhythm of cortisol secretion is an early and consistent finding in patients with CS.

Measuring cortisol at nadir (which occurs between 11 p.m. and midnight in individuals who sleep during the night) is of significant diagnostic value (Raff and Findling, 2003; Findling and Raff, 2006; Nieman et al., 2008; Nunes et al., 2009). Of the various tests developed, late night salivary cortisol has become the most popular as a corollary to its reliability and simplicity. Salivary specimens are collected between 11 p.m. and midnight using a clean tube (either by chewing on a clean swab or drooling directly in the container). Patients are counseled to avoid specimen contamination by blood (no tooth brushing or flossing before collection) or exogenous glucocorticoids (no use of creams, cosmetics or mouthwashes before collection) (Nieman et al., 2008). In addition, tobacco use, including chewing, is best avoided. Late night salivary cortisol is not a suitable test in third shift workers, who are likely to have an altered diurnal rhythm of cortisol secretion. Salivary cortisol is stable at room temperature for several days, allowing for specimen transportation via regular mail to the laboratory for analysis. Salivary cortisol can be assayed using enzyme-linked immunosorbent assay (ELISA) or LC/MS-MS. The latter method can also detect contamination by synthetic glucocorticoids or even exogenous hydrocortisone (by measuring cortisol to cortisone ratio) (Raff, 2012). Collection of a minimum of two separate specimens is recommended in order to ensure adequate diagnostic accuracy (92–100% sensitivity and specificity) (Trilck et al., 2005; Nieman et al., 2008). The test has been gaining popularity, but may still not be universally available around the world.

Midnight serum cortisol can also be measured in sleeping or awake individuals (using different cutpoints) as an index of nadir cortisol (Papanicolaou et al., 1998; Nieman et al., 2008). Several studies suggest that this test can be accurate in the diagnosis of CS, but estimates vary (90–100% sensitivity and 20–87% specificity). To avoid raising serum cortisol levels as a result of stress induced by hospitalization or venipuncture, cortisol has been measured in sleeping patients after a 48–hour hospital stay (Papanicolaou et al., 1998; Nieman et al., 2008). This protocol has not been validated in acutely ill patients and may not be practical in several countries and healthcare systems, including the US. These considerations have limited the use of this test in the US, where hospitalization for routine laboratory testing is not reimbursed.

A variety of DSTs have been used in the diagnosis of CS for several decades (Findling and Raff, 2006; Nieman et al., 2008). They all aim at identifying the absence of feedback inhibition of the HPA axis by low doses of dexamethasone, which generally occurs in patients with CS. The overnight 1 mg DST is often used as a screening test because of its simplicity (Findling and Raff, 2006; Nieman et al., 2008). Dexamethasone (1 mg) is taken orally between 11 p.m. and midnight and blood is drawn the following morning between 8 and 9 a.m. Adequate patient instruction is essential. Of note, some pituitary corticotroph adenomas maintain some degree of sensitivity to feedback inhibition, necessitating a very low diagnostic cutpoint (1.8 μg/dL (50 nmol/L)) for optimal test sensitivity. In addition, patients with abnormalities in dexamethasone clearance (including patients taking medications that affect dexamethasone metabolism) may have erroneous (falsely positive or falsely negative) results in this test, as a result of either inadequate or excessive exposure to dexamethasone, respectively (Findling and Raff, 2006; Nieman et al., 2008). Measuring serum dexamethasone levels in the same serum specimen is prudent in order to assure adequate exposure, but this test may not be widely available. In addition, patients with conditions raising corticosteroid binding globulin (CBG) levels, including pregnancy or oral contraceptive use, may have a falsely positive test result on DST (Findling and Raff, 2006; Nieman et al., 2008). It is advisable to discontinue potentially offending medications at least 6 weeks before the test. The 1 mg DST is considered to be a screening test (95% sensitivity and 80% specificity) for CS (Pecori Giraldi et al., 2007; Nieman et al., 2008).

The traditional, 2 day "low dose" DST (2 mg/day in divided doses administered every 6 hours) is in use in some practices and countries (Findling et al., 2004). It is cumbersome and requires adequate patient instruction in order to be performed correctly. The diagnostic accuracy of this test is high (approximately 95%) if serum cortisol (rather than 24 hour UFC) is used as the endpoint (Nieman and Ilias, 2005; Nieman et al., 2008). More recently, the 2 day DST has been combined with peripheral CRH stimulation in the dexamethasone-suppressed CRH stimulation test (Yanovski et al., 1993a, 1998). Originally reported to distinguish between patients with CS and individuals with pseudo-Cushing's states with perfect sensitivity and specificity, it has more recently been noted to have somewhat lower diagnostic accuracy (92–93% sensitivity and specificity), which is further decreased in patients on medications affecting dexamethasone metabolism (Erickson et al., 2007; Valassi et al., 2009). The test is also cumbersome to conduct and requires use of a cortisol assay with high analytic sensitivity and accuracy at very low levels of serum cortisol. In addition, CRH is currently not available for diagnostic use in the US. A proposed algorithm for the evaluation of patients with suspected CS is shown in Figure 15.1.

After the diagnosis of presumptive CS is established, additional laboratory testing as well as imaging studies are needed in order to identify the underlying cause. Measuring plasma ACTH levels in morning specimens helps to distinguish between ACTH-dependent causes

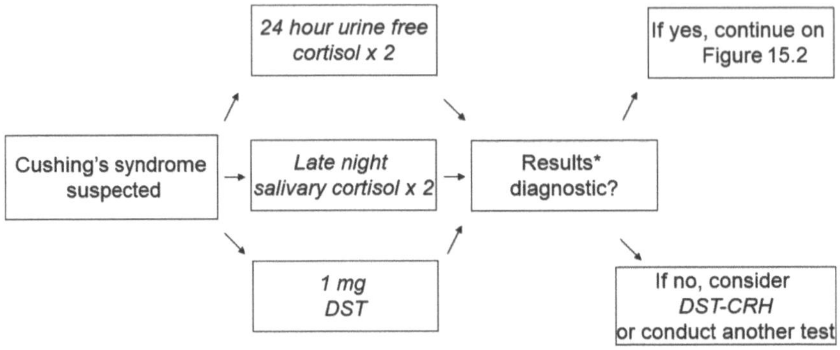

Fig. 15.1. Diagnostic approach to patients with suspected Cushing's syndrome. CRH, corticotropin-releasing hormone; DST, dexamethasone suppression test. *As the 1 mg DST is a screening test, it is advisable to confirm abnormal results with another test (e.g., 24 hour urinary free cortisol or late night salivary cortisol).

(pituitary or ectopic ACTH syndrome or, rarely, ectopic CRH secretion) and ACTH-independent causes (Raff and Findling, 2003; Biller et al., 2008). Patients with ACTH-dependent CS generally have morning plasma ACTH levels exceeding 20 pg/mL (4.4 pmol/L), and patients with ACTH-independent CS usually have morning plasma ACTH levels less than 5 pg/mL (1.1 pmol/L) (Newell-Price et al., 2006). In the latter case, adrenal imaging using computed tomography (CT) will usually identify the underlying lesion (adenoma, carcinoma, or hyperplasia), with the exception of primary pigmented nodular adrenal disease, where the adrenals may appear radiographically normal. If CRH is available, a peripheral CRH stimulation test may be performed in the minority of patients with repeated intermediate (5–20 pg/mL (1.1–4.4 pmol/L)) morning plasma ACTH levels, and may typically demonstrate a rise in ACTH levels in patients with ACTH-dependent CS, but not in those with ACTH-independent CS (Raff and Findling, 2003; Biller et al., 2008).

Once the diagnosis of ACTH-dependent CS is established, the next step involves distinguishing between a pituitary and an ectopic source. Pituitary magnetic resonance imaging (MRI), using a dedicated, high-resolution protocol, is recommended (Raff and Findling, 2003; Nieman and Ilias, 2005). Corticotropin-secreting pituitary adenomas are generally small, only 6–10% being macroadenomas (Bonneville et al., 2002). Only 60–70% of corticotroph adenomas are detectable on MRI (Doppman et al., 1988; Raff and Findling, 2003; Nieman and Ilias, 2005). In addition, the prevalence of incidental pituitary microadenomas is substantial in the general population (approximately 10%) (Hall et al., 1994). Based on these considerations, patients with larger sellar lesions (> 6–10 mm in diameter, depending on local practice) can be reasonably presumed to have CD.

In contrast, patients with smaller sellar masses need additional evaluation to distinguish between a pituitary and ectopic origin of ACTH excess.

The most accurate test to help make this distinction is bilateral inferior petrosal sinus sampling (BIPSS) (Oldfield et al., 1985, 1991). The procedure can be safely performed in outpatients but requires considerable expertise for successful catheterization of both inferior petrosal sinuses via the femoral veins. Blood is then sampled concurrently from both centrally placed catheters as well as a peripheral vein, and plasma specimens are assayed to measure ACTH levels. Several additional samples are then obtained after peripheral administration of CRH or desmopressin (Castinetti et al., 2007a; Machado et al., 2007).

The presence of a central to peripheral vein ACTH gradient (2:1 or higher before CRH or desmopressin, and 3:1 or higher after CRH or desmopressin) predicts the presence of a pituitary source of ACTH excess with high (approximately 94%) sensitivity and specificity (Bonelli et al., 2000; Raff and Findling, 2003; Nieman and Ilias, 2005). Use of fluoroscopy is advised in order to verify accurate catheter placement during BIPSS. The accuracy of BIPSS may be increased by simultaneous measurement of serum prolactin levels in the same blood specimens (Mulligan et al., 2011, 2012). Measuring 24 hour UFC on the day before BIPSS is advisable in order to assure that hypercortisolism is present, which is needed in order to suppress ACTH secretion by normal corticotrophs during BIPSS. Of note, healthy individuals may have ACTH gradients on BIPSS that overlap with those of patients with CD as a result of ACTH secretion by normal corticotrophs (Yanovski et al., 1993b). When performed by expert interventional radiologists, BIPSS is generally safe. Occasionally, deep venous thrombosis may occur. In addition, symptoms suggestive of brainstem ischemia have, rarely, occurred in the past and were associated with the use of microcatheters which are no longer employed during BIPSS (Seyer et al., 1994).

Patients who are predicted to have an ectopic lesion based on BIPSS may then undergo imaging of the neck, chest, abdomen, and pelvis (CT, MRI) in order to detect

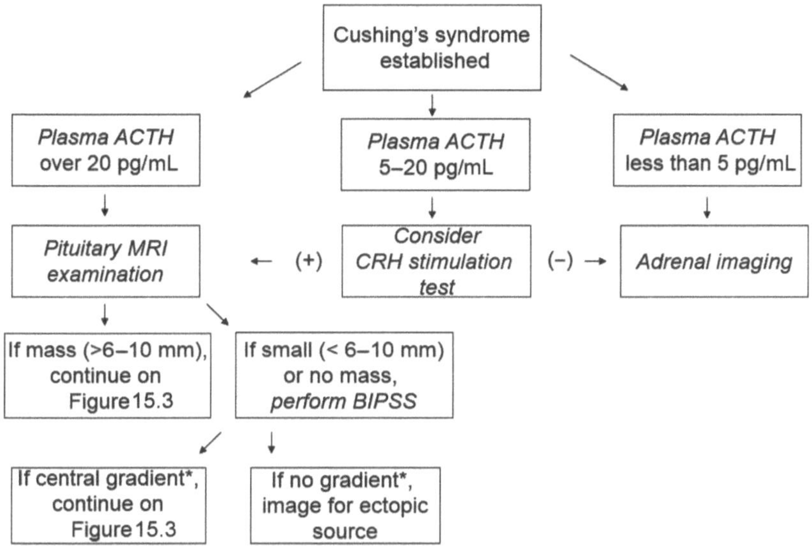

Fig. 15.2. Diagnostic approach to the identification of the cause of Cushing's syndrome. *A central gradient (peak ratio between either right or left inferior petrosal sinus ACTH to peripheral ACTH level) > 2 before CRH (or desmopressin) administration, or > 3 after CRH (or desmopressin) administration is considered consistent with a pituitary source of ACTH excess. If neither of these criteria is met, the results are consistent with an ectopic source of ACTH excess. ACTH, corticotropin; BIPSS, bilateral inferior petrosal sinus sampling; CRH, corticotropin-releasing hormone; MRI, magnetic resonance imaging.

the underlying tumor (Findling and Raff, 2006). In some cases, somatostatin receptor scintigraphy or positron emission tomography (PET) in combination with CT may be helpful in detecting an occult cctopic source (Ilias et al., 2005). Some laboratory tests may help to identify an ectopic tumor, including serum calcitonin (medullary thyroid carcinoma), fractionated plasma metanephrines (pheochromocytoma), chromogranin A (wide variety of neuroendocrine tumors), or serotonin levels (some carcinoids) (Ilias et al., 2005). Despite extensive testing, an ectopic tumor may not be detectable in approximately 19% of patients (Ilias et al., 2005).

Before BIPSS became available, two other tests were widely used to distinguish between a pituitary and an ectopic tumor, including the high-dose DST and the peripheral CRH stimulation test (Nieman et al., 1986, 1993). The principle of both tests is that pituitary tumors often maintain some degree of physiologic response to feedback inhibition by dexamethasone in high doses (8 mg) or stimulation by exogenous CRH, whereas ectopic tumors generally (but not always) lack such responses. Unfortunately, both tests lack sufficiently high sensitivity and specificity for CD (8 mg DST: sensitivity: 82% and specificity: 67–79%; peripheral CRH stimulation test: sensitivity: 70–93% and specificity: 88%) (Nieman and Ilias, 2005). Among patients with ACTH-dependent CS, the pretest probability of CD is 85% or higher (especially in young women), suggesting that neither the 8 mg DST nor the peripheral CRH stimulation test are sufficiently accurate in most cases (Findling and Raff, 2006). A proposed algorithm for the identification of the cause of CS is shown in Figure 15.2.

MANAGEMENT

In patients with CS/CD, goals of therapy include normalization of cortisol secretion, adequate control of the underlying tumor, prevention and management of comorbidities associated with CS, while restoring life expectancy and quality of life to those in the general population (Biller et al., 2008). A multidisciplinary approach is important in optimizing patient care. Pituitary surgery is currently the treatment of choice for most patients with CD. Patients who are acutely ill may require preoperative medical therapy in order to improve their surgical risk in preparation for surgery. Preoperative medical therapy is also advised if surgery will be delayed. A proposed algorithm for the management of patients with CD is shown in Figure 15.3.

Pituitary surgery

Patients with CD should be referred to an experienced pituitary neurosurgeon (Barker et al., 2003). Transsphenoidal pituitary tumor resection (TSS) leads to remission of CD in 70–90% of patients (Swearingen et al., 1999a, b). Preoperative and postoperative images of the pituitary of a patient with CD are shown in Figure 15.4.

Several factors may predict remission, including tumor size and location, remission being more

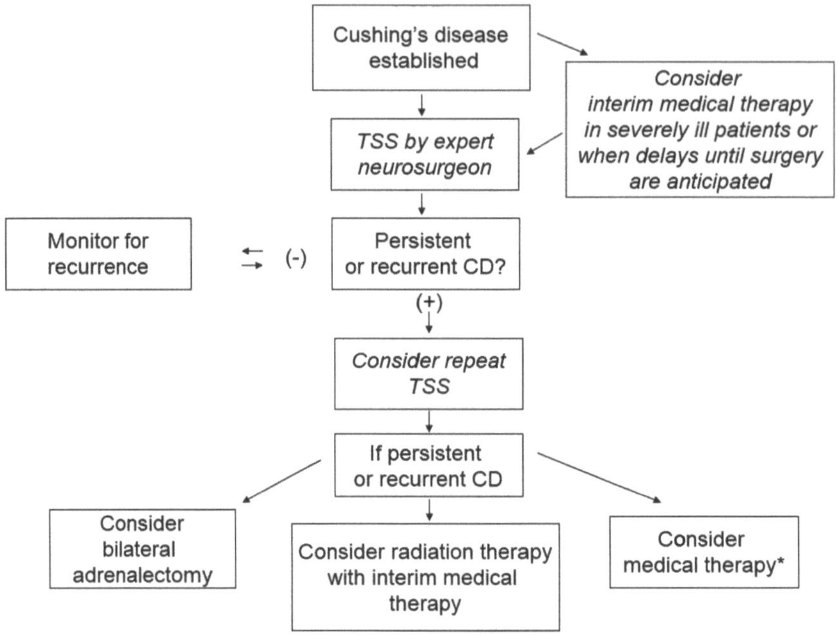

Fig. 15.3. Management of Cushing's disease. CD, Cushing's disease; TSS, transsphenoidal surgery. *In patients with recurrent CD, medical therapy is typically not advised as definitive therapy, but used as a "bridge" to other interventions. Medical therapy alone might be used in some patients who are not candidates for, or decline, other interventions.

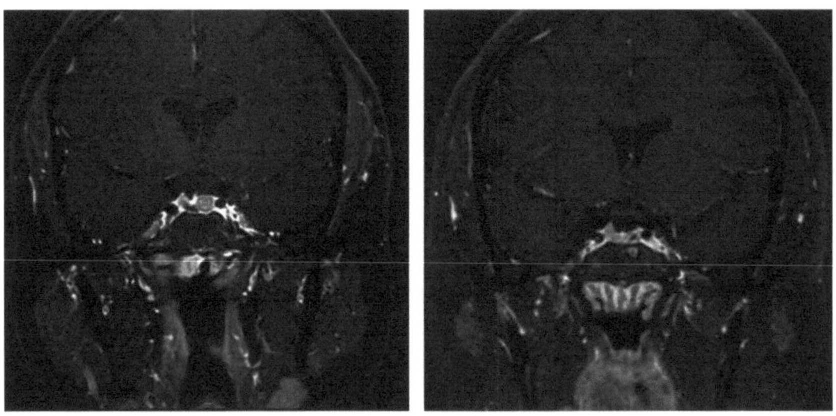

Fig. 15.4. Coronal T1-weighted, contrast-enhanced magnetic resonance images of the pituitary of a patient with Cushing's disease, showing a 5 × 5 mm mass on the left side of the pituitary (left panel). Postoperative appearance of the pituitary at 6 weeks after transsphenoidal pituitary adenomectomy (right panel).

frequent in patients with microadenomas (Blevins et al., 1998; Guilhaume et al., 1988; Swearingen et al., 1999a, b; Shimon et al., 2002; Hammer et al., 2004; Rollin et al., 2004). Other factors suggested as being predictive of remission in some studies include preoperative tumor visualization on MRI, intraoperative tumor visualization by the surgeon, and identification of a corticotroph adenoma on pathology (Guilhaume et al., 1988; Bochicchio et al., 1995; Invitti et al., 1999; Chee et al., 2001; Rees et al., 2002; Hammer et al., 2004; Rollin et al., 2004). It may be noted, however, that some patients may achieve remission even if no tumor is found on pathology, presumably reflecting the fragility of some of these minute specimens, which may not survive tissue handling and processing.

Experienced pituitary neurosurgeons can perform TSS with very low perioperative mortality (ranging between zero and 1.5%) and low complication rates (Barker et al., 2003). Even in experienced hands, cerebrospinal fluid leak (up to 8%), bleeding or hematomas (up to 6%), epistaxis (up to 6%), venous thromboembolism (up to 4%), infection (meningitis, up to 3%) may occur (Semple and Laws, 1999). In addition, diabetes insipidus (3–9%) may occur, which is generally transient, as well as hyponatremia (10–25%) caused by the syndrome of inappropriate antidiuretic hormone secretion (SIADH). Anterior hypopituitarism (2–40%) may also occur (Semple and Laws, 1999).

Patients with CD entering remission are typically anticipated to develop central hypoadrenalism, usually lasting for 6–18 months postoperatively (Rollin et al., 2004). While the exact remission cutpoints for morning serum cortisol and 24 hour UFC vary between institutions, morning serum cortisol lower than 2–5 μg/dL and 24 hour UFC less than 20 μg/24 hours (at nadir) usually indicate remission postoperatively (Swearingen et al., 1999a, b). Glucocorticoid replacement is needed to prevent symptomatic adrenal insufficiency and alleviate glucocorticoid withdrawal symptoms, until recovery of the HPA axis occurs. Regular clinical evaluation is recommended to detect glucocorticoid withdrawal symptoms that might emerge during subsequent glucocorticoid tapering. In addition, periodic laboratory reassessment of early morning serum cortisol levels is advised in order to monitor the recovery of the HPA axis, which eventually occurs in the large majority of patients. When morning basal (or cosyntropin stimulated) serum cortisol levels exceed 18 μg/dL (500 nmol/L), recovery of the HPA axis is likely. In general, glucocorticoid replacement may then be discontinued.

Patients in remission should be periodically monitored for recurrence of CD, which may occur in up to 20–25% of patients on long-term follow-up (up to 20 years) (Sonino et al., 1996; Barbetta et al., 2001; Patil et al., 2008). While it is not possible to predict recurrence of CD in individual patients with certainty, there are published data on possible predictors. Very low or undetectable morning serum cortisol levels in the early postoperative period have been associated with a lower risk of recurrence (Trainer et al., 1993; Yap et al., 2002; Pereira et al., 2003; Esposito et al., 2006; Patil et al., 2008). However, recurrence of CD is possible even in patients with very low serum cortisol levels postoperatively. Low plasma ACTH levels in the postoperative period have also been associated with a lower risk of recurrence (Invitti et al., 1999). Prolonged requirement for glucocorticoid replacement (exceeding 1 year after TSS) has also been associated with a lower recurrence risk (Bochicchio et al., 1995; Estrada et al., 2001). In contrast, normal (rather than low) serum cortisol levels in the postoperative period, inadequate suppression of serum cortisol on dexamethasone or loperamide suppression testing or excessive stimulation on CRH, desmopressin, dexamethasone-suppressed desmopressin or metyrapone testing may predict a higher risk of recurrence (Avgerinos et al., 1987; Trainer et al., 1993; van Aken et al., 1997; Barbetta et al., 2001; Chen et al., 2003; Esposito et al., 2006; Castinetti et al., 2009b). In addition, patients with delayed remission of CD after TSS appear to be at increased risk of recurrence (Valassi et al., 2010a). As none of these predictors is completely reliable, all patients should be followed for the possibility of recurrence.

Management of recurrent Cushing's disease

Treatment options for patients with recurrent CD may include a second pituitary operation, radiation therapy to the pituitary with interim medical therapy, possibly medical therapy alone, or occasionally bilateral adrenalectomy (Biller et al., 2008; Tritos et al., 2011a). Repeat TSS may lead to remission of CD in approximately 50–60% of patients, but is associated with increased risk of complications (including cerebrospinal fluid leak and anterior hypopituitarism) in comparison with the first TSS (Ram et al., 1994; Biller et al., 2008; Tritos et al., 2011a).

Radiation therapy can be administered as either conventional photon radiation therapy or stereotactic (photon or proton beam) radiation therapy (Littley et al., 1990; Jagannathan et al., 2007; Petit et al., 2008). Stereotactic photon radiation therapy can be delivered using a Gamma Knife device or a linear accelerator, and proton beam radiation therapy is delivered using a cyclotron (Witt et al., 1998; Sheehan et al., 2000; Castinetti et al., 2007b; Petit et al., 2008). Stereotactic radiation therapy can be delivered in a single fraction (stereotactic "radiosurgery") in patients whose tumors do not abut the optic apparatus.

Remission of hypercortisolism occurs in up to 86% and tumor control is achieved in 90–100% of patients after radiation therapy (Witt et al., 1998; Sheehan et al., 2000; Petit et al., 2008). However, there is a considerable delay between administration of radiation therapy and remission of hypercortisolism, ranging between months and years. Stereotactic radiosurgery may lead to remission sooner than conventional radiation therapy. Medical therapy is generally administered until radiation therapy takes effect (which is typically determined based on results of 24 hour UFC and late night salivary cortisol tests).

Stereotactic radiation therapy spares the normal brain from low level radiation exposure in comparison with conventional radiation therapy (Witt et al., 1998; Petit et al., 2008). However, the risk of anterior hypopituitarism is considerable on follow-up (40% at 5 years and likely higher in the long-term), necessitating periodic reevaluation of pituitary function in irradiated patients. Other uncommon adverse events in patients who have received radiation therapy include optic neuropathy, additional cranial neuropathies, secondary tumor formation, temporal lobe necrosis, and possibly cerebrovascular disease as a result of radiation-induced angiopathy (Witt et al., 1998; Petit et al., 2008). Use of stereotactic techniques for delivery of radiation therapy, careful planning of treatment sessions, and dose fractionation (for tumors near the optic apparatus) may help to minimize these risks. However, this has not yet been established with certainty.

In patients with CD, bilateral adrenalectomy is generally reserved for patients with recurrent disease, who are either not candidates or have failed repeat TSS, and have either failed or prefer not to have radiation therapy (Biller et al., 2008). At present, bilateral adrenalectomy is typically performed laparoscopically, leading to improved postoperative morbidity and decreased hospital stay in comparison with open adrenalectomy (Thompson et al., 1997, 2007). Possible perioperative complications may occur in up to 12% of patients, including need for conversion to open adrenalectomy, bleeding, infection, organ injury, persistent pain, hernia formation, and deep vein thrombosis (Vella et al., 2001; Zada et al., 2010).

Bilateral adrenalectomy will cure CD in almost all patients. However, primary adrenal insufficiency will also occur as a consequence of adrenalectomy, necessitating lifelong glucocorticoid and mineralocorticoid replacement. In addition, corticotroph tumor progression (Nelson's syndrome) may occur in up to 50% of patients with CD who underwent bilateral adrenalectomy (Nelson et al., 1958; Assie et al., 2007). The reader is referred to the chapter on Nelson's syndrome (Ch. 22) for details. Briefly, this complication is characterized by growth of the corticotroph pituitary tumor, which might exert mass effect, and progressive rise in plasma ACTH levels, leading to skin (and possibly mucosal) hyperpigmentation. Pituitary surgery, either alone or in combination with radiation therapy, may be needed in order to control tumor growth in patients with Nelson's syndrome (Nelson et al., 1958; Assie et al., 2007; Vik-Mo et al., 2009; Tritos et al., 2011b).

Medical therapy

Medical therapy may be used preoperatively in severely hypercortisolemic or symptomatic patients (suffering with infection, psychosis, severe muscle weakness, or hypokalemia) in order to improve their overall status in preparation for surgery. In addition, medical therapy may be used if substantial delays until surgery are anticipated. More often, medical therapy is used as adjunctive treatment in patients who have failed surgery, including patients who have received radiation therapy to the sella, until the salutary effects of radiotherapy lead to control of hypercortisolism (Nieman, 2002; Tritos et al., 2011a; Tritos and Biller, 2012).

Medications used to treat CD include steroidogenesis inhibitors, centrally acting agents, and a glucocorticoid receptor antagonist (mifepristone). All medications should be carefully titrated and monitored to avoid precipitating hypoadrenalism. Using steroidogenesis inhibitors, a "block and replace" regimen may also be employed, involving complete suppression of endogenous cortisol secretion and add-on glucocorticoid replacement (Nieman, 2002; Biller et al., 2008).

Steroidogenesis inhibitors currently in use include ketoconazole, metyrapone, mitotane, and etomidate, which are all used "off-label" as therapies for CD. Ketoconazole is an imidazole antifungal agent with a rapid onset of action, inhibiting multiple enzymatic steps in the adrenal cortex (including cholesterol side chain cleavage, 17α-hydroxylase and 17,20 lyase) (Sonino et al., 1991; Tabarin et al., 1991; Engelhardt and Weber, 1994). Additional, direct inhibitory effects of ketoconazole on pituitary corticotrophs have been proposed. Taken in divided doses (200–600 mg by mouth two to three times daily), ketoconazole decreases cortisol secretion in approximately 70% of patients with CD. Side-effects may include nausea, vomiting, reversible transaminitis (in 10% of patients), or rare idiosyncratic severe hepatotoxicity (in ~1:15 000 patients) (McCance et al., 1987). Monitoring of liver enzymes is advised in patients receiving ketoconazole. Other side-effects include headache, hypogonadism, rash, and sedation. Ketoconazole potently inhibits several P450 enzymes (including CYP3A4, CYP2C9, and CYP1A2), leading to potentially significant interactions with other medications metabolized by these enzymes. Ketoconazole may interfere with masculinization of a male fetus and is best avoided during pregnancy.

Metyrapone inhibits 11β-hydroxylase and suppresses hypercortisolism in up to 75% of patients with CD, taken in doses ranging between 250 and 1500 mg by mouth four times daily (Thoren et al., 1985; Verhelst et al., 1991). It has been effective when used during pregnancy (Gormley et al., 1982; Lindsay et al., 2005). Escape from its effects may occur as a result of increased ACTH output by the pituitary tumor. Side-effects may include nausea and dizziness. In addition, metyrapone generally leads to increased secretion of precursor steroids with mineralocorticoid and androgenic activity, resulting in hypertension, hypokalemia, edema, and hirsutism (the latter in women).

Mitotane is an inhibitor of multiple enzymes in the adrenal cortex, including cholesterol side chain cleavage enzyme, 3-β-hydroxysteroid dehydrogenase, 11β-hydroxylase and aldosterone synthase (Luton et al., 1979; Terzolo et al., 2007). It has a delayed onset of action exceeding 2 weeks. Mitotane suppresses hypercortisolism in up to 83% of patients with CD. It also has adrenolytic properties when taken in higher doses (greater than 3 g/day) for a prolonged period. Mitotane is most frequently recommended in patients with adrenocortical carcinoma (Luton et al., 1979; Terzolo et al., 2007). Mitotane has been used in a dose ranging between 1 and 12 g/day, though dose-related gastrointestinal (nausea, diarrhea) and neurologic (dizziness, vertigo, ataxia, confusion) side-effects frequently limit mitotane dose. The medication is teratogenic and extensively stored in adipose tissue. Pregnancy should be avoided for up to 5 years after mitotane discontinuation. In addition, dyslipidemia may occur in patients on mitotane. Glucocorticoid replacement dose requirements are higher in these patients, as a result of accelerated glucocorticoid clearance (Robinson et al., 1987).

Etomidate is an intravenous anesthetic agent with potent inhibitory effects on 11β-hydroxylase (as well as weaker inhibitory effects on 17,20 lyase activity) and is effective in suppressing hypercortisolism in up to 100% of patients, even in subhypnotic doses (loading dose: 0.03 mg/kg bolus intravenously, followed by 0.1 mg/kg/hour) (Allolio et al., 1988; Drake et al., 1998). Etomidate is particularly helpful in acutely ill patients and can help bridge them to surgery. Careful monitoring is required to avoid excessive sedation.

Aminoglutethimide is another steroidogenesis inhibitor, which is no longer available (Thoren et al., 1985). Trilostane is a weakly active agent which is not available for human use. A novel steroidogenesis inhibitor, LCI699, inhibits 11β-hydroxylase and aldosterone synthase (Calhoun et al., 2011). It is currently under development as potential therapy in patients with CS.

Several centrally acting agents have been evaluated as potential therapies in CD. Bromocriptine, octreotide, rosiglitazone, pioglitazone, valproate, and cyproheptadine have very limited effectiveness in CD (Nieman, 2002). Cabergoline, a selective dopamine receptor (D2) agonist, has been extensively used in patients with hyperprolactinemia and prolactinoma (Klibanski, 2010). Cabergoline has been used "off-label" in patients with CD in doses ranging between 1 and 7 mg/week (which are higher than those recommended in most hyperprolactinemic patients). Cabergoline suppresses hypercortisolism in 50–70% of patients with CD in the short term (12 months) (Pivonello et al., 1999, 2009; Godbout et al., 2010). However, only 30–40% of patients show a sustained response to cabergoline after 24–36 months of therapy. Of note, cabergoline is effective in suppressing ACTH levels and decreasing tumor size in some patients with Nelson's syndrome (Pivonello et al., 1999; Tritos et al., 2011b). Cabergoline is well-tolerated in most patients. However, nausea, vomiting, dizziness, headache, constipation and occasionally psychiatric manifestations have been associated with cabergoline use. In addition, high dose cabergoline therapy has been associated with cardiac valvulopathy in patients with Parkinson's disease (Schade et al., 2007; Zanettini et al., 2007; Valassi et al., 2010b).

Pasireotide (SOM 230) is a somatostatin receptor (sst) agonist which has substantial binding activity to several sst isoforms, including sst1, sst2, sst3, sst5 (Ben-Shlomo et al., 2009; Boscaro et al., 2009). In a recent phase III study, 88% of patients with CD who were administered pasireotide showed a decline in 24 hour UFC (Colao et al., 2012). Complete normalization of 24 hour UFC was achieved in up to 26% of patients on pasireotide. Adverse events include gastrointestinal toxicity (diarrhea, constipation, bloating, abdominal pain, gallstones), sinus bradycardia, QT prolongation, and hyperglycemia (73% of patients). In addition, diabetes mellitus of new onset occurred in 34% of patients. Pasireotide is currently investigational in the US but has been granted approval by the European Medicines Agency (EMA). Preliminary data suggest that combination therapy with ketoconazole, cabergoline, and pasireotide may be of value in patients showing an insufficient response to one agent alone (Feelders et al., 2010). Although octreotide, an older somatostatin receptor agonist engaging sst2 and, to a lesser extent, sst5, is not effective in most patients with CS, it can be helpful in decreasing ACTH levels and inducing tumor responses in some patients with Nelson's syndrome (Lamberts et al., 1989).

Temozolomide is an orally active alkylating agent that has significant activity in patients with locally aggressive pituitary adenomas, including a response rate of 60% in aggressive corticotroph tumors (McCormack et al., 2011). Other centrally acting agents under development include bexarotene, a retinoic acid receptor (RXR) agonist, and lapatinib, an inhibitor of the epidermal growth factor (EGF)/erb2 receptor tyrosine kinase (Farol and Hymes, 2004; Schteingart, 2009; Fukuoka et al., 2011). Preclinical data suggest that R-roscovitine, a cyclin-dependent kinase 2 (CDK2)/cyclin E inhibitor, may suppress corticotroph tumor growth and ACTH secretion (Liu et al., 2011).

Mifepristone is a type 2 glucocorticoid receptor antagonist, which was recently approved by the US Food and Drug Administration (FDA) for use in patients with CS and hyperglycemia who are not surgical candidates (Castinetti et al., 2009a). Mifepristone was evaluated in a recently published phase III open label study, including 50 patients with CS of diverse causes (Fleseriu et al., 2012). Administered in doses ranging between 300 mg and 1200 mg orally daily, mifepristone led to improvement in glycemia in 60% and improvement in diastolic blood pressure in 38% of patients in predefined study subgroups. Overall clinical status improved in 87% of patients. Mifepristone must be titrated and monitored based on clinical criteria, as serum and 24 hour UFC rise on mifepristone therapy in patients with CD (Castinetti et al., 2009a; Fleseriu et al., 2012). For the same reason, hypoadrenalism has to be recognized on clinical grounds alone. If suspected, adrenal insufficiency can be treated by temporarily holding and decreasing mifepristone dose, while administering dexamethasone as rescue therapy. Of note, the medication antagonizes progesterone receptors, resulting in endometrial thickening and irregular vaginal bleeding. In addition, hypertension and hypokalemia may occur as a result of unopposed interaction of endogenous cortisol with the mineralocorticoid receptor, which is not blocked by mifepristone. Serum potassium levels should be monitored and hypokalemia corrected with potassium replacement and/or spironolactone (or eplerenone) administration. Skin rash, dyslipidemia, and abnormalities in thyroid function tests (including rise in serum thyrotropin) may also occur. Mifepristone is metabolized through the CYP3A4 pathway and can interact with other medications that utilize, inhibit, or induce this enzyme, raising the potential for significant drug interactions.

Management and prevention of complications of CD is an essential part of patient care, and may include treatment of hypertension, hyperglycemia, dyslipidemia, bone loss, correction of electrolyte abnormalities, prevention and treatment of thromboembolic disease, coronary heart disease, or infection (Boscaro et al., 2002; Arnaldi et al., 2003).

CONCLUSION

Over the past 100 years, significant advances have been made in the diagnosis and management of CD, translating into improvements in prognosis, especially for patients in remission (Swearingen et al., 1999a, b; Lindholm et al., 2001; Hassan-Smith et al., 2012). However, CD remains a challenging condition to diagnose and treat. Delays in diagnosis are still common. A high index of suspicion and an organized, stepwise diagnostic approach is essential in the evaluation of these patients. Once the diagnosis of CS is made, further efforts should be directed at establishing the underlying cause (including CD). Effective management generally requires the combined expertise of pituitary neurosurgeons, endocrinologists, and neuropathologists. Recurrent CD is still a major clinical problem and can be managed with a second pituitary operation, radiation therapy with interim medical therapy, or, in some cases, bilateral adrenalectomy. Quality of life remains impaired in many patients with CD, including some of those who are in biochemical remission (Lindholm et al., 2001). It is anticipated that better understanding of the currently obscure pathogenesis of CD may lead to more accurate diagnostic tests and more efficacious therapies for this condition.

NOTES ADDED IN PROOF

1. Pasireotide was approved by the FDA as therapy for Cushing's disease in December 2012, including patients who are not surgical candidates or those not cured by surgery. 2. CRH is now available in the US.

References

Alexandraki KI, Kaltsas GA, Isidori AM et al. (2009). The prevalence and characteristic features of cyclicity and variability in Cushing's disease. Eur J Endocrinol 160: 1011–1018.

Allolio B, Schulte HM, Kaulen D et al. (1988). Nonhypnotic low-dose etomidate for rapid correction of hypercortisolaemia in Cushing's syndrome. Klin Wochenschr 66: 361–364.

Arnaldi G, Angeli A, Atkinson AB et al. (2003). Diagnosis and complications of Cushing's syndrome: a consensus statement. J Clin Endocrinol Metab 88: 5593–5602.

Assie G, Bahurel H, Coste J et al. (2007). Corticotroph tumor progression after adrenalectomy in Cushing's disease: a reappraisal of Nelson's syndrome. J Clin Endocrinol Metab 92: 172–179.

Avgerinos PC, Chrousos GP, Nieman LK et al. (1987). The corticotropin-releasing hormone test in the postoperative evaluation of patients with Cushing's syndrome. J Clin Endocrinol Metab 65: 906–913.

Azziz R (2005). Diagnostic criteria for polycystic ovary syndrome: a reappraisal. Fertil Steril 83: 1343–1346.

Bals-Pratsch M, Hanker JP, Hellhammer DH et al. (1996). Intermittent Cushing's disease in hirsute women. Horm Metab Res 28: 105–110.

Barbetta L, Dall'Asta C, Tomei G et al. (2001). Assessment of cure and recurrence after pituitary surgery for Cushing's disease. Acta Neurochir (Wien) 143: 477–481, discussion 481–482.

Barker 2nd FG, Klibanski A, Swearingen B (2003). Transsphenoidal surgery for pituitary tumors in the United States, 1996–2000: mortality, morbidity, and the effects of hospital and surgeon volume. J Clin Endocrinol Metab 88: 4709–4719.

Ben-Shlomo A, Schmid H, Wawrowsky K et al. (2009). Differential ligand-mediated pituitary somatostatin receptor subtype signaling: implications for corticotroph tumor therapy. J Clin Endocrinol Metab 94: 4342–4350.

Biller BM, Grossman AB, Stewart PM et al. (2008). Treatment of adrenocorticotropin-dependent Cushing's syndrome: a consensus statement. J Clin Endocrinol Metab 93: 2454–2462.

Blevins Jr LS, Christy JH, Khajavi M et al. (1998). Outcomes of therapy for Cushing's disease due to adrenocorticotropin-secreting pituitary macroadenomas. J Clin Endocrinol Metab 83: 63–67.

Bochicchio D, Losa M, Buchfelder M (1995). Factors influencing the immediate and late outcome of Cushing's disease treated by transsphenoidal surgery: a retrospective study by the European Cushing's Disease Survey Group. J Clin Endocrinol Metab 80: 3114–3120.

Bonelli FS, Huston 3rd J, Carpenter PC et al. (2000). Adrenocorticotropic hormone-dependent Cushing's syndrome: sensitivity and specificity of inferior petrosal sinus sampling. AJNR Am J Neuroradiol 21: 690–696.

Bonneville JF, Cattin F, Bonneville F et al. (2002). Pituitary gland imaging in Cushing's disease. Neurochirurgie 48: 173–185.

Boscaro M, Sonino N, Scarda A et al. (2002). Anticoagulant prophylaxis markedly reduces thromboembolic complications in Cushing's syndrome. J Clin Endocrinol Metab 87: 3662–3666.

Boscaro M, Ludlam WH, Atkinson B et al. (2009). Treatment of pituitary-dependent Cushing's disease with the multireceptor ligand somatostatin analog pasireotide (SOM230): a multicenter, phase II trial. J Clin Endocrinol Metab 94: 115–122.

Calhoun DA, White WB, Krum H et al. (2011). Effects of a novel aldosterone synthase inhibitor for treatment of primary hypertension: results of a randomized, double-blind, placebo- and active-controlled phase 2 trial. Circulation 124: 1945–1955.

Castinetti F, Morange I, Dufour H et al. (2007a). Desmopressin test during petrosal sinus sampling: a valuable tool to discriminate pituitary or ectopic ACTH-dependent Cushing's syndrome. Eur J Endocrinol 157: 271–277.

Castinetti F, Nagai M, Dufour H et al. (2007b). Gamma knife radiosurgery is a successful adjunctive treatment in Cushing's disease. Eur J Endocrinol 156: 91–98.

Castinetti F, Fassnacht M, Johanssen S et al. (2009a). Merits and pitfalls of mifepristone in Cushing's syndrome. Eur J Endocrinol 160: 1003–1010.

Castinetti F, Martinie M, Morange I et al. (2009b). A combined dexamethasone desmopressin test as an early marker of postsurgical recurrence in Cushing's disease. J Clin Endocrinol Metab 94: 1897–1903.

Catargi B, Rigalleau V, Poussin A et al. (2003). Occult Cushing's syndrome in type-2 diabetes. J Clin Endocrinol Metab 88: 5808–5813.

Cavagnini F, Pecori Giraldi F (2001). Epidemiology and follow-up of Cushing's disease. Ann Endocrinol (Paris) 62: 168–172.

Chee GH, Mathias DB, James RA et al. (2001). Transsphenoidal pituitary surgery in Cushing's disease: can we predict outcome? Clin Endocrinol (Oxf) 54: 617–626.

Chen JC, Amar AP, Choi S et al. (2003). Transsphenoidal microsurgical treatment of Cushing disease: postoperative assessment of surgical efficacy by application of an overnight low-dose dexamethasone suppression test. J Neurosurg 98: 967–973.

Colao A, Petersenn S, Newell-Price J et al. (2012). A 12-month phase 3 study of pasireotide in Cushing's disease. N Engl J Med 366: 914–924.

Cushing H (1932). The basophil adenomas of the pituitary body and their clinical manifestations (pituitary basophilism). Bull Johns Hopkins Hosp 50: 137–195.

Cushing H (1969). The basophil adenomas of the pituitary body. Ann R Coll Surg Engl 44: 180–181.

Doppman JL, Frank JA, Dwyer AJ et al. (1988). Gadolinium DTPA enhanced MR imaging of ACTH-secreting microadenomas of the pituitary gland. J Comput Assist Tomogr 12: 728–735.

Drake WM, Perry LA, Hinds CJ et al. (1998). Emergency and prolonged use of intravenous etomidate to control hypercortisolemia in a patient with Cushing's syndrome and peritonitis. J Clin Endocrinol Metab 83: 3542–3544.

Elamin MB, Murad MH, Mullan R et al. (2008). Accuracy of diagnostic tests for Cushing's syndrome: a systematic review and metaanalyses. J Clin Endocrinol Metab 93: 1553–1562.

Engelhardt D, Weber MM (1994). Therapy of Cushing's syndrome with steroid biosynthesis inhibitors. J Steroid Biochem Mol Biol 49: 261–267.

Erickson D, Natt N, Nippoldt T et al. (2007). Dexamethasone-suppressed corticotropin-releasing hormone stimulation test for diagnosis of mild hypercortisolism. J Clin Endocrinol Metab 92: 2972–2976.

Esposito F, Dusick JR, Cohan P et al. (2006). Clinical review. Early morning cortisol levels as a predictor of remission after transsphenoidal surgery for Cushing's disease. J Clin Endocrinol Metab 91: 7–13.

Estrada J, Garcia-Uria J, Lamas C et al. (2001). The complete normalization of the adrenocortical function as the criterion of cure after transsphenoidal surgery for Cushing's disease. J Clin Endocrinol Metab 86: 5695–5699.

Farol LT, Hymes KB (2004). Bexarotene: a clinical review. Expert Rev Anticancer Ther 4: 180–188.

Feelders RA, de Bruin C, Pereira AM et al. (2010). Pasireotide alone or with cabergoline and ketoconazole in Cushing's disease. N Engl J Med 362: 1846–1848.

Findling JW, Raff H (2006). Cushing's syndrome: important issues in diagnosis and management. J Clin Endocrinol Metab 91: 3746–3753.

Findling JW, Raff H, Aron DC (2004). The low-dose dexamethasone suppression test: a reevaluation in patients with Cushing's syndrome. J Clin Endocrinol Metab 89: 1222–1226.

Fleseriu M, Biller BM, Findling JW et al. (2012). Mifepristone, a glucocorticoid receptor antagonist, produces clinical and metabolic benefits in patients with Cushing's syndrome. J Clin Endocrinol Metab 97: 2039–2049.

Fukuoka H, Cooper O, Mizutani J et al. (2011). HER2/ErbB2 receptor signaling in rat and human prolactinoma cells: strategy for targeted prolactinoma therapy. Mol Endocrinol 25: 92–103.

Godbout A, Manavela M, Danilowicz K et al. (2010). Cabergoline monotherapy in the long-term treatment of Cushing's disease. Eur J Endocrinol 163: 709–716.

Gormley MJ, Hadden DR, Kennedy TL et al. (1982). Cushing's syndrome in pregnancy – treatment with metyrapone. Clin Endocrinol (Oxf) 16: 283–293.

Guilhaume B, Bertagna X, Thomsen M et al. (1988). Transsphenoidal pituitary surgery for the treatment of Cushing's disease: results in 64 patients and long term follow-up studies. J Clin Endocrinol Metab 66: 1056–1064.

Hall WA, Luciano MG, Doppman JL et al. (1994). Pituitary magnetic resonance imaging in normal human volunteers: occult adenomas in the general population. Ann Intern Med 120: 817–820.

Hammer GD, Tyrrell JB, Lamborn KR et al. (2004). Transsphenoidal microsurgery for Cushing's disease: initial outcome and long-term results. J Clin Endocrinol Metab 89: 6348–6357.

Hassan-Smith ZK, Sherlock M, Reulen RC et al. (2012). Outcome of Cushing's disease following transsphenoidal surgery in a single center over 20 years. J Clin Endocrinol Metab 97: 1194–1201.

Ilias I, Torpy DJ, Pacak K et al. (2005). Cushing's syndrome due to ectopic corticotropin secretion: twenty years' experience at the National Institutes of Health. J Clin Endocrinol Metab 90: 4955–4962.

Invitti C, Pecori Giraldi F, de Martin M et al. (1999). Diagnosis and management of Cushing's syndrome: results of an Italian multicentre study Study Group of the Italian Society of Endocrinology on the Pathophysiology of the Hypothalamic–Pituitary–Adrenal axis. J Clin Endocrinol Metab 84: 440–448.

Jagannathan J, Sheehan JP, Pouratian N et al. (2007). Gamma Knife surgery for Cushing's disease. J Neurosurg 106: 980–987.

Kidambi S, Raff H, Findling JW (2007). Limitations of nocturnal salivary cortisol and urine free cortisol in the diagnosis of mild Cushing's syndrome. Eur J Endocrinol 157: 725–731.

Klibanski A (2010). Clinical practice. Prolactinomas. N Engl J Med 362: 1219–1226.

Lamberts SW, Uitterlinden P, Klijn JM (1989). The effect of the long-acting somatostatin analogue SMS 201–995 on ACTH secretion in Nelson's syndrome and Cushing's disease. Acta Endocrinol (Copenh) 120: 760–766.

Lindholm J, Juul S, Jorgensen JO et al. (2001). Incidence and late prognosis of Cushing's syndrome: a population-based study. J Clin Endocrinol Metab 86: 117–123.

Lindsay JR, Jonklaas J, Oldfield EH et al. (2005). Cushing's syndrome during pregnancy: personal experience and review of the literature. J Clin Endocrinol Metab 90: 3077–3083.

Littley MD, Shalet SM, Beardwell CG et al. (1990). Long-term follow-up of low-dose external pituitary irradiation for Cushing's disease. Clin Endocrinol (Oxf) 33: 445–455.

Liu NA, Jiang H, Ben-Shlomo A et al. (2011). Targeting zebrafish and murine pituitary corticotroph tumors with a cyclin-dependent kinase (CDK) inhibitor. Proc Natl Acad Sci U S A 108: 8414–8419.

Luton JP, Mahoudeau JA, Bouchard P et al. (1979). Treatment of Cushing's disease by O, p′DDD. Survey of 62 cases. N Engl J Med 300: 459–464.

Machado MC, de Sa SV, Domenice S et al. (2007). The role of desmopressin in bilateral and simultaneous inferior petrosal sinus sampling for differential diagnosis of ACTH-dependent Cushing's syndrome. Clin Endocrinol (Oxf) 66: 136–142.

McCance DR, Ritchie CM, Sheridan B et al. (1987). Acute hypoadrenalism and hepatotoxicity after treatment with ketoconazole. Lancet 1: 573.

McCormack AI, Wass JA, Grossman AB (2011). Aggressive pituitary tumours: the role of temozolomide and the assessment of MGMT status. Eur J Clin Invest 41: 1133–1148.

Mulligan GB, Eray E, Faiman C et al. (2011). Reduction of false-negative results in inferior petrosal sinus sampling with simultaneous prolactin and corticotropin measurement. Endocr Pract 17: 33–40.

Mulligan GB, Faiman C, Gupta M et al. (2012). Prolactin measurement during inferior petrosal sinus sampling improves the localization of pituitary adenomas in Cushing's disease. Clin Endocrinol (Oxf) 77: 268–274.

Nelson DH, Meakin JW, Dealy Jr JB et al. (1958). ACTH-producing tumor of the pituitary gland. N Engl J Med 259: 161–164.

Newell-Price J, Bertagna X, Grossman AB et al. (2006). Cushing's syndrome. Lancet 367: 1605–1617.

Nieman LK (2002). Medical therapy of Cushing's disease. Pituitary 5: 77–82.

Nieman LK, Ilias I (2005). Evaluation and treatment of Cushing's syndrome. Am J Med 118: 1340–1346.

Nieman LK, Chrousos GP, Oldfield EH et al. (1986). The ovine corticotropin-releasing hormone stimulation test and the dexamethasone suppression test in the differential diagnosis of Cushing's syndrome. Ann Intern Med 105: 862–867.

Nieman LK, Oldfield EH, Wesley R et al. (1993). A simplified morning ovine corticotropin-releasing hormone stimulation test for the differential diagnosis of adrenocorticotropin-dependent Cushing's syndrome. J Clin Endocrinol Metab 77: 1308–1312.

Nieman LK, Biller BM, Findling JW et al. (2008). The diagnosis of Cushing's syndrome: an Endocrine Society clinical practice guideline. J Clin Endocrinol Metab 93: 1526–1540.

Nunes ML, Vattaut S, Corcuff JB et al. (2009). Late-night salivary cortisol for diagnosis of overt and subclinical Cushing's syndrome in hospitalized and ambulatory patients. J Clin Endocrinol Metab 94: 456–462.

Oldfield EH, Chrousos GP, Schulte HM et al. (1985). Preoperative lateralization of ACTH-secreting pituitary microadenomas by bilateral and simultaneous inferior petrosal venous sinus sampling. N Engl J Med 312: 100–103.

Oldfield EH, Doppman JL, Nieman LK et al. (1991). Petrosal sinus sampling with and without corticotropin-releasing hormone for the differential diagnosis of Cushing's syndrome. N Engl J Med 325: 897–905.

Papanicolaou DA, Yanovski JA, Cutler Jr GB et al. (1998). A single midnight serum cortisol measurement distinguishes Cushing's syndrome from pseudo-Cushing states. J Clin Endocrinol Metab 83: 1163–1167.

Patil CG, Prevedello DM, Lad SP et al. (2008). Late recurrences of Cushing's disease after initial successful transsphenoidal surgery. J Clin Endocrinol Metab 93: 358–362.

Pecori Giraldi F, Ambrogio AG, De Martin M et al. (2007). Specificity of first-line tests for the diagnosis of Cushing's syndrome: assessment in a large series. J Clin Endocrinol Metab 92: 4123–4129.

Pereira AM, van Aken MO, van Dulken H et al. (2003). Long-term predictive value of postsurgical cortisol concentrations for cure and risk of recurrence in Cushing's disease. J Clin Endocrinol Metab 88: 5858–5864.

Petit JH, Biller BM, Yock TI et al. (2008). Proton stereotactic radiotherapy for persistent adrenocorticotropin-producing adenomas. J Clin Endocrinol Metab 93: 393–399.

Pivonello R, Faggiano A, Di Salle F et al. (1999). Complete remission of Nelson's syndrome after 1-year treatment with cabergoline. J Endocrinol Invest 22: 860–865.

Pivonello R, De Martino MC, Cappabianca P et al. (2009). The medical treatment of Cushing's disease: effectiveness of chronic treatment with the dopamine agonist cabergoline in patients unsuccessfully treated by surgery. J Clin Endocrinol Metab 94: 223–230.

Raff H (2012). Cushing's syndrome: diagnosis and surveillance using salivary cortisol. Pituitary 15: 64–70.

Raff H, Findling JW (2003). A physiologic approach to diagnosis of the Cushing syndrome. Ann Intern Med 138: 980–991.

Ram Z, Nieman LK, Cutler Jr GB et al. (1994). Early repeat surgery for persistent Cushing's disease. J Neurosurg 80: 37–45.

Rees DA, Hanna FW, Davies JS et al. (2002). Long-term follow-up results of transsphenoidal surgery for Cushing's disease in a single centre using strict criteria for remission. Clin Endocrinol (Oxf) 56: 541–551.

Robinson BG, Hales IB, Henniker AJ et al. (1987). The effect of o, p′-DDD on adrenal steroid replacement therapy requirements. Clin Endocrinol (Oxf) 27: 437–444.

Rollin GA, Ferreira NP, Junges M et al. (2004). Dynamics of serum cortisol levels after transsphenoidal surgery in a cohort of patients with Cushing's disease. J Clin Endocrinol Metab 89: 1131–1139.

Schade R, Andersohn F, Suissa S et al. (2007). Dopamine agonists and the risk of cardiac-valve regurgitation. N Engl J Med 356: 29–38.

Schteingart DE (2009). Drugs in the medical treatment of Cushing's syndrome. Expert Opin Emerg Drugs 14: 661–671.

Semple PL, Laws Jr ER (1999). Complications in a contemporary series of patients who underwent transsphenoidal surgery for Cushing's disease. J Neurosurg 91: 175–179.

Seyer H, Honegger J, Schott W et al. (1994). Raymond's syndrome following petrosal sinus sampling. Acta Neurochir (Wien) 131: 157–159.

Sheehan JM, Vance ML, Sheehan JP et al. (2000). Radiosurgery for Cushing's disease after failed transsphenoidal surgery. J Neurosurg 93: 738–742.

Shimon I, Ram Z, Cohen ZR et al. (2002). Transsphenoidal surgery for Cushing's disease: endocrinological follow-up monitoring of 82 patients. Neurosurgery 51: 57–61, discussion 61–62.

Sonino N, Boscaro M, Paoletta A et al. (1991). Ketoconazole treatment in Cushing's syndrome: experience in 34 patients. Clin Endocrinol (Oxf) 35: 347–352.

Sonino N, Zielezny M, Fava GA et al. (1996). Risk factors and long-term outcome in pituitary-dependent Cushing's disease. J Clin Endocrinol Metab 81: 2647–2652.

Swearingen B, Barker 2nd FG, Zervas NT (1999a). The management of pituitary adenomas: the MGH experience. Clin Neurosurg 45: 48–56.

Swearingen B, Biller BM, Barker 2nd FG et al. (1999b). Long-term mortality after transsphenoidal surgery for Cushing disease. Ann Intern Med 130: 821–824.

Tabarin A, Navarranne A, Guerin J et al. (1991). Use of ketoconazole in the treatment of Cushing's disease and ectopic ACTH syndrome. Clin Endocrinol (Oxf) 34: 63–69.

Terzolo M, Angeli A, Fassnacht M et al. (2007). Adjuvant mitotane treatment for adrenocortical carcinoma. N Engl J Med 356: 2372–2380.

Thompson GB, Grant CS, van Heerden JA et al. (1997). Laparoscopic versus open posterior adrenalectomy: a case-control study of 100 patients. Surgery 122: 1132–1136.

Thompson SK, Hayman AV, Ludlam WH et al. (2007). Improved quality of life after bilateral laparoscopic adrenalectomy for Cushing's disease: a 10-year experience. Ann Surg 245: 790–794.

Thoren M, Adamson U, Sjoberg HE (1985). Aminoglutethimide and metyrapone in the management of Cushing's syndrome. Acta Endocrinol (Copenh) 109: 451–457.

Trainer PJ, Lawrie HS, Verhelst J et al. (1993). Transsphenoidal resection in Cushing's disease: undetectable serum cortisol as the definition of successful treatment. Clin Endocrinol (Oxf) 38: 73–78.

Trilck M, Flitsch J, Ludecke DK et al. (2005). Salivary cortisol measurement – a reliable method for the diagnosis of Cushing's syndrome. Exp Clin Endocrinol Diabetes 113: 225–230.

Tritos NA, Biller BM (2012). Advances in medical therapies for Cushing's syndrome. Discov Med 13: 171–179.

Tritos NA, Biller BM, Swearingen B (2011a). Management of Cushing disease. Nat Rev Endocrinol 7: 279–289.

Tritos NA, Schaefer PW, Stein TD (2011b). Case records of the Massachusetts General Hospital Case 40-2011. A 52-year-old man with weakness, infections, and enlarged adrenal glands. N Engl J Med 365: 2520–2530.

Valassi E, Swearingen B, Lee H et al. (2009). Concomitant medication use can confound interpretation of the combined dexamethasone-corticotropin releasing hormone test in Cushing's syndrome. J Clin Endocrinol Metab 94: 4851–4859.

Valassi E, Biller BM, Swearingen B et al. (2010a). Delayed remission after transsphenoidal surgery in patients with Cushing's disease. J Clin Endocrinol Metab 95: 601–610.

Valassi E, Klibanski A, Biller BM (2010b). Clinical review. Potential cardiac valve effects of dopamine agonists in hyperprolactinemia. J Clin Endocrinol Metab 95: 1025–1033.

van Aken MO, de Herder WW, van der Lely AJ et al. (1997). Postoperative metyrapone test in the early assessment of outcome of pituitary surgery for Cushing's disease. Clin Endocrinol (Oxf) 47: 145–149.

Vella A, Thompson GB, Grant CS et al. (2001). Laparoscopic adrenalectomy for adrenocorticotropin-dependent Cushing's syndrome. J Clin Endocrinol Metab 86: 1596–1599.

Verhelst JA, Trainer PJ, Howlett TA et al. (1991). Short and long-term responses to metyrapone in the medical management of 91 patients with Cushing's syndrome. Clin Endocrinol (Oxf) 35: 169–178.

Vik-Mo EO, Oksnes M, Pedersen PH et al. (2009). Gamma Knife stereotactic radiosurgery of Nelson syndrome. Eur J Endocrinol 160: 143–148.

Witt TC, Kondziolka D, Flickinger JC (1998). Gamma Knife radiosurgery for pituitary tumors. In: LD Lunsford, D Kondziolka, J Flickinger (Eds.), In: Gamma Knife Brain Surgery. Progress in Neurological Surgery. Vol. 14. Karger, Basel, pp. 114–127.

Yanovski JA, Cutler Jr GB, Chrousos GP et al. (1993a). Corticotropin-releasing hormone stimulation following low-dose dexamethasone administration. A new test to distinguish Cushing's syndrome from pseudo-Cushing's states. JAMA 269: 2232–2238.

Yanovski JA, Cutler Jr GB, Doppman JL et al. (1993b). The limited ability of inferior petrosal sinus sampling with corticotropin-releasing hormone to distinguish Cushing's disease from pseudo-Cushing states or normal physiology. J Clin Endocrinol Metab 77: 503–509.

Yanovski JA, Cutler Jr GB, Chrousos GP et al. (1998). The dexamethasone-suppressed corticotropin-releasing hormone stimulation test differentiates mild Cushing's disease from normal physiology. J Clin Endocrinol Metab 83: 348–352.

Yap LB, Turner HE, Adams CB et al. (2002). Undetectable postoperative cortisol does not always predict long-term remission in Cushing's disease: a single centre audit. Clin Endocrinol (Oxf) 56: 25–31.

Zada G, Governale LS, Laws Jr ER (2010). Intraoperative conversion from endoscopic to microscopic approach for the management of sellar pathology: incidence and rationale in a contemporary series. World Neurosurg 73: 334–337.

Zanettini R, Antonini A, Gatto G et al. (2007). Valvular heart disease and the use of dopamine agonists for Parkinson's disease. N Engl J Med 356: 39–46.

Handbook of Clinical Neurology, Vol. 124 (3rd series)
Clinical Neuroendocrinology
E. Fliers, M. Korbonits, and J.A. Romijn, Editors

Chapter 16

Craniopharyngioma

HERMANN L. MÜLLER*
Department of Pediatrics, Klinikum Oldenburg, Medical Campus University Oldenburg, Oldenburg, Germany

INTRODUCTION

Craniopharyngiomas are rare embryogenic malformations of the sellar and parasellar area with low-grade histological malignancy. Despite high survival rates (87–95% in recent series), quality of life is frequently impaired in long-term survivors due to sequelae caused by the anatomic proximity of craniopharyngiomas to the optic nerve and hypothalamic–pituitary axes (Karavitaki et al., 2006; Müller, 2008, 2010a, b; Wisoff, 2008). Any meaningful improvement in the prognosis of craniopharyngioma patients will require development of risk-adapted neurosurgical and radio-oncological treatment strategies in a multidisciplinary setting that provides medical as well as psychological support for these patients. Due to its rare occurrence, high survival rate, and adverse quality of life effects, recent multicenter cooperation has already led to beneficial results (Wisoff, 2008; Müller, 2010a).

EPIDEMIOLOGY AND PATHOLOGY

Craniopharyngioma is a nonglial intracranial tumor derived from a malformation of embryonal tissue (Garrè and Cama, 2007). There are controversial hypotheses on its embryonal origin: originating from ectodermal remnants of Rathke's pouch versus originating from residual embryonal epithelium of the anterior pituitary gland and of the anterior infundibulum (Garrè and Cama, 2007).

Craniopharyngioma incidence is 0.5–2 cases per million persons per year, with 30–50% of all cases presenting during childhood and adolescence (Bunin et al., 1998; Nielsen et al., 2011). Craniopharyngioma represents 1.2–4% of all childhood intracranial tumors. In childhood and adolescence, its histologic type is usually adamantinomatous with cyst formation (Müller-Scholden et al., 2000; Rushing et al., 2007). Typical incidence of adult onset craniopharyngioma is between the ages of 50 and 75 years, usually presenting with a squamous-papillary histologic type. More than 70% of the predominantly childhood craniopharyngioma adamantinomatous type bear a mutation of the β-catenin gene, which is not detectable in the adult papillary type of craniopharyngioma (Holsken et al., 2010). Recently, a new mouse model of craniopharyngioma due to an activation of Wnt signalling has been published (Gaston-Massuet et al., 2011).

The German Pediatric Cancer Registry (DKKR) systematically documents cases of craniopharyngioma according to international guidelines (Müller et al., 2003b). Their data from 1980 to 2007 obtained for 496 craniopharyngioma patients diagnosed ≤ 18 years reveals most (451; 91%) were younger than 15 years of age at diagnosis, with a 1:1 sex ratio and median age at diagnosis of 8.8 years. The 1980–2007 (contemporary) survival rate is 97% after 3 years from diagnosis, 96% after 5 years, and 93% after 10 years. Patients who developed the disease in the 1980s had a lower survival rate than those diagnosed in the 1990s (survival at 5 years, 91% versus 98%) (Sherlock et al., 2010) (Müller et al., 2003b).

CLINICAL MANIFESTATIONS AT THE TIME OF DIAGNOSIS

The diagnosis of craniopharyngioma is often made late – sometimes years after the initial appearance of symptoms (Müller et al., 2003b) – with the clinical picture at time of diagnosis often dominated by nonspecific manifestations of intracranial pressure (e.g., headache and nausea). Further primary manifestations are visual impairment (62–84%) and endocrine deficits (52–87%) (Fig. 16.1). Endocrine deficits are caused by disturbances to the hypothalamic–pituitary axis and affect growth

*Correspondence to: Hermann L. Müller, MD, Department of Pediatrics, Klinikum Oldenburg, Medical Campus University Oldenburg, Rahel-Straus-Strasse 10, 26133 Oldenburg, Germany. Email: mueller.hermann@klinikum-oldenburg.de, homepage: www.kraniopharyngeom.net

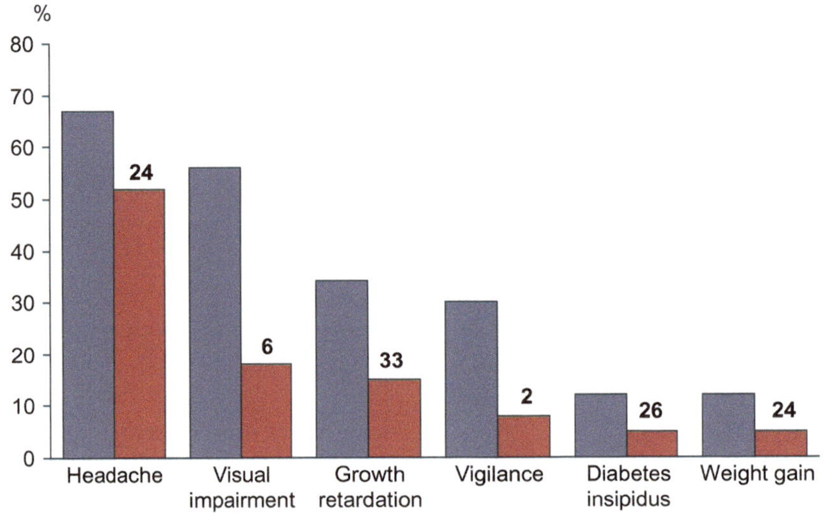

Fig. 16.1. Manifestations before diagnosis of craniopharyngioma in children and adolescents. Frequency of occurrence of each manifestation before diagnosis (blue) and frequency of occurrence as the initial manifestation (red). The median time (months) from the appearance of each initial manifestation until diagnosis is indicated above each red column. In the overall group, the median time from the initial manifestation of disease until diagnosis was 12 months, with a range of 0.01–96 months. (Modified from Müller et al., 2003b, with the kind permission of Springer.)

hormone secretion (75%), gonadotropins (40%), adrenocorticotropic hormone (ACTH) (25%), and thyroid-stimulating hormone (TSH) (25%). At diagnosis, 40–87% of patients present with at least one hormonal deficit (Hoffman et al., 1992; Caldarelli et al., 2005; Müller, 2008) and other endocrine symptoms such as neurohormonal diabetes insipidus present preoperatively in 17–27% of patients (Hoffman et al., 1992; Müller, 2008; Elliott and Wisoff, 2010). An analysis of anthropometric data obtained in routine checkups before the diagnosis of craniopharyngioma in 90 children (Müller et al., 2004) revealed that a pathologically reduced growth rate – an early manifestation of the disease – presents in patients as young as 12 months, but that an increase in weight, predictive of hypothalamic obesity, tends to occur as a later manifestation, shortly before diagnosis. Recent literature has documented that the clinical combination of headache, visual impairment, decreased growth rate, and polydipsia/polyuria should arouse suspicion of craniopharyngioma in the differential diagnosis (Müller, 2010a).

IMAGING STUDIES

Both computed tomography (CT) and magnetic resonance imaging (MRI) reveal that childhood craniopharyngioma is typically a cystic tumor of the intra- and/or suprasellar region. The most common localization is suprasellar, with an intrasellar portion; only 20% are exclusively suprasellar and even fewer (5%) exclusively intrasellar (Warmuth-Metz et al., 2004). CT is the only way to definitively detect or exclude calcification, which is found in approximately 90% of these tumors. The signal intensity of craniopharyngioma in MRI is highly variable because it depends on the protein content of the cysts. Solid tumor portions and cyst membranes appear isointense in T1-weighted images, often with a mildly heterogeneous structure (Fig. 16.2). The combination of solid, cystic, and calcified tumor components is an important radiologic clue to the diagnosis. MRI before and after gadolinium application is the standard imaging for detection of craniopharyngioma, further imaged by unenhanced CT to detect calcifications (Warmuth-Metz et al., 2004). After preoperative detection of calcifications and complete resection confirmed by postoperative MRI, postsurgical unenhanced CT of the sellar/parasellar area is recommended for definitive confirmation of complete resection (Warmuth-Metz et al., 2004).

TREATMENT STRATEGIES

Neurosurgery: strategies and effects

Before surgery, a tumor-related impairment of cerebrospinal fluid flow often causes hydrocephalus, which can be of varying severity (Choux and Lena, 1979). Resection of the tumor is the preferred treatment for restoring normal cerebrospinal fluid flow, but a pre-resection shunt operation may also be required. For large cystic craniopharyngioma, particularly in infancy or early childhood, the stereotactic or open implantation of an intracystic catheter is a valuable treatment option, both for the relief of pressure and, in some cases, for a possible intracystic

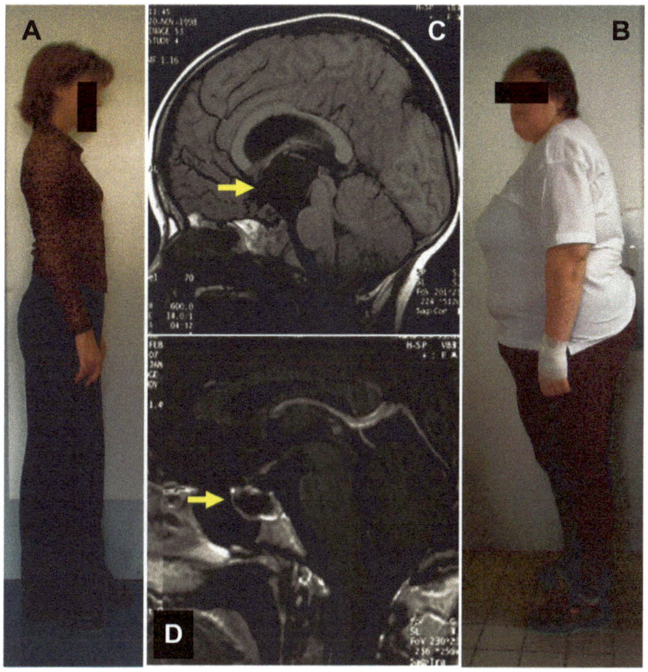

20 y. 32 y.

Fig. 16.2. Degree of obesity in relation to the location of childhood craniopharyngioma. In both patients craniopharyngioma (as indicated by arrow on magnetic resonance imaging before surgery) could be completely resected. Both patients had complete hypopituitarism after surgery requiring endocrine substitution of all hypothalamic–pituitary axes. The patient depicted in (**B**) developed severe obesity due to hypothalamic lesions of suprasellar parts of craniopharyngioma (**C**). The patient depicted in (**A**) presented with a small tumor confined to the sellar region (**D**). After complete resection she kept normal weight without any eating disorders. (Modified from Müller et al., 2003b, with the kind permission of Springer.)

instillation of sclerosing substances. The stereotactic or open surgical implantation of an intracystic catheter with a subcutaneous reservoir can also be useful in reducing cyst volume while waiting for optimal timing of radiotherapy or surgical resection. For patients with large cysts whose pressure causes preoperative visual impairment, the recommended two-staged approach is cyst drainage to relieve pressure and improve vision, followed by resection (Fahlbusch et al., 1999). The surgical approach is dictated by localization and extent of the tumor, the standard being a right frontotemporal approach. Because pediatric craniopharyngiomas typically extend to the suprasellar area, these are best removed through a transcranial approach. Intrasellar tumors can be operated via the transsphenoidal route, which for topographic-anatomic reasons has the critical advantage of preserving hypothalamic functions (Zona and Spaziante, 2006). Neuroendoscopic routes to craniopharyngiomas such as transnasal-transsphenoidal, transventricular, and the supraorbital approach can only be used for smaller, primarily intrasellar tumors (Elliott et al., 2011).

For favorably localized tumors, the treatment of choice is an attempt at complete resection with preservation of visual, hypothalamic, and pituitary function (Choux and Lena, 1979). For unfavorably localized tumors too close to or too entangled with the optic nerve and/or the hypothalamus, controversy exists over whether complete resection should still be attempted or whether a planned limited resection (biopsy, partial/subtotal resection) should be performed. Many authors take a critical view of planned radical resection in these cases because of the risk of surgically induced deficits (mainly hypothalamic) and the high rate of recurrence in infants and small children despite apparently complete resection (Choux and Lena, 1979). Whereas following incomplete resection the residual tumor shows progression in 71–90% of patients, the rate of progression after incomplete resection followed by radiotherapy is 21% (Becker et al., 1999). Ultimately, whatever the preoperative intention, the final decision regarding the extent of surgical resection can only be made intraoperatively by the neurosurgeon.

The published literature to date (Hetelekidis et al., 1993; Rajan et al., 1993; Merchant et al., 2002; Karavitaki et al., 2005; Müller, 2006; Puget et al. 2006; Vinchon et al., 2009; Elliott et al., 2011) has not settled the controversy over the best treatment strategy for craniopharyngioma in children and adolescents (intended primary radical resection versus biopsy/partial resection followed by radiotherapy). Therapeutic consequences of irradiation and surgery also remain controversial. Above all, effects of treatment sequence (immediate irradiation versus progression-contingent irradiation of residual tumor) on quality of life are unclear in the retrospective data published to date. A retrospective analysis cited primary therapy FSIQ (full scale IQ) losses of 9.8 points after a single complete resection compared to a loss of 1.25 points after limited resection followed by irradiation (Merchant et al., 2002). For a repeat surgical intervention carried out following a relapse, the loss was 13.1 points, statistically suggesting that radical and/or repeated surgeries seem to generate negative influences on neurocognitive functions compared to limited surgical intervention plus immediate radiation treatment. The fact of the matter is that only very limited available retrospective data exist.

The German multicenter, prospective surveillance study of children and adolescents with craniopharyngioma KRANIOPHARYNGEOM 2000 was designed to address these issues. Newly diagnosed patients (117; 2001–2006) from Germany, Austria, and Switzerland were entered into the prospective evaluation. An analysis of 3 year, event-free survival rates (EFS) revealed frequent early events (i.e., tumor progression after incomplete resection and tumor relapse after complete resection in the first 3 years after primary surgery) (Fig. 16.3). In a multivariate analysis on risk factors for the observed low EFS rates, the authors found that the risk for relapses after complete resection was 80% lower compared to incomplete resection. The risk for progression was 88% lower in irradiated patients when compared with patients without or before irradiation (Müller et al., 2010).

Any discussion of treatment and post-treatment strategies must take into account the late sequelae and quality of life experienced by the patients after treatment. A follow-up study of quality of life in children after complete resection of craniopharyngioma revealed that quality of life depends to a statistically significant extent on the experience of the operating neurosurgeon (Sanford, 1994).

Irradiation

The site and rate of progression of the tumor, as well as the patient's age, are important considerations when deciding whether reoperation and/or radiotherapy should be performed. Craniopharyngioma tumors are sharply bordered in the imaging. In contrast to primary brain tumors, they tend towards less infiltrative growth, permitting a small safety margin of 5 mm maximum. These biological characteristics also allow the option of using high-precision, three-dimensional conformation technology. A conventional, fractionated irradiation target (total) volume dose of 54 Gy has been established worldwide (Becker et al., 1999; Habrand et al., 2006; Merchant et al., 2006; Minniti et al., 2009).

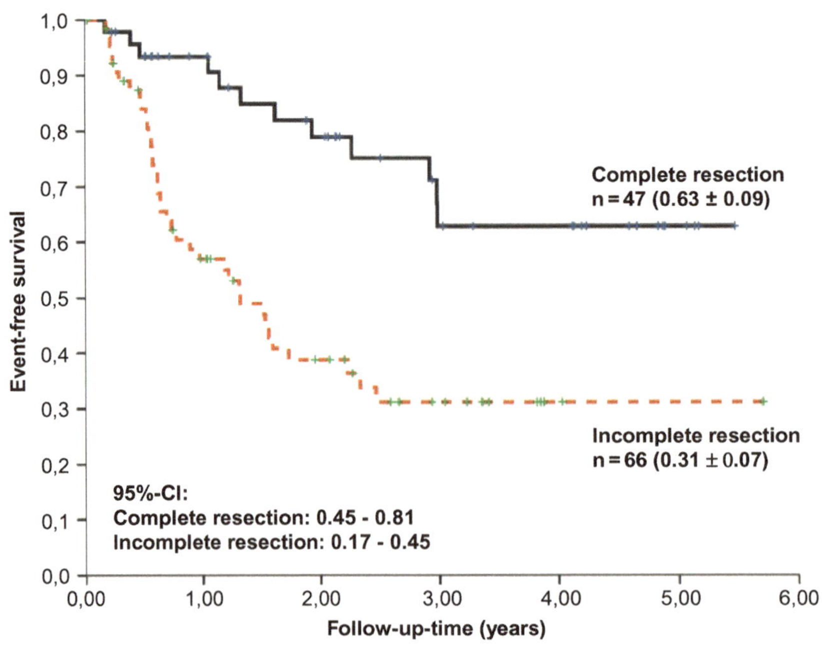

Fig. 16.3. Kaplan–Meier analyses of event-free survival rates (EFS) depending on the extent of resection among the 117 craniopharyngioma patients recruited in the KRANIOPHARYNGEOM 2000 trial. (Modified from Müller et al., 2006c, with the kind permission of Karger.)

Proton beam therapy

Proton beams have an "inverse dose profile" across the tissues, whereby the dose released by the particles increases with penetration depth until reaching a maximum at the end of the particle range (Bragg peak). Beyond the Bragg peak, practically no dose is deposited.

Fitzek and colleagues published the first series of 15 craniopharyngioma patients treated with combined proton–photon irradiation for residual or recurrent disease (Fitzek et al., 2006). Actuarial 5- and 10-year local control rates were 93% and 85%, respectively, with 10 year survival expectancy in 72% of patients. No treatment-related neurocognitive deficits were recorded; functional status, academic skills and professional abilities were unaltered after proton beam therapy. Luu and colleagues published a preliminary report on 16 patients treated with proton beam therapy (Luu et al., 2006). Local tumor control was achieved in 14 of 16 (87.5%) patients. During follow-up (12–121 months), late sequelae included newly diagnosed panhypopituitarism, a cerebrovascular accident, and an out-of-proton-field posterior fossa meningioma (59 months following proton beam therapy administered to patient who previously received photon radiotherapy). One study of proton beam therapy in craniopharyngioma assessed cyst growth during the treatment course: 24% of patients demonstrated cyst enlargement and 5% cyst reduction requiring modification of the treatment plan, while one patient required cyst drainage during treatment (Winkfield et al., 2009).

Clinical outcome data are very limited for assessing the value of proton beam therapy compared to modern photon therapy, as the technique is available in only a few centers. However, proton beam therapy is being refined with spot scanning technology in order to deliver intensity-modulated proton therapy (IMPT) with potential advantages of better conformation of dose to the target volume, sparing of critical structures, reduced integral dose, and lower dose of secondary neutrons, which should reduce the risk of secondary malignancies (Beltran et al., 2012; Boehling et al., 2012). Data should become available in due course with sufficient follow-up time and as more centers are able to deliver this promising therapy.

Radiosurgery

The most frequently used system for delivery of single fraction radiotherapy is the Gamma Knife (Lee et al., 2008). The Gamma Knife requires the patient to be immobilized using a stereotactic fixed frame and delivers the treatment in a single radiosurgery session. Generally, patients treated with radiosurgery had small (<3 cm), mainly solid tumors, which were well circumscribed on imaging and sited > 3 mm away from critical structures such as brainstem, optic chiasm, and optic nerves. Dose constraints for radiosurgery applied to the optic chiasm and brainstem were 8–9 Gy and 12–14 Gy, respectively (Mokry, 1999; Chung et al., 2000; Amendola et al., 2003; Kobayashi, 2009). In published Gamma Knife series, tumor control rates ranged from 67 to 94%. Rates of complications directly attributable to Gamma Knife radiosurgery ranged from 0 to 38%, including visual deterioration (0–38%, endocrine morbidity (0–19%), and neurologic complications (0–2%). No treatment-related mortality has been reported (Mokry, 1999; Amendola et al., 2003; Kobayashi, 2009).

Hypofractionated stereotactic radiotherapy: CyberKnife

The CyberKnife is a compact 6 MV pencil beam linear accelerator mounted on a robotic arm with 6 degrees of freedom of movement and is equipped with an image-guided system for target tracking. Data on the efficacy and tolerability of CyberKnife in craniopharyngioma are limited (Lee et al., 2008). Tumor shrinkage was observed in seven patients and stabilized disease in three. No deterioration of vision or neuroendocrine function was recorded within the limited follow-up period (Lee et al., 2008).

Intracavitary β irradiation

Stereotactic instillation of radioisotopes has been discussed as an alternative therapeutic option, mainly for monocystic craniopharyngioma recurrences. Nevertheless, this treatment method is restricted to cystic craniopharyngioma and should be considered only for postoperative recurrences and after percutaneous irradiation (Julow et al., 1989; Becker et al., 1999; Szeifert et al., 2007).

Instillation of sclerosing substances for cystic recurrent tumors

The implantation of an intracystic catheter with a subcutaneous reservoir enables the possibility of repeated decompression of the cyst. Even though it often relieves pressure only transiently, it is a useful therapeutic method for cystic recurrent tumors whose anatomic configuration and localization makes them difficult to resect (Cavalheiro et al., 1996; Kim et al., 2007). The instillation of sclerosing substances such as bleomycin in craniopharyngioma cysts, using an intracystic catheter implanted by a stereotactic or open procedure, was used in such cases (Schubert et al., 2009). Severe neurotoxic side effects were observed in cases due to cystic leakage of bleomycin into cerebrospinal fluid (Lafay-Cousin et al., 2007). Recent reports on the effect and tolerability of intracystic instillation of interferon α are promising (Cavalheiro et al., 2010).

SEQUELAE

Pituitary deficiencies

Pituitary hormone deficiencies are common in childhood craniopharyngioma. At diagnosis, 40–87% of patients (Hoffman et al., 1992; Caldarelli et al., 2005; Müller, 2008) present with at least one hormonal deficit and 17–27% (Hoffman et al., 1992; Müller, 2008; Elliott and Wisoff, 2010) have been reported to have diabetes insipidus. The rate of postsurgical pituitary hormone deficiencies increases due to the tumor's proximity or even entanglement with the hypothalamic–pituitary axes (Hoffman et al., 1992; De Vile et al., 1996a; Merchant et al., 2002; Müller et al., 2004; Caldarelli et al., 2005; Elliott and Wisoff, 2010; Jung et al., 2010; Steno et al., 2011), transient diabetes insipidus occurring in 80–100% cases in some series after surgery (Poretti et al., 2004; Caldarelli et al., 2005). Permanent diabetes insipidus after surgery ranges between 40 and 93% (Hoffman et al., 1992; De Vile et al., 1996a; Merchant et al., 2002; Poretti et al., 2004; Caldarelli et al., 2005; Ahmet et al., 2006; Elliott and Wisoff, 2010; Elliott et al., 2011; Müller et al., 2011a).

Growth hormone deficiency has been described at diagnosis in 26–75% of childhood craniopharyngiomas (Ahmet et al., 2006; Müller, 2008), and impaired growth may occur years before diagnosis (Müller et al., 2004). Growth hormone deficiency following treatment for childhood craniopharyngioma is found in about 70–92% of patients (Müller et al., 2004; Halac and Zimmerman, 2005; Crom et al., 2010; Müller et al., 2011a). A positive response to growth hormone treatment is seen in most cases (Geffner et al., 2004).

That said, normal growth in childhood craniopharyngioma without growth hormone is reported in the literature (Srinivasan et al., 2004). In fact, childhood patients with hypothalamic involvement were found to achieve normal adult height more often than those without hypothalamic involvement (Müller et al., 2004). Even though this phenomenon of "growth without growth hormone" was described in childhood craniopharyngioma almost five decades ago (Matson, 1964), the physiology of growth in these cases is not fully understood, although insulin and/or leptin is suspected to play a compensating role. Both have been hypothesized to induce growth in the fetus and in obese children (Costin et al., 1976; Geffner, 1996; Phillip et al., 2002), with leptin reported to function as a bone growth factor acting directly at the level of bone growth centers, independently of growth hormone (Phillip et al., 2002). Mechanisms by which insulin stimulates growth include its known anabolic effects. At high levels it may bind to the type 1 insulin-like growth factor (IGF) receptor and induce growth, mediated by its actions to decrease IGF-binding protein 1 levels, resulting in increased levels of free IGF-1 (Phillip et al., 2002). In support of this theory, obese patients with childhood craniopharyngioma were found to present with higher height Standard deviation score (SDS) at diagnosis and at last follow-up with no difference in hormonal substitution, including growth hormone (Müller et al., 2001). In contrast, another study found that children who were growing despite growth hormone deficiency were not different from those requiring growth hormone substitution in terms of anthropometric measures, body composition, and metabolic indexes, including insulin levels (Srinivasan et al., 2004). Because these are pediatric patients, sex hormones may also play a role in inducing growth (Phillip et al., 2002).

Neurologic and visual outcomes

Due to frequent suprasellar tumor localization, visual deficits (both visual acuity and visual fields) in childhood craniopharyngioma are relatively common. Visual impairment was found as an initial manifestation of childhood craniopharyngioma in more than half of the patients (Müller, 2008), with some postsurgical improvement of vision in 41–48% of patients (Caldarelli et al., 2005; Elliott et al., 2011). Risk factors for postsurgical visual impairment include severe presurgical visual deficits and tumor localization in the prechiasmatic area (Caldarelli et al., 2005; Steno et al., 2011). Improved results were found in cases treated using the transsphenoidal approach (Elliott et al., 2011), but such an approach is limited to resection of intrasellar tumors. Because pediatric craniopharyngiomas typically extend to the suprasellar area, these are best removed through a transcranial approach.

Neurologic abnormalities include hemiparesis, cranial nerve deficits, cerebrovascular disease manifestations, epilepsy, learning problems, and headaches (Merchant et al., 2002; Crom et al., 2010; Elliott and Wisoff, 2010). A significant part of these are transient and the total prevalence of long-term neurologic complications is reported to be 8% (Caldarelli et al., 2005); however, this rises to 36% for large-sized tumors (Elliott and Wisoff, 2010) and 30% when including both visual and neurologic complications (Poretti et al., 2004).

Hypothalamic dysfunction

Symptoms related to hypothalamic dysfunction, such as obesity, daytime sleepiness, disturbed circadian rhythm and sleep irregularities, behavioral changes, and imbalances in regulation of thirst, body temperature, heart rate and/or blood pressure have been found at diagnosis in 35% of childhood craniopharyngioma patients (Elliott and Wisoff, 2010). The rate of hypothalamic dysfunction dramatically increases following treatment, in some series up to 65–80% (Poretti et al., 2004; Elliott and Wisoff, 2010). Even though presurgical evaluation of hypothalamic damage has proved difficult, both clinically and radiologically (Steno et al., 2011), tumor involvement of the third ventricle or obstructive hydrocephalus are suggestive findings (Caldarelli et al., 2005). A three-level clinical grading system for hypothalamic dysfunction has been suggested based on the degree of obesity and existence of other manifestations (de Vile et al., 1996b).

Associated with high morbidity, suprachiasmatic lesions are difficult to treat. Surgical removal of tumor tissue beyond the mammillary bodies endangers hypothalamic structures and may cause hypothalamic obesity (Müller, 2011; Müller et al., 2011a). With the aid of imaging studies, recent reports have indicated that the degree of obesity of affected patients is positively correlated with the degree and extent of hypothalamic damage (de Vile et al., 1996a; Holmer et al., 2009, 2010; Müller et al., 2011a). Taking these considerations into account, a novel classification of presurgical involvement and postsurgical lesions of hypothalamic structures based on magnetic resonance imaging was recently published (Flitsch et al., 2011). The classification might help to establish more risk-adapted surgical strategies (Fig. 16.4).

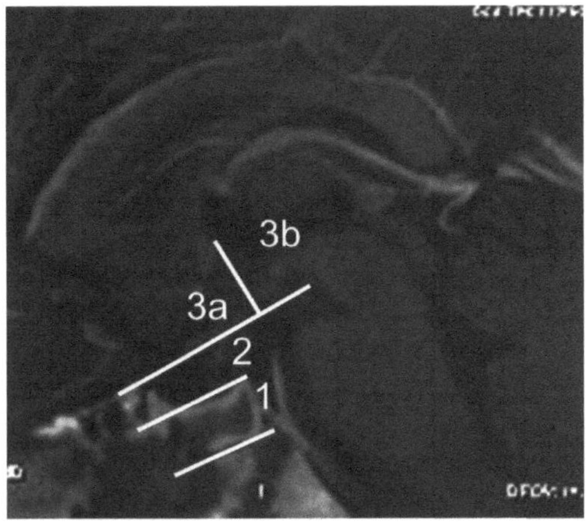

Fig. 16.4. Sagittal magnetic resonance imaging (MRI) of the midline. Suggestion of a classification system of craniopharyngiomas by preoperative MRI criteria. The intra- and suprasellar region can be divided into three sections. Section 1 is limited by the diaphragma sella; section 2 is below the optic chiasm and the mammillary bodies; and section 3 is above the chiasm and mammillary bodies, subdivided into an area anterior and posterior to the mammillary bodies. In this particular patient, a transsphenoidal surgery of a type 1 craniopharyngioma had been performed previously. Section 1 is usually reached by the transsphenoidal route, whereas sections 3a and 3b are mostly reserved for transcranial procedures. Depending on the tumor extension, section 2 can be reached by transcranial as well as transsphenoidal procedures. (Modified from Flitsch et al., 2011, with the kind permission of the authors and Frontiers Endocrinology.)

Obesity and eating disorders

Of the manifestations of hypothalamic damage in childhood craniopharyngioma, rapid weight gain is one of the most perplexing complications. Weight gain in childhood craniopharyngioma patients often occurs years before diagnosis (Müller et al., 2004), with 12–19% of patients reported to be obese at diagnosis (Hoffman et al., 1992; Poretti et al., 2004; Ahmet et al., 2006; Müller, 2008). Weight gain occurs despite adequate endocrine replacement of pituitary hormone deficiencies. The hypothalamic disturbance in energy management contributes to obesity and is exacerbated by factors limiting physical activity such as marked daytime sleepiness and disturbances of circadian rhythms (Müller et al., 2002). The degree of obesity frequently increases early after treatment and rapid weight gain occurs the first 6–12 months after treatment (Müller et al., 2001; Ahmet et al., 2006; Holmer et al., 2010). Following treatment, the prevalence of obesity is higher, reaching up to 55% (Hoffman et al., 1992; Müller et al., 2001; Müller et al., 2003a; Poretti et al., 2004; Srinivasan et al., 2004; Ahmet et al., 2006; Lek et al., 2010; O'Gorman et al., 2010; Elliott et al., 2011). Obesity and eating disorders result in increased risks of metabolic syndrome (Srinivasan et al., 2004) and cardiovascular disease (Holmer et al., 2009), including sudden death events (Mong et al., 2008), multisystem morbidity (Pereira et al., 2005) and mortality (Scott et al., 1994; Fisher et al., 1998; Khafaga et al., 1998; Kalapurakal et al., 2003; Poretti et al., 2004; Tomita and Bowman, 2005; Habrand et al., 2006; Lin et al., 2008; Visser et al., 2010).

Although the relation of obesity with hypothalamic damage is obvious in childhood craniopharyngioma (Lustig et al., 2003a; Holmer et al., 2009, 2010), the mechanisms responsible for increased prevalence of cardiometabolic complications in these patients are unclear. It is likely that in case of suprasellar extension, hypothalamic function will be compromised and will remain compromised to a certain extent when treated surgically or with irradiation. Although it is a relatively small structure of only 4 mL, the hypothalamus contains many groups of nerve cell bodies forming distinct nuclei, which have highly diverse molecular, structural, and functional organizations (Swaab et al., 1992). The hypothalamus plays a major role in keeping the internal environment stable by synchronizing circadian rhythms and biological clock mechanisms. Recent data indicate that a proper balance of the autonomic nervous system is crucial for metabolism. It is well known that adipose tissue is richly innervated by sympathetic nerve fibers that control lipolysis. It now appears that lipogenesis is also controlled by parasympathetic innervation of adipose tissue originating from separate sympathetic and parasympathetic neurons in the periventricular nucleus and suprachiasmatic nucleus (Kreier et al., 2002). Such a high level of differentiation puts the suprachiasmatic nucleus in a key position to balance circadian activity of both branches of the autonomous nervous system. Considering the large proportion of patients with damage to suprasellar structures, it is likely that craniopharyngiomas and/or the effects of treatment damage the suprachiasmatic hypothalamic nucleus. This in turn affects regulation of central clock mechanisms, which predisposes to alterations in metabolism. Clearly, surgical strategies to preserve hypothalamic integrity are mandatory for the prevention of sequelae such as severe obesity owing to hypothalamic lesions.

When elevated serum leptin levels relative to body mass index (BMI) were found in childhood craniopharyngioma patients with a suprasellar tumor extension (Roth et al., 1998), researchers suggested that normal appetite inhibition failed to occur in these patients due to disruption of hypothalamic receptors that regulates negative feedback loop in which leptin, formed in adipocytes, binds to hypothalamic leptin receptors. However, a study involving self-assessment by nutritional diaries revealed that hypothalamic obesity can occur in patients with childhood craniopharyngioma even when their caloric intake is similar to controls matched for BMI (Harz et al., 2003).

Physical activity and energy expenditure

An analysis of physical activity by accelerometric measurements showed that patients with childhood craniopharyngioma had a markedly lower level of physical activity than healthy controls matched for BMI (Harz et al., 2003). Marked daytime sleepiness and disturbances of circadian rhythms have been demonstrated in patients with childhood craniopharyngioma and obesity (Müller et al., 2002). Daytime sleepiness and obesity in these patients were both correlated with low nocturnal and early morning melatonin levels in saliva. The proposed pathogenic mechanism involves impaired hypothalamic regulation of circadian melatonin rhythms in patients with craniopharyngioma extending to the suprasellar area. Initial experiences with melatonin substitution in patients with childhood craniopharyngioma were promising: melatonin levels normalized and daytime sleepiness and physical activity improved (Müller et al., 2006b). However, data on the long-term effect of melatonin substitution on weight development have not yet been published.

Polysomnographic studies in patients with childhood craniopharyngioma and severe daytime sleepiness have revealed sleeping patterns typical for hypersomnia and secondary narcolepsy (Müller et al., 2006a; Müller, 2010c; O'Gorman et al., 2010). Treatment with central stimulating agents (methylphenidate, modafinil) had a significantly beneficial effect on daytime sleepiness in these patients (Müller et al., 2006a). Secondary narcolepsy should be taken into consideration as a pathogenic factor in severely obese children and adolescents with craniopharyngioma. Mason et al. (2002) treated five patients with childhood craniopharyngioma and hypothalamic obesity (age range: 6.0–9.8 years) with the central stimulating agent dexamfetamine, for the purpose of weight reduction. Dexamfetamine therapy stabilized patient BMI and parents reported noticeable improvements in their child's physical activity and alertness.

A decreased metabolic rate, in terms of both resting and total energy expenditure, has been suggested to contribute to weight gain in this population. Adults and pediatric patients with childhood onset craniopharyngioma were found to have a lower resting energy expenditure (REE) compared to controls (Shaikh et al., 2008; Holmer et al., 2010; Kim et al., 2010) that was not explained by differences in body composition. This energy intake/REE ratio was significantly lower in those with tumors involving the third ventricle (Holmer et al., 2010). Decreased physical activity might contribute to overall lowering of total energy expenditure (Harz et al., 2003; Shaikh et al., 2008; Holmer et al., 2009, 2010). Further factors that could potentially contribute to decreased physical activity are neurologic and visual deficits, increased daytime sleepiness, and psychosocial difficulties.

Autonomous nervous system

Lustig and colleagues (2003b) have postulated that hypothalamic disinhibition of vagal output is a cause of increased β cell stimulation in patients with childhood craniopharyngioma, and that this disinhibition leads to hyperinsulinism and severe obesity. They therefore studied treatment with the somatostatin analog octreotide, which suppresses β cell activity. Roth and colleagues (2007) hypothesized that decreased physical activity and severe obesity in patients with childhood craniopharyngioma could be related to impaired central sympathetic output, based on reduced urine concentrations of catecholamine metabolites correlating with the degree of obesity and the level of physical activity.

Appetite regulation

Roth and colleagues (2011) recently analyzed the gastrointestinal hormones ghrelin and peptide YY and their effect on satiety in patients with childhood craniopharyngioma and obesity. Their findings support the hypothesis that reduced ghrelin secretion and reduced postprandial suppression of ghrelin in patients with craniopharyngioma and severe obesity leads to disturbed regulation of appetite. Peptide YY levels did not differ between normal weight, obese, and very obese patients with childhood craniopharyngioma. A possible pathogenic role of peripheral α-melanocyte-stimulating hormone in childhood craniopharyngioma obesity has also been reported (Roth et al., 2010).

Pharmacologic treatment of hypothalamic obesity

Due to the above-reported disturbances in energy expenditure, central sympathetic output, and appetite regulation, craniopharyngioma patients with hypothalamic obesity typically develop morbid obesity that is mainly unresponsive to conventional lifestyle modifications (diet and exercise). Based on impairment of sympathoadrenal activation and epinephrine production manifesting as a reduced hormonal response to hypoglycemia, treating this disorder with amfetamine derivates has been suggested (Schofl et al., 2002; Coutant et al., 2003). Use of dexamfetamine started at 10 months post surgical intervention for craniopharyngioma and lasting for 24 months was shown to diminish continuous weight gain and stabilize BMI (Mason et al., 2002); importantly, spontaneous physical activity increased significantly. Even shorter periods of dexamfetamine treatment have caused a subjective improvement in daytime sleepiness (Ismail et al., 2006).

Sibutramine is a neurotransmitter reuptake inhibitor that reduces the reuptake of serotonin, norepinephrine, and dopamine, thereby increasing the levels of these substances in synaptic clefts and helping enhance satiety. Sibutramine has been widely used to treat obesity, leading to a weight loss of 7–10% when combined with a regulated diet. As part of a group with several obesity syndromes, sibutramine was tested in a randomized placebo-controlled cross over trial (Danielsson et al., 2007). While the drug was well tolerated and safe, the weight loss response in patients with hypothalamic obesity was less pronounced in comparison to other participants (such as trisomy 21 and Prader–Willi syndrome). While the effect on BMI was promising, the drug has been taken off the market and further clinical trials are not expected.

Childhood craniopharyngioma patients with hypothalamic obesity have a "parasympathetic predominance" of the autonomic nervous system induced by vagal activation and manifesting as daytime sleepiness, and reduced body temperature and heart rate (Lustig, 2008). Parasympathetic stimulation causes insulin secretion by way of direct activation of β cells as well as promoting adipogenesis. As insulin is an anabolic hormone, it has been suggested as an important driver of weight gain in hypothalamic obesity. Octreotide is a somatostatin analog and thus causes reduction in insulin secretion. Lustig and colleagues used octreotide in a double-blind randomized controlled study in children with hypothalamic obesity and demonstrated moderate reductions in weight gain (Lustig et al., 2003b) and showed that insulin levels during a proof-of-concept oral glucose tolerance test decreased without leading to major changes in glucose tolerance. This study was followed by a larger trial performed using octreotide LAR in 60 patients with cranial surgical interventions that led to hypothalamic obesity (http://clinicaltrials.gov/ct2/show/NCT00076362). This 6 month intervention showed no efficacy at all in changing BMI and the open label segment of this study was terminated earlier than planned due to an increased risk of gallstone formation.

Bariatric treatment of hypothalamic obesity

Initial experiences with bariatric surgery in severely obese childhood craniopharyngioma patients achieved sufficient tolerability and short-term weight reduction (Inge et al., 2007; Müller et al., 2007). An instant improvement of binge-eating behavior in patients with childhood craniopharyngioma immediately after laparoscopic adjustable gastric banding (LAGB) was observed, but failed in long-term weight reduction. Nevertheless, weight stabilization could be achieved with regular follow-up monitoring (Müller et al., 2011b). Treatment with invasive, non-reversible bariatric methods such as gastric bypass is controversial in the pediatric population because of medical, ethical, and legal considerations (Rottembourg et al., 2009; Schultes et al., 2009; Müller et al., 2011b).

Despite the availability of these promising therapeutic approaches, it must be emphasized that currently no generally accepted (pharmacologic or bariatric) therapy for hypothalamic obesity in craniopharyngioma has been shown to be effective in randomized studies.

Quality of life, neurocognitive outcome, and psychosocial functioning

Quality of life in children can be affected by both the tumor itself and the treatment received. Reports assessing psychosocial and physical functioning show variable results, ranging from excellent in a majority of subjects to impaired function in almost half of the patients (Hoffman et al., 1992; Van Effenterre and Boch, 2002; Poretti et al., 2004). The most common areas of difficulty reported include social and emotional functioning, with patients rating their psychosocial health to be lower than their physical health (Poretti et al., 2004). Other challenges included somatic complaints such as pain, mobility, and self-care (Merchant et al., 2002; Poretti et al., 2004). Behavioral questionnaires indicate more frequent presence of psychopathologic symptoms, including depression, anxiety, and withdrawal. The most frequent problems in children's everyday functioning include inability to control emotions, difficulties in learning, unsatisfactory peer relationships, and concerns regarding physical appearance (Crom et al., 2010; Ondruch et al., 2011).

Factors associated with worsening quality of life outcomes as well as psychosocial and neurocognitive functioning include younger age at diagnosis and preoperative functional impairment, and tumor characteristics including larger size, hypothalamic and third ventricle involvement at presentation. Treatment strategies have also been implicated, with worse outcomes for surgery alone compared to limited surgery and irradiation and for multiple operations for tumor recurrence. Neurologic, ophthalmologic, and endocrine sequelae all adversely affect quality of life outcome (Hoffman et al., 1992; Merchant et al., 2002; Van Effenterre and Boch, 2002; Müller et al., 2003c; Poretti et al., 2004; Elliott and Wisoff, 2010; Steno et al., 2011). Hypothalamic dysfunction was found to have the most important negative effect on physical ability, social functioning, and body image (Müller et al., 2001, 2003c; Poretti et al., 2004).

Long-term neurocognitive complications following treatment for childhood craniopharyngioma include cognitive problems, particularly those affecting attention, executive function, working memory, and episodic memory (Cavazzuti et al., 1983; Riva et al., 1998; Carpentieri et al., 2001; Poretti et al., 2004; Sands et al., 2005; Kiehna et al., 2006; Crom et al., 2010; Ondruch et al., 2011; Özyurt et al., 2014). Also, long-term survivors of childhood craniopharyngioma treated primarily with irradiation and subtotal surgical resection were found to have psychological and educational deficits (Crom et al., 2010). Neurocognitive deficits include memory disturbances, slower cognitive speed, attention problems, and behavioral instability (Cavazzuti et al., 1983; Colangelo et al., 1990; Carpentieri et al., 2001; Kiehna et al., 2006; Crom et al., 2010; Ondruch et al., 2011). While intact intellectual functioning has been reported in up to 82% of patients, visual memory is reduced despite normal visual-spatial abilities (Crom et al., 2010; Ondruch et al., 2011). The acquired deficits in higher cognitive processing such as attention problems are considered precursors to poor academic achievement and vocational failure.

Reports on the effectiveness of rehabilitation efforts to treat neurocognitive deficiencies and psychosocial handicaps are preliminary but encouraging (Anderson et al., 1997; Riva et al., 1998; Poretti et al., 2004; Metzler-Baddeley and Jones, 2010; Hammond and Hall, 2011).

Survival and late mortality

Survival rates in children treated for craniopharyngioma are generally high (Müller et al., 2003b). However, disease-related mortality can still occur many years after treatment. Data regarding survival includes primarily surgically treated patients. The reported postsurgical 5 year overall survival is 88–94% (De Vile et al., 1996a; Van Effenterre and Boch, 2002; Müller et al., 2001, 2006c), and the reported 10 year overall survival is 70–92% (Hoffman et al., 1992; Van Effenterre and Boch, 2002; Poretti et al., 2004; Elliott and Wisoff, 2010; Visser et al., 2010), with a 20 year survival of 76%. Causes of late mortality include those directly related to the tumor or treatment, such as progressive disease with multiple recurrences, hormonal deficiencies, chronic hypothalamic insufficiency, cerebrovascular disease, and seizures (De Vile et al., 1996a; Poretti et al., 2004; Elliott and Wisoff, 2010; Visser et al., 2010; Steno et al., 2011). Other causes have been described, including decreased mineral bone density and nonalcoholic steatohepatitis leading to liver cirrhosis in some cases (Basenau et al., 1994; Altuntas et al., 2002; Müller et al., 2003a; Poretti et al., 2004; Caldarelli et al., 2005; Visser et al., 2010; Holmer et al., 2011).

Cerebrovascular morbidity

Radiation-induced vasculopathies are an uncommon consequence of radiation therapy for craniopharyngioma. In patients irradiated for craniopharyngioma, moyamoya syndrome (a radiation-induced cerebrovascular condition predisposing to stroke) has been described (Hetelekidis et al., 1993; Liu et al., 2009). A retrospective estimate was that 27% of 22 patients treated with irradiation and some combination of surgery and intracystic chemotherapy with a median radiation dose of 52.2 Gy developed some kind of vasculopathy, only half of which were symptomatic (Liu et al., 2009). No association was found between age, radiation prescription dose, and maximum or mean dose to the internal carotid arteries with the presence of vascular abnormalities. Regine and colleagues reported a 13.7% rate of cerebrovascular events, all in cases receiving over 61 Gy (Regine and Kramer, 1992). No cerebrovascular clinical events were reported in any other series of conventionally fractionated radiotherapy for craniopharyngioma, including those with large patient numbers (Rajan et al., 1993; Stripp et al., 2004; Karavitaki et al., 2005; Moon et al., 2005; Pemberton et al., 2005; Merchant et al., 2006; Combs et al., 2007).

Second malignant neoplasms

In the largest reported series no second malignancies were seen in 173 treated patients with a median follow-up of 12 years (Rajan et al., 1993). Overall only four cases of second malignant neoplasm were reported (Hetelekidis et al., 1993; Regine et al., 1993; Habrand et al., 1999; Winkfield et al., 2009) comprising two in-field glioblastomas (Regine et al., 1993; Habrand et al., 1999), one in-field glioma with unspecified grade of malignancy (Hetelekidis et al., 1993), and a posterior fossa meningioma (Winkfield et al., 2009).

ADULT-ONSET CRANIOPHARYNGIOMA

Age-dependent differences between childhood and adult onset craniopharyngioma are related to histologic diagnosis, biological behavior, clinical manifestations, treatment options, and follow-up (Koranyi et al., 2001; Attanasio et al., 2002; Kendall-Taylor et al., 2005; Gautier et al., 2012). The Kendall-Taylor et al. study compared childhood craniopharyngioma with adult-onset craniopharyngioma and reported a poor state of health and quality of life in both cohorts. The majority of childhood and adult-onset craniopharyngioma patients displayed pituitary insufficiency, with 60% suffering from diabetes insipidus. Nearly all patients were overweight or obese, reporting a poor quality of life.

QUESTIONS AND TREATMENT PERSPECTIVES

Surgical treatment strategies: degree of resection

One of the biggest challenges in treating craniopharyngioma is identifying the best candidates for a radical versus conservative treatment approach. Experiential expertise in large centers has increased the likelihood of safe gross total resection, evidenced by two reports representing historically different attitudes: the first at Necker Hospital, Paris, France (Puget et al., 2006), which is more surgically oriented; and the second in North America (Merchant et al., 2006), which is more oriented towards a conservative approach. The North American experience shows that most recent cases now receive moderate to aggressive surgery and only 42% have limited surgery before irradiation. The Necker authors (Puget et al., 2007) show in a contemporary series that 96% of their recent cases achieve complete (23%) or subtotal resection (73%), and that radiotherapy is performed in 50% of cases after subtotal resection. It appears there is a trend towards radiotherapy in centers with past predominantly surgical approaches, and towards more radical surgical treatment strategies in centers historically conservative-oriented. There are current prospective studies underway at a national and multinational level to adopt strategies tailored to risk factors for morbidity and quality of life (Müller et al., 2006d; Puget et al., 2007; Trivin et al., 2009).

Controversy over time point of irradiation

In clinical practice, the timing of postoperative residual tumor irradiation is both unclear and inconsistently regarded (Sung, 1982; Regine and Kramer, 1992; Stripp et al., 2004; Moon et al., 2005; Müller et al., 2005, 2006c; Tomita and Bowman, 2005; Trivin et al., 2009). Some favor immediate postoperative irradiation in the event of life-impairing clinical conditions, proactively preventing tumor progression (Merchant et al., 2002). On the other hand, some favor a "wait-and-see" procedure, delaying irradiation in order to reduce both its necessity and the negative consequences associated with radiation therapy. Without a doubt, immediate postoperative irradiation significantly delays tumor progression (Stripp et al., 2004). However, progression-contingent irradiation has proven effective, as overall survival is statistically unaffected by this wait-and-see strategy (Regine and Kramer, 1992). Three recent series retrospectively compared the immediate postoperative irradiation strategy with progression-contingent deployment (Stripp et al., 2004; Moon et al., 2005; Tomita and Bowman, 2005).

No differences in overall and progression-free survival were detected between immediate irradiation and progression-contingent treatment in the series evaluated by Moon and colleagues (2005). Relapse-free overall survival rates were 83% and 70% after 5 and 10 years, respectively, in the series analyzed by Tomita and colleagues (Tomita and Bowman, 2005). The corresponding numbers for patients after incomplete resection followed by immediate irradiation were 71% and 36% after 5 and 10 years, respectively. After incomplete resection without radiation therapy, the relapse-free survival rate after 5 years was only 9%. Progression-contingent irradiation achieved similar final overall survival- and progression-free survival rates of 90% and 70%, respectively, meaning progression-contingent irradiation in this series was highly effective. However, it has to be emphasized that the interpretation of these studies is difficult due to confounding factors in terms of indication for treatment, and unknown factors regarding medical staff experience and state of facilities. Regardless of concerns over quality and completeness of data, the relevant endpoint in analysis of childhood craniopharyngioma treatment should be patients' quality of life and morbidity.

KRANIOPHARYNGEOM 2007, a German prospective childhood craniopharygioma multinational trial (Müller, 2006, 2008, 2010a, b; Müller et al., 2006c, d), is currently evaluating craniopharyngioma patients' prognoses (quality of life, event-free survival, overall survival) following defined therapeutic strategies. A stratified randomization of two treatment arms is conducted with respect to timing of postoperative irradiation (immediate radiotherapy versus radiotherapy at the time of progression) for the subgroup of patients ≥ 5 years of age at the time of incomplete resection. The schedule of prospective data collection and the set and definition of parameters in KRANIOPHARYNGEOM 2007 are based on a multinational European consensus (Müller et al., 2006d) (Fig. 16.5).

Expertise

There are only a few studies analyzing the outcome of patients with craniopharyngioma in relation to the neurosurgeons' experience. Sanford (1994) and Boop (2007) report a marked difference in outcome according to the neurosurgeons' experience with the condition.

The degree of obesity and quality of life based on reference assessment of tumor localization and postsurgical hypothalamic lesions were analyzed in a recent report (Müller et al., 2011a) (Fig. 16.6). Treatment was also analyzed regarding neurosurgical strategy and the neurosurgical center sizes based on patient load. Surgical lesions of anterior and posterior hypothalamic areas were associated with postsurgical obesity, negatively impacting quality of life in patients with posterior hypothalamic lesions. Treatment in large centers was less radical, the rates of complete resection and hypothalamic surgical lesions lower than those of middle-sized and small centers. However, a multivariable analysis showed that preoperative hypothalamic involvement was the only independent risk factor for severe obesity (Müller et al., 2011a).

Based on the most current literature presenting factors affecting the genesis, pathology, treatment, and quality of life effects of childhood craniopharyngioma, it is advisable to have a multidisciplinary team able to discuss diagnostic, mitigating, and treatment strategies, adopting the most sophisticated approaches feasible based on sufficient in-house surgical, radio-oncological, and sociopsychological experience for treating patients with childhood craniopharyngioma.

Risk-adapted strategies/treatment algorithms for craniopharyngioma

Recently published risk-adapted treatment strategies are focusing on the main goals: (1) relief of increased intracranial pressure, (2) reversal of visual compression symptoms, (3) substitution of pituitary hormone deficits plus all other supplement-supportive measures, and (4) prevention of tumor recurrence or progression, while minimizing acute and long-term morbidity and mortality. With regard to the last goal, treatment strategies aiming at gross total resection in craniopharyngioma patients with hypothalamic tumor involvement are not recommended. Accordingly, several recommendations and treatment algorithms reflecting this hypothalamus-sparing aspect

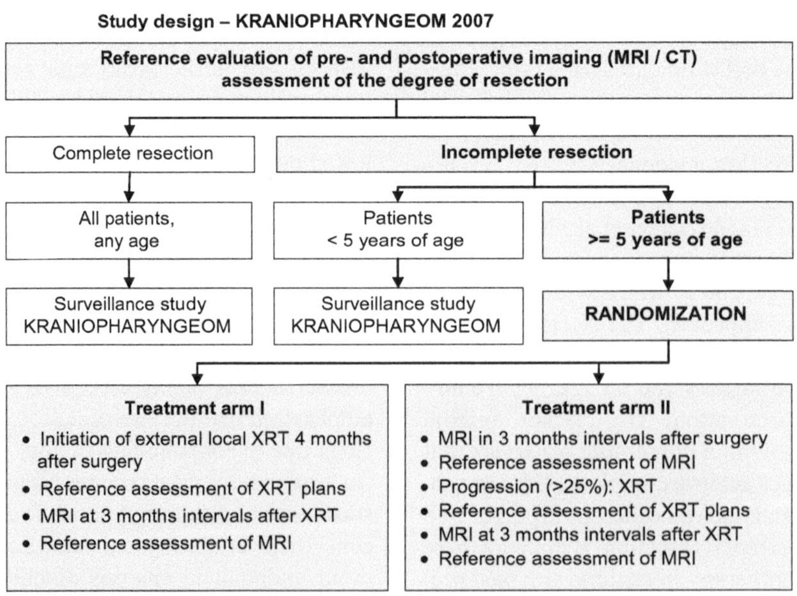

Fig. 16.5. Design of the KRANIOPHARYNGEOM 2007 study (www.kraniopharyngeom.net). (Modified from Müller et al., 2006c, with the kind permission of Karger.)

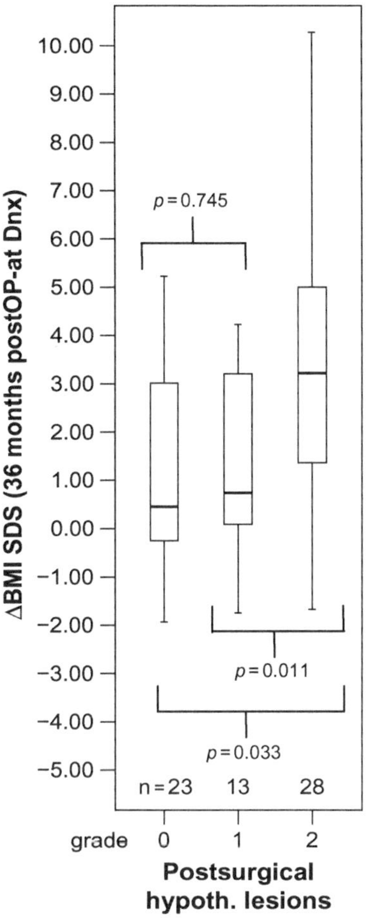

Fig. 16.6. Changes in body mass index (BMI Standard deviation score (SDS)) during the first 36 months after diagnosis of 117 childhood craniopharyngioma patients recruited in the KRANIOPHARYNGEOM 2000 trial relative to the extent of surgical hypothalamic (hypoth.) lesions (grade 0 to 2). The horizontal line in the middle of the box depicts the median. The edges of the box mark the 25th and 75th percentiles. Whiskers indicate the range of values that fall within 1.5 box lengths. (Modified from Müller et al., 2011a, with the kind permission of Bioscientifica.)

(Table 16.1) have been published recently (Spoudeas et al., 2006; Garrè and Cama, 2007; Flitsch et al., 2011; Müller et al., 2011a, 2012; Mallucci et al., 2012; Puget, 2012; Elowe-Gruau et al., 2013; Fjalldal et al., 2013).

Puget et al. (2007) recently published an algorithm for neurosurgical treatment of childhood craniopharyngioma patients, suggesting a hypothalamus-sparing surgical strategy based on the grading of hypothalamic tumor involvement in preoperative magnetic resonance imaging. The authors observed (Elowe-Gruau et al., 2013) that patients treated according to this algorithm using a hypothalamus-sparing surgical strategy had similar recurrence rates and a lower prevalence of severe obesity than patients treated by gross total resection (28% versus 54%, respectively). The study by Elowe-Gruau et al. is the first report in the literature (Elowe-Gruau et al., 2013) proving the efficacy and tolerability of a hypothalamus-sparing strategy by comparing cohorts treated by the same experienced surgical team at a single institution (Müller, 2013), and thus eliminating the bias of surgical experience on outcome analyses.

Garrè et al. (Garrè and Cama, 2007) used the grading system of hypothalamic involvement published by Puget et al. (Puget et al., 2007) and proposed a modified algorithm for risk-adapted treatment strategy in childhood craniopharyngioma. The authors emphasized the treatment by experienced multidisciplinary and neurosurgical teams and suggested proton beam therapy, especially for young patients (<5 years of age at the time of diagnosis) after limited hypothalamus-sparing surgery.

Müller et al. (2011a) suggested a risk-adapted treatment strategy based on pre- and postsurgical grading of hypothalamic tumor involvement/damage in magnetic resonance imaging. The assessment of tumor extension and location of surgical lesions in relation to the mammillary bodies is essential for the grading in terms of anterior or posterior hypothalamic involvement/lesion. According to Müller et al. (2011a), patients with postsurgical lesions of the posterior hypothalamic area presented with a higher BMI and lower self-assessed quality of life during prospective follow-up (Fig. 16.6). Flitsch et al. (2011) modified this grading system (Müller et al., 2011a) and published an algorithm suggesting different, risk-adapted surgical approaches (transsphenoidal versus transcranial) based on preoperative magnetic resonance imaging (Fig. 16.4).

Fjalldal et al. (2013) published a novel grading system on hypothalamic involvement in childhood craniopharyngioma patients and demonstrated that radical surgical resection in craniopharyngioma with extension into the third ventricle results in lower cognitive performance and psychosocial health during long-term follow-up of craniopharyngioma patients.

Spoudeas et al. (2006) published a hypothalamus-sparing surgical strategy based on a grading system that includes hypothalamic involvement, tumor size of craniopharyngioma, and the presence of hydrocephalus. Based on this grading system, Mallucci et al. (2012) published a treatment algorithm, suggesting a two-staged neurosurgical approach with initial relief of cystic pressure and thereby down-staging the risk grade in appropriate cases.

All of the above-mentioned treatment strategies and algorithms recommend that (1) in case of hypothalamic tumor involvement, limited hypothalamus-sparing surgical approaches and postoperative radiotherapy are advisable, and (2) treatment of craniopharyngioma should be confined to experienced multidisciplinary teams.

Table 16.1

Novel grading systems and treatment algorithms for childhood craniopharyngioma patients based on magnetic resonance imaging

Author and year*	n	Follow-up (years)	Grade 0	Grade I	Grade II	Treatment recommendation	Outcome parameters
Müller (2011a)	120	3	No HI	HI/lesion of the anterior hypothalamus not involving the MB and the hypothalamic area beyond MB	HI/lesion of the anterior + posterior hypothalamic area, i.e., involving the MB and the area beyond MB	Grade 0: GTR Grade I: attempt at GTR; if not achieved: XRT Grade II: subtotal resection with hypothalamic preservation + XRT	Higher BMI and lower QoL in the II° cohort treated by GTR resulting in posterior hypothalamic lesions
Flitsch (2011)	n.a.	n.a.	No HI	According to Müller et al. (2011a), specifying sections below and above the diaphragma sellae	According to Müller et al. (2011a)	Grade 0: GTR Grade I: attempt at GTR – transsphenoidal approach; if not achieved: XRT Grade II: subtotal resection with hypothalamic preservation – transcranial approach, followed by XRT	n.a.
Puget et al. (2007)	65	3	No HI	HI (distortion/elevation) with negligible hypothalamic lesions, the hypothalamus is still visible	Tumor involvement of hypothalamic structures, which are no longer identifiable	Grade 0: GTR Grade I: attempt at GTR; if not achieved: second surgery ± XRT Grade II: subtotal resection with hypothalamic preservation + XRT	Lower BMI and similar recurrence rate in a prospective cohort treated according to algorithm compared with historical cohort
Garre (2007)	n.a.	n.a.	No HI	According to Puget et al. (2007)	According to Puget et al. (2007)	Grade 0 + I: attempt at GTR by experienced surgeon; if not achieved: XRT Grade II: cyst drainage ± XRT (proton beam therapy at age < 5 years)	n.a.
Fjalldal (2013)	42	20	No HI	Suprasellar growth, not towards or into the third ventricle (non-TGTV)	Suprasellar growth towards or into the third ventricle (TGTV)	Non-TGTV: GTR TGTV: subtotal resection with hypothalamic preservation + XRT	Lower cognitive performance in TGTV patients treated by GTR
Spoudeas (2006)	n.a.	n.a.	No HI	Tumor size (≤2–4 cm), no hydrocephalus, no hypothalamic syndrome, no breech third ventricle	Retrochiasmatic tumor, (>4 cm), hydrocephalus, hypothalamic syndrome ±breech third ventricle	Grade 0: GTR Grade I: consider GTR Grade II: limited resection + XRT	n.a.

*First named author and year of publication.

n, size of cohort; HI, hypothalamic involvement; n.a., not analyzed; MB, mamillary bodies; GTR, gross total resection; XRT, irradiation; BMI, body mass index; TGTV, growth towards third ventricle.

CONCLUSIONS

Risk-adapted surgical strategies at initial diagnosis should aim at a maximal degree of resection keenly focused on respecting the integrity of optic and hypothalamic structures to prevent severe sequelae and therein minimize consequences that could negatively impact patients' quality of life. Because initial hypothalamic tumor involvement has an *a priori* effect on the clinical course (Müller et al., 2004, 2011a), childhood craniopharyngioma should be recognized as a chronic disease requiring constant monitoring of the consequences and medical resources for treatment in order to provide not only optimal quality of life for patients, but also to garner additional information with the intent of minimizing what at present are severe consequences of both the disease and its treatment (Müller, 2011).

NOTE

This chapter was composed in the absence of any commercial or financial relationships that could be construed as a potential conflict of interest.

ACKNOWLEDGMENTS

The author is grateful to Mrs Neff-Heinrich (Göttingen, Germany) for help in proofreading and editing the manuscript.

References

Ahmet A, Blaser D, Stephens D et al. (2006). Weight gain in craniopharyngioma – a model for hypothalamic obesity. J Pediatr Endocrinol Metab 19: 121–127.

Attanasio AF, Howell S, Bates PC et al. (2002). Body composition, IGF-I and IGFBP-3 concentrations as outcome measures in severely GH-deficient (GHD) patients after childhood GH treatment: a comparison with adult onset GHD patients. J Clin Endocrinol Metab 87: 3368–3372.

Altuntas B, Ozcakar B, Bideci A et al. (2002). Cirrhotic outcome in patients with craniopharyngioma. J Pediatr Endocrinol Metab 15: 1057–1058.

Amendola BE, Wolf A, Coy SR et al. (2003). Role of radiosurgery in craniopharyngiomas: a preliminary report. Med Pediatr Oncol 41: 123–127.

Anderson CA, Wilkening GN, Filley CM et al. (1997). Neurobehavioral outcome in pediatric craniopharyngioma. Pediatr Neurosurg 26: 255–260.

Basenau D, Stehphani U, Fischer G (1994). Development of complete liver cirrhosis in hyperphagia-induced fatty liver. Klin Padiatr 206: 62–64.

Becker G, Kortmann RD, Skalej M et al. (1999). The role of radiotherapy in the treatment of craniopharyngioma – indications, results, side effects. Front Radiat Ther Oncol 33: 100–113.

Beltran C, Roca M, Merchant TE (2012). On the benefits and risks of proton therapy in pediatric craniopharyngioma. Int J Radiat Oncol Biol Phys 82: e281–e287.

Boehling NS, Grosshans DR, Bluett JB (2012). Dosimetric comparison of three-dimensional conformal proton radiotherapy, intensity-modulated proton therapy, and intensity-modulated radiotherapy for treatment of pediatric craniopharyngiomas. Int J Radiat Oncol Biol Phys 82: 643–652.

Boop FA (2007). Craniopharyngioma. J Neurosurg 106 (1 Suppl): 1–2, discussion 2.

Bunin GR, Surawicz TS, Witman PA (1998). The descriptive epidemiology of craniopharyngioma. J Neurosurg 89: 547–551.

Caldarelli M, Massimi L, Tamburrini G et al. (2005). Long-term results of the surgical treatment of craniopharyngioma: the experience at the Policlinico Gemelli, Catholic University, Rome. Childs Nerv Syst 21: 747–757.

Carpentieri SC, Waber DP, Scott RM et al. (2001). Memory deficits among children with craniopharyngioma. Neurosurgery 49: 1053–1058.

Cavalheiro S, Sparapani FV, Franco JO et al. (1996). Use of bleomycin in intratumoral chemotherapy for cystic craniopharyngioma. Case report. J Neurosurg 84: 124–126.

Cavalheiro S, Di Rocco C, Valenzuela S et al. (2010). Craniopharyngiomas: intratumoral chemotherapy with interferon-alpha: a multicenter preliminary study with 60 cases. Neurosurg Focus 28: E12.

Cavazzuti V, Fischer EG, Welch K et al. (1983). Neurological and psychophysiological sequelae following different treatments of craniopharyngioma in children. J Neurosurg 59: 409–417.

Choux M, Lena G (1979). Bases of surgical management of craniopharyngioma in children. proceedingsActa Neurochir Suppl (Wien) 28: 348.

Chung WY, Pan DH, Shiau CY (2000). Gamma Knife radiosurgery for craniopharyngiomas. J Neurosurg 93 (Suppl 3): 47–56.

Colangelo M, Ambrosio A, Ambrosio C (1990). Neurological and behavioral sequelae following different approaches to craniopharyngioma. Long-term follow-up review and therapeutic guidelines. Childs Nerv Syst 6: 379–382.

Combs SE, Thilmann C, Huber PE et al. (2007). Achievement of long-term local control in patients with craniopharyngiomas using high precision stereotactic radiotherapy. Cancer 109: 2308–2314.

Costin G, Kogut MD, Phillips LS et al. (1976). Craniopharyngioma: the role of insulin in promoting postoperative growth. J Clin Endocrinol Metab 42: 370–379.

Coutant R, Maurey H, Rouleau S et al. (2003). Defect in epinephrine production in children with craniopharyngioma: functional or organic origin? J Clin Endocrinol Metab 88: 5969–5975.

Crom DB, Smith D, Xiong Z et al. (2010). Health status in long-term survivors of pediatric craniopharyngiomas. J Neurosci Nurs 42: 323–328.

Danielsson P, Janson A, Norgren S et al. (2007). Impact sibutramine therapy in children with hypothalamic obesity or obesity with aggravating syndromes. J Clin Endocrinol Metab 92: 4101–4106.

De Vile CJ, Grant DB, Kendall BE et al. (1996a). Management of childhood craniopharyngioma: can the morbidity of radical surgery be predicted? J Neurosurg 85: 73–81.

de Vile CJ, Grant DB, Hayward RD et al. (1996b). Obesity in childhood craniopharyngioma: relation to post-operative hypothalamic damage shown by magnetic resonance imaging. J Clin Endocrinol Metab 81: 2734–2737.

Elliott RE, Wisoff JH (2010). Surgical management of giant pediatric craniopharyngiomas. J Neurosurg Pediatr 6: 403–416.

Elliott RE, Jane Jr JA, Wisoff JH (2011). Surgical management of craniopharyngiomas in children: meta-analysis and comparison of transcranial and transsphenoidal approaches. Neurosurgery 69: 630–643, discussion 643.

Elowe-Gruau E, Beltrand J, Brauner R et al. (2013). Childhood craniopharyngioma: hypothalamus-sparing surgery decreases the risk of obesity. J Clin Endocrinol Metab 98: 2376–2382.

Fahlbusch R, Honegger J, Paulus W et al. (1999). Surgical treatment of craniopharyngiomas: experience with 168 patients. J Neurosurg 90: 237–250.

Fisher PG, Jenab J, Gopldthwaite PT et al. (1998). Outcomes and failure patterns in childhood craniopharyngiomas. Childs Nerv Syst 14: 558–563.

Fitzek MM, Linggood RM, Adams J et al. (2006). Combined proton and photon irradiation for craniopharyngioma: long-term results of the early cohort of patients treated at Harvard Cyclotron Laboratory and Massachusetts General Hospital. Int J Radiat Oncol Biol Phys 64: 1348–1354.

Fjalldal S, Holmer H, Rylander L et al. (2013). Hypothalamic involvement predicts cognitive performance and psychosocial health in long-term survivors of childhood craniopharyngioma. J Clin Endocrinol Metab 98: 3253–3262.

Flitsch J, Müller HL, Burkhardt T (2011). Surgical strategies in childhood craniopharyngioma. Front Endocrinol (Lausanne) 2: 96.

Garrè ML, Cama A (2007). Craniopharyngioma: modern concepts in pathogenesis and treatment. Curr Opin Pediatr 19: 471–479.

Gaston-Massuet C, Andoniadou CL, Signore M et al. (2011). Increased Wingless (Wnt) signaling in pituitary progenitor/stem cells gives rise to pituitary tumors in mice and humans. Proc Natl Acad Sci U S A 108: 11482–11487.

Gautier A, Godbout A, Grosheny C et al. (2012). Markers of recurrence and long-term morbidity in craniopharyngioma: a systematic analysis of 171 patients. J Clin Endocrinol Metab 97: 1258–1267.

Geffner ME (1996). The growth without growth hormone syndrome. Endocrinol Metab Clin North Am 25: 649–663.

Geffner M, Lundberg M, Koltowska-Haggstrom M et al. (2004). Changes in height, weight, and body mass index in children with craniopharyngioma after three years of growth hormone therapy: analysis of KIGS (Pfizer International Growth Database). J Clin Endocrinol Metab 89: 5435–5440.

Habrand JL, Ganry O, Couanet D et al. (1999). The role of radiation therapy in the management of craniopharyngioma: a 25-year experience and review of the literature. Int J Radiat Oncol Biol Phys 44: 255–263.

Habrand JL, Saran F, Alapetite C et al. (2006). Radiation therapy in the management of craniopharyngioma: current concepts and future developments. J Pediatr Endocrinol Metab 19 (Suppl 1): 389–394.

Halac I, Zimmerman D (2005). Endocrine manifestations of craniopharyngioma. Childs Nerv Syst 21: 640–648.

Hammond JL, Hall SS (2011). Functional analysis and treatment of aggressive behavior following resection of a craniopharyngioma. Dev Med Child Neurol 53: 369–374.

Harz KJ, Müller HL, Waldeck E et al. (2003). Obesity in patients with craniopharyngioma: assessment of food intake and movement counts indicating physical activity. J Clin Endocrinol Metab 88: 5227–5231.

Hetelekidis S, Barnes PD, Tao ML (1993). 20-year experience in childhood craniopharyngioma. Int J Radiat Oncol Biol Phys 27: 189–195.

Hoffman HJ, De Silva M, Humphreys RP et al. (1992). Aggressive surgical management of craniopharyngiomas in children. J Neurosurg 76: 47–52.

Holmer H, Ekman B, Bjork J et al. (2009). Hypothalamic involvement predicts cardiovascular risk in adults with childhood onset craniopharyngioma on long-term GH therapy. Eur J Endocrinol 161: 671–679.

Holmer H, Pozarek G, Wirfalt E et al. (2010). Reduced energy expenditure and impaired feeding-related signals but not high energy intake reinforces hypothalamic obesity in adults with childhood onset craniopharyngioma. J Clin Endocrinol Metab 95: 5395–5402.

Holmer H, Popovic V, Ekman B et al. (2011). Hypothalamic involvement and insufficient sex steroid supplementation are associated with low bone mineral density in women with childhood onset craniopharyngioma. Eur J Endocrinol 165: 25–31.

Holsken A, Buchfelder M, Fahlbusch R et al. (2010). Tumour cell migration in adamantinomatous craniopharyngiomas is promoted by activated Wnt-signalling. Acta Neuropathol 119: 631–639.

Inge TH, Pfluger P, Zeller M et al. (2007). Gastric bypass surgery for treatment of hypothalamic obesity after craniopharyngioma therapy. Nat Clin Pract Endocrinol Metab 3: 606–609.

Ismail D, O'Connell MA, Zacharin MR (2006). Dexamphetamine use for management of obesity and hypersomnolence following hypothalamic injury. J Pediatr Endocrinol Metab 19: 129–134.

Julow J, Lanyi F, Hajda M et al. (1989).Intracystic instillation of yttrium 90 silicate colloid in cystic craniopharyngioma Orv Hetil 130: 1367–1368, 1371–1375.

Jung TY, Jung S, Moon KS et al. (2010). Endocrinological outcomes of pediatric craniopharyngiomas with anatomical pituitary stalk preservation: preliminary study. Pediatr Neurosurg 46: 205–212.

Kalapurakal JA, Goldman S, Hsieh YC et al. (2003). Clinical outcome in children with craniopharyngioma treated with primary surgery and radiotherapy deferred until relapse. Med Pediatr Oncol 40: 214–218.

Karavitaki N, Brufani C, Warner JT et al. (2005). Craniopharyngiomas in children and adults: systematic analysis of 121 cases with long-term follow-up. Clin Endocrinol (Oxf) 62: 397–409.
Karavitaki N, Cudlip S, Adams CB et al. (2006). Craniopharyngiomas. Endocr Rev 27: 371–397.
Kendall-Taylor P, Jönsson PJ, Abs R et al. (2005). The clinical, metabolic and endocrine features and the quality of life in adults with childhood-onset craniopharyngioma compared with adult-onset craniopharyngioma. Eur J Endocrinol 152: 557–567.
Khafaga Y, Jenkin D, Kanaan I et al. (1998). Craniopharyngioma in children. Int J Radiat Oncol Biol Phys 42: 601–606.
Kiehna EN, Mulhern R, Li C et al. (2006). Changes in attentional performance of children and young adults with localized primary brain tumors after conformal radiation therapy. J Clin Oncol 24: 5283–5290.
Kim SD, Park JY, Park J et al. (2007). Radiological findings following postsurgical intratumoral bleomycin injection for cystic craniopharyngioma. Clin Neurol Neurosurg 109: 236–241.
Kim RJ, Shah R, Tershakovec AM et al. (2010). Energy expenditure in obesity associated with craniopharyngioma. Childs Nerv Syst 26: 913–917.
Kobayashi T (2009). Long-term results of Gamma Knife radiosurgery for 100 consecutive cases of craniopharyngioma and a treatment strategy. Prog Neurol Surg 22: 63–76.
Koranyi J, Svensson J, Götherström G et al. (2001). Baseline characteristics and the effects of five years of GH replacement therapy in adults with GH deficiency of childhood or adulthood onset: a comparative, prospective study. J Clin Endocrinol Metab 86: 4693–4699.
Kreier F, Fliers E, Voshol PJ et al. (2002). Selective parasympathetic innervation of subcutaneous and intra-abdominal fat – functional implications. J Clin Invest 110: 1243–1250.
Lafay-Cousin L, Bartels U, Raybaud C et al. (2007). Neuroradiological findings of bleomycin leakage in cystic craniopharyngioma. Report of three cases. J Neurosurg 107 (4 Suppl): 318–323.
Lee M, Kalani MY, Cheshier S et al. (2008). Radiation therapy and CyberKnife radiosurgery in the management of craniopharyngiomas. Neurosurg Focus 24: E4.
Lek N, Prentice P, Williams RM et al. (2010). Risk factors for obesity in childhood survivors of suprasellar brain tumours: a retrospective study. Acta Paediatr 99: 1522–1526.
Lin LL, El Naqa I, Leonard JR et al. (2008). Long-term outcome in children treated for craniopharyngioma with and without radiotherapy. J Neurosurg Pediatr 1: 126–130.
Liu AK, Bagrosky B, Fenton LZ et al. (2009). Vascular abnormalities in pediatric craniopharyngioma patients treated with radiation therapy. Pediatr Blood Cancer 52: 227–230.
Lustig RH (2008). Hypothalamic obesity: causes, consequences, treatment. Pediatr Endocrinol Rev 6: 220–227.
Lustig RH, Post SR, Srivannaboon K et al. (2003a). Risk factors for the development of obesity in children surviving brain tumors. J Clin Endocrinol Metab 88: 611–616.
Lustig RH, Hinds PS, Ringwald-Smith K et al. (2003b). Octreotide therapy of pediatric hypothalamic obesity: a double-blind, placebo-controlled trial. J Clin Endocrinol Metab 88: 2586–2592.
Luu QT, Loredo LN, Archambeau JO (2006). Fractionated proton radiation treatment for pediatric craniopharyngioma: preliminary report. Cancer J 12: 155–159.
Mallucci C, Pizer B, Blair J et al. (2012). Management of craniopharyngioma: the Liverpool experience following the introduction of the CCLG guidelines. Introducing a new risk assessment grading system. Childs Nerv Syst 28: 1181–1192.
Mason PW, Krawiecki N, Meacham LR (2002). The use of dextroamphetamine to treat obesity and hyperphagia in children treated for craniopharyngioma. Arch Pediatr Adolesc Med 156: 887–892.
Matson DD (1964). Craniopharyngioma. Clin Neurosurg 10: 14.
Merchant TE, Kiehna EN, Sanford RA et al. (2002). Craniopharyngioma: the St Jude Children's Research Hospital experience 1984–2001. Int J Radiat Oncol Biol Phys 53: 533–542.
Merchant TE, Kiehna EN, Kun LE et al. (2006). Phase II trial of conformal radiation therapy for pediatric patients with craniopharyngioma and correlation of surgical factors and radiation dosimetry with change in cognitive function. J Neurosurg 104 (2 Suppl): 94–102.
Metzler-Baddeley C, Jones RW (2010). Brief communication. Cognitive rehabilitation of executive functioning in a case of craniopharyngioma. Appl Neuropsychol 17: 299–304.
Minniti G, Esposito V, Amichetti M et al. (2009). The role of fractionated radiotherapy and radiosurgery in the management of patients with craniopharyngioma. Neurosurg Rev 32: 125–132, discussion 132.
Mokry M (1999). Craniopharyngiomas: a six year experience with Gamma Knife radiosurgery. Stereotact Funct Neurosurg 72 (Suppl 1): 140–149.
Mong S, Pomeroy SL, Cecchin F et al. (2008). Cardiac risk after craniopharyngioma therapy. Pediatr Neurol 38: 256–260.
Moon SH, Kim IH, Park SW et al. (2005). Early adjuvant radiotherapy toward long-term survival and better quality of life for craniopharyngiomas – a study in single institute. Childs Nerv Syst 21: 799–807.
Müller HL (2006). More or less? Treatment strategies in childhood craniopharyngioma. Childs Nerv Syst 22: 156–157.
Müller HL (2008). Childhood craniopharyngioma. Recent advances in diagnosis, treatment and follow-up. Horm Res 69: 193–202.
Müller HL (2010a). Childhood craniopharyngioma – current concepts in diagnosis, therapy and follow-up. Nat Rev Endocrinol 6: 609–618.
Müller HL (2010b). Childhood craniopharyngioma: current controversies on management in diagnostics, treatment and follow-up. Expert Rev Neurother 10: 515–524.
Müller HL (2010c). Increased daytime sleepiness in patients with childhood craniopharyngioma and hypothalamic tumor involvement: review of the literature and perspectives. Int J Endocrinol 2010: 519607.

Müller HL (2011). Consequences of craniopharyngioma surgery in children. J Clin Endocrinol Metab 96: 1981–1991.

Müller HL (2013). Paediatrics: surgical strategy and quality of life in craniopharyngioma. Nat Rev Endocrinol 9: 447–449.

Müller HL, Bueb K, Bartels U et al. (2001). Obesity after childhood craniopharyngioma – German multicenter study on pre-operative risk factors and quality of life. Klin Padiatr 213: 244–249.

Müller HL, Handwerker G, Wollny B et al. (2002). Melatonin secretion and increased daytime sleepiness in childhood craniopharyngioma patients. J Clin Endocrinol Metab 87: 3993–3996.

Müller HL, Heinrich M, Bueb K et al. (2003a). Perioperative dexamethasone treatment in childhood craniopharyngioma – influence on short-term and long-term weight gain. Exp Clin Endocrinol Diabetes 111: 330–334.

Müller HL, Kaatsch P, Warmuth-Metz M et al. (2003b). Childhood craniopharyngioma – diagnostic and therapeutic strategies. Monatsschr Kinderheilkd 151: 1056–1063.

Müller HL, Faldum A, Etavard-Gorris N et al. (2003c). Functional capacity, obesity and hypothalamic involvement: cross-sectional study on 212 patients with childhood craniopharyngioma. Klin Padiatr 215: 310–314.

Müller HL, Emser A, Faldum A et al. (2004). Longitudinal study on growth and body mass index before and after diagnosis of childhood craniopharyngioma. J Clin Endocrinol Metab 89: 3298–3305.

Müller HL, Gebhardt U, Etavard-Gorris N et al. (2005). Current strategies in diagnostics and endocrine treatment of patients with childhood craniopharyngioma during follow-up – recommendations in KRANIOPHARYNGEOM 2000. Onkologie 28: 150–156.

Müller HL, Müller-Stover S, Gebhardt U et al. (2006a). Secondary narcolepsy may be a causative factor of increased daytime sleepiness in obese childhood craniopharyngioma patients. J Pediatr Endocrinol Metab 19 (Suppl 1): 423–429.

Müller HL, Handwerker G, Gebhardt U et al. (2006b). Melatonin treatment in obese patients with childhood craniopharyngioma and increased daytime sleepiness. Cancer Causes Control 17: 583–589.

Müller HL, Gebhardt U, Pohl F et al. (2006c). Relapse pattern after complete resection and early progression after incomplete resection of childhood craniopharyngioma. Klin Padiatr 218: 315–320.

Müller HL, Albanese A, Calaminus G et al. (2006d). Consensus and perspectives on treatment strategies in childhood craniopharyngioma: results of a meeting of the Craniopharyngioma Study Group (SIOP), Genova, 2004. J Pediatr Endocrinol Metab 19 (Suppl 1): 453–454.

Müller HL, Gebhardt U, Wessel V et al. (2007). First experiences with laparoscopic adjustable gastric banding (LAGB) in the treatment of patients with childhood craniopharyngioma and morbid obesity. Klin Padiatr 219: 323–325.

Müller HL, Gebhardt U, Schroder S et al. (2010). Analyses of treatment variables for patients with childhood craniopharyngioma – results of the multicenter prospective trial KRANIOPHARYNGEOM 2000 after three years of follow-up. Horm Res Paediatr 73: 175–180.

Müller HL, Gebhardt U, Teske C et al. (2011a). Post-operative hypothalamic lesions and obesity in childhood craniopharyngioma: results of the multinational prospective trial KRANIOPHARYNGEOM 2000 after 3-year follow-up. Eur J Endocrinol 165: 17–24.

Müller HL, Gebhardt U, Maroske J et al. (2011b). Long-term follow-up of morbidly obese patients with childhood craniopharyngioma after laparoscopic adjustable gastric banding (LAGB). Klin Padiatr 223: 372–373.

Müller HL, Gebhardt U, Faldum A et al. (2012). Xanthogranuloma, Rathke's cyst, and childhood craniopharyngioma: results of prospective multinational studies of children and adolescents with rare sellar malformations. J Clin Endocrinol Metab 97: 3935–3943.

Müller-Scholden J, Lehrnbecher T, Müller HL et al. (2000). Radical surgery in a neonate with craniopharyngioma report of a case. Pediatr Neurosurg 33: 265–269.

Nielsen EH, Feldt-Rasmussen U, Poulsgaard L et al. (2011). Incidence of craniopharyngioma in Denmark (n = 189) and estimated world incidence of craniopharyngioma in children and adults. J Neurooncol 104: 755–763.

O'Gorman CS, Simoneau-Roy J, Pencharz P et al. (2010). Sleep-disordered breathing is increased in obese adolescents with craniopharyngioma compared with obese controls. J Clin Endocrinol Metab 95: 2211–2218.

Ondruch A, Maryniak A, Kropiwnicki T et al. (2011). Cognitive and social functioning in children and adolescents after the removal of craniopharyngioma. Childs Nerv Syst 27: 391–397.

Özyurt J, Thiel CM, Lorenzen A et al. (2014). Neuropsychological outcome in patients with childhood craniopharyngioma and hypothalamic involvement. J Pediatr 164: 876–881.

Pemberton LS, Dougal M, Magee B et al. (2005). Experience of external beam radiotherapy given adjuvantly or at relapse following surgery for craniopharyngioma. Radiother Oncol 77: 99–104.

Pereira AM, Schmid EM, Schutte PJ et al. (2005). High prevalence of long-term cardiovascular, neurological and psychosocial morbidity after treatment for craniopharyngioma. Clin Endocrinol (Oxf) 62: 197–204.

Phillip M, Moran O, Lazar L (2002). Growth without growth hormone. J Pediatr Endocrinol Metab 15 (Suppl 5): 1267–1272.

Poretti A, Grotzer MA, Ribi K et al. (2004). Outcome of craniopharyngioma in children: long-term complications and quality of life. Dev Med Child Neurol 46: 220–229.

Puget S (2012). Treatment strategies in childhood craniopharyngioma. Front Endocrinol (Lausanne) 3: 64.

Puget S, Grill J, Habrand JL et al. (2006). Multimodal treatment of craniopharyngioma: defining a risk-adapted strategy. J Pediatr Endocrinol Metab 19 (Suppl 1): 367–370.

Puget S, Garnett M, Wray A et al. (2007). Pediatric craniopharyngiomas: classification and treatment according to the degree of hypothalamic involvement. J Neurosurg 106: 3–12.

Rajan B, Ashley S, Gorman C et al. (1993). Craniopharyngioma – a long-term results following limited surgery and radiotherapy. Radiother Oncol 26: 1–10.

Regine WF, Kramer S (1992). Pediatric craniopharyngiomas: long term results of combined treatment with surgery and radiation. Int J Radiat Oncol Biol Phys 24: 611–617.

Regine WF, Mohiuddin M, Kramer S (1993). Long-term results of pediatric and adult craniopharyngiomas treated with combined surgery and radiation. Radiother Oncol 27: 13–21.

Riva D, Pantaleoni C, Devoti M et al. (1998). Late neuropsychological and behavioral outcome of children surgically treated for craniopharyngioma. Childs Nerv Syst 14: 179–184.

Roth C, Wilken B, Hanefeld F et al. (1998). Hyperphagia in children with craniopharyngioma is associated with hyperleptinaemia and a failure in the downregulation of appetite. Eur J Endocrinol 138: 89–91.

Roth CL, Hunneman DH, Gebhardt U et al. (2007). Reduced sympathetic metabolites in urine of obese patients with craniopharyngioma. Pediatr Res 61: 496–501.

Roth CL, Enriori PJ, Gebhardt U et al. (2010). Changes of peripheral alpha-melanocyte-stimulating hormone in childhood obesity. Metabolism 59: 186–194.

Roth CL, Gebhardt U, Müller HL (2011). Appetite-regulating hormone changes in patients with craniopharyngioma. Obesity (Silver Spring) 19: 36–42.

Rottembourg D, O'Gorman CS, Urbach S et al. (2009). Outcome after bariatric surgery in two adolescents with hypothalamic obesity following treatment of craniopharyngioma. J Pediatr Endocrinol Metab 22: 867–872.

Rushing EJ, Giangaspero F, Paulus W et al. (2007). Craniopharyngioma. In: DN Louis, H Ohgaki, OD Wiestler, KC Webster (Eds.), WHO Classification of Tumours of the Central Nervous System. 3rd edn. WHO PRESS, Geneva.

Sands SA, Milner JS, Goldberg J et al. (2005). Quality of life and behavioral follow-up study of pediatric survivors of cranipharyngioma. J Neurosurg Pediatr 103: 302–311.

Sanford RA (1994). Craniopharyngioma: results of survey of the American Society of Pediatric Neurosurgery. Pediatr Neurosurg 21 (Suppl 1): 39–43.

Schofl C, Schleth A, Berger D et al. (2002). Sympathoadrenal counterregulation in patients with hypothalamic craniopharyngioma. J Clin Endocrinol Metab 87: 624–629.

Schubert T, Trippel M, Tacke U et al. (2009). Neurosurgical treatment strategies in childhood craniopharyngiomas: is less more? Childs Nerv Syst 25: 1419–1427.

Schultes B, Ernst B, Schmid F et al. (2009). Distal gastric bypass surgery for the treatment of hypothalamic obesity after childhood craniopharyngioma. Eur J Endocrinol 161: 201–206.

Scott RM, Hetelekidis S, Barnes PD et al. (1994). Surgery, radiation, and combination therapy in the treatment of childhood craniopharyngioma – a 20-year experience. Pediatr Neurosurg 21 (Suppl 1): 75–81.

Shaikh MG, Grundy RG, Kirk JM (2008). Reductions in basal metabolic rate and physical activity contribute to hypothalamic obesity. J Clin Endocrinol Metab 93: 2588–2593.

Sherlock M, Ayuk J, Tomlinson JW et al. (2010). Mortality in patients with pituitary disease. Endocr Rev 31: 301–342.

Spoudeas HA, Saran F, Pizer B (2006). A multimodality approach to the treatment of craniopharyngiomas avoiding hypothalamic morbidity: a UK perspective. J Pediatr Endocrinol Metab 19: 447–451.

Srinivasan S, Ogle GD, Garnett SP et al. (2004). Features of the metabolic syndrome after childhood craniopharyngioma. J Clin Endocrinol Metab 89: 81–86.

Steno J, Bizik I, Steno A et al. (2011). Craniopharyngiomas in children: how radical should the surgeon be? Childs Nerv Syst 27: 41–54.

Stripp DC, Maity A, Janss AJ et al. (2004). Surgery with or without radiation therapy in the management of craniopharyngiomas in children and young adults. Int J Radiat Oncol Biol Phys 58: 714–720.

Sung DI (1982). Suprasellar tumors in children: a review of clinical manifestations and managements. Cancer 50: 1420–1425.

Swaab DF, Gooren LJ, Hofman MA (1992). The human hypothalamus in relation to gender and sexual orientation. Prog Brain Res 93: 205–217, discussion 217–219.

Szeifert GT, Balint K, Sipos L et al. (2007). Pathological findings in cystic craniopharyngiomas after stereotactic intracavitary irradiation with yttrium-90 isotope. Prog Neurol Surg 20: 297–302.

Tomita T, Bowman RM (2005). Craniopharyngiomas in children: surgical experience at Children's Memorial Hospital. Childs Nerv Syst 21: 729–746.

Trivin C, Busiah K, Mahlaoui N et al. (2009). Childhood craniopharyngioma: greater hypothalamic involvement before surgery is associated with higher homeostasis model insulin resistance index. BMC Pediatr 9: 24.

Van Effenterre R, Boch AL (2002). Craniopharyngioma in adults and children: a study of 122 surgical cases. J Neurosurg 97: 3–11.

Vinchon M, Weill J, Delestret I et al. (2009). Craniopharyngioma and hypothalamic obesity in children. Childs Nerv Syst 25: 347–352.

Visser J, Hukin J, Sargent M et al. (2010). Late mortality in pediatric patients with craniopharyngioma. J Neurooncol 100: 105–111.

Warmuth-Metz M, Gnekow AK, Müller H et al. (2004). Differential diagnosis of suprasellar tumors in children. Klin Padiatr 216: 323–330.

Winkfield KM, Linsenmeier C, Yock TI et al. (2009). Surveillance of craniopharyngioma cyst growth in children treated with proton radiotherapy. Int J Radiat Oncol Biol Phys 73: 716–721.

Wisoff JH (2008). Craniopharyngioma. J Neurosurg Pediatr 1: 124–125, discussion 125.

Zona G, Spaziante R (2006). Management of cystic craniopharyngiomas in childhood by a transsphenoidal approach. J Pediatr Endocrinol Metab 19 (Suppl 1): 381–388.

Handbook of Clinical Neurology, Vol. 124 (3rd series)
Clinical Neuroendocrinology
E. Fliers, M. Korbonits, and J.A. Romijn, Editors

Chapter 17

Rathke's cleft cyst

SARAH LARKIN[1], NIKI KARAVITAKI[2], AND OLAF ANSORGE[1,*]

[1]*Department of Neuropathology, Nuffield Department of Clinical Neurosciences, University of Oxford, John Radcliffe Hospital, Oxford, UK*

[2]*Oxford Centre for Diabetes, Endocrinology and Metabolism, Churchill Hospital, Oxford, UK*

INTRODUCTION

Rathke's cleft cysts are sellar or suprasellar nonneoplastic cystic lesions arising from remnants of Rathke's pouch. Symptomatic cases are relatively rare and the majority are discovered incidentally (11.3% of unselected postmortem cases) (Teramoto et al., 1994). The prevalence of these lesions is uncertain, but they are detected with increasing frequency as imaging techniques improve and neuroimaging becomes more commonplace. Generally found in adults, the natural history and pathogenesis of these lesions is unclear. Rathke's cleft cysts become symptomatic when they are large enough to cause pressure effects on surrounding structures. The most common presenting symptoms include headaches (33–81% of patients) (Isono et al., 2001; Kim et al., 2004), visual disturbance (12–58% of patients) (Ross et al., 1992; Trifanescu et al., 2011), and pituitary hormone abnormalities (one or more axes affected in 19–81% of patients) (Shin et al., 1999; Madhok et al., 2010). Treatment is almost exclusively surgical with the aim of draining the cyst contents and removal of as much of the surrounding capsule as possible. This usually results in symptomatic improvement (resolution or improvement of headache in 40–100% of patients and visual disturbance in 33–100%) (Midha et al., 1991; Shin et al., 1999; Isono et al., 2001; Benveniste et al., 2004; Billeci et al., 2004; Kim et al., 2004; Aho et al., 2005; Sade et al., 2005; Nishioka et al., 2006a, b; Zada et al., 2009; Madhok et al., 2010; Lillehei et al., 2011; Kasperbauer et al., 2002; Trifanescu et al., 2011). Surgical intervention is not indicated in incidentally discovered asymptomatic lesions. Reported recurrence rates vary and few studies describe observation periods longer than 2 years. Here we discuss the clinical, radiologic, and pathologic characteristics of these lesions along with their epidemiology, natural history, and pathogenesis. We also summarize the current literature on their treatment and risk factors for recurrence.

EPIDEMIOLOGY

Symptomatic Rathke's cleft cysts are rare and so the prevalence of this lesion is uncertain. In a series of 1000 unselected pituitaries from routine postmortem examinations, subclinical Rathke's cleft cysts were found in 113 cases (11.3%). Of these, 33 were located centrally and four were located in the lateral wings of the pituitary. The largest lesion reported was 8 mm in diameter and there were 37 cases with lesions larger than 2 mm (33%) (Teramoto et al., 1994). Another study of 2598 patients undergoing pituitary MRI detected a Rathke's cleft cyst in 3.4% of cases (Famini et al., 2011), although this study was conducted at a tertiary pituitary centre and so does not reflect the general population. As imaging techniques advance and scans become more frequent, a more accurate estimate of the prevalence of this lesion will emerge.

Rathke's cleft cysts have been reported in all age groups, but are more common in adults, with a peak incidence at 30–50 years and mean ages ranging from 34 to 44 years (Voelker et al., 1991; Teramoto et al., 1994; Isono et al., 2001; Kim et al., 2004; Raper and Besser, 2009; Madhok et al., 2010; Lillehei et al., 2011). Pediatric cases are rare (Zada et al., 2009; Jahangiri et al., 2011). Pituitary cysts suggestive of Rathke's cleft cysts were observed in 1.2% of patients between 1 and 4 years old who

*Correspondence to: Olaf Ansorge, MD, FRCPath, Nuffield Department of Clinical Neurosciences, University of Oxford, Department of Neuropathology, John Radcliffe Hospital, Headley Way, Oxford, OX3 9DU, UK. Tel: +44-1865-231434, Fax: +44-1865-231157, E-mail: olaf.ansorge@ndcn.ox.ac.uk

underwent MRI following presentation with epilepsy, psychomotor retardation, or headache (Takanashi et al., 2005). The tendency of symptomatic lesions to occur in adulthood is suggestive of a slow-growing lesion that develops over time and remains subclinical until later in life. No variation in prevalence of Rathke's cleft cyst has been reported between different ethnic groups. A female preponderance is observed in clinical studies and female:male ratios of between 1 and 5 have been reported (Voelker et al., 1991; Teramoto et al., 1994; Isono et al., 2001; Kim et al., 2004; Raper and Besser, 2009; Madhok et al., 2010; Jahangiri et al., 2011; Lillehei et al., 2011). However, the female:male ratio observed postmortem was 7:30 (Teramoto et al., 1994), suggesting that Rathke's cleft cysts in males may be underreported. It has been proposed that the greater clinical incidence in female patients is due to earlier detection as a result of menstrual irregularity reflecting pituitary dysfunction (Shin et al., 1999); however, the same female preponderance is seen in the pediatric population. In pediatric populations undergoing surgery for Rathke's cleft cyst, female:male ratios of 3.7 (Jahangiri et al., 2011) and 1.5 (Zada et al., 2009) have been reported.

PATHOLOGY

Rathke's cleft cysts are benign, cystic, predominantly intrasellar lesions originating from Rathke's cleft in the vestigial pars intermedia of the pituitary. Occasionally, suprasellar location has been reported. These lesions vary in size from a few mm up to 40 mm (Shin et al., 1999; Kim et al., 2004; Nishioka et al., 2006b; Chuang et al., 2010), with the majority between 10 and 20 mm in diameter (Shin et al., 1999; Kim et al., 2004).

Macroscopically, the cyst is encapsulated in a delicate membrane and the contents are generally thick, mucoid or gelatinous material, although CSF-like, motor-oil, and milky contents have been reported (Eguchi et al., 1994; Sade et al., 2005). Microscopically, the cyst wall may be simple or pseudostratified cuboidal or columnar epithelium; cilia and mucous-secreting goblet cells may be present (Fig. 17.1). Squamous metaplasia is common (20–40%) (Kim et al., 2004; Le et al., 2007) and may be extensive, resembling papillary craniopharyngioma, making differential diagnosis difficult. The presence of squamous metaplasia has been associated with increased risk of recurrence after surgery (Kim et al., 2004; Aho et al., 2005) and these metaplastic regions have been shown to have a higher Ki-67 proliferation index than the remainder of the cyst (Ikeda and Yoshimoto, 2002). Cyst contents are composed of cholesterol and protein that can form nodules of mucinous material that may or may not be connected to the cyst wall (Brassier et al., 1999; Byun et al., 2000; Binning et al., 2005). An inflammatory reaction has been reported in the cyst wall and adjacent tissue and is most visible at sites of cyst rupture. It has been proposed that

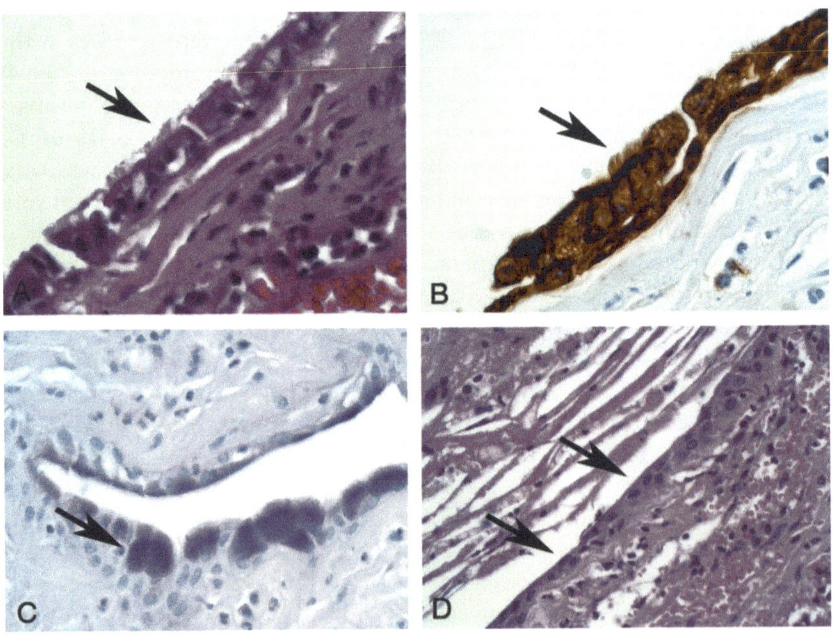

Fig. 17.1. The typical epithelial lining consists of tall epithelial cells with cilia (**A**, arrow). The epithelium expresses cytokeratins (**B**, Cam5.2 antibody, brown reaction product, arrow) and may be mucin-rich (**C**, PAS, arrow). Long-standing cyst epithelium may show pressure atrophy (**D**, arrows), resulting in flat epithelium without cilia (note clefted cyst contents). All magnifications ×600. (Reproduced from Trifanescu et al., 2012.)

this inflammation is a foreign body reaction to cyst contents. Changes in the cyst wall epithelium from simple to stratified correlate with the degree of inflammation and the inflammatory response has been suggested as the cause of stratification of the cyst wall epithelium (Hama et al., 2002). Pituitary deficit is often associated with Rathke's cleft cyst and may not completely recover after surgery. As well as the compressive effects of the cyst, inflammation has been suggested to contribute to destruction of the adjacent pituitary gland and the loss of pituitary function (Hama et al., 2002; Nishioka et al., 2006b). Other features that are uncommon in Rathke's cleft cysts include keratin nodules, hemosiderin, and cholesterol clefts (Kim et al., 2004).

Differential diagnosis of Rathke's cleft cyst and other cystic sellar lesions is often challenging due to surgical sampling bias: specimens often contain only cyst contents and very little epithelium. Overlapping features of Rathke's cleft cyst, craniopharyngioma, and epidermoid cyst, as well as collision lesions of Rathke's cleft cyst and pituitary adenoma may be encountered (Trokoudes et al., 1978; Arita et al., 1994; Bader et al., 2004; Karavitaki et al., 2008; Koutourousiou et al., 2010). In small biopsy samples, the capacity for Rathke's cleft cyst to undergo squamous metaplasia makes distinction from papillary craniopharyngioma particularly difficult. Table 17.1 summarizes the pathologic features of these lesions.

PATHOGENESIS

Formation of Rathke's pouch and pituitary organogenesis

The pituitary is an organ of dual origin. The anterior lobe – the adenohypophysis – is derived from oral ectoderm and is epithelial in origin, whereas the posterior lobe – the neurohypophysis – derives from the neural ectoderm. The composite nature of the pituitary requires that the neural and oral ectoderm interact physically and biochemically. Precise spatial and temporal coordination and regulation of signaling molecule release from both sites and the appropriate response by target cells is critical for pituitary formation and differentiation of the various hormone-producing cell types in the mature anterior lobe. The expression of transcription factors that control pituitary organogenesis, cell proliferation, lineage commitment, and terminal differentiation in the developing anterior lobe must be precisely regulated. Disruption of this regulation can lead to numerous developmental disorders.

Pituitary organogenesis begins during weeks 3–4 of fetal development. Under the control of the transcription factors Hesx1, expressed in the oral ectoderm, and Bmp4, expressed in the ventral diencephalon, a thickening of cells in the oral ectoderm forms the hypophyseal placode. This gives rise to Rathke's pouch, an upward outfolding of the oral ectoderm that extends towards the ventral diencephalon. Mutations in Hesx1 have been found in patients with combined pituitary hormone deficiency (Carvalho et al., 2003; Tajima et al., 2003), isolated growth hormone deficiency and septo-optic dysplasia (Wales and Quarrell, 1996; Dattani et al., 1998). Along with Hesx1, members of the Wnt, Notch, and Hedgehog pathways, Otx2 Lhx3 Hes1, Sox2 Bmp2 and Six family proteins are expressed in the developing Rathke's pouch (Ericson et al., 1998; Schmitt et al., 2000; Kioussi et al., 2002; Li et al., 2003; Raetzman et al., 2004; Davis and Camper, 2007; Gaston-Massuet et al., 2008; Kelberman et al., 2008; Gorbenko Del Blanco et al., 2012; for a detailed review, see Kelberman et al., 2009).

At the same time as the hypophyseal placode forms in the oral ectoderm, a downward extension of the ventral diencephalon develops that will eventually form the posterior lobe of the pituitary. Expression of Bmp4, FGF 8/10/18 Sox2/3 and Wnt 5a among others in the ventral diencephalon regulates this process and signals to the developing Rathke's pouch, resulting in connection of the two nascent lobes to create the composite structure of the adult pituitary (Takuma et al., 1998; Treier et al., 1998, 2001). Rathke's pouch constricts at its base and eventually separates altogether from the oral ectoderm during week 6–8 (Fig. 17.2). The cells of the anterior wall of Rathke's pouch undergo extensive proliferation to form the anterior lobe while the posterior wall proliferates more slowly to form the vestigial intermediate lobe. A switch in expression of transcription factors from those that promote cell proliferation to those governing cell patterning occurs within the anterior lobe, facilitating the formation of the five principal specialized endocrine cell types of the pituitary gland. Cell lineage commitment and terminal differentiation then occurs, regulated by transcription factors GATA-2 and Nr5a1 (gonadotrophs) (Ingraham et al., 1994), Tbx19 and NeuroD (corticotrophs) (Poulin et al., 2000; Pulichino et al., 2003), and POU1F1 (thyrotrophs, somatotrophs, and lactotrophs) (Rhodes et al., 1994). The cleft that remains between the anterior and posterior lobes is termed Rathke's cleft and persists into adulthood.

Pathogenesis of Rathke's cleft cyst and other cystic sellar lesions

Rathke's cleft cysts are not common and are seldom symptomatic. Consequently, the pathogenesis of Rathke's cleft cysts is not well understood. Several ideas have been proposed to explain their origin, including neuroepithelial derivation (Shuangshoti et al., 1970), metaplasia of anterior pituitary cells, and an endodermal

Table 17.1

Pathologic features of craniopharyngioma, Rathke's cleft cyst, and epidermoid cyst

	Adamantinomatous craniopharyngioma	Papillary craniopharyngioma	Rathke's cleft cyst	Epidermoid cyst
Macroscopic features				
Site	Suprasellar with infrequent intrasellar component and extension into neighboring structures	Suprasellar or within third ventricle	Intrasellar, occasional suprasellar extension	Cerebellopontine angle most common, but may occur anywhere in the intracranial cavity. Occasionally sellar or parasellar (Verkijk and Bots, 1980; Boggan et al., 1983; Oge and Ozgen, 1991; Tatagiba et al., 2000; Iaconetta et al., 2001; Sani et al., 2005). Usually off midline
Boundary	Lobular structure with sharp, irregular interface. Invasive and adherent to surrounding structures	Encapsulated. Discrete and often solid	Encapsulated in delicate membrane	Smooth, often translucent cyst wall with mother-of-pearl sheen. Irregular nodular surface, frequent incorporation of adjacent nerves or vessels
Cyst features/ contents	Dark, "motor-oil"-like fluid, often with cholesterol crystals	When cystic, clear contents	Mostly mucoid or gelatinous, comprising cholesterol and protein. Occasionally "motor-oil", milky or clear, CSF-like contents	Flaky, keratinous material
Microscopic features				
Epithelium	Peripheral, pallisading epithelium containing whorls of closely packed epithelial cells that often appear near the infiltrating edge of the tumor	Squamous, well-differentiated and nonkeratinizing	Simple or pseudostratified, ciliated, columnar epithelium. Some goblet cells. Squamous metaplasia common	Mature, keratinizing squamous epithelium
Stroma	Loosely aggregated stellate cells (stellate reticulum) containing nodules of anuclear "ghost cells" or wet keratin	Fibrovascular core with no stellate reticulum	N/A	N/A
Cyst contents	Dark, "motor-oil" appearance. Cholesterol crystals	N/A	Nodules of cholesterol and protein that may be connected to the cyst wall	Flaky, keratinous material containing anuclear, squamous "ghost cells"
Mitoses	Fewer in whorls		More frequent in squamous epithelium than simple or pseudostratified (Ikeda and Yoshimoto, 2002)	Absent or rare

N/A, not applicable; CSF, cerebrospinal fluid.

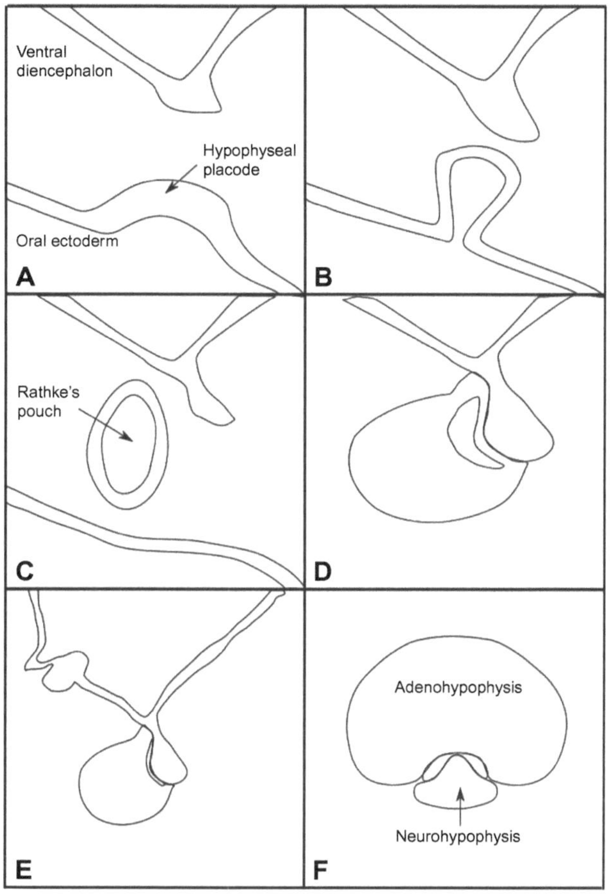

Fig. 17.2. Pituitary development. (**A**) Cells of the oral ectoderm proliferate to form the hypophyseal placode. (**B**) An upward outfolding of the oral ectoderm and a downward extension of the ventral diencephalon form the nascent Rathke's pouch. (**C**) Rathke's pouch constricts at its base and migrates along the craniopharyngeal canal to contact the ventral diencephalon. (**D**) Cells in the anterior wall of Rathke's pouch proliferate to form the adenohypophysis, while those in the posterior wall form the vestigial pars intermedia. (**E**) The composite structure of the developed pituitary gland in sagittal and (**F**) horizontal section.

origin (Hirano and Ghatak, 1974). These theories have largely been superseded by the idea that Rathke's cleft cysts originate from epithelial cells of the developing Rathke's pouch, and that these cells surrounding Rathke's cleft have the potential to give rise to various cystic lesions, including Rathke's cleft cyst, epidermoid cysts, and craniopharyngiomas. This is an attractive hypothesis as these lesions share many features. They have a similar location in the sellar and suprasellar regions and may show overlapping histopathologic features (Table 17.1). A series of observations of the similarities of these lesions has led to the hypothesis that Rathke's cleft cysts fall on a histopathologic continuum that includes epithelial, epidermoid, and dermoid cysts, and both papillary and adamantinomatous craniopharyngiomas (Harrison et al., 1994; Berx et al., 1995; Crotty et al., 1995; Mao et al., 2001; Hofmann et al., 2006). Although experimental evidence is lacking there are reports of transitional lesions that lend support to this idea. Several authors have speculated that papillary craniopharyngiomas represent an intermediate entity between Rathke's cleft cysts and adamantinomatous craniopharyngiomas, as these lesions have been found to contain ciliated epithelial cells and goblet cells characteristic of Rathke's cleft cysts (Goodrich et al., 1985; Harrison et al., 1994; Crotty et al., 1995; Oka et al., 1997).

A survey of patients diagnosed with Rathke's cleft cysts by Ikeda et al. showed a spectrum of lesions that contained features ranging from the characteristic ciliated columnar epithelium of Rathke's cleft cyst through ciliated squamous epithelium to squamous epithelium that is more commonly a feature of papillary craniopharyngioma (Ikeda and Yoshimoto, 2002). A study of the Ki-67 proliferation index in lesions from this cohort showed that it was significantly higher in regions of the lesion containing squamous epithelium compared to those containing simple epithelium. These findings raise the possibility of pathologic progression from Rathke's cleft cysts that contain large areas of squamous epithelium to papillary craniopharyngioma. Furthermore, in a large series of cases (n = 118), Aho et al. found that the presence of squamous metaplasia in the cyst wall was a significant predictor of recurrence of Rathke's cleft cyst after surgery, suggesting a more aggressive lesion (Aho et al., 2005). A study of 12 Rathke's cleft cysts with squamous metaplasia and features deemed "transitional" between classic Rathke's cleft cyst and craniopharyngioma by Ogawa et al. revealed nuclear and cytosolic β-catenin, a feature typically seen only in adamantinomatous craniopharyngioma (Hassanein et al., 2003; Buslei et al., 2005; Oikonomou et al., 2005; Hofmann et al., 2006; Holsken et al., 2009; Andoniadou et al., 2012; Ogawa et al., 2013). β-catenin was found in the nucleus and cytosol of 83% of patients who experienced postoperative re-enlargement of the cyst and all of the patients with nuclear and cytosolic β-catenin experienced postoperative re-enlargement. The authors propose that this pattern of β-catenin staining may be associated with a more aggressive clinical course. However, these "transitional" lesions are morphologically not well defined, and a much larger study including 30 classic Rathke's cleft cysts, nine of which showed squamous metaplasia, did not show any evidence of nuclear β-catenin translocation (Crotty et al., 1995). In adamantinomatous craniopharyngiomas, mutations in β-catenin are associated with cytosolic and nuclear accumulation due to disruption of sites that are phosphorylated by the β-catenin destruction complex. Mutation in β-catenin has been implicated in the tumorigenesis of adamantinomatous craniopharyngioma (Gaston-Massuet et al., 2008), although the relationship between β-catenin phosphorylation and mutation remains to be established (Sekine et al., 2003; Kato et al., 2004; Holsken et al., 2009). There have been case reports of craniopharyngiomas, particularly the papillary form, that have arisen after treatment for Rathke's cleft cysts which may indicate progression from one lesion to the other, although the possibility of coexisting lesions cannot be excluded (Sato et al., 2006; Park et al., 2009; Wolfe and Heros, 2010).

The rarity and largely asymptomatic nature of Rathke's cleft cyst has meant that research into the pathogenesis of this lesion is scant. Much further study is required to determine the molecular pathologic events underlying Rathke's cleft cyst and other cystic sellar lesions and to establish whether the idea of a continuum of cystic lesions is tenable.

PRESENTING MANIFESTATIONS

Rathke's cleft cysts are mostly asymptomatic and become clinically apparent if they are large enough to compress adjacent structures or rupture. In this case, the most common chronic symptoms at presentation are headache, visual disturbance, and pituitary dysfunction. Occasionally, acute symptoms resembling pituitary apoplexy may be present. The mean duration of clinical symptoms until diagnosis varies greatly, from a few days to years, but is generally between 9 and 24 months (Midha et al., 1991; Mukherjee et al., 1997; Shin et al., 1999; Billeci et al., 2004; Kim et al., 2004; Chaiban et al., 2011; Trifanescu et al., 2011).

Headaches

The most common manifestation of Rathke's cleft cyst at presentation is headache. In up to 40% of cases, this may be the only symptom (Nishioka et al., 2006a, b). Headache has been reported in between 33% and 81% of patients and it may be chronic or continuous in 60% (Mukherjee et al., 1997; Shin et al., 1999; Isono et al., 2001; Kasperbauer et al., 2002; Kim et al., 2004; Aho et al., 2005; Nishioka et al., 2006b; Madhok et al., 2010; Wait et al., 2010; Trifanescu et al., 2011; Xie et al., 2011). Headaches associated with Rathke's cleft cyst are most commonly episodic and nonpulsating frontal, bilateral, or deep retro-orbital pain and are occasionally associated with nausea and vomiting. However, they have also been reported to affect the occipital or temporal regions or the whole head (Nishioka et al., 2006a). A study of headaches associated with Rathke's cleft cyst by Nishioka et al. found no correlation between headache symptoms and cyst size or location or the presence of pituitary dysfunction. However, headache was more frequently observed in Rathke's cleft cyst with hyper- or isointense cyst contents on T1-weighted magnetic resonance imaging (MRI) than in hypo-intense T1. This hyperintesity is associated with a cyst that has a higher protein content – usually mucinous. No relationship has been reported between headache and cyst size or location, but they are associated with hyper- or isointensity on T1-weighted MRI and with mucinous cyst content or cyst wall inflammation. Although the presence of pituitary dysfunction *per se* does not correlate with headache, cyst wall inflammation, which can lead to irreversible pituitary damage and subsequent endocrine dysfunction, is associated with episodic headache and may suggest a surgical approach (Nishioka et al., 2006a).

Visual field disturbance

Between 12% and 75% of patients reported visual disturbance at presentation, including disturbances in visual acuity and visual field (Eguchi et al., 1994; Mukherjee et al., 1997; Shin et al., 1999; Isono et al., 2001; Kim et al., 2004; Aho et al., 2005; Nishioka et al., 2006a, b; Wait et al., 2010; Trifanescu et al., 2011; Xie et al., 2011). Visual field abnormalities have been found to correlate significantly with cyst size and compression of optic nerve or chiasm (Nishioka et al., 2006b).

Endocrine dysfunction

Anterior pituitary hormone deficits in one or more axes associated with Rathke's cleft cyst have been reported in between 19% and 81% of patients. The most frequently reported abnormalities were hypogonadism with amenorrhea or impotence/low libido and galactorrhea as a result of hyperprolactinemia (Shin et al., 1999; Isono et al., 2001; Kim et al., 2004; Aho et al., 2005; Sade et al., 2005; Raper and Besser, 2009). However, the diagnostic tests used to determine endocrine function varied between studies (Table 17.2). No correlation between hypopitutarism and cyst size was observed (Nishioka et al., 2006b).

Diabetes insipidus

The prevalence of diabetes insipidus (DI) at presentation has been variably reported between 0% and 19% (el-Mahdy and Powell, 1998; Shin et al., 1999; Kim et al., 2004; Aho et al., 2005; Sade et al., 2005; Nishioka et al., 2006a, b; Raper and Besser, 2009; Trifanescu et al., 2011). There is an association between a hyper- or isointense T1 MRI signal and DI at presentation (Nishioka et al., 2006b). Some authors suggest that T1 hyperintensity is a characteristic of lesions that show increased inflammatory reaction to cyst contents and associated damage to the pituitary (Nishioka et al., 2006a).

Apoplexy

Apoplexy associated with Rathke's cleft cyst is uncommon and may or may not be associated with intracystic hemorrhage. Symptoms include sudden onset of or increase in severity of headache, with visual disturbance, nausea and vomiting, meningismus, occulomotor palsies, diplopia, and impairment of pituitary function (Wen et al., 2010; Chaiban et al., 2011). Alterations in mental status may also be observed (Chaiban et al., 2011).

Other presenting manifestations

Rupture of the cyst and leakage of contents into the subarachnoid space along with extension of inflammation into surrounding structures has been proposed to explain the uncommon presentation of aseptic meningitis and subsequent visual loss (Benveniste et al., 2004; Nishioka et al., 2006a; Zada et al., 2009). Other reports of rare presenting manifestations include sphenoid sinusitis, syncope, seizures (Benveniste et al., 2004; Lillehei et al., 2011), ataxia (Cao et al., 2008), mood disturbance (Cao et al., 2008; Lillehei et al., 2011), and precocious puberty.

LOCATION AND IMAGING FEATURES

Rathke's cleft cysts are intrasellar and suprasellar lesions. Exclusively suprasellar location is rare. The most common location is the pars intermedia and cysts are medial (rarely lateral), well-circumscribed and ovoid or dumb-bell shaped. The cyst may displace the pituitary in any direction and the gland is often located around or under the cyst. Estimations of the rate of suprasellar extension of Rathke's cleft cysts vary from 16% to 97% and some reports describe extension into the sphenoid or cavernous sinus or frontal area, but this is very rare. Generally the maximum diameter of these lesions is 5–50 mm (mean 10–20 mm).

Table 17.2

Endocrine disturbances in Rathke's cleft cyst at presentation

Study	Number of patients	LH/FSH	GH	TSH	ACTH/cortisol	PRL
Eguchi et al.	19	31.6/84.2	78.9	26.3	68.4	52.6%
Aho et al.	118	62	66	8	7	
Kim et al.	40	25	15	22.5	40	20
Raper and Besser	12	33/33	25	16.7	16.7 / 25	50
Shin et al.	18	43	48	35	57	39
Trifanescu et al., 2011	33	60		36	36	
Benveniste et al.	62	45	12	34	35	42
Mean ± SEM	43 ± 14	43 ± 5.4	41 ± 11.4	26 ± 4.0	37 ± 8.0	41 ± 5.7

LH, luteinizing hormone; FSH, follicle-stimulating hormone; GH, growth hormone; TSH, thyroid-stimulating hormone; ACTH, adrenocorticotropic hormone; PRL, prolactin; SEM, standard error of mean.

Computed tomography

Computed tomography (CT) reveals homogenous lesions that may appear hypodense, isodense or slightly hyperdense compared to brain parenchyma. Rathke's cleft cysts may or may not show enhancement, but when present, enhancement is ring-like or capsular.

Magnetic resonance imaging

Rathke's cleft cysts appear as well demarcated, homogenous lesions with occasional enhancement of the cyst wall, which has been attributed variously to an inflammatory reaction, hemosiderin deposition, cholesterol crystals, squamous metaplasia of the cyst wall, or a displaced pituitary gland. The intensity of signal is highly variable and dependent on the cyst contents (Fig. 17.3). A cyst containing clear, CSF-like fluid will appear hypointense on T1- and hyperintense on T2-weighted imaging, whereas a cyst containing thick, mucoid material will appear hyperintense on T1- and isointense on T2-weighted imaging. A cyst containing blood will give a hyperintense signal on both T1 and T2.

Although the more common finding is hyperintensity on T2 imaging (about 70% of cases), a cyst that gives a hypointense T2-weighted signal can be useful in differential diagnosis from other lesions that more commonly show hyperintense T2 signal, such as chordomas, chondroid tumors, abscesses, mucoceles, and lipomas (Bonneville et al., 2006, 2007). A feature that is almost pathognomonic for Rathke's cleft cyst is an intracystic, nonenhancing nodule, showing low intensity on T2- and high intensity on T1-weighted imaging compared to cyst fluid. The various imaging features of Rathke's cleft cyst and related lesions can be useful in preoperative differential diagnosis (Table 17.3).

NATURAL HISTORY

The natural history of Rathke's cleft cysts is not well understood. There are few studies that report large series with long follow-up times. In a study by Aho et al. of pituitary incidentalomas, 61 patients with presumed Rathke's cleft cyst were followed for a period of 9 years. Cyst growth, visual loss or impairment, or endocrinopathy were observed in 19 patients (31%), who were subsequently treated surgically. The remaining patients showed no endocrine dysfunction or cyst growth during follow-up (Aho et al., 2005). Other studies have estimated the incidence of cyst growth at 5.3% (mean follow-up period 27 months) (Sanno et al., 2003). Spontaneous involution of Rathke's cleft cyst in patients receiving no surgical intervention has also been reported in 9 of 29 patients (31%) associated with resolution of headache in seven cases (Amhaz et al., 2010). The mechanism of regression of the cyst in these patients is not known, but absorption of cyst fluid or repeated rupture have been proposed to explain this phenomenon (Nishio et al., 2001).

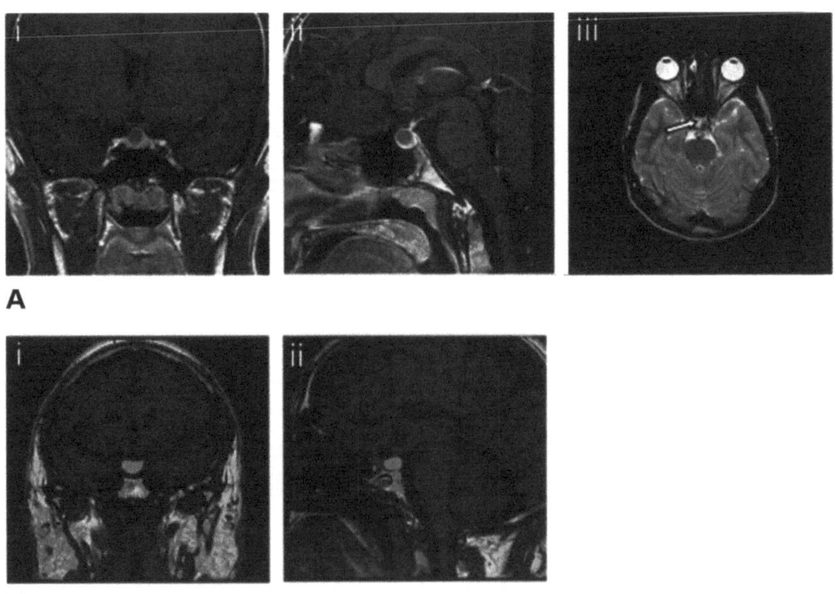

Fig. 17.3. MRIs of RCCs. (**A**) Coronal (i) and sagittal (ii) T1-weighted postgadolinium images showing low signal intensity and rim enhancement. The same cyst demonstrated hyperintense signal on T2-weighted images (shown with arrow) (iii). (**B**) Coronal (i) and sagittal (ii) T1-weighted images showing hyperintense signal.

Table 17.3

Imaging features of Rathke's cleft cyst and similar intra- and suprasellar masses†

			MRI		CT		
	Shape	Solid/ cystic	Intensity	Contrast enhancement	Attenuation	Contrast enhancement	Other
Rathke's cleft cyst (Mukherjee et al., 1997; Binning et al., 2005; Nishioka et al., 2006b)	Ovoid, small	Cystic	Variable depending on cyst contents	Little or no wall enhancement	Variable depending on cyst contents	Generally none. Occasionally seen in cyst wall	
Pituitary adenoma (Davis et al., 1987; Nishioka et al., 2006b; Choi et al., 2007; Rennert and Doerfler, 2007)	"Snowman"	Solid		Homogenous in sold portion	Similar to brain in solid regions	Moderate in solid regions	
Craniopharyngioma	Superiorly lobulated, larger	Mixed		Reticular enhancement of solid portion			Compression of third ventricle or regions of calcification common
Dermoid cyst	Round, lobulated	Cystic	Hyperintense	None	Hypointense	None	Foci of calcification
Epidermoid cyst	Lobulated	Cystic	Similar to CSF	None	Similar to CSF		Occasional calcification, restricted diffusion on diffusion-weighted imaging
Arachnoid cyst	Distinct border	Cystic	Similar to CSF	None	Similar to CSF	None	No calcification
Abscess			Hyperintense T1 Hyperintense T2	Significant wall enhancement	Centrally hypodense with ring of increased density	Marked	

†Adapted from Trifanescu et al., 2012.
MRI, magnetic resonance imaging; CT, computed tomography; CSF, cerebrospinal fluid.

TREATMENT

Treatment strategies

Small, asymptomatic cysts do not require surgery. When cysts are large enough to manifest as clinical symptoms, surgery is indicated with the aim of draining the cyst content and removing as much of the capsule as safely possible. The most common approach is via the endonasal or sublabial transsphenoidal route with subsequent fenestration and drainage of the cyst (Midha et al., 1991; Zada et al., 2003, 2009; Benveniste et al., 2004; Aho et al., 2005; Madhok et al., 2010). An endonasal endoscopic transsphenoidal approach has also been reported (Frank et al., 2005; Xie et al., 2011) and some larger or suprasellar cysts have required craniotomy (Benveniste et al., 2004). Resolution or improvement of pressure effects, particularly headaches and visual impairment, is generally achieved with this approach. Several studies have reported the use of ethanol cauterization in cases where the arachnoid membrane was uncompromised with the intention of preventing cyst regrowth by treating any microscopic rests of epithelium. This treatment showed no significant effect on relapse rates and a single case report has described significant neurologic complications from ethanol administration (Hsu et al., 2004). Radiotherapy is very uncommon, but may be of value especially for recurrent cysts (Mukherjee et al., 1997).

Complications

Postoperative cerebrospinal fluid (CSF) leak has been reported in up to 25% of patients (Kim et al., 2004; Nishioka et al., 2006b; Raper and Besser, 2009; Xie et al., 2011) and is more common in patients undergoing more aggressive surgeries with the aim of gross total resection as opposed to cyst drainage (Higgins et al., 2011). Diabetes insipidus has also been reported and associated with more aggressive resections (Higgins et al., 2011). This can be transient (up to 67%) (Raper and Besser, 2009) or permanent (up to 20%) (Schreckinger et al., 2013). Other postoperative complications include anterior pituitary hormone deficits, hyponatremia, meningitis, sinusitis, and infection (up to 12%) (Benveniste et al., 2004; Tate et al., 2010; Lillehei et al., 2011). There are no reports of perioperative mortality (Shin et al., 1999; Benveniste et al., 2004; Aho et al., 2005; Nishioka et al., 2006b; Raper and Besser, 2009; Lillehei et al., 2011).

Outcomes

Visual field disturbances and headaches improve or resolve after surgery, with between 40% and 100% showing improvement or resolution of headaches (Midha et al., 1991; Shin et al., 1999; Isono et al., 2001; Benveniste et al., 2004; Billeci et al., 2004; Kim et al., 2004; Aho et al., 2005; Sade et al., 2005; Nishioka et al., 2006a, b; Madhok et al., 2010; Lillehei et al., 2011) and 33–100% showing improvement or resolution of visual field defects (Midha et al., 1991; Shin et al., 1999; Isono et al., 2001; Kasperbauer et al., 2002; Benveniste et al., 2004; Billeci et al., 2004; Kim et al., 2004; Aho et al., 2005; Sade et al., 2005; Nishioka et al., 2006a, b; Madhok et al., 2010; Lillehei et al., 2011; Trifanescu et al., 2011). Recovery of pituitary function is more variable. In cases of partial hypopituitarism, between 14% and 50% of patients experience postsurgical recovery of pituitary function (Midha et al., 1991; Shin et al., 1999; Aho et al., 2005; Nishioka et al., 2006b; Zada et al., 2009; Madhok et al., 2010; Lillehei et al., 2011), with the most commonly resolved abnormality being hyperprolactinemia (Voelker et al., 1991; Eguchi et al., 1994; Shin et al., 1999; Kasperbauer et al., 2002; Billeci et al., 2004; Kim et al., 2004; Aho et al., 2005; Nishioka et al., 2006b; Sade et al., 2005; Lillehei et al., 2011). In contrast, patients with panhypopituitarism rarely recover pituitary function postsurgery (Isono et al., 2001; Billeci et al., 2004; Sade et al., 2005). This poor recovery of function has been linked to chronic inflammation that causes damage to the pituitary (Hama et al., 2002; Benveniste et al., 2004; Sade et al., 2005; Nishioka et al., 2006a, b). Mucinous cyst contents that give a high signal intensity on T1-weighted MRI are more likely to be associated with an inflammatory reaction visible on histology and poor recovery of pituitary function (Nishioka et al., 2006a).

RECURRENCE

Relapse rates

Reports on the rate of recurrence of Rathke's cleft cyst and associated risk factors have relatively short follow-up periods (mean between 19 and 48 months), leading to a probable underestimation of the relapse rate (Mukherjee et al., 1997; Shin et al., 1999; Isono et al., 2001; Kasperbauer et al., 2002; Benveniste et al., 2004; Billeci et al., 2004; Kim et al., 2004; Aho et al., 2005; Sade et al., 2005; Nishioka et al., 2006a; Koutourousiou et al., 2009; Zada et al., 2009; Madhok et al., 2010; Wait et al., 2010; Trifanescu et al., 2011; Xie et al., 2011) which ranges from 0% to 48%. Most relapses are reported within 5–6 years, although varying criteria for the definition of relapse have led to a lack of consensus in the literature (intervals between 1 month and 24 years have been reported) (Mukherjee et al., 1997; Shin et al., 1999; Isono et al., 2001; Kasperbauer et al., 2002; Benveniste et al., 2004; Billeci et al., 2004; Kim et al., 2004). Nevertheless, these studies suggest that follow-up for at least 5 years after surgery is needed.

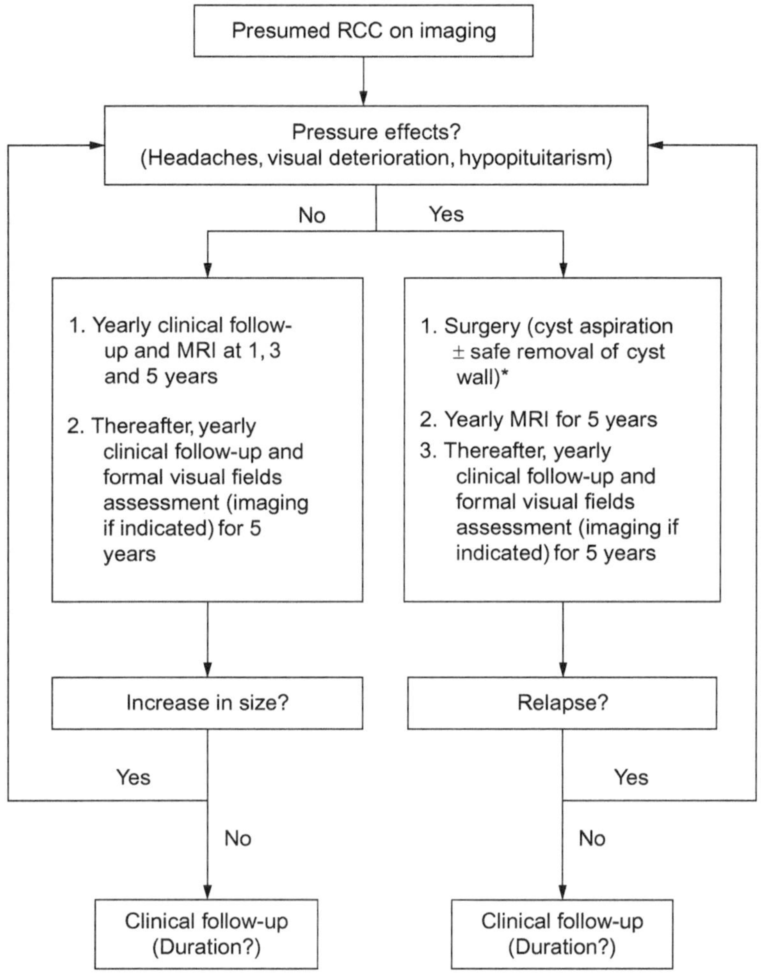

* In case of multiple relapses external irradiation may be considered

Fig. 17.4. Management algorithm for Rathke's cleft cysts. (Modified from Trifanescu et al., 2012.)

Multiple relapses have been reported, but these are rare (Benveniste et al., 2004; Billeci et al., 2004; Raper and Besser, 2009; Lillehei et al., 2011; Trifanescu et al., 2011). Based on the likelihood that recurrence rates have been underestimated, Trifanescu et al. present a proposed management algorithm for Rathke's cleft cyst (Fig. 17.4) (Trifanescu et al., 2012).

Risk factors for relapse

Risk factors that are reported to be associated with increased recurrence include cyst size (also significantly associated with shorter time to recurrence) (Shin et al., 1999), presence of squamous metaplasia of the cyst wall (Benveniste et al., 2004; Kim et al., 2004; Aho et al., 2005; Wait et al., 2010; Zada et al., 2010; Lillehei et al., 2011) (possibly as a consequence of an inflammatory reaction to cyst contents), incomplete resection as evidenced by residual cyst on postoperative MRI (Kim et al., 2004; Raper and Besser, 2009), enhancement of the cyst on MRI (Benveniste et al., 2004; Kim et al., 2004), intraoperative CSF leak (Aho et al., 2005; Lillehei et al., 2011), and the need for an abdominal fat graft for sellar packing (Aho et al., 2005), presence of inflammation (Benveniste et al., 2004; Kim et al., 2004) and infection (Tate et al., 2010). Antibiotic treatment reduces the rate of recurrence of an infected Rathke's cleft cyst to near that of an uninfected lesion (13% and 9% respectively) (Tate et al., 2010). Increased risk of relapse has been linked with a more aggressive surgical approach and the use of intraoperative ethanol cauterization in some studies (Benveniste et al., 2004; Kim et al., 2004), but there is no clear consensus on whether the extent of resection is related to increased risk of recurrence. A more aggressive surgical approach is, however, linked to a higher incidence of new postoperative pituitary hormone deficiencies, especially diabetes insipidus (Benveniste et al., 2004; Kim et al., 2004).

Optimal treatment strategies for recurrent, symptomatic Rathke's cleft cyst are unclear. Reported approaches have included aspiration and cyst wall removal (Mukherjee et al., 1997; Trifanescu et al., 2011) and external beam pituitary radiotherapy, but the efficacy of this treatment in preventing further recurrence is undetermined (Mukherjee et al., 1997).

ACKNOWLEDGMENTS

Supported by the National Institute for Health Research (NIHR) Oxford Biomedical Research Centre based at Oxford University Hospitals NHS Trust and University of Oxford, UK. The views expressed are those of the author(s) and not necessarily those of the UK National Health Service, the NIHR, or the UK Department of Health.

References

Aho CJ, Liu C, Zelman V et al. (2005). Surgical outcomes in 118 patients with Rathke cleft cysts. J Neurosurg 102: 189–193.

Amhaz HH, Chamoun RB, Waguespack SG et al. (2010). Spontaneous involution of Rathke cleft cysts: is it rare or just underreported? J Neurosurg 112: 1327–1332.

Andoniadou CL, Gaston-Massuet C, Reddy R et al. (2012). Identification of novel pathways involved in the pathogenesis of human adamantinomatous craniopharyngioma. Acta Neuropathol 124: 259–271.

Arita K, Uozumi T, Takechi A et al. (1994). A case of Cushing's disease accompanied by Rathke's cleft cyst: the usefulness of cavernous sinus sampling in the localization of microadenoma. Surg Neurol 42: 112–116.

Bader LJ, Carter KD, Latchaw RE et al. (2004). Simultaneous symptomatic Rathke's cleft cyst and GH secreting pituitary adenoma: a case report. Pituitary 7: 39–44.

Benveniste RJ, King WA, Walsh J et al. (2004). Surgery for Rathke cleft cysts: technical considerations and outcomes. J Neurosurg 101: 577–584.

Berx G, Cleton-Jansen AM, Nollet F et al. (1995). E-cadherin is a tumour/invasion suppressor gene mutated in human lobular breast cancers. EMBO J 14: 6107–6115.

Billeci D, Marton E, Tripodi M et al. (2004). Symptomatic Rathke's cleft cysts: a radiological surgical and pathological review. Pituitary 7: 131–137.

Binning MJ, Gottfried ON, Osborn AG et al. (2005). Rathke cleft cyst intracystic nodule: a characteristic magnetic resonance imaging finding. J Neurosurg 103: 837–840.

Boggan JE, Davis RL, Zorman G et al. (1983). Intrasellar epidermoid cyst. Case report. J Neurosurg 58: 411–415.

Bonneville F, Cattin F, Marsot-Dupuch K et al. (2006). T1 signal hyperintensity in the sellar region: spectrum of findings. Radiographics 26: 93–113.

Bonneville F, Chiras J, Cattin F et al. (2007). T2 hypointense signal of Rathke cleft cyst. AJNR Am J Neuroradiol 28: 397.

Brassier G, Morandi X, Tayiar E et al. (1999). Rathke's cleft cysts: surgical-MRI correlation in 16 symptomatic cases. J Neuroradiol 26: 162–171.

Buslei R, Nolde M, Hofmann B et al. (2005). Common mutations of beta-catenin in adamantinomatous craniopharyngiomas but not in other tumours originating from the sellar region. Acta Neuropathol 109: 589–597.

Byun WM, Kim OL, Kim D (2000). MR imaging findings of Rathke's cleft cysts: significance of intracystic nodules. AJNR Am J Neuroradiol 21: 485–488.

Cao Z, Lv J, Ding Z et al. (2008). Pathological laughter in a patient with Rathke cleft cyst. J Clin Neurosci 15: 1279–1282.

Carvalho LR, Woods KS, Mendonca BB et al. (2003). A homozygous mutation in HESX1 is associated with evolving hypopituitarism due to impaired repressor-corepressor interaction. J Clin Invest 112: 1192–1201.

Chaiban JT, Abdelmannan D, Cohen M et al. (2011). Rathke cleft cyst apoplexy: a newly characterized distinct clinical entity. J Neurosurg 114: 318–324.

Choi SH, Kwon BJ, Na DG et al. (2007). Pituitary adenoma craniopharyngioma and Rathke cleft cyst involving both intrasellar and suprasellar regions: differentiation using MRI. Clin Radiol 62: 453–462.

Chuang CC, Chen YL, Jung SM et al. (2010). A giant retroclival Rathke's cleft cyst. J Clin Neurosci 17: 1189–1191.

Crotty TB, Scheithauer BW, Young Jr WF et al. (1995). Papillary craniopharyngioma: a clinicopathological study of 48 cases. J Neurosurg 83: 206–214.

Dattani MT, Martinez-Barbera JP, Thomas PQ et al. (1998). Mutations in the homeobox gene HESX1/Hesx1 associated with septo-optic dysplasia in human and mouse. Nat Genet 19: 125–133.

Davis SW, Camper SA (2007). Noggin regulates Bmp4 activity during pituitary induction. Dev Biol 305: 145–160.

Davis PC, Hoffman Jr JC, Spencer T et al. (1987). MR imaging of pituitary adenoma: CT clinical and surgical correlation. AJR Am J Roentgenol 148: 797–802.

Eguchi K, Uozumi T, Arita K et al. (1994). Pituitary function in patients with Rathke's cleft cyst: significance of surgical management. Endocr J 41: 535–540.

el-Mahdy W, Powell M (1998). Transsphenoidal management of 28 symptomatic Rathke's cleft cysts with special reference to visual and hormonal recovery. Neurosurgery 42: 7–16, discussion 16–17.

Ericson J, Norlin S, Jessell TM et al. (1998). Integrated FGF and BMP signaling controls the progression of progenitor cell differentiation and the emergence of pattern in the embryonic anterior pituitary. Development 125: 1005–1015.

Famini P, Maya MM, Melmed S (2011). Pituitary magnetic resonance imaging for sellar and parasellar masses: ten-year experience in 2598 patients. J Clin Endocrinol Metab 96: 1633–1641.

Frank G, Sciarretta V, Mazzatenta D et al. (2005). Transsphenoidal endoscopic approach in the treatment of Rathke's cleft cyst. Neurosurgery 56: 124–128, discussion 129.

Gaston-Massuet C, Andoniadou CL, Signore M et al. (2008). Genetic interaction between the homeobox transcription factors HESX1 and SIX3 is required for normal pituitary development. Dev Biol 324: 322–333.

Goodrich JT, Post KD, Duffy P (1985). Ciliated craniopharyngioma. Surg Neurol 24: 105–111.

Gorbenko Del Blanco D, Romero CJ, Diaczok D et al. (2012). A novel OTX2 mutation in a patient with combined pituitary hormone deficiency pituitary malformation and an underdeveloped left optic nerve. Eur J Endocrinol 167: 441–452.

Hama S, Arita K, Nishisaka T et al. (2002). Changes in the epithelium of Rathke cleft cyst associated with inflammation. J Neurosurg 96: 209–216.

Harrison MJ, Morgello S, Post KD (1994). Epithelial cystic lesions of the sellar and parasellar region: a continuum of ectodermal derivatives? J Neurosurg 80: 1018–1025.

Hassanein AM, Glanz SM, Kessler HP et al. (2003). beta-Catenin is expressed aberrantly in tumors expressing shadow cells. Pilomatricoma craniopharyngioma and calcifying odontogenic cyst. Am J Clin Pathol 120: 732–736.

Higgins DM, Van Gompel JJ, Nippoldt TB et al. (2011). Symptomatic Rathke cleft cysts: extent of resection and surgical complications. Neurosurg Focus 31: E2.

Hirano A, Ghatak NR (1974). The fine structure of colloid cysts of the third ventricle. J Neuropathol Exp Neurol 33: 333–341.

Hofmann BM, Kreutzer J, Saeger W et al. (2006). Nuclear beta-catenin accumulation as reliable marker for the differentiation between cystic craniopharyngiomas and Rathke cleft cysts: a clinico-pathologic approach. Am J Surg Pathol 30: 1595–1603.

Holsken A, Kreutzer J, Hofmann BM et al. (2009). Target gene activation of the Wnt signaling pathway in nuclear beta-catenin accumulating cells of adamantinomatous craniopharyngiomas. Brain Pathol 19: 357–364.

Hsu HY, Piva A, Sadun AA (2004). Devastating complications from alcohol cauterization of recurrent Rathke cleft cyst. Case report. J Neurosurg 100: 1087–1090.

Iaconetta G, Carvalho GA, Vorkapic P et al. (2001). Intracerebral epidermoid tumor: a case report and review of the literature. Surg Neurol 55: 218–222.

Ikeda H, Yoshimoto T (2002). Clinicopathological study of Rathke's cleft cysts. Clin Neuropathol 21: 82–91.

Ingraham HA, Lala DS, Ikeda Y et al. (1994). The nuclear receptor steroidogenic factor 1 acts at multiple levels of the reproductive axis. Genes Dev 8: 2302–2312.

Isono M, Kamida T, Kobayashi H et al. (2001). Clinical features of symptomatic Rathke's cleft cyst. Clin Neurol Neurosurg 103: 96–100.

Jahangiri A, Molinaro AM, Tarapore PE et al. (2011). Rathke cleft cysts in pediatric patients: presentation surgical management and postoperative outcomes. Neurosurg Focus 31: E3.

Karavitaki N, Scheithauer BW, Watt J et al. (2008). Collision lesions of the sella: co-existence of craniopharyngioma with gonadotroph adenoma and of Rathke's cleft cyst with corticotroph adenoma. Pituitary 11: 317–323.

Kasperbauer JL, Orvidas LJ, Atkinson JL et al. (2002). Rathke cleft cyst: diagnostic and therapeutic considerations. Laryngoscope 112: 1836–1839.

Kato K, Nakatani Y, Kanno H et al. (2004). Possible linkage between specific histological structures and aberrant reactivation of the Wnt pathway in adamantinomatous craniopharyngioma. J Pathol 203: 814–821.

Kelberman D, de Castro SC, Huang S et al. (2008). SOX2 plays a critical role in the pituitary forebrain and eye during human embryonic development. J Clin Endocrinol Metab 93: 1865–1873.

Kelberman D, Rizzoti K, Lovell-Badge R et al. (2009). Genetic regulation of pituitary gland development in human and mouse. Endocr Rev 30: 790–829.

Kim JE, Kim JH, Kim OL et al. (2004). Surgical treatment of symptomatic Rathke cleft cysts: clinical features and results with special attention to recurrence. J Neurosurg 100: 33–40.

Kioussi C, Briata P, Baek SH et al. (2002). Identification of a Wnt/Dvl/beta-Catenin –>Pitx2 pathway mediating cell-type-specific proliferation during development. Cell 111: 673–685.

Koutourousiou M, Grotenhuis A, Kontogeorgos G et al. (2009). Treatment of Rathke's cleft cysts: experience at a single centre. J Clin Neurosci 16: 900–903.

Koutourousiou M, Kontogeorgos G, Wesseling P et al. (2010). Collision sellar lesions: experience with eight cases and review of the literature. Pituitary 13: 8–17.

Le BH, Towfighi J, Kapadia SB et al. (2007). Comparative immunohistochemical assessment of craniopharyngioma and related lesions. Endocr Pathol 18: 23–30.

Li X, Oghi KA, Zhang J et al. (2003). Eya protein phosphatase activity regulates Six1-Dach-Eya transcriptional effects in mammalian organogenesis. Nature 426: 247–254.

Lillehei KO, Widdel L, Astete CA et al. (2011). Transsphenoidal resection of 82 Rathke cleft cysts: limited value of alcohol cauterization in reducing recurrence rates. J Neurosurg 114: 310–317.

Madhok R, Prevedello DM, Gardner P et al. (2010). Endoscopic endonasal resection of Rathke cleft cysts: clinical outcomes and surgical nuances. J Neurosurg 112: 1333–1339.

Mao J, Wang J, Liu B et al. (2001). Low-density lipoprotein receptor-related protein-5 binds to Axin and regulates the canonical Wnt signaling pathway. Mol Cell 7: 801–809.

Midha R, Jay V, Smyth HS (1991). Transsphenoidal management of Rathke's cleft cysts. A clinicopathological review of 10 cases. Surg Neurol 35: 446–454.

Mukherjee JJ, Islam N, Kaltsas G et al. (1997). Clinical radiological and pathological features of patients with Rathke's cleft cysts: tumors that may recur. J Clin Endocrinol Metab 82: 2357–2362.

Nishio S, Morioka T, Suzuki S et al. (2001). Spontaneous regression of a pituitary cyst: report of two cases. Clin Imaging 25: 15–17.

Nishioka H, Haraoka J, Izawa H et al. (2006a). Headaches associated with Rathke's cleft cyst. Headache 46: 1580–1586.

Nishioka H, Haraoka J, Izawa H et al. (2006b). Magnetic resonance imaging clinical manifestations and management of Rathke's cleft cyst. Clin Endocrinol (Oxf) 64: 184–188.

Ogawa Y, Watanabe M, Tominaga T (2013). Rathke's cleft cysts with significant squamous metaplasia-high risk of

postoperative deterioration and close origins to craniopharyngioma. Acta Neurochir (Wien) 155: 1069–1075.
Oge K, Ozgen T (1991). Transsphenoidal removal of an intra- and suprasellar epidermoid cyst. Neurochirurgia 34: 94–96.
Oikonomou E, Barreto DC, Soares B et al. (2005). Beta-catenin mutations in craniopharyngiomas and pituitary adenomas. J Neurooncol 73: 205–209.
Oka H, Kawano N, Yagishita S et al. (1997). Ciliated craniopharyngioma indicates histogenetic relationship to Rathke cleft epithelium. Clin Neuropathol 16: 103–106.
Park YS, Ahn JY, Kim DS et al. (2009). Late development of craniopharyngioma following surgery for Rathke's cleft cyst. Clin Neuropathol 28: 177–181.
Poulin G, Lebel M, Chamberland M et al. (2000). Specific protein-protein interaction between basic helix-loop-helix transcription factors and homeoproteins of the Pitx family. Mol Cell Biol 20: 4826–4837.
Pulichino AM, Vallette-Kasic S, Tsai JP et al. (2003). Tpit determines alternate fates during pituitary cell differentiation. Genes Dev 17: 738–747.
Raetzman LT, Ross SA, Cook S et al. (2004). Developmental regulation of Notch signaling genes in the embryonic pituitary: Prop1 deficiency affects Notch2 expression. Dev Biol 265: 329–340.
Raper DM, Besser M (2009). Clinical features management and recurrence of symptomatic Rathke's cleft cyst. J Clin Neurosci 16: 385–389.
Rennert J, Doerfler A (2007). Imaging of sellar and parasellar lesions. Clin Neurol Neurosurg 109: 111–124.
Rhodes SJ, DiMattia GE, Rosenfeld MG (1994). Transcriptional mechanisms in anterior pituitary cell differentiation. Curr Opin Genet Dev 4: 709–717.
Ross DA, Norman D, Wilson CB (1992). Radiologic characteristics and results of surgical management of Rathke's cysts in 43 patients. Neurosurgery 30: 173–178, discussion 178–179.
Sade B, Albrecht S, Assimakopoulos P et al. (2005). Management of Rathke's cleft cysts. Surg Neurol 63: 459–466, discussion 466.
Sani S, Smith A, Leppla DC et al. (2005). Epidermoid cyst of the sphenoid sinus with extension into the sella turcica presenting as pituitary apoplexy: case report. Surg Neurol 63: 394–397, discussion 397.
Sanno N, Oyama K, Tahara S et al. (2003). A survey of pituitary incidentaloma in Japan. Eur J Endocrinol 149: 123–127.
Sato K, Oka H, Utsuki S et al. (2006). Ciliated craniopharyngioma may arise from Rathke cleft cyst. Clin Neuropathol 25: 25–28.
Schmitt S, Biason-Lauber A, Betts D et al. (2000). Genomic structure chromosomal localization and expression pattern of the human LIM-homeobox3 (LHX 3) gene. Biochem Biophys Res Commun 274: 49–56.
Schreckinger M, Walker B, Knepper J et al. (2013). Post-operative diabetes insipidus after endoscopic transsphenoidal surgery. Pituitary 16: 445–451.
Sekine S, Sato S, Takata T et al. (2003). Beta-catenin mutations are frequent in calcifying odontogenic cysts but rare in ameloblastomas. Am J Pathol 163: 1707–1712.
Shin JL, Asa SL, Woodhouse LJ et al. (1999). Cystic lesions of the pituitary: clinicopathological features distinguishing craniopharyngioma Rathke's cleft cyst and arachnoid cyst. J Clin Endocrinol Metab 84: 3972–3982.
Shuangshoti S, Netsky MG, Nashold Jr BS (1970). Epithelial cysts related to sella turcica. Proposed origin from neuroepithelium. Arch Pathol 90: 444–450.
Tajima T, Hattorri T, Nakajima T et al. (2003). Sporadic heterozygous frameshift mutation of HESX1 causing pituitary and optic nerve hypoplasia and combined pituitary hormone deficiency in a Japanese patient. J Clin Endocrinol Metab 88: 45–50.
Takanashi J, Tada H, Barkovich AJ et al. (2005). Pituitary cysts in childhood evaluated by MR imaging. AJNR Am J Neuroradiol 26: 2144–2147.
Takuma N, Sheng HZ, Furuta Y et al. (1998). Formation of Rathke's pouch requires dual induction from the diencephalon. Development 125: 4835–4840.
Tatagiba M, Iaconetta G, Samii M (2000). Epidermoid cyst of the cavernous sinus: clinical features pathogenesis and treatment. Br J Neurosurg 14: 571–575.
Tate MC, Jahangiri A, Blevins L et al. (2010). Infected Rathke cleft cysts: distinguishing factors and factors predicting recurrence. Neurosurgery 67: 762–769, discussion 769.
Teramoto A, Hirakawa K, Sanno N et al. (1994). Incidental pituitary lesions in 1000 unselected autopsy specimens. Radiology 193: 161–164.
Treier M, Gleiberman AS, O'Connell SM et al. (1998). Multistep signaling requirements for pituitary organogenesis in vivo. Genes Dev 12: 1691–1704.
Treier M, O'Connell S, Gleiberman A et al. (2001). Hedgehog signaling is required for pituitary gland development. Development 128: 377–386.
Trifanescu R, Stavrinides V, Plaha P et al. (2011). Outcome in surgically treated Rathke's cleft cysts: long-term monitoring needed. Eur J Endocrinol 165: 33–37.
Trifanescu R, Ansorge O, Wass JA et al. (2012). Rathke's cleft cysts. Clin Endocrinol (Oxf) 76: 151–160.
Trokoudes KM, Walfish PG, Holgate RC et al. (1978). Sellar enlargement with hyperprolactinemia and a Rathke's pouch cyst. JAMA 240: 471–473.
Verkijk A, Bots GT (1980). An intrasellar cyst with both Rathke's cleft and epidermoid characteristics. Acta Neurochir (Wien) 51: 203–207.
Voelker JL, Campbell RL, Muller J (1991). Clinical radiographic and pathological features of symptomatic Rathke's cleft cysts. J Neurosurg 74: 535–544.
Wait SD, Garrett MP, Little AS et al. (2010). Endocrinopathy vision headache and recurrence after transsphenoidal surgery for Rathke cleft cysts. Neurosurgery 67: 837–843, discussion 843.
Wales JK, Quarrell OW (1996). Evidence for possible Mendelian inheritance of septo-optic dysplasia. Acta Paediatr 85: 391–392.
Wen L, Hu LB, Feng XY et al. (2010). Rathke's cleft cyst: clinicopathological and MRI findings in 22 patients. Clin Radiol 65: 47–55.
Wolfe SQ, Heros RC (2010). A Rathke cleft cyst to craniopharyngioma: is there a spectrum? J Neurosurg 112: 1322–1323, discussion 1323.

Xie T, Hu F, Yu Y et al. (2011). Endoscopic endonasal resection of symptomatic Rathke cleft cysts. J Clin Neurosci 18: 760–762.

Zada G, Kelly DF, Cohan P et al. (2003). Endonasal transsphenoidal approach for pituitary adenomas and other sellar lesions: an assessment of efficacy safety and patient impressions. J Neurosurg 98: 350–358.

Zada G, Ditty B, McNatt SA et al. (2009). Surgical treatment of Rathke cleft cysts in children. Neurosurgery 64: 1132–1137, author reply 1037–1038.

Zada G, Lin N, Ojerholm E et al. (2010). Craniopharyngioma and other cystic epithelial lesions of the sellar region: a review of clinical imaging and histopathological relationships. Neurosurg Focus 28: E4.

Handbook of Clinical Neurology, Vol. 124 (3rd series)
Clinical Neuroendocrinology
E. Fliers, M. Korbonits, and J.A. Romijn, Editors

Chapter 18

Alternative causes of hypopituitarism: traumatic brain injury, cranial irradiation, and infections

SANDRA PEKIC AND VERA POPOVIC*
Faculty of Medicine, University of Belgrade, and Clinic of Endocrinology, Diabetes and Metabolic Diseases, University Clinical Center Belgrade, Belgrade, Serbia

INTRODUCTION

A wide spectrum of congenital disorders and structural lesions in the hypothalamic/pituitary region may cause insufficiency of one or more anterior pituitary hormones (hypopituitarism) (Vance, 1994; Lamberts et al., 1998; Ho, 2007; Schneider et al., 2007a. Acquired hypopituitarism in adults is most commonly due to hypothalamic/pituitary tumors. Generally, growth hormone (GH) is the first hormone to be lost, followed by the gonatotropins (FSH/LH), adrenocorticotropic hormone (ACTH), and thyroid-stimulating hormone (TSH) (Growh Hormone Research Society, 1998). Data that have emerged since 2000 demonstrate the relevance of traumatic brain injury (TBI) as a cause of hypopituitarism, while postpartum pituitary necrosis, lymphocytic hypophysitis, infections, and neoplastic lesions in the sella remain rarer causes.

TRAUMATIC BRAIN INJURY

Traumatic brain injury (TBI) is an important public health problem in the Western world. Each year, TBI contribute to substantial number of deaths and cases of permanent disability. The incidence rate of TBI has been estimated to be between 150 and 250 cases per 100000 population per year (Leon-Carrion et al., 2005). About 10% are severe injuries, another 10% are moderate, and 80% are mild. In the US, TBI affects 1.7 million people annually (CDC, 2010). Of these 52000 die, 275000 are hospitalized, and 1.365 million (nearly 80%) are treated and released from an emergency department. TBI is a contributing factor to a third (30.5%) of all injury-related deaths in the US. Children aged 0–4 years, older adolescents aged 15–19 years, and adults aged 65 years and older are most likely to sustain a TBI. In every age group, TBI rates are higher for males than for females. Falls are the leading cause of TBI (more than half a million emergency department visits), the highest rates for children aged 0–4 years and for adults aged 75 years and older. Motor vehicle–traffic injuries are the leading cause of TBI-related death; the highest rates are for adults aged 20–24 years. From 2002 to 2006 there was an increase in TBI-related emergency department visits (14.4%) and hospitalizations (19.5%) among every age group (CDC, 2010). These data suggest that incidence of TBI is higher than that of other heavily funded conditions (breast cancer, HIV, spinal cord injuries, multiple sclerosis). The severity of TBI may range from "mild" (a brief change in mental status or consciousness) to "severe" (an extended period of unconsciousness, coma, or amnesia after the injury). Cognitive impairment after TBI is often diagnosed as post-traumatic dementia. Hypopituitarism secondary to TBI may not be readily apparent and the contribution of pituitary dysfunction to cognitive impairment remains unknown. Millions are currently living with a traumatic brain injury-related disabilities.

Historical background of hypopituitarism after traumatic brain injury

Endocrine failure after TBI has been recognized for more than a century with most reports being clinical case presentations (Kahler, 1886; Maranon and Pintos, 1917; Cyran, 1918; Edwards and Clark, 1986). The first patients with post-traumatic hypopituitarism were described at autopsy (Cyran, 1918; Daniel et al., 1959; Daniel and Treip, 1966). After a systematic study in 1942 that

*Correspondence to: Professor dr Vera Popovic, MD, PhD, FRCP, Clinic of Endocrinology, Dr Subotica 13, 11000 Belgrade, Serbia. Tel: +381-11-3639702, Fax: +381-11-2685357, E-mail: popver@eunet.rs

identified a low prevalence of hypopituitarism after TBI (0.7%), it was thought to be rare event (Escamilla and Lisser, 1942). In the 1970s a few pathologic reports were concerned with changes in the pituitary and hypothalamus following fatal head injury (Treip, 1970). A diagram to illustrate possible mechanisms of hypothalamic and pituitary damage is presented in Fig. 18.1 (Treip, 1970). This report showed that hypothalamic damage which may be less immediately fatal may permit deficiency syndromes to develop later. In 1971, Escamilla et al. published the case of a pituitary dwarf who was treated sucessfully with growth hormone in childhood, in whom the cause of panhypopituitarism was a skull fracture and coma for 8 days at age 3 years (Escamilla and Forsham, 1971). This patient was possibly the first patient with hypopituitarism due to head trauma who received human growth hormone along with other pituitary hormones.

Renewed interest in hypopituitarism secondary to head trauma arose following two systematic studies in the year 2000. In a survey, Benvenga et al. (2000) reported 367 cases of hypopituitarism due to head trauma of differing causes. In another study, Kelly et al. (2000) found hypopituitarism in 40% of patients with moderate or severe head injury, with growth hormone and gonadotropin deficiencies being most common. These studies initiated many others confirming that pituitary dysfunction following head injury is not so rare, raising the awareness that a majority of these patients are not diagnosed (Lieberman et al., 2001; Kokshoorn et al., 2010) (Table 18.1). Although there has been considerable advancement in the last 10 years in the understanding of the underlying pathophysiology, natural history, risk factors for developing hypopituitarism, and potential benefits from treatment, conflicting results in the prevalence of hypopituitarism allowed some skepticism in the endocrine community for the the need to screen such a large population of TBI patients.

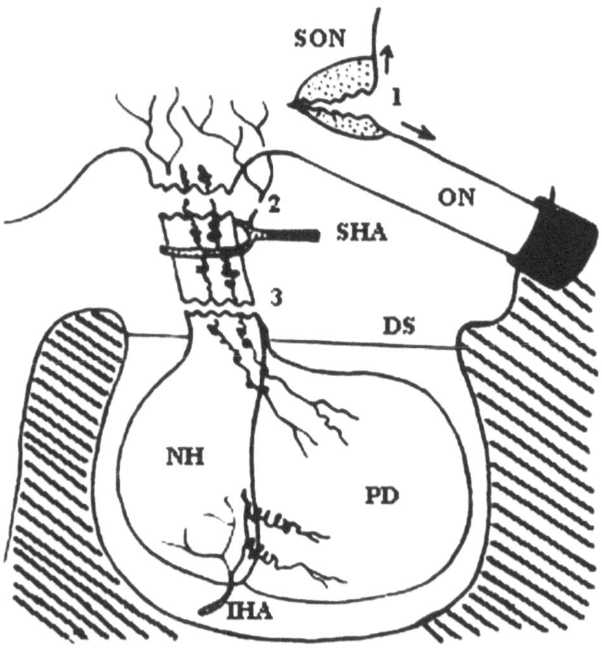

Fig. 18.1. Diagram to illustrate possible mechanisms of hypothalamic and pituitary damage. (1) Dural tethering of the optic nerve at the optic foramen results in tearing of the supraoptic nucleus with sudden movements of the brain. (2) High transection of the pituitary stalk results in rupture of the tuberal and infundibular vessels, the portal vessels to the pars distalis escaping injury. (3) Low stalk transection results in rupture of the long portal vessels to the pars distalis. ON, optic nerve; DS, diaphragma sellae; IHA, inferior hypophyseal artery; NH, neurohypophysis; PD, pars distalis; SHA, superior hypophyseal artery; SON, supraoptic nucleus. (Reproduced with kind permission from BMJ Publishing Group Ltd. © Treip CS (1970) J Clin Pathol 23(Suppl): 178–186.)

Neuroendocrine dysfunction after traumatic brain injury

Although diabetes insipidus has been recognized to occur particularly in the acute phase of TBI and usually recovers, partial or complete hypopituitarism may develop months and years after TBI and may escape detection for a long period of time (Popovic et al., 2005; Herrmann et al., 2006; Schneider et al., 2006; Hannon et al., 2011; Kreitschmann-Andermahr et al., 2011; Schneider et al., 2011).

In a systematic review of the literature analyzing published data of 931 patients in the chronic phase after TBI, the prevalence of pituitary hormone deficiency was 15–90% (Table 18.1) (Kokshoorn et al., 2010). This striking difference in the prevalence of any pituitary deficiency is thought to be due to the use of different dynamic tests and different normative data. In most individuals a single pituitary axis was affected. Growth hormone (GH) deficiency was the most common (Popovic, 2005; Casanueva et al., 2008). Studies showed that 8–15% of the adult patients with TBI had severe growth hormone deficiency (GHD) (Agha et al., 2004a; Aimaretti et al., 2004; Popovic et al., 2004). GH deficiency is followed by gonadotropin, adrenocorticotropic (ACTH), and thyrotropic (TSH) deficiency (Fig. 18.2). Why is GHD so common? Although not yet clear, it seems that GH-secreting cells are anatomically more vulnerable to damage. GH cells are located in the exposed lateral wings of the pituitary which are dependent on vascular input from the portal system alone (Urban, 2006). Gonadotrope cells (FSH/LH) are distributed throughout the anterior pituitary. Corticotrope (ACTH) and thyrotrope cells (TSH) are located in the protected median wedge area of the anterior pituitary and are supplied with

Table 18.1

Studies on traumatic brain injury and pituitary deficiency

Study	Year of publication	Number of patients	Time of testing post TBI (months (median))	Trauma severity (GCS)	BMI (kg/m^2)	Any pituitary deficiency (%)
Kelly et al.	2000	22	3–276 (median 26)	3–15	25.1±6.5	37
Lieberman et al.	2001	70	1–276 (median 13)	3–15 84% GCS ≤8	NR	69
Bondanelli et al.	2004	50	12–64	3–15 54% GCS ≤8	24.6±0.4	54
Agha et al.	2004a	102	6–36 (median 17)	3–13 56% GCS ≤8	NR	28
Popovic et al.	2004	67	12–264	3–13	24.8±0.5	34
Aimaretti et al.	2005	70	12	3–15 21% GCS ≤8	23.8±0.4	23
Leal-Cerro et al.	2005	99	>12	≤8	25.2±3.0 (n=44)	25
Schneider et al.	2006	70	12	3–15	23.8±3.2	36
Tanriverdi et al.	2006	52	12	3–15 25% GCS ≤8	NR	51
Herrmann et al.	2006	76	5–47	≤8	25.8±4.2	24
Bushnik et al.	2007	64	>12	NR	NR	90
Klose et al.	2007a	104	10–27 (median 13)	3–15 38% GCS ≤8	25† (17–39)	15
Tanriverdi et al.	2008a	30	36	3–15 16.7% GCS ≤8	NR	30
Wachter et al.	2009	55	NR	3–15 17% GCS ≤8	NR	25
Total no. of patients		931				

(Reproduced by the kind permission of the Society of the European Journal of Endocrinology © Kokshoorn NE et al., 2010, Eur J Endocrinol 162: 11–18.)

BMI, body mass index reported as mean ± SEM; GCS, Glasgow Coma Scale score; TBI, traumatic brain injury; NR, not reported.

†Reported as median (range).

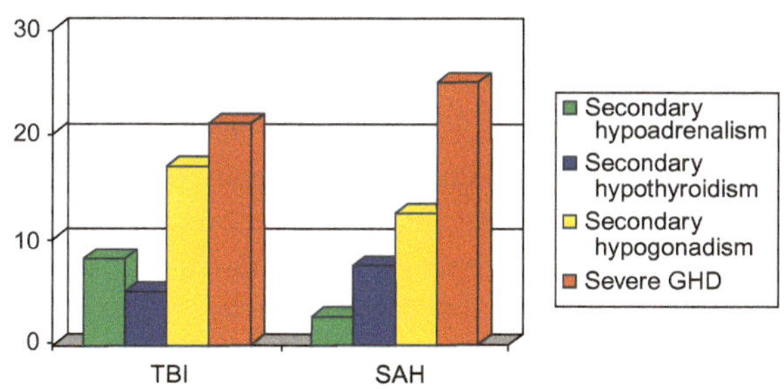

Fig. 18.2. Percentage of single pituitary deficits in patients with traumatic brain injury (TBI) and subarachnoid hemorrhage (SAH), 3 months after the pathologic event. (Reproduced with kind permission from John Wiley and Sons © Aimaretti G et al. (2004) Clin Endocrinol 61: 320–326.)

blood via both the long hypophyseal portal vessels and the inferior hypophyseal artery. Thus these cells are anatomically less vulnerable to injury than gonadotrophs and somatotrophs and this may explain the lower prevalence of ACTH and TSH deficiencies compared to GH and gonadotropin deficiencies.

In patients in the acute phase of TBI, the most important consideration is to detect glucocorticoid deficiency as a potential life-threatening condition. These patients develop hypoglycemia, hyponatremia, or refractory hypotension after injury with dramatic improvement after glucocorticoid replacement therapy (Agha et al., 2005a). In one study, 16% of TBI patients developed acute ACTH deficiency after moderate to severe TBI (Agha et al., 2004b).

Screening tests for hypopituitarism should not be performed earlier than 1 year after TBI due to the possibility of its reversal (Casanueva et al., 2009; Tanriverdi et al., 2011). A confirmation test is necessary since 20–30% of screening results are false-positive. Thus a much lower prevalence of anterior pituitary dysfunction after TBI may be found. In the studies in which extensive endocrine evaluation with confirmation tests were performed, anterior pituitary dysfunction was diagnosed in < 1% of patients (Van der Eerden et al., 2010). Kokshoorn et al. (2011) reported that hypopituitarism was diagnosed in 5.4% of patients after TBI. In conclusion, identification of predictors for neuroendocrine dysfunction after TBI is of utmost importance in order to define a cost-effective screening strategy for these patients.

Predictors of neuroendocrine dysfunction after traumatic brain injury

The possible predictors of neuroendocrine dysfunction after TBI are: early hormonal abnormalities, the type of brain injury, and the severity of the trauma (Klose and Feldt-Rasmussen, 2008; Klose et al., 2007a).

Hormonal changes during the acute phase after brain injury represent important neuroendocrine adaptive responses, rather than true deficiencies. In most patients the physiologic responses to stress include hormonal alterations mimicking central hypothyroidism, central hypogonadism, elevation of prolactin, ACTH, cortisol, GH, and vasopressin, decrease of cortisol-binding globulin, and increased tissue sensitivity to glucocorticoids (Van den Berghe et al., 1998). Neuroendocrine dysfunction is frequent in the acute phase of moderate and severe TBI (Agha et al., 2005b; Tanriverdi et al., 2006). After 1 year of follow-up these abnormalities recover in most patients. Hypocorticism during the acute phase may be a life-threatening condition with major impact for the recovery of the patient and it is very important adequately to assess adrenal function. A normal Synacthen test in the acute phase after TBI does not exclude hypocorticism since adrenal atrophy has not yet developed. The insulin tolerance test (ITT) is relatively contraindicated in these situations; glucagon testing could be an alternative but needs further verification.

The predictive value of trauma-related parameters (severity or type of trauma) has been suggested by some, but not all authors (Kelly et al., 2000; Agha et al., 2004a; Bondanelli et al., 2004; Aimaretti et al., 2005; Leal-Cerro et al., 2005; Klose et al., 2007a; Klose and Feldt-Rasmussen, 2008; Schneider et al., 2008; Tanriverdi et al., 2010a). The severity of TBI was defined by a low score on the Glasgow Coma Scale (GCS), increased intracranial pressure, and the number of days of intubation and hospitalization (Klose et al., 2007a). Other risk factors include diffuse axonal/neuronal injury, basilar skull fracture, diffuse brain swelling, intracranial hematoma, and multiple contusions of the brain in the acute phase (Medic-Stojanoska, 2009).

Sport

Although road traffic accidents are the most common cause of TBI, responsible for more than half of the head injuries, recent data clearly demonstrate that sports-related chronic repetitive head trauma may also result in hypopituitarism. Concussion is the main type of TBI and is associated with a number of sports including boxing, kickboxing, football, and ice hockey. Sports-related repetitive head trauma of low intensity increases the risk for isolated GH deficiency (Kelestimur et al., 2004; Tanriverdi et al., 2007, 2008a, 2010a). In one of the studies, 45% of professional boxers were diagnosed with isolated GH deficiency (Kelestimur et al., 2004). Retired boxers with GHD had significantly lower pituitary volume on MRI than retired boxers with normal GH (Tanriverdi et al., 2008b).

Modern military operations: blast-related traumatic brain injury

Soldiers in military operations suffer from blast-related mild TBI, an extremely common injury in modern wars. In 2010, the injuries in 80% of over 30 000 US military service members medically diagnosed with TBI were classified as mild TBI (Military Health System, 2011). About 10–20% of returnees report having experienced at least one blast concussion. A recent report showed that these soldiers are at risk for post-traumatic hypopituitarism, in particular GH deficiency and central hypogonadism (Wilkinson et al., 2012). Male US veterans of combat in Iraq and Afghanistan with and without blast concussion had baseline hormone measurements and 42% of participants with blast concussion had abnormal hormone levels in at least one axis, while none of veterans who did not experience blast trauma had evidence of pituitary dysfunction.

Traumatic brain injury in children and adolescents

There are insufficient data in the literature about the epidemiology, pathology, prevalence, diagnosis, and treatment of TBI-induced hypopituitarism in children and adolescents. As in adults, TBI in children and adolescents may be complicated with hypopituitarism in the acute or chronic phase (Medic-Stojanoska, 2009). Einaudi et al. (2006) reported endocrine failure in 23% of children during the acute phase of TBI. In a study by Poomthavorn et al. (2008), hypothalamopituitary dysfunction was identified in 16.6% of childhood survivors from TBI. However, in a recent longitudinal study of 198 survivors of TBI in early childhood no cases of hypopituitarism were reported (Heather et al., 2012). In conclusion, pediatricians should be aware that in cases of slow growth rate and or disordered puberty, TBI should be considered as a cause.

Pathophysiologic mechanisms of neuroendocrine dysfunction due to traumatic brain injury

Several mechanisms have been suggested for the pathogenesis of neuroendocrine dysfunction after head trauma. Head trauma may trigger a cascade of vascular changes, histopathologic changes, inflammatory reactions, and activation of the immune system, which all contribute to late neuroendocrine dysfunction. Inflammatory mediators (cytokines, in particular interleukin 6, free radicals, amino acids, and nitric oxide) may lead to accelerated neuronal cell necrosis (Gaetz, 2004). Kasturi and Stein (2009) investigated the effects of cortical contusion injury on growth hormone (GH) secretion in adult male rats and described pathologic changes in the hypothalamopituitary region. They found that rats with cortical contusion injury had significantly lower levels of GH in serum and anterior pituitary gland 2 months after TBI. They demostrated upregulation of the proinflammatory marker (ILβ) in the hypothalamus with upregulation of the marker for proliferation/expansion of glial cells (glial fibrillary acidic protein, GFAP) in the hypothalamus and pituitary gland. These data are in accordance with the findings of Holmin and Mathiesen (1999), who demonstrated the persistence of inflammation and astrocytosis at the site of injury at 3 months after TBI. Others described the persistence and spread of inflammatory factors at the site of injury which in turn resulted in progressive demyelination, secondary necrosis, and apoptosis of subcortical brain tissue (Smith et al., 1997; Nonaka et al., 1999). Toxic factors secreted by injured cells may affect distant brain structures by volume transmission–diffusion through the extracellular space or the cerebral ventricles to distal neurons, causing systemic inflammation and neuronal loss (Bach-y-Rita, 2003). It is possible that TBI initiates proinflammatory response which in turn initiates degenerative processes in the brain, explaining the evolution of hypopituitarism after TBI with time.

Pituitary function improves over time in a considerable number of patients, but it may also worsen (Aimaretti et al., 2005; Tanriverdi et al., 2006). The mechanisms of these dynamic changes in the pituitary function after TBI remains to be clarified. Studies in animals demonstrate the presence of autoantibodies against neurons in the brains of rats with cortical lesions (Stein et al., 2002). Recent studies by Tanriverdi et al. suggest that autoimmunity may contribute to the pathogenesis of TBI-induced hypopituitarism in humans also (Tanriverdi et al., 2008c, 2010b). They showed the presence of antipituitary antibodies (APA) in 44.8% of TBI patients several years after head trauma (Tanriverdi et al., 2008c). Pituitary dysfunction was associated with APA positivity and APA titers were associated with low GH response to the stimulatory test. In another study, antihypothalamic antibodies (AHA) were detected in 21.3% of boxers and APA in 22.9% of boxers (Tanriverdi et al., 2010b). These authors proposed that sequestered pituitary antigens released from necrotic tissue may trigger pituitary autoimmunity with consequent post-traumatic hypopituitarism. An appropriate study using a double immunofluorescence method is needed to clarify which cells are the targets of these antibodies at hypothalamic level.

There are some data suggesting a genetic predisposition to developing pituitary dysfunction after TBI (Tanriverdi et al., 2008d). Apolipoprotein E (APO E) is the primary apolipoprotein synthesized within the brain, including the hypothalamopituitary region, and it is upregulated after brain injury (Nishida et al., 2005). APO E reduces the neuroinflammatory response *in vitro* and *in vivo* in an isoform-specific fashion, APO E3 isoform being more effective than APO E4 isoform (Laskowitz et al., 1997). It is demonstrated that individuals with the APO E3 genotype have a decreased risk for pituitary dysfunction after TBI when compared with individuals with other genotypes (Tanriverdi et al., 2008d).

The pathologic reports from the 1970s already mentioned, as well as more recent reports, demostrate vascular events that underlie the pathogenesis of hypothalamic/pituitary dysfunction. Infarctions and hemorrhages in the hypothalamus have been demonstrated in half of patients who died after a motor vehicle accident (Fig. 18.3A, B) (Treip, 1970; Salehi et al., 2007). Vascular damage is the most likely explanation for the early pituitary abnormalities in TBI patients.

Since GH deficiency is the most common pituitary deficiency in TBI, the mechanisms seem to be structural and functional alterations in the median eminence, the structure supporting secretions of hypothalamic hypophysiotropic factors, and this was shown in an animal model (Osterstock et al., personal communication).

Diagnosis of neuroendocrine dysfunction after brain injury

The differences in the prevalence of hypopituitarism in clinical studies can be partly explained by the timing of evaluation, the methods used in evaluating neuroendocrine function, different criteria for diagnosis of hypopituitarism, and the presence of confounding factors, such as obesity (Hannon et al., 2011; Kokshoorn et al., 2011; Gasco et al., 2012).

Since the most affected axis is the somatotropic axis, various tests have been used to diagnose GH deficiency (Growth Hormone Research Society, 1998, 2000; Ho, 2000). Although insulin-induced hypoglycemia is the gold standard for diagnosing GHD in adults, it may

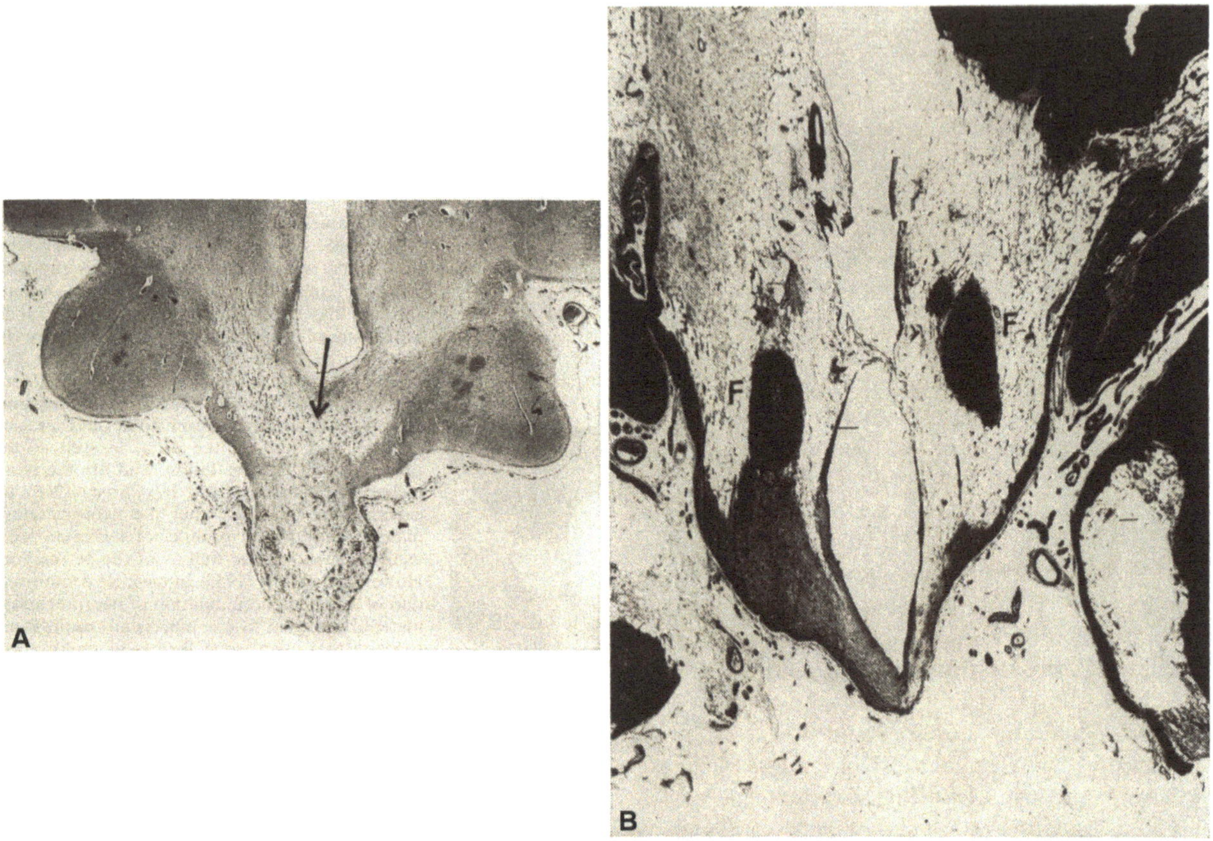

Fig. 18.3. (**A**) Acute infundibular infarction immediately above pituitary stalk (arrow), haematoxylin and eosin × 6 (male, age 41 years, survived 27 days, severe hypernatremia). (Reproduced with kind permission from BMJ Publishing Group Ltd. © Treip CS (1970) J Clin Pathol 23(Suppl): 178–186.) (**B**) Old infarction (pale) of most of hypothalamus, including the wall of the third ventricle, but sparing the fornices (F), Loyez × 5.5 (male, aged 65, survived 110 days, persistent hypothermia). (Reproduced with kind permission from BMJ Publishing Group Ltd. © Treip CS (1970) J Clin Pathol 23 Suppl: 178–186.)

sometimes be contraindicated in patients who have survived severe brain injury. Alternative provocative tests are: glucagon, L-dopa, arginine, GHRH + arginine test, or GHRH + GHRP-6 test. The criterion of 3 μg/L for diagnosing severe GHD is recommended by the Growth Hormone Research Society for the insulin tolerance test and provides a specificity of more than 96% (Growth Hormone Research Society, 2000). Measurement of IGF-1 is not able accurately to diagnose GHD because there is substantial overlap of IGF-1 levels between healthy subjects and those with GHD (Hoffman et al., 1994; Mukherjee and Shalet, 2009). However, despite the low diagnostic sensitivity of this parameter, very low IGF-I levels in a patient with multiple pituitary hormone deficiencies is diagnostic for GHD.

The diagnosis of secondary hypogonadism is challenging and no consensus exists. Most studies define hypogonadism in males as low or normal gonadotropin levels with a low total testosterone level, or less frequently only a low free testosterone level. Hypogonadotropic hypogonadism in postmenopausal women is defined as inappropriately low gonadotropins for the patient's age. In premenopausal women it is defined as the presence of amenorrhea or oligomenorrhea, associated with low estradiol and inappropriately low gonadotropins.

Unrecognized ACTH deficiency can be a life-threatening condition. In some studies hypocorticism was defined as low serum cortisol level or a decreased level of 24 hour urinary free cortisol. However, most authors diagnose hypocorticism with a stimulation test (insulin tolerance test, glucagon stimulation test, short ACTH stimulation test).

Hypothyroidism is mostly defined by low free T_4 and low or normal TSH levels.

According to all published results, routine screening for hormone disturbances in unselected patients after TBI is not cost-effective. However, screening should be advised in all patients with symptoms and signs of hypopituitarism, with a history of TBI, and with some predictors of risk. Screening should probably be advised in patients with severe TBI.

Cognitive impairments after traumatic brain injury

Abnormal anterior pituitary hormone levels after TBI may contribute to other traumatic brain injury-related disabilities, such as cognitive impairments (Klose et al., 2007b). It is well known that hypopituitarism from other causes is associated with cognitive impairment, depression, anxiety, fatigue, apathy, and impaired quality of life that improves with GH treatment (Deijen et al., 1996, 1998; Van Nieuwpoort and Drent, 2008). The GH-deficient TBI patients have decreased energy levels, decreased energy performance, are emotionally more labile, with depression and anxiety, and some are diagnosed as having post-traumatic dementia (Springer and Chollet, 2001; Maric et al., 2010, Pavlovic et al., 2010). Some authors advocate pituitary function screening in all patients after TBI if there is severe cognitive impairment (Agha et al., 2005b; Schneider et al., 2007b). An open dilemma exists as to whether cognitive impairment in TBI survivors is caused by GH deficiency or by brain injury itself, and whether GH replacement therapy would improve cognitive ability. GH receptors are found in many regions in the CNS vulnerable to trauma (the choroid plexus, thalamus, hypothalamus, pituitary, putamen, and hippocampus) (Nyberg and Burman, 1996; Nyberg, 2000). IGF-1 receptors are also found in the hippocampus, parahippocampal areas, cerebellum, frontal and other neocortical areas, and caudate nucleus (Adem et al., 1989). The hippocampus is involved in cognition, memory, and spatial functions. GH replacement therapy has been shown to induce some changes in the levels of neurotransmitters in the cerebrospinal fluid (CSF), which may influence cognitive and psychiatric functions (Burman et al., 1993; Johansson et al., 1995; Burman et al., 1996). Hypopituitary patients on GH therapy have a decrease in CSF concentrations of the dopamine metabolite homovanillic acid (HVA) similar to depressed patients after successful treatment. The level of aspartate is increased in CSF during GH therapy (Burman et al., 1993; Johansson et al., 1995; Burman et al., 1996). Aspartate is an excitatory amino acid, a ligand for N-methyl-D-aspartate (NMDA), receptor which is involved in memory function and attentional performance.

In the few case report studies, GH hormone replacement therapy in GHD head-injured patients resulted in neurobehavioral improvements (High et al., 2010; Maric et al., 2010; Tanriverdi et al., 2010c; Beca et al., 2012). Some studies did not find any evidence for the association of neuropsychological changes and GH secretion in adult patients tested more than 1 year after TBI (Pavlovic et al., 2010). In conclusion, randomized placebo-controlled clinical trials on the effect of GH replacement on neuropsychological functions in TBI-induced hypopituitary patients are still needed.

CRANIAL IRRADIATION

Introduction

It is well documented that patients may develop frank or subtle abnormalities in the hypothalamic–pituitary axis several years after prophylactic or therapeutic cranial irradiation (Shalet et al., 1976a, b; Constine et al., 1993; Ogilvy-Stuart et al., 1994a; Schmiegelow et al., 2000; Oeffinger et al., 2006; Darzy and Shalet, 2009a, b; Fernandez et al., 2009; Appelman-Dijkstra et al., 2011). Patients at risk for developing hypopituitarism are children with brain tumors who received cranial irradiation, patients treated for head and neck tumors, sellar and parasellar tumors, and whole body irradiation for hematologic malignancies (Shalet et al., 1976a, b, 1977; Bhandare et al., 2008; Klose et al., 2009; Appelman-Dijkstra et al., 2011). According to data from the Childhood Cancer Survivor Study (CCSS), 43% of children treated for brain tumors had at least one endocrinopathy (Gurney et al., 2003). The severity and frequency of endocrine sequelae after cranial irradiation correlate with the total radiation dose delivered to the hypothalamo–pituitary axis, radiation schedules, and the time that has elapsed since treatment (Darzy and Shalet, 2009a, b). If the radiation dose is administered over a short period it will induce more damage to the hypothalamo–pituitary axis than if the same dose is administered over a longer period. Hypothalamo–pituitary dysfunction after cranial radiation is progressive and irreversible. Treatment of brain tumors in childhood often affects growth, causing growth retardation, and also affects sexal development in children. Thus regular endocrine testing is recommended in all long-term survivors who have been cranially irradiated in childhood. In adults the symptoms of endocrine dysfunction caused by cranial radiation are usually masked by the sequelae of the primary disease. Periodic endocrine assessment is recommended in adult patients exposed to cranial radiotherapy.

Neuroendocrine dysfunction after cranial irradiation

The GH axis is the most vulnerable to radiation damage and isolated GH deficiency may occur with low radiation dose of 18 Gy (Darzy and Shalet, 2003, 2005). If the radiation dose is <30 Gy, often used in patients with leukemia or brain tumors, isolated GH deficiency is present in 30% of patients (Littley et al., 1989; Lam et al., 1991; Constine et al., 1993). With the increase in the radiation dose from 30 to 50 Gy, the incidence of GH deficiency

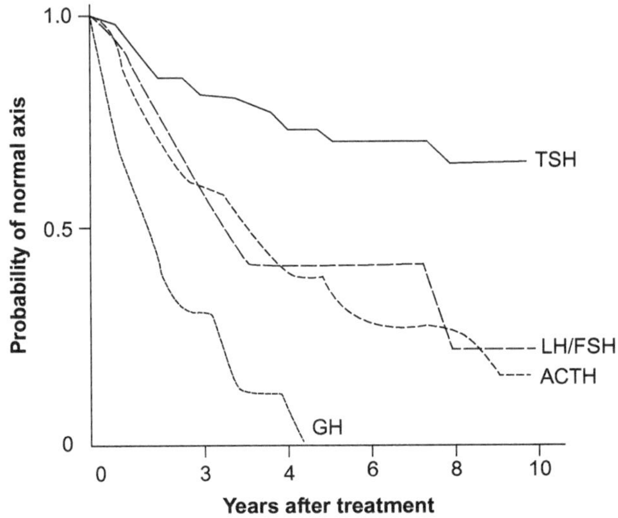

Fig. 18.4. The probability of hypothalamic–pituitary axis dysfunction after irradiation. Life-table analysis indicating the probabilities of initially normal hypothalamic–pituitary hormonal axes remaining normal after conventional radiotherapy for pituitary adenomas. ACTH, adrenocorticotropic hormone; FSH, follicle-stimulating hormone; GH, growth hormone; LH, luteinizing hormone. (Reproduced with kind permission from Oxford University Press © Littley MD et al. (1989) QJ Med 70: 145–160.)

increases to 50–100% of patients and other pituitary hormone deficiencies develop more frequently (Fig. 18.4) (Samaan et al., 1982, 1987; Chen et al., 1989; Lam et al., 1991). In these patients, with time, gonadotropin deficiency occurs in 20–30% of patients, TSH deficiency in 3–9%, and ACTH deficiency in 3–6% (Darzy and Shalet, 2009a, b). The occurence of neuroendocrine dysfunction as a late complication of cranial radiotherapy is related to duration of follow-up after radiotherapy. After a follow-up period of 10 years, multiple pituitary hormone deficiencies occur in 30–60% of patients (Samaan et al., 1987; Lam et al., 1991). Agha et al. (2005c) showed that 41% of patients irradiated for brain tumors distant from the hypothalamo–pituitary axis developed hypopituitarism. In a large study of 4110 patients with organic adult-onset GH deficiency from the Pfizer International Metabolic Database (KIMS), a history of irradiation was recorded in 36% of patients with isolated GH deficiency and in 37% of patients with multiple pituitary hormone deficiencies (Klose et al., 2009). A recent meta-analysis of 18 studies, with a total of 813 patients, reported that approximately two-thirds of all patients previously treated with cranial irradiation developed some degree of hypopituitarism, irrespective of the underlying disease (Appelman-Dijkstra et al., 2011).

Stereotactic techniques for irradiation have been developed with the aim of improving the efficacy of irradiation, delivering more localized irradiation, and minimizing the long-term consequences of irradiation. Stereotactic irradiation can be given in a single dose as stereotactic radiosurgery (SRS) or in multiple doses as fractionated stereotactic radiotherapy. SRS has been used to treat recurrent pituitary adenomas, small brain tumors, or vascular malformations (aneurysms). In patients with pituitary adenomas, SRS is usually offered to patients with relatively small adenomas (less than 3 cm) situated more than 2–3 mm from the optic chiasm. Gamma Knife surgery (GKS) is the most widely performed radiosurgical technique. The reported overall rate of serious complications after SRS is low (Minniti et al., 2012). The most common complication following GKS for pituitary adenomas is delayed hypopituitarism. In the analysis of the literature on SRS for GH-secreting pituitary adenomas, the incidence of hypopituitarism was between 0% and 69% (Minniti et al., 2012). In a study of 95 acromegalic patients treated with GKS, hypopituitarism was reported in 34% of patients (Jagganathan et al., 2008). In other series of 39 patients with acromegaly treated with SRS, hypopituitarism occurred in one-third of patients (Inoue et al., 1999). In this study, 10% of patients developed new pituitary deficit at 2 years and 33% at 5 years. Similar results are in patients with nonfunctioning pituitary adenomas treated with GKS. It is demonstrated that 30.3% of these patients develop hypopituitarism after GKS (Starke et al., 2012). In the retrospective study of 130 patients (68 with nonfunctioning pituitary adenoma and 62 with secreting pituitary adenoma), 16 patients (12.3%) showed a new pituitary deficit in one or more axes (Sicignano et al., 2012). The incidence of new hormonal deficits after GKS for pituitary adenoma was dose-dependent and the mean dose to the stalk/pituitary and the amount of healthy tissue within the high dose region were strong predictors of pituitary dysfunction.

Diagnosis of impaired growth hormone secretion after cranial irradiation

After cranial irradiation, spontaneous endogenous 24 hour GH secretion may be impaired, but recently there has been dispute as to whether radiation-induced GH neurosecretory dysfunction exists (Darzy et al., 2007). Spontaneous 24 hour GH secretion was found to be normal in many. There is a 40% reduction in the peak GH responsiveness to stimulation tests, suggesting reduced somatotroph reserve. GH response is stimulus-dependent (Lissett et al., 2001; Popovic et al., 2002). Two stimulation tests for estimating GH secretion are required in the case of isolated GHD. Interpretation of results to the GH stimulatory tests may be complicated if the results of the two tests are discordant (ITT and arginine + GHRH test) (Darzy and Shalet, 2003; Darzy, 2009; Darzy et al., 2009). Failing to pass a hypoglycemia test (ITT) is more common after radiation compared to other stimulatory tests (Ahmed et al., 1986; Lissett et al., 2001). In a study of 166 patients with radiation damage to the somatotropic axis, patients who failed the ITT passed the alternative tests such as the arginine + GHRH stimulation test or GHRH + GHRP-6 (Lissett et al., 2001; Popovic et al., 2002). The ITT, arginine + GHRH, and GHRP-6 + GHRH stimulation tests have divergent underlying mechanisms governing GH responsiveness. Arginine induces GH release by somatostatin inhibition in the hypothalamus, GHRP-6 stimulates GHRH release in the hypothalamus, while the mechanisms of insulin-induced hypoglycemia (ITT) for GH secretion are complex and not yet fully elucidated. It is proposed that α_2-adrenergic pathways, basomedial hypothalamus, somatostatin suppression, and GHRH release are involved. It has been suggested that radiation doses less than 40 Gy predominantly cause hypothalamic damage with GHRH deficiency causing somatotrope atrophy. Hypothalamus seems to be more vulnerable to radiation damage than the pituitary and the ITT is the preferred test (Shalet, 1997; Growth Hormone Research Society, 1998; Ho, 2000; Darzy et al., 2007). If cranially irradiated patients have low IGF-1 and an impaired GH response to the ITT, then GH deficiency is highly likely despite a possible normal GH response to the other stimulation tests (Popovic et al., 2002; Darzy, 2009; Darzy et al., 2009). If cranially irradiated patients fail to pass the GHRH + arginine test or GHRH + GHRP-6 it is always indicative of GH deficiency.

Abnormalities in other pituitary hormones

Abnormalities in gonadotropin secretion are dose-dependent. Low dose irradiation (<25 Gy) in prepubertal children can cause precocious puberty mostly in girls, and in both sexes with a radiation dose of 25–50 Gy (Leiper et al., 1987; Ogilvy-Stuart et al., 1994b; Darzy and Shalet, 2009a, b). The possible mechanism of premature activation of the hypothalamo–pituitary–gonadal axis after cranial irradiation might be the result of damage of inhibitory neurons (γ-aminobutryic acid), with resultant desinhibition and premature activation of hypothalamic GnRH neurons (Roth et al., 2001). Subsequently central gonadotropin deficiency and delayed puberty occur with high radiation doses and are usually a late complication of cranial radiotherapy with a cumulative incidence of 20–50% on long-term follow-up (Samaan et al., 1982; Littley et al., 1989; Lam et al., 1991; Constine et al., 1993; Agha et al., 2005c). Gonadotropin deficiency after radiotherapy may be subtle, subclinical (detected only by GnRH test), or more severe with normal or descreased LH/FSH levels, decreased sex hormone levels, and impaired fertility. A recent meta-analysis of neuroendocrine dysfunction after cranial radiation (18 studies analyzed) showed that hypogonadotropic hypogonadism was present in 30% of irradiated patients (Appelman-Dijkstra et al., 2011).

Hyperprolactinemia is a pathologic finding in some cranially irradiated patients and indicates hypothalamic damage which reduces the normal inhibitory action of dopamine on prolactin release (Constine et al., 1993; Mahajan and Lightman, 2000; Popovic et al., 2002; Darzy and Shalet, 2009a, b). Hyperprolactinemia is mostly seen in young females after high-dose cranial irradiation (above 40 Gy), is usually subclinical, and is noted in 20–50% of these patients (Samaan et al., 1982; Constine et al., 1987; Lam et al., 1991; Agha et al., 2005c). The meta-analysis of neuroendocrine dysfunction after cranial radiation showed that hyperprolactinemia was present in 29% of patients (Appelman-Dijkstra et al., 2011). Hyperprolactinema after radiotherapy may be subclinical, without clinical significance, or more severe with impaired gonadotropin secretion, pubertal delay or arrest in children, decreased libido and impotence in adult men, or galactorrhea and ovarian dysfunction in adult women (Samaan et al., 1982). With time, in cranially irradiated patients elevated prolactin levels decline and even normalize, suggesting progressive reduction of the pituitary lactrotroph pool by radiotherapy (Littley et al., 1989).

The hypothalamo–pituitary–adrenal axis (HPA) is more radioresistant than GH and gonadotropin axes. ACTH deficiency occurs after a large dose of cranial radiation (>50 Gy), with a long-term cumulative frequency of 3–6% (30–60% after 10 years) (Darzy and Shalet, 2009a, b). Central hypocorticism is subtle, subclinical, and only detected by stimulatory tests: ITT, glucagon, Synacthen test. Central hypocorticism is usually a late complication of cranial radiotherapy with doses above 50 Gy, with reported rates of 27–35% up to 15 years after irradiation (Samaan et al., 1982; Lam et al., 1991; Constine et al., 1993). In patients treated with conventional irradiation for pituitary tumors the reported incidence of ACTH deficiency was 31–60% (Littley et al., 1989). In the mentioned meta-analysis of neuroendocrine dysfunction after cranial radiation (18 studies analyzed), adrenal insufficiency occurred in 0–50% of patients with nasopharyngeal tumors and in 3–62% of the patients with intracerebral tumors (Appelman-Dijkstra et al., 2011).

The hypothalamo–pituitary–thyroid axis is also more radioresistant than GH and gonadotropin axes. TSH deficiency occurs after a large dose of cranial radiation (>50 Gy) used for nasopharyngeal carcinomas and tumors of the skull base or conventional radiotherapy (30–50 Gy) or stereotactic radiotherapy for pituitary tumors, with a long-term cumulative frequency of 3–6% (30–60% after 10 years) (Darzy and Shalet, 2009a, b). TSH deficiency after radiotherapy may be subtle and detected by stimulatory TRH test. If it is more severe then normal or low TSH levels with low FT4 levels are diagnostic of central hypothyroidism which is usually a late complication of cranial radiotherapy with doses above 50 Gy (Chen et al., 1989; Lam et al., 1991; Constine et al., 1993). In adult patients treated with conventional irradiation for nonpituitary brain tumors the reported rate of TSH deficiency was 9% (Agha et al., 2005c). A meta-analysis of endocrine dysfunction after cranial radiation which included 18 studies showed thyroid dysfunction in 26% of patients (Appelman-Dijkstra et al., 2011).

In conclusion, the underlying pathophysiology of the radiation-induced damage to the hypothalamus remains ill understood. The damage certainly depends on the total radiation dose, the fraction size, and the time between fractions for tissue repair. It is hypothesized that radiotherapy can cause direct injury to hypothalamic neurons and reduce cerebral flow (vascular damage, fibrosis) (Chieng et al., 1991; Shalet, 1997). Radiotherapy could also cause secondary hypothalamic dysfunction due to altered neurotransmitter input from other brain centers injured by radiotherapy (Jorgensen et al., 1993).

INFECTIONS IN THE HYPOTHALAMIC–PITUITARY REGION

Despite numerous case reports, the incidence of hypothalamic–pituitary dysfunction following infectious diseases particularly of the central nervous system (CNS) has been underestimated. Hypopituitarism usually relates to the severity of the disease, type of causative agent, and primary localization of the infection. Infectious lesions in the hypothalamic–pituitary region include: pituitary abscess, tuberculosis, fungal, viral, and parasitic infections. Hypopituitarism may be misinterpreted as postencephalitic syndrome (Schaefer et al., 2008). On the other hand, the presence of a sellar mass with suprasellar extension may be misdiagnosed as pituitary macroadenoma in a patient with pituitary abscess, which is potentially a life-threatening disease.

Pituitary infections may be primary, without an identifiable source, or secondary in origin.

Sources of infections spreading to the hypothalamic–pituitary region

1. Hematogenous spread in immunocompromised hosts
2. Direct extension from adjacent anatomic sites (meningeal infection, sphenoid sinus, cavernous sinus, skull)
3. Previous infectious diseases of the CNS of different etiologies
4. Iatrogenic in the setting of multiple surgical interventions for sellar and suprasellar lesions.

Predisposing factors for pituitary infections

1. Diabetes mellitus
2. Tuberculosis
3. Renal transplant
4. Human immunodeficiency virus (HIV) infection
5. Non-Hodgkin lymphoma and chemotherapy
6. Cushing's syndrome.

Clinical features of pituitary infections

Neurologic symptoms

1. Headache – most common complaint
2. Visual disturbances – optic chiasm compression
3. Cranial nerve palsy III, IV or VI.

Endocrine dysfunction

1. Hyponatremia
2. Complete hypopituitarism
3. Hypogonadotropic hypogonadism
4. Hyperprolactinemia
5. Central diabetes insipidus (DI).

Pituitary abscess

Pituitary abscess is a rare disease. More than 200 cases of pituitary abscess have been reported in the literature, mostly in the form of isolated case reports. In children, pituitary abscess is exceptionally rare and some 20 cases have been reported mostly due to infection of Rathke's

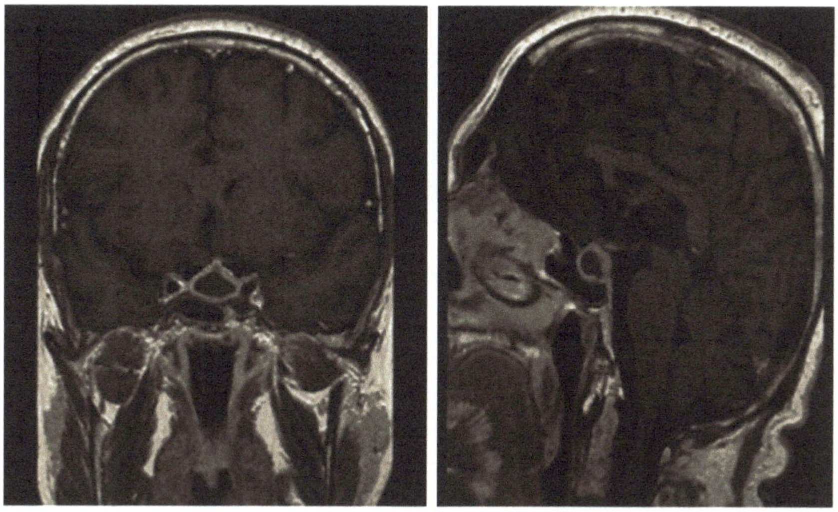

Fig. 18.5. Pituitary abscess: gadolinium-enhanced T1-weighted MRI scan showing sellar and suprasellar mass with peripheral contrast enhancement in a 56-year-old male patient. (Courtesy of Professor dr Milica Medic, Endocrinology, Cinical Center, University Novi Sad.)

cleft cyst (Pepene et al., 2010). The delay in diagnosis is due to its indolent course. It is frequently misdiagnosed as pituitary adenoma. Pituitary abscess on MRI presents as a heterogeneous intrasellar lesion with hyperintense peripheral capsule mimicking apoplexy (Fig. 18.5) (Vates et al., 2001; Ciappetta et al., 2008; Dalan and Leow, 2008). Transsphenoidal decompression is the most effective treatment and intraoperatively pus is found (Zhang et al., 2012). Gram-staining and cultures should be performed in order to reveal the responsible microorganism and treatment commenced with appropriate antibiotics in order to avoid recurrences. However, in most instances no organism is isolated. Hypopituitarism including diabetes insipidus rarely recovers and in most patients remains permanent.

Nonspecific inflammation of the cavernous sinus: the Tolosa–Hunt syndrome

Tolosa–Hunt syndrome presents with throbbing headache and painful ophthalmoplegia, cranial nerve palsies (III, IV, VI), fever, and leukocytosis. Cavernous sinus is located on either side of sella turcica. Cavernous sinuses communicate not only with cerebral veins but also with facial allowing extracranial infections to pass intracranially. The patient with Tolosa–Hunt syndrome presents with granulomatous inflammation within the cavernous sinus causing acute unilateral orbital pain. Ophthalmoparesis and disordered eye movement occur when the third, fourth and sixth cranial nerves are damaged by inflammation. Tolosa–Hunt syndrome is usually unilateral and its diagnosis is made by exclusion. MRI and CT scans of the brain and orbit are particularly helpful in

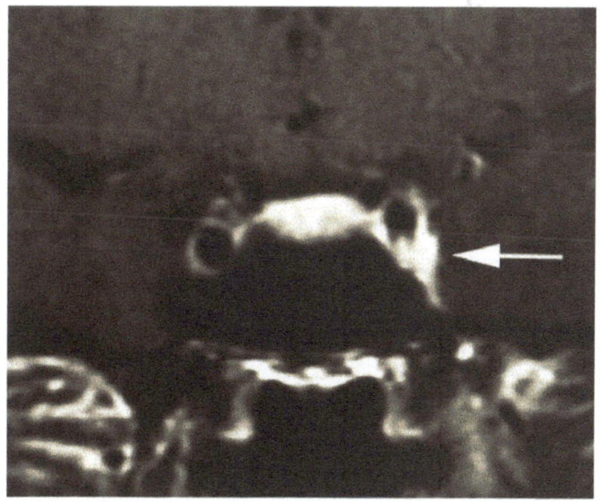

Fig. 18.6. Tolosa–Hunt syndrome: MRI scan of the sellar region of a 50-year-old male with severe periorbital pain, ophthalmoplegia and hyponatremia due to hypopituitarism.

detecting inflammation in the cavernous sinus and rarely in the sella. MRI imaging enhancement demonstrates nonspecific fullness involving the left cavernous sinus and the pituitary (Fig. 18.6). Treatment consists of antibiotics and corticosteroids. For refractory cases immunosupressants (methotrexate or azathioprine) may be used along with radiotherapy.

Hypothalamic–pituitary tuberculosis

Tuberculosis is common in the developing world and accompanies a rise in the incidence of AIDS in developed countries as well. Although tubercular infections of the CNS are frequent, pituitary tuberculosis is rare (Husain et al., 2008). Tuberculous hypophysitis has been described in a few case reports. In most cases there is a past history of pulmonary tuberculosis, tuberculous meningitis, or spondylitis of the spine. The prevalence of hypopituitarism in tuberculous meningitis in childhood has been studied in 49 patients (mean age 23 years) who had tuberculous meningitis in childhood at a mean age of 6 years. Hypopituitarism was documented in 20% of patient-years after recovery from tuberculous meningitis (Lam et al., 1993). The cause appeared to be tuberculous lesions affecting the hypothalamus, the pituitary stalk, and directly the pituitary. In a series of 75 patients with tubercular meningitis in India, the most common hormonal abnormalities were hyperprolactinemia (49.3%) followed by cortisol insufficiency (42.7%) and central hypothyroidism (30.7%) (Dhanwal et al., 2010). Pituitary MRI usually shows thickening of the pituitary stalk which is considered one of the signs but this finding is not specific since it can occur in other inflammatory conditions such as autoimmune hypophysitis (Fig. 18.7). Surgery is not indicated for tuberculous hypophysitis but if performed then histopathology in some case reports shows central areas of necrosis surrounded by giant Langhans cells. Acid-fast bacilli are usually not demonstrable but polymerase chain reaction (PCR) for tuberculosis in DNA extracted from tissue may be helpful. Usually paranasal sinuses are involved as well. Hypophyseal tuberculomas are rare lesions and may mimic pituitary adenomas. The largest reported series is one with 18 cases which were histologically proven (Sharma et al., 2000; Arunkumar and Rajshekhar, 2001).

Fungal infections

The majority of pituitary abscesses are caused by bacterial infections. Fungal pituitary abscess is extremely rare. Only a few cases have been reported and fungal infections usually occur in immune compromised patients. We regularly inhale spores of diverse fungal species, yet fungal disease is uncommon. Fungal organisms within nasal mucin are present in up to 93% of patients with chronic sinusitis and even in healthy persons without sinonasal disease. Aspergilli are ubiquitous fungi found in soil and organic materials. They are saprophytes and grow within the respiratory tract. The organism can become pathogenic and CNS aspergillosis presents with meningitis, encephalitis, brain abscess, mycotic arteritis, and pituitary abscess (Iplikcioglu et al., 2004; Hao et al., 2008). MRI findings are not specific and are similar with the images of pituitary abscess. Treatment consists of transsphenoidal drainage followed by liposomal amphotericin B, itraconazole, and corticosteroids according to the established protocols for treating serious fungal infections when aspergillosis has been confirmed (Liu et al., 2010).

There is also a report of fungal sinusitis that occurred in an immunocompetent patient in whom the chronic inflammatory infiltrate from the sphenoid sinus

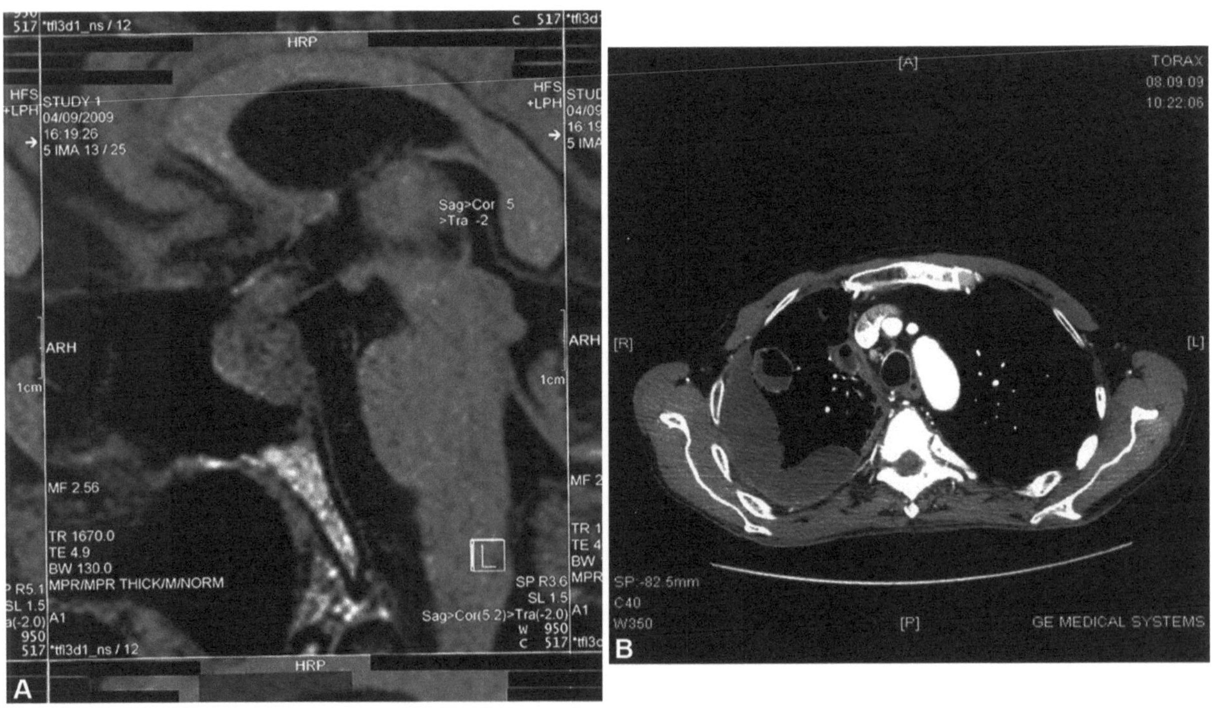

Fig. 18.7. MRI scan of the sella showing tubercular hypophysitis with thickening of the stalk in a 64-year-old male patient (**A**) with pulmonary tuberculous pleuritis and cavern (CT scan (**B**)).

progressed through a bone erosion in the sellar floor and caused compression of pituitary stalk (Fig. 18.8A, B) (Pekic et al., 2010).

Viral infections affecting the hypothalmus and/or pituitary

Pituitary insufficiency might occur in the acute phase or in the late stage of meningitis/encephalitis (Schaefer et al., 2008). Increased incidence of pituitary insufficiency following infectious meningitis has been reported (Tanriverdi et al., 2008e; Tsiakalos et al., 2010).

Causative agents of central nervous system infections include:

- meningitis (herpes, varicella, enterovirus, uknown origin)
- meninogoencephalitis
  - encephalitis (tick-borne, herpes simplex, cytomegalovirus)
- neuroborreliosis.

In a study of hypothalamic–pituitary insufficiency following infectious diseases of the CNS, patients were investigated at least 2 years after mild-moderate infection with good neurologic recovery; a substantial number of patients were found to have isolated corticotroph deficiency (Schaefer et al., 2008). In that study no DI or other abnormalities were found.

Hemorrhagic fever with renal syndrome (HFRS) caused by hantaviruses is a severe systemic infection, with acute shock, vascular leakage, hypotension, and acute renal failure. Hantaviruses are RNA viruses of the family Bunyaviridiae with serotypes *Hantaan*, *Seoul*, *Puumala* and *Dobrava-Belgrade*. Rodents are carriers of *Hantavirus* and infection occurs via aerosolized rodent urine, feces, saliva, and rodent bites. HFRS is an infection that has influenced the course of wars (Finland, Korea). Soldiers and farmers bear an occupational risk of infection. Eurosurveillance 2005 reported an outbreak in Western Europe with 1114 confirmed cases (Heyman et al., 2007). HFRS is endemic in the Balkans. Many outbreaks have been recorded, with a mortality rate of 6.6%. Pituitary hemorrhage and necrosis underlie the cause of hypopituitarism. HFRS as a cause of hypopituitarism has been sporadically reported (Lim et al., 1986; Forslund et al., 1992; Settergren et al., 1992; Suh et al., 1995; Park and Pyo, 1996; Kim et al., 2001; Pekic et al., 2005). Autopsy reports suggest a slightly enlarged pituitary with hemorrhage and necrosis (Klebanov, 1990; Valtonen et al., 1995; Hautala et al., 2002). Direct viral invasion was confirmed in the pituitary causing viral hypophysitis (Fig. 18.9). A MRI scan of the sella in an HFRS survivor with hypopituitarism reveals pituitary atrophy and secondary empty sella (Fig. 18.10) (Pekic et al., 2005). In a study of 60 adults who recovered from HFRS, in 8 patients (13.3%) growth hormone deficiency was confirmed, in five patients with multiple pituitary hormone deficiencies, and isolated in three patients (Stojanovic et al., 2008). The identified high prevalence of hypopituitarism after

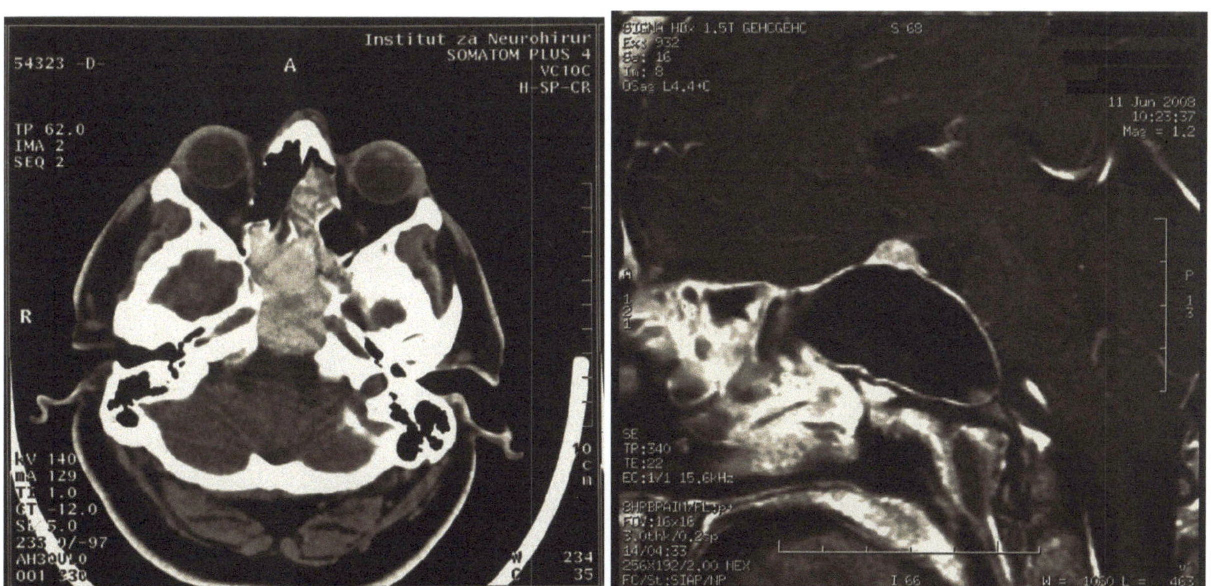

Fig. 18.8. Fungal infections in the sella: CT scan of the sinuses showed opacification of the sinuses, erosion in sellar floor, and propagation of pathologic process, a large sellar mass (**A**), and MRI scan of the sella confirmed a giant hypointense lesion in the sellar region (**B**).

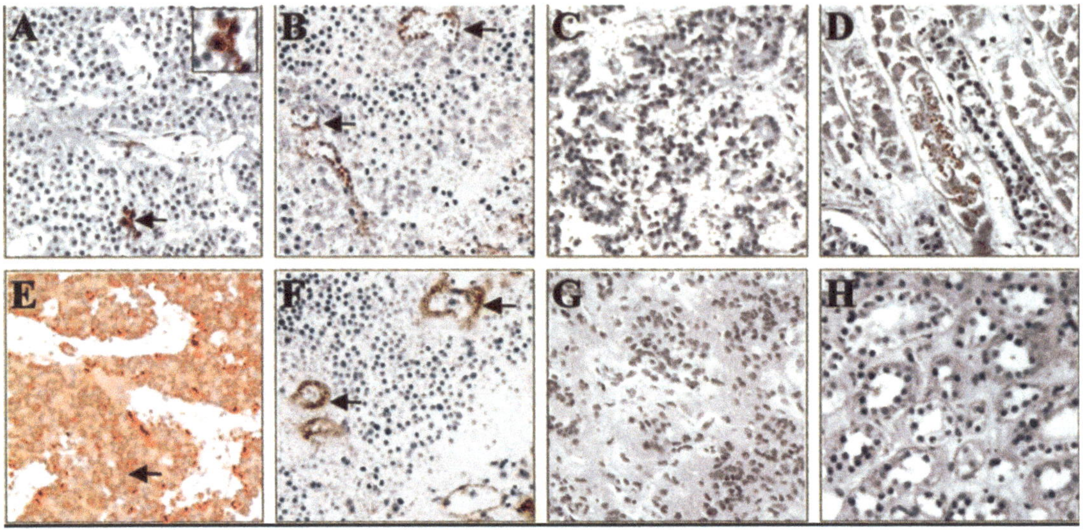

Fig. 18.9. Immunohistochemical detection of *Puumala virus* nucleocapsid protein in the hypophysis (**A–C**, **E–G**) and kidney (**D**, **H**), and characterization of the infected cells in the hypophysis. The antigen is detected with a *Puumala virus* N-specific immune serum and is visualized with 3,3′-diaminobenzidine (for hypophysis) or fast red (for kidney samples). An example of a *Puumala virus* N-protein-positive cell is seen in hypophysis of patient 1 (**A**); the infected cell was identified as a chromogranin-positive neuroendocrine cell (**E**). The arrows are pointing to parallel locations in serial sections in both (**A**) and (**E**), and also in (**B**) and (**F**). Some CD31-positive vascular endothelial cells (**F**) were also positive for *Puumala virus* N-protein in the hypophysis (**B**). The same tissue stained with preimmune serum was negative (**C**), as was a control hypophysis tissue sample incubated with the *Puumala virus* N-specific serum (**G**). Granular specific antibody binding is detected in kidney sample obtained from patient 1 (**D**), whereas a control kidney sample (**H**) shows no staining. (Reproduced with kind permission from Oxford University Press © Hautala T et al. (2002) Clin Inf Dis 35: 96–101.)

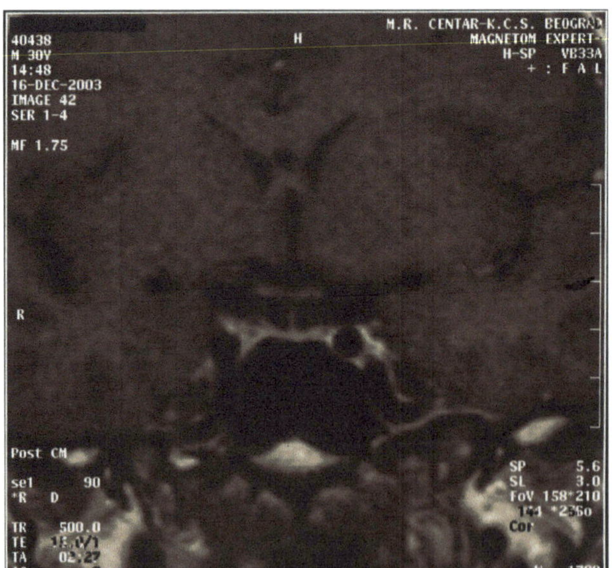

Fig. 18.10. MRI scan of the sella in HFRS survivor with hypopituitarism revealed pituitary atrophy and secondary empty sella. HFRS, hemorrhagic fever with renal syndrome. (Reproduced with kind permission from Springer © Pekic S et al. (2005) Endocrine 26: 79–82.)

recovery from HFRS raises the awareness of the possibility of viral hypophysitis.

Major causes of central diabetes insipidus (DI) are neoplastic or infiltrative lesions of the hypothalamus or pituitary, severe head injuries, and surgery. Central DI caused by viral infections has been reported rarely, mostly in immunosuppressed patients (AIDS or Cushing's syndrome) (Torres et al., 2000; Moses et al., 2003; Scheinpflug et al., 2006). The causative agents in these patients were herpes infection (herpes encephalitis) or cytomegalovirus infection (CMV). Hypothalamic areas producing AVP (vasopressin/oxytocin) infected with CMV have been documented in a case report (Moses et al., 2003).

PARASITES IN THE PITUITARY: *TOXOPLASMA GONDII*

Toxoplasma gondii is a protozoan parasite that is globally spread and infects approximately 30% of the world's population, but causes overt clinical symptoms in only a small proportion of people. *T. gondii* forms cysts that are located at various anatomic sites including the brain during chronic infection. The involvement of the pituitary in cases with toxoplasmosis is extremely rare and includes patients with congenital toxoplasmosis and in patients with central nervous system toxoplasmosis with AIDS (Milligan et al., 1984). Recently two cases of pituitary adenoma (prolactinomas) were found to be associated with *T. gondii* infection (Zhang et al., 2002). *T. gondii* cysts were found among the tumor cells (Fig. 18.11). The association of *T. gondii* infection and prolactin is interesting. Recent studies suggest that prolactin may be a regulator of antiparasitic immune activity by stimulating IFN-γ and IL-2 by Th1 lymphocytes (De Bellis et al., 2005; Plociński et al., 2007).

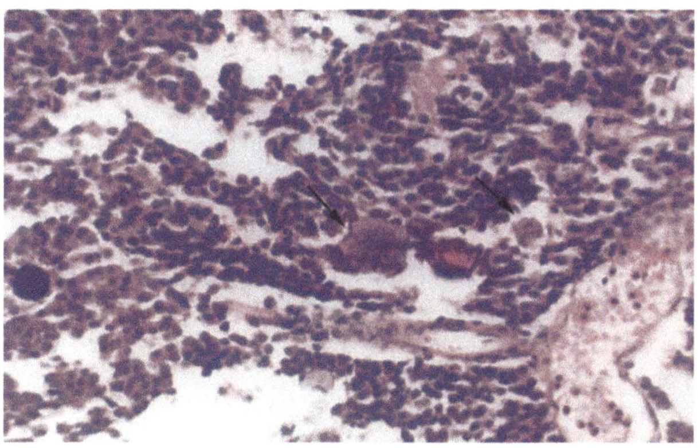

Fig. 18.11. *Toxoplasma gondii* cysts (arrows) in the tumor tissue consisted of homogeneous cells (haematoxylin and eosin staining). (Reproduced with kind permission from BMJ Publishing Group Ltd. © Zhang X et al. (2002) J Clin Pathol 55: 965–966.)

SUMMARY

Traumatic brain injury (TBI), cranial irradiation, and infectious diseases of the CNS are recognized as risk factors for hypopituitarism. The neuroendocrine dysfunction which affects a significant proportion of patients with these alternative causes of hypopituitarism may be mistakenly ascribed to depression, cognitive impairment, and in the elderly to neurologic causes. Hypopituitarism may progress (as is the case after cranial irradiation) or reverse with time (after TBI, or infections of the CNS). The incidence and prevalence of hypothalamic–pituitary dysfunction due to these alternative causes are currently unknwon and certainly underestimated.

ACKNOWLEDGMENT

This study was supported by a grant from the Ministry of Science of Republic of Serbia (Project 175033).

References

Adem A, Jossan SS, d'Argy R et al. (1989). Insulin-like growth factor I (IGF-1) receptors in the human brain: quantitative autoradiographic localization. Brain Res 503: 299–303.

Agha A, Rogers B, Sherlock M et al. (2004a). Anterior pituitary dysfunction in survivors of traumatic brain injury. J Clin Endocrinol Metab 89: 4929–4936.

Agha A, Rogers B, Mylotte D et al. (2004b). Neuroendocrine dysfunction in the acute phase of traumatic brain injury. Clin Endocrinol (Oxf) 60: 584–591.

Agha A, Sherlock M, Thompson CJ (2005a). Post-traumatic hyponatraemia due to acute hypopituitarism. QJM 98: 463–464.

Agha A, Phillips J, O'Kelly P et al. (2005b). The natural history of post-traumatic hypopituitarism: implications for assessment and treatment. Am J Med 118: 1416e1–1416e7.

Agha A, Sherlock M, Brennan S et al. (2005c). Hypothalamic-pituitary dysfunction after irradiation of nonpituitary brain tumors in adults. J Clin Endocrinol Metab 90: 6355–6360.

Ahmed SR, Shalet SM, Beardwell CG (1986). The effects of cranial irradiation on growth hormone secretion. Acta Paediatr Scand 75: 255–260.

Aimaretti G, Ambrosio M, Di Somma C et al. (2004). Traumatic brain injury and subarachnoid haemorrhage are conditions at high risk for hypopituitarism: screening study at 3 months after the brain injury. Clin Endocrinol (Oxf) 61: 320–326.

Aimaretti G, Ambrosio MR, Di Somma C et al. (2005). Residual pituitary function after brain injury-induced hypopituitarism: a prospective 12-month study. J Clin Endocrinol Metab 90: 6085–6092.

Appelman-Dijkstra NM, Kokshoorn NE, Dekkers OM et al. (2011). Pituitary dysfunction in adult patients after cranial radiotherapy: systematic review and meta-analysis. J Clin Endocrinol Metab 96: 2330–2340.

Arunkumar MJ, Rajshekhar V (2001). Intrasellar tuberculoma presenting as pituitary apoplexy. Neurol India 49: 407–410.

Bach-y-Rita P (2003). Theoretical basis for brain plasticity after a TBI. Brain Inj 17: 643–651.

Beca SG, High Jr WM, Masel BE et al. (2012). What are critical outcome measures for patients receiving pituitary replacement following brain injury? Pituitary 15: 10–19.

Benvenga S, Campenni A, Ruggeri R et al. (2000). Hypopituitarism secondary to head trauma. J Clin Endocrinol Metab 85: 1353–1361.

Bhandare N, Kennedy L, Malyapa RS et al. (2008). Hypopituitarism after radiotherapy for extracranial head and neck cancers. Head Neck 30: 1182–1192.

Bondanelli M, De Marinis L, Ambrosio MR et al. (2004). Occurrence of pituitary dysfunction following traumatic brain injury. J Neurotrauma 21: 685–696.

Burman P, Hetta J, Karlssonn A (1993). Effect of growth hormone on brain neurotransmitters. Lancet 342: 1492–1493.

Burman P, Hetta J, Wide L et al. (1996). Growth hormone treatment affects brain neurotransmitters and thyroxone. Clin Endocrinol (Oxf) 44: 319–324.

Bushnik T, Englander J, Katznelson L (2007). Fatigue after TBI: association with neuroendocrine abnormalities. Brain Inj 21: 559–566.

Casanueva FF, Leal A, Kotlowska-Haggstrom M et al. (2008). Traumatic brain injury as a relevant cause of growth hormone deficiency in adults: a KIMS-based study. Arch Phys Med Rehabil 86: 463–468.

Casanueva FF, Catro AI, Micic D et al. (2009). New guidelines for the diagnosis of growth hormone deficiency in adults. Horm Res 71 (suppl 1): 112–115.

CDC (2010). Traumatic brain injury in the United States. Emergency department visits, hospitalizations and deaths 2002–2006. Available at, http:/www.cdc.gov/TraumaticBrainInjury (accessed July 11, 2012).

Chen MS, Lin FJ, Huang MJ et al. (1989). Prospective hormone study of hypothalamic-pituitary function in patients with nasopharyngeal carcinoma after high dose irradiation. Jpn J Clin Oncol 19: 265–270.

Chieng PU, Huang TS, Chang C et al. (1991). Reduced hypothalamic blood flow after radiation treatment of nasopharyngeal cancer: SPECT studies in 34 patients. AJNR Am J Neuroradiol 12: 661–665.

Ciappetta P, Calace A, D'Urso PI et al. (2008). Endoscopic treatment of pituitary abscess: two case reports and literature review. Neurosurg Rev 31: 237–246.

Constine LS, Rubin P, Woolf PD et al. (1987). Hyperprolactinemia and hypothyroidism following cytotoxic therapy for central nervous system malignancies. J Clin Oncol 5: 1841–1851.

Constine LS, Woolf PD, Cann D et al. (1993). Hypothalamic-pituitary dysfunction after radiation for brain tumors. N Engl J Med 328: 87–94.

Cyran E (1918). Hypophysenschadigung durch Schadelbasisfraktur. Dtsch Med Wochenschr 44: 1261.

Dalan R, Leow MK (2008). Pituitary abscess: our experience with a case and a review of the literature. Pituitary 11: 299–306.

Daniel P, Prichard M, Treip C (1959). Traumatic infarction of the anterior lobe of the pituitary gland. Lancet 2: 927–930.

Daniel PM, Treip CS (1966). Lesions of the pituitary gland associated with head injuries. In: GW Harris, BT Donovan (Eds.), The Pituitary Gland. Vol. 2. Butterworth, London, pp. 519–534.

Darzy KH, Shalet SM (2003). Radiation-induced growth hormone deficiency. Horm Res 59 (Suppl 1): 1–11.

Darzy KH, Shalet SM (2005). Hypopituitarism as a consequence of brain tumours and radiotherapy. Pituitary 8: 203–211.

Darzy KH, Pezzoli SS, Thorner MO et al. (2007). Cranial irradiation and growth hormone neurosecretory dysfunction: a critical appraisal. J Clin Endocrinol Metab 92: 1666–1672.

Darzy KH, Shalet SM (2009a). Hypopituitarism following radiotherapy. Pituitary 12: 40–50.

Darzy KH, Shalet SM (2009b). Hypopituitarism following radiotherapy revisited. Endocr Dev 15: 1–24.

Darzy KH, Thorner MO, Shalet SM (2009). Cranially irradiated adult cancer survivors may have normal spontaneous GH secretion in the presence of discordant peak GH responses to stimulation tests (compensated GH deficiency). Clin Endocrinol (Oxf) 70: 287–293.

Darzy KH (2009). Radiation-induced hypopituitarism after cancer therapy: who, how and when to test. Nat Clin Pract Endocrinol Metab 5: 88–99.

De Bellis A, Bizzarro A, Pivonello R et al. (2005). Prolactin and autoimmunity. Pituitary 8: 25–30.

Deijen J, de Boer H, Blok GJ et al. (1996). Cognitive impairments and mood disturbances in growth hormone deficient men. Psychoneuroendocrinology 21: 313–322.

Deijen JB, de Boer H, van der Veen EA (1998). Cognitive changes during growth hormone replacement in adult men. Psychoneuroendocrinology 23: 45–55.

Dhanwal DK, Vyas A, Sharma A et al. (2010). Hypothalamic pituitary abnormalities in tubercular meningitis at the time of diagnosis. Pituitary 13: 304–310.

Edwards OM, Clark JD (1986). Post-traumatic hypopituitarism. Six cases and a review of the literature. Medicine 65: 281–290.

Einaudi S, Matarazzo P, Peretta P et al. (2006). Hypothalamo-hypophyseal dysfunction after traumatic brain injury in children and adolescents: a preliminary retrospective and prospective study. J Pediatr Endocrinol Metab 19: 691–703.

Escamilla R, Lisser H (1942). Simmonds disease. J Clin Endocrinol Metab 2: 65–96.

Escamilla RF, Forsham PH (1971). Treatment with human growth hormone (Li) for overt eight years-effects of long-term therapy in a pituitary dwarf. Calif Med 115: 72–76.

Fernandez A, Brada M, Zabuliene L et al. (2009). Radiation-induced hypopituitarism. Endocr Relat Cancer 16: 733–772.

Forslund T, Saltevo J, Anttinen J et al. (1992). Complications of nephropathia epidemica: three cases. J Intern Med 232: 87–90.

Gaetz M (2004). The neurophysiology of brain injury. Clin Neurophysiol 115: 4–18.

Gasco V, Prodam F, Pagano L et al. (2012). Hypopituitarism following brain injury: when does it occur and how best to test? Pituitary 15: 20–24.

Growth Hormone Research Society (1998). Consensus guidelines for the diagnosis and treatment of adults with GH deficiency: summary statement of the Growth Hormone Research Society workshop on adult growth hormone deficiency. J Clin Endocrinol Metab 83: 379–381.

Growth Hormone Research Society (2000). Consensus guidelines for the diagnosis and treatment of growth hormone (GH) deficiency in childhood and adolescence: summary statement of the GH Research Society. J Clin Endocrinol Metab 85: 3990–3993.

Gurney JG, Kadan-Lottick NS, Packer RJ et al. (2003). Endocrine and cardiovascular late effects among adult survivors of childhood brain tumors. Childhood Cancer Survivor Study. Cancer 97: 663–673.

Hannon MJ, Sherlock M, Thompson CJ (2011). Pituitary dysfunction following traumatic brain injury or subarachnoid haemorrhage – in "Endocrine Management in the Intensive Care Unit". Best Pract Res Clin Endocrinol Metab 25: 783–798.

Hao L, Jing C, Bowen C et al. (2008). Aspergillus sellar abscess: case report and review of the literature. Neurol India 56: 186–188.

Hautala T, Sironen T, Vapalahti O et al. (2002). Hypophyseal hemorrhage and panhypopituitarism during Puumala virus infection: magnetic resonance imaging and detection of viral antigen in the hypophysis. Clin Infect Dis 35: 96–101.

Heather NL, Jefferies C, Hofman PL et al. (2012). Permanent hypopituitarism is rare after structural traumatic brain injury in early childhood. J Clin Endocrinol Metab 97: 599–604.

Herrmann BL, Rehder J, Kahlke S et al. (2006). Hypopituitarism following severe traumatic brain injury. Exp Clin Endocrinol Diabetes 114: 316–321.

Heyman P, Cochez C, Ducoffre G et al. (2007). Haemorrhagic fever with renal syndrome: an analysis of the outbreaks in Belgium, the Netherlands and Luxembourg in 2005. Euro Surveill 12: E15–E16.

High WM, Briones-Galang M, Clark JA et al. (2010). Effect of growth hormone replacement therapy on cognition after traumatic brain injury. J Neurotrauma 27: 1565–1575.

Ho KKY (2000). Diagnosis of adult GH deficiency. Lancet 356: 1125.

Ho KK (2007). GH Deficiency Consensus Workshop Participants. Consensus guidelines for the diagnosis and treatment of adults with GH deficiency II: a statement of the GH Research Society in association with the European Society for Pediatric Endocrinology, Lawson Wilkins Society, European Society of Endocrinology, Japan Endocrine Society, and Endocrine Society of Australia. Eur J Endocrinol 157: 695–700.

Hoffman DM, O'Sullivan AJ, Baxter RC et al. (1994). Diagnosis of growth hormone-deficiency in adults. Lancet 343: 1064–1068.

Holmin S, Mathiesen T (1999). Long-term intracerebral inflammatory response after experimental focal brain injury in rat. Neuroreport 10: 1889–1891.

Husain N, Husain M, Rao P (2008). Pituitary tuberculosis mimicking idiopathic granulomatous hypophysitis. Pituitary 11: 313–315.

Inoue HK, Kohga H, Hirato M et al. (1999). Pituitary adenomas treated by microsurgery with or without gamma-knife surgery: experience in 122 cases. Stereotact Funct Neurosurg 72 (Supplement 1): 125–131.

Iplikcioglu AC, Bek S, Bikmaz K et al. (2004). Aspergillus pituitary abscess. Acta Neurochir 146: 521–524.

Jagganathan J, Sheehan JP, Pouratian N et al. (2008). Gamma-knife radiosurgery for acromegaly: outcomes after failed transsphenoidal surgery. Neurosurgery 62: 1262–1269.

Johansson J, Larson G, Andersson M et al. (1995). Treatment of growth hormone-deficient adults with recombinant human growth hormone increases concentration of growth hormone in cerebrospinal fluid and affects neurotransmitters. Neuroendocrinology 61: 57–66.

Jorgensen EV, Schwartz ID, Hvizdala E et al. (1993). Neurotransmitter control of growth hormone secretion in children after cranial radiation therapy. J Pediatr Endocrinol 6: 131–142.

Kahler O (1886). Die dauernde polyurie als cerebrales herdsymptom. Z Heilk 7: 105–220.

Kasturi BS, Stein DG (2009). Traumatic brain injury causes long-term reduction in serum growth hormone and persistent astrocytosis in the cortico-hypothalamo-pituitary axis of adult male rats. J Neurotrauma 26: 1315–1324.

Kelestimur F, Tanriverdi F, Atmaca H et al. (2004). Boxing as a sport activity associated with isolated GH. J Endocrinol Invest 27: RC28–RC32.

Kelly DF, Gonzalo IT, Cohan P et al. (2000). Hypopituitarism following traumatic brain injury and aneurismal subarachnoid hemorrhage: a preliminary report. J Neurosurg 93: 743–752.

Kim NH, Cho JG, Ahn YK et al. (2001). A case of torsade de pointes associated with hypopituitarism due to hemorrhagic fever with renal syndrome. J Korean Med Sci 16: 355–359.

Klebanov IuA (1990). Analysis of fatal outcome in hemorrhagic fever with renal syndrome. Klin Med 68: 64–67.

Klose M, Juul A, Poulsgaard L et al. (2007a). Prevalence and predictive factors of post-traumatic hypopituitarism. Clin Endocrinol (Oxf) 67: 193–201.

Klose M, Watt T, Brennum J et al. (2007b). Posttraumatic hypopituitarism is associated with an unfavorable body composition and lipid profile, and decreased quality of life 12 months after injury. J Clin Endocrinol Metab 92: 3861–3868.

Klose M, Feldt-Rasmussen U (2008). Does the type and severity of brain injury predict hypothalamo-pituitary dysfunction? Does post-traumatic hypopituitarism predict worse outcome? Pituitary 11: 255–261.

Klose M, Jonsson B, Abs R et al. (2009). From isolated GH deficiency to multiple pituitary hormone deficiency: an evolving continuum-a KIMS analysis. Eur J Endocrinol 161: S75–S83.

Kokshoorn NE, Wassenaar MJE, Biermasz NR et al. (2010). Hypopituitarism following traumatic brain injury: prevalence is affected by the use of different dynamic tests and different normal values. Eur J Endocrinol 162: 11–18.

Kokshoorn NE, Smit JWA, Nieuwlaat WA et al. (2011). Low prevalence of hypopituitarism after traumatic brain injury: a multicenter study. Eur J Endocrinol 165: 225–231.

Kreitschmann-Andermahr I, Hartmann Y, Poll E et al. (2011). The German database on hypopituitarism after traumatic brain injury and aneurysmal subarachnoid hemorrhage – description, objectives and design. Exp Clin Endocrinol Diabetes 119: 15–20.

Lam KS, Tse VK, Wang C et al. (1991). Effects of cranial irradiation on hypothalamic-pituitary function – a 5-year longitudinal study in patients with nasopharyngeal carcinoma. Q J Med 78: 165–176.

Lam KS, Sham MM, Tam SC et al. (1993). Hypopituitarism after tuberculous meningitis in childhood. Ann Intern Med 118: 701–706.

Lamberts SWJ, de Herder WW, van der Lely AJ (1998). Pituitary insufficiency. Lancet 352: 127–134.

Laskowitz DT, Goel S, Bennett ER et al. (1997). Apolipoprotein E suppresses glial cell secretion of TNF alpha. J Neuroimmunol 76: 70–74.

Leal-Cerro A, Flores JM, Rincon M et al. (2005). Prevalence of hypopituitarism and growth hormone deficiency in adults long-term after severe traumatic brain injury. Clin Endocrinol (Oxf) 62: 525–532.

Leiper AD, Stanhope R, Kitching P et al. (1987). Precocious and premature puberty associated with treatment of acute lymphoblastic leukamia. Arch Dis Child 62: 1107–1112.

Leon-Carrion J, Dominguez-Morales M del R, Barroso y Martin JM et al. (2005). Epidemiology of traumatic brain injury and subarachnoid hemorrhage. Pituitary 8: 197–202.

Lieberman SA, Oberoi AL, Gilkinson CR et al. (2001). Prevalence of neuroendocrine dysfunction in patients recovering from traumatic brain injury. J Clin Endocrinol Metab 86: 2752–2756.

Lim TH, Chang KH, Han MC et al. (1986). Pituitary atrophy in Korean (epidemic) hemorrhagic fever: CT correlation with pituitary function and visual field. AJNR Am J Neuroradiol 7: 633–637.

Lissett CA, Saleem S, Rahim A et al. (2001). The impact of irradiation on growth hormone responsiveness to provocative agents is stimulus dependent: results in 161 individuals with radiation damage to the somatotropic axis. J Clin Endocrinol Metab 86: 663–668.

Littley MD, Shalet SM, Beardwell CG et al. (1989). Hypopituitarism following exernal radiotherapy for pituitary tumours in adults. Q J Med 70: 145–160.

Liu W, Chen H, Cai B et al. (2010). Successful treatment of sellar aspergillus abscess. J Clin Neurosci 17: 1587–1589.

Mahajan T, Lightman SL (2000). A simple test for growth hormone deficiency in adults. J Clin Endocrinol Metab 85: 1473–1476.

Maranon G, Pintos G (1917). Lésion traumatique pure de l'hypophyse. Syndrome adiposo-génital et diabète insipide. Nouv Iconogr Salpet 28: 185–195.

Maric NP, Doknic M, Pavlovic D et al. (2010). Psychiatric and neuropsychological changes in growth hormone-deficient patients after traumatic brain injury in response to growth hormone therapy. J Endocrinol Invest 33: 770–775.

Medic-Stojanoska M (2009). Traumatic brain injury induced hypopituitarism in children and adolescents. Pediatr Helath 3: 1–9.

Military Health System (2011). DoD Worldwide Numbers for Traumatic Brain Injury. Available at, http:www.health.mil/Research/TBI_Numbers.aspx (accessed May 12, 2014).

Milligan SA, Katz MS, Craven PC et al. (1984). Toxoplasmosis presenting as panhypopituitarism in a patient with the acquired immune deficiency syndrome. Am J Med 77: 760–764.

Minniti G, Scaringi C, Amelio D et al. (2012). Stereotactic irradiation of GH-secreting pituitary adenomas. Int J Endocrinol 2012: 482861.

Moses AM, Thomas DG, Canfield MC et al. (2003). Central diabetes insipidus due to cytomegalovirus infection of the hypothalamus in a patient with acquired immunodeficiency syndrome: a clinical, pathological, and immunohistochemical case study. J Clin Endocrinol Metab 88: 51–54.

Mukherjee A, Shalet SM (2009). The value of IGF1 estimation in adults with GH deficiency. Eur J Endocrinol 161: S33–S39.

Nishida Y, Yoshioka M, Armand J (2005). The top 10 most abundant transcripts are sufficient to characterize the organ's functional specificity: evidence from the cortex, hypothalamus and pituitary gland. Gene 334: 133–141.

Nonaka M, Chen XN, Pierce JE et al. (1999). Prolonged activation of NF-kappaB following traumatic brain injury in rats. J Neurotrauma 16: 1023–1034.

Nyberg F, Burman P (1996). Growth hormone and its receptors in the central nervous system – location and functional significance. Horm Res 45: 18–22.

Nyberg F (2000). Growth hormone in the brain: characteristics of specific brain targets for the hormone and their functional significance. Front Neuroendocrinol 21: 300–348.

Oeffinger KC, Mertens AC, Sklar CA et al. (2006). Chronic health conditions in adult survivors of childhood cancer. N Engl J Med 355: 1572–1582.

Ogilvy-Stuart A, Wallace W, Shalet S (1994a). Radiation and neuroregulatory control of growth hormone secretion. Clin Endocrinol (Oxf) 41: 163–168.

Ogilvy-Stuart AL, Clayton PE, Shalet SM (1994b). Cranial irradiation and early puberty. J Clin Endocrinol Metab 78: 1282–1286.

Park JE, Pyo HJ (1996). Delayed onset of diuresis in a patient with acute renal failure due to hemorrhagic fever with renal syndrome who also developed anterior hypopituitarism. Clin Nephrol 46: 141–145.

Pavlovic D, Pekic S, Stojanovic M et al. (2010). Chronic cognitive sequelae after traumatic brain injury are not related to growth hormone deficiency in adults. Eur J Neurol 17: 696–702.

Pekic S, Cvijovic G, Stojanovic M et al. (2005). Hypopituitarism as a late complication of hemorrhagic fever. Endocrine 26: 79–82.

Pekic S, Arsenijevic VA, Gazibara MS et al. (2010). What lurks in the sellar? Lancet 375: 432.

Pepene CE, Ilie I, Mihu D et al. (2010). Primary pituitary abscess followed by empty sella syndrome in an adolescent girl. Pituitary 13: 385–389.

Płociński P, Dzitko K, Długońska H (2007). Prolactin as a modulator of antiparasitic immunity. Wiad Parazytol 53: 263–270.

Poomthavorn P, Maixner W, Zacharin M (2008). Pituitary function in pediatric survivors of severe traumatic brain injury. Arch Dis Child 93: 133–137.

Popovic V, Pekic S, Golubicic I et al. (2002). The impact of cranial irradiation on GH responsiveness to GHRH plus GH-releasing peptide-6. J Clin Endocrinol Metab 87: 2095–2099.

Popovic V, Pekic S, Pavlovic D et al. (2004). Hypopituitarism as a consequence of traumatic brain injury (TBI) and its possible relation with cognitive disabilities and mental distress. J Endocrinol Invest 11: 1048–1054.

Popovic V, Aimaretti G, Casanueva FF et al. (2005). Hypopituitarism following traumatic brain injury. Growth Horm IGF Res 15: 177–184.

Popovic V (2005). GH deficiency as the most common pituitary defect after TBI: clinical implications. Pituitary 8: 239–243.

Roth C, Schmidberger H, Lakomek M et al. (2001). Reduction of gamma-aminobutryic acid-ergic neurotransmission as a putative mechanism of radiation induced activation of the gonadotropin releasing-hormone-pulse generator leading to precocious puberty in female rats. Neurosci Lett 297: 45–48.

Salehi F, Kovacs K, Scheithauer BW et al. (2007). Histologic study of the human pituitary gland in acute traumatic brain injury. Brain Inj 21: 651–656.

Samaan NA, Vieto R, Schultz PN et al. (1982). Hypothalamic, pituitary and thyroid dysfunction after radiotherapy to the head and neck. Int J Radiat Oncol Biol Phys 8: 1857–1867.

Samaan NA, Schultz PN, Yang KP et al. (1987). Endocrine complications after radiotherapy for tumors of the head and neck. J Lab Clin Med 109: 364–372.

Schaefer S, Boegershausen N, Meyer S et al. (2008). Hypothalamic-pituitary insufficiency following infectious diseases of the central nervous system. Eur J Endocrinol 158: 3–9.

Scheinpflug K, Schalk E, Reschke K et al. (2006). Diabetes insipidus due to herpes encephalitis in a patient with diffuse large cell lymphoma. A case report. Exp Clin Endocrinol Diabetes 114: 31–34.

Schmiegelow M, Lasen S, Poulsen HS et al. (2000). Cranial radiotherapy of childhood brain tumours: growth hormone deficiency and its relation to the biological effective dose of irradiation in a large population based study. Clin Endocrinol (Oxf) 53: 191–197.

Schneider HJ, Schneider M, Saller B et al. (2006). Prevalence of anterior pituitary insufficiency 3 and 12 months after traumatic brain injury. Eur J Endocrinol 154: 259–265.

Schneider HJ, Aimaretti G, Kreitschmann-Andermahr I et al. (2007a). Hypopituitarism. Lancet 369: 1461–1470.

Schneider HJ, Kreitschmann-Andermahr I, Ghigo E et al. (2007b). Hypothalamopituitary dysfunction following traumatic brain injury and aneurismal subarachnoid hemorrhage. JAMA 298: 1429–1438.

Schneider M, Schneider HJ, Yassouridis A et al. (2008). Predictors of anterior pituitary insufficiency after traumatic brain injury. Clin Endocrinol (Oxf) 68: 206–212.

Schneider HJ, Schneider M, Kreitschmann-Andermahr I et al. (2011). Structured assessment of hypopituitarism after traumatic brain injury and aneurysmal subarachnoid hemorrhage in 1242 patients: the German Interdisciplinary Database. J Neurotrauma 28: 1693–1698.

Settergren B, Boman J, Linderholm M et al. (1992). A case of nephropathia epidemica associated with panhypopituitarism and nephrotic syndrome. Nephron 61: 234–235.

Shalet SM, Beardwell CG, Jones PH et al. (1976a). Growth hormone deficiency after treatment of acute leukaemia in children. Arch Dis Child 51: 489–493.

Shalet SM, Beardwell CG, Morris-Jones P et al. (1976b). Growth hormone deficiency in children with brain tumors. Cancer 37: 1144–1148.

Shalet SM, Beardwell CG, MacFarlane IA et al. (1977). Endocrine morbidity in adults treated with cerebral irradiation for brain tumours during childhood. Acta Endocrinol 84: 673–680.

Shalet S (1997). Cytotoxic endocrinopathy: a legacy of insults. J R Soc Med 90: 192–199.

Sharma MC, Arora R, Mahapatra AK et al. (2000). Intrasellar tuberculoma – an enigmatic pituitary infection: a series of 18 cases. Clin Neurol Neurosurg 102: 72–77.

Sicignano G, Losa M, del Vecchio A et al. (2012). Dosimetric factors associated with pituitary function after Gamma Knife surgery (GKS) of pituitary adenomas. Radiother Oncol 104: 119–124.

Smith DH, Chen XN, Pierce JE et al. (1997). Progressive atrophy and neuron death for one year following brain trauma in the rat. J Neurotrauma 14: 715–727.

Springer J, Chollet A (2001). A traumatic car crash. Lancet 357: 1848.

Starke RM, Williams BJ, Jane Jr JA et al. (2012). Gamma Knife surgery for patients with nonfunctioning pituitary macroadenomas: predictors of tumor control, neurological deficits and hypopituitarism. J Neurosurg 117: 129–135.

Stein TD, Fedynyshyn JP, Kalil RE (2002). Circulating autoantibodies recognize and bind dying neurons following injury to the brain. J Neuropathol Exp Neurol 61: 1100–1108.

Stojanovic M, Pekic S, Cvijovic G et al. (2008). High risk of hypopituitarism in patients who recovered from hemorrhagic fever with renal syndrome. J Clin Endocrinol Metab 93: 2722–2728.

Suh DC, Park JS, Park SK et al. (1995). Pituitary hemorrhage as a complication of hantaviral disease. AJNR Am J Neuroradiol 16: 175–178.

Tanriverdi F, Senyurek H, Uhluhizarci K et al. (2006). High risk of hypopituitarism after traumatic brain injury: a prospective investigation of anterior pituitary function in the acute phase and 12 months after trauma. J Clin Endocrinol Metab 91: 2105–2111.

Tanriverdi F, Unluhizarci K, Coksevim B et al. (2007). Kickboxing sport as a new cause of traumatic brain injury-mediated hypopituitarism. Clin Endocrinol (Oxf) 66: 360–366.

Tanriverdi F, Ulutabanca H, Unluhizarci K et al. (2008a). Three years prospective investigation of anterior pituitary function after traumatic brain injury: a pilot study. Clin Endocrinol (Oxf) 68: 573–579.

Tanriverdi F, Unluhizarci K, Kocyigit I et al. (2008b). Brief communication: pituitary volume and function in competing and retired male boxers. Ann Intern Med 148: 827–831.

Tanriverdi F, De Bellis A, Bizzaro A et al. (2008c). Antipituitary antibodies after traumatic brain injury: is head trauma-induced pituitary dysfunction associated with autoimmunity? Eur J Endocrinol 159: 7–13.

Tanriverdi F, Taheri S, Ulutabanca H et al. (2008d). Apolipoprotein E3/E3 genotype decrease the risk of pituitary dysfunction after traumatic brain injury due to various causes: preliminary data. J Neurotrauma 25: 1071–1077.

Tanriverdi F, Alp E, Demiraslan H et al. (2008e). Investigation of pituitary functions in patients with acute meningitis: a pilot study. J Endocrinol Invest 31: 489–491.

Tanriverdi F, Unluhizarci K, Kelestimur F (2010a). Pituitary function in subjects with mild tramatic brain injury: a review of literature and proposal of a screening strategy. Pituitary 13: 146–153.

Tanriverdi F, De Bellis A, Bataglia M et al. (2010b). Investigation of antihypothalamus and antipituitary antibodies in amateur boxers: is chronic repetitive head trauma-induced pituitary dysfunction associated with autoimmunity? Eur J Endocrinol 162: 861–867.

Tanriverdi F, Unluhizarci K, Karaca Z et al. (2010c). Hypopituitarism due to sports related head trauma and the effects of growth hormone replacement in retired amateur boxers. Pituitary 13: 111–114.

Tanriverdi F, Agha A, Aimaretti G et al. (2011). Manifesto for the current understanding and management of traumatic brain injury-induced hypopituitarism. J Endocrinol Invest 34: 541–543.

Torres AM, Kazee AM, Simon H et al. (2000). Central diabetes insipidus due to herpes simplex in a patient immunosuppressed by Cushing's syndrome. Endocr Pract 6: 26–28.

Treip CS (1970). Hypothalamic and pituitary injury. J Clin Pathol 23 (Suppl): 178–186.

Tsiakalos A, Xynos ID, Sipsas NV et al. (2010). Pituitary insufficiency after infectious meningitis: a prospective study. J Clin Endocrinol Metab 957: 3277–3281.

Urban RJ (2006). Hypopituitarism after acute brain injury. Growth Horm IGF Res 16 (Suppl A): S25–S29.

Valtonen M, Kauppila M, Kotilainen P et al. (1995). Four fatal cases of nephropathia epidemica. Scand J Infect Dis 27: 515–517.

Van den Berghe G, de Zegher F, Bouillon R (1998). Clinical review 95: acute and prolonged critical illness as different neuroendocrine paradigms. J Clin Endocrinol Metab 83: 1827–1834.

Van der Eerden A, Twickler M, Sweep F et al. (2010). Should anterior pituitary function be tested during follow-up of all patients presenting at the emergency department because of traumatic brain injury? Eur J Endocrinol 162: 19–28.

Van Nieuwpoort IC, Drent ML (2008). Cognition in the adult with childhood-onsed GH deficiency. Eur J Endocrinol 159: S53–S57.

Vance ML (1994). Hypopituitarism. N Engl J Med 330: 1651–1662.

Vates GE, Berger MS, Wilson CB (2001). Diagnosis and management of pituitary abscess: a review of twenty-four cases. J Neurosurg 95: 233–241.

Wachter D, Gundling K, Oertel MF et al. (2009). Pituitary insufficiency after traumatic brain injury. J Clin Neurosci 16: 202–208.

Wilkinson CW, Pagulayan KF, Petrie EC et al. (2012). High prevalence of chronic pituitary and target-organ hormone abnormalities after blast-related mild traumatic brain injury. Front Neurol 3: 1–12.

Zhang X, Li Q, Hu P et al. (2002). Two case reports of pituitary adenoma associated with *Toxoplasma gondii* infection. J Clin Pathol 55: 965–966.

Zhang X, Sun J, Shen M et al. (2012). Diagnosis and minimally invasive surgery for the pituitary abscess: a review of twenty nine cases. Clin Neurol Neurosurg 114: 957–961.

Handbook of Clinical Neurology, Vol. 124 (3rd series)
Clinical Neuroendocrinology
E. Fliers, M. Korbonits, and J.A. Romijn, Editors

Chapter 19

Surgical approach to pituitary tumors

DOMENICO SOLARI, LUIGI MARIA CAVALLO, AND PAOLO CAPPABIANCA*
Department of Neurological Sciences, Division of Neurosurgery, Università degli Studi di Napoli Federico II, Naples, Italy

INTRODUCTION

Pituitary surgery is a continuously evolving specialty of neurosurgery that requires precise anatomic knowledge, technical skills, and an integrated appreciation of pituitary pathophysiology. Thus, it cannot be considered to be solely a surgical procedure, but rather the result of a close cooperation between different specialists, i.e., the endocrinologist, neurosurgeon, neuroradiologist, pathologist, ophthalmologist, and others. In this teamwork environment each member plays a well-defined role, offering a contribution to the final result, specifically tailored to single patients. It is currently possible to manage many of the different pituitary syndromes with more than one option, including medical, surgical, and radiotherapeutic options, either alone or in various combinations. Pituitary surgery, perhaps more than other area of neurosurgery, requires careful and specific postoperative management and long-term patient follow-up; these can make the difference between a satisfactory result and a poor one. A patient may be operated on successfully, but the outcome may not be as brilliant as the surgical procedure promises if there is no mutual exchange among the members of the team or if their work is not implemented: each participant contributes to the final outcome for the patient while promoting growth of the other components, which calls for further work and better allocation of competencies and effectiveness: A virtuous circle develops.

It is in such a context that pituitary surgery should exist today, where the neurosurgeon dealing with techniques, indications, and results is playing a refined role as a member of an expert team: he or she should know detailed anatomy, be experienced in neuroimaging, know pathophysiology and the natural history of pituitary disease, and be familiar with all the different therapeutic options (McLaughlin et al., 2012). The neurosurgeon plays a crucial role, fully informed about current therapeutic possibilities in the interest of the patient and of the institution where the operation is done.

HISTORICAL BACKGROUND

The first operation on a pituitary tumor was performed by Horsley in 1889; in 1906 he published the results obtained on a series of 10 patients first by means of a frontal craniotomy and later using a temporal approach (Handelsmann and Horsley, 1911). The first surgeon reporting on an operation specifically for a pituitary tumor was a British general surgeon, Paul, who, in 1893, performed a temporal decompression in an acromegalic patient without actually reaching the tumor (Caton, 1893). The next milestone was the first transsphenoidal approach, performed by the Viennese surgeon Schloffer in 1907 (Schloffer, 1907), based on the anatomic studies of the Italian physician Giordano (Giordano, 1911; Artico et al., 1998), chief surgeon of the hospital in Venice.

The first totally endonasal procedure without complete dislocation of the nose was achieved in 1910 by Hirsch, a Viennese rhinologist, who was the first to incorporate a nasal speculum (Hirsch, 1910). Cushing performed his first transsphenoidal procedure in 1909 (Cushing, 1909, 1981), a classic sublabial, transseptal, transsphenoidal approach combining the evolution of his own technique and aspects of the different methods reported so far. He abandoned this procedure because of difficulties in achieving hemostasis and completeness of tumor removal in large suprasellar tumors, as compared to the transcranial procedure (Cushing, 1932; Rosegay, 1981; Lanzino and Laws, 2001). In 1918 the American neurosurgeon Dandy offered his view that “the nasal route is impractical and can never be otherwise”,

*Correspondence to: Paolo Cappabianca, MD, Division of Neurosurgery, Department of Neurological Sciences, Università degli Studi di Napoli Federico II, via S. Pansini 5, 80131, Naples, Italy. Tel: +39-081-7462559, Fax: +39-081-19560905, E-mail: paolo.cappabianca@unina.it

presenting his experience of 20 cases operated on by an intracranial intradural approach to the chiasm, according to a frontotemporal route to the pituitary (Dandy, 1934). The two main transcranial options, i.e., the subfrontal and the frontotemporal, are still used today together with more recent skull base approaches.

The only pupil of Cushing who did not abandon the transsphenoidal method was Dott, a neurosurgeon at the Royal Infirmary of Edinburgh (Lanzino and Laws, 2001; Liu et al., 2001); he kept the procedure alive, improving the technique, and taught the method to the French neurosurgeon Guiot during the latter's visit to the Royal Infirmary in 1956. Guiot at the Hôpital Foch in Paris, and his trainee Hardy in Montreal, deserve the credit for the "transsphenoidal renaissance" in the 1960s (Kanter et al., 2005). Modern transsphenoidal surgery takes advantage of the innovations of intraoperative image intensification and fluoroscopy, introduced by Guiot, and of the operating microscope as used by Hardy (1969, 1971), who introduced the concept of microadenoma and selective microsurgical resection.

It was Gerard Guiot (Guiot et al., 1963; Guiot, 1973; Lanzino and Laws, 2001; Liu et al., 2001; Cappabianca et al., 2003; Kanter et al., 2005) who first proposed the use of the endoscope during a classic transsphenoidal transnasorhinoseptal approach in order to explore the sellar contents. Nevertheless, it was not until the 1990s that, embracing the experience of otorhinolaryngologists in functional endoscopic sinus surgery (FESS) (Kennedy, 1985; Stammberger and Posawetz, 1990; Jankowski et al., 1992), the "pure" endoscopic endonasal approach to the sellar area, with the endoscope as the sole visualizing tool during the whole procedure, was defined by the Pittsburgh duo of an otorhinolaryngologist and a neurosurgeon, namely Carrau and Jho (Carrau et al., 1996; Jho et al., 1996a); they were followed by our group in Naples (Cappabianca et al., 1998; de Divitiis et al., 2003; Cappabianca et al., 2004).

SURGERY

Pituitary surgery performed by means of a transsphenoidal or transcranial approach has been developing through advances in medical science and technological progress; the surgical procedure for the removal of pituitary adenomas is, however, targeted at achieving multiple goals (Laws, 1993b; Laws and Lanzino, 2010):

1. Normalization of excess hormone secretion
2. Preservation or restoration of normal pituitary function
3. Relief of mass effect
4. Preservation or restoration of normal neurologic function, usually visual acuity or visual field (or both)
5. Prevention of tumor recurrence
6. Achievement of a complete histologic diagnosis
7. Obtaining tissue for scientific studies.

It should be remembered that biologically, endocrinologically, and pathologically, pituitary tumors represent a heterogeneous group of lesions, so that the role of surgery will be different for different pituitary tumor subtypes. Indications for surgery have changed through time, due to the refinement of surgical techniques and the evaluation of results and experiences, the development of knowledge about the molecular biology of diseases, and the use of effective new pharmacologic agents and radiation techniques (Cappabianca et al., 2010).

Nonetheless, the primary role of surgery has been established (Laws and Ebersold, 1982; Cardoso and Peterson, 1984; Bevan et al., 1992; Bills et al., 1993; Davis et al., 1993; Brisman et al., 1996; Fahlbusch et al., 1996; Colao et al., 1997, 2003; Abosch et al., 1998; Swearingen et al., 1998; Shomali and Katznelson, 1999; Jane et al., 2001; Lohmann et al., 2001; Simmons et al., 2001; Chen et al., 2003; Casanueva et al., 2006; Esposito et al., 2006; Losa et al., 2006) for the following:

- Nonfunctioning pituitary tumors
- Pituitary apoplexy
- Progressive mass effect, producing compression of the surrounding neurovascular structures, regardless hormonal status
- Cushing's disease, because of the present inadequacy of pharmacologic agents
- Acromegaly, in combination with medical treatment (preoperatively and postoperatively, if necessary)
- Secondary hyperthyroidism.

The role of surgery in prolactinoma is secondary (Molitch et al., 1985; Bevan et al., 1992; Colao et al., 2003), but still necessary in selected conditions. Indications for surgery also include:

- Failure of, or resistance to, medical treatment; intolerable side-effects of medical therapy
- Complications of medical therapy such as cerebrospinal fluid (CSF) leakage due to tumor shrinkage, or apoplexy (e.g., in prolonged and massive cabergoline treatment)
- Recurrences, in combination or in association with the other therapeutic options, medical and/or radiotherapeutic
- Patient choice.

The surgical approach, with respect to the basic principles for resecting pituitary adenomas, can be performed via two main routes, each of them amenable to several different approaches:

1. Transsphenoidal
   a. microsurgical
      (i) transnasal
      (ii) sublabial
      (iii) endonasal
   b. endoscopic.
2. Transcranial
   a. subfrontal unilateral
   b. frontolateral or pterional
   c. subfrontal bilateral interhemispheric.

In recent decades, this field of surgery has been taking advantage of evolving ideas and surgical tools in the attempt to attain the lowest possible rates of morbidity and mortality in a safe, feasible, and practical way. The transsphenoidal midline route, however, has become the standard approach to the pituitary area, being the less traumatic, direct route to the sella, avoiding brain retraction, and providing excellent visualization of the pituitary gland and adjacent pathology, with a lower morbidity and mortality rate as compared with transcranial procedures (Perneczky et al., 1999; Leonhard et al., 2003; Cappabianca and de Divitiis, 2004; Doglietto et al., 2005; Kanter et al., 2005). Transsphenoidal surgery today is used in over 95% of surgical procedures to the sellar area and in about 97% of all surgery for the treatment of pituitary adenomas. Absolute indications were established about 30 years ago and are still valid today:

- elevated surgical risk of the transcranial route
- in the elderly
- in longstanding compression of the chiasm
- in case of acute endosellar hypertension
- in most cases of pituitary apoplexy
- in paninvasive not radically removable adenomas
- in cases of adenoma with downward development
- in cases of microadenoma
- in non-neoplastic intrasellar cysts (Baskin and Wilson, 1984; Ross et al., 1992; el-Mahdy and Powell, 1998; Cavallo et al., 2008)
- in craniopharyngiomas, especially cystic, extra-arachnoidal, and infradiaphragmatic (Guiot and Derome, 1972), with an enlarged sella (Laws, 1980; Abe and Ludecke, 1999).

To these classic guidelines for the transsphenoidal option, in more recent decades (Zada and Cappabianca, 2010) the following can be added:

- The extended *transplanum-transtuberculum* approach (Weiss, 1987; Kelley et al., 1996; Kato et al., 1998; Kim et al., 2000; Kouri et al., 2000; Maira et al., 2004; Laws et al., 2005; Locatelli et al., 2006; Castelnuovo et al., 2007; Kitano and Taneda, 2009), for the removal of suprasellar craniopharyngiomas, Rathke's cleft cysts, some tuberculum sellae meningiomas, and anterior cranial base CSF leaks; the extended approach to the clival area and to the parasellar compartment for invasive adenomas (Fraioli et al., 1995; Alfieri and Jho, 2001; Kitano et al., 2008; Dallan et al., 2011; Di Maio et al., 2011) and chordomas (Hardy and Vezina, 1976; Laws, 1993a; Jho et al., 1996b; Maira et al., 1996; Jho et al., 1997; Jho and Ha, 2004; Stippler et al., 2009; Fraser et al., 2010; Komotar et al., 2011).
- A multistaged transsphenoidal approach, for the removal of intrasuprasellar adenomas, as an intentionally two-stage transsphenoidal operation, in order to favor the descent of a suprasellar remnant of the adenoma and limit the risks of a brisk decompression of huge lesions (Saito et al., 1995).

Nevertheless, there are conditions that limit and sometimes contraindicate the choice of the transsphenoidal approach in favor of the transcranial, either related to the anatomy of the surgical route or to the inner features of the lesion itself (Zada et al., 2011). The size of the sella, its degree of ossification, the size and the pneumatization of the sphenoid sinus and/or carotid arteries, position and shape can increase the difficulty of the transsphenoidal procedure.

Indications for transcranial surgery include the following (Wilson, 1990; Yasargil, 1996; Thapar and Laws, 2001; Powell and Pollock, 2003):

- Tumors with extensive intracranial invasion, with asymmetric lateral development, into the anterior cranial fossa or lateral or posterior extension into the middle and posterior cranial fossa, particularly if major vessel involvement is present
- Suprasellar tumors not completely resectable through the transsphenoidal route
- Recurrent or residual pituitary tumors in patients who have already had unsuccessful transsphenoidal surgery.

TRANSSPHENOIDAL APPROACHES

The transsphenoidal approach represents a minimally traumatic corridor of surgical access to the sella, providing direct and superior visualization of the pituitary gland and adjacent pathology (Laws, 1993b; Perneczky et al., 1999; Elias and Laws, 2000; Leonhard et al., 2003; Cappabianca and de Divitiis, 2004; Doglietto et al., 2005; Kanter et al., 2005). It has been performed since the 1960s by means of the operating microscope, through transnasal transseptal, sublabial transseptal, or endonasal procedures (microsurgical transsphenoidal procedures) (Kanter et al., 2005). Recently, the endoscope has been introduced in transsphenoidal surgery as the sole visualizing tool during the entire surgical procedure, defining the "pure" endoscopic endonasal transsphenoidal approach (Doglietto et al., 2005). The combined use of the microscope and the endoscope during the same approach defines the procedure of endoscope-assisted microsurgery.

Microsurgical transsphenoidal approaches

There are three basic microsurgical transsphenoidal approaches to pituitary tumors: (1) the transnasal transseptal transsphenoidal approach, (2) the sublabial transseptal transsphenoidal approach, and (3) the endonasal transsphenoidal approach, performed with an operating microscope for visualization, illumination, and magnification of the surgical field.

Microsurgical transnasal transseptal transsphenoidal approach

In the right nostril, the columella is retracted to expose the anterior edge of the septal cartilage; the cartilaginous septum is then dissected and freed from the bony septum. Posterior submucosal tunnels are created along both sides of the bony septum, which is partially removed to facilitate the introduction of a self-retaining transsphenoidal retractor. Care must be taken to avoid mucosal perforation during these maneuvers.

Microsurgical sublabial transseptal transsphenoidal approach

The upper lip is retracted, and an incision is made along the buccogingival junction, between the two canine fossae; the upper lip and the periosteum are elevated to expose the anterior nasal spine and the inferior border of the pyriform aperture of the nasal cavities. The mucosa of the floor of the nose is elevated first on both sides; the inferior and posterior portion of the cartilaginous septum is dissected from the bony nasal septum and is deflected laterally. The self-retaining nasal speculum is introduced and widely opened; indeed, a more anterior trajectory is provided as compared to the transnasal option.

Microsurgical endonasal transsphenoidal approach

A hand-held speculum is inserted into the nostril along the middle turbinate, and a vertical mucosal incision is performed at the junction of the keel of the sphenoid bone and the posterior nasal septum; the septum, with its intact mucosa, is pushed off the midline by the medial blade of the speculum (Griffith and Veerapen, 1987). Bilateral mucosal flaps over the keel of the sphenoid bone are reflected laterally.

When the anterior wall of the sphenoid sinus has been reached by one of the aforementioned three routes, microdrill and/or bone punches are used to make a large opening. One or more septa can be identified, dividing the sphenoid sinus into concamerations; the removal or the flattening of septa allows the exposure of all the useful anatomic keypoints inside the sphenoid cavity, especially on the posterior wall. When those landmarks are not clearly visible, C-arm fluoroscopy or, more recently, a neuronavigation system could be helpful to provide surgical orientation.

It is crucial to realize an adequate bony exposure of the sellar floor for the success of the approach. This latter is opened with a microdrill or bone punches or both; then the dura is incised in a midline position, in a linear or cross fashion, taking care, especially in the case of microadenomas, to avoid damaging a possibly ectopic carotid artery within the sella, which is very likely in acromegalic patients.

For removal of a microadenoma, if it is visible on the surface of the gland, a cleavage plane between the microadenoma and the residual anterior pituitary should be found; when the microadenoma is not superficial a small incision can be made in the normal pituitary gland on the same side of the lesion, which can be removed with the help of small ring curettes.

For removal of macroadenomas, the inferior and lateral components of the lesion are removed before the superior aspect. Indeed, the removal of the superior part first will prematurely allow the suprasellar cistern and the redundant diaphragma to fall into the operative field, thus reducing the ability to expose and remove the lateral portions of the lesion. Nevertheless, if the descent of the suprasellar portion of the lesion is not observed, a Valsalva maneuver can be useful, causing the protrusion of the suprasellar cistern into the sellar cavity.

At the end of the procedure, the speculum is retracted and nasal structures are placed back in their primitive position. Nasal packing is placed for 24 hours in selected cases, but is not routinely employed.

Endoscopic endonasal transsphenoidal approach

The endoscopic endonasal approach is performed by means of the endoscope as a stand-alone visualizing instrument, without the need for the transsphenoidal retractor; it has the same indications as the conventional microsurgical technique. It requires specific endoscopic skills and is based on a different concept because the endoscopic view that the surgeon receives on the video monitor is not the transposition of the real image, as it would be looking through the eyepiece of a microscope, but is the result of a microprocessor's elaboration

(Jho, 2000; de Divitiis and Cappabianca, 2002; Cappabianca et al., 2004; Cappabianca et al., 2008a; Shahlaie et al., 2010).

The procedure consists of three main aspects: exposure of the lesion, management of the relevant pathology, and reconstruction of the sella, that go through three different steps, the nasal, the sphenoid, and the sellar phases. In the first two steps, the corridor to the lesion and the room to work comfortably are identified and adapted to the need of each single case, while in the sellar phase the lesion is removed and the tailored reconstruction of the sellar area is realized.

When the endoscope (18 cm in length, 4 mm in diameter) is introduced, it is possible to identify the main anatomic landmarks, such as the inferior turbinate laterally and the nasal septum medially. Cottonoid pledgets soaked with diluted adrenaline (2:100 000) or with xylometazoline hydrochloride are positioned between the middle turbinate and the nasal septum; then, the middle turbinate is gently pushed laterally. If more space is required, middle turbinectomy on one side as well as a posterior bilateral ethmoidectomy can be carried out.

Coagulation of the sphenoethmoid recess and the area around the sphenoid ostium is performed in order to avoid arterial bleeding from septal branches of the sphenopalatine artery. The nasal septum is detached from the sphenoid rostrum by means of a microdrill and the anterior wall of the sphenoid sinus is opened wide; the posterior nasal septum is then removed with a retrograde bone punch. The removal of all the sphenoid septa is required to expose the anatomic landmarks inside the sphenoid cavity. After such a maneuver, the posterior and lateral walls of the sphenoid sinus, with the sellar floor at the center, the sphenoethmoid planum above it, and the clival indentation below, become visible.

From this point on, the surgeon performs a bimanual dissection while a coworker holds the endoscope dynamically, allowing the comfortable introduction of two instruments through one or both nostrils, without coming into conflict with it – the so-called "3-4 hands technique" (Castelnuovo et al., 2006); this requires a high level of collaboration between two surgeons.

The endoscope is held by the assistant in the patient's right nostril; this is stretched upward (at 12 o'clock) by means of another instrument, usually a suction tube held by the first surgeon, in the most inferior position in the same nostril (at 6 o'clock). The main instrument is held in the left nostril by the primary surgeon, using his or her dominant hand.

The sellar phase of the procedure follows the same rules as the microsurgical transsphenoidal approach. Lesions involving the medial wall of the cavernous sinus can also be removed under endoscopic control; in case of microadenoma, it is easier to dissect tumor pseudocapsule from pituitary gland tissue, in order to achieve an "en bloc" removal. Finally, after lesion removal, an endoscopic exploration of the tumor cavity, by means of a 0-degree and/or angled scope, is performed to assess for the presence of any tumor remnants.

It should be noted that some pituitary adenomas present some features (e.g., dumb-bell shape, supra- or parasellar extension, and/or fibrous or rubbery consistency, which is high likely in the case of a recurrent tumor) that may hinder such a route. In these cases an extended, purely endoscopic endonasal technique (Cavallo et al., 2005b; Kassam et al., 2005; de Divitiis et al., 2007; Cappabianca et al., 2008b), as described for suprasellar lesions, may be suitable for the removal of these selected pituitary adenomas. This technique allows the use of two surgical corridors, the conventional endosellar extra-arachnoidal and a suprasellar transarachnoidal (Di Maio et al., 2011). The suprasellar aspect of the lesion is debulked and its capsule is dissected from the surrounding neurovascular structures through an arachnoid plane, using microscissors and sharp dissection, as in a conventional open microsurgical technique.

In cases where an adenoma extends into the cavernous sinus, it should be noted that two different surgical corridors have been described to gain access to different areas of the cavernous sinus, according to the relationship with the intracavernous carotid artery (ICA); one corridor permits access to a compartment medial to the ICA, while the other allows access to a cavernous sinus compartment lateral to it (Frank and Pasquini, 2002; Cavallo et al., 2005a).

The first approach is indicated for pituitary adenomas projecting through the medial wall of the cavernous sinus, without extension into the lateral compartment. The tumor itself enlarges the C-shaped parasellar segment of the internal carotid artery, thus making easier the suctioning and the curettage through this corridor. Conversely the approach to the lateral compartment of the cavernous sinus is indicated in the case of tumors involving the entire cavernous sinus.

The tumor removal proceeds from the extracavernous to the intracavernous portion. In the case of tumors occupying mainly the lateral compartment of the cavernous sinus, the growth of the lesion usually displaces the ICA medially and pushes the cranial nerves laterally. Delicate maneuvers of curettage and suction usually allow the removal of the parasellar portion of the lesion, in the same way as the intrasellar portion.

After lesion removal, closure of the sellar floor is required. Various techniques could be adopted for sellar repair (intra- and/or extradural closure of the sella and packing of the sella with or without packing of the sphenoid sinus), depending on the size of osteodural defect and of the "dead space" inside the sella. Usually autologous or heterologous materials, either resorbable or not, if necessary, are used, taking care to avoid overpacking to prevent compression of the optic apparatus (Cappabianca et al., 2002b). If an extended approach has been used, especially in the suprasellar area, a consistent intraoperative CSF leakage occurs, due to intentional arachnoid opening, and thus an accurate reconstruction of the skull base defect is mandatory (Cavallo et al., 2007, Esposito et al., 2007).

At the end of the procedure, hemostasis is obtained and the middle turbinate is gently restored in position. Packing of the nasal cavity is not commonly considered necessary.

TRANSCRANIAL APPROACHES

There are different standard transcranial or alternative skull base approaches routinely used for the resection of pituitary tumors with extensive suprasellar and parasellar extension, depending on the direction of the extrasellar growth of the lesion: the unilateral subfrontal approach, the pterional approach, and the bilateral subfrontal interhemispheric approach. Depending on the particular compartment where the tumor is located, the size of the opening must be commensurate with the best and the safest removal of the tumor – "as small as possible, as large as necessary, but cosmetically optimal" (Yasargil, 1984, 1996).

The unilateral subfrontal approach is indicated mainly for large suprasellar adenomas with an asymmetric supraparasellar extension and when the tumor has expanded into the upper prepontine cistern. It gives excellent bilateral access to the optic nerves and the chiasm (Powell and Pollock, 2003). Usually a bicoronal skin incision is adopted and craniotomy has a quadrangular shape, with the basal cut as low as possible. Preoperatively, the surgeon should have a clear idea of the size of the frontal sinuses: when opened, it should be stripped of mucosa and packed, then covered with galea capitis, temporalis fascia, or dural substitute. After the dural opening, the olfactory nerve is microscopically freed and the frontal lobe is retracted gently. Once the lesion has been identified, bipolar coagulation and incision of the capsule permit debulking and removal of the adenoma, between the optic nerves, preserving the pituitary stalk. During retrochiasmatic removal, care must be taken to minimize the manipulations, avoiding damage to the optic pathways.

The frontolateral craniotomy, also known as the pterional approach, is a versatile craniotomy that gives good exposure of the inferolateral portion of the frontal lobe and the anterior temporal lobe. The pterional approach provides a short distance to the suprasellar region and is the craniotomy of choice for adenomas with unilateral extrasellar parasellar extension, when there is the need to expose the compartment between the optic nerve and the ICA or the ICA and the third cranial nerve. It may be useful when cavernous sinus area invasion or a significant retrochiasmatic component is present (Dolenc, 1997). The skin incision designs a large radius arc to terminate at the midline, posterior to the hairline; the bone flap resembles a relatively circular shape, centered on the *pterion*. The dura is opened in a curvilinear fashion and the sylvian fissure is opened to ease frontal lobe retraction in order to realize an adequate corridor. The carotid cistern is completely opened, revealing the carotid artery and its branches. The tumor can be found in the optochiasmatic cistern, the interpeduncular cistern, and the cistern of lamina terminalis. In the case of intraventricular extension, the adenoma can be reached through the translamina terminalis corridor (King, 1979; Maira et al., 2000). The optimal technique is to decompress the tumor between the optic nerves and to mobilize it from the opticocarotid space to the interoptic space. When a large pituitary tumor invades the parasellar-cavernous sinus area and the adjacent central skull base regions, Dolenc's variations (Dolenc et al., 1987; Dolenc, 1997) of this technique could be adopted.

The bilateral interhemispheric subfrontal approach is not used today as frequently as the other two options, being mostly indicated for the treatment of large lesions, above all craniopharyngiomas, with retrochiasmatic extension. It offers a wide exposure – the craniotomy and dural opening are extended bilaterally – of the anterior cranial base with a good overview of the sellar, suprasellar, and parasellar areas. It affords an excellent midline orientation and may be used for the treatment of huge pituitary adenomas with large bilateral suprasellar extension.

COMPLICATIONS

Complications of pituitary surgery depend on the surgical route employed to reach the sella; microsurgical transsphenoidal surgery offers a lower mortality and morbidity rate, and furthermore does not leave visible scars, when compared with the conventional transcranial approaches, resulting in it being more appealing to both patients and physicians. Serious complications of transsphenoidal surgery are uncommon and seem to be mainly related to the size of the tumor and the experience of the surgeon. Nevertheless, even if the mortality rate is low (usually $< 1\%$), complications still occur (Ciric et al., 1997; Laws and Kern, 1982; Black et al., 1987; Cappabianca et al., 2002a; Kassam et al., 2011). Major morbidity (CSF leak, meningitis, stroke, intracranial hemorrhage, visual loss) occurs in 3.4% of cases, whereas minor complications (sinus disease, nasal septal perforations) are present in approximately 4.6% of procedures. Therefore, according to the anatomic compartments and relative structures involved during the different steps of the procedure, the possible complications can be divided into the following categories: (1) oronasofacial (mostly related to the approach itself); (2) sphenoid sinus (bleeding from the sphenopalatine artery, sinusitis); (3) sellar (CSF leak); (4) parasellar and suprasellar (including central nervous system injuries, cranial nerve damage – optic, olphactory, abducens, etc., vascular problems – carotid artery, basilar artery, cavernous sinus, etc.); and (5) endocrine complications (Laws and Kern, 1982; Cappabianca et al., 2002a).

Nasofacial complications, including nasal septal perforations, bleeding from the mucosal branches of the sphenopalatine artery, injury or fracture of the cribriform plate with subsequent CSF leak, anesthesia of the upper lip and of the anterior maxillary teeth, saddle nose, anosmia caused by undue superior nasal septum dissection, diastasis of the maxilla, or fracture of the hard palate due to overspreading of the speculum, have been reported. The occurrence of postoperative disorders such as numbness of the upper lip and/or the anterior maxillary teeth, nasal septum perforations, diastasis of the maxilla or fracture of the hard palate caused by overspreading of the speculum, fracture of the orbit and, moreover, injury or fracture of the cribriform plate and subsequent CSF leak, could be definitely addressed to the microsurgical transsphendoidal technique.

Sphenoid sinus complications more frequently occurring are sinusitis and mucocele, a rare and usually late-onset disorder, caused by obstruction of the airflow at the osteomeatal complex. Fracture of the sphenoid body with injury to the optic nerves and the carotid arteries, mostly related to the use of a transsphenoidal retractor and/or to its overspread, sometimes due to thin or absent bone, is exceptional, but must be kept in mind.

Complications reported for the sellar phase of the procedure account essentially for the CSF leak, due to violation of the arachnoid membrane, subarachnoid hemorrhage, vasospasm, and tension pneumocephalus. A wide but infrequent range of suprasesllar and parasellar complications have been reported: hypothalamic injury, visual damage, hemorrhage or ischemia, vascular injury to one of the vessels of the circle of Willis, meningitis related to a CSF leak or to contamination; cavernous sinus injury (ICA; sixth, thirth, and fourth cranial nerve injury), when dealing with lesions extending into the parasellar area, and, finally, brainstem injury due to a misdirected approach toward the clivus. Conversely, the endocrine sequelae are the most frequent complications, namely loss of one or more anterior pituitary functional axes occurs in approximately 3% of microadenomas, while in macroadenomas it occurs in about 5% of cases. Permanent diabetes insipidus occurs in 3% of cases (Laws and Kern, 1982; Thapar and Laws, 2001).

The endoscopic approach is endonasal, so that rare nasal phase complications such as anesthesia of the upper lip and of the anterior maxillary teeth, nasal septal perforations, and saddle nose are almost absent. The lack of the nasal speculum avoids the development of other rare complications, such as diastasis of the maxilla or fracture of the hard palate due to overspreading of the speculum, fracture of the orbit, and injury or fracture of the cribriform plate and subsequent CSF leak.

Nevertheless, even though the corridor is completely endonasal and no incision of nasal mucosa is required, the insertion and movement of the endoscope and blunt and/or, above all, sharp instruments could cause a direct mucosal tearing. In addition, during the nasal step of an extended endonasal approach, it should be remembered that the extensive manipulations necessary when performing maneuvers such as middle turbinectomy, posterior septectomy, ethmoidectomy and/or the harvesting of the mucosal nasoseptal flap could increase the risk of bleeding from sphenopalatine artery branches.

Series of endoscopic operations show an overall decreased incidence of complications compared with historical microsurgical transsphenoidal series (Ciric et al., 1997).

The morbidity and mortality of transcranial approaches have consistently decreased in the microsurgical era. A direct comparison of the complications between the two groups is not possible because the respective inclusion criteria have changed over the years. One aspect that should not be underestimated is that nowadays transcranial surgery is usually employed for giant and invasive pituitary adenomas or adenomas invading the parasellar compartment or the central skull base, representing a cohort of subjects with the most difficult surgical management and intricate surgical problems. Despite these considerations, a surprisingly high total tumor resection rate (63–96%) has been reported more recently.

Specific complications are those events common to any supratentorial craniotomy and related to traction on the frontal and temporal lobes, dissection of major or perforating vessels, and manipulation of the optic or oculomotor nerves. The most common postoperative complication is diabetes insipidus, which can be immediate, delayed (4–5 days), or triphasic, and either transient (31.8%) or permanent (21.1%); the next most common is hemiparesis, either transient (33.3%), or permanent (9.1%). Loss of vision can occur, most commonly due to disruption of the blood supply to the chiasm or the optic nerves. Other complications, such as worsening of anterior pituitary function, epilepsy, infection, and CSF leak, can also occur. The strict operative mortality (5%) or mortality from disease-related complications (11.3%) is not negligible (Yasargil, 1996).

FINAL REMARKS

Surgery, either transsphenoidal or transcranial, should accomplish the goal of a total removal of the lesion during the first operation, if possible, for the patient's best chance of "cure". Only a reasonable risk can be borne by the patient in terms of complications and postoperative morbidity; the surgeon must always attempt a complete and radical result, but at the same time should remember that a wide variety of different options – medical, surgical, and radiotherapeutic – are now effective treatment in terms of long-term results. What is crucial, regardless of the surgical option selected for a single case, whether transsphenoidal or transcranial, is to relate the goal of surgery to the patient's needs, selecting the best option for the actual condition of the patient from among all the options available, surgical or otherwise.

References

Abe T, Ludecke DK (1999). Transnasal surgery for infradiaphragmatic craniopharyngiomas in pediatric patients. Neurosurgery 44: 957–964, discussion 964–966.

Abosch A, Tyrrell JB, Lamborn KR et al. (1998). Transsphenoidal microsurgery for growth hormone-secreting pituitary adenomas: initial outcome and long-term results. J Clin Endocrinol Metab 83: 3411–3418.

Alfieri A, Jho HD (2001). Endoscopic endonasal approaches to the cavernous sinus: surgical approaches. Neurosurgery 49: 354–360, discussion 360–362.

Artico M, Pastore FS, Fraioli B et al. (1998). The contribution of Davide Giordano (1864–1954) to pituitary surgery: the transglabellar-nasal approach. Neurosurgery 42: 909–911, discussion 911–912.

Baskin DS, Wilson CB (1984). Transsphenoidal treatment of non-neoplastic intrasellar cysts. A report of 38 cases. J Neurosurg 60: 8–13.

Bevan JS, Webster J, Burke CW et al. (1992). Dopamine agonists and pituitary tumor shrinkage. Endocr Rev 13: 220–240.

Bills DC, Meyer FB, Laws Jr ER et al. (1993). A retrospective analysis of pituitary apoplexy. Neurosurgery 33: 602–608, discussion 608–609.

Black PM, Zervas NT, Candia GL (1987). Incidence and management of complications of transsphenoidal operation for pituitary adenomas. Neurosurgery 20: 920–924.

Brisman MH, Katz G, Post KD (1996). Symptoms of pituitary apoplexy rapidly reversed with bromocriptine. Case report. J Neurosurg 85: 1153–1155.

Cappabianca P, de Divitiis E (2004). Endoscopy and transsphenoidal surgery. Neurosurgery 54: 1043–1048, discussion 1048–1050.

Cappabianca P, Alfieri A, de Divitiis E (1998). Endoscopic endonasal transsphenoidal approach to the sella: towards functional endoscopic pituitary surgery (FEPS). Minim Invasive Neurosurg 41: 66–73.

Cappabianca P, Cavallo LM, Colao A et al. (2002a). Surgical complications associated with the endoscopic endonasal transsphenoidal approach for pituitary adenomas. J Neurosurg 97: 293–298.

Cappabianca P, Cavallo LM, Esposito F et al. (2002b). Sellar repair in endoscopic endonasal transsphenoidal surgery: results of 170 cases. Neurosurgery 51: 1365–1371, discussion 1371–1372.

Cappabianca P, de Divitiis O, Maiuri F (2003). Evolution of transsphenoidal surgery. In: E de Divitiis, P Cappabianca (Eds.), Endoscopic Endonasal Transsphenoidal Surgery, Springer, Vienna and New York, pp. 1–7.

Cappabianca P, Cavallo LM, de Divitiis E (2004). Endoscopic endonasal transsphenoidal surgery. Neurosurgery 55: 933–940, discussion 940–941.

Cappabianca P, Cavallo LM, de Divitiis O et al. (2008a). Endoscopic pituitary surgery. Pituitary 11: 385–390.

Cappabianca P, Cavallo LM, Esposito F et al. (2008b). Extended endoscopic endonasal approach to the midline skull base: the evolving role of transsphenoidal surgery. In: JD Pickard, N Akalan, C Di Rocco et al. (Eds.), Advances and Technical Standards in Neurosurgery, Springer, Vienna and New York, pp. 152–199.

Cappabianca P, Cavallo LM, de Divitiis O et al. (2010). Pituitary surgery. In: LJ DeGroot, JL Jameson (Eds.), Endocrinology, 6th edn. Saunders Elsevier, Philadelphia, pp. 358–376.

Cardoso ER, Peterson EW (1984). Pituitary apoplexy: a review. Neurosurgery 14: 363–373.

Carrau RL, Jho HD, Ko Y (1996). Transnasal-transsphenoidal endoscopic surgery of the pituitary gland. Laryngoscope 106: 914–918.

Casanueva FF, Molitch ME, Schlechte JA et al. (2006). Guidelines of the Pituitary Society for the diagnosis and management of prolactinomas. Clin Endocrinol (Oxf) 65: 265–273.

Castelnuovo P, Pistochini A, Locatelli D (2006). Different surgical approaches to the sellar region: focusing on the "two nostrils four hands technique". Rhinology 44: 2–7.

Castelnuovo P, Dallan I, Pistochini A et al. (2007). Endonasal endoscopic repair of Sternberg's canal cerebrospinal fluid leaks. Laryngoscope 117: 345–349.

Caton R (1893). Notes of a case of acromegaly treated by operation. Br Med J 2: 1421–1423.

Cavallo LM, Cappabianca P, Galzio R et al. (2005a). Endoscopic transnasal approach to the cavernous sinus versus transcranial route: anatomic study. Neurosurgery 56: ONS379–ONS389.

Cavallo LM, Messina A, Cappabianca P et al. (2005b). Endoscopic endonasal surgery of the midline skull base: anatomical study and clinical considerations. Neurosurg Focus 19: E2.

Cavallo LM, Messina A, Esposito F et al. (2007). Skull base reconstruction in the extended endoscopic transsphenoidal approach for suprasellar lesions. J Neurosurg 107: 713–720.

Cavallo LM, Prevedello D, Esposito F et al. (2008). The role of the endoscope in the transsphenoidal management of cystic lesions of the sellar region. Neurosurg Rev 31: 55–64, discussion 64.

Chen JC, Amar AP, Choi S et al. (2003). Transsphenoidal microsurgical treatment of Cushing disease: postoperative assessment of surgical efficacy by application of an overnight low-dose dexamethasone suppression test. J Neurosurg 98: 967–973.

Ciric I, Ragin A, Baumgartner C et al. (1997). Complications of transsphenoidal surgery: results of a national survey review of the literature and personal experience. Neurosurgery 40: 225–236, discussion 236–237.

Colao A, Ferone D, Cappabianca P et al. (1997). Effect of octreotide pretreatment on surgical outcome in acromegaly. J Clin Endocrinol Metab 82: 3308–3314.

Colao A, Di Sarno A, Cappabianca P et al. (2003). Withdrawal of long-term cabergoline therapy for tumoral and nontumoral hyperprolactinemia. N Engl J Med 349: 2023–2033.

Cushing H (1909). III. Partial hypophysectomy for acromegaly: with remarks on the function of the hypophysis. Ann Surg 50: 1002–1017.

Cushing H (1932). Intracranial tumors. Note upon a series of two thousands verified cases with surgical-mortality percentages pertaining thereto, Charles C, Thomas, Springfield, MA.

Cushing H (1981). Medical Classic. The functions of the pituitary body: Harvey Cushing. Am J Med Sci 281: 70–78.

Dallan I, Seccia V, Muscatello L et al. (2011). Transoral endoscopic anatomy of the parapharyngeal space: a step-by-step logical approach with surgical considerations. Head Neck 33: 557–561.

Dandy WE (1934). The brain. In: D Lewis (Ed.), Practice of Surgery, WF Prior, Hagerstown, pp. 556–605.

Davis DH, Laws Jr ER, Ilstrup DM et al. (1993). Results of surgical treatment for growth hormone-secreting pituitary adenomas. J Neurosurg 79: 70–75.

de Divitiis E, Cappabianca P (2002). Endoscopic endonasal transsphenoidal surgery. In: JD Pickard (Ed.), Advances and Technical Standards in Neurosurgery, Springer-Verlag, Vienna and New York, pp. 173–177.

de Divitiis E, Cappabianca P, Cavallo LM (2003). Endoscopic endonasal transsphenoidal approach to the sellar region. In: E de Divitiis, P Cappabianca (Eds.), Endoscopic Endonasal Transsphenoidal Surgery, Springer, Vienna and New York, pp. 91–130.

de Divitiis E, Cavallo LM, Cappabianca P et al. (2007). Extended endoscopic endonasal transsphenoidal approach for the removal of suprasellar tumors: Part 2. Neurosurgery 60: 46–58, discussion 58–59.

Di Maio S, Cavallo LM, Esposito F et al. (2011). Extended endoscopic endonasal approach for selected pituitary adenomas: early experience. J Neurosurg 114: 345–353.

Doglietto F, Prevedello DM, Jane Jr JA et al. (2005). Brief history of endoscopic transsphenoidal surgery – from Philipp Bozzini to the First World Congress of Endoscopic Skull Base Surgery. Neurosurg Focus 19: E3.

Dolenc VV (1997). Transcranial epidural approach to pituitary tumors extending beyond the sella. Neurosurgery 41: 542–550, discussion 551–552.

Dolenc VV, Kregar T, Ferluga M et al. (1987). Treatment of tumors invading the cavernous sinus. In: VV Dolenc (Ed.), The Cavernous Sinus, Springer-Verlag, New York, pp. 377–391.

Elias WJ, Laws ER (2000). Transsphenoidal approach to lesion of the sella. In: HH Schmideck (Ed.), Schmideck and Sweet, 4th edn, WB Saunders, Philadelphia, Operative Neurosurgical Technique. Indications Methods and Results, pp 373–384.

el-Mahdy W, Powell M (1998). Transsphenoidal management of 28 symptomatic Rathke's cleft cysts with special reference to visual and hormonal recovery. Neurosurgery 42: 7–16, discussion 16–17.

Esposito F, Dusick JR, Cohan P et al. (2006). Clinical review. Early morning cortisol levels as a predictor of remission after transsphenoidal surgery for Cushing's disease. J Clin Endocrinol Metab 91: 7–13.

Esposito F, Dusick JR, Fatemi N et al. (2007). Graded repair of cranial base defects and cerebrospinal fluid leaks in transsphenoidal surgery. Neurosurgery 60: 295–303, discussion 303–304.

Fahlbusch R, Honegger J, Buchfelder M (1996). Acromegaly – the place of the neurosurgeon. Metabolism 45: 65–66.

Fraioli B, Esposito V, Santoro A et al. (1995). Transmaxillosphenoidal approach to tumors invading the medial compartment of the cavernous sinus. J Neurosurg 82: 63–69.

Frank G, Pasquini E (2002). Endoscopic endonasal approaches to the cavernous sinus: surgical approaches. Neurosurgery 50: 675.

Fraser JF, Nyquist GG, Moore N et al. (2010). Endoscopic endonasal transclival resection of chordomas: operative technique clinical outcome and review of the literature. J Neurosurg 112: 1061–1069.

Giordano D (1911). Compendio di chirurgia operativa italiana, UTET, Turin.

Griffith HB, Veerapen R (1987). A direct transnasal approach to the sphenoid sinus. Technical note. J Neurosurg 66: 140–142.

Guiot G (1973). Transsphenoidal approach in surgical treatment of pituitary adenomas: general principles and indications in non-functioning adenomas. In: PO Kohler, GT Ross (Eds.), Diagnosis and Treatment of Pituitary Adenomas, Excerpta Medica, Amsterdam, pp. 159–178.

Guiot G, Derome P (1972). Indications for trans-sphenoid approach in neurosurgery. 521 cases. Ann Med Interne (Paris) 123: 703–712.

Guiot G, Rougerie J, Fourestier M et al. (1963). Intracranial endoscopic explorations [in French]. Presse Med 71: 1225–1228.

Handelsmann Horsley V (1911). Preliminary note on experimental investigations on the pituitary body. Br Med J 2: 1150–1151.

Hardy J (1969). Transphenoidal microsurgery of the normal and pathological pituitary. Clin Neurosurg 16: 185–217.

Hardy J (1971). Transsphenoidal hypophysectomy. J Neurosurg 34: 582–594.

Hardy J, Vezina JL (1976). Transsphenoidal neurosurgery of intracranial neoplasm. Adv Neurol 15: 261–273.

Hirsch O (1910). Endonasal method of removal of hypofiseal tumors. With a report of two successful cases. JAMA 55: 772–774.
Jane Jr JA, Thapar K, Laws Jr ER (2001). Acromegaly: historical perspectives and current therapy. J Neurooncol 54: 129–137.
Jankowski R, Auque J, Simon C et al. (1992). Endoscopic pituitary tumor surgery. Laryngoscope 102: 198–202.
Jho HD (2000). Endoscopic transsphenoidal surgery. In: HH Schmidek (Ed.), Schmidek and Sweet, Operative Neurosurgical Techniques. WB Saunders, Philadelphia, pp. 385–397.
Jho HD, Ha HG (2004). Endoscopic endonasal skull base surgery. Part 3. The clivus and posterior fossa. Minim Invasive Neurosurg 47: 16–23.
Jho HD, Carrau RL, Ko Y (1996a). Endoscopic pituitary surgery. In: H Wilkins, S Rengachary (Eds.), Neurosurgical Operative Atlas, American Association of Neurological Surgeons, Park Ridge, IL, pp. 1–12.
Jho HD, Carrau RL, McLaughlin ML et al. (1996b). Endoscopic transsphenoidal resection of a large chordoma in the posterior fossa. Case report. Neurosurg Focus 1: e3, discussion one page following e3.
Jho HD, Carrau RL, McLaughlin MR et al. (1997). Endoscopic transsphenoidal resection of a large chordoma in the posterior fossa. Acta Neurochir (Wien) 139: 343–347, discussion 347–348.
Kanter AS, Dumont AS, Asthagiri AR et al. (2005). The transsphenoidal approach. A historical perspective. Neurosurg Focus 18: e6.
Kassam A, Snyderman CH, Mintz A et al. (2005). Expanded endonasal approach: the rostrocaudal axis. Part I. Crista galli to the sella turcica. Neurosurg Focus 19: 1–12, E3.
Kassam AB, Prevedello DM, Carrau RL et al. (2011). Endoscopic endonasal skull base surgery: analysis of complications in the authors' initial 800 patients. J Neurosurg 114: 1544–1568.
Kato T, Sawamura Y, Abe H et al. (1998). Transsphenoidal-transtuberculum sellae approach for supradiaphragmatic tumours: technical note. Acta Neurochir (Wien) 140: 715–719.
Kelley TF, Stankiewicz JA, Chow JM et al. (1996). Endoscopic closure of postsurgical anterior cranial fossa cerebrospinal fluid leaks. Neurosurgery 39: 743–746.
Kennedy DW (1985). Functional endoscopic sinus surgery. Technique Arch Otolaryngol 111: 643–649.
Kim J, Choe I, Bak K et al. (2000). Transsphenoidal supradiaphragmatic intradural approach: technical note. Minim Invasive Neurosurg 43: 33–37.
King TT (1979). Removal of intraventricular craniopharyngiomas through the lamina terminalis. Acta Neurochir (Wien) 45: 277–286.
Kitano M, Taneda M (2009). Extended transsphenoidal surgery for suprasellar craniopharyngiomas: infrachiasmatic radical resection combined with or without a suprachiasmatic trans-lamina terminalis approach. Surg Neurol 71: 290–298.
Kitano M, Taneda M, Shimono T et al. (2008). Extended transsphenoidal approach for surgical management of pituitary adenomas invading the cavernous sinus. J Neurosurg 108: 26–36.
Komotar RJ, Starke RM, Raper DM et al. (2011). The endoscope-assisted ventral approach compared with open microscope-assisted surgery for clival chordomas. World Neurosurg 76: 318–327, discussion 259–262.
Kouri JG, Chen MY, Watson JC et al. (2000). Resection of suprasellar tumors by using a modified transsphenoidal approach. Report of four cases. J Neurosurg 92: 1028–1035.
Lanzino G, Laws Jr ER (2001). Pioneers in the development of transsphenoidal surgery: Theodor Kocher Oskar Hirsch and Norman Dott. J Neurosurg 95: 1097–1103.
Laws Jr ER (1980). Transsphenoidal microsurgery in the management of craniopharyngioma. J Neurosurg 52: 661–666.
Laws E (1993a). Clivus chordomas. In: LN Sekhar, IP Janecka (Eds.), Surgery of Cranial Base Tumors, Raven Press, New York, pp. 679–685.
Laws ERJ (1993b). Transsphenoidal surgery. In: MLJ Apuzzo (Ed.), Brain Surgery: Complications, Avoidance and Management, Churchill Livingstone, New York, pp. 357–362.
Laws Jr ER, Ebersold MJ (1982). Pituitary apoplexy – an endocrine emergency. World J Surg 6: 686–688.
Laws ER, Kern EB (1982). Complications of transsphenoidal surgery. In: ER Laws, RV Randall, EB Kern et al. (Eds.), Management of Pituitary Adenomas and Related Lesions with Emphasis on Transsphenoidal Microsurgery, Appleton-Century-Crofts, New York, pp. 329–346.
Laws ER, Lanzino G (Eds.), (2010). Transsphenoidal Surgery, Saunders Elsevier, Philadelphia.
Laws ER, Kanter AS, Jane Jr JA et al. (2005). Extended transsphenoidal approach. J Neurosurg 102: 825–827, discussion 827–828.
Leonhard M, Cappabianca P, de Divitiis E (2003). The endoscope, endoscopic equipment and instrumentation. In: E de Divitiis, P Cappabianca (Eds.), Endoscopic Endonasal Transsphenoidal Surgery, Springer-Verlag, Vienna and New York, pp. 9–19.
Liu JK, Das K, Weiss MH et al. (2001). The history and evolution of transsphenoidal surgery. J Neurosurg 95: 1083–1096.
Locatelli D, Rampa F, Acchiardi I et al. (2006). Endoscopic endonasal approaches for repair of cerebrospinal fluid leaks: nine-year experience. Neurosurgery 58: ONS-246–256, discussion ONS-256–257..
Lohmann T, Trantakis C, Biesold M et al. (2001). Minor tumour shrinkage in nonfunctioning pituitary adenomas by long-term treatment with the dopamine agonist cabergoline. Pituitary 4: 173–178.
Losa M, Mortini P, Urbaz L et al. (2006). Presurgical treatment with somatostatin analogs in patients with acromegaly: effects on the remission and complication rates. J Neurosurg 104: 899–906.
Maira G, Pallini R, Anile C et al. (1996). Surgical treatment of clival chordomas: the transsphenoidal approach revisited. J Neurosurg 85: 784–792.
Maira G, Anile C, Colosimo C et al. (2000). Craniopharyngiomas of the third ventricle: trans-lamina terminalis approach. Neurosurgery 47: 857–863, discussion 863–865.

Maira G, Anile C, Albanese A et al. (2004). The role of transsphenoidal surgery in the treatment of craniopharyngiomas. J Neurosurg 100: 445–451.

McLaughlin N, Laws ER, Oyesiku NM et al. (2012). Pituitary centers of excellence. Neurosurgery 71: 916–924, discussion 924–926.

Molitch ME, Elton RL, Blackwell RE et al. (1985). Bromocriptine as primary therapy for prolactin-secreting macroadenomas: results of a prospective multicenter study. J Clin Endocrinol Metab 60: 698–705.

Perneczky A, Muller-Forell W, van Lindert E et al. (1999). Keyhole Concept in Neurosurgery, Thieme, Stuttgart and New York.

Powell MP, Pollock JR (2003). Transcranial surgery. In: MP Powell, SL Lightman, ERJ Laws (Eds.), Management of Pituitary Tumors, Humana Press, Totowa, NJ, pp. 147–159.

Rosegay H (1981). Cushing's legacy to transsphenoidal surgery. J Neurosurg 54: 448–454.

Ross DA, Norman D, Wilson CB (1992). Radiologic characteristics and results of surgical management of Rathke's cysts in 43 patients. Neurosurgery 30: 173–178, discussion 178–179.

Saito K, Kuwayama A, Yamamoto N et al. (1995). The transsphenoidal removal of nonfunctioning pituitary adenomas with suprasellar extensions: the open sella method and intentionally staged operation. Neurosurgery 36: 668–675, discussion 675–676.

Schloffer H (1907). Erforgleiche operationen eines hypophysentumors auf nasalem wege. Wien Klin Wochenschr 20: 621–624.

Shahlaie K, McLaughlin N, Kassam AB et al. (2010). The role of outcomes data for assessing the expertise of a pituitary surgeon. Curr Opin Endocrinol Diabetes Obes 17: 369–376.

Shomali ME, Katznelson L (1999). Medical therapy for gonadotroph and thyrotroph tumors. Endocrinol Metab Clin North Am 28: 223–240, viii.

Simmons NE, Alden TD, Thorner MO et al. (2001). Serum cortisol response to transsphenoidal surgery for Cushing disease. J Neurosurg 95: 1–8.

Stammberger H, Posawetz W (1990). Functional endoscopic sinus surgery. Concept indications and results of the Messerklinger technique. Eur Arch Otorhinolaryngol 247: 63–76.

Stippler M, Gardner PA, Snyderman CH et al. (2009). Endoscopic endonasal approach for clival chordomas. Neurosurgery 64: 268–277, discussion 277–278.

Swearingen B, Barker 2nd FG, Katznelson L et al. (1998). Long-term mortality after transsphenoidal surgery and adjunctive therapy for acromegaly. J Clin Endocrinol Metab 83: 3419–3426.

Thapar K, Laws ERJ (2001). Pituitary tumors. In: AH Kaye, ER Laws Jr (Eds.), Brain Tumors, Churchill Livingstone, London, pp. 804–854.

Weiss MH (1987). The transnasal transsphenoidal approach. In: MLJ Apuzzo (Ed.), Surgery of the Third Ventricle, Williams and Wilkins, Baltimore, pp. 476–494.

Wilson CB (1990). Role of surgery in the management of pituitary tumors. Neurosurg Clin N Am 1: 139–159.

Yasargil MG (1984). General operative techniques. In: MG Yasargil (Ed.), Microsurgical Anatomy of the Basal Cisterns and Vessels of the Brain, Diagnostic Studies, General Operative Techniques and Pathological Considerations of the Intracranial Aneurysms, Georg Thieme Verlag, New York, pp. 215–233.

Yasargil MG (1996). Transcranial surgery for large pituitary adenomas. In: MG Yasargil (Ed.), Microneurosurgery: Microneurosurgery of CNS Tumors, Georg Thieme Verlag, Stuttgart, pp. 200–204.

Zada G, Cappabianca P (2010). Raising the bar in transsphenoidal pituitary surgery. World Neurosurg 74: 452–454.

Zada G, Du R, Laws Jr ER (2011). Defining the "edge of the envelope": patient selection in treating complex sellar-based neoplasms via transsphenoidal versus open craniotomy. J Neurosurg 114: 286–300.

Handbook of Clinical Neurology, Vol. 124 (3rd series)
Clinical Neuroendocrinology
E. Fliers, M. Korbonits, and J.A. Romijn, Editors

Chapter 20

Medical approach to pituitary tumors

S.J.C.M.M. NEGGERS* AND A.J. VAN DER LELY
Section of Endocrinology, Department of Medicine, Erasmus University Medical Center, Rotterdam, The Netherlands

INTRODUCTION

Medical and surgical treatment are the two cornerstones in the management of pituitary adenoma. However, the type of adenoma does influence the reported efficacy of the two approaches and therefore affects the choice between medical and surgical options. There are adenomas such as prolactinomas in which surgical treatment has been more or less replaced by medical treatment. However, in Cushing's disease or nonfunctional adenoma surgery is still the first step in treatment. In this chapter the possible medical options for the different adenomas will be discussed according to type of adenoma.

NONFUNCTIONAL ADENOMA

Gonadotroph adenomas constitute about 80% of the nonfunctional adenomas (NFA) as their secretory products usually do not cause recognizable clinical symptoms or a syndrome and therefore they are also considered nonfunctional adenomas. Patients with NFA usually first seek medical attention when the adenoma's size results in pressure on surrounding tissues, such as, e.g., the optic nerve, causing symptoms such as headaches, visual field defects, diplopia, and symptoms related to hypopituitarism.

Medical treatment

The main aim in the treatment of NFA is a decrease in tumor size, since the majority of these adenomas do not cause a clinical syndrome. To date, no medical treatment modality has been found that has demonstrated reliable reduction in tumor size of NFAs (Greenman, 2007).

There were attempts to use dopamine agonists to address tumor size. However, the small nonrandomized studies carried out mainly found a stabilization of tumor size after surgery (Pivonello et al., 2004; Greenman et al., 2005). The lack of tumor size response has been attributed to a low density of membrane-bound dopamine receptors (Bevan et al., 1992; Pivonello et al., 2004; Greenman, 2007). A trial, preferably large, randomized, double-blind, and placebo-controlled, could end the speculation on the efficacy of dopamine agonists in NFA.

In our opinion there is a limited role, if any, for dopamine agonists in NFA. They are potentially useful only after surgery, and in those rare cases in whom radiotherapy and a second surgical intervention are less attractive.

Side-effects of medical treatment

Dopamine agonists have side-effects which are described below, under Prolactinoma.

PROLACTINOMA

Prolactinomas, or lactotroph adenomas, are the most common pituitary secreting adenomas. More than any other type of pituitary adenoma, prolactinomas are treated primarily medically. To date, dopamine agonists have been used as first-line treatment as they usually rapidly normalize prolactin levels and reverse galactorrhea, they restore fertility, and they also cause tumor shrinkage in most patients (Verhelst et al., 1999). In macroadenomas, treatment is a necessity, while in micradenomas it is not. However, when patients seek medical attention due to infertility, hypogonadism, gynecomastia, galactorrhea, or when there is growth of the (micro)adenoma, pharmacologic intervention is indicated.

Other treatment approaches should only be considered in the minority of prolactinomas that do not respond to dopamine agonists. Resistant prolactinomas can be treated by surgery and/or radiotherapy. This definition of pharmacologic resistance is empirically

*Correspondence to: S.J.C.M.M. Neggers, Section of Endocrinology, Department of Medicine, Erasmus University, Erasmus MC, PO Box 2040, 3000 CA Rotterdam, The Netherlands. E-mail: s.neggers@erasmusmc.nl

defined; the most commonly used version is a failure to reach normal prolactin serum levels and/or reduce tumor volume by at least 50% during dopamine agonist treatment (Molitch, 2005; Gillam et al., 2006).

Medical treatment

Dopamine agonists reduce tumor size and decrease prolactin hypersecretion of the lactotroph adenoma in more than 90% of patients. Thereby they decrease signs and symptoms caused by the hyperprolactemia. The lactotroph adenoma cells express dopamine receptors (D_1 and D_2 subtypes of dopamine receptors). By binding these cell-surface $D_{1/2}$ receptors, dopamine agonists reduce synthesis and secretion of prolactin and also adenoma cell size (Verhelst et al., 1999). Usually, a decrease in serum prolactin levels is accompanied by tumor size reduction. Decrease in prolactin levels can be expected within 2–3 weeks after initiation of pharmacologic treatment and normally it precedes the reduction in tumor size. Tumor reduction can be observed from well before 6 weeks to 6 months after the initiation of dopamine agonist treatment (Molitch et al., 1985). In general the magnitude of decrease in serum prolactin levels correlates well with the decrease in tumor size.

Several dopamine agonists are used in prolactinoma patients. Depending on the patient's characteristics, therapeutic expectation, and possible side-effects, the optimal dopamine agonist should be chosen to increase efficacy and patient compliance. Bromocriptine is one of the oldest ergot derivatives and has been used for approximately 30 years. To reach an optimal therapeutic effect a twice-daily dose is needed (Vance et al., 1984). Cabergoline is an ergot dopamine agonist that can be administered once or twice weekly and is more effective than bromocriptine (Verhelst et al., 1999). Cabergoline also seems to have a lower tendency to cause nausea than bromocriptine (Biller et al., 1996).

Pergolide, another ergot derivative, was used primarily in high doses of > 3 mg/day for the treatment of Parkinson's disease (Kleinberg et al., 1983). In prolactinoma patients typical doses range from 0.05 to 1.0 mg/day. In high doses, as used for the treatment of Parkinson's disease, pergolide is associated with an increased risk of valvular heart disease (Zanettini et al., 2007). In 2007, pergolide was withdrawn from the US market due to this risk.

Quinagolide is the only once daily administered nonergot dopamine; however, it is not available in some countries (van der Lely et al., 1991; Barlier and Jaquet, 2006).

Efficacy

Cabergoline seems to be superior to bromocriptine in decreasing serum prolactin levels (Verhelst et al., 1999). Normalization of prolactin levels with an efficacy rate of 80–90% can be expected. A higher efficacy rate > 95% was observed in patients treated with high doses of cabergoline, up to 12 mg weekly (Ono et al., 2008).

Quinagolide has more or less the same efficacy as cabergoline although a few studies suggest superiority to cabergoline (Barlier and Jaquet, 2006). The great advantage is the nonergot nature of the drug. This would imply an absence of the increased reported risk of the development of valvular heart disease, which can be seen during high-dose ergot derivate treatment.

Withdrawal of dopamine agonists

Remission rates after discontinuation of a dopamine agonist differ greatly between studies. For macroadenomas rates of 16–64% and for microadenomas rates of 21–69% are reported. In a recent meta-analysis, an average remission rate of 21% for microadenomas and 16% for macroadenomas was reported (Dekkers et al., 2010). Higher remission rates were seen in studies in which cabergoline was used, the duration of the treatment was longer, and when shrinkage of the adenoma was > 50%. The highest probability of remission can be expected when prolactin serum levels are normal for a longer period of time and no visible tumor can be seen on MRI for at least 2 years.

Adverse effects

The most common side-effects of dopamine agonist drugs are nausea, orthostatic hypotension, and mental disturbances. The gastrointestinal and orthostatic hypotension can be minimized when dopamine agonists are initiated at a low dose and slowly increased. Generally, cabergoline should be started at 0.25 mg once or twice a week, taken with food and/or at bedtime. Bromocriptine more frequently causes nausea than cabergoline. The usual bromocriptine starting dose is 1.25 mg twice daily. Although low doses decrease side-effects, some patients seem to be intolerant of dopamine agonists. For women, intravaginal administration of bromocriptine is reported to decrease the incidence of nausea (Kletzky and Vermesh, 1989). Less frequent side-effects are depression, constipation, Raynaud's phenomenon, and alcohol intolerance.

When patients with macrolactrotroph adenomas that have infiltrated the base of the skull are treated with a dopamine agonist, cerebrospinal fluid (CSF) leakage as rhinorrhea may occur due to rapid tumor shrinkage (Leong et al., 2000). Early recognition of this complication is important as there is a potential risk of bacterial meningitis.

Valvular heart disease

Ergot derivatives such as cabergoline and pergolide, when given in high doses, are associated with valvular heart disease in, e.g., Parkinson's patients (Zanettini et al., 2007). As mentioned above, this association appears to be a dose-dependent effect. In Parkinson's disease doses usually exceed those used in endocrine disorders.

After the initial reports on cardiac valvulopathy in Parkinson's patients during cabergoline use, a number of studies used cardiac ultrasonography in patients taking cabergoline for hyperprolactinemia (Valassi et al., 2010) to detecet valvulopathy. One study reported a higher frequency of moderate tricuspid regurgitation than in age- and gender-matched subjects. This was an exception; no other studies found any valvular regurgitation during long-term cabergoline treatment (Valassi et al., 2010). The majority of prolactinoma patients in these studies used standard doses of cabergoline of 0.5–1.5 mg/week. However, if patients use higher doses for longer periods there might be a risk of them developing valvular disease.

Therefore, cardiac ultrasonography approximately every 2 years in patients using cabergoline > 2 mg per week should be advocated. Also, dose adaptation to the lowest dose of cabergoline that is necessary to lower prolactin to the normal range is desirable.

ACROMEGALY

Acromegaly is the second most common secreting pituitary adenoma and acromegalic patients have clinical signs and symptoms that are related to the pathologic growth hormone (GH) secretion. In more than 95% of the patients with acromegaly the condition is caused by a pituitary adenoma (Melmed, 2006). More than 75% of these pituitary adenomas are macroadenomas, which often extend dorsally of the suprasellar region or laterally into the cavernal sinus (Melmed, 2006). Clinically, acromegaly is characterized by soft tissue swelling, excessive skeletal growth, reduced life expectancy, and a reduced quality of life (QoL) (Melmed, 2006; Neggers and van der Lely, 2009). The cornerstone of the biochemical diagnosis of acromegaly consists of insufficient GH suppression after an oral glucose load and elevated insulin-like growth factor 1 (IGF-1) serum levels. Treatment should aim for a reduction in signs and symptoms, an improvement in life expectancy, and optimization of QoL.

Treatment objectives of active acromegaly are therefore threefold: management of GH hypersecretion in order to control the hypersecretion of IGF-1, control of tumor size, and optimized QoL. Biochemical control is reserved for subjects in whom a normalized serum GH and IGF-1 level within the age-adjusted limits of healthy individuals have been reached by whatever treatment modality. Normalization of GH excess will result in a decrease in signs and symptoms, a decrease in morbidity, and a normal life expectancy (Neggers and van der Lely, 2011). With regard to long-term treatment, morbidity and especially mortality rates, both the biochemical parameters GH and IGF-1 seem to be important. The reported mortality rates are mainly assessed in acromegaly patients treated with pituitary surgery and sometimes with radiotherapy (RT). To date, there are hardly any data available on mortality rates during (primary) medical treatment. Medical treatment modalities have high reported efficacy rates and therefore it might be expected that future data will become available on a possible reduction in mortality and morbidity (Neggers et al., 2012).

Important biochemical predictors of mortality in acromegaly are GH (Bates et al., 1993; Orme et al., 1998; Ayuk et al., 2004) and IGF-1 (Swearingen et al., 1998; Biermasz et al., 2004; Holdaway et al., 2004). The latest GH correlates better with an increased mortality risk than the latest IGF-1 (Ayuk and Sheppard, 2008; Dekkers et al., 2008). GH, and sex- and age-matched IGF-1 concentrations are not only important biochemical parameters for mortality but are also at present the most widely accepted parameters for monitoring treatment response. There is an ongoing discussion on normal values of GH and IGF-1 (Pokrajac et al., 2007; Neggers et al., 2012) but this is beyond the scope of this chapter.

Medical treatment

To date, over 90% of acromegaly patients are initially treated by transsphenoidal surgery. However, reported remission rates vary from 80% (Giustina et al., 2000; Holdaway, 2004) to less than 40% for microadenomas (Bates et al., 2008). Remission can be achieved in less than 50% of patients with macroadenomas, even in highly experienced neurosurgical centers (Melmed, 2006). Reported remission rates for macroadenomas are between 20% and 30% (Bates et al., 2008). Furthermore, recurrence rates after initial curative surgery are reported to be between 3% and 10% (Abosch et al., 1998; Swearingen et al., 1998; Kreutzer et al., 2001; Barker et al., 2003; Dekkers et al., 2008). If we compare surgery with long-acting somatostatin analogs (LA-SRIFS), with an efficacy rate of 44–77% in normalizing IGF-1 levels, remission rates for surgery of 30–80% are within the same range. However, if we compare surgery with reported efficacy rates of medical treatment with the GH receptor antagonist pegvisomant (PEG-V), with or without LA-SRIFS co-treatment, surgical efficacy rates in acromegaly patients with macroadenomas are considerably lower. It is worth noting that successful surgery is followed by a rapid fall in GH and IGF-1 levels, and the costs relative to long-term drug therapy are probably much lower.

Somatostatin analogs

Somatotroph adenomas predominately express somatostatin receptor subtypes 2 (SSTR2) and 5 (SSTR5) (Murray and Melmed, 2008). The two commercially available LA-SRIFS, octreotide LAR and lanreotide autosolution, predominantly bind to the SSTR2 receptor, with similar affinity (Murray and Melmed, 2008). LA-SRIFS suppress GH secretion by the pituitary adenoma but also inhibit tumor growth via stimulation of apoptosis and inhibition of angiogenesis (Feelders et al., 2009).

For obvious reasons, there are more long-term efficacy and safety data available on LA-SRIFS than on PEG-V, since LA-SRIFS were introduced more than 20 years ago while PEG-V only became available around 2003.

The studies that have been conducted on long-term treatment with LA-SRIFS were biased as they frequently included patients who were responsive to LA-SRIFS before the start of the study (Melmed, 2006). In one of the few available prospective studies, LA-SRIFS reduced signs and symptoms and induced tumor shrinkage in about 42% of the patients (Bevan, 2005; Feelders et al., 2009). In theory, preoperative tumor shrinkage might improve surgical efficacy rates. Recently, a randomized controlled trial indeed observed improved normalization rate of IGF-1, but not of GH, in patients with macroadenomas who were pretreated for 6 months with octreotide LAR (Carlsen et al., 2008).

Efficacy

Octreotide LAR and lanreotide autosolution both inhibit GH secretion of the pituitary adenoma. Octreotide is available in depot preparations of 10, 20, and 30 mg, while lanreotide preparations include 60, 90, and 120 mg (Feelders et al., 2009). Both formulations are usually administrated once every 4 weeks and have more or less the same efficacy rates (Murray and Melmed, 2008; Feelders et al., 2009). Normalization of IGF-1 and GH (<2.5 μg/L) is observed in about 44% of the patients, for both adjuvant and primary medical therapy (Holdaway, 2004). However, the efficacy rates could be higher in older studies because patients were selected. This bias could explain the fact that more recent studies find lower efficacy rates. The efficacy rates of LA-SRIFS can possibly be increased by gross debulking of the macroadenoma, as was reported by, e.g., Petrossians et al. (2005). However, this could not be confirmed by Sherlock et al. (2009b), although they did observe that previous RT decreased the efficacy of the LA-SRIFS (Sherlock et al., 2009a). The exact reason for this observation is not clear. Both acute and long-term responses to LA-SRIFS are associated with tumoral expression of SSTR2a as assessed by immunohistochemistry (Fougner et al., 2008). Another predictor for LA-SRIFS responsiveness is the granulation pattern of somatotroph adenomas. The densely granulated adenomas have a higher responsiveness to SRIFS than sparsely granulated adenomas. However, densely granulated adenomas also have a higher SSTR2a expression (Fougner et al., 2012). The alternative for somatostatin receptor 2a expression is SSTR2A protein expression. However many studies did not find a relation between LA-SRIFS responsiveness and SSTR2a protein expression, mainly in patients pretreated with LA-SRIF. Recently, a group of LA-SRIF-naïve operated patients had a good correlation between high SSTR2a protein expression and good long-term response to adjuvant LA-SRIFS treatment (Gatto et al., 2013). It seems that pretreatment of LA-SRIFS affects the expression of SSTR2a protein expression and that therefore SSTR2a protein expression is only useful as a marker for treatment responsiveness in LA-SRIFS naïve adenomas (Gatto et al., 2013). In addition, the GH response in the acute octreotide test may predict long-term GH control by LA-SRIF, although this test is not recommended for decision-making on prolonged treatment with LA-SRIF (de Herder et al., 2005).

Pegvisomant

About a decade ago PEG-V, a growth hormone receptor (GHR) antagonist, was introduced. To date, PEG-V is the most effective therapy for acromegaly (Clark et al., 1996; Kopchick et al., 2002; Neggers et al., 2009a, 2012; Neggers and van der Lely, 2011; van der Lely et al., 2011, 2012). PEG-V is a recombinant growth-hormone analog with a Gly120Arg substitution in GH binding site 2 and eight amino acid substitutions in GH binding site 1, which results in an enhanced affinity for the GHR on the one hand but a lack of functional GHR signaling on the other (Kopchick et al., 2002). Due to its pegylation (that is, covalent attachment of polyethylene glycol polymer chains), PEG-V is a stable 42–46 kD compound with decreased immunogenicity and prolonged half-life (Clark et al., 1996; Drake, 2001). PEG-V treatment induces a fast reduction in serum IGF-1 levels which results, due to the negative feedback, in a rise in GH serum levels in a proportion of patients (van der Lely et al., 2001a; Neggers and van der Lely, 2011). However, because PEG-V is a mutated GH molecule, most commercial GH assays cannot differentiate between endogenous GH and PEG-V due to cross-reactivity. For these two reasons GH cannot be assessed in patients treated with PEG-V, unless specific assays are used. The key biomarker during the treatment of PEG-V is therefore IGF-1 along with clinical signs and symptoms.

Efficacy

During long-term treatment with PEG-V in the registration studies, more than 90% of patients with acromegaly have a normalization of IGF-1 levels (van der Lely et al., 2001a). However, in the observational registries ACROSTUDY[tm] or the German Pegvisomant Observational Study (GPOS), lower efficacy rates are observed (van der Lely et al., 2012). After 5 years, ACROSTUDY reported an efficacy of 62% with a weekly dose of, respectively, 113 mg and 106 mg for controlled and uncontrolled subjects (Trainer, 2009). GPOS efficacy rates started at 71%, but declined to 58% after 5 years (Buchfelder et al., 2009). The majority of patients included in the available registries are treated with daily PEG-V. PEG-V was combined with LA-SRIF in 23%, with dopamine agonist in 6%, and a combination of the three agents in 4% of patients (Trainer, 2009). The mean weekly dose in van der Lely's study was around 130 mg, with an efficacy rate of > 90% (van der Lely et al., 2001a). The mean weekly doses of the observational registries are 106 mg in patients with a normal IGF-1, and 113 mg in those with an elevated IGF-1 (van der Lely et al., 2012). The lower efficacy of 62% might be explained by the relatively low dose of PEG-V administered in the registries. To achieve efficacy rates of above 90% by PEG-V monotherapy, the average expected weekly dose will rise, probably above 120–130 mg.

Combination therapy with pegvisomant and somatostatin analogs

Efficacy rates of PEG-V as single agent and in combination therapy of LA-SRIFS and PEG-V are comparable. However, the necessary weekly PEG-V dose during combined treatment with LA-SRIFS appears to be much lower (around 50%) and can still be accompanied by efficacy rates of over 90% (Neggers et al., 2009a; Neggers and van der Lely, 2011; van der Lely et al., 2011). The mean cumulative weekly dose was 77 mg PEG-V during combined treatment of LA-SRIF and PEG-V (Neggers et al., 2009a; Neggers and van der Lely, 2011). In another report, two patients who were controlled with high-dose PEG-V monotherapy were converted to lanreotide autogel 120 mg monthly and PEG-V weekly (Neggers et al., 2011a). After conversion, a weekly dose reduction of 80 and 150 mg could be achieved, without a change in IGF-1 levels (Neggers et al., 2011a). Van der Lely et al. reported a dose reduction of 51% during coadministration of lanreotide autogel 120 mg (van der Lely et al., 2011). These data suggest that a significant dose reduction during combined treatment can be expected. However, the magnitude of dose reduction in PEG-V can differ significantly between patients (Neggers et al., 2011a). The implication of these studies is that for those patients requiring high-dose PEG-V monotherapy, a significant economic gain can be made when their dose can be reduced by 50%, concomitantly increasing the cost-effectiveness of treatment.

A direct comparison between PEG-V alone and combined treatment has been studied but did not show a significant difference in efficacy (Trainer et al., 2009). However, this study is difficult to interpret as the IGF-1 assay that was used for assessing efficacy and dose titration had to be changed during the study period. Also, the study was not optimally designed to detect a dose difference between the combined use of LA-SRIF and PEG-V and PEG-V monotherapy (Trainer et al., 2009). Nevertheless, the pharmacologic properties of PEG-V more or less guarantee an efficacy close to 100%, provided that patients are treated with the proper dose. Therefore, any study that compares the efficacy of different treatment strategies which incorporate PEG-V in all stratification arms are more or less meaningless, since any observed efficacy less than close to 100%, in whatever treatment arm, simply indicates that the dose of PEG-V was suboptimal, and not that the medication used in that specific treatment arm was less effective.

The reason why the PEG-V dose can be reduced during combination treatment with LA-SRIF might be partially explained by the 20% increase in PEG-V serum levels compared to monotherapy with an equal dose (van der Lely et al., 2001b; Jorgensen et al., 2005). Since PEG-V is a competitive blocker of GH, competing endogenous GH concentrations are very relevant. GH levels increase during PEG-V treatment (van der Lely et al., 2001b), but when PEG-V is combined with LA-SRIF, lower GH serum concentrations are observed (van der Lely et al., 2001b; Jorgensen et al., 2005). PEG-V meets less competition of endogenous GH around the GHR, resulting in a lower necessary dose of PEG-V to block all GHRs, during combination therapy (Feenstra et al., 2005; Neggers et al., 2007, 2009b; Hodish and Barkan, 2008; Neggers and van der Lely, 2009, 2011; van der Lely et al., 2011). Additionally, LA-SRIF in rodent studies showed that the number of GHRs expressed in the liver is reduced as a result of decreased portal insulin concentration (Wurzburger et al., 1993; Shishko et al., 1994; Leung et al., 2000). Therefore,

PEG-V (being a competitive GHR blocker) has less GH to compete with and less GHRs to block. In addition, LA-SRIFS may directly inhibit the production of IGF-1 by hepatocytes (Murray et al., 2004). These mechanisms together are the basis for the observed dose reduction during combined use of LA-SRIF and PEG-V (Feenstra et al., 2005; Neggers et al., 2007; Hodish and Barkan, 2008; Neggers et al., 2009a; Neggers and van der Lely, 2009, 2011; van der Lely et al., 2011).

The PEG-V dose required to normalize IGF-1 was positively correlated with baseline IGF-1 levels, corrected for age and gender ($r=0.48$; $p=0.006$) (Neggers et al., 2007). A similar result was observed during coadministration of lanreotide autogel and PEG-V (van der Lely et al., 2011). However, gross debulking of the pituitary ademoma or RT did not reduce the necessary dose of PEG-V to normalize IGF-1 levels during combined treatment (Neggers et al., 2007).

Quality of life aspects of combination therapy

From a patient's perspective, QoL is an important parameter of disease control (Neggers et al., 2008a). From a clinician's perspective the main parameter seems to be normalization of both IGF-1 and GH, which has been shown to correlate with normalization of the elevated long-term mortality rates of patients with acromegaly (Swearingen et al., 1998; Beauregard et al., 2003; Holdaway et al., 2003, 2004, 2008). As mentioned above, biochemical normalization does not guarantee complete relief from acromegalic symptoms (Hua et al., 2006; Neggers et al., 2008a). These residual symptoms appear to result in a measurable impairment in QoL in patients with acromegaly (Biermasz et al., 2004, 2005; Bonapart et al., 2005). QoL can be quantified by the Acromegaly Quality of Life Questionnaire (AcroQoL™) (Webb et al., 2002) and symptoms by questionnaires such as the Patient-assessed Acromegaly Symptom Questionnaire (PASQ™) (van der Lely et al., 2001a). In a prospective double-blind, placebo-controlled crossover study, QoL was assessed by AcroQoL and symptoms by PASQ, with or without the addition of PEG-V to LA-SRIF. All patients in that study had IGF-1 levels within the age- and sex-adjusted normal limits and GH levels of less than 2.5 μg/L (Neggers et al., 2008a). During the period of 16 weeks with 40 mg PEG-V weekly, QoL improved in these so-called controlled acromegalic subjects, as indicated by an increase in AcroQoL total score and especially AcroQoL physical dimension. This was accompanied by a reduction in total PASQ score and the single PASQ questions, perspiration, soft tissue swelling, and overall health status. The observation of the improvement in QoL and signs and symptoms was not accompanied by a significant decrease in IGF-1. No correlation between change in IGF-1 and improvement in QoL was observed, but body weight correlated with the improvement in AcroQoL physical, although the absolute decrease in body weight was not significant (Neggers et al., 2008a). Noteworthy is that this improvement in QoL after the addition of PEG-V in controlled acromegaly patients was not reconfirmed by Madsen et al. (2011). Madsen's study design, however, was different from the design of Neggers's study. In the Danish study the dose of LA-SRIF was changed when PEG-V was added and the AcroQoL was not used. Furthermore, patients were not used as their own control, which decreased the power of their study. These differences in study design might explain the contradictory results.

The mode of action of LA-SRIFS might also explain why the addition of PEG-V might improve QoL (Neggers et al., 2008b; Neggers and van der Lely, 2011). As already mentioned above, LA-SRIF treatment reduces portal insulin concentrations and therefore the number of available GHRs in the liver (Leung et al., 2000; Neggers et al., 2011b). LA-SRIFS can also directly inhibit IGF-1 production by hepatocytes (Murray et al., 2004). These mechanisms suggest that whereas the liver becomes relatively resistant to GH during LA-SRIF treatment, GH activity in other organs and tissues of the body is still too high (Neggers et al., 2008b, 2011b; Rubeck et al., 2010). Recently, in a Danish study, acromegaly patients with a normalized IGF-1 during LA-SRIF treatment still had higher nadir GH levels as compared with patients with a normal IGF-1 after surgery and had a reduced disease-specific health status (Rubeck et al., 2010). In line with this, we recently introduced the concept of "extrahepatic acromegaly" (Neggers et al., 2011b) (Fig. 20.1). One might expect that treatment of this "extrahepatic acromegaly" with low-dose, weekly PEG-V would improve the GH-dependent signs and symptoms and the patient's QoL (Neggers et al., 2008a).

New developments

As stated before, the SSTR pattern of GH-producing pituitary adenomas is characterized by predominant expression of SSTR2 and SSTR5 (Thodou et al., 2006; Murray and Melmed, 2008). *In vitro* studies suggest a synergistic effect between SSTR2 and SSTR5 activation in the control of GH secretion (Ren et al., 2003). Pasireotide (SOM230) is a so-called universal LA-SRIF that binds with high affinity to all SSTR except for SSTR4 (Schmid, 2008). Pasireotide has a particularly high affinity for SSTR5 and could thus be a potent GH inhibitor in octreotide-resistant adenomas. Pasireotide potently

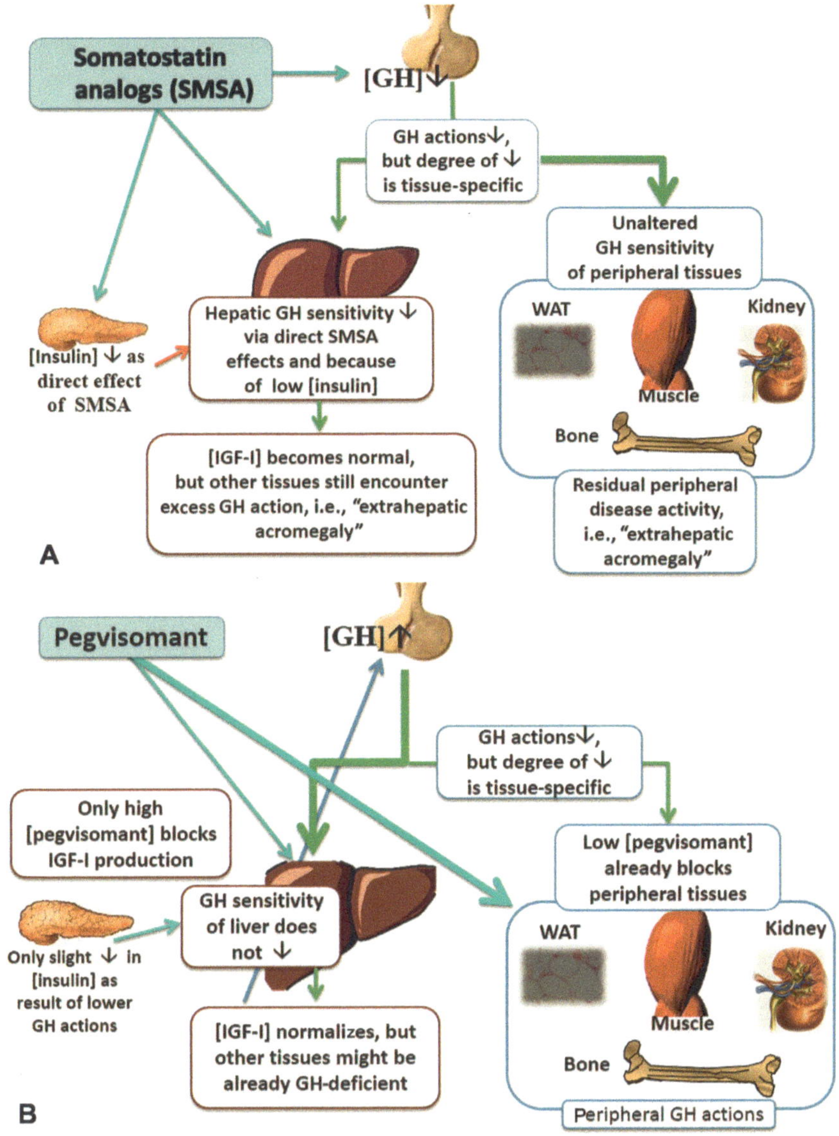

Fig. 20.1. (**A**) Effects of somatostatin analogs (SMSA) in SMSA-sensitive acromegalic subjects. Red arrows indicate inhibitory effects; green arrows indicate stimulatory effects; while thickness of arrow indicates level of inhibition. (**B**) Effects of pegvisomant in acromegalic subjects. Red arrows indicate inhibitory effects; green arrows indicate stimulatory effects; while thickness of arrow indicates level of inhibition. WAT, white adipose tissue; GH, growth hormone; IGF-1, insulin-like growth factor 1. (Reproduced from Neggers et al., 2011b, with permission from Bioscientifica.)

suppressed IGF-1 levels in rats (Weckbecker et al., 2002). In a single-dose proof-of-concept study in acromegalic patients, pasireotide (100 and 250 μg) and octreotide (100 μg) were equipotent in GH suppression in 8/12 patients, but in 3/12 patients pasireotide was more efficacious than octreotide. Adenomatous tissue from one of these octreotide-resistant patients could be assessed for SSTR expression and showed a relatively high SSTR5 expression (van der Hoek et al., 2004). In a phase II study, pasireotide at dose between 200 and 600 μg induced a complete biochemical response, i.e., normalization of GH and IGF-1 levels after 3 months in 27% of patients (Petersenn et al., 2010). Studies evaluating the efficacy and safety of long-acting pasireotide are ongoing.

The majority of GH-producing adenomas simultaneously express SSTR2 and dopamine receptor subtype 2 (Da2) and a functional interaction between SSTR2 and Da2 via heterodimerization has been shown *in vitro* (Rocheville et al., 2000). This may in part explain the synergistic effects of combined treatment with a somatostatin analog and dopamine agonist in acromegaly (Flogstad et al., 1994). Based on these observations, chimeric compounds have been developed that target both somatostatin and dopamine receptors. For instance BIM-23A760, which binds to SSTR2, SSTR5, and Da2, strongly suppresses GH production *in vitro* (Jaquet et al., 2005). This compound was, however, not potent enough during use in early studies in humans.

Adverse effects

Somatostatin analogs have relatively few side-effects. Transient complaints of loose stools, abdominal discomfort, nausea, flatulence, and fat malabsorption during the first weeks of treatment have been reported in about two-thirds of patients. Most of these side-effects resolve within 10 weeks (Lamberts et al., 1996). In the first 18 months of treatment, the development of asymptomatic gallstones or sludge in about 25% of the patients has been reported, most likely due to the reduced postprandial gallbladder contractility and emptying (Freda, 2002).

There are three important side-effects of PEG-V. Usually these side-effects are quite mild, transient, and self-limiting. Lipohypertrophy has been described in several reports during PEG-V monotherapy and in combination treatment with LA-SRIF, with a low prevalence (Maffei et al., 2006; Bonert et al., 2008; Neggers and van der Lely, 2009; van der Lely et al., 2012). When lipohypertrophy does occur, frequent rotation of the injection site can reverse local lipohypertrophy, although it can remain detectable for more than 8 months (Neggers et al., 2009b). In some cases it remains present so that PEG-V therapy needs to be discontinued. It is probably caused by a severe local GH deficiency due to very high PEG-V tissue levels at the injection site in the presence of insulin. This disbalance may lead to accumulation of adipose tissue around the injection sites (Hodish and Barkan, 2008; Neggers and van der Lely, 2009).

The second side-effect involves hepatocellular liver enzyme disturbances. Elevations in liver enzyme levels are mainly transient and self-limiting and occur both during PEG-V monotherapy and when used in combination with LA-SRIF. There are two types of hepatic enzyme disturbance: hepatocellular and cholestatic. Cholestatic disturbances are most often related to current or previous treatment with LA-SRIFS (Shi et al., 1993; Biering et al., 2006; Neggers et al., 2007; Neggers and van der Lely, 2009). A significant proportion of these patients had asymptomatic bile stones and active biliary disease is rare during LA-SRIF treatment (Shi et al., 1993). When LA-SRIFS are discontinued, contractility of the biliary gland normalizes and when sludge or stones are present, symptoms of biliary obstruction or active biliary disease can occur. Hepatocellular liver enzyme disturbances are probably directly related to the use of PEG-V. They seem to occur almost exclusively within the first year of treatment and there is no dose dependency (Biering et al., 2006; Neggers et al., 2007, 2009b; Hodish and Barken, 2008; Trainer, 2009; Trainer et al., 2009; Neggers and van der Lely, 2011). The incidence of elevated transaminases of more than three times the upper limit of normal ($\times$ULN), is higher during combined treatment ($\leq$15%) (Neggers and van der Lely, 2009, 2011; Madsen et al., 2011) than during PEG-V monotherapy (5.2%) (Biering et al., 2006; van der Lely et al., 2012). A recent combination study observed a 10% prevalence of elevated transaminases of more than 2 $\times$ULN (van der Lely et al., 2011). The difference in follow up of the patients in various studies could explain the differences in the prevalence of elevated transaminases, because the GPOS (Biering et al., 2006; Schreiber et al., 2007) resembles everyday practice, whereas the combined studies from Rotterdam have a more systematic follow-up (Feenstra et al., 2005, 2006; Neggers et al., 2007, 2008a, 2009b; Neggers and van der Lely, 2009, 2011; van der Lely et al., 2011). When the intervals between outpatient visits are long enough, many episodes of transient elevated transaminases will occur unnoticed. It seems that a specific group of acromegaly patients might have an increased risk of developing these elevated transaminases with a common polymorphism associated with Gilbert's syndrome, UGT1A1*28 and male sex (Bernabeu et al., 2010). The incidence of homozygous and heterozygous genotypes of UGT1A1*28 in acromegaly patients was 54% (Bernabeu et al., 2010). In some studies it has been reported that diabetic acromegaly patients have a 2.3 times higher risk (Neggers et al., 2007, 2009b). However, the impact of diabetes mellitus on elevated transaminases over 3 $\times$ULN seems to be less visible in larger follow-up studies in cohorts of patients (Neggers and van der Lely, 2009; Neggers et al., 2009b). In other studies, no relationship between diabetes and elevated transaminases was reported (Biering et al., 2006; Schreiber et al., 2007; van der Lely et al., 2011). Nor does (cumulative) PEG-V dose or concomitant medication seem to be related to the occurrence of these hepatocellular liver enzyme disturbances (Neggers et al., 2007; Neggers and van der Lely, 2009).

In the recent past, some concerns were raised that PEG-V might induce pituitary tumor growth. However, despite a few reports of an increase in tumor size during PEG-V therapy, there is no unequivocal evidence that PEG-V directly promotes tumor growth (Buchfelder et al., 2009; van der Lely et al., 2011, 2012). It might be more correct to conclude that PEG-V is unable to prevent tumor growth. Tumor growth can also occur during long-term LA-SRIF, in which it seems to be present in 2.6% of the treated patients (Bevan, 2005). In the GPOS database, tumor size increase was carefully and systematically reviewed in over 300 patients. In this systematic review, only three of the eight patients with an initially reported increase in tumor size had a real, but minor increase after PEG-V treatment was initiated (Buchfelder et al., 2009). In another three of the eight subjects, the initially reported increase had already started before the initiation of PEG-V treatment. In two subjects a detectable rebound of tumor size was encountered after cessation of LA-SRIF therapy (Buchfelder et al., 2009). In 75 patients in a Spanish survey, five (6.7%) acromegaly patients were identified with an increase in pituitary tumor size (Marazuela et al., 2011). All of these patients were pretreated with LA-SRIF analogs and then switched to PEG-V monotherapy. The reference MRI was made just after LA-SRIF was discontinued. The patients with tumor size increase were pretreated with LA-SRIFS for a shorter period of time and had not been treated before with radiotherapy (Marazuela et al., 2011). Therefore the tumor size increase in this study could again be due to the rebound phenomenon after cessation of the LA-SRIF treatment.

When LA-SRIFS are continued and PEG-V added, no increase in tumor size was observed in about 99 patients (Neggers and van der Lely, 2009). Only in the study by Jorgensen and co-workers was an increase in tumor size observed in one of the 11 patients studied (Jorgensen et al., 2005). Some of the patients in this study, however, received a high dose of octreotide LAR (30 mg every 2 weeks) prior to study entry. There were no data on the tumor size increase prior to study entry and the study also included a period with PEG-V monotherapy. Therefore, it is impossible to determine whether the increase in tumor size was caused by a rebound effect after discontinuation or decrease in frequency of the LA-SRIFS or was continued growth of the adenoma that had already started prior to study entry. Moreover, during combination treatment, in about 19% of the subjects a decrease in tumor size was observed, suggesting that LA-SRIF treatment can still control tumor size, even in the presence of PEG-V (Neggers and van der Lely, 2009; Neggers et al., 2009a).

CUSHING'S DISEASE

Cushing's disease (CD) is caused by an adenoma that secrets corticotropin (ACTH), which results in a glucocorticoid excess. This excess gives rise to the clinical signs and symptoms of Cushing's syndrome. This results in many signs and symptoms that can be seen as part of the metabolic syndrome, such as central obesity, hypertension, dyslipidemia, and diabetes mellitus, as well as others such as hirsutism, proximal muscle weakness, easy bruising, and osteoporosis. CD has the highest mortality rate of all pituitary adenomas (Dekkers et al., 2007).

Primary treatment is transsphenoidal surgery and, if necessary, additional pituitary radiation. After this first approach, persistent disease can be treated either by a second transsphenoidal surgery or with medical therapy, radiotherapy, or bilateral adrenalectomy in very selected cases. The medical therapy of CD can be used as a bridging, preoperative, or primary medical treatment when surgery is contraindicated. Unlike prolactinoma and acromegaly patients, primary medical treatment in CD is a rare exception. Clearly there is an unmet need for highly efficacious and safe pharmacologic treatment in CD.

Medical treatment

The main aim in the treatment of CD is to reduce the clinical syndrome. There are several types of medical treatment modalities. The first type of pharmacologic agents target the adrenal cortisol production. They inhibit the steroidogenesis by steroid enzyme inhibition or by both inhibition and cytotoxic effects on the adrenocortical cells (adrenolytic medication).

Steroid enzyme inhibitors are aminoglutethimide, etomidate, fluconazole, ketoconazole, metyrapone, and trilostane. They block one or more enzymes involved in the synthesis of cortisol (Tritos et al., 2011). Ketoconazole and metyrapone are the most commonly used. Ketoconazole inhibits several steroidogenic enzymes and thereby decreases cortisol production in about 70–80% of the patients (Tritos et al., 2011). The doses necessary to achieve this efficacy range from 200 to 1200 mg daily, divided into two administrations and starting with a twice or thrice daily dose of 200 mg. Metyrapone inhibits 11β hydrolase and is used as a monotherapy. It results in a similar efficacy to ketoconazole with a decrease in cortisol synthesis in 75–80% of patients (Tritos et al., 2011). The daily dose ranges from 750 to 6000 mg, divided over two to three daily administrations, starting with a twice or thrice daily dose of 250 mg. The combination of metyrapone with other steroid inhibitors may increase efficacy and limit the side-effects. Metyrapone is the most frequently used inhibitor during pregnancy, although there are some safety concerns (Tritos et al., 2011). Unlike metyrapone and ketoconazole, etomidate is not an oral formulation. It is primarily used as a parenteral anesthetic drug. Etomidate administration results in a rapid control of cortisol excess but also leads to unwanted sedation. Doses used to control cortisol excess are < 0.1 mg/kg/h.

Mitotane is an adrenocorticolytic drug in high doses, but in low doses inhibits steriodogenesis. It is not the preferred first-line medication for CD, since it has a slow onset and an accumulation in fat that results in detectable plasma levels long after the treatment has been terminated (Hogan et al., 1978). Mitotane should be primarily reserved for the treatment of adrenal carcinoma.

Mifepristone is mainly an antiprogestin. However, at higher doses it acts as a glucocorticoid receptor antagonist. Some of the CD symptoms decrease rapidly and in some patients it may even cause symptoms of hypocortisolism. Monitoring this potential hypocortisolism is cumbersome, since the levels of cortisol no longer reflect cortisol action when the glucocorticoid receptor is blocked. Doses needed to control cortisol excess range between 300 and 1200 mg daily (Fleseriu et al., 2012). Starting doses of mifepristone are usually 200–400 mg daily.

The other type of medical intervention in CD targets the pathologic ACTH hypersecretion of the pituitary adenoma.

Centrally acting agents are the dopamine agonist cabergoline and the somatostatin analog pasireotide (SOM230). Currently available dopamine agonists and somatostatin analogs that have been tested in Cushing's disease have not been very effective. Chronic use of cabergoline can decrease urinary free cortisol (UFC) up to 125% of the ULN in about 29% of patients with CD (Godbout et al., 2010; Vilar et al., 2010). The dose required ranged from 1 to 7 mg weekly. Pasireotide is a multiligand somatostatin analog. It is an agonist of SSTR subtypes 1, 2, 3, and 5. Subtype 5 is frequently expressed in pituitary adenomas of CD patients, and when serum levels of cortisol normalize, subtype 2 receptor expression increases (de Bruin et al., 2009). Pasireotide can normalize UFC levels in 16% of CD patients. However, in a stepwise combination with cabergoline and ketoconazole its efficacy can reach 66–88% (Feelders et al., 2010; Tritos et al., 2011). At present only pasireotide has been registered for the treatment of Cushing's disease.

Side-effects of medical treatment

Ketoconazole can cause transaminitis, and even severe hepatotoxity, for which dose adjustment or discontinuation is necessary. Transient transaminitis can be observed in 10% of treated patients and serious liver injury in < 0.01% of patients (Tritos et al., 2011). Ketoconazole can decrease sex-steroid levels in both men and women. In men it causes gynecomastia and testosterone replacement may be required. In females, low estrogens do not cause clinical symptoms, since most female Cushing's patients already have oligomenorrhea or amenorrhea.

The inhibition of steroidogenesis by metyrapone causes an increase in androgens that can lead to hirsutism. The increase in deoxycorticosterone can lead to hypertension and salt retention. Combinations of other steroid-inhibiting agents with metyrapone may decrease the side-effects caused by the accumulation of the precursors.

Gastrointestinal side-effects are possible during treatment with steroid-inhibiting agents and pasireotide. Pasireotide has the typical gastrointestinal side-effects seen during treatment with the well-known and presently available somatostatin analogs octreotide and lanreotide (see side-effects of acromegaly). The main side-effect of pasireotide is hyperglycemia; this can be observed in 73% of patients, with 63% requiring additional glucose-lowering medication (Colao et al., 2012). In about 70% of patients an IGF-1 level below the lower limit of normal was observed.

Cabergoline's side-effects are discussed above in the section on prolactinoma, although doses in CD are sometimes significantly higher.

Mitotane can cause dyslipidemia and because of its accumulation in fatty tissue, women should not become pregnant within 5 years after discontinuation of this medication.

Most of this medication can be titrated until cortisol levels are within the normal range, although a so-called "block and replace" treatment can be another solution.

We recommend that medical treatment in CD is reserved for specific groups of patients, mainly as a possible presurgical intervention or as bridging treatment, or as a last resort.

THYROTROPIN-SECRECTING ADENOMA

Thyrotropin-secrecting adenomas are very rare and account for < 1% of all functional pituitary adenomas. The excess in thyrotropin-releasing hormone (TSH) gives rise to the clinical signs and symptoms of a classic hyperthyroidism. Surgical resection has a poor reported result. Remission can be achieved in less than a third of patients, and in another third it leads to an improved biochemical profile (Beck-Peccoz et al., 1996). Medical treatment, therefore, is required in the majority of patients.

Medical treatment

Dopamine agonists

Bromocriptine and cabergoline can be efficacious in these patients, usually in those that present with co-secretion of prolactin (Beck-Peccoz et al., 1996). Bromocriptine doses of 10–20 mg are required, and for cabergoline doses of 0.25–0.5 mg twice weekly are necessary.

Somatostatin analogs

Octreotide and lanreotide have more or less the same efficacy in the treatment of acromegaly. It can be expected that the same is true in the treatment of thyrotropin-secrecting adenomas. However, there are more data available on the efficacy of lanreotide than octreotide. In 13 of 16 patients, TSH and thyroid hormones normalized (Kuhn et al., 2000).

Antithyroid treatment

Any type of antithyroid treatment primarily targeting the thyroid should be avoided. The decrease in thyroid hormone levels could lead to a further increase in TSH secretion and therefore further stimulate the growth of the pituitary adenoma. Hyperthyroidism symptoms can be controlled with β-blockers such as propranolol or atenolol.

Side-effects

The side-effects of dopamine agonists are presented above, in the section on prolactinomas. The side-effects of somatostatin analogs are presented in the section on acromegaly.

References

Abosch A, Tyrrell JB, Lamborn KR et al. (1998). Transsphenoidal microsurgery for growth hormone-secreting pituitary adenomas: initial outcome and long-term results. J Clin Endocrinol Metab 83: 3411–3418.

Ayuk J, Sheppard MC (2008). Does acromegaly enhance mortality? Rev Endocr Metab Disord 9: 33–39.

Ayuk J, Clayton RN, Holder G et al. (2004). Growth hormone and pituitary radiotherapy but not serum insulin-like growth factor-I concentrations predict excess mortality in patients with acromegaly. J Clin Endocrinol Metab 89: 1613–1617.

Barker 2nd FG, Klibanski A, Swearingen B (2003). Transsphenoidal surgery for pituitary tumors in the United States 1996–2000: mortality, morbidity and the effects of hospital and surgeon volume. J Clin Endocrinol Metab 88: 4709–4719.

Barlier A, Jaquet P (2006). Quinagolide – a valuable treatment option for hyperprolactinaemia. Eur J Endocrinol 154: 187–195.

Bates AS, Van't Hoff W, Jones JM et al. (1993). An audit of outcome of treatment in acromegaly. Q J Med 86: 293–299.

Bates PR, Carson MN, Trainer PJ et al. (2008). Wide variation in surgical outcomes for acromegaly in the UK. Clin Endocrinol (Oxf) 68: 136–142.

Beauregard C, Truong U, Hardy J et al. (2003). Long-term outcome and mortality after transsphenoidal adenomectomy for acromegaly. Clin Endocrinol (Oxf) 58: 86–91.

Beck-Peccoz P, Brucker-Davis F, Persani LS et al. (1996). Thyrotropin-secreting pituitary tumors. Endocr Rev 17: 610–638.

Bernabeu I, Marazuela M, Lucas T et al. (2010). Pegvisomant-induced liver injury is related to the ugt1a1*28 polymorphism of Gilbert's syndrome. J Clin Endocrinol Metab 95: 2147–2154.

Bevan JS (2005). Clinical review. The antitumoral effects of somatostatin analog therapy in acromegaly. J Clin Endocrinol Metab 90: 1856–1863.

Bevan JS, Webster J, Burke CW et al. (1992). Dopamine agonists and pituitary tumor shrinkage. Endocr Rev 13: 220–240.

Biering H, Saller B, Bauditz J et al. (2006). Elevated transaminases during medical treatment of acromegaly: a review of the German pegvisomant surveillance experience and a report of a patient with histologically proven chronic mild active hepatitis. Eur J Endocrinol 154: 213–220.

Biermasz NR, Van Thiel SW, Pereira AM et al. (2004). Decreased quality of life in patients with acromegaly despite long-term cure of growth hormone excess. J Clin Endocrinol Metab 89: 5369–5376.

Biermasz NR, Pereira AM, Smit JW et al. (2005). Morbidity after long-term remission for acromegaly: persisting joint-related complaints cause reduced quality of life. J Clin Endocrinol Metab 90: 2731–2739.

Biller BM, Molitch ME, Vance ML et al. (1996). Treatment of prolactin-secreting macroadenomas with the once-weekly dopamine agonist cabergoline. J Clin Endocrinol Metab 81: 2338–2343.

Bonapart IE, van Domburg R, ten Have SM et al. (2005). The "bio-assay" quality of life might be a better marker of disease activity in acromegalic patients than serum total IGF-I concentrations. Eur J Endocrinol 152: 217–224.

Bonert VS, Kennedy L, Petersenn S et al. (2008). Lipodystrophy in patients with acromegaly receiving pegvisomant. J Clin Endocrinol Metab 93: 3515–3518.

Buchfelder M, Schlaffer S, Droste M et al. (2009). The German ACROSTUDY: past and present. Eur J Endocrinol 161 (Suppl 1): s3–s10.

Carlsen SM, Lund-Johansen M, Schreiner T et al. (2008). Preoperative octreotide treatment in newly diagnosed acromegalic patients with macroadenomas increases cure short-term postoperative rates: a prospective randomized trial. J Clin Endocrinol Metab 93: 2984–2990.

Clark R, Olson K, Fuh G et al. (1996). Long-acting growth hormones produced by conjugation with polyethylene glycol. J Biol Chem 271: 21969–21977.

Colao A, Petersenn S, Newell-Price J et al. (2012). A 12-month phase 3 study of pasireotide in Cushing's disease. N Engl J Med 366: 914–924.

de Bruin C, Feelders RA, Lamberts SW et al. (2009). Somatostatin and dopamine receptors as targets for medical treatment of Cushing's syndrome. Rev Endocr Metab Disord 10: 91–102.

de Herder WW, Taal HR, Uitterlinden P et al. (2005). Limited predictive value of an acute test with subcutaneous octreotide for long-term IGF-I normalization with sandostatin LAR in acromegaly. Eur J Endocrinol 153: 67–71.

Dekkers OM, Biermasz NR, Pereira AM et al. (2007). Mortality in patients treated for Cushing's disease is increased compared with patients treated for nonfunctioning pituitary macroadenoma. J Clin Endocrinol Metab 92: 976–981.

Dekkers OM, Biermasz NR, Pereira AM et al. (2008). Mortality in acromegaly: a metaanalysis. J Clin Endocrinol Metab 93: 61–67.

Dekkers OM, Lagro J, Burman P et al. (2010). Recurrence of hyperprolactinemia after withdrawal of dopamine agonists: systematic review and meta-analysis. J Clin Endocrinol Metab 95: 43–51.

Drake WM (2001). Experience with pegvisomant in the treatment of acromegaly. Growth Horm IGF Res 11 (Suppl A): S111–S114.

Feelders RA, Hofland LJ, van Aken MO et al. (2009). Medical therapy of acromegaly efficacy and safety of somatostatin analogues. Drugs 69: 2207–2226.

Feelders RA, de Bruin C, Pereira AM et al. (2010). Pasireotide alone or with cabergoline and ketoconazole in Cushing's disease. N Engl J Med 362: 1846–1848.

Feenstra J, de Herder WW, ten Have SM et al. (2005). Combined therapy with somatostatin analogues and weekly pegvisomant in active acromegaly. Lancet 365: 1644–1646.

Feenstra J, van Aken MO, de Herder WW et al. (2006). Drug-induced hepatitis in an acromegalic patient during combined treatment with pegvisomant and octreotide long-acting repeatable attributed to the use of pegvisomant. Eur J Endocrinol 154: 805–806.

Fleseriu M, Biller BM, Findling JW et al. (2012). Mifepristone, a glucocorticoid receptor antagonist, produces clinical and metabolic benefits in patients with Cushing's syndrome. J Clin Endocrinol Metab 97: 2039–2049.

Flogstad AK, Halse J, Grass P et al. (1994). A comparison of octreotide bromocriptine or a combination of both drugs in acromegaly. J Clin Endocrinol Metab 79: 461–465.

Fougner SL, Borota OC, Berg JP et al. (2008). The clinical response to somatostatin analogues in acromegaly correlates to the somatostatin receptor subtype 2a protein expression of the adenoma. Clin Endocrinol (Oxf) 68: 458–465.

Fougner SL, Casar-Borota O, Heck A et al. (2012). Adenoma granulation pattern correlates with clinical variables and effect of somatostatin analogue treatment in a large series of patients with acromegaly. Clin Endocrinol (Oxf) 76: 96–102.

Freda PU (2002). Somatostatin analogs in acromegaly. J Clin Endocrinol Metab 87: 3013–3018.

Gatto F, Feelders RA, van der Pas R et al. (2013). Immunoreactivity score using an anti-sst2a receptor monoclonal antibody strongly predicts the biochemical response to adjuvant treatment with somatostatin analogs in acromegaly. J Clin Endocrinol Metab 98: E66–E71. http://dx.doi.org/10.1210/jc.2012-2609.

Gillam MP, Molitch ME, Lombardi G et al. (2006). Advances in the treatment of prolactinomas. Endocr Rev 27: 485–534.

Giustina A, Barkan A, Casanueva FF et al. (2000). Criteria for cure of acromegaly: a consensus statement. J Clin Endocrinol Metab 85: 526–529.

Godbout A, Manavela M, Danilowicz K et al. (2010). Cabergoline monotherapy in the long-term treatment of Cushing's disease. Eur J Endocrinol 163: 709–716.

Greenman Y (2007). Dopaminergic treatment of nonfunctioning pituitary adenomas. Nat Clin Pract Endocrinol Metab 3: 554–555.

Greenman Y, Tordjman K, Osher E et al. (2005). Postoperative treatment of clinically nonfunctioning pituitary adenomas with dopamine agonists decreases tumour remnant growth. Clin Endocrinol (Oxf) 63: 39–44.

Hodish I, Barkan A (2008). Long-term effects of pegvisomant in patients with acromegaly. Nat Clin Pract Endocrinol Metabol 4: 324–332.

Hogan TF, Citrin DL, Johnson BM et al. (1978). Op′-ddd (mitotane) therapy of adrenal cortical carcinoma: observations on drug dosage toxicity and steroid replacement. Cancer 42: 2177–2181.

Holdaway IM (2004). Treatment of acromegaly. Horm Res 62 (suppl 3): 79–92.

Holdaway IM, Rajasoorya CR, Gamble GD et al. (2003). Long-term treatment outcome in acromegaly. Growth Horm IGF Res 13: 185–192.

Holdaway IM, Rajasoorya RC, Gamble GD (2004). Factors influencing mortality in acromegaly. J Clin Endocrinol Metab 89: 667–674.

Holdaway IM, Bolland MJ, Gamble GD (2008). A meta-analysis of the effect of lowering serum levels of GH and IGF-I on mortality in acromegaly. Eur J Endocrinol 159: 89–95.

Hua SC, Yan YH, Chang TC (2006). Associations of remission status and lanreotide treatment with quality of life in patients with treated acromegaly. Eur J Endocrinol 155: 831–837.

Jaquet P, Saveanu A, Barlier A (2005). New SRIF analogs in the control of human pituitary adenomas: perspectives. J Endocrinol Invest 28: 14–18.

Jorgensen JO, Feldt-Rasmussen U, Frystyk J et al. (2005). Cotreatment of acromegaly with a somatostatin analog and a growth hormone receptor antagonist. J Clin Endocrinol Metab 90: 5627–5631.

Kleinberg DL, Boyd 3rd AE, Wardlaw S et al. (1983). Pergolide for the treatment of pituitary tumors secreting prolactin or growth hormone. N Engl J Med 309: 704–709.

Kletzky OA, Vermesh M (1989). Effectiveness of vaginal bromocriptine in treating women with hyperprolactinemia. Fertil Steril 51: 269–272.

Kopchick JJ, Parkinson C, Stevens EC et al. (2002). Growth hormone receptor antagonists: discovery, development and use in patients with acromegaly. Endocr Rev 23: 623–646.

Kreutzer J, Vance ML, Lopes MB et al. (2001). Surgical management of GH-secreting pituitary adenomas: an outcome study using modern remission criteria. J Clin Endocrinol Metab 86: 4072–4077.

Kuhn JM, Arlot S, Lefebvre H et al. (2000). Evaluation of the treatment of thyrotropin-secreting pituitary adenomas with a slow release formulation of the somatostatin analog lanreotide. J Clin Endocrinol Metab 85: 1487–1491.

Lamberts SW, van der Lely AJ, de Herder WW et al. (1996). Octreotide. N Engl J Med 334: 246–254.

Leong KS, Foy PM, Swift AC et al. (2000). CSF rhinorrhoea following treatment with dopamine agonists for massive invasive prolactinomas. Clin Endocrinol (Oxf) 52: 43–49.

Leung KC, Doyle N, Ballesteros M et al. (2000). Insulin regulation of human hepatic growth hormone receptors: divergent effects on biosynthesis and surface translocation. J Clin Endocrinol Metab 85: 4712–4720.

Madsen M, Poulsen PL, Orskov H et al. (2011). Cotreatment with pegvisomant and a somatostatin analog (SA) in SA-responsive acromegalic patients. J Clin Endocrinol Metab 96: 2405–2413.

Maffei P, Martini C, Pagano C et al. (2006). Lipohypertrophy in acromegaly induced by the new growth hormone receptor antagonist pegvisomant. Ann Intern Med 145: 310–312.

Marazuela M, Paniagua AE, Gahete MD et al. (2011). Somatotroph tumor progression during pegvisomant therapy: a clinical and molecular study. J Clin Endocrinol Metab 96: e251–e259.

Melmed S (2006). Medical progress: acromegaly. N Engl J Med 355: 2558–2573.

Molitch ME (2005). Pharmacologic resistance in prolactinoma patients. Pituitary 8: 43–52.

Molitch ME, Elton RL, Blackwell RE et al. (1985). Bromocriptine as primary therapy for prolactin-secreting macroadenomas: results of a prospective multicenter study. J Clin Endocrinol Metab 60: 698–705.

Murray RD, Melmed S (2008). A critical analysis of clinically available somatostatin analog formulations for therapy of acromegaly. J Clin Endocrinol Metab 93: 2957–2968.

Murray RD, Kim K, Ren SG et al. (2004). Central and peripheral actions of somatostatin on the growth hormone-IGF-I axis. J Clin Invest 114: 349–356.

Neggers SJ, van der Lely AJ (2009). Somatostatin analog and pegvisomant combination therapy for acromegaly. Nat Rev Endocrinol 5: 546–552.

Neggers S, van der Lely AJ (2011). Combination treatment with somatostatin analogues and pegvisomant in acromegaly. Growth Horm IGF Res 21: 129–133.

Neggers S, van Aken MO, Janssen J et al. (2007). Long-term efficacy and safety of combined treatment of somatostatin analogs and pegvisomant in acromegaly. J Clin Endocrinol Metab 92: 4598–4601.

Neggers S, van Aken MO, de Herder WW et al. (2008a). Quality of life in acromegalic patients during long-term somatostatin analog treatment with and without pegvisomant. J Clin Endocrinol Metab 93: 3853–3859.

Neggers SJ, van Aken MO, de Herder WW et al. (2008b). Quality of life in acromegalic patients during long-term somatostatin analog treatment with and without pegvisomant. J Clin Endocrinol Metab 93: 3853–3859.

Neggers S, de Herder WW, Janssen J et al. (2009a). Combined treatment for acromegaly with long-acting somatostatin analogs and pegvisomant: long-term safety for up to 45 years (median 22 years) of follow-up in 86 patients. Eur J Endocrinol 160: 529–533.

Neggers SJ, de Herder WW, Janssen JA et al. (2009b). Combined treatment for acromegaly with long-acting somatostatin analogs and pegvisomant: long-term safety for up to 45 years (median 22 years) of follow-up in 86 patients. Eur J Endocrinol 160: 529–533.

Neggers S, de Herder WW, Feelders RA et al. (2011a). Conversion of daily pegvisomant to weekly pegvisomant combined with long-acting somatostatin analogs in controlled acromegaly patients. Pituitary 14: 253–258.

Neggers SJ, Kopchick JJ, Jorgensen JOL et al. (2011b). Hypothesis: extra-hepatic acromegaly: a new paradigm? Eur J Endocrinol 164: 11–16.

Neggers SJ, Biermasz NR, van der Lely AJ (2012). What is active acromegaly and which parameters do we have? Clin Endocrinol (Oxf) 76: 609–614.

Ono M, Miki N, Kawamata T et al. (2008). Prospective study of high-dose cabergoline treatment of prolactinomas in 150 patients. J Clin Endocrinol Metab 93: 4721–4727.

Orme SM, Mcnally RJ, Cartwright RA et al. (1998). Mortality and cancer incidence in acromegaly: a retrospective cohort study. United Kingdom Acromegaly Study Group. J Clin Endocrinol Metab 83: 2730–2734.

Petersenn S, Schopohl J, Barkan A et al. (2010). Pasireotide (SOM230) demonstrates efficacy and safety in patients with acromegaly: a randomized multicenter phase II trial. J Clin Endocrinol Metab 95: 2781–2789.

Petrossians P, Borges-Martins L, Espinoza C et al. (2005). Gross total resection or debulking of pituitary adenomas improves hormonal control of acromegaly by somatostatin analogs. Eur J Endocrinol 152: 61–66.

Pivonello R, Matrone C, Filippella M et al. (2004). Dopamine receptor expression and function in clinically nonfunctioning pituitary tumors: comparison with the effectiveness of cabergoline treatment. J Clin Endocrinol Metab 89: 1674–1683.

Pokrajac A, Wark G, Ellis AR et al. (2007). Variation in GH and IGF-I assays limits the applicability of international consensus criteria to local practice. Clin Endocrinol (Oxf) 67: 65–70.

Ren SG, Kim S, Taylor J et al. (2003). Suppression of rat and human growth hormone and prolactin secretion by a novel somatostatin/dopaminergic chimeric ligand. J Clin Endocrinol Metab 88: 5414–5421.

Rocheville M, Lange DC, Kumar U et al. (2000). Receptors for dopamine and somatostatin: formation of hetero-oligomers with enhanced functional activity. Science 288: 154–157.

Rubeck KZ, Madsen M, Andreasen CM et al. (2010). Conventional and novel biomarkers of treatment outcome in patients with acromegaly: discordant results after somatostatin analog treatment compared with surgery. Eur J Endocrinol 163: 717–726.

Schmid HA (2008). Pasireotide (SOM230): development mechanism of action and potential applications. Mol Cell Endocrinol 286: 69–74.

Schreiber I, Buchfelder M, Droste M et al. (2007). Treatment of acromegaly with the GH receptor antagonist pegvisomant in clinical practice: safety and efficacy evaluation from the German pegvisomant observational study. Eur J Endocrinol 156: 75–82.

Sherlock M, Aragon Alonso A, Reulen RC et al. (2009a). Monitoring disease activity using GH and IGF-I in the follow-up of 501 patients with acromegaly. Clin Endocrinol (Oxf) 71: 74–81.

Sherlock M, Fernandez-Rodriguez E, Alonso AA et al. (2009b). Medical therapy in patients with acromegaly: predictors of response and comparison of efficacy of dopamine agonists and somatostatin analogues. J Clin Endocrinol Metab 94: 1255–1263.

Shi YF, Zhu XF, Harris AG et al. (1993). Prospective study of the long-term effects of somatostatin analog (octreotide) on gallbladder function and gallstone formation in Chinese acromegalic patients. J Clin Endocrinol Metab 76: 32–37.

Shishko PI, Dreval AV, Abugova IA et al. (1994). Insulin-like growth factors and binding proteins in patients with recent-onset type 1 (insulin-dependent) diabetes mellitus: influence of diabetes control and intraportal insulin infusion. Diabetes Res Clin Pract 25: 1–12.

Swearingen B, Barker 2nd FG, Katznelson L et al. (1998). Long-term mortality after transsphenoidal surgery and adjunctive therapy for acromegaly. J Clin Endocrinol Metab 83: 3419–3426.

Thodou E, Kontogeorgos G, Theodossiou D et al. (2006). Mapping of somatostatin receptor types in GH or/and PRL producing pituitary adenomas. J Clin Pathol 59: 274–279.

Trainer PJ (2009). ACROSTUDY: the first 5 years. Eur J Endocrinol 161 (Suppl 1): s19–s24.

Trainer PJ, Ezzat S, D'Souza GA et al. (2009). A randomized controlled multicentre trial comparing pegvisomant alone with combination therapy of pegvisomant and long-acting octreotide in patients with acromegaly. Clin Endocrinol (Oxf) 71: 549–557.

Tritos NA, Biller BM, Swearingen B (2011). Management of Cushing disease. Nat Rev Endocrinol 7: 279–289.

Valassi E, Klibanski A, Biller BM (2010). Clinical review. Potential cardiac valve effects of dopamine agonists in hyperprolactinemia. J Clin Endocrinol Metab 95: 1025–1033.

van der Hoek J, de Herder WW, Feelders RA et al. (2004). A single-dose comparison of the acute effects between the new somatostatin analog SOM230 and octreotide in acromegalic patients. J Clin Endocrinol Metab 89: 638–645.

van der Lely AJ, Brownell J, Lamberts SW (1991). The efficacy and tolerability of CV 205-502 (a nonergot dopaminergic drug) in macroprolactinoma patients and in prolactinoma patients intolerant to bromocriptine. J Clin Endocrinol Metab 72: 1136–1141.

van der Lely AJ, Hutson RK, Trainer PJ et al. (2001a). Long-term treatment of acromegaly with pegvisomant, a growth hormone receptor antagonist. Lancet 358: 1754–1759.

van der Lely AJ, Muller A, Janssen JA et al. (2001b). Control of tumor size and disease activity during cotreatment with octreotide and the growth hormone receptor antagonist pegvisomant in an acromegalic patient. J Clin Endocrinol Metab 86: 478–481.

van der Lely AJ, Bernabeu I, Cap J et al. (2011). Coadministration of lanreotide autogel and pegvisomant normalizes IGF1 levels and is well tolerated in patients with acromegaly partially controlled by somatostatin analogs alone. Eur J Endocrinol 164: 325–333.

van der Lely AJ, Biller BM, Brue T et al. (2012). Long-term safety of pegvisomant in patients with acromegaly: comprehensive review of 1288 subjects in ACROSTUDY. J Clin Endocrinol Metab 97: 1589–1597.

Vance ML, Evans WS, Thorner MO (1984). Drugs five years later. Bromocriptine. Ann Intern Med 100: 78–91.

Verhelst J, Abs R, Maiter D et al. (1999). Cabergoline in the treatment of hyperprolactinemia: a study in 455 patients. J Clin Endocrinol Metab 84: 2518–2522.

Vilar L, Naves LA, Azevedo MF et al. (2010). Effectiveness of cabergoline in monotherapy and combined with ketoconazole in the management of Cushing's disease. Pituitary 13: 123–129.

Webb SM, Prieto L, Badia X et al. (2002). Acromegaly quality of life questionnaire (AcroQoL) a new health-related quality of life questionnaire for patients with acromegaly: development and psychometric properties. Clin Endocrinol (Oxf) 57: 251–258.

Weckbecker G, Briner U, Lewis I et al. (2002). SOM230: a new somatostatin peptidomimetic with potent inhibitory effects on the growth hormone/insulin-like growth factor-I axis in rats, primates and dogs. Endocrinology 143: 4123–4130.

Wurzburger MI, Prelevic GM, Sonksen PH et al. (1993). The effect of recombinant human growth hormone on regulation of growth hormone secretion and blood glucose in insulin-dependent diabetes. J Clin Endocrinol Metab 77: 267–272.

Zanettini R, Antonini A, Gatto G et al. (2007). Valvular heart disease and the use of dopamine agonists for Parkinson's disease. N Engl J Med 356: 39–46.

Handbook of Clinical Neurology, Vol. 124 (3rd series)
Clinical Neuroendocrinology
E. Fliers, M. Korbonits, and J.A. Romijn, Editors

Chapter 21

Radiation therapy in the management of pituitary adenomas

ITAI PASHTAN[1], KEVIN S. OH[2], AND JAY S. LOEFFLER[2]*
[1]*Harvard Radiation Oncology Program, Boston, MA, USA*
[2]*Department of Radiation Oncology, Massachusetts General Hospital, Boston, MA, USA*

INTRODUCTION

Radiation therapy has an important role to play in the management of pituitary adenomas that are incompletely resected, that have recurred or persist biochemically or radiographically, those considered a high risk for recurrence despite surgical resection, or in the rare medically inoperable patient. This chapter reviews decision-making, technology, clinical outcomes, and late sequelae of radiation therapy in the management of pituitary adenomas.

BACKGROUND ON FRACTIONATED RADIATION THERAPY AND SINGLE-FRACTION RADIOSURGERY

"Fractionation" refers to the number of radiation treatments, which when multiplied by the dose per fraction yields the total dose delivered. "Conventional fractionation" refers to the use of 1.8–2 Gy per day. Radiobiologic effects on both benign and malignant tumors are heavily influenced by both dose per fraction and total dose. For example, 50 Gy in 2 Gy fractions is felt to have a similar tumorical effect as 12–14 Gy delivered as a single fraction. This phenomenon is felt to be due to interfraction cell repopulation or repair of sublethal damage that occurs in a more protracted schedule. The benefit of fractionating radiation into small daily doses is the potential to decrease late normal tissue complications, but at the expense of inconvenience. Fractionated doses for pituitary adenomas range from 45 to 54 Gy, which implies 25–30 fractions, or 5–6 weeks of daily treatment.

Stereotactic radiosurgery (SRS) is the delivery of a single high dose of radiation under conditions of extreme precision. Depending on histology, typical SRS doses for intracranial tumors range from 12 to 35 Gy × 1 dose. There are several commercially developed and linear accelerator-based (Linac) systems that are used to achieve the same goals of submillimeter precision via rigid immobilization and meticulous setup verification. Gamma Knife (GK; Elekta, Stockholm, Sweden) and CyberKnife (Accuray, Sunnyvale, CA, USA) are two of the most common commercially developed systems. In GK, the patient is rigidly immobilized in a metal collimator helmet, which comprises a hemispheric distribution of > 200 cobalt-60 sources behind bores that converge on a single point (the isocenter). These bores are strategically removed to form dose distributions, and several overlapping isocenters may be used for irregularly shaped targets. GK systems are single use, meaning that they cannot be used for non-SRS purposes. In contrast, linear accelerator-based systems are used to deliver SRS with the same multipurpose machine used for most other radiation therapy applications. Linac-based systems can achieve rigid immobilization in a variety of ways, such as (1) invasive frame secured to the calvarium by four pins above the level of the brow, (2) fixed bite piece that relies on the patient's dental impression, or (3) less rigid mask and high-fidelity image-guidance system. In Linac-based systems, the beam is shaped by either cones or multileaf collimators (MLCs). A steep dose gradient between the target and normal tissue can be achieved by several strategies, such as multiple noncoplanar stationary fields or arc therapy in which the beam is on while the treatment unit is rotating around the patient. All SRS systems exploit the steepest portion of the beam profile, which is often the 50% isodose line (IDL) for GK or the 80–90% IDL for Linac-based SRS.

*Correspondence to: Jay S. Loeffler, MD, 100 Blossom Street, Cox 3, Boston, MA 02114, USA. Tel: +1-617-724-1548, E-mail: jloeffler@partners.org

Deciding between fractionated radiotherapy and single-fraction radiosurgery

When safely achievable, stereotactic radiosurgery is often preferred over fractionated radiation therapy. The advantages of SRS in the management of pituitary adenomas are twofold: (1) convenience of single day treatment and (2) the potential for faster biochemical normalization in secretory tumors (Landolt et al., 1998; Mitsumori et al., 1998; Sheehan et al., 2005). However, many patients are not candidates for SRS because of unfavorable tumor size, irregular geometry, or tumor location near critical normal tissues such as the optic apparatus and brainstem. For intracranial targets, the safety of SRS generally declines for tumors > 3 cm, but this is dependent on the shape and cystic make-up. The location of the target is critically important in deciding whether a particular SRS dose can be safely achieved. The distance required between the target and critical normal tissues is entirely dependent on (1) the prescribed dose to the tumor, (2) the safe allowable dose to the normal tissue, (3) the gradient of dose fall-off per millimeter. For example, to keep the optic chiasm and nerves from receiving < 8–10 Gy at typical SRS doses of 12–15 Gy, the target should generally be located at least 3–5 mm away from the optic apparatus. However, a target requiring 16–30 Gy, such as a functioning adenoma or malignancy, will require further separation.

Therefore, fractionated radiation therapy is often required for pituitary adenomas with considerable extrasellar extension, of bulky size, or located near the optic apparatus. When safety is in doubt, fractionated RT should be considered the standard of care to minimize the risk of late complications (Mitsumori et al., 1998).

Radiation treatment planning

Target definition is the first step in the radiation planning process. The gross tumor volume (GTV) is the grossly visible pituitary adenoma by magnetic resonance imaging (MRI) and computed tomography (CT). Parasellar extension of gross tumor is common for nonfunctioning pituitary adenomas in particular and may extend into the cavernous sinus, sphenoid sinus, or intracranial parenchyma. The clinical target volume (CTV) includes both gross disease and subclinical extent. For certain functioning pituitary adenomas, this may include the inferior or entire sella turcica and medial walls of the left and right cavernous sinuses. The planning target volume (PTV) includes the CTV with an additional margin to account for uncertainties from setup and physiologic motion. Depending on the immobilization and verification system used, the PTV margin may range from 0 to 5 mm. For example, a standard thermoplastic mask may require 3–5 mm of PTV margin for uncertainty. However, a stereotactic setup can often eliminate the need for a PTV expansion altogether. Certain critical normal tissues to be avoided ("organs at risk", OAR) should also be defined and should include the brainstem, optic apparatus, temporal lobes, and cochlea.

The next step in treatment planning is the choice of beam characteristics and arrangement. Multiple noncoplanar stationary fields or arcs generally achieve the sharpest gradient between PTV and OARs and is commonly used for stereotactic radiosurgery. For fractionated cases, three-dimensional conformal radiation therapy (3D-CRT) is often employed and uses at least three unopposed beams to reduce heterogeneity ("hot spots") and minimize dose in surrounding uninvolved tissue. This may include laterals, obliques, and/or a vertex field, which has been a standard approach for many years. More recently, inverse-planned intensity-modulated radiation therapy (IMRT) is used to create complex and concave dose distributions which are particularly useful for tumors with a close relationship to the optic apparatus or other critical normal tissues. There are several methodologies for IMRT, but all require the use of optimization software to iterate and evaluate fluence through small beamlets to achieve the planned objectives.

Normal tissue tolerances

The optic apparatus, including the chiasm and optic nerves, are the primary normal tissues that require sparing in radiation planning for pituitary adenomas. The historical dose limit to these structures, as described by Emami et al., is 50–54 Gy, with the belief that the 5 year risk of visual deficits are 5% at 50 Gy and 50% at 65 Gy with standard fractionation (Emami et al., 1991). However, these guidelines were based on consensus agreement. In 2010, the Quantitative Analysis of Normal Tissue Effects in the Clinic guidelines (QUANTEC) provided recommendations for safe irradiation of the optic nerve and chiasm based on available clinical data. These guidelines suggest that radiation-induced optic neuropathy (RION) is extremely rare at < 55 Gy with fraction sizes of 1.8–2 Gy per day. The risk increases to 3–7% at doses of 55–60 Gy and > 7–20% at >60 Gy (Mayo et al., 2010). However, rare complications have been reported at doses as low as 46 Gy at 1.8 Gy/fraction (Aristizabal et al., 1977; van den Bergh et al., 2003; Mackley et al., 2007). Single-fraction dose limits to the optic apparatus are an entirely different range. Several series have shown that RION is very rare at < 8–10 Gy, has increasing frequency from 10–12 Gy, and reaches > 10% at 12–15 Gy (Tishler et al., 1993; Mayo et al., 2010). In summary, a reasonable range for optic chiasm and nerve tolerances are < 54–55 Gy with standard fractionation except for patients at high risk for visual sequelae and < 8–10 Gy with SRS. The brainstem tolerance is thought to be 12 Gy in a single fraction for 1 cc or more of tissue.

Proton therapy

Proton therapy and other positively charged heavy particles are forms of radiation therapy that can deliver the intended dose to the target while minimizing the exit dose. In doing so, the nearby uninvolved tissues can be spared any unnecessary low and medium scatter dose (Fig. 21.1). The physical properties of protons are such that their velocity in tissue rapidly decreases near their end of range. As this occurs, the energy transferred exponentially increases until the particles come to a complete rest, forming what is called a "Bragg peak." Multiple Bragg peaks of varying energy can be superimposed to create a "spread out Bragg peak." The clinical benefit of sparing brain parenchyma and soft tissues is the minimization of neurocognitive deficits and second malignancies. However, prospective clinical studies to confirm this advantage are sparse at present, although currently enrolling at the Massachusetts General Hospital (MGH) and other likeminded institutions. The widespread use of proton therapy is limited by the expense associated with such facilities, although the last decade has seen a rapid rise in the number of proton centers.

CLINICAL OUTCOMES OF RADIATION THERAPY IN PITUITARY ADENOMAS

Nonfunctioning adenomas

Radiation therapy is commonly used for nonfunctioning adenomas (NFAs) that are unresectable or subtotally resected and felt to be at high risk for impending symptoms with the goal of radiographic/clinical stabilization. The risks of late sequelae in combination with the indolent growth pattern of NFAs suggest that radiation therapy should be employed judiciously and with a high threshold. In cases of gross recurrence, repeat surgical debulking is often used as a strategy to delay RT (Lillehei et al., 1998; Park et al., 2004). When needed, both fractionated and SRS are associated with very high rates of radiographic local control (95–100%).

Conventional fractionated radiation at fairly low doses of 45–50.4 Gy in 1.8 Gy fractions has a long history in the treatment of NFAs. A retrospective experience of two hospitals in the UK reviewed 126 patients with NFAs. One hospital routinely offered postoperative RT to 45 Gy to 63 patients, while the other did not. The 15 year progression-free survival (PFS) significantly favored the subgroup who received radiation therapy (93% versus 33%, $p < 0.05$) (Gittoes et al., 1998). Similarly, van den Bergh et al. compared 76 patients who received postoperative RT ranging from 45 to 55.8 Gy with 28 patients who did not receive immediate radiation (van den Bergh et al., 2007). The 10 year local control rate heavily

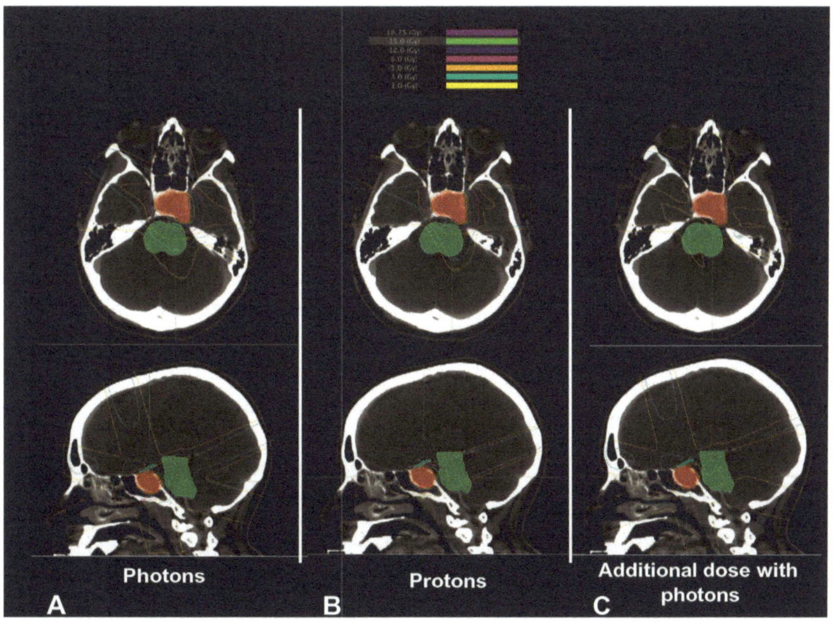

Fig. 21.1. Radiosurgery plans for a 31-year-old woman with a GH-secreting pituitary adenoma. The prescription dose is 15 Gy × 1 (green line). (**A**) Photon-based plan, (**B**) proton-based plan, and (**C**) proton plan subtracted from photon plan representing the additional low-dose of radiation (~1–3 Gy) given with photon therapy.

favored the RT group (95% versus 22%). One review of 63 patients with NFAs who were treated with fractionated stereotactic RT to 50.4 Gy at 1.8 Gy per fraction reported local control of 100% with median follow-up of 6.8 years (Colin et al., 2005). The fractionated data with proton therapy is limited, but also encouraging. In the Loma Linda, California, experience, 24 patients with NFAs were 100% locally controlled after median of 54 cobalt gray equivalents (CGE) and median follow-up of 3.9 years (Ronson et al., 2006). After gross or near total resection, postoperative radiation therapy is generally not required for NFAs because the risk of recurrence is very low (Lillehei et al., 1998; Park et al., 2004; Alameda et al., 2005; Dekkers et al., 2006).

Stereotactic radiosurgery can be used to successfully treat NFAs when their size, geometry, and location are favorable and normal tissue tolerances can be met. Pollock et al. retrospectively reported 3 and 7 year local control rates of 95% in 62 patients who received GK radiosurgery with a median marginal dose of 16 Gy and follow-up of 5.3 years (Pollock et al., 2008). Voges et al. prospectively evaluated 37 patients with NFAs treated with SRS at a median dose of 13.4 Gy and reported 100% local control at mean follow-up of 4.7 years (Voges et al., 2006).

Functioning adenomas

While radiation therapy provides excellent radiographic local control for functioning adenomas, the actuarial rate of biochemical normalization is far less predictable and heavily depends on histology (Zierhut et al., 1995; Estrada et al., 1997; Minniti et al., 2007). To achieve biochemical remission, functioning adenomas are felt to require higher doses than nonfunctioning adenomas. The recommended fractionated RT dose is 45–54 Gy in 1.8–2 Gy fractions for prolactinomas, Cushing's disease, acromegaly, and prolactinoma. Thyroid-stimulating hormone (TSH)-producing adenomas are often treated with slightly higher dose (e.g., 54 Gy) because they are felt to be locally aggressive and less radioresponsive. SRS doses range widely from 18 to 35 Gy in a single fraction.

During the interval between radiation and biochemical normalization (i.e., "latency period"), medical suppression is required as this period can last for several years or decades. Several clinical reports suggest that SRS achieves biochemical normalization with a shorter latency period when compared to fractionated RT. For example, a comparison of SRS versus standard fractionated RT in 29 hormonally active adenomas reported latency periods of 8.5 versus 18 months in favor or SRS (Mitsumori et al., 1998).

Prolactinomas

Radiation therapy is not commonly required for prolactin-secreting adenomas because the combination of medical therapy (such as dopamine agonists) and surgery achieves biochemical remission and radiographic stability in the vast majority of cases. Biochemical remission of prolactinomas is often defined as normal serum prolactin of $<$ 20–25 ng/mL depending on sex.

When radiation therapy is required for refractory prolactinomas, its success in achieving remission appears to be quite poor compared to other adenoma subtypes. Biochemical remission occurs in only 15–30% of cases with radiation alone and often requires a latency period of many years. For example, Littley et al. reviewed 58 patients who received standard fractionated RT (20–42.5 Gy in 8–15 fractions) for prolactinoma and reported only 21% biochemical normalization when dopamine agonists were discontinued (Littley et al., 1991). The same efficacy seems to be true of SRS as well. In a comprehensive systematic review of SRS in pituitary adenomas, Sheehan et al. included 22 series and 393 prolactinomas (Sheehan et al., 2005). Remission rates ranged from 15% to 30% in those studies that clearly defined their endpoints. The largest series included 128 patients treated with GK to a marginal dose of 9–35 Gy and a median follow-up of 2.8 years (Pan et al., 2000). Only 13% of evaluable patients had durable biochemical normalization off bromocriptine at 2 years.

Similar to other functional adenomas, there are retrospective data in prolactinomas that suggest that concurrent use of pharmacotherapy while receiving radiation therapy leads to poorer rates of biochemical response. In a small GK series of 20 patients, all five patients who achieved a complete biochemical remission were not on a concurrent dopamine agonist. Conversely, none of the nine patients using pharmacotherapy achieved a complete remission (Landolt and Lomax, 2000).

Based on these data, radiation therapy for prolactinomas may be thought of as adjunctive therapy to be used in combination with dopamine agonists and/or surgical resection, as it rarely achieves biochemical normalization when used alone. These data should be interpreted with caution given the low patient numbers in the retrospective literature.

Adrenocorticotropic hormone-secreting adenomas

In corticotropin (adrenocorticotropic hormone, ACTH)-secreting adenomas, indications for RT include: radiographic residual disease, biochemical or radiographic recurrence after transsphenoidal surgery, and excessive hormone production despite lack of gross disease. The definition of biochemical remission varies widely in the literature, but is typically defined as normalization of urinary free cortisol (UFC) and serum ACTH. Both standard fractionated RT and SRS appear to achieve biochemical remission rates of 50–80%, although SRS seems to have a shorter latency period.

In a larger series of Cushing's patients treated with fractionated RT, Estrada et al. included 30 patients treated with 48–54 Gy and reported actuarial remission rates of 44% at 1 year and 83% at 3 years (Estrada et al., 1997). Another study reviewed 40 patients with Cushing's disease treated with 45–100 Gy and reported actuarial 10 year PFS rate of 59% (Hughes et al., 1993). In the SRS literature, remission rates are 35–63% with shorter-term follow-up when compared to literature for standard fractionated RT. Sheehan et al. treated 43 patients with refractory Cushing's disease with GK to a mean marginal dose of 20 Gy, and achieved 63% biochemical normalization at a median follow-up of 44 months (Sheehan et al., 2000). The experience with heavy particles seems to be comparable to photon-based radiation therapy. The Massachusetts General Hospital reported its experience of 33 patients with ACTH-producing adenomas treated with proton SRS and concluded 52% complete radiographic and biochemical response off medical therapy after a median dose of 20 CGE at a median follow-up of 62 months (Petit et al., 2008). There are older data with helium ions from the Lawrence Berkeley Laboratory, California, in which 83 patients with Cushing's disease were treated with 30–150 Gy in three to four fractions and achieved an 85% biochemical cure rate (Levy et al., 1991).

Growth hormone-secreting adenomas

Similar to other functioning adenomas, RT is often used for patients with acromegaly who have unresectable disease or biochemical recurrence/persistence after surgery and medical therapy. The most commonly accepted criterion for biochemical remission is normalization of IGF-1 (matched for age and gender) and growth hormone (GH) level < 1 ng/mL after glucose challenge. While radiographic local control is often achieved (Milker-Zabel et al., 2001; Minniti et al., 2005), biochemical and clinical normalization after RT typically requires many years, or even decades. Predictors of biochemical response to RT are not well defined, but there is suggestion that lower baseline IGF-1 and GH levels are associated with higher rates of normalization (Littley et al., 1990; Attanasio et al., 2003; Castinetti et al., 2005). Similar to other functioning adenomas, there is some concern that use of octreotide at the time of RT is associated with a longer latency period to biochemical response (Landolt et al., 2000), although this is not a universal finding (Castinetti et al., 2005).

There is wide variation in the reported biochemical remission rates after fractionated RT for acromegaly, but this likely reflects inconsistency in definition of endpoints and the long follow-up required to document response. The largest retrospective experience of fractionated RT included 884 patients at 14 centers in the UK (Jenkins et al., 2006). Some 63% achieved normal IGF-1 levels at 10 years. GH levels fell to < 2.5 ng/mL in 22% at 2 years, 60% at 10 years, and 77% at 20 years. Barrande et al. reported a single-institution experience of 128 patients with acromegaly treated with fractionated radiation (Barrande et al., 2000). At last follow-up, 79% had achieved IGF-1 normalization. At median follow-up of 11.5 years, basal levels of GH < 2.5 ng/mL were achieved in 7% at 2 years, 53% at 10 years, and 66% at 15 years. These experiences illustrate the extremely long latency period often required for normalization.

Biochemical remission rates in large series of SRS range from 40% to 96%, but these low rates are likely related to shorter follow-up and poorer endpoint definition when compared to the literature governing fractionated RT. The largest single series included 82 patients treated with GK to a marginal dose of 12–40 Gy (Castinetti et al., 2005). Seventeen percent achieved a biochemical remission, defined as GH < 2 ng/mL and normalized IGF-1 after discontinuation of somatostatin agonists for at least 3 months. The mean follow-up was 50 months. Kobayashi et al. reviewed 67 patients treated with GK to a mean marginal dose of 18.9 Gy (Kobayashi et al., 2005). At a mean follow-up of 63 months, GH levels significantly decreased (<5 ng/mL) in 41%. Serum IGF-1 levels also significantly decreased (<400 ng/mL) in 41% of cases. A study by Zhang et al. included 68 acromegalic patients treated with GK to a mean marginal dose of 31 Gy (Zhang et al., 2000). Biochemical remission (generously defined as GH < 12 ng/mL) was achieved by 40% at 1 year and 96% at 2 years. Regarding proton therapy, the MGH reported 22 patients treated with proton SRS for refractory acromegaly after transphenoidal resection (Petit et al., 2007). After delivery of a median dose of 20 cobalt gray equivalents (CGE) and 6.3 years of follow-up, 59% achieved biochemical complete response (CR) defined as sustained IGF-1 normalization without medical therapy. In these patients, the median time to CR was 3.5 years.

Thyroid-stimulating hormone-secreting adenomas

TSH-secreting adenomas are rare and account for < 1% of all functioning pituitary adenomas, and therefore the data on radiation therapy for TSH-secreting adenomas are sparse. TSH-secreting adenomas often have an aggressive natural history and are considered more resistant to all types of therapy. Most require a combination of transsphenoidal resection, pharmacotherapy such as octreotide or bromocriptine, and radiation therapy. Surgery alone results in cure in only one-third of patients. Many would advocate radiation therapy to a higher dose (such as 54 Gy as fractionated therapy) in all cases.

Pituitary carcinomas

Pituitary carcinomas are extremely rare and comprise approximately 0.2% of pituitary tumors (Ragel and Couldwell, 2004). They are defined as pituitary tumors with metastasis outside the central nervous system (CNS) or occurring as separate foci within the CNS (Kaltsas et al., 2005). Approximately 75% are hormone secreting and the remaining 25% are nonfunctioning (Pernicone et al., 1997). Given their rarity, there is no consensus for the optimal therapy for pituitary carcinomas. The experience with fractionated involved field radiation therapy has been limited to case reports (Landman et al., 2002; Ragel and Couldwell, 2004). Outcomes are poor even after trimodality treatment including surgery, radiation, and chemotherapy.

SEQUELAE AFTER RADIATION THERAPY

The risk of late sequelae after radiation is related to the adjacent normal tissue that receives clinically relevant dose, which for pituitary adenomas includes the hypophysis itself, optic chiasm and nerves, carotid vessels, and the circle of Willis. Although some brain parenchyma and cranial nerves within the cavernous sinus often receive radiation therapy, it can be argued that the volume and dose used for pituitary adenomas are not clinically relevant. For example, radionecrosis within uninvolved brain parenchyma is extremely uncommon with 45–54 Gy in 1.8–2 Gy fractions with a reported risk of < 1% (Becker et al., 2002; Sheehan et al., 2005). Several technologic advances allow for improved sparing of uninvolved normal tissues, including CT-based three-dimensional planning, IMRT, stereotactic strategies, and proton therapy.

Hypopituitarism is a common late effect of radiation therapy when the prescribed dose to the sella is in the 45–54 Gy range used to treat most benign neoplasms. Because the sella is often the target, the risk of hypopituitarism is unavoidable. The risk of new hypopituitarism affecting at least one axis is 20–60% at 5 years (Estrada et al., 1997; Sheehan et al., 2000; Hoybye et al., 2001; Minniti et al., 2007). In an experience with Cushing's disease, Estrada et al. reported 57% risk of new growth hormone deficiency after fractionated RT (Estrada et al., 1997). Minniti reported new hypopituitarism in 85% at 15 years (Minniti et al., 2007). Several dosimetric studies have attempted to clarify dose thresholds to specific anatomic structures as predictors for hypopituitarism, including the entire pituitary (Vladyka et al., 2003), infundibulum (Feigl et al., 2002), and hypothalamus (Pai et al., 2001). Many would argue that with sufficiently long follow-up the risk of hypopituitarism is nearly certain.

Visual deficits after radiation therapy for pituitary adenomas are rare at the recommended tolerance doses of < 54–55 Gy in 1.8–2 Gy fractions and < 8–10 Gy in a single SRS fraction. In 2010 the Quantitative Analysis of Normal Tissue Effects in the Clinic (QUANTEC) provided recommendations for safe irradiation of the optic nerve and chiasm based on available clinical data. These guidelines suggest that radiation-induced optic neuropathy (RION) is extremely rare at < 55 Gy with fraction sizes of 1.8–2 Gy per day. Specific to pituitary adenomas, the Royal Marsden Hospital (RMH) in the UK reviewed 411 patients treated with 45–50 Gy in 25–30 fractions and found 1.5% incidence of visual deterioration at 20 years (Brada et al., 1993). A visual risk of 1% was reported by a compilation of SRS experiences for pituitary adenomas which included >1600 patients (Sheehan et al., 2005).

The development of a treatment-induced malignancy or benign tumor is a feared late effect after radiation therapy. Radiation-induced tumors are most commonly meningiomas, gliomas, and sarcomas. The risk of second malignancy from radiation therapy is difficult to predict and is dependent on dose, treatment volume, and follow-up interval. A large review of SRS series including 1621 patients reported no radiation-induced malignancies (Sheehan et al., 2005). However, with longer-term follow-up the risk of second tumors after standard fractionated RT seems to be 1–3% at 20 years (Brada et al., 1993; Tsang et al., 1993; Becker et al., 2002). Some may consider this risk concerning given that it is in the context of treating benign pituitary tumors. Advances in radiation therapy including 3D treatment planning and proton therapy may theoretically reduce the risk of radiation-induced tumor formation by reducing the volume of normal tissues exposed to low doses. However, the clinical confirmation of this benefit will require decades of follow-up.

The risk of vascular events after pituitary irradiation is low, but increases over decades. When targeting the sella turcica, it is often difficult to spare dose to the nearby carotid vessels and/or circle of Willis. In one review of 331 patients with pituitary adenomas irradiated at the RMH between 1962 and 1986, the actuarial incidences for cerebrovascular accidents were 4%, 11%, and 21% at 5, 10, and 20 years, respectively (Brada et al., 1999). This was estimated to represent a relative risk increase of 4.1 when compared to the general population. However, this study did not necessarily reflect modern radiation technologies and strategies.

CONCLUSION

Fractionated radiation therapy and radiosurgery play an important role in the management of pituitary adenomas. For nonfunctioning adenomas, RT leads to extremely high rates of local control. However, RT is less predictable in offering biochemical normalization for refractory or recurrent functioning adenomas. The use of more modern technology holds the potential to minimize dose to normal tissues and its consequential sequelae. In all cases of benign tumors, the decision to offer radiation therapy is often a delicate balance between risk and benefit in a population who otherwise would have a normal length and quality of life. These decisions are best made in a thoughtful multidisciplinary setting.

REFERENCES

Alameda C, Lucas T, Pineda E et al. (2005). Experience in management of 51 non-functioning pituitary adenomas: indications for post-operative radiotherapy. J Endocrinol Invest 28: 18–22.

Aristizabal S, Caldwell WL, Avila J (1977). The relationship of time–dose fractionation factors to complications in the treatment of pituitary tumors by irradiation. Int J Radiat Oncol Biol Phys 2: 667–673.

Attanasio R, Epaminonda P, Motti E et al. (2003). Gamma-knife radiosurgery in acromegaly: a 4-year follow-up study. J Clin Endocrinol Metab 88: 3105–3112.

Barrande G, Pittino-Lungo M, Coste J et al. (2000). Hormonal and metabolic effects of radiotherapy in acromegaly: long-term results in 128 patients followed in a single center. J Clin Endocrinol Metab 85: 3779–3785.

Becker G, Kocher M, Kortmann RD et al. (2002). Radiation therapy in the multimodal treatment approach of pituitary adenoma. Strahlenther Onkol 178: 173–186.

Brada M, Rajan B, Traish D et al. (1993). The long-term efficacy of conservative surgery and radiotherapy in the control of pituitary adenomas. Clin Endocrinol (Oxf) 38: 571–578.

Brada M, Burchell L, Ashley S et al. (1999). The incidence of cerebrovascular accidents in patients with pituitary adenoma. Int J Radiat Oncol Biol Phys 45: 693–698.

Castinetti F, Taieb D, Kuhn JM et al. (2005). Outcome of gamma knife radiosurgery in 82 patients with acromegaly: correlation with initial hypersecretion. J Clin Endocrinol Metab 90: 4483–4488.

Colin P, Jovenin N, Delemer B et al. (2005). Treatment of pituitary adenomas by fractionated stereotactic radiotherapy: a prospective study of 110 patients. Int J Radiat Oncol Biol Phys 62: 333–341.

Dekkers OM, Pereira AM, Roelfsema F et al. (2006). Observation alone after transsphenoidal surgery for nonfunctioning pituitary macroadenoma. J Clin Endocrinol Metab 91: 1796–1801.

Emami B, Lyman J, Brown A et al. (1991). Tolerance of normal tissue to therapeutic irradiation. Int J Radiat Oncol Biol Phys 21: 109–122.

Estrada J, Boronat M, Mielgo M et al. (1997). The long-term outcome of pituitary irradiation after unsuccessful transsphenoidal surgery in Cushing's disease. N Engl J Med 336: 172–177.

Feigl GC, Bonelli CM, Berghold A et al. (2002). Effects of gamma knife radiosurgery of pituitary adenomas on pituitary function. J Neurosurg 97 (5 Suppl): 415–421.

Gittoes NJ, Bates AS, Tse W et al. (1998). Radiotherapy for non-function pituitary tumours. Clin Endocrinol (Oxf) 48: 331–337.

Hoybye C, Grenback E, Rahn T et al. (2001). Adrenocorticotropic hormone-producing pituitary tumors: 12- to 22-year follow-up after treatment with stereotactic radiosurgery. Neurosurgery 49: 284–291, discussion 291–292.

Hughes MN, Llamas KJ, Yelland ME et al. (1993). Pituitary adenomas: long-term results for radiotherapy alone and post-operative radiotherapy. Int J Radiat Oncol Biol Phys 27: 1035–1043.

Jenkins PJ, Bates P, Carson MN et al. (2006). Conventional pituitary irradiation is effective in lowering serum growth hormone and insulin-like growth factor-I in patients with acromegaly. J Clin Endocrinol Metab 91: 1239–1245.

Kaltsas GA, Nomikos P, Kontogeorgos G et al. (2005). Clinical review. Diagnosis and management of pituitary carcinomas. J Clin Endocrinol Metab 90: 3089–3099.

Kobayashi T, Mori Y, Uchiyama Y et al. (2005). Long-term results of gamma knife surgery for growth hormone-producing pituitary adenoma: is the disease difficult to cure? J Neurosurg 102 (Suppl): 119–123.

Landman RE, Horwith M, Peterson RE et al. (2002). Long-term survival with ACTH-secreting carcinoma of the pituitary: a case report and review of the literature. J Clin Endocrinol Metab 87: 3084–3089.

Landolt AM, Lomax N (2000). Gamma knife radiosurgery for prolactinomas. J Neurosurg 93 (Suppl 3): 14–18.

Landolt AM, Haller D, Lomax N et al. (1998). Stereotactic radiosurgery for recurrent surgically treated acromegaly: comparison with fractionated radiotherapy. J Neurosurg 88: 1002–1008.

Landolt AM, Haller D, Lomax N et al. (2000). Octreotide may act as a radioprotective agent in acromegaly. J Clin Endocrinol Metab 85: 1287–1289.

Levy RP, Fabrikant JI, Frankel KA et al. (1991). Heavy-charged-particle radiosurgery of the pituitary gland: clinical results of 840 patients. Stereotact Funct Neurosurg 57: 22–35.

Lillehei KO, Kirschman DL, Kleinschmidt-DeMasters BK et al. (1998). Reassessment of the role of radiation therapy in the treatment of endocrine-inactive pituitary macroadenomas. Neurosurgery 43: 432–438, discussion 438–439.

Littley MD, Shalet SM, Swindell R et al. (1990). Low-dose pituitary irradiation for acromegaly. Clin Endocrinol (Oxf) 32: 261–270.

Littley MD, Shalet SM, Reid H et al. (1991). The effect of external pituitary irradiation on elevated serum prolactin levels in patients with pituitary macroadenomas. Q J Med 81: 985–998.

Mackley HB, Reddy CA, Lee SY et al. (2007). Intensity-modulated radiotherapy for pituitary adenomas: the preliminary report of the Cleveland Clinic experience. Int J Radiat Oncol Biol Phys 67: 232–239.

Mayo C, Martel MK, Marks LB et al. (2010). Radiation dose-volume effects of optic nerves and chiasm. Int J Radiat Oncol Biol Phys 76 (3 Suppl): S28–S35.

Milker-Zabel S, Debus J, Thilmann C et al. (2001). Fractionated stereotactically guided radiotherapy and radiosurgery in the treatment of functional and nonfunctional adenomas of the pituitary gland. Int J Radiat Oncol Biol Phys 50: 1279–1286.

Minniti G, Jaffrain-Rea ML, Osti M et al. (2005). The long-term efficacy of conventional radiotherapy in patients with GH-secreting pituitary adenomas. Clin Endocrinol (Oxf) 62: 210–216.

Minniti G, Osti M, Jaffrain-Rea ML et al. (2007). Long-term follow-up results of postoperative radiation therapy for Cushing's disease. J Neurooncol 84: 79–84.

Mitsumori M, Shrieve DC, Alexander 3rd E et al. (1998). Initial clinical results of LINAC-based stereotactic radiosurgery and stereotactic radiotherapy for pituitary adenomas. Int J Radiat Oncol Biol Phys 42: 573–580.

Pai HH, Thornton A, Katznelson L et al. (2001). Hypothalamic/pituitary function following high-dose conformal radiotherapy to the base of skull: demonstration of a dose-effect relationship using dose-volume histogram analysis. Int J Radiat Oncol Biol Phys 49: 1079–1092.

Pan L, Zhang N, Wang EM et al. (2000). Gamma knife radiosurgery as a primary treatment for prolactinomas. J Neurosurg 93 (Suppl 3): 10–13.

Park P, Chandler WF, Barkan AL et al. (2004). The role of radiation therapy after surgical resection of nonfunctional pituitary macroadenomas. Neurosurgery 55: 100–106, discussion 106–107.

Pernicone PJ, Scheithauer BW, Sebo TJ et al. (1997). Pituitary carcinoma: a clinicopathologic study of 15 cases. Cancer 79: 804–812.

Petit JH, Biller BM, Coen JJ et al. (2007). Proton stereotactic radiosurgery in management of persistent acromegaly. Endocr Pract 13: 726–734.

Petit JH, Biller BM, Yock TI et al. (2008). Proton stereotactic radiotherapy for persistent adrenocorticotropin-producing adenomas. J Clin Endocrinol Metab 93: 393–399.

Pollock BE, Cochran J, Natt N et al. (2008). Gamma knife radiosurgery for patients with nonfunctioning pituitary adenomas: results from a 15-year experience. Int J Radiat Oncol Biol Phys 70: 1325–1329.

Ragel BT, Couldwell WT (2004). Pituitary carcinoma: a review of the literature. Neurosurg Focus 16: E7.

Ronson BB, Schulte RW, Han KP et al. (2006). Fractionated proton beam irradiation of pituitary adenomas. Int J Radiat Oncol Biol Phys 64: 425–434.

Sheehan JM, Vance ML, Sheehan JP et al. (2000). Radiosurgery for Cushing's disease after failed transsphenoidal surgery. J Neurosurg 93: 738–742.

Sheehan JP, Niranjan A, Sheehan JM et al. (2005). Stereotactic radiosurgery for pituitary adenomas: an intermediate review of its safety, efficacy, and role in the neurosurgical treatment armamentarium. J Neurosurg 102: 678–691.

Tishler RB, Loeffler JS, Lunsford LD et al. (1993). Tolerance of cranial nerves of the cavernous sinus to radiosurgery. Int J Radiat Oncol Biol Phys 27: 215–221.

Tsang RW, Laperriere NJ, Simpson WJ et al. (1993). Glioma arising after radiation therapy for pituitary adenoma. A report of four patients and estimation of risk. Cancer 72: 2227–2233.

van den Bergh AC, Dullaart RP, Hoving MA et al. (2003). Radiation optic neuropathy after external beam radiation therapy for acromegaly. Radiother Oncol 68: 95–100.

van den Bergh AC, van den Berg G, Schoorl MA et al. (2007). Immediate postoperative radiotherapy in residual nonfunctioning pituitary adenoma: beneficial effect on local control without additional negative impact on pituitary function and life expectancy. Int J Radiat Oncol Biol Phys 67: 863–869.

Vladyka V, Liscak R, Novotny Jr J et al. (2003). Radiation tolerance of functioning pituitary tissue in gamma knife surgery for pituitary adenomas. Neurosurgery 52: 309–316, discussion 316–317.

Voges J, Kocher M, Runge M et al. (2006). Linear accelerator radiosurgery for pituitary macroadenomas: a 7-year follow-up study. Cancer 107: 1355–1364.

Zhang N, Pan L, Dai J et al. (2000). Gamma Knife radiosurgery as a primary surgical treatment for hypersecreting pituitary adenomas. Stereotact Funct Neurosurg 75: 123–128.

Zierhut D, Flentje M, Adolph J et al. (1995). External radiotherapy of pituitary adenomas. Int J Radiat Oncol Biol Phys 33: 307–314.

Section 3

Controversial issues and hot topics

Handbook of Clinical Neurology, Vol. 124 (3rd series)
Clinical Neuroendocrinology
E. Fliers, M. Korbonits, and J.A. Romijn, Editors

Chapter 22

Nelson syndrome: definition and management

T.M. BARBER[1], E. ADAMS[2], AND J.A.H. WASS[2]*

[1]*Division of Metabolic and Vascular Health, Warwick Medical School, University of Warwick, Coventry, UK*

[2]*Department of Endocrinology, Oxford Centre for Diabetes, Endocrinology and Metabolism, Churchill Hospital, Oxford, UK*

INTRODUCTION

Overall, Nelson syndrome is encountered infrequently in clinical practice. However, it is a relatively common complication of treatment for refractory Cushing's disease, occurrence rates in such patients being 8–43% for adults (Nagesser et al., 2000) and 25–66% for children (Hopwood and Kenny, 1977; Thomas et al., 1984; Wright-Pascoe et al., 2001), although its presentation has been reported to be delayed up to 24 years post-total bilateral adrenalectomy (TBA) and mean presentation 15 years post-TBA (Nagesser et al., 2000). The first case of Nelson syndrome was described in 1958; the patient was a 33-year-old woman who, 3 years previously, had undergone TBA for refractory Cushing's disease. The presenting features in this case included skin hyperpigmentation, visual field defects, raised plasma adrenocorticotropic hormone (ACTH) levels, and a large sellar mass (a pituitary corticotropinoma) demonstrated on an X-ray of the skull (Nelson et al., 1958). Surgical removal of the tumor resulted in resolution of her symptoms (Nelson et al., 1958).

Effective modern management of corticotropinomas that are refractory to transsphenoidal surgery (TSS) remain a clinical challenge. Options include repeating the TSS procedure, the use of drugs that block cortisol production, the administration of pituitary irradiation, and finally, TBA (Kelly, 2007). The scenarios in which TBA tends to be implemented include the presence of an undetectable or surgically unresectable corticotropinoma, or the recurrence of a corticotropinoma following TSS (Porterfield et al., 2008). In such cases, TBA can be associated with an 85–100% success rate at controlling hypercortisolemia depending on the successful removal of all adrenal tissue (Kelly, 2007; Nagesser et al., 2000), particularly when life-threatening manifestations of hypercortisolism occur (with immediate control of hypercortisolemia) (Kelly et al., 1983). However, TBA can result in the development of Nelson syndrome (Newell-Price et al., 2006).

Nelson syndrome is a potentially life-threatening condition (mortality rate 12% in the first reported case series (Sprague, 1953)) and can be associated with significant morbidity. However, the mortality rate associated with this condition has diminished significantly in recent times, and mortality was absent in two more recently reported case series (McCance et al., 1993; Jenkins et al., 1995). The improved mortality of Nelson syndrome is attributable at least in part to its earlier diagnosis and effective management. This in turn has resulted from improved awareness and screening for the condition with advancements in radiologic techniques and biochemical assays (Barber et al., 2010). A central feature of Nelson syndrome is the enlargement of an underlying pituitary corticotropinoma (with positive immunostaining for ACTH and ACTH secretion being associated with more aggressive and invasive tumors) (Fig. 22.1) (Bradley et al., 2003; De Tommasi et al., 2005). Before computed tomography (CT) and magnetic resonance imaging (MRI) scans were available, detection of Nelson tumors usually resulted from the late clinical manifestations of such tumor invasion and compression of surrounding structures. The introduction of MRI, with its highly resolved images, facilitated the early detection of enlarging corticotropinomas post-TBA and therefore enabled earlier diagnosis and management of Nelson syndrome.

Aspects of Nelson syndrome remain poorly understood, including its pathophysiology and the factors that are predictive for its onset and progression. In this

*Correspondence to: Professor J.A.H. Wass, Department of Endocrinology, Oxford Centre for Diabetes, Endocrinology and Metabolism, Churchill Hospital, Old Road, Headington, Oxford, OX3 7LJ, UK. Tel: +44-7973841508, E-mail: john.wass@noc.nhs.uk

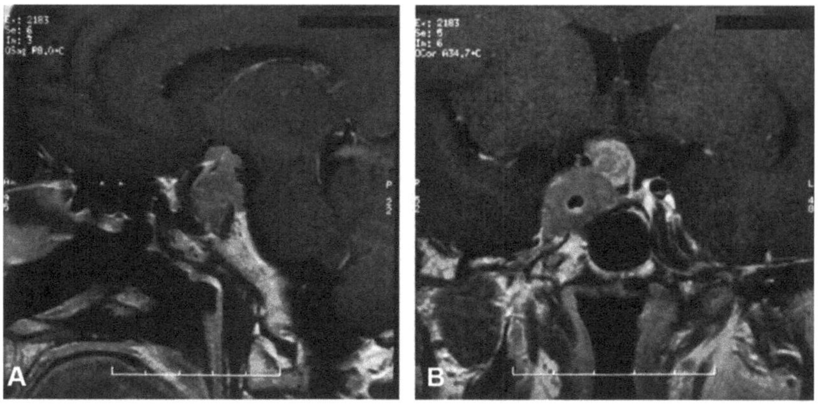

Fig. 22.1. Sagittal (**A**) and coronal (**B**) gadolinium-enhanced T1-weighted magnetic resonance images (MRI) showing an invasive corticotropinoma in a patient with Nelson syndrome. (Reproduced from Barber et al., 2010, copyright BioScientifica).

chapter, we discuss key aspects of Nelson syndrome relating to its diagnosis and effective management (summarized in Table 22.1):

1. Effective diagnosis of Nelson syndrome
2. Predictive factors for the onset and progression of Nelson syndrome
3. Pathophysiology of Nelson syndrome
4. Pathologic features of corticotropinomas in Nelson syndrome
5. Effective management of Nelson syndrome.

EFFECTIVE DIAGNOSIS OF NELSON SYNDROME

Clinical, biochemical, and radiologic features

Prior to the introduction of modern imaging techniques and sensitive assays, Nelson syndrome often presented with late clinical features resulting from compressive effects and invasion of an enlarging pituitary corticotropinoma into surrounding structures, metastases being unusual (Moore et al., 1976; Pernicone et al., 1997; Garcia et al., 2007). Late manifestations also include visual field defects (Ernest and Ekman, 1972; Moore et al., 1976; Assie et al., 2007) and cranial nerve palsies (McCance et al., 1993). The highly resolved imaging techniques of CT and MRI and the availability of sensitive ACTH assays has resulted in earlier diagnosis of Nelson syndrome and substantially reduced the frequency with which it presents with these late features. In modern practice, Nelson syndrome more typically presents clinically with hyperpigmentation of skin and mucous membranes (extensor surfaces, flexures, scars, areolae), with a frequency of up to 42% (Kelly et al., 2002), although visual field defects have been reported in some series in 10–57% (Banasiak and Malek, 2007). Other less frequent manifestations of Nelson syndrome include headaches, diabetes insipidus (Rees and Zilva, 1959), pituitary apoplexy (Ernest and Ekman, 1972; Kasperlik-Zaluska et al., 1983), panhypopituitarism, paratesticular tumors (Kasperlik-Zaluska et al., 1983), presenting with testicular pain and oligospermia (Johnson and Scheithauer, 1982), and paraovarian tumors (Baranetsky et al., 1979). In men, adrenal tissue may be present anywhere along the route of descent of the testis, and particularly in the paratesticular area. In women, adrenal rest cells can occur within the ovaries, their likely origin being from mesonephric remnants. The gonadal features of Nelson syndrome result from hyperstimulation and hyperplasia of gonadal adrenal rest cells (Shekarriz et al., 1996), and can be mistaken for a relapse of Cushing's disease (Johnson and Scheithauer, 1982).

Biochemically, Nelson syndrome is characterized by marked elevation of plasma ACTH levels following adrenalectomy, with basal and pulsatile volumes of ACTH secretions being up to six- and ninefold larger respectively in patients with Nelson syndrome compared to those with Cushing's disease (van Aken et al., 2004). Given that ACTH levels are affected by exogenously administered steroids, it is important that attention be given to the timing of the blood sample which should be taken at 08:00 hours, prior to the morning dose of glucocorticoid replacement and ideally 20 hours following the last dose of glucocorticoid tablet (Munir and Newell-Price, 2007). A further sample for ACTH should be taken at 2 hours following the morning dose of glucocorticoid tablet.

Radiologically, Nelson syndrome is characterized by an enlarging pituitary mass when compared with images taken post-TSS (often regrowth of pituitary remnant tissue) (Banasiak and Malek, 2007). Use of high-resolution MRI that is capable of identifying tumors as small as 3 mm in diameter enables identification of early tumor progression so that prompt therapeutic intervention can be instigated (Banasiak and Malek, 2007). Whilst many patients develop Nelson syndrome early following TBA for refractory Cushing's disease (20% in the first

Table 22.1

Summary of the clinicopathologic features of Nelson syndrome and its effective management

Clinical features	Hypothesized predictive factors	Hypothesized pathophysiology	Management
Hyperpigmentation of skin/ mucous membranes	Residual pituitary tumor	Released negative feedback hypothesis	Pituitary surgery
Visual field defects	ACTH levels in the first postoperative year	Aggressive corticotropinoma hypothesis	Adjuvant pituitary radiotherapy
Raised ACTH level	Neoadjuvant radiotherapy post-TBA		Stereotactic radiosurgery
Enlarging sellar mass	Duration of Cushing's disease pre-TBA		Selective somatostatin analogs
Cranial nerve palsies	Residual adrenal remnant post-TBA		PPAR γ agonists
Headaches	Age		Sodium valproate
Diabetes insipidus	Raised urinary cortisol		Dopamine agonists
Pituitary apoplexy	Insufficient exogenous steroid post-TBA		Temozolomide
Panhypopituitarism	Lack of cortisol suppressibility on high-dose dexamethasone pre-TBA		
Paratesticular tumors (testicular pain and oligospermia)			
Paraovarian tumors			

ACTH, adrenocorticotropic hormone; TBA, total bilateral adrenalectomy; PPAR γ, peroxisome proliferator-activated receptor gamma.

year and 35% in the first 2 years post-TBA (Assie et al., 2007)), the development of Nelson syndrome has also been reported up to 24 years post-TBA (Assie et al., 2007). It is therefore imperative that all such patients undergo lifelong pituitary imaging, with the first MRI scan at 3 months post-TBA (Banasiak and Malek, 2007), followed by repeat scans at 6-monthly intervals for 2 years and then yearly scans thereafter (lifelong). In this scenario, any increase in tumor size is significant (Barber et al., 2010).

Diagnostic criteria

Unfortunately, there is no clear consensus in the literature for a set of diagnostic criteria for Nelson syndrome. This has adverse implications for both clinical management of the condition and direct comparisons between studies that employ different diagnostic criteria. The utility of hyperpigmentation as a diagnostic criterion for Nelson syndrome poses problems in that this may lack sensitivity as a marker of ACTH levels (Kimura et al., 1996). There is clearly a need for a single set of diagnostic criteria.

Based on the existing literature, we published a proposal for a new set of diagnostic criteria for Nelson syndrome, to improve how we approach this condition in both clinical and research settings (Barber et al., 2010). In our proposal, a prerequisite for diagnosis of Nelson syndrome is previous TBA surgery and pre-existing Cushing's disease. Other criteria include: an expanding pituitary mass lesion; elevated ACTH levels (08:00 hours presteroid level > 500 ng/L based on ACTH levels in those patients who develop Nelson syndrome in previous case series) (Kasperlik-Zaluska et al., 1983; Pereira et al., 1998; Nagesser et al., 2000; Assie et al., 2007; Banasiak and Malek, 2007; Gil-Cardenas et al., 2007); and progressive elevations of ACTH levels on at least three consecutive and separate time-points (a rise of ACTH by > 30% of the initial post-TBA sample) (Barber et al., 2010). Our proposal, based on observations from current literature, is an attempt to unify the diagnostic criteria for Nelson syndrome, to facilitate diagnosis and screening in the clinical setting, and to enable direct comparisons between future studies on patients with this condition.

PREDICTIVE FACTORS FOR THE ONSET AND PROGRESSION OF NELSON SYNDROME

It is difficult to predict which patients with a history of Cushing's disease will subsequently develop Nelson syndrome following TBA. This underlies the importance of effective screening for Nelson syndrome in all such patients. However, there are some indicators that may predict future development of this condition:

Residual pituitary tumor shown on imaging prior to total bilateral adrenalectomy

It is clear from a number of studies that the presence of residual pituitary tumor following TSS treatment for Cushing's disease is a risk factor for future development of Nelson syndrome post-TBA (Jenkins et al., 1995; Sonino et al., 1996; Pereira et al., 1998; Hawn et al., 2002). In one study on 30 such patients, it was reported that in those who developed Nelson syndrome, 30% had radiologic evidence of residual tumor at the time of TBA, the majority (55%) being macroscopically visible (Jenkins et al., 1995). In contrast, only 17% of those patients who did not develop Nelson syndrome had radiologic evidence of residual tumor at the time of TBA, and only 33% of these were macroscopically visible (Jenkins et al., 1995). In a further series, it was reported that the development of Nelson syndrome occurred in 30% with versus 26% without residual pituitary tumor post-TSS (Gil-Cardenas et al., 2007). The utility of residual pituitary tumor as a predictive factor for the future development of Nelson syndrome is dependent upon the resolution of the imaging technique employed (usually MRI), and direct comparison of post-TBA pituitary images with those taken post-TSS.

Adrenocorticotropic hormone levels in the first postoperative year

Elevated levels of ACTH following TBA may be associated with future development of Nelson syndrome (Barnett et al., 1983; McCance et al., 1993; Pereira et al., 1998) through association with corticotropinoma progression (Assie et al., 2007). However, it has been observed in long-term follow-up studies that persistent elevations of ACTH levels may occur in up to 42% of post-TBA patients (McCance et al., 1993), with only some of these patients developing Nelson syndrome. Although a rise in plasma ACTH of $>$ 100 ng/L in the first year post-TBA may help to predict the future development of Nelson syndrome (Assie et al., 2007), this assertion requires further validation through long-term studies before it can be usefully adopted as a clinical tool. Other possible predictors including plasma ACTH levels pre-TBA and following metyrapone suppression should also be fully assessed in future studies.

Administration of neoadjuvant radiotherapy post-total bilateral adrenalectomy surgery

In patients who undergo TBA for refractory Cushing's disease, neoadjuvant pituitary radiotherapy (administered at the time of TBA or soon after this procedure) is often considered. There is some controversy in the literature regarding the efficacy of pituitary neoadjuvant radiotherapy administered to patients with Cushing's disease at the time of TBA in the prevention of the subsequent development of Nelson syndrome. In a long-term study over 15 years following TBA in 39 patients with Cushing's disease, none of those who received neoadjuvant radiotherapy (versus 50% of those who did not) had subsequent development of Nelson syndrome (Gil-Cardenas et al., 2007). Another study has also demonstrated a potential benefit of neoadjuvant radiotherapy, with development of Nelson syndrome in 25% of those receiving radiotherapy versus 50% of those who did not (Jenkins et al., 1995). However, other studies do not show that lack of pituitary radiotherapy at the time of TBA is associated with future development of Nelson syndrome (Moore et al., 1976; Manolas et al., 1984; McCance et al., 1993; Sonino et al., 1996). Clearly, pituitary radiotherapy is associated with future development of adverse sequelae and these potential risks need to be balanced with potential benefits of this prophylactic treatment. Given the risks associated with residual pituitary tumor outlined above, a pragmatic approach is to administer prophylactic neoadjuvant pituitary radiotherapy to those patients with the presence of pituitary remnant tissue prior to TBA surgery, but not in those without residual tissue. The role of radiosurgery in this scenario should be explored in future studies.

Duration of Cushing's disease prior to total bilateral adrenalectomy

In one study involving 43 patients who were followed up for a median of 10 years post-TBA, those who developed pituitary enlargement had had symptoms of Cushing's disease prior to TBA for twice as long as those who developed no pituitary enlargement (Kelly et al., 1983). In a smaller study with seven patients with Cushing's disease who had undergone TBA, there was no association between the duration of Cushing's disease and the development of Nelson syndrome (Moreira et al., 1993). There is some controversy in the literature therefore, and further studies are required to clarify the potential role of Cushing's disease duration pre-TBA as a potential predictor of subsequent Nelson syndrome development.

Residual adrenal remnant after total bilateral adrenalectomy

The presence of a small adrenal remnant post-TBA does not appear to protect against the subsequent development of Nelson syndrome (Manolas et al., 1984). Furthermore, there is a risk of recurrence of hypercortisolemia (resulting from raised ACTH levels) if adrenal remnant tissue is left *in situ* (Barber et al., 2010). This was demonstrated in a large study in which of those patients with Cushing's disease who had been treated with TBA and in whom adrenal remnants had been left *in situ* (n = 12 representing 27% of the sample), two patients developed early recurrence of Cushing's disease from hyperfunctioning adrenal remnant tissue (Nagesser et al., 2000).

Age

Age may be a predictive factor for subsequent development of Nelson syndrome, with younger patients – particularly children (Hopwood and Kenny, 1977; Thomas et al., 1984) – at the time of TBA being at higher risk (Moore et al., 1976; Kemink et al., 1994; Imai et al., 1996). However, once again controversy exists, with not all studies showing age at TBA to be a risk factor for Nelson syndrome development (Assie et al., 2007). It is possible that children may develop more aggressive subtypes of corticotropinomas, although other age-related factors may exist (Leinung and Zimmerman, 1994).

High urinary cortisol

For some macrocorticotropinomas, urinary cortisol levels pre-TBA may reflect tumor size and functionality, and thereby predict subsequent development of Nelson syndrome post-TBA (Boscaro et al., 2009; Nagesser et al., 2000; Sonino et al., 1996). However, such an association between pre-TBA urinary cortisol levels and risk of Nelson syndrome development post-TBA has not been demonstrated in some studies (Barnett et al., 1983; Kelly et al., 1983; Pereira et al., 1998), precluding adoption of urinary cortisol levels as a reliable predictive factor in clinical practice.

Insufficient exogenous steroid replacement therapy post-total bilateral adrenalectomy surgery

Although it is possible that suboptimal or absent steroid replacement therapy may increase the risk of Nelson syndrome development post-TBA (Pollock and Young, 2002; Kasperlik-Zaluska et al., 2006), or promote the development of an aggressive tumor subtype (Banasiak and Malek, 2007; Nagesser et al., 2000), most studies have not shown this to be the case (Barnett et al., 1983; Kelly et al., 1983; Nagesser et al., 2000).

Lack of cortisol suppression on high-dose dexamethasone pre-total bilateral adrenalectomy

It is possible that the minority (20%) of patients who lack suppression of cortisol on high-dose dexamethasone pre-TSS may be at higher risk of developing Nelson syndrome (Barber et al., 2010). However, existing evidence is lacking: in a study of 48 patients with Cushing's disease who underwent TBA (eight of whom subsequently developed Nelson syndrome), the suppressibility of cortisol following administration of dexamethasone (8 mg and 16 mg) prior to TBA did not appear to predict the subsequent development of Nelson syndrome (Kemink et al., 1994). Given the unpredictability of which patients go on to develop Nelson syndrome following TBA, further larger studies are required to explore pre-TSS and pre-TBA cortisol suppressibility as a possible clinical predictor of this outcome.

It is clear that there is much controversy in the current literature regarding the clinical utility of a number of potential predictors for the development of Nelson syndrome post-TBA. It is therefore imperative that all patients with a history of Cushing's disease who undergo TBA surgery also undergo close long-term follow-up to enable early identification and management of incipient Nelson syndrome.

PATHOPHYSIOLOGY OF NELSON SYNDROME

Although the pathophysiology of Nelson syndrome is incompletely understood, various hypotheses exist. One such hypothesis is that following TBA, a drop in the previously elevated cortisol levels results in reduced negative feedback on corticotroph cells and restoration of hypothalamic corticotropin-releasing hormone (CRH) production (Barber et al., 2010), which, in turn, may drive corticotroph neoplasia. A similar mechanism may occur rarely with the development of neoplasia in occasional patients with Addison's disease (Sugiyama et al., 1996). Data from rodent studies appear to support this hypothesis, with elevations in CRH levels (Carey et al., 1984), POMC gene products, and corticotroph cell hyperplasia (Wynn et al., 1985) observed following adrenalectomy in rats, and increased hypothalamic arginine vasopressin (AVP) production (McNicol and Carbajo-Perez, 1999), which may induce corticotroph proliferation (van Wijk et al., 1995) following adrenalectomy in mice. Furthermore, it has been demonstrated in rodents that chronic CRH infusion can result in corticotroph cell hyperplasia (Gertz et al., 1987). Finally, cortisol lowers the proliferation rate of human pituitary adenoma cells *in vitro* (Resetic et al., 1990).

Although the "released negative-feedback" hypothesis is attractive, there are some important inconsistencies from both clinical and laboratory observations. Not all patients with a history of Cushing's disease develop Nelson syndrome during long-term follow-up following TBA. Furthermore, Nelson syndrome develops mostly in those post-TBA patients who receive adequate steroid replacement therapy exogenously (Barber et al., 2010). A further inconsistent observation is that corticotrophs from patients with Nelson syndrome demonstrate attenuated negative-feedback responses to glucocorticoids *in vivo* (Wolfsen and Odell, 1980). Furthermore, secretion of POMC-derived peptides from Cushing's disease patients does not appear to be attenuated by glucocorticoids (Cook et al., 1976).

An alternative explanation to the released negative-feedback hypothesis for the pathophysiology of Nelson syndome is that this condition tends to develop in the subgroup of patients with Cushing's disease with aggressive forms of corticotropinomas (Assie et al., 2007). Such patients also have a higher chance of relapse following TSS (7–12% of patients with Cushing's disease treated with TSS have a relapse within one year following the procedure (Lanzi et al., 1998)) and subsequent need for TBA. Consistent with this hypothesis is the observation that compared with patients with less aggressive forms of pituitary tumor, patients with invasive corticotropinomas at the time of TSS are also at greater risk of subsequently developing Nelson syndrome at an earlier stage following TBA (Nagesser et al., 2000; Gil-Cardenas et al., 2007). The *de novo* development of a new corticotropinoma that is clonally distinct from the original tumor is also possible in some patients, although there is little evidence to support this alternative theory.

A more complete understanding of the pathophysiology of Nelson syndrome (including molecular and histopathologic features of Nelson tumors) would facilitate future therapeutic developments and would also assist in the development of novel predictive factors. In one small study on 12 patients with Cushing's disease, it was shown that mitotic figures or Ki-67-expressing nuclei in corticotroph adenomas at baseline was not predictive of adenoma progression following TBA (Assie et al., 2007). Further cell-based studies in larger numbers of subjects are required to define the molecular characteristics of those corticotropinomas that subsequently develop into Nelson tumors. These data could then be used for predictive purposes in clinical practice, based on histopathologic and immunocytochemical features (including percentage of mitoses and Ki-67-immunopositive nuclei) of tumor derived at TSS.

PATHOLOGIC FEATURES OF CORTICOTROPINOMAS IN NELSON SYNDROME

There are many similarities (both histologic and molecular) between the corticotropinomas that develop in Cushing's disease and those that develop in Nelson syndrome (Bertagna, 1992; Assie et al., 2004). The tumors in both conditions manifest monoclonality, originating from the same corticotroph cells (Herman et al., 1990). Furthermore, cells from these tumors maintain some normal functional activity including POMC processing (Raffin-Sanson et al., 2003), expression of functional CRH and vasopressin V3 receptors (de Keyzer et al., 1997), and usually expression of the two isoforms of glucocorticoid receptor (Dahia et al., 1997). Morphologically, Nelson tumor cells can appear distinct from corticotropinomas in Cushing's disease, including unique ultrastructural features (inconspicuous type 1 cytokeratin filaments (Herman et al., 1990)) and larger cells with significant pleomorphism (Machado et al., 2005; Nagesser et al., 2000).

Despite the similarities outlined above, and as alluded to earlier, compared to those that develop in Cushing's disease, corticotropinomas that develop in Nelson syndrome tend to be macroadenomas, manifest higher proliferation (1.1% versus 0.5%) and lower p27 labeling indices (13% versus 28%) and are more invasive (Scheithauer et al., 2006). The aggressiveness of Nelson tumors is further supported by clinical observations. In one study, four out of seven corticotroph carcinomas developed in the context of Nelson syndrome (average interval 15.3 years (Pernicone et al., 1997)). In a further study involving 33 ACTH-producing pituitary carcinomas, 15 (45%) developed in the context of Nelson syndrome (Landman et al., 2002). Although the actual prevalence of pituitary carcinoma developing in the context of Nelson syndrome is not clear (with case series often not reporting on such an outcome), even in the context of Nelson syndrome the development of pituitary carcinoma is rare (Landman et al., 2002).

EFFECTIVE MANAGEMENT OF NELSON SYNDROME

Pituitary surgery

Wherever possible, pituitary surgery should be the first-line treatment option for Nelson syndrome, with success rates that range between 10% and 70% (Wolfsen and

Odell, 1980; Kemink et al., 2001; Kelly et al., 2002; De Tommasi et al., 2005). Most Nelson tumors are amenable to a transsphenoidal procedure, although in 33% of cases (particularly those with extrasellar extension) a transcranial approach is required (De Tommasi et al., 2005). Perhaps not surprisingly, long-term remission and complication rates are related to tumor size, with those confined to the sellar region having the most successful outcome (Kemink et al., 2001). Although mortality following the surgical management of Nelson syndrome is low at 5% (Kelly et al., 2002; Xing et al., 2002), morbidity rates are high, with up to 69% of patients developing postoperative panhypopituitarism (Banasiak and Malek, 2007), 15% developing a CSF leak, 8% developing meningitis, and 5% acquiring a cranial nerve palsy (Kelly et al., 2002; Xing et al., 2002).

Adjuvant radiotherapy

Despite surgical intervention, Nelson tumors may still subsequently progress in some patients (Kelly et al., 2002), with adjuvant radiotherapy being required in between 20% and 30% of such patients (Ludecke et al., 1982; Wislawski et al., 1985; Kemink et al., 2001; Kelly et al., 2002). As alluded to earlier in this chapter, a pragmatic approach would be to administer adjuvant radiotherapy as a preventive strategy at the time of TBA to those patients with Nelson syndrome who also have remnant corticotroph tumor tissue. Administration of this form of therapy should be balanced with its potential complications.

Stereotactic radiosurgery

The role of Gamma Knife surgery (GKS) in Nelson syndrome is unclear (Levy et al., 1991; Ganz et al., 1993; Wolffenbuttel et al., 1998; Mauermann et al., 2007), although this therapy appears to be most effective when administered soon after TBA (Vik-Mo et al., 2009). It is also known that GKS is most effective when the anatomic target is clear and discrete. Therefore, GKS may be less effective following surgical management when the tumor border may become indistinct (Ganz et al., 1993; Mauermann et al., 2007). Consideration should also be given to the adverse effects of GKS that include panhypopituitarism (Degerblad et al., 1986) and cranial nerve palsies, which precludes this form of therapy for tumors that are adjacent to the optic apparatus and those that invade the cavernous sinus. The current literature on the effectiveness of GKS in Nelson syndrome is limited and conflicting, with one study showing no tumor regrowth at 7 years post-GKS therapy (Vik-Mo et al., 2009) and another showing remission rates to be only 14% (Jane et al., 2003). Clearly additional studies are required to explore further the role of GKS in the management of Nelson syndrome.

Selective somatostatin analogs

It is hypothesized that somatostatin analogs (SSAs) act through reducing plasma ACTH levels (Kasperlik-Zaluska et al., 2005) and tumor volume (Petrini et al., 1994), although human evidence to support the use of SSAs in patients with Nelson syndrome is lacking. There is some evidence to suggest that pasireotide (SOM 230, having affinity to somatostatin receptors 1, 2, 3 and 5) inhibits ACTH secretion *in vitro* from human corticotroph cells from patients with Cushing's disease (Hofland et al., 2005), inhibits ACTH release *in vivo* (Schmid, 2008) and reduces the proliferative rate of human corticotroph cells (Batista et al., 2006). Pasireotide has also been shown to provide long-term efficacy in patients with Cushing's disease (Colao et al., 2012). Again, however, further studies are required to explore the possible future role of SSAs in the effective management of patients with Nelson syndrome.

Peroxisome proliferator-activated receptor γ agonists

The withdrawal of rosiglitazone has limited somewhat the prospects of peroxisome proliferator-activated receptor (PPAR) γ agonist drugs as future therapies for Nelson syndrome, particularly given that much of the (limited) current evidence to support the use of PPAR γ agonists in this context relates to rosiglitazone. Furthermore, PPAR γ agonists remain unlicensed for use in Nelson syndrome. However, there is some evidence to suggest that this class of drug may be useful in Nelson syndrome. It is known that expression of PPAR γ receptors occurs in normal corticotrophs, and especially in corticotroph adenoma cells (Heaney et al., 2002). Furthermore, transcription of POMC mRNA from murine corticotrophs cultured *in vitro* is reduced fourfold when exposed to rosiglitazone (Heaney et al., 2002). This drug has also been demonstrated to cause cell cycle arrest, apoptosis, and reduced ACTH secretion from corticotroph cells in a mouse model of Cushing's disease (Heaney et al., 2002). However, human data have been disappointing, with one reported study on the use of rosiglitazone in patients with Nelson syndrome ($n=7$) demonstrating no lowering of ACTH levels (Mullan et al., 2006). In support of this, two more recent studies on the use of rosiglitazone in Nelson syndrome also reported on its ineffectiveness as a treatment in this condition (Kreutzer et al., 2009; Munir et al., 2007).

Sodium valproate

It is possible that use of sodium valproate, which inhibits the hypothalamic reuptake of GABA, thereby reducing CRH release, may be effective in patients with Nelson syndrome (Batista et al., 2006). This hypothesis has not been supported by data from patients with Nelson syndrome (Kasperlik-Zaluska et al., 2005), with no beneficial effects on ACTH levels (Kasperlik-Zaluska et al., 2005; Batista et al., 2006), or corticotropinoma growth (Batista et al., 2006).

Dopamine agonists

It is known that dopamine receptors are expressed ubiquitously within pituitary adenomas. Support for the efficacy of dopamine agonists in Nelson syndrome include the observation that ACTH levels are reduced by administration of bromocriptine (Batista et al., 2006) and even that remission of Nelson syndrome (Pivonello et al., 1999; Shraga-Slutzky et al., 2006) and tumor resolution (Casulari et al., 2004) can occur in response to the use of cabergoline therapy. Future directions include the identification of dopamine receptor subclass in Nelson tumors to enable novel therapeutic developments (Barber et al., 2010).

Temozolomide

Although more data are required, the existing literature reveals that the alkylating agent temozolomide represents a promising therapy for patients with Nelson syndrome, especially those with aggressive tumors. One case report demonstrated an excellent response to temozolomide therapy with a significant reduction in ACTH level and regression of the underlying tumor following just four cycles of treatment in a patient with an aggressive Nelson tumor that had previously failed to respond to surgery, radiotherapy, and Gamma Knife treatment modalities (Moyes et al., 2009). It has been suggested that in patients with aggressive pituitary tumors, immunoexpression of the DNA repair protein 0(6)-methylguanine DNA methyl transferase (MGMT) is predictive of the response to temozolomide therapy (McCormack et al., 2009), low expression of MGMT being correlated with tumor responsiveness to temozolomide in patients with regrowth of nonfunctioning pituitary adenomas (Widhalm et al., 2009). The efficacy and placement of temozolomide therapy in Nelson syndrome should be a focus for future research.

CONCLUSIONS

To the nonspecialist, Nelson syndrome is encountered infrequently in the clinical arena. However, its associated morbidity and mortality, and the importance of early diagnosis and instigation of effective therapy, promotes Nelson syndrome as a condition of high importance and worthy of cognizance amongst all clinicians. The diagnosis and management of Nelson syndrome, due at least in part to the frequent aggressiveness of the underlying corticotropinoma, can be a challenge. It is imperative that all patients at risk of developing this condition (patients with a history of Cushing's disease who have been treated with TBA) are followed up long-term and undergo close screening with ACTH levels and MRI scans of the pituitary at regular intervals. Unfortunately, the current literature in the field of Nelson syndrome is littered with controversy due partly to the rarity of the condition, but also to factors such as lack of consensus regarding diagnostic criteria, for example. It is our hope that widespread acceptance and adoption of the diagnostic criteria for Nelson syndrome set out in this chapter will enable the generation of a robust and reliable evidence base on which to generate appropriate guidance regarding management of this condition. Other future directions include further exploration of the molecular aspects of Nelson tumors, including quantification of CRH-receptor and glucocorticoid-receptor densities, and associations of these molecular features with clinicopathologic outcomes. Further research on the predictive factors for and pathophysiology of Nelson syndrome will enable a more focused approach towards screening and development of future novel therapeutic agents, ultimately benefiting the patients who develop this important condition.

ACKNOWLEDGMENTS

We acknowledge all the patients, relatives, nurses, and physicians who contributed to the ascertainment of the various clinical samples reported on in this chapter.

References

Assie G, Bahurel H, Bertherat J et al. (2004). The Nelson's syndrome revisited. Pituitary 7: 209–215.

Assie G, Bahurel H, Coste J et al. (2007). Corticotroph tumor progression after adrenalectomy in Cushing's Disease: A reappraisal of Nelson's Syndrome. J Clin Endocrinol Metab 92: 172–179.

Banasiak MJ, Malek AR (2007). Nelson syndrome: comprehensive review of pathophysiology diagnosis and management. Neurosurg Focus 23: E13.

Baranetsky NGRD, Zipser U, Goebelsmann (1979). Adrenocorticotropin–dependent virilizing paraovarian tumors in Nelson's syndrome. J Clin Endocrinol Metab 49: 381–386.

Barber TM, Adams E, Ansorge O et al. (2010). Nelson's syndrome. Eur J Endocrinol 163: 495–507.

Barnett AH, Livesey JH, Friday K et al. (1983). Comparison of preoperative and postoperative ACTH concentrations after bilateral adrenalectomy in Cushing's disease. Clin Endocrinol (Oxf) 18: 301–305.

Batista DL, Zhang X, Gejman R et al. (2006). The effects of SOM230 on cell proliferation and adrenocorticotropin secretion in human corticotroph pituitary adenomas. J Clin Endocrinol Metab 91: 4482–4488.

Bertagna X (1992). Unrestrained production of proopiomelanocortin (POMC) and its peptide fragments by pituitary corticotroph adenomas in Cushing's disease. J Steroid Biochem Mol Biol 43: 379–384.

Boscaro M, Ludlam WH, Atkinson B et al. (2009). Treatment of pituitary–dependent Cushing's disease with the multireceptor ligand somatostatin analog pasireotide (SOM230): a multicenter phase II trial. J Clin Endocrinol Metab 94: 115–122.

Bradley KJ, Wass JA, Turner HE (2003). Non–functioning pituitary adenomas with positive immunoreactivity for ACTH behave more aggressively than ACTH immunonegative tumours but do not recur more frequently. Clin Endocrinol (Oxf) 58: 59–64.

Carey RMSK, Varma CR, Jr Drake et al. (1984). Ectopic secretion of corticotropin–releasing factor as a cause of Cushing's syndrome A clinical morphologic and biochemical study. N Engl J Med 311: 13–20.

Casulari LA, Naves LA, Mello PA et al. (2004). Nelson's syndrome: complete remission with cabergoline but not with bromocriptine or cyproheptadine treatment. Horm Res 62: 300–305.

Colao A, Petersenn S, Newell–Price J et al. and Pasireotide B2305 Study Group (2012). A 12–month phase 3 study of pasireotide in Cushing's disease. N Engl J Med 366: 914–924.

Cook DM, Kendall JW, Allen JP et al. (1976). Nyctohemeral variation and suppressibility of plasma ACTH in various stages of Cushing's disease. Clin Endocrinol (Oxf) 5: 303–312.

Dahia PL, Honegger J, Reincke M et al. (1997). Expression of glucocorticoid receptor gene isoforms in corticotropin–secreting tumors. J Clin Endocrinol Metab 82: 1088–1093.

de Keyzer Y, Rene P, Lenne F et al. (1997). V3 vasopressin receptor and corticotropic phenotype in pituitary and non-pituitary tumors. Horm Res 47 (4–6): 259–262.

De Tommasi C, Vance ML, Okonkwo DO et al. (2005). Surgical management of adrenocorticotropic hormone-secreting macroadenomas: outcome and challenges in patients with Cushing's disease or Nelson's syndrome. J Neurosurg 103: 825–830.

Degerblad M, Rahn T, Bergstrand G et al. (1986). Long–term results of stereotactic radiosurgery to the pituitary gland in Cushing's disease. Acta Endocrinol (Copenh) 112: 310–314.

Ernest I, Ekman H (1972). Adrenalectomy in Cushing's disease A long-term follow-up. Acta Endocrinol Suppl (Copenh) 160: 3–41.

Ganz JC, Backlund EO, Thorsen FA (1993). The effects of Gamma Knife surgery of pituitary adenomas on tumor growth and endocrinopathies. Stereotact Funct Neurosurg 61 (Suppl 1): 30–37.

Garcia C, Bordier L, Garcia–Hejl C et al. (2007). Nelson's syndrome management: current knowledge. Rev Med Interne 28: 766–769.

Gertz BJ, Contreras LN, McComb DJ et al. (1987). Chronic administration of corticotropin-releasing factor increases pituitary corticotroph number. Endocrinology 120: 381–388.

Gil-Cardenas A, Herrera MF, Diaz-Polanco A et al. (2007). Nelson's syndrome after bilateral adrenalectomy for Cushing's disease. Surgery 141: 147–151, discussion 151–2.

Hawn MT, Cook D, Deveney C et al. (2002). Quality of life after laparoscopic bilateral adrenalectomy for Cushing's disease. Surgery 132: 1064–1068, discussion 1068–9.

Heaney AP, Fernando M, Yong WH et al. (2002). Functional PPAR–gamma receptor is a novel therapeutic target for ACTH–secreting pituitary adenomas. Nat Med 8: 1281–1287.

Herman V, Fagin J, Gonsky R et al. (1990). Clonal origin of pituitary adenomas. J Clin Endocrinol Metab 71: 1427–1433.

Hofland LJ, Hoek J van der, Feelders R et al. (2005). The multi–ligand somatostatin analogue SOM230 inhibits ACTH secretion by cultured human corticotroph adenomas via somatostatin receptor type 5 Eur. J Endocrinol 152: 645–654.

Hopwood NJ, Kenny FM (1977). Incidence of Nelson's syndrome after adrenalectomy for Cushing's disease in children: results of a nationwide survey. Am J Dis Child 131: 1353–1356.

Imai T, Funahashi H, Tanaka Y et al. (1996). Adrenalectomy for treatment of Cushing syndrome: results in 122 patients and long–term follow–up studies. World J Surg 20: 781–786, discussion 786–7.

Jane Jr JA, Vance ML, Woodburn CJ et al. (2003). Stereotactic radiosurgery for hypersecreting pituitary tumors: part of a multimodality approach. Neurosurg Focus 14: e12.

Jenkins PJ, Trainer PJ, Plowman PN et al. (1995). The long–term outcome after adrenalectomy and prophylactic pituitary radiotherapy in adrenocorticotropin–dependent Cushing's syndrome. J Clin Endocrinol Metab 80: 165–171.

Johnson RE, Scheithauer B (1982). Massive hyperplasia of testicular adrenal rests in a patient with Nelson's syndrome. Am J Clin Pathol 77: 501–507.

Kasperlik-Zaluska AA, Bonicki W, Jeske W et al. (2006). Nelson's syndrome – 46 years later: clinical experience with 37 patients. Zentralbl Neurochir 67: 14–20.

Kasperlik-Zaluska A, Nielubowicz AJ, Wislawski J et al. (1983). Nelson's syndrome: incidence and prognosis. Clin Endocrinol (Oxf) 19: 693–698.

Kasperlik-Zaluska A, Zgliczynski AW, Jeske W et al. (2005). ACTH responses to somatostatin valproic acid and dexamethasone in Nelson's syndrome. Neuro Endocrinol Lett 26: 709–712.

Kelly DF (2007). Transsphenoidal surgery for Cushing's disease: a review of success rates remission predictors management of failed surgery and Nelson's Syndrome. Neurosurg Focus 23: E5.

Kelly PA, Samandouras G, Grossman AB et al. (2002). Neurosurgical treatment of Nelson's syndrome. J Clin Endocrinol Metab 87: 5465–5469.

Kelly WF, MacFarlane IA, Longson D et al. (1983). Cushing's disease treated by total adrenalectomy: long–term observations of 43 patients. Q J Med 52 (206): 224–231.

Kemink L, Pieters G, Hermus A et al. (1994). Patient's age is a simple predictive factor for the development of Nelson's syndrome after total adrenalectomy for Cushing's disease. J Clin Endocrinol Metab 79: 887–889.

Kemink SA, Grotenhuis JA, De Vries J et al. (2001). Management of Nelson's syndrome: observations in fifteen patients. Clin Endocrinol (Oxf) 54: 45–52.

Kimura N, Ishikawa T, Sasaki Y et al. (1996). Expression of prohormone convertase PC2 in adrenocorticotropin–producing thymic carcinoid with elevated plasma corticotropin-releasing hormone. J Clin Endocrinol Metab 81: 390–395.

Kreutzer J, Jeske I, Hofmann B et al. (2009). No effect of the PPAR-gamma agonist rosiglitazone on ACTH or cortisol secretion in Nelson's syndrome and Cushing's disease in vitro and in vivo. Clin Neuropathol 28: 430–439.

Landman RE, Horwith M, Peterson RE et al. (2002). Long-term survival with ACTH-secreting carcinoma of the pituitary: a case report and review of the literature. J Clin Endocrinol Metab 87: 3084–3089.

Lanzi R, Montorsi F, Losa M et al. (1998). Laparoscopic bilateral adrenalectomy for persistent Cushing's disease after transsphenoidal surgery. Surgery 123: 144–150.

Leinung MC, Zimmerman D (1994). Cushing's disease in children. Endocrinol Metab Clin North Am 23: 629–639.

Levy RP, Fabrikant JI, Frankel KA et al. (1991). Heavy–charged–particle radiosurgery of the pituitary gland: clinical results of 840 patients. Stereotact Funct Neurosurg 57 (1–2): 22–35.

Ludecke DK, Breustedt HJ, Bramswig J et al. (1982). Evaluation of surgically treated Nelson's syndrome. Acta Neurochir (Wien) 65 (1–2): 3–13.

Machado AL, Nomikos P, Kiesewetter F et al. (2005). DNA–flow cytometry of 207 pituitary adenomas: ploidy proliferation and prognosis. J Endocrinol Invest 28: 795–801.

Manolas KJ, Farmer HM, Wilson HK et al. (1984). The pituitary before and after adrenalectomy for Cushing's syndrome. World J Surg 8: 374–387.

Mauermann WJ, Sheehan JP, Chernavvsky DR et al. (2007). Gamma Knife surgery for adrenocorticotropic hormone–producing pituitary adenomas after bilateral adrenalectomy. J Neurosurg 106: 988–993.

McCance DR, Russell CF, Kennedy TL et al. (1993). Bilateral adrenalectomy: low mortality and morbidity in Cushing's disease. Clin Endocrinol (Oxf) 39: 315–321.

McCormack AIKL, McDonald AJ, Gill S et al. (2009). Low O6–methylguanine–DNA methyltransferase (MGMT) expression and response to temozolomide in aggressive pituitary tumours. Clin Endocrinol (Oxf) 71: 226–233.

McNicol AM, Carbajo–Perez E (1999). Aspects of anterior pituitary growth with special reference to corticotrophs. Pituitary 1 (3–4): 257–268.

Moore TJ, Dluhy RG, Williams GH et al. (1976). Nelson's syndrome: frequency prognosis and effect of prior pituitary irradiation. Ann Intern Med 85: 731–734.

Moreira AC, Castro M, Machado HR (1993). Longitudinal evaluation of adrenocorticotrophin and beta–lipotrophin plasma levels following bilateral adrenalectomy in patients with Cushing's disease. Clin Endocrinol (Oxf) 39: 91–96.

Moyes VJ, Alusi G, Sabin HI et al. (2009). Treatment of Nelson's syndrome with temozolomide. Eur J Endocrinol 160: 115–119.

Mullan KR, Leslie H, McCance DR et al. (2006). The PPAR–gamma activator rosiglitazone fails to lower plasma ACTH levels in patients with Nelson's syndrome. Clin Endocrinol (Oxf) 64: 519–522.

Munir A, Newell–Price J (2007). Nelson's Syndrome. Arq Bras Endocrinol Metabol 51: 1392–1396.

Munir A, Song F, Ince P et al. (2007). Ineffectiveness of rosiglitazone therapy in Nelson's syndrome. J Clin Endocrinol Metab 92: 1758–1763.

Nagesser SK, van Seters AP, Kievit J et al. (2000). Long–term results of total adrenalectomy for Cushing's disease. World J Surg 24: 108–113.

Nelson DHJW, Meakin JB, Jr Dealy et al. (1958). ACTH–producing tumor of the pituitary gland. N Engl J Med 259: 161–164.

Newell-Price J, Bertagna X, Grossman AB et al. (2006). Cushing's syndrome. Lancet 367 (9522): 1605–1617.

Pereira MA, Halpern A, Salgado LR et al. (1998). A study of patients with Nelson's syndrome. Clin Endocrinol (Oxf) 49: 533–539.

Pernicone PJ, Scheithauer BW, Sebo TJ et al. (1997). Pituitary carcinoma: a clinicopathologic study of 15 cases. Cancer 79: 804–812.

Petrini L, Gasperi M, Pilosu R et al. (1994). Long-term treatment of Nelson's syndrome by octreotide: a case report. J Endocrinol Invest 17: 135–139.

Pivonello R, Faggiano A, Di Salle F et al. (1999). Complete remission of Nelson's syndrome after 1-year treatment with cabergoline. J Endocrinol Invest 22: 860–865.

Pollock BE, Young Jr WF (2002). Stereotactic radiosurgery for patients with ACTH-producing pituitary adenomas after prior adrenalectomy. Int J Radiat Oncol Biol Phys 54: 839–841.

Porterfield JRGB, Thompson WF, Jr Young et al. (2008). Surgery for Cushing's syndrome: an historical review and recent ten–year experience. World J Surg 32: 659–677.

Raffin-Sanson ML, de Keyzer Y, Bertagna X (2003). Proopiomelanocortin a polypeptide precursor with multiple functions: from physiology to pathological conditions. Eur J Endocrinol 149: 79–90.

Rees JR, Zilva JF (1959). Diabetes insipidus complicating total adrenalectomy. J Clin Pathol 12: 530–534.

Resetic J, Reiner Z, Ludecke D et al. (1990). The effects of cortisol 11-epicortisol and lysine vasopressin on DNA and RNA synthesis in isolated human adrenocorticotropic hormone-secreting pituitary tumor cells. Steroids 55: 98–100.

Scheithauer BW, Gaffey TA, Lloyd RV et al. (2006). Pathobiology of pituitary adenomas and carcinomas. Neurosurgery 59: 341–353, discussion 341–53.

Schmid HA (2008). Pasireotide (SOM230): development mechanism of action and potential applications. Mol Cell Endocrinol 286: 69–74.

Shekarriz M, Schneider C, Sabanegh E et al. (1996). Excessive testosterone production in a patient with Nelson syndrome and bilateral testicular tumors. Urol Int 56: 200–203.

Shraga-Slutzky I, Shimon I, Weinshtein R (2006). Clinical and biochemical stabilization of Nelson's syndrome with long–term low–dose cabergoline treatment. Pituitary 9: 151–154.

Sonino N, Zielezny M, Fava GA et al. (1996). Risk factors and long–term outcome in pituitary-dependent Cushing's disease. J Clin Endocrinol Metab 81: 2647–2652.

Sprague RG (1953). Cushing's syndrome with special reference to bilateral adrenalectomy. Proc R Soc Med 46: 1070–1077.

Sugiyama K, Kimura M, Abe T et al. (1996). Hyper–adrenocorticotropinemia in a patient with Addison's disease after treatment with corticosteroids. Intern Med 35: 555–559.

Thomas Jr CG, Smith AT, Benson M et al. (1984). Nelson's syndrome after Cushing's disease in childhood: a continuing problem. Surgery 96: 1067–1077.

van Aken MO, Pereira AM, Berg G van den et al. (2004). Profound amplification of secretory–burst mass and anomalous regularity of ACTH secretory process in patients with Nelson's syndrome compared with Cushing's disease. Clin Endocrinol (Oxf) 60: 765–772.

van Wijk PA, van Neck JW, Rijnberk A et al. (1995). Proliferation of the murine corticotropic tumour cell line AtT20 is affected by hypophysiotrophic hormones growth factors and glucocorticoids. Mol Cell Endocrinol 111: 13–19.

Vik-Mo EO, Oksnes M, Pedersen PH et al. (2009). Gamma knife stereotactic radiosurgery of Nelson syndrome. Eur J Endocrinol 160: 143–148.

Widhalm G, Wolfsberger S, Preusser M et al. (2009). O(6)–methylguanine DNA methyltransferase immunoexpression in nonfunctioning pituitary adenomas: are progressive tumors potential candidates for temozolomide treatment? Cancer 115: 1070–1080.

Wislawski JA, Kasperlik-Zaluska A, Jeske W et al. (1985). Results of neurosurgical treatment by a transsphenoidal approach in 10 patients with Nelson's syndrome. J Neurosurg 62: 68–71.

Wolffenbuttel BH, Kitz K, Beuls EM (1998). Beneficial gamma-knife radiosurgery in a patient with Nelson's syndrome. Clin Neurol Neurosurg 100: 60–63.

Wolfsen AR, Odell WD (1980). The dose–response relationship of ACTH and cortisol in Cushing's disease. Clin Endocrinol (Oxf) 12: 557–568.

Wright-Pascoe R, Charles CF, Richards R et al. (2001). A clinico-pathological study of Cushing's syndrome at the University Hospital of the West Indies and a review of the literature. West Indian Med J 50: 55–61.

Wynn PC, Harwood JP, Catt KJ et al. (1985). Regulation of corticotropin-releasing factor (CRF) receptors in the rat pituitary gland: effects of adrenalectomy on CRF receptors and corticotroph responses. Endocrinology 116: 1653–1659.

Xing B, Ren Z, Su C et al. (2002). Microsurgical treatment of Nelson's syndrome. Chin Med J (Engl) 115: 1150–1152.

Handbook of Clinical Neurology, Vol. 124 (3rd series)
Clinical Neuroendocrinology
E. Fliers, M. Korbonits, and J.A. Romijn, Editors

Chapter 23

Familial pituitary tumors

NEDA ALBAND AND MÁRTA KORBONITS*

Department of Endocrinology, Barts and the London School of Medicine, Queen Mary University of London, London, UK

INTRODUCTION

Pituitary tumors are common benign monoclonal neoplasms accounting for approximately 10–25% of all primary intracranial tumors (Newey et al., 2013). A meta-analysis by Ezzat et al. in 2004 showed that pituitary adenomas occur with a frequency of 14.4% and 22.5% in autopsy and radiologic studies respectively (Ezzat et al., 2004). However, only a small fraction of these adenomas cause clinically significant disease. Two recent cross-sectional studies showed that clinically relevant pituitary tumors occur in about 1:1000 in the population (Daly et al., 2006a; Fernandez et al., 2010). The benign nature of these tumors means that most cancer registration systems do not record pituitary adenomas, resulting in limited availability of data on these tumors from various countries (Hemminki et al., 2007). Although histologically benign, pituitary tumors can cause significant morbidity due to mass effect, and/or inappropriate pituitary hormone secretion (Melmed, 2011).

Approximately two-thirds of pituitary adenomas produce an excess of endogenous hormones, while a third are clinically nonfunctioning and present with local compressive symptoms (Evans et al., 2001). These include headache, visual disturbances, and/or altered hormone expression due to pituitary stalk disruption with compromised hypothalamic hormone access, and pituitary failure due to compression of normal pituitary tissue (Ben-Shlomo and Melmed, 2008). The most common pituitary tumor types are prolactinomas (40–50%), nonfunctioning pituitary adenomas (NFPAs) (24–27%), which are usually of gonadotropic origin, growth hormone (GH)-secreting adenomas (16–21%), adrenocorticotropic hormone (ACTH)-secreting adenomas (4.7–16%), thyroid-stimulating hormone (TSH)-secreting adenomas (0.4%) and luteinizing hormone (LH)/follicle-stimulating hormone (FSH)-secreting adenomas (0.9%) (Yamada, 2001).

The vast majority of pituitary adenomas are sporadic tumors while a minority of pituitary adenomas can be seen in hereditary syndromes such as multiple endocrine neoplasia type 1 (MEN1), Carney complex (CNC), and familial isolated pituitary adenoma (FIPA), caused by mutations in tumor suppressor genes (TSG) such as MEN1, PRKAR1a, and AIP respectively. Data from previous studies suggest that pituitary adenomas that occur in a familial setting account for about 5% of all pituitary adenomas (Beckers and Daly, 2007; Leontiou et al., 2008; Tichomirowa et al., 2009; Vandeva et al., 2010).

PITUITARY TUMORIGENESIS

Pituitary tumors are widely considered to be monoclonal in origin (Herman et al., 1990). Several factors are thought to be involved in the pathogenesis of pituitary tumors; these include: genetic mutations, epigenetic dysregulation of cell cycle regulators, local growth factors, and possibly hypothalamic dysregulation (Hanahan and Weinberg 2000; Korbonits et al., 2004; Asa and Ezzat, 2005; Yu et al., 2006; Beckers and Daly, 2007; Fedele et al., 2010). Two key mechanisms may be involved in the tumorigenic process: oncogene activation and TSG inactivation. These can occur either independently or in combination with one another (Yu et al., 2006). Oncogenes cause tumorigenesis through gain of function. The most important oncogene involved in sporadic pituitary tumorigenesis is the "gsp" oncogene. Gsp is a mutated variant of the α-subunit of the Gs signaling protein, GNAS, which transmits, among others, the effects of the growth hormone-releasing hormone (GHRH) receptor on somatotroph cells (Farfel et al., 1999; Adams et al., 2000; Arafah and Nasrallah, 2001). The gsp mutation leads to the loss of the GTP-ase

*Correspondence to: Professor Márta Korbonits, Department of Endocrinology, Barts and the London School of Medicine, Queen Mary University of London, Charterhouse Square, London, EC1A 6BQ, UK. Tel: +44-20-7882 6238, Fax: +44-20-7882-6197, E-mail: m.korbonits@qmul.ac.uk

activity of Gsα and therefore keeps Gsα in a constitutively active state. Up to 40% of somatotropinomas have a somatic mutation of this paternally imprinted gene on the maternal allele (Hayward et al., 2001). It is not fully understood how these mutations results in pituitary tumorigenesis but it has been suggested that Ser133 phosphorylated cAMP response element-binding protein (CREB) enhances mitogenic signaling in GH-secreting cells (Bertherat et al., 1995; Landis et al., 1989; Vallar et al., 1987).

In contrast to oncogene activation, tumors resulting from TSG inactivation usually require both alleles to be lost, according to the Knudson's "two-hit" hypothesis (Knudson, 1971; Shimon and Melmed, 1997; Heaney and Melmed, 2000; Arafah and Nasrallah, 2001). The first hit may be an inherited germline mutation, a somatic mutation, or loss of one allele of a TSG. The second hit turns off the other allele of the TSG. The second hit most commonly is a partial chromosome deletion leading to loss of heterozygosity (LOH) of common polymorphisms around the TSG in the tumor tissue, or it could be methylation of the TSG promoter. A recently described second hit mechanism involves the upregulation of a microRNA which then turns off the remaining TSG allele. This mechanism has been described in parathyroid tumor samples of MEN1 patients, where the oncomir mir24 which targets the MEN1 gene is somatically upregulated and therefore downregulates the expression of menin protein from the nonmutated copy of the MEN1 (Luzi et al., 2012).

In sporadic pituitary adenomas LOH has been described in approximately 20% of cases on chromosome 9, 11q13, and 13. The oncogenes and tumor suppressor genes implicated in pituitary tumorigenesis are listed in Table 23.1.

PITUITARY ADENOMAS OF GENETIC ORIGIN

(Fig. 23.1)

In addition to the classic inherited syndromes of MEN1 and CNC, pituitary adenomas can occur due to postzygotic mosaic mutations in the *GNAS* gene (i.e. the same gene affected in somatic gsp mutations) in McCune–Albright syndrome. Familial pituitary adenomas are also seen in kindreds without abnormalities in other endocrine glands or other organs; this is referred to as familial isolated pituitary adenoma (FIPA) (Daly et al., 2006b). Additionally, preliminary description of a pituitary tumor, a pituitary blastoma associated with germline *DICER1* mutation, has been recently described (Wildi-Runge et al., 2011). Several cases of pituitary adenomas have also been identified in patients harboring SDH mutation (Brahma et al., 2009; Denes et al., 2012; Varsavsky et al., 2012; Xekouki et al., 2012).

MULTIPLE ENDOCRINE NEOPLASIA TYPE 1 (MEN1) OMIM #131100

At the beginning of the past century, autopsy findings of a patient with acromegaly and four enlarged parathyroid glands were reported (Falchetti et al., 2009; Syro et al., 2012). However, it was not until 1953 that Underdahl et al. first published a review of MEN1 syndrome describing 14 literature cases and eight cases from the Mayo Clinic (Underdahl et al., 1953; Syro et al., 2012). In 1954, Wermer suggested an autosomal dominant inheritance pattern for this syndrome, referred to as MEN1 syndrome. The MEN1 clinical phenotype was fully characterized in the 1960s (Wermer, 1954; Syro et al., 2012).

MEN1 is an autosomal dominant disease with high penetrance characterized by presence of several endocrine tumors, in particular, pituitary, parathyroid, and pancreatic islet cells (Thakker, 2001). In addition to these tumors, carcinoid tumors, adrenocortical tumors, facial angiofibromas, lipomatous tumors, and collagenomas have also been identified in patients with MEN1 (Thakker, 2010) (Fig. 23.2).

Clinical features of MEN1

The incidence of MEN1 is estimated to be around 0.25% from postmortem studies, approximately affecting 1 in 30 000 individuals. About 10% of the mutations are *de novo*, resulting in sporadic presentation; the remainder are in a familial setting (Lips et al., 1984; Thakker, 2010). MEN1 is estimated to have an incidence of 1–18% in patients with primary hyperparathyroidism (Brandi et al., 1987), 16–38% among patients with gastrinomas (Thakker, 2010); and less than 3% in patients with pituitary tumors (Corbetta et al., 1997).

MEN1 affects both sexes equally but the pituitary manifestations have a female preponderance. The reported age range is 5–81 years (Thakker, 2010). The clinical manifestations of the disorder are present in over 95% of patients by the fifth decade (Thakker, 2010).

Parathyroid tumors

Primary hyperparathyroidism is the most common clinical manifestation of MEN1 in approximately 95% of MEN1 patients (Trump et al., 1996; Brandi et al., 2001; Thakker, 2010). Patients with primary hyperparathyroidism can present with asymptomatic hypercalcemia or with symptoms associated with hypercalcemia including nephrolithiasis, osteitis fibrosa cystica, polyuria, polydipsia, abdominal pain, malaise, constipation, and occasionally peptic ulcers. MEN1 patients with primary hyperparathyroidism on average have earlier age of onset (20–25 years) compared to non-MEN1 patients with primary hyperparathyroidism (55 years) (Thakker, 2010).

Table 23.1

Oncogenes and tumor-suppressor genes involved in pituitary tumorigenesis and their chromosomal location

Oncogenes	Defect
Cyclin D1 (*CCND1*) Chromosome 11q13	Important in regulation of cell progression through the G1 phase of the cell cycle. Overexpression in nonfunctioning adenomas and somatotropinomas (Wang et al., 2001)
Gsp (*GNAS*) Chromosome 20q13	Biallelically expressed in tumors. Somatic activating mutations in up to 40% of somatotropinomas and mosaicism in McCune–Albright syndrome (Landis et al., 1989; Hayward et al., 2001)
RAS Chromosome 5p13	Somatic activating mutations in pituitary carcinomas (Pei et al., 1994)
PTTG-1 Chromosome 5q33	Increased expression in invasive pituitary tumors (Zhang et al., 1999)
HMGA2 Chromosome 12q14	HMGA2 overexpression is common in both early- and late-stage high-grade ovarian serous papillary carcinoma. Overexpression in mixed growth hormone/prolactin secreting pituitary adenomas
Ptd-FGFR4 Chromosome 5q35	Alternative transcription initiation associated with more invasive somatotropinomas (Ezzat et al., 2002; Morita et al., 2008)
Tumor suppressor genes	
AIP Chromosome 11q13	Germline mutations in some FIPA families, particularly in families with somatotropinomas, somatomammotropinomas or a mixture of prolactinomas and somatotropinomas. Mutation in young-onset sporadic adenomas (Leontiou et al., 2008; Newey et al., 2013)
p27Kip1 (CDKN1B) Chromosome 12p13	Germline heterozygous nonsense mutation in MEN4 (a rare MEN1-like syndrome). Reduced protein expression in sporadic adenomas, especially ACTH-secreting ones, but no somatic mutations identified (Dahia et al., 1998; Lidhar et al., 1999; Georgitsi et al., 2007)
p16INK4A (CDKN2A) Chromosome 9p21	Hypermethylation of promoter region in pituitary adenoma development (Ruebel et al., 2001)
p18INK4C (CDKN2C) Chromosome 1p32	Hypermethylation of promoter region in pituitary adenoma development (Kirsch et al., 2009)
GADD45 gamma Chromosome 9p22	Growth suppressor controlling pituitary cell proliferation. Promoter methylation in nonfunctioning adenomas, prolactinomas, and somatotropinomas (Zhang et al., 2002)
PKA (PRKAR1A) Chromosome 17q24	Truncating mutations in Carney complex leading to somatomammotroph hyperplasia and adenomas (Veugelers et al., 2004)
WIF 1 Chromosome 12q14	Hypermethylation of promoter region in pituitary adenomas, especially in non functioning adenomas. Inhibitors of the Wnt pathway (WIF1, SFRP2, frizzled B = SFRP3 (FZDB), SFRP4) were all downregulated in pituitary tumors compared to normal pituitaries (Elston et al., 2008)
p53 (*TP53*) Chromosome 17p13	Somatic inactivating mutations (very rare) or overexpression in a subset of pituitary carcinomas (Thapar et al., 1996; Tanizaki et al., 2007)
ZAC1 (*PLAGL1/LOT1*) Chromosome 6q24-25	Hypermethylation of promoter region in pituitary adenomas, especially in nonfunctioning adenomas (Theodoropoulou et al., 2006)
Retinoblastoma (*RB1*) Chromosome 13q14	Hypermethylation of promoter region in pituitary adenomas and rare cases of pituitary carcinomas (Bates et al., 1997; Simpson et al., 2000)
MEN1 Chromosome 11q13	Inactivating germline mutations in all pituitary tumor types (Scheithauer et al., 1987)
MEG3a Chromosome 14q32	Loss of expression as a result of promoter region hypermethylation found in nonfunctioning adenomas and gonadotropinomas (Zhang et al., 2003; Zhao et al., 2005)
DAP1 Chromosome 5p15	Loss of DAP kinase expression in invasive adenomas
Wee1 Chromosome 11p15	Wee1 kinase is a nuclear protein that delays mitosis by Cdk1 phosphorylation. Reduced Wee1 protein expression in NFAs and GH-producing tumors (Butz et al., 2010)
PTAG Chromosome 22q12	Pituitary tumor apoptosis by CpG island methylation and loss of transcription (Bahar et al., 2004)

Diagnosis of primary hyperparathyroidism is by measuring serum calcium which is raised usually in association with a raised primary hyperparathyroid hormone (PTH) level. Imaging with neck ultrasound and Tc99m sestamibi parathyroid scintigraphy could identify multiglandular disease, but if MEN1 diagnosis is known preoperatively, neck exploration is indicated irrespective of imaging studies.

Fig. 23.1. Familial young-onset acromegaly in two French brothers. The Hugo brothers, who were 226 cm and 231 cm tall, lived at the beginning of the 20th century. They are shown in this picture with their siblings and their family tree. In the family tree, squares represent males, circles females, and filled symbols affected subjects.

Pancreatic tumors

Pancreatic islet cell tumors have a prevalence of 30–80% amongst patients with MEN1. These tumors include gastrinomas, insulinomas, glucagonomas, VIPomas and somatostatinomas producing usually high levels of hormones (Thakker, 2001). A small number of pancreatic islet cell tumors do not produce hormones and remain nonfunctional. MEN1 patients with pancreatic islet cell tumors on average have an earlier age of onset compared to non-MEN1 patients with similar lesions.

Gastrinomas are the most common pancreatic islet cell tumors in MEN1 patients, accounting for over half of all pancreatic islet cell tumors seen in these patients. The majority of MEN1 gastrinomas are malignant and will have disseminated by the time of presentation. Gastrinomas produce high levels of gastrin causing recurrent severe peptic ulceration (Zollinger-Ellison syndrome), diarrhea, abdominal pain, and steatorrhea. Gastrinomas are the major cause of morbidity and mortality in MEN1 patients. Diagnosis of gastrinoma is by demonstrating raised fasting serum gastrin concentration in association with increased gastric acid secretion. MRI or somatostatin receptor scintigraphy is subsequently used to localize the gastrinoma.

Insulinomas account for nearly 10–30% of all pancreatic islet cell tumors and can occur in association with gastrinomas in 10% of MEN1 patients. Clinical features of insulinoma are symptoms of hypoglycemia after a period of starvation or during exertion which resolves following glucose intake. Symptoms of hypoglycemia include hunger, headache, anxiety, sweating, tremor, convulsions, and loss of consciousness (Thakker, 2010). A raised plasma insulin concentration and symptoms of hypoglycemia during a supervised 72 hour fast confirm the diagnosis. Elevated concentrations of C-peptide and proinsulin are also diagnostic.

Glucagonomas are seen in less than 3% of MEN1 patients. The characteristic presentation of glucagonoma includes diarrhea, weight loss, anemia, necrolytic migratory erythema, and stomatitis. Glucose intolerance and hyperglucagonemia is commonly seen in patients with glucagonoma (Brandi et al., 2001).

VIP-secreting pancreatic tumors (VIPoma) occur rarely in MEN1 patients. The clinical syndrome commonly referred to as Verner–Morrison or VIPoma syndrome results in diarrhea, hypokalemia and achlorhydria. Diagnosis of VIPoma is by demonstrating a stool volume in excess of 0.5–1.01 L per day during a fast, in association with elevated plasma VIP and in the absence of laxative or diuretic abuse (Thakker, 2010).

Somatostatinomas and GHRHomas have been reported in some MEN1 patients.

Pituitary tumors

Pituitary tumors have an incidence of 15–90% amongst MEN1 patients in different series (Thakker, 2010; Syro et al., 2012). A pituitary adenoma is the first manifestation of MEN1 in ~15% of patients (Verges et al., 2002) (ranging from 10% to 25%). Pediatric patients are suggested to be screened for MEN1 as well as AIP mutations (Cuny et al., 2013). The majority of these tumors are functional with approximately 60% secreting prolactin, 25% growth hormone and somatomammotroph, and 5% adrenocorticotropic hormone. The remainder are nonfunctional (Horvath and Stratakis, 2008; Thakker, 2010). Approximately 65–85% of pituitary tumors in MEN1 syndrome are macroadenomas (Syro et al., 2012). Multihormonal and mixed tumors as well as

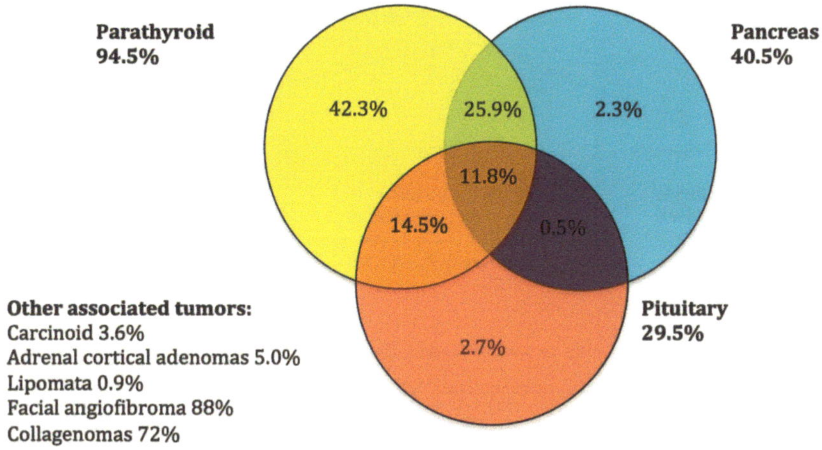

Fig. 23.2. MEN1 tumor distribution in 220 patients with MEN1. The diagram is a schematic representation of 384 MEN1 tumors found in 220 patients with MEN1. The proportion of patients with parathyroid, pituitary, pancreatic, and other associated tumors is clearly shown in each box. The overlap in the diagram shows the combination of tumors found in the 220 MEN1 patients. For example, 37.7% (25.9% + 11.8%) of patients had both a parathyroid and a pancreatic tumor, whereas 2.3% of patients had only a pancreatic tumor. Parathyroid tumor is by far the most common tumor type in MEN1 (95%) followed by pancreatic (40%) and pituitary tumors (30%). Other tumors including carcinoid tumors, adrenal cortical tumors, lipomata, angiofibromas, and collagenomas make up a small proportion of tumors in MEN1. (Reproduced from Trump et al.,1996, with permission from Oxford University Press.)

multiple adenomas also occur more frequently than monohormonal adenomas in MEN1 patients (Trouillas et al., 2008; Syro et al., 2012).

The clinical manifestation of these tumors in MEN1 patients occur at a younger age compared to sporadic adenomas (Thakker, 2010; Syro et al., 2012). Patients may present with typical symptoms of hyperprolactinemia (galactorrhea, amenorrhea, and infertility in women and decreased libido and impotence in men), acromegaly, or Cushing's disease, or symptoms due to mass effect from enlarging pituitary tumor compressing adjacent structures such as optic chiasm causing headache and bitemporal hemianopia as well as hypopituitarism. MEN1-related adenomas are larger and more invasive compared to sporadic counterparts and their response to therapy is also reduced (Thakker, 2010; Syro et al., 2012). There is a preponderance of female patients with MEN1-related pituitary adenomas compared to male patients.

Even though pituitary tumors are reported to be larger and more often invasive in patients with MEN1 syndrome than in sporadic tumors, malignant transformations were not more frequent. However, three recent cases of MEN1-associated pituitary carcinomas have been reported, which is intriguing considering the low frequency of pituitary adenomas in general, which represent about 0.1–0.2% of all pituitary tumors (Benito et al., 2005; Gordon et al., 2007; Scheithauer et al., 2009; Syro et al., 2012).

A summary of the diagnosis of MEN1 according to the latest guidelines is shown in Figure 23.3.

Genetics of MEN1 syndrome

The genetic background of MEN1 syndrome is heterogeneous. While the majority (~80%) of cases are due to a mutation in the *MEN1* gene, around 1.5% of patients have a mutation in the coding region or upstream open reading frame of the *CDKN1B* gene, coding for the cell cycle inhibitor p27 (Lemos and Thakker, 2008; Occhi et al., 2013). A few MEN1 cases have been described with mutations in genes coding for other cell cycle inhibitors: p15, p18, and p21. The remaining patients (5–25%) with MEN1-associated tumors are reported not to harbor a mutation. This could be due to either other, currently unknown disease-causing genes, phenocopies, or lack of full assessment of the known disease-causing genes, for example, lack of testing for large deletions.

The MEN1 susceptibility gene was initially linked to a locus on chromosome 11q13 in 1988 by Larsson et al. MEN1 gene was subsequently cloned in 1997 by two independent groups (Chandrasekharappa et al., 1997; Lemmens et al., 1997). The MEN1 gene has 10 exons of which exons 2–10 encode a 610 amino acid nuclear protein, menin, whose functions are still being elucidated. Menin appears to be located mostly in the nucleus, where it has multiple binding partners, including junD and members of histone methyltransferase complexes. Menin potentially interacts with promoter regions of many genes, indicating its wide transcriptional regulatory role (Lemos and Thakker, 2008; Tichomirowa et al., 2009). It has been long hypothesized that the tumor suppressive actions of menin is mediated by regulating

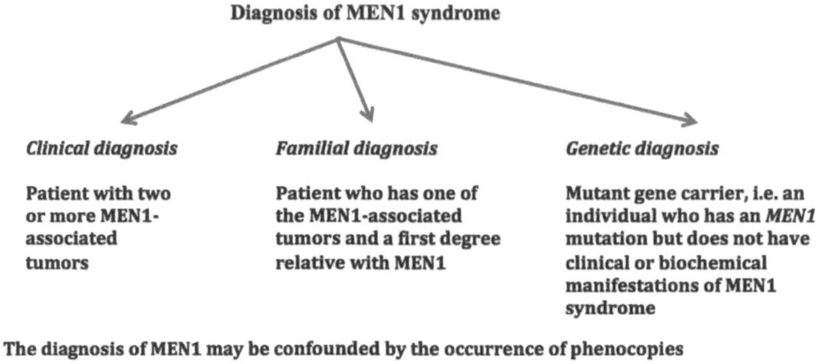

Fig. 23.3. Diagnosis of MEN1 according to the latest guidelines (Thakker, 2010).

histone methylation in promoters of *p27* and *HOX* genes and possibly other CKD inhibitors (Karnik et al., 2005). Expression of p27 and p18 as well as H3 k4 methylation were shown to be dependent on menin in pancreatic islet cells while knockout (KO) mice with loss of both p27 and p18 were noted to develop tumors in the pituitary, parathyroid, thyroid, pancreas, and stomach, similar to human MEN1 and MEN2 patients (Chen et al., 2006). MEN1 KO mice have severe developmental defects and die at embryonic age (Bertolino et al., 2003a). Mice with heterozygous *MEN1* deletion show a phenotype similar to humans. Prolactin or GH-positive pituitary tumors were noted in 19% of mice at 13–18 months and 36.6% at 19–26 months (Bertolino et al., 2003b).

A detailed review of MEN1 in 2008, by Lemos and Thakker, showed 1336 reported sequence abnormalities in the MEN1 gene (Lemos and Thakker, 2008). Seventy-five percent of MEN1 mutations are inactivating and are consistent with those expected in a tumor suppressor gene. The mutations are widely scattered throughout the 1830 base pair coding region of the MEN1 gene.

Somatic MEN1 mutations are commonly found in sporadic parathyroid (20%) and pancreatic neuroendocrine tumors (NET) (30%). These mutations are extremely rare in sporadic pituitary adenomas. LOH in 11q13 has been described in 30% of sporadic pituitary adenomas although MEN1 mRNA is not downregulated in these tumors (Satta et al., 1999; Thakker et al., 2012).

Management of pituitary disease in MEN1

Pituitary tumors in MEN1 patients appear to be larger and more aggressive/invasive than in patients without MEN1, with macroadenomas being present in 85% of MEN1 patients compared to 42% of sporadic cases. Prolactinomas are the most common pituitary tumor type in MEN1-associated pituitary adenomas (60%), followed by GH (25%), NFPA, and ACTH-secreting adenomas. Plurihormonal adenomas have also been observed in patients with MEN1-associated pituitary tumors (Thakker et al., 2012). In patients with MEN1 mutation and clinical signs of acromegaly, the possibility of GHRH-secreting NET should be considered and GHRH level should be assessed (Garby et al., 2012; Saleem et al., 2012).

The clinical management of MEN1-associated pituitary adenomas in general is not different to that of the sporadic cases. Dopamine agonists, such as cabergoline, are the treatment of choice in MEN1 patients with prolactin-secreting tumors, and somatostatin analogs for GH-secreting adenomas (Brandi et al., 2001). Medical therapy is less successful in MEN1-positive cases (Thakker et al., 2012). Selective adenomectomy, lack of prolactinoma shrinkage in response to dopamine agonist therapy, pressure on optic structures, or debulking could be indications for transsphenoidal surgery. Radiotherapy may be necessary if tumor growth or hormone release cannot be controlled.

Regular clinical assessment and biochemical follow-up for pituitary hormones combined with MRI scans (every 3 years) is indicated in MEN1 mutation carrier subjects (Thakker et al., 2012).

All MEN1 syndrome patients are likely to have a mutation in the *MEN1* gene. The MEN1 germline mutation test is recommended for MEN1 carrier identification in patients with the clinical diagnosis of MEN1 or in their unaffected "at risk" family members. MEN1 mutation testing can be considered in patients with pituitary adenoma and no other MEN1 manifestation in childhood or young-onset cases. In 15% of MEN1 cases pituitary adenoma can be the first manifestation, while 1% of unselected pituitary adenoma patients harbor a germline MEN1 mutation. Prenatal diagnosis for high-risk pregnancies can also be carried out if the disease-causing mutation in a family is known. In patients with MEN1 phenotype but negative *MEN1* test, screening for CDKN1B gene mutations could be considered, although not routinely recommended (Owens et al., 2009; Tichomirowa et al., 2009).

FAMILIAL ISOLATED PITUITARY ADENOMA (FIPA): RELATED OMIM ENTRIES: PITUITARY ADENOMA, GROWTH HORMONE-SECRETING #102200 AND AIP*605555

Although pituitary adenomas usually occur as a sporadic disease, a few families have been previously described with familial pituitary adenomas but no other associated physical abnormality. These families show an autosomal dominant inheritance with incomplete penetrance, and no clinical or genetic features of the MEN1 syndrome or CNC. This condition was classified in various publications as isolated familial somatotropinoma, familial isolated pituitary adenoma (FIPA), or pituitary adenoma predisposition, and since 2006 several hundred families have been described (Soares and Frohman, 2004; Vierimaa et al., 2006; Horvath and Stratakis, 2008; Newey et al., 2013). The genetic background of this disease is heterogeneous. In a large Finnish family with acromegaly and prolactinomas, linkage studies identified a mutation in a novel gene, AIP (aryl hydrocarbon receptor interacting protein) (Vierimaa et al., 2006). Current data suggest that this gene is responsible for about 20% of FIPA cases, while the disease-causing gene in the majority of the families with FIPA is currently unknown (Daly et al., 2010; Beckers et al., 2013).

Clinical features of FIPA

Approximately 20% of all FIPA families and 40% of somatotroph adenoma families harbor a mutation in the *AIP* gene, while the rest of the families probably have a mutation in a currently unknown gene (or genes). Patients with AIP mutations have a distinct phenotype. The two most characteristic features of AIP mutation-positive patients are the prevalence of somatotroph or somatolactotroph adenomas (~80% of cases) and the young onset of the disease (mean age of onset is around 20–24 years) (Chahal et al., 2010; Daly et al., 2010; Igreja et al., 2010). AIP mutation-positive patients usually present with larger tumors (88% macroadenomas) and have reduced response to treatment with somatostatin analogs (Daly et al., 2010; Chahal et al., 2012; Gadelha et al., 2013). There is a male preponderance in affected subjects (Daly et al., 2010). Penetrance is incomplete and can be variable with data from large families suggesting ~30% penetrance rate (Naves et al., 2007; Chahal et al., 2011). Apparently sporadic, young-onset pituitary macroadenoma patients have also been identified with germline AIP mutations. In apparently sporadic pituitary adenoma patients, 20% of childhood-onset and 11% of acromegaly patients less than 30 years old harbor an AIP mutation. As the prevalence of *de novo* mutations is very low, the lack of known family history is probably due to a combination of low penetrance, no information on other family members, and lack of diagnosis in family members in previous generations.

Pituitary tumors in FIPA families can be either homogeneous, displaying tumors of the same type, or heterogeneous, displaying different tumor types (Vierimaa et al., 2006; Beckers and Daly, 2007; Georgitsi et al., 2008; Leontiou et al., 2008; Newey et al., 2013). In AIP mutation-negative families the pituitary tumors are also predominantly macroadenomas (71%). The observed tumor types are also dominated by prolactin and growth hormone-secreting tumors, but NFPA and rarely corticotroph adenomas are also described (Igreja et al., 2010; Newey et al., 2013). Male/female ratio is equal and age of onset is more similar to sporadic pituitary adenoma patients. Penetrance is probably slightly lower than in AIP-positive families (Igreja et al., 2010).

Genetics of FIPA

The *AIP* gene is located on chromosome 11q13 in the vicinity (3 Mb distal) of the *MEN1* gene. The *AIP* gene contains six exons and encodes a 330 amino acid co-chaperone protein which is a well-conserved protein throughout evolution (Trivellin and Korbonits, 2011). The amino-terminus of the AIP protein has an immunophilin-like domain, with significant homology to immunophilins FKBP12 and FKBP52. However, it differs from other immunophilins by not sharing the ability to bind to immunosuppressant drugs such as ciclosporin or rapamycin (Carver et al., 1998; Newey et al., 2013). The carboxy-terminus contains seven α helices: three 34 amino acid structures (tetratricopeptide repeat (TPR) domains), each with 2 helices, and a final seventh α helix (Fig. 23.4). These helices are necessary for

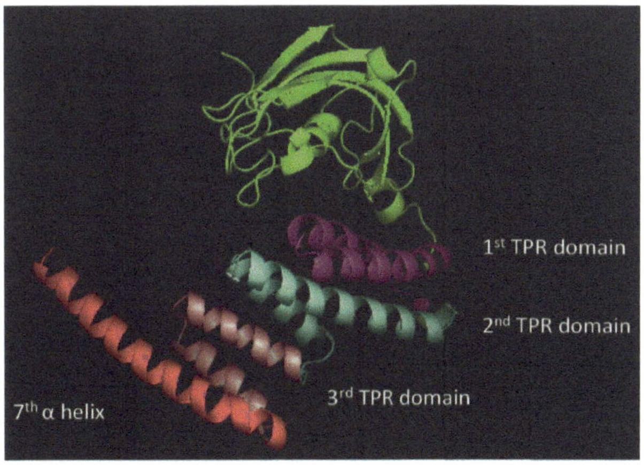

Fig. 23.4. Human AIP structure based on the crystal structure of the N-terminal (Linnert et al., 2012) and C terminal part (Morgan et al., 2012) of human AIP.

protein–protein interactions of AIP (Carver et al., 1998; Meyer and Perdew, 1999; Bell and Poland, 2000). The crystal structure of the human AIP gene has now been identified (Linnert et al., 2012; Morgan et al., 2012).

Over 70 different AIP mutations have been identified to date (Fig. 23.5), these include deletions, insertions, segmental duplications, nonsense, missense mutations and deletions of whole exons or the whole gene. Most of the pathogenic missense mutations directly affect the TPR domains or the C-terminal α-helix. Two-thirds of the AIP mutations lead to protein truncations, which remove segments of the TPR domains and/or carboxy-terminal end, and therefore lead to loss of function of the protein (Meyer and Perdew, 1999; Bell and Poland, 2000; Chahal et al., 2010). A common genetic "hotspot" for mutations in the AIP protein is the 304 residue (R304X and R304Q) which affects a CpG sequence and has been shown to be present in several independent families from different parts of the world. Other potential hotspots include the 271 and the 81 locus (Daly et al., 2007, 2009; Tichomirowa et al., 2009; Chahal et al., 2010, 2011).

It has been previously suggested that *AIP* functions as a TSG. Recent data indeed demonstrate a tumor suppressor role for AIP. Transient overexpression of wild-type AIP in human fibroblast cell lines (TIG3 and HEK293) and in GH3 cells (rodent somatomammotroph pituitary cell line) causes reduced cell proliferation, while cells transfected with mutant AIP do not demonstrate a reduction in cell proliferation (Leontiou et al., 2008). On the other hand, inhibition of AIP expression with siRNA causes increased cell proliferation (Heliovaara et al., 2009; Chahal et al., 2012).

AIP has numerous binding partners, such as viral proteins (HBV X and EBNA-3), AhR, Hsp90, Hsc70, PDEs,

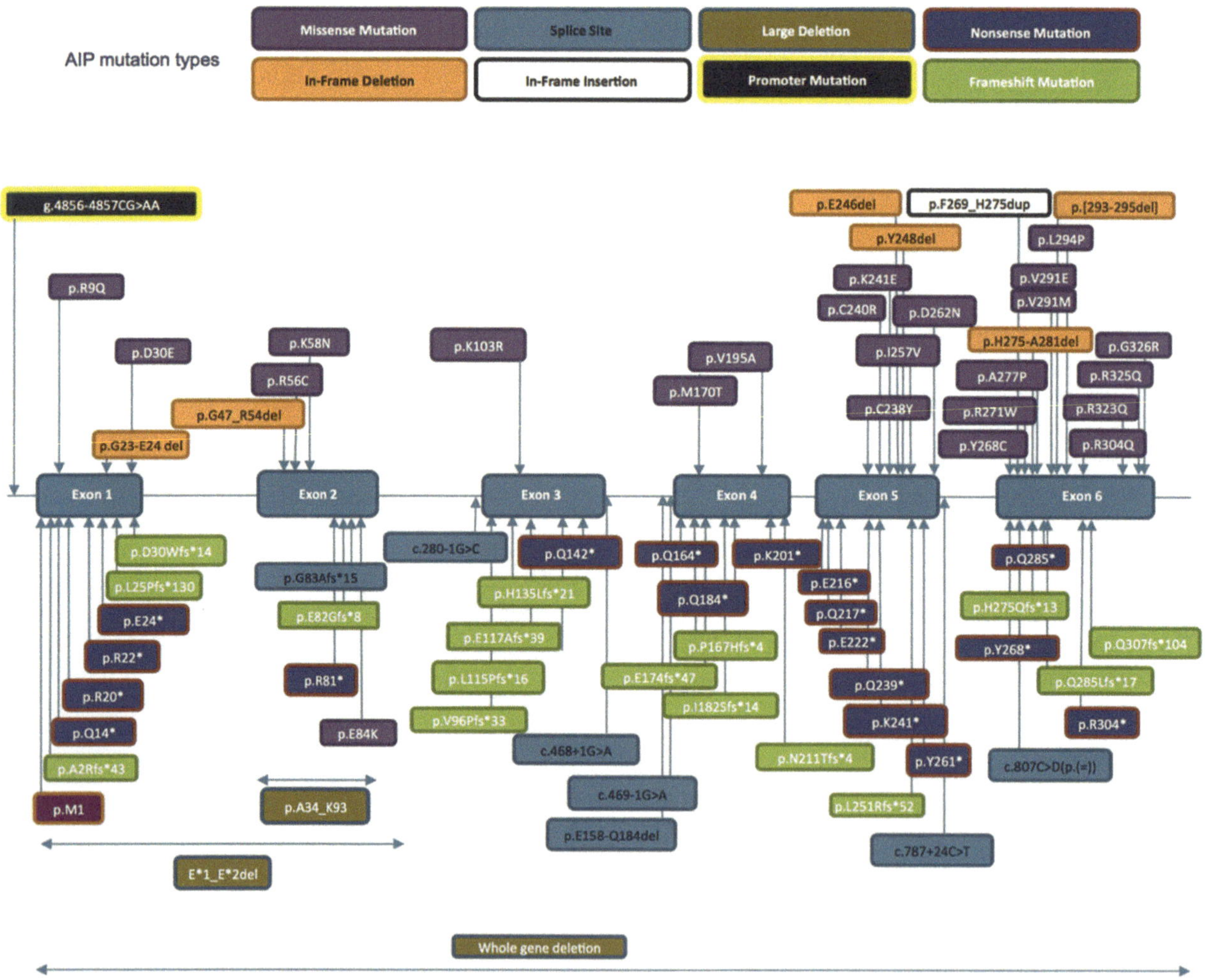

Fig. 23.5. Mutations on AIP gene. Mutations currently known in the AIP gene color-coded according to mutation types. The six exons of the AIP gene are shown. Over 70 different AIP mutations have been identified; these include deletions, insertions, segmental duplications, nonsense and missense mutations, and large deletions. Mutations resulting in complete disruption of the AIP protein (for example stop, deletion, or frameshift mutations) are scattered over the entire length of the gene, whereas the vast majority of point mutations affect only the C-terminal, known to be important for the biological function of the AIP protein.

nuclear and transmembrane receptors, G proteins, TOMM20, survivin, TNNI3K, and cytoskeletal proteins, but the exact mechanism by which AIP exerts its tumor suppressive action in the pituitary is not yet understood (Trivellin and Korbonits, 2011). A recent study has demonstrated that upregulation of *AIP* in the liver of transgenic mice increases the expression of a TSG named *ZAC/PLAGL1/Lot-1* (Hollingshead et al., 2006). This might be the mechanism by which AIP exerts its tumor suppressor effects in the pituitary. A recent study demonstrated that ZAC mRNA expression was significantly increased in GH3 cells (rat somatomammotroph cell line) transiently transfected with wild- type AIP compared to the empty vector and to those transfected with mutant forms of AIP (C238Y, R304X). This suggests that AIP may exert its tumor suppressor role in the pituitary by upregulating ZAC mRNA expression (Chahal et al., 2012). New data from *in vitro* experiments on mouse embryonic fibroblast (MEF) and pituitary adenoma cell lines demonstrate that AIP deficiency results in increased levels of cAMP through defective Gαi signaling. This results in subsequent downregulation of phosphorylated extracellular signal-regulated kinases 1/2 (p-ERK1/2) and cAMP response element binding protein (p-CREB).

This new evidence suggests that defective Gαi signaling is potentially a major contributor to the development of GH-secreting pituitary adenomas in AIP mutation carriers (Tuominen et al., 2014).

Management of pituitary disease in FIPA

Overall, management of pituitary tumors in the FIPA setting does not differ from the management of sporadic cases.

FIPA patients, particularly those with AIP mutations, tend to have more aggressive disease, and hence treatment of these patients can be challenging, especially if diagnosed late in the disease process. It is therefore vital to provide genetic counseling and AIP mutation testing in all patients with a family history of pituitary adenoma as well as providing genetic screening to "at risk" family members and clinical screening to carrier subjects (Tichomirowa et al., 2009; Chahal et al., 2011).

AIP mutation screening includes testing for single base-pair or small sequence changes as well as large deletions. This is now available in various laboratories (Korbonits and Kumar, 2012). Based on the age of onset of pituitary disease in some AIP mutation-positive patients, genetic testing is suggested to be performed at the age of 4 years or earlier. AIP mutation carriers identified through screening should undergo baseline assessments including magnetic resonance imaging (MRI) as well as clinical and biochemical tests followed by regular check-ups.

The screening and follow-up of AIP mutation-negative families is more difficult as gene carrier status cannot be currently established, age of onset has a significantly wider range, and penetrance is lower (Igreja et al., 2010). Patient education is important but policy about clinical screening should be discussed with individual patients outside the research setting. It is also important to emphasize that pituitary incidentalomas are common findings on MRI imaging in the general population.

CARNEY COMPLEX SYNDROME (CNC): RELATED OMIM ENTRIES: CNC1 #160980, PRKAR1A *188830

Carney complex syndrome (CNC) also known as LAMB syndrome (lentigines, atrial myxomas, blue nevi) and NAME syndrome (nevi, atrial myxomas, ephelides), is a rare autosomal dominant condition with variable penetrance, characterized by various endocrine and other tumor types including pituitary hyperplasia or adenoma (Stratakis et al., 2001; Horvath and Stratakis, 2008, 2009) (Figs. 23.6–23.8). CNC is a rare condition and has been described in about 500 people to date (Boikos and Stratakis, 2007). One of the first cases of CNC was

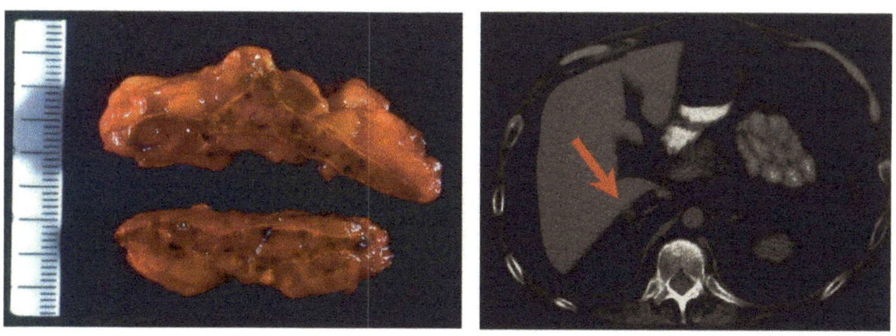

A B

Fig. 23.6. Macroscopic and CT findings in primary pigmented nodular adrenocortical disease (PPNAD). (**A**) Macroscopic appearance of the multiple pigmented micronodules of adrenal gland in PPNAD. The periadrenal fat is also visible around the adrenal capsule (Bertherat, 2006). (**B**) Adrenal CT-scan in PPNAD showing a micronodule on the external part of the left adrenal (red arrow). (Reproduced from Bertherat, 2006, with permission from BioMed Central.)

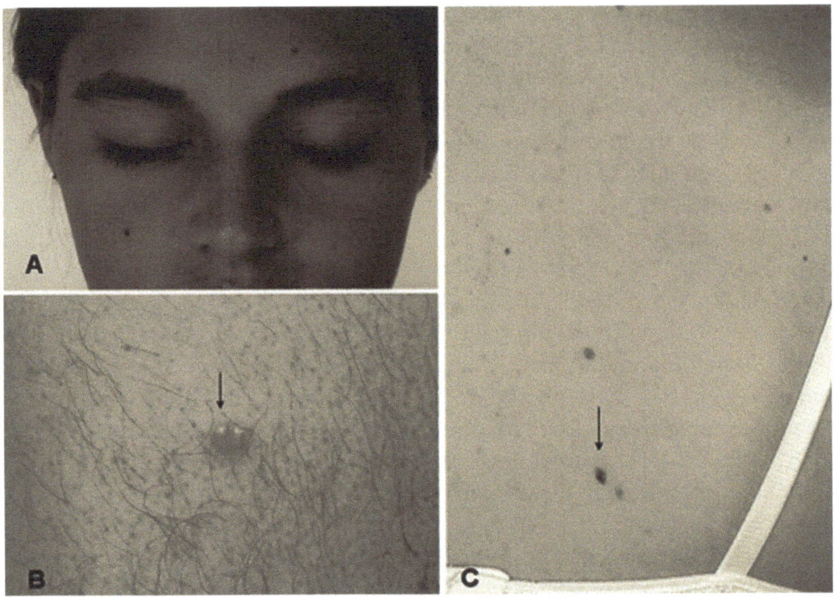

Fig. 23.7. Cutaneous lesions in Carney complex (CNC). Spotty skin pigmentations noted in a patient with CNC: (**A**) lentigines on the nose, around the eye, cheeks and vermilion border of the lips; (**B**) slow-growing subcutaneous nodules-myxomas (arrow); (**C**) pigmented cutaneous lesions such as blue nevi (arrow). (Reproduced from Boikos and Stratakis, 2006, with permission from Springer Science.)

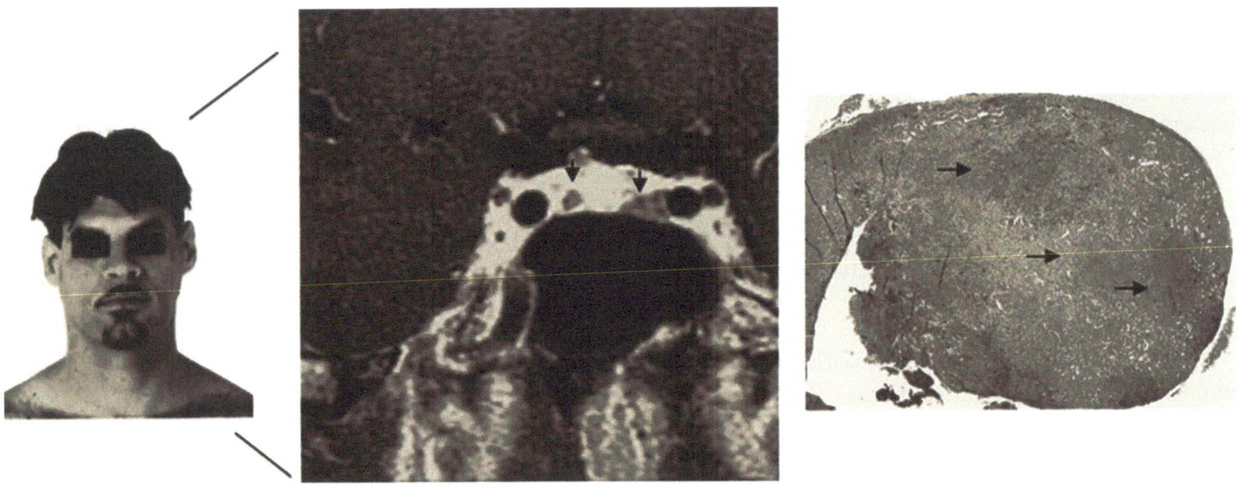

Fig. 23.8. Carney complex (CNC) and acromegaly. A patient with CNC and acromegaly. MRI brain shows multiple microadenomas (arrows). The pituitary gland of this patient was removed due to the presence of multiple small tumors on a background of hyperplasia. (Reproduced from Boikos and Stratakis, 2006, with permission from Springer Science.)

described in 1981 in a patient with both acromegaly and Cushing syndrome, two rare endocrine conditions. However, this condition was not properly described until almost a decade later by Dr Carney who reported the association of myxomas, spotty skin pigmentation (lentigines), and endocrine overactivity in four patients of a total of 40 cases in his original series (Carney et al., 1985). CNC is an umbrella term used to describe patients previously diagnosed with LAMB and NAME syndrome. CNC is familial in 70% of cases with slight female predominance (Stratakis et al., 2001).

Clinical features of CNC

The manifestations of CNC can be variable and may appear over many years. The median age of detection of the disease is 20 years with various tumors associated with the syndrome presenting at different age groups. The main endocrine features of CNC are primary pigmented nodular adrenocortical disease (PPNAD), testicular tumors (large cell calcifying Sertoli cell tumor (LCCSCT), Leydig cell tumors, etc.), thyroid tumors and increased levels of GH and prolactin due to hyperplasia or adenoma of somatotrophs and mammotrophs (Stratakis et al., 2001; Stergiopoulos and Stratakis, 2003). Nonendocrine features include myxomas (heart, skin, and breast), skin pigmentations (multiple skin lentigines and blue nevi), schwannomas, and other nonendocrine manifestations.

Clinically evident acromegaly is uncommon in CNC and only affects around 10% of CNC patients; however, approximately 75% of patients with CNC exhibit asymptomatic elevations of GH, IGF-1, or prolactin, and show abnormal response to pituitary dynamic testing. A histologic analysis of pituitary tumors in CNC patients with acromegaly demonstrated that all tumors were positive for GH and prolactin while a minority also stained positively for LH, TSH, or α-subunit (Pack et al., 2000). Most patients with CNC do not have an aggressive pituitary tumor profile. Abnormal hormone levels and true acromegaly usually develop insidiously and may arise from multifocal hyperplasia which may lead to formation of GH/prolactin-secreting adenoma. These zones of hyperplasia are poorly demarcated and demonstrate altered reticulin staining (Kurtkaya-Yapicier et al., 2002). Multifocal somatomammotroph cell hyperplasia does not appear to be present in MEN1 pituitary tumors (Horvath and Stratakis, 2008) or in FIPA patients, although it has been described in one family (Villa et al., 2011). The mean age of onset of acromegaly in CNC patients was reported as 35.8 years in a recent study (Pack et al., 2005; Boikos and Stratakis, 2006). Similar to MEN1 and FIPA, sporadic pituitary tumors usually do not exhibit somatic mutations in PRKAR1A (Kaltsas et al., 2002; Sandrini et al., 2002a).

Adrenocorticotropic hormone-independent Cushing syndrome is commonly seen in 25–30% of patients with CNC. Overall, PPNAD is the most common clinically significant endocrine lesion in CNC. Patients with PPND commonly present in the second and third decade of life, although some patients are also diagnosed in the first 2–3 years of life. Interesting is the possible constellation of Cushing syndrome and acromegaly as the symptoms of one can counteract the typical manifestations of the other. Testicular and thyroid nodules also commonly appear in the first decade of life and can lead to malignancy in later years (Stratakis and Horvath, 1993; Horvath and Stratakis, 2009; Jaffrain-Rea et al., 2011). Abnormal skin pigmentations are also common findings in CNC patients at birth. However, lentigines usually appear shortly before, and during puberty. Lentigines are small brown to black macules typically located around the upper and lower lips, on the eyelids, the ears, and the genital area (Horvath and Stratakis, 2009).

CNC-associated cardiac myxomas are the most common noncutaneous lesions in CNC patients. These tumors can occur in any cardiac chamber and may be multiple and recur frequently and can manifest as intracardiac obstruction of blood flow, embolism, and heart failure. Cardiac myxomas are responsible for more than 50% of the disease-specific mortality among CNC patients (Stratakis and Horvath, 1993). Large-cell calcifying Sertoli cell tumors are observed in one-third of affected males within the first decade and in almost all adult males (Stratakis and Horvath, 1993; Horvath and Stratakis, 2008). Psammomatous melanotic schwannoma, a rare tumor of the nerve sheath, occurs in an estimated 10% of affected individuals. The median age of diagnosis is 20 years. It rarely occurs as a sporadic tumor. Syndromes associated with schwannomas are CNC, neurofibromatosis, and familial schwannomatosis; however, in CNC these tumors are always pigmented, due to heavy melanin deposition. Schawannomas can occur anywhere in the central and peripheral nervous system but frequently affect the gastrointestinal tract and the paraspinal sympathetic chain (Stratakis and Horvath, 1993; Horvath and Stratakis, 2009).

Genetics of CNC

CNC is genetically heterogeneous. Three disease gene loci have been identified in CNC patients, these are 17q22-24, 1p31.1 and 2p16. The gene located on 2p16 has not been identified yet while a single patient has recently been described with a duplication event on 1.p31.1 (Forlino et al., 2014). Over 60% of families with CNC have a germline-inactivating mutation in the cAMP-dependent protein kinase A type I-a regulatory subunit (*PRKAR1A*) gene, located on chromosome 17q24 (Stratakis et al., 1996; Veugelers et al., 2004; Horvath and Stratakis, 2008). PKA is a second messenger-dependent enzyme implicated in a wide range of cellular processes including hormone release, transcriptional regulation, cell cycle progression, and apoptosis. PKA consists naturally of two homodimers of regulatory (R) and two catalytic (C) subunits. The R subunit exists in two forms, R-I (coded by PRKAR1A) and R-II, and produces two alternative enzymes PKA-I or PKA-II (Horvath and Stratakis, 2008; Stratakis et al., 2010). Modulators of PKA activity include factors that either activate or inhibit adenylate cyclase, resulting in an increase or decrease in cAMP levels. Loss of regulatory activity of PKA results in enhanced activity of PKA and therefore enhances the response to cAMP signaling in affected tissues. For example in somatotroph cells, there is increased activity of the GHRH-induced signal transduction pathway. Functional data suggest that a slight alteration of PRKAR1A function is sufficient for increasing PKA activity leading to tumorigenesis and CNC (Groussin et al., 2002).

PRKAR1A comprises 11 exons covering a genomic region of approximately 21 kb with a coding region of

1143 base pairs. The reported mutations are spread over the entire gene and consist of single base pair substitutions, insertions, small deletions or combined rearrangements of less than 15 base pairs (Kirschner et al., 2000a). Since the identification of PRKAR1A as the cause for CNC, a least 40 different mutations have been identified. The majority of these mutations generate a direct or frameshift premature stop codon (Kirschner et al., 2000a, b; Groussin et al., 2002). A two base-pair deletion in exon 5 of the gene is the most frequently seen mutation in CNC patients. This genetic defect is found *de novo* in approximately 30% of CNC cases, identifying a hotspot for mutations in the PRKAR1A gene (Stratakis and Horvath, 1993; Horvath and Stratakis, 2008). Approximately 70% of individuals diagnosed with CNC have an affected parent (Stratakis and Horvath, 1993).

Truncating PRK*AR1A* mutations result in mRNA instability due to the nonsense-mediated decay mechanism. LOH at 17q22-24 and loss of the normal allele have been demonstrated in CNC tumors. Somatic mutations have not been identified in pituitary adenomas in the PRKAR1A gene (Kaltsas et al., 2002), but have been rarely described in sporadic thyroid tumors (Sandrini et al., 2002b).

Studies of R-Ia knockout in mice have demonstrated that homozygous mice die as embryos due to failed cardiac morphogenesis. However, transgenic mice carrying an antisense transgene for Prkar1a exon 2 (X2AS) under the control of a tetracycline responsive promoter developed many characteristics of CNC patients including thyroid follicular hyperplasia/adenomas, adrenocortical hyperplasia, and other features of PPNAD, including late-onset weight gain, visceral adiposity, and hypercorticosteronemia not responsive to dexamethasone (Griffin et al., 2004; Kirschner et al., 2005).

Management of pituitary disease in CNC

Diagnosis of CNC depends on the patient having at least two or more of the typical manifestations (Horvath and Stratakis, 2008; Jaffrain-Rea et al., 2011). Patients who meet the criteria for CNC should then undergo PRKAR1A mutation screening. Mutation carriers should then have annual clinical, biochemical (measurement of urinary free cortisol and serum IGF-1) assessment and imaging (MRI, echocardiogram, thyroid ultrasound, testicular ultrasound in males and transabdominal pelvic ultrasound in females) for early detection of CNC. In patients who meet these criteria, germline DNA sequencing for mutations in the *PRKAR1A* gene should be carried out (Tichomirowa et al., 2009; Jaffrain-Rea et al., 2011). In PRKAR1A mutation-positive patients screening can be done as young as 6 months of age. In these patients biochemical and imaging tests should be undertaken at least yearly for manifestations of CNC (Horvath and Stratakis, 2008; Tichomirowa et al., 2009). In PRKAR1A mutation-negative patients further screening for lager genomic deletions/duplication in the *PRKAR1A* gene should be performed (Tichomirowa et al., 2009).

The treatment of individual manifestations of CNC is similar to that of the sporadic cases. Cardiac myxomas should be removed surgically. The main treatment of PPNAD is bilateral adrenalectomy, although ketokonazole or mitotane may be used for medical adrenolysis under certain circumstances (Stratakis and Horvath, 1993). Treatment of schwannomas is rather challenging due to the critical location of these neoplasms usually in or around nerve roots and also metastasis to brain, liver, and lung early in the disease process. There is no effective medical or surgical treatment for metastatic schwannomas (Stratakis and Horvath, 1993). Growth hormone-producing pituitary tumors may be excised surgically if large in size. Somatostatin analogs are also frequently used, either as a primary treatment or to shrink the tumors prior to surgery.

MCCUNE–ALBRIGHT SYNDROME: OMIM 174800

McCune–Albright syndrome (MAS) was first described in 1937 by Donavan James McCune and Fuller Albright. This a rare, sporadically occurring condition characterized by mosaic mutations in the gene encoding the adenylate cyclase-stimulating G α protein (GNAS, guanine nucleotide binding α-subunit gene) (Weinstein et al., 1991; Schwindinger et al., 1992; Horvath and Stratakis, 2008). This disease is not inherited but has a genetic origin (Fig. 23.9).

Clinical features of McCune–Albright syndrome

Diagnosis of MAS is established on clinical grounds with patients having at least two features of the triad of polyostotic fibrous dysplasia (FD), café-au-lait skin pigmentation, and autonomous endocrine hyperfunction, including precocious puberty, thyrotoxicosis, pituitary gigantism, and Cushing syndrome as well as renal phosphate wasting (Dumitrescu and Collins, 2008; Horvath and Stratakis, 2008; Keil and Stratakis, 2008). The most common forms of autonomous endocrine hyperfunction in this syndrome are gonadotropin-independent precocious puberty, thyrotoxicosis, pituitary gigantism, and Cushing syndrome (Frohman and Eguchi, 2004; Horvath and Stratakis, 2008).

In girls, precocious puberty usually presents with vaginal bleeding or spotting, accompanied by development of breast tissue, usually without the development of pubic hair. A recent study has demonstrated that gonadal pathology in MAS is common not just in females but also in males, showing equal incidence in the two sexes

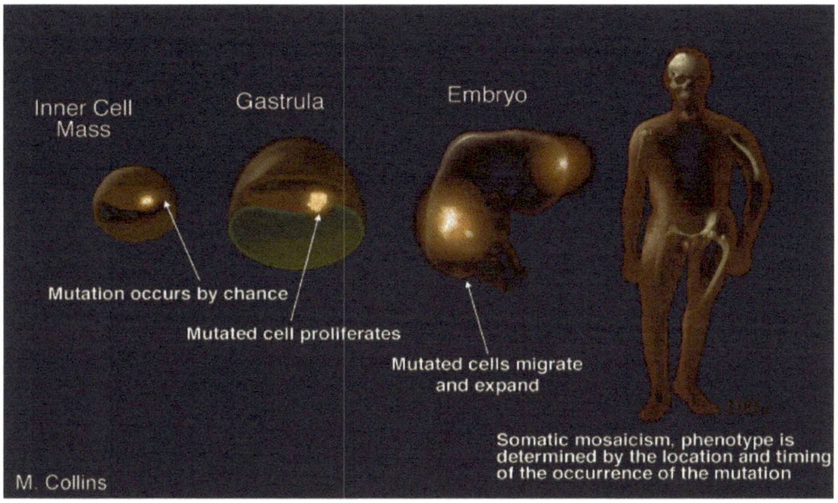

Fig. 23.9. Molecular and developmental defects in McCune–Albright syndrome (MAS). A sporadic mutation occurs in a single cell (bright spot) at some point early in the development. If a mutation occurs at the inner cell mass stage, tissues from all three germ layers will be affected. As the cells derived from this mutated clone are dispersed throughout the organism, the final phenotype emerges, MAS (Dumitrescu and Collins, 2008). The diagram demonstrates the mechanism by which GNAS1 mutations cause a rather heterogeneous disease encompassing a broad spectrum of phenotypic manifestations, ranging from a relatively benign disease to a severe condition with a significant impact on development and quality of life. (Reproduced from Dumitrescu and Collins, 2008, with permission from BioMed Central.)

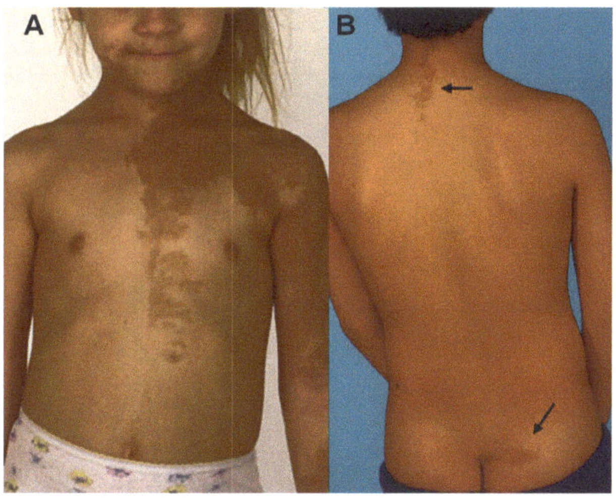

Fig. 23.10. Café-au-lait skin pigmentation. (**A**) Typical café-au-lait spots on the face, chest, and arm of a 5-year-old girl with McCune–Albright syndrome which demonstrates jagged "coast of Maine" borders. Café-au-lait spots are typically associated with the midline. (**B**) Typical lesions that are often found on the nape of the neck and crease of the buttocks are shown (arrows). (Reproduced from Dumitrescu and Collins, 2008, with permission from BioMed Central.)

(Boyce et al., 2012). In boys, testicular and penile enlargement, pubic and axillary hair, and early-onset sexual behavior are the manifestations of precocious puberty (Dumitrescu and Collins, 2008), due to predominantly Leydig cell hyperplasia on histologic studies, which is associated with low risk of malignant transformation (Boyce et al., 2012).

Café-au-lait spots are commonly the first manifestation in MAS and usually appear at birth or shortly thereafter (Fig. 23.10). However, it is most often precocious puberty or fibrous dysplasia that brings the child to medical attention and the café-au-lait spots are often missed. Café-au-lait spots in MAS have been classically described as having a "coast of Maine" border, which refers to the jagged appearance of the Maine coastline as it appears on maps (Collins et al., 2012). Café-au-lait spots found in MAS are usually associated with the midline, though frequent exceptions occur. This is in contrast to café-au-lait spots seen in neurofibromatosis which typically have a smooth edge and are classically described as having a "coast of California" border (Collins et al., 2012).

Pituitary disease in MAS, similar to CNC, often manifests as GH- and PRL-producing cell hyperplasia (Fig. 23.11). GH excess and lack of GH suppression after OGTT is present in up to 20% of MAS patients; however, detectable pituitary tumors are noted in approximately 50% of MAS cases (Kovacs et al., 1984; Cuttler et al., 1989; Pack et al., 2000; Horvath and Stratakis, 2008). Patients with MAS should be investigated for subclinical GH/IGF-1 hypersecretion as this negatively impacts bone disease especially craniofacial FD resulting in hearing and visual defects in approximately a third of affected individuals (Akintoye et al., 2002). Elevated GH levels in MAS have also been implicated in sarcomatous transformation of FD (Kaushik et al., 2002). Elevated GH or IGF-1 levels may be seen in MAS patients as early as infancy and can result in gigantism (Akintoye et al., 2002). Diagnosis of GH excess can be challenging in MAS patients. In children with MAS, rapid linear growth which could be as a result of GH excess is often attributed to precocious puberty, which is a common finding in patients with MAS. In addition, characteristic features of acromegaly such as coarsening of the face, frontal bossing, and prognathism not only develop insidiously but can be wrongly attributed to fibrous dysplasia of the skull which can result in dysmorphic features (Akintoye et al., 2006).

Fibrous dysplasia (FD) is characterized by the lack of differentiation and proliferation of bone-forming stromal cells leading to replacement of normal bone and marrow by fibrous tissue. FD most commonly behaves as a slow and indolent growing mass lesion. Depending on the type and location of FD, the signs and symptoms vary and include facial deformity, visual and hearing impairment, nasal congestion and/or obstruction, paresthesia, and pain. Occasionally diagnosis of FD is made when a family member, friend, or healthcare provider notices asymmetry in facial features. Incidental findings of FD have also been noted on dental X-rays or head and neck computed tomogram (Lee et al., 2012). The areas most commonly involved are the proximal femur, the craniofacial bones, and the ribs. FD in the appendicular skeleton usually presents with a limp and/or pain, or occasionally with a pathologic fracture (Dumitrescu and Collins, 2008). Ninety percent of the total body skeletal disease burden is usually established by the age of 15 and the progression of the lesions appears to diminish after puberty; however, the course of FD in craniofacial disease is less clear (Hart et al., 2007; Dumitrescu and Collins, 2008). Extraskeletal

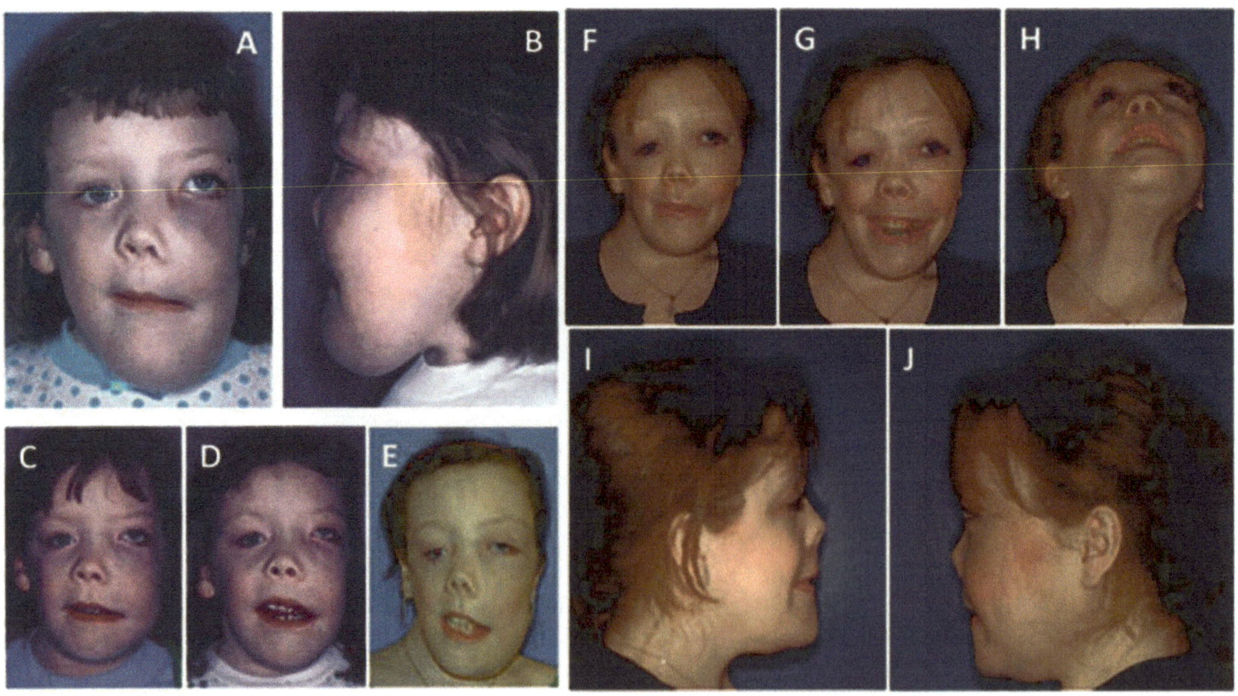

Fig. 23.11. McCune–Albright syndrome (MAS) with extensive fibrous dysplasia (FD) complicated by growth hormone excess. Serial images of a woman who initially presented at the age of 9 with MAS and extensive FD complicated by growth hormone excess. (**A, B**) Initial presentation; extensive FD of craniofacial bones resulting in blindness in left eye, displacement of tongue and gross facial deformity. (**C–E**) After first surgery; marked improvement in facial contours. Lesions continued to grow but stabilized by age 17 years. (**F–J**) Images taken 5 years after the second surgery. General improvement in facial contours but with remaining orbital asymmetry. (Reproduced from Lee et al., 2012, with permission from BioMed Central.)

manifestations of FD/MAS are established early in the disease process (Collins et al., 2012).

FD lesions may demonstrate rapid growth resulting in extensive bone deformity and pain. In some patients, this is associated with other pathologic lesions such as aneurysmal bone cysts or mucoceles (Lee et al., 2012) (Figs. 23.12 and 23.13).

Malignancies associated with MAS are distinctly rare occurrences. Malignant transformation of FD lesions occurs in probably less than 1% of the cases (Dumitrescu and Collins, 2008). In addition, breast and thyroid cancers are also rare occurrences (Tanabeu et al., 1998; Collins et al., 2003; Dumitrescu and Collins, 2008).

Genetics of McCune–Albright syndrome

McCune–Albright syndrome is due to postzygotic activating mutations in the GNAS1 gene. GNAS maps on chromosome 20q13 and encodes the ubiquitously expressed stimulatory (Gsα) subunit of the G protein (Figs. 23.14 and 23.15). Activating missense mutations result in substitution of normal arginine at position 201 (R201) with either a cysteine or a histidine, or rarely with serine, leucine, or glycine. Interestingly, while the typical somatic gsp mutations in somatotroph adenomas can affect both the 201 and 227 positions, patients with McCune–Albright syndrome only have 201 locus alterations (Landis et al., 1989). One possible explanation is that these mutations are more activating than R201 mutations (Landis et al., 1989). *GNAS1* mutations are found in approximately 90% of affected tissues in MAS, with the exception of skin lesions (Lumbroso et al., 2004; Jaffrain-Rea et al., 2011). The GNAS1 locus is under parent-of-origin control (imprinting). In MAS patients with acromegaly the GNAS1 mutation is almost always on the maternal allele (Mantovani et al., 2004), similar to sporadic GH-secreting adenoma cases harboring gsp mutations (Hayward et al., 2001). Other endocrine organ manifestations of MAS do not show imprinting characteristics (Mantovani et al., 2004).

Management of pituitary disease in McCune–Albright syndrome

Management of pituitary hypersecretion in MAS patients may be challenging. Somatostatin analogs and/or dopamine agonists are the mainstay of the treatment in MAS. However, GH-producing tumors in MAS show a consistent but inadequate response to treatment with somatostatin analogs, with only 50% of patients

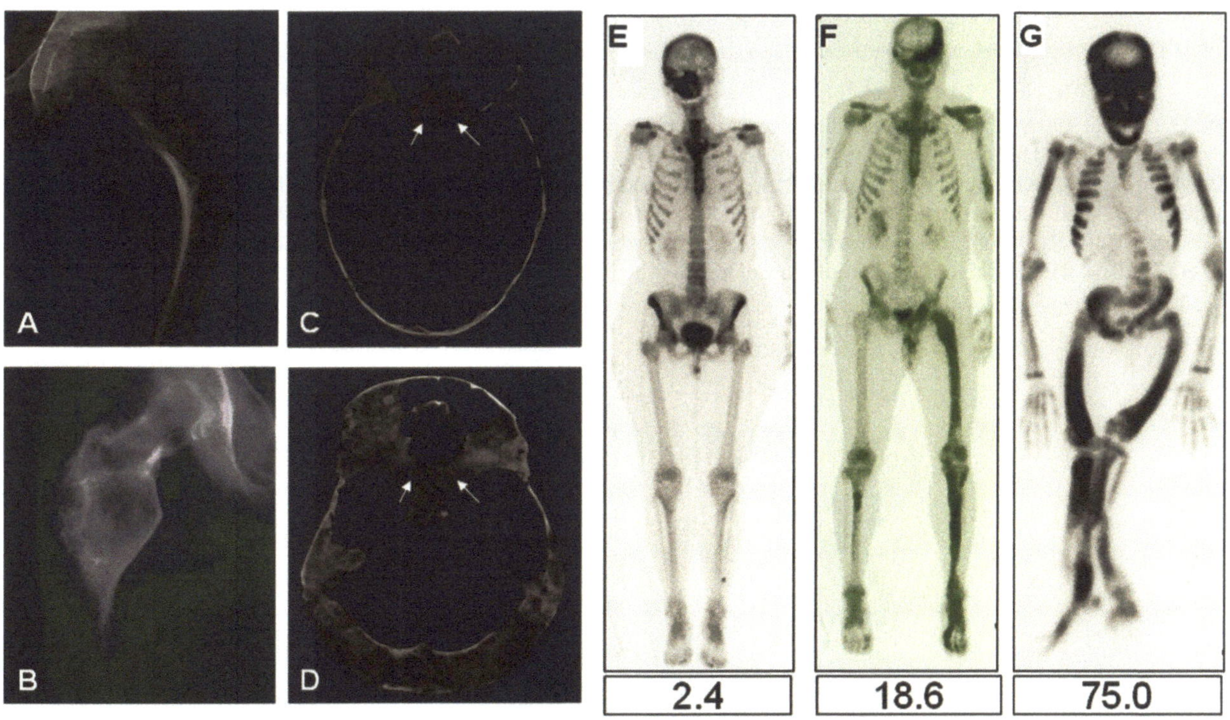

Fig. 23.12. Radiographic appearance of fibrous dysplasia (FD). (**A**) Proximal femur with typical ground glass appearance and shepherd's crook deformity in a 10-year-old child. (**B**) Sclerotic FD lesions in the femur of an untreated 40-year-old man. (**C**) The typical ground glass appearance of FD in the craniofacial region on a CT image of a 10-year-old child; optic nerves (arrows) are typically encased with FD. (**D**) Mixed solid and "cystic" lesions of FD on CT brain of a 40-year-old. (**E–F**) Craniofacial fibrous dysplasia is shown (**G**). A 16-year-old boy with McCune–Albright syndrome and involvement of virtually all skeletal sites (panostotic) is shown. (Reproduced from Dumitrescu and Collins, 2008, with permission from BioMed Central.)

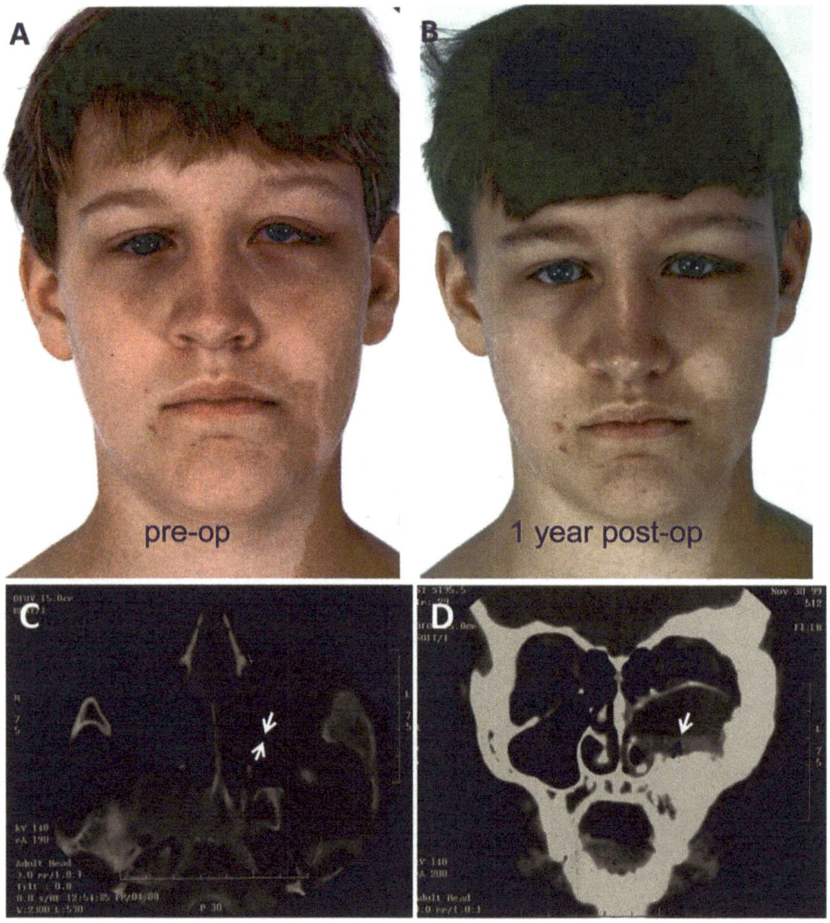

Fig. 23.13. Fibrous dysplasia with secondary aneurysmal bone cyst (ABC). (**A**) Before surgery: patient with McCune–Albright syndrome (MAS) and history of worsening vision and asymmetry of the left eye and face. He was found to have a rapidly growing ABC within FD. (**B**) After surgery: patient had resection and decompression of the ABC. Improved facial symmetry is noted post-operatively. Classic café-au-lait spots associated with MAS are seen on the face and neck. (**C, D**) Preoperative CT images of the patient in (**A**) showing the FD lesion and associated ABC (arrows). The fluid/fluid level is diagnostic of an ABC. (Reproduced from Lee et al., 2012, with permission from BioMed Central.)

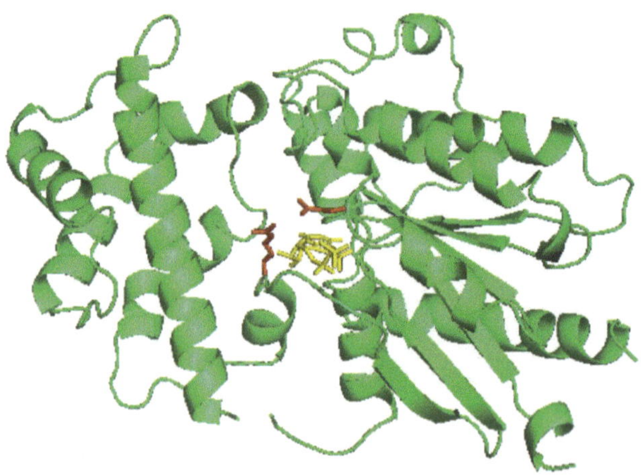

Fig. 23.14. G protein α-subunit. The diagram shows G protein α-subunit in its GTP-bound (GTP-γ-S on the model) form. Mutational replacements of red residues R201 and Q227 impair GTP hydrolysis. Bound GTP nucleotide is depicted as yellow. We used the model of G protein αs-subunit to prepare this model (PDB (protein database) code 1AZT) (Sunahara et al., 1997). (Figure kindly created by Dr Chrisostomos Prodromou, University of Sussex, Brighton, UK.)

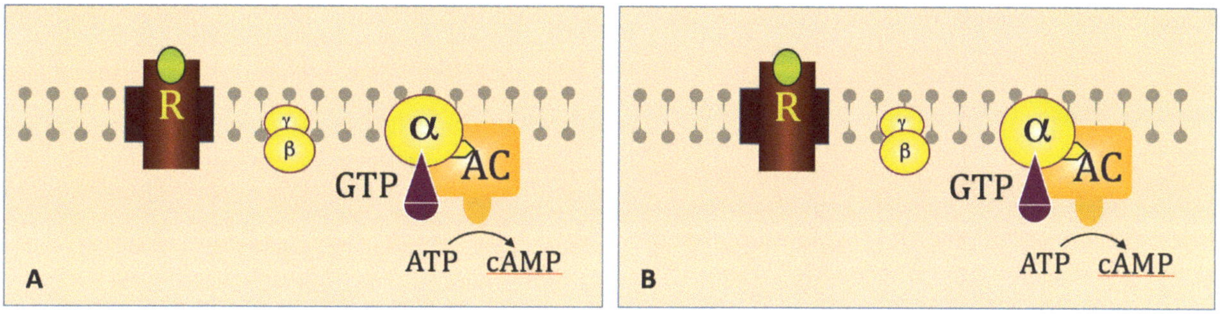

Fig. 23.15. Gsα subunit. The α-subunits comprise of four families: Gαs, Gαq/Gα11, Gα12/13 and Gαi/Gαo which determine the identity of G protein. (**A**) Resting state: free receptors do not interact with Gs-protein. (**B**) Active state: binding of the ligand (H) to the receptor (R) activates the receptor which in turn binds to the trimeric G-complex. G-α becomes active and GDP will change to GTP. G-α-GTP dissociates from the β-γ complex. G-α-GTP interacts with adenylyl cyclase (AC), which turns ATP into cAMP which acts as second messenger. The G-α-GTP complex is short-lived as the G-α subunit has an intrinsic GTPase activity. The gsp mutation destroys this GTPase activity therefore leaving the G-α-GTP complex active to stimulate AC.

having adequate GH/IGF-1 control (Akintoye et al., 2002). The use of transsphenoidal surgery is limited in MAS patients as PFD frequently involves the skull base. Surgical adenomectomy in MAS patients also rarely eliminates GH excess as the underlying cause of acromegaly in MAS is somatotroph hyperplasia involving the entire pituitary gland, with or without development of somatotroph adenoma (Vortmeyer et al., 2012). The GH-receptor antagonist pegvisomant has recently been proposed as an effective medical agent for uncontrolled MAS pituitary tumors or for simple hypersomatotropinemia without a visible adenoma. Pegvisomant has been proven effective in normalizing IGF-1 levels in MAS patients (Akintoye et al., 2006). Although pegvisomant was shown to effectively reduce IGF-1 and IGFBP-3 levels in gsp-mediated GH excess, it had no effect on FD (Akintoye et al., 2006). Radiotherapy has a limited efficiency in MAS and carries a potential risk of sarcomatous transformation of PFD, but has been used in cases where no other measure can influence hormone levels (Akintoye et al., 2006).

Genetic testing in MAS may be challenging since the chance to detect the *GNAS1* mutation on leukocyte DNA is only 45–59% in a patient with classic manifestations of the disease (Lumbroso et al., 2004). This decreases significantly in subjects who exhibit fewer features of MAS. Genetic testing in MAS is therefore not absolutely necessary as there is no genotype–phenotype correlation and no vertical transmission (Horvath and Stratakis, 2008; Jaffrain-Rea et al., 2011).

Familial hyperprolactinemia

Recently a heterozygous mutation has been described in the prolactinoma receptor in a family with familial mild to moderate hyperprolactinemia and normal pituitary gland on MRI imaging (Newey et al., 2013). Some family members had fertility problems while others did not. *In vitro* studies suggested a loss of function and possible dominant negative effect of this mutation on the intracellular signaling pathways of the prolactin receptor.

CONCLUSION

Pituitary adenomas usually occur sporadically and only a small percentage have mosaic (McCune–Albright syndrome) or germline (MEN1, CNC, and FIPA) genetic origin. While in the majority, but not all, MEN1 syndrome patients the disease-causing gene has been identified, in 30–40% of CNC and 80% of FIPA cases novel genes are yet to be discovered. Further studies in familial pituitary disease may lead to early diagnosis and prevention of severe complications in family members and better understanding of the pathophysiologic mechanisms that can help research into novel therapies.

ABBREVIATIONS

ACTH, adrenocorticotropic hormone; AIP, aryl hydrocarbon receptor interacting protein; CNC, Carney complex; CDK, cyclin-dependent kinase; CpG, cytosine preceded guanine; FIPA, familial isolated pituitary adenoma; FSH, follicle-stimulating hormone; GAPDH, glyceraldehyde 3-phosphate dehydrogenase; GH, growth hormone; GHR, growth hormone receptor; GHRH, growth hormone releasing hormone; GNAS, guanine nucleotide binding protein α-subunit; Gsα, guanine protein α-subunit; GTP, guanosine triphosphate; HEK293, human embryonic kidney 293 cell; HMGA2, high-mobility group AT-hook 2; IGF-1, insulin-like growth factor 1; KO, knockout; LCCSCT, large cell calcifying Sertoli cell tumor; LH, luteinizing hormone; LOH, loss of heterozygosity; Lot1, lost on transformation; MAS, McCune–Albright syndrome; MEG3a, maternally expressed gene 3a; MEN-1, multiple endocrine neoplasia type I; MRI, magnetic resonance imaging; mRNA, messenger ribonucleic acid; NET, neuroendocrine tumor; NFPA, nonfunctioning pituitary adenomas; OGTT, oral glucose tolerance test; PDE, phosphodiesterase; PFD, polyostotic fibrous dysplasia; PI3K, phosphatidylinositol 3-kinase; PKA, protein kinase A; PLAGL1, pleiomorphic adenoma gene-like1; PMS, psammomatous melanotic schwannoma; PPNAD, primary pigmented nodular adrenocortical disease; PRKAR1a, (gene name for the regulatory subunit 1a of protein kinase A); PRL, prolactin; PTAG, pituitary tumor apoptosis gene; PTTG, pituitary tumour transforming gene; Rb, retinoblastoma; Tc99, technetium-99; TGFα, transforming growth factor α; TPR, tetratricopeptide repeat; TRH, thyrotropin-releasing hormone; TSG, tumour suppressor gene; TSH, thyroid stimulating hormone; WIF1, Wnt inhibitory factor 1; XAP2, hepatitis B virus X-associated protein.

References

Adams EF, Symowski H, Buchfelder M et al. (2000). A polymorphism in the growth hormone (GH)-releasing hormone (GHRH) receptor gene is associated with elevated response to GHRH by human pituitary somatotrophinomas in vitro. Biochem Biophys Res Commun 275: 33–36.

Akintoye SO, Chebli C, Booher S et al. (2002). Characterization of gsp-mediated growth hormone excess in the context of McCune–Albright syndrome. J Clin Endocrinol Metab 87: 5104–5112.

Akintoye SO, Kelly MH, Brillante B et al. (2006). Pegvisomant for the treatment of gsp-mediated growth hormone excess in patients with McCune–Albright syndrome. J Clin Endocrinol Metab 91: 2960–2966.

Arafah BM, Nasrallah MP (2001). Pituitary tumors: pathophysiology, clinical manifestations and management. Endocr Relat Cancer 8: 287–305.

Asa SL, Ezzat S (2005). Genetics and proteomics of pituitary tumors. Endocrine 28: 43–47.

Bahar A, Simpson DJ, Cutty SJ et al. (2004). Isolation and characterization of a novel pituitary tumor apoptosis gene. Mol Endocrinol 18: 1827–1839.

Bates AS, Farrell WE, Bicknell EJ et al. (1997). Allelic deletion in pituitary adenomas reflects aggressive biological activity and has potential value as a prognostic marker. J Clin Endocrinol Metab 82: 818–824.

Beckers A, Daly AF (2007). The clinical, pathological, and genetic features of familial isolated pituitary adenomas. Eur J Endocrinol 157: 371–382.

Beckers A, Aaltonen LA, Daly AF et al. (2013). Familial isolated pituitary adenomas (FIPA) and the pituitary adenoma predisposition due to mutations in the aryl hydrocarbon receptor interacting protein (AIP) gene. Endocr Rev 34: 239–277.

Bell DR, Poland A (2000). Binding of aryl hydrocarbon receptor (AhR) to AhR-interacting protein. The role of hsp90. J Biol Chem 275: 36407–36414.

Benito M, Asa SL, Livolsi VA et al. (2005). Gonadotroph tumor associated with multiple endocrine neoplasia type 1. J Clin Endocrinol Metab 90: 570–574.

Ben-Shlomo A, Melmed S (2008). Acromegaly. Endocrinol Metab Clin North Am 37: 101–122, viii.

Bertherat J (2006). Carney complex (CNC). Orphanet J Rare Dis 1: 21.

Bertherat J, Chanson P, Montminy M (1995). The cyclic adenosine 3′,5′-monophospate-responsive factor CREB is constitutively activated in human somatotroph adenomas. Mol Endocrinol 9: 777–783.

Bertolino P, Radovanovic, Casse H et al. (2003a). Genetic ablation of the tumor suppressor menin causes lethality at mid-gestation with defects in multiple organs. Mech Dev 120: 549–560.

Bertolino P, Tong WM, Galendo D et al. (2003b). Heterozygous Men1 mutant mice develop a range of endocrine tumors mimicking multiple endocrine neoplasia type 1. Mol Endocrinol 17: 1880–1892.

Boikos SA, Stratakis CA (2006). Pituitary pathology in patients with Carney complex: growth-hormone producing hyperplasia or tumors and their association with other abnormalities. Pituitary 9: 203–209.

Boikos SA, Stratakis CA (2007). Carney complex: the first 20 years. Curr Opin Oncol 19: 24–29.

Boyce AM, Chong WH, Shawker TH et al. (2012). Characterization and management of testicular pathology in McCune–Albright syndrome. J Clin Endocrinol Metab 97: E1782–E1790.

Brahma A, Heyburn P, Swords F (2009). Familial prolactinoma occuring in association with SDHB mutation positive paraganglioma. Endocrine Abstracts 19: 239.

Brandi ML, Marx SJ, Aurbach GD et al. (1987). Familial multiple endocrine neoplasia type I: a new look at pathophysiology. Endocr Rev 8: 391–405.

Brandi ML, Gagel RF, Angeli A et al. (2001). Guidelines for diagnosis and therapy of MEN type 1 and type 2. J Clin Endocrinol Metab 86: 5658–5671.

Butz H, Liko I, Czirjak S et al. (2010). Down-regulation of Wee1 kinase by a specific subset of microRNA in human sporadic pituitary adenomas. J Clin Endocrinol Metab 95: E181–E191.

Carney JA, Gordon H, Carpenter PC et al. (1985). The complex of myxomas, spotty pigmentation, and endocrine overactivity. Medicine (Baltimore) 64: 270–283.

Carver LA, LaPres JJ, Jain S et al. (1998). Characterization of the Ah receptor-associated protein, ARA9. J Biol Chem 273: 33580–33587.

Chahal HS, Chapple JP, Frohman LA et al. (2010). Clinical, genetic and molecular characterization of patients with familial isolated pituitary adenomas (FIPA). Trends Endocrinol Metab 21: 419–427.

Chahal HS, Stals K, Unterlander M et al. (2011). AIP mutation in pituitary adenomas in the 18th century and today. N Engl J Med 364: 43–50.

Chahal HS, Trivellin G, Leontiou CA et al. (2012). Somatostatin analogs modulate AIP in somatotroph adenomas: the role of the ZAC1 pathway. J Clin Endocrinol Metab 97: E1411–E1420.

Chandrasekharappa SC, Guru SC, Manickam P et al. (1997). Positional cloning of the gene for multiple endocrine neoplasia-type 1. Science 276: 404–407.

Chen YX, Yan J, Keeshan K et al. (2006). The tumor suppressor menin regulates hematopoiesis and myeloid transformation by influencing Hox gene expression. Proc Natl Acad Sci U S A 103: 1018–1023.

Collins MT, Sarlis NJ, Merino MJ et al. (2003). Thyroid carcinoma in the McCune–Albright syndrome: contributory role of activating Gs alpha mutations. J Clin Endocrinol Metab 88: 4413–4417.

Collins MT, Singer FR, Eugster E (2012). McCune–Albright syndrome and the extraskeletal manifestations of fibrous dysplasia. Orphanet J Rare Dis 7 (Suppl 1): S4.

Corbetta S, Pizzocaro A, Peracchi M et al. (1997). Multiple endocrine neoplasia type 1 in patients with recognized pituitary tumours of different types. Clin Endocrinol (Oxf) 47: 507–512.

Cuny T, Pertuit M, Sahnoun-Fathallah M et al. (2013). Genetic analysis in young patients with sporadic pituitary macroadenomas: besides AIP don't forget MEN1 genetic analysis. Eur J Endocrinol 168: 533–541.

Cuttler L, Jackson JA, Saeed uz-Zafar M et al. (1989). Hypersecretion of growth hormone and prolactin in McCune–Albright syndrome. J Clin Endocrinol Metab 68: 1148–1154.

Dahia PL, Aguiar RC, Honegger J et al. (1998). Mutation and expression analysis of the p27/kip1 gene in corticotrophin-secreting tumours. Oncogene 16: 69–76.

Daly AF, Rixhon M, Adam C et al. (2006a). High prevalence of pituitary adenomas: a cross-sectional study in the province of Liege, Belgium. J Clin Endocrinol Metab 91: 4769–4775.

Daly AF, Jaffrain-Rea ML, Ciccarelli A et al. (2006b). Clinical characterization of familial isolated pituitary adenomas. J Clin Endocrinol Metab 91: 3316–3323.

Daly AF, Vanbellinghen JF, Khoo SK et al. (2007). Aryl hydrocarbon receptor-interacting protein gene mutations in familial isolated pituitary adenomas: analysis in 73 families. J Clin Endocrinol Metab 92: 1891–1896.

Daly AF, Tichomirowa MA, Beckers A et al. (2009). The epidemiology and genetics of pituitary adenomas. Best Pract Res Clin Endocrinol Metab 23: 543–554.

Daly AF, Tichomirowa MA, Petrossians P et al. (2010). Clinical characteristics and therapeutic responses in patients with germ-line AIP mutations and pituitary adenomas: an international collaborative study. J Clin Endocrinol Metab 95: E373–E383.

Denes J, Swords F, Xekouki P et al. (2012). Familial pituitary adenoma and paraganglioma syndrome – a novel type of multiple endocrine neoplasia. Endocr Rev. Endocrine Society Annual Meeting 33: OR41–2.

Dumitrescu CE, Collins MT (2008). McCune–Albright syndrome. Orphanet J Rare Dis 3: 12.

Elston MS, Gill AJ, Conaglen JV et al. (2008). Wnt pathway inhibitors are strongly down-regulated in pituitary tumors. Endocrinology 149: 1235–1242.

Evans CO, Young AN, Brown MR et al. (2001). Novel patterns of gene expression in pituitary adenomas identified by complementary deoxyribonucleic acid microarrays and quantitative reverse transcription-polymerase chain reaction. J Clin Endocrinol Metab 86: 3097–3107.

Ezzat S, Zheng L, Zhu XF et al. (2002). Targeted expression of a human pituitary tumor-derived isoform of FGF receptor-4 recapitulates pituitary tumorigenesis. J Clin Invest 109: 69–78.

Ezzat S, Asa SL, Couldwell WT et al. (2004). The prevalence of pituitary adenomas: a systematic review. Cancer 101: 613–619.

Falchetti A, Marini F, Luzi E et al. (2009). Multiple endocrine neoplasia type 1 (MEN1): not only inherited endocrine tumors. Genet Med 11: 825–835.

Farfel Z, Bourne HR, Iiri T (1999). The expanding spectrum of G protein diseases. N Engl J Med 340: 1012–1020.

Fedele M, Palmieri D, Fusco A (2010). HMGA2: a pituitary tumour subtype-specific oncogene? Mol Cell Endocrinol 326: 19–24.

Fernandez A, Karavitaki N, Wass JA (2010). Prevalence of pituitary adenomas: a community-based, cross-sectional study in Banbury (Oxfordshire, UK). Clin Endocrinol (Oxf) 72: 377–382.

Forlino A, Vetro A, Garavelli L et al. (2014). PRKACB and Carney complex. N Eng J Med 370: 1065–1067.

Frohman LA, Eguchi K (2004). Familial acromegaly. Growth Horm IGF Res 14 (Suppl A): S90–S96.

Gadelha MR, Kasuki L, Korbonits M (2013). Novel pathway for somatostatin analogs in patients with acromegaly. Trends Endocrinol Metab 24: 238–246.

Garby L, Caron P, Claustrat F et al. (2012). Clinical characteristics and outcome of acromegaly induced by ectopic secretion of growth hormone-releasing hormone (GHRH): a French nationwide series of 21 cases. J Clin Endocrinol Metab 97: 2093–2104.

Georgitsi M, Raitila A, Karhu A et al. (2007). Germline CDKN1B/p27Kip1 mutation in multiple endocrine neoplasia. J Clin Endocrinol Metab 92: 3321–3325.

Georgitsi M, De Menis E, Cannavo S et al. (2008). Aryl hydrocarbon receptor interacting protein (AIP) gene mutation analysis in children and adolescents with sporadic pituitary adenomas. Clin Endocrinol (Oxf) 69: 621–627.

Gordon MV, Varma D, McLean CA et al. (2007). Metastatic prolactinoma presenting as a cervical spinal cord tumour in multiple endocrine neoplasia type one (MEN-1). Clin Endocrinol (Oxf) 66: 150–152.

Griffin KJ, Kirschner LS, Matyakhina L et al. (2004). A transgenic mouse bearing an antisense construct of regulatory subunit type 1A of protein kinase A develops endocrine and other tumours: comparison with Carney complex and other PRKAR1A induced lesions. J Med Genet 41: 923–931.

Groussin L, Kirschner LS, Vincent-Dejean C et al. (2002). Molecular analysis of the cyclic AMP-dependent protein kinase A (PKA) regulatory subunit 1A (PRKAR1A) gene in patients with Carney complex and primary pigmented nodular adrenocortical disease (PPNAD) reveals novel mutations and clues for pathophysiology: augmented PKA signaling is associated with adrenal tumorigenesis in PPNAD. Am J Hum Genet 71: 1433–1442.

Hanahan D, Weinberg RA (2000). The hallmarks of cancer. Cell 100: 57–70.

Hart ES, Kelly MH, Brillante B et al. (2007). Onset, progression, and plateau of skeletal lesions in fibrous dysplasia and

the relationship to functional outcome. J Bone Miner Res 22: 1468–1474.
Hayward BE, Barlier A, Korbonits M et al. (2001). Imprinting of the G(s)alpha gene GNAS1 in the pathogenesis of acromegaly. J Clin Invest 107: R31–R36.
Heaney AP, Melmed S (2000). New pituitary oncogenes. Endocr Relat Cancer 7: 3–15.
Heliovaara E, Raitila A, Launonen V et al. (2009). The expression of AIP-related molecules in elucidation of cellular pathways in pituitary adenomas. Am J Pathol 175: 2501–2507.
Hemminki K, Forsti A, Ji J (2007). Incidence and familial risks in pituitary adenoma and associated tumors. Endocr Relat Cancer 14: 103–109.
Herman V, Fagin J, Gonsky R et al. (1990). Clonal origin of pituitary adenomas. J Clin Endocrinol Metab 71: 1427–1433.
Hollingshead BD, Patel RD, Perdew GH (2006). Endogenous hepatic expression of the hepatitis B virus X-associated protein 2 is adequate for maximal association with aryl hydrocarbon receptor-90-kDa heat shock protein complexes. Mol Pharmacol 70: 2096–2107.
Horvath A, Stratakis CA (2008). Clinical and molecular genetics of acromegaly: MEN1, Carney complex, McCune–Albright syndrome, familial acromegaly and genetic defects in sporadic tumors. Rev Endocr Metab Disord 9: 1–11.
Horvath A, Stratakis CA (2009). Carney complex and lentiginosis. Pigment Cell Melanoma Res 22: 580–587.
Igreja S, Chahal HS, King P et al. (2010). Characterization of aryl hydrocarbon receptor interacting protein (AIP) mutations in familial isolated pituitary adenoma families. Hum Mutat 31: 950–960.
Jaffrain-Rea ML, Daly AF, Angelini M et al. (2011). Genetic susceptibility in pituitary adenomas: from pathogenesis to clinical implications. Exp Rev Endocrinol Metab 6: 195–214.
Kaltsas GA, Kola B, Borboli N et al. (2002). Sequence analysis of the PRKAR1A gene in sporadic somatotroph and other pituitary tumours. Clin Endocrinol (Oxf) 57: 443–448.
Karnik SK, Hughes CM, Gu X et al. (2005). Menin regulates pancreatic islet growth by promoting histone methylation and expression of genes encoding p27Kip1 and p18INK4c. Proc Natl Acad Sci U S A 102: 14659–14664.
Kaushik S, Smoker WR, Frable WJ (2002). Malignant transformation of fibrous dysplasia into chondroblastic osteosarcoma. Skeletal Radiol 31: 103–106.
Keil MF, Stratakis CA (2008). Pituitary tumors in childhood: update of diagnosis, treatment and molecular genetics. Expert Rev Neurother 8: 563–574.
Kirsch M, Morz M, Pinzer T et al. (2009). Frequent loss of the CDKN2C (p18INK4c) gene product in pituitary adenomas. Genes Chromosomes Cancer 48: 143–154.
Kirschner LS, Sandrini F, Monbo J et al. (2000a). Genetic heterogeneity and spectrum of mutations of the PRKAR1A gene in patients with the carney complex. Hum Mol Genet 9: 3037–3046.
Kirschner LS, Carney JA, Pack SD et al. (2000b). Mutations of the gene encoding the protein kinase A type I-alpha regulatory subunit in patients with the Carney complex. Nat Genet 26: 89–92.
Kirschner LS, Kusewitt DF, Matyakhina L et al. (2005). A mouse model for the Carney complex tumor syndrome develops neoplasia in cyclic AMP-responsive tissues. Cancer Res 65: 4506–4514.
Knudson Jr AG (1971). Mutation and cancer: statistical study of retinoblastoma. Proc Natl Acad Sci U S A 68: 820–823.
Korbonits M, Kumar AV (2012). AIP-related familial isolated pituitary adenomas. In: RA Pagon, MP Adam, HH Ardinger et al. (Eds.), GeneReviews® [Internet]. University of Washington, Seattle (WA), Seattle; 1993–2014.
Korbonits M, Morris DG, Nanzer A et al. (2004). Role of regulatory factors in pituitary tumour formation. Front Horm Res 32: 63–95.
Kovacs K, Horvath E, Thorner MO et al. (1984). Mammosomatotroph hyperplasia associated with acromegaly and hyperprolactinemia in a patient with the McCune–Albright syndrome. A histologic, immunocytologic and ultrastructural study of the surgically-removed adenohypophysis. Virchows Arch A Pathol Anat Histopathol 403: 77–86.
Kurtkaya-Yapicier O, Scheithauer BW, Carney JA et al. (2002). Pituitary adenoma in Carney complex: an immunohistochemical, ultrastructural, and immunoelectron microscopic study. Ultrastruct Pathol 26: 345–353.
Landis CA, Masters SB, Spada A et al. (1989). GTPase inhibiting mutations activate the α chain of G_s and stimulate adenylyl cyclase in human pituitary tumours. Nature 340: 692–696.
Lee JS, FitzGibbon EJ, Chen YR et al. (2012). Clinical guidelines for the management of craniofacial fibrous dysplasia. Orphanet J Rare Dis 7 (Suppl 1): S2.
Lemmens I, Van de Ven WJ, Kas K et al. (1997). Identification of the multiple endocrine neoplasia type 1 (MEN1) gene. The European Consortium on MEN1. Hum Mol Genet 6: 1177–1183.
Lemos MC, Thakker RV (2008). Multiple endocrine neoplasia type 1 (MEN1): analysis of 1336 mutations reported in the first decade following identification of the gene. Hum Mutat 29: 22–32.
Leontiou CA, Gueorguiev M, van der Spuy J et al. (2008). The role of the aryl hydrocarbon receptor-interacting protein gene in familial and sporadic pituitary adenomas. J Clin Endocrinol Metab 93: 2390–2401.
Lidhar K, Korbonits M, Jordan S et al. (1999). Low expression of the cell cycle inhibitor p27Kip1 in normal corticotroph cells, corticotroph tumors, and malignant pituitary tumors. J Clin Endocrinol Metab 84: 3823–3830.
Linnert M, Haupt K, Lin YJ et al. (2012). NMR assignments of the FKBP-type PPIase domain of the human aryl-hydrocarbon receptor-interacting protein (AIP). Biomol NMR Assign 6: 209–212.
Lips CJ, Vasen HF, Lamers CB (1984). Multiple endocrine neoplasia syndromes. Crit Rev Oncol Hematol 2: 117–184.
Lumbroso S, Paris F, Sultan C (2004). Activating Gsalpha mutations: analysis of 113 patients with signs of McCune–Albright syndrome – a European collaborative study. J Clin Endocrinol Metab 89: 2107–2113.

Luzi E, Marini F, Giusti F et al. (2012). The negative feedback-loop between the oncomir Mir-24-1 and menin modulates the Men1 tumorigenesis by mimicking the "Knudson's second hit". PLoS One 7: e39767.

Mantovani G, Bondioni S, Lania AG et al. (2004). Parental origin of Gsalpha mutations in the McCune–Albright syndrome and in isolated endocrine tumors. J Clin Endocrinol Metab 89: 3007–3009.

Melmed S (2011). Pathogenesis of pituitary tumors. Nat Rev Endocrinol 7: 257–266.

Meyer BK, Perdew GH (1999). Characterization of the AhR-hsp90-XAP2 core complex and the role of the immunophilin-related protein XAP2 in AhR stabilization. Biochemistry 38: 8907–8917.

Morgan RM, Hernández-Ramírez LC, Trivellin G et al. (2012). Structure of the TPR domain of AIP: lack of client protein interaction with the C-terminal alpha-7 helix of the TPR domain of AIP is sufficient for pituitary adenoma predisposition. PLoS One 7: e53339.

Morita K, Takano K, Yasufuku-Takano J et al. (2008). Expression of pituitary tumour-derived, N-terminally truncated isoform of fibroblast growth factor receptor 4 (ptd-FGFR4) correlates with tumour invasiveness but not with G-protein alpha subunit (gsp) mutation in human GH-secreting pituitary adenomas. Clin Endocrinol (Oxf) 68: 435–441.

Naves LA, Daly AF, Vanbellinghen JF et al. (2007). Variable pathological and clinical features of a large Brazilian family harboring a mutation in the aryl hydrocarbon receptor-interacting protein gene. Eur J Endocrinol 157: 383–391.

Newey PJ, Gorvin CM, Cleland SJ et al. (2013). Mutant prolactin receptor and familial hyperprolactinemia. N Engl J Med 369: 2012–2020.

Occhi G, Regazzo D, Trivellin G et al. (2013). A novel mutation in the upstream open reading frame of the CDKN1B gene causes a MEN4 phenotype. PLoS Genet 9: e1003350.

Owens M, Stals K, Ellard S et al. (2009). Germline mutations in the CDKN1B gene encoding p27 Kip1 are a rare cause of multiple endocrine neoplasia type 1. Clin Endocrinol (Oxf) 70: 499–500.

Pack SD, Kirschner LS, Pak E et al. (2000). Genetic and histologic studies of somatomammotropic pituitary tumors in patients with the "complex of spotty skin pigmentation, myxomas, endocrine overactivity and schwannomas" (Carney complex). J Clin Endocrinol Metab 85: 3860–3865.

Pack SD, Qin LX, Pak E et al. (2005). Common genetic changes in hereditary and sporadic pituitary adenomas detected by comparative genomic hybridization. Genes Chromosomes Cancer 43: 72–82.

Pei L, Melmed S, Scheithauer B et al. (1994). H-ras mutations in human pituitary carcinoma metastases. J Clin Endocrinol Metab 78: 842–846.

Ruebel KH, Jin L, Zhang S et al. (2001). Inactivation of the p16 gene in human pituitary nonfunctioning tumors by hypermethylation is more common in null cell adenomas. Endocr Pathol 12: 281–289.

Saleem TF, Santhanam P, Hamoudeh E et al. (2012). Acromegaly caused by growth hormone releasing hormone (GHRH) secreting tumor in multiple endocrine neoplasia (MEN-1). W V Med J 108: 26–30.

Sandrini F, Kirschner LS, Bei T et al. (2002a). PRKAR1A, one of the Carney complex genes, and its locus (17q22-24) are rarely altered in pituitary tumours outside the Carney complex. J Med Genet 39: e78.

Sandrini F, Matyakhina L, Sarlis NJ et al. (2002b). Regulatory subunit type I-alpha of protein kinase A (PRKAR1A): a tumor-suppressor gene for sporadic thyroid cancer. Genes Chromosomes Cancer 35: 182–192.

Satta MA, Korbonits M, Jacobs RA et al. (1999). Expression of menin gene mRNA in pituitary tumours. Eur J Endocrinol 140: 358–361.

Scheithauer BW, Laws Jr ER, Kovacs K et al. (1987). Pituitary adenomas of the multiple endocrine neoplasia type I syndrome. Semin Diagn Pathol 4: 205–211.

Scheithauer BW, Kovacs K, Nose V et al. (2009). Multiple endocrine neoplasia type 1-associated thyrotropin-producing pituitary carcinoma: report of a probable de novo example. Hum Pathol 40: 270–278.

Schwindinger WF, Francomano CA, Levine MA (1992). Identification of a mutation in the gene encoding the alpha subunit of the stimulatory G protein of adenylyl cyclase in McCune–Albright syndrome. Proc Natl Acad Sci U S A 89: 5152–5156.

Shimon I, Melmed S (1997). Genetic basis of endocrine disease: pituitary tumor pathogenesis. J Clin Endocrinol Metab 82: 1675–1681.

Simpson DJ, Hibberts NA, McNicol AM et al. (2000). Loss of pRb expression in pituitary adenomas is associated with methylation of the RB1 CpG island. Cancer Res 60: 1211–1216.

Soares BS, Frohman LA (2004). Isolated familial somatotropinoma. Pituitary 7: 95–101.

Stergiopoulos SG, Stratakis CA (2003). Human tumors associated with Carney complex and germline PRKAR1A mutations: a protein kinase A disease! FEBS Lett 546: 59–64.

Stratakis CA, Horvath A (1993). Carney complex. In: RA Pagon, TD Bird, CR Dolan et al. (Eds.), Gene reviews. University of Washington, Seattle.

Stratakis CA, Carney JA, Lin JP et al. (1996). Carney complex, a familial multiple neoplasia and lentiginosis syndrome. Analysis of 11 kindreds and linkage to the short arm of chromosome 2. J Clin Invest 97: 699–705.

Stratakis CA, Kirschner LS, Carney JA (2001). Clinical and molecular features of the Carney complex: diagnostic criteria and recommendations for patient evaluation. J Clin Endocrinol Metab 86: 4041–4046.

Stratakis CA, Tichomirowa MA, Boikos S et al. (2010). The role of germline AIP, MEN1, PRKAR1A, CDKN1B and CDKN2C mutations in causing pituitary adenomas in a large cohort of children, adolescents, and patients with genetic syndromes. Clin Genet 78: 457–463.

Sunahara RK, Tesmer JJ, Gilman AG et al. (1997). Crystal structure of the adenylyl cyclase activator Gsalpha. Science 278: 1943–1947.

Syro LV, Scheithauer BW, Kovacs K et al. (2012). Pituitary tumors in patients with MEN1 syndrome. Clinics (Sao Paulo) 67 (Suppl 1): 43–48.

Tanabeu Y, Nakahara S, Mitsuyama S et al. (1998). Breast cancer in a patient with McCune–Albright syndrome. Breast Cancer 5: 175–178.

Tanizaki Y, Jin L, Scheithauer BW et al. (2007). P53 gene mutations in pituitary carcinomas. Endocr Pathol 18: 217–222.

Thakker RV (2001). Molecular genetics and patient management of multiple endocrine neoplasia type 1. Best Pract Res Clin Endocrinol Metab 15: 189–212.

Thakker RV (2010). Multiple endocrine neoplasia type 1 (MEN1). Best Pract Res Clin Endocrinol Metab 24: 355–370.

Thakker RV, Newey PJ, Walls GV et al. (2012). Clinical practice guidelines for multiple endocrine neoplasia type 1 (MEN1). J Clin Endocrinol Metab 97: 2990–3011.

Thapar K, Scheithauer BW, Kovacs K et al. (1996). p53 expression in pituitary adenomas and carcinomas: correlation with invasiveness and tumor growth fractions. Neurosurgery 38: 765–770, discussion 770–771.

Theodoropoulou M, Zhang J, Laupheimer S et al. (2006). Octreotide, a somatostatin analogue, mediates its antiproliferative action in pituitary tumor cells by altering phosphatidylinositol 3-kinase signaling and inducing Zac1 expression. Cancer Res 66: 1576–1582.

Tichomirowa MA, Daly AF, Beckers A (2009). Familial pituitary adenomas. J Intern Med 266: 5–18.

Trivellin G, Korbonits M (2011). AIP and its interacting partners. J Endocrinol 210: 137–155.

Trouillas J, Labat-Moleur F, Sturm N et al. (2008). Pituitary tumors and hyperplasia in multiple endocrine neoplasia type 1 syndrome (MEN1): a case-control study in a series of 77 patients versus 2509 non-MEN1 patients. Am J Surg Pathol 32: 534–543.

Trump D, Farren B, Wooding C et al. (1996). Clinical studies of multiple endocrine neoplasia type 1 (MEN1). QJM 89: 653–669.

Tuominen I, Heliövaara E, Raitila A et al. (2014). AIP inactivation leads to pituitary tumorigenesis through defective Gαi-cAMP signalling. Oncogene. http://dx.doi.org/10.1038/onc.2014.50.

Underdahl LO, Woolner LB, Black BM (1953). Multiple endocrine adenomas: report of 8 cases in which the parathyroids, pituitary and pancreatic islets were involved. J Clin Endocrinol Metab 13: 20–47.

Vallar L, Spada A, Giannattasio G (1987). Altered G_s and adenylate cyclase activity in human GH-secreting pituitary adenomas. Nature 330: 566–568.

Vandeva S, Jaffrain-Rea ML, Daly AF et al. (2010). The genetics of pituitary adenomas. Best Pract Res Clin Endocrinol Metab 24: 461–476.

Varsavsky M, Sebastian-Ochoa A, Torres Vela E (2012). Coexistence of a pituitary macroadenoma and multicentric paraganglioma: a strange coincidence. Endocrinol Nutr 60: 154–156.

Verges B, Boureille F, Goudet P et al. (2002). Pituitary disease in MEN type 1 (MEN1): data from the France–Belgium MEN1 multicenter study. J Clin Endocrinol Metab 87: 457–465.

Veugelers M, Wilkes D, Burton K et al. (2004). Comparative PRKAR1A genotype-phenotype analyses in humans with Carney complex and prkar1a haploinsufficient mice. Proc Natl Acad Sci U S A 101: 14222–14227.

Vierimaa O, Georgitsi M, Lehtonen R et al. (2006). Pituitary adenoma predisposition caused by germline mutations in the AIP gene. Science 312: 1228–1230.

Villa C, Lagonigro MS, Magri F et al. (2011). Hyperplasia-adenoma sequence in pituitary tumorigenesis related to aryl hydrocarbon receptor interacting protein gene mutation. Endocr Relat Cancer 18: 347–356.

Vortmeyer AO, Glasker S, Mehta GU et al. (2012). Somatic GNAS mutation causes widespread and diffuse pituitary disease in acromegalic patients with McCune–Albright syndrome. J Clin Endocrinol Metab 97: 2404–2413.

Wang Z, Yu R, Melmed S (2001). Mice lacking pituitary tumor transforming gene show testicular and splenic hypoplasia, thymic hyperplasia, thrombocytopenia, aberrant cell cycle progression, and premature centromere division. Mol Endocrinol 15: 1870–1879.

Weinstein LS, Shenker A, Gejman PV et al. (1991). Activating mutations of the stimulatory G protein in the McCune–Albright syndrome. N Engl J Med 325: 1688–1695.

Wermer P (1954). Genetic aspects of adenomatosis of endocrine glands. Am J Med 16: 363–371.

Wildi-Runge S, Bahubeshi A, Carret AS et al. (2011). New phenotype in familial DICER1 tumor syndrome: pituitary blastoma presenting at age 9 months. Endocr Rev 32: 1–777.

Xekouki P, Pacak K, Almeida M et al. (2012). Succinate dehydrogenase (SDH) D subunit (SDHD) inactivation in a growth-hormone-producing pituitary tumor: a new association for SDH? J Clin Endocrinol Metab 97: E357–E366.

Yamada S (2001). Epidemiology of pituitary tumors. In: Diagnosis and Management of Pituitary Tumors, Humana Press, Totawa.

Yu R, Bonert V, Saporta I et al. (2006). Aryl hydrocarbon receptor interacting protein variants in sporadic pituitary adenomas. J Clin Endocrinol Metab 91: 5126–5129.

Zhang X, Horwitz GA, Heaney AP et al. (1999). Pituitary tumor transforming gene (PTTG) expression in pituitary adenomas. J Clin Endocrinol Metab 84: 761–767.

Zhang X, Sun H, Danila DC et al. (2002). Loss of expression of GADD45 gamma, a growth inhibitory gene, in human pituitary adenomas: implications for tumorigenesis. J Clin Endocrinol Metab 87: 1262–1267.

Zhang X, Zhou Y, Mehta KR et al. (2003). A pituitary-derived MEG3 isoform functions as a growth suppressor in tumor cells. J Clin Endocrinol Metab 88: 5119–5126.

Zhao J, Dahle D, Zhou Y et al. (2005). Hypermethylation of the promoter region is associated with the loss of MEG3 gene expression in human pituitary tumors. J Clin Endocrinol Metab 90: 2179–2186.

Handbook of Clinical Neurology, Vol. 124 (3rd series)
Clinical Neuroendocrinology
E. Fliers, M. Korbonits, and J.A. Romijn, Editors

Chapter 24

Long-term effects of treatment of pituitary adenomas

ALBERTO M. PEREIRA*
Department of Endocrinology and Center for Endocrine Tumors, Leiden University Medical Center, Leiden, The Netherlands

TREATMENT OF PITUITARY ADENOMAS: THE HISTORICAL PERSPECTIVE

Pituitary adenomas are benign neuroendocrine tumors that can be treated effectively in the vast majority of cases. However, pituitary adenomas can cause serious morbidity due to overproduction of pituitary hormones and to local mass effects resulting in pituitary insufficiency and optic chiasm compression. This chapter will discuss the long-term effects of these tumors and of their treatment. The treatment of pituitary adenomas includes transsphenoidal surgery, medical therapy (e.g., with somatostatin analogs, growth hormone receptor antagonists, or dopamine agonists), and/or radiotherapy (Shimon and Melmed, 1998; Melmed, 2008). Remission rates induced by transsphenoidal surgery alone for macroadenomas do not exceed 50–70% and are between 80% and 90% for microadenomas in referral centers (Roelfsema et al., 2012). The most important side-effect of surgical treatment is *de novo* pituitary insufficiency, developing in 10–15% of patients. These intrinsic imperfections of transsphenoidal surgery (limited remission rates and new-onset pituitary insufficiency) preclude optimal recovery for all patients. However, many patients (up to 90%) with nonfunctioning macroadenomas (NFMA) and 10–50% of patients with functioning adenomas already had deficits preoperatively, some of which can resolve after onset of cure (i.e., hypogonadism) (Dekkers et al., 2006; Burgers et al., 2011; Honegger et al., 2012).

When surgery does not lead to remission, radiotherapy and medical treatment (with dopamine agonists, somatostatin analogs, or pegvisomant) are available for functioning tumors. For prolactinomas, dopamine agonists are the first choice. For acromegaly, somatostatin analogs are either first or second choice, followed by pegvisomant if needed, with radiotherapy being reserved for selected cases. For Cushing's disease, either reoperation or irradiations are secondary treatment options after transsphenoidal surgery. New developments suggest that medical treatment with pasireotide (Colao et al., 2012), possibly combined with cabergoline and ketoconazole (Feelders et al., 2010), but also monotherapy with mifepristone (Fleseriu et al., 2012) may represent potential novel treatment options for Cushing's disease.

Multimodality treatment – a combination of surgery followed by radiotherapy or medical treatment – will result in long-term remission rates of more than 90% of patients (Pereira et al., 2003; Kars et al., 2008; Albarel et al., 2013). However, despite the curative treatment of these adenomas *per se*, multiple physical and psychological complaints may persist, even when remission has been present for many years (Webb, 2011). It is likely that the causes of persistent morbidity are multifactorial, and over the last decade an accumulating number of data have emerged indicating that, amongst others, intrinsic imperfections of surgical treatment and endocrine replacement therapy, but also persistent effects of hormone excess on the central nervous system, may affect personality, behavior, and metabolism. This chapter will focus on mortality and disease-specific morbidity in patients with pituitary disease, particularly based on observations in patients who were considered to be cured because they were enjoying long-term remission after treatment.

MORTALITY

One of the most objective parameters for the long-term efficacy of treatment is normalization of increased mortality rates. If remission in the long term equals cure, mortality should be normal, as well as disease-related morbidity. All available data from historically treated

*Correspondence to: Alberto M. Pereira, Department of Endocrinology, Building 1, Room C-04-082, Leiden University Medical Center, 2333 ZA Leiden, The Netherlands. Tel: +31-71-526-3793, E-mail: A.M.Pereira@lumc.nl

cohorts of patients with pituitary tumors, both functioning and nonfunctioning, indicate increased mortality in these patients. For patients with acromegaly, the weighted mean standardized mortality ratio (SMR) in a meta-analysis was 1.72 (95% confidence interval 1.62–1.83). Interestingly, in studies that used transsphenoidal surgery as the primary therapy, the weighted mean SMR was lower (1.32, 95% confidence interval 1.12–1.56), but not normal (Dekkers et al., 2008). As with acromegaly, in Cushing's disease mortality has been evaluated in only a few small studies. The reported SMRs in these studies ranged from 1.7 to 4.8 with a consistently higher mortality risk in the patients with persistent hypercortisolemia (Feelders et al., 2012). As with acromegaly, transsphenoidal surgery as a first-line treatment for Cushing's disease has been an important advance, as high remission rates after initial surgery have been accompanied by mortality rates that in some, but not all, studies have even mirrored those observed in the general population (Swearingen et al., 1999; Hammer et al., 2004). In the Leiden series of pituitary patients who were treated by the same neurosurgeon with a single neurosurgical procedure for either acromegaly, Cushing's disease or nonfunctioning adenoma, the SMR after treatment was 1.24 for nonfunctioning adenoma, indicating a 24% persistent increased mortality rate. For acromegaly, the SMR was 1.32, while for Cushing's disease the increase in mortality was even higher, at 80% (Fig. 24.1). These observations point towards long-lasting hormone-specific effects despite long-term remission, especially of cortisol overexposure, on mortality. It is tempting to speculate on the potential mechanism. Recent studies in mice have shown that chronic high glucocorticoid exposure induces increased expression and decreased DNA methylation of fkbp5, a gene encoding a co-chaperone of the glucocorticoid receptor, thereby enhancing the response to glucocorticoids (Lee et al., 2010). In agreement, the degree of fkbp5 DNA methylation was strongly correlated with the percentage of visceral fat mass found in mice exposed to supraphysiologic doses of glucocorticoids (Karatsoreos et al., 2010).

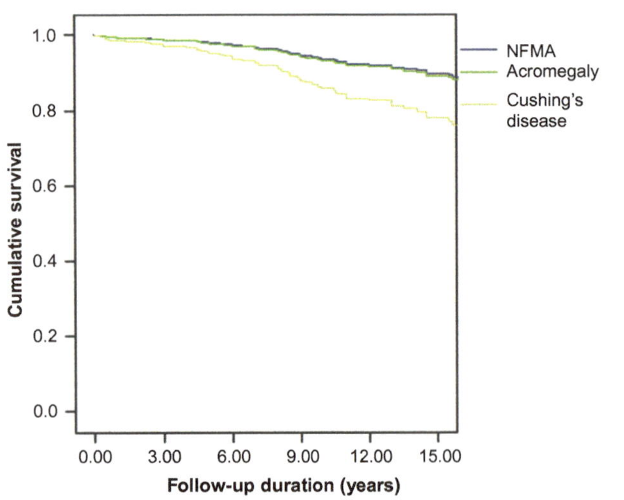

Fig. 24.1. Mortality in patients treated for nonfunctioning macroadenoma, acromegaly, and Cushing's disease (Cox model, corrected for age and gender).

HYPOPITUITARISM AND MORTALITY

The presence of hypopituitarism *per se* in patients with pituitary tumors is associated with increased mortality: data from a limited number of historically treated cohorts (including a total of 2211 patients) indicate that mortality risk is approximately twofold increased in patients with hypopituitarism (Isley, 2002). In the largest cohort, the West Midlands study, the SMR for patients with hypopituitarism was 1.87 and the excess mortality was predominantly from vascular and respiratory disease (Tomlinson et al., 2001). Age at diagnosis, female sex, and above all, craniopharyngioma were significant independent risk factors for increased mortality. Specific endocrine axis deficiency, with the exception of untreated gonadotropin deficiency, did not seem to affect mortality. However, the vast majority (90%) of the patients were not tested for growth hormone deficiency and one-third of the patients received cranial irradiation. The effect of long-term recombinant human GH (rGH) replacement on mortality has been reported recently after evaluation of the Dutch National Registry of Growth Hormone Treatment in Adults (retrospectively monitored between 1985 and 2009) (van Bunderen et al., 2011). Mortality in the treated group (when compared to normative data from the Dutch population) was slightly increased (SMR: 1.27, 95% CI 1.04–1.56). Mortality was higher in women than in men, and after exclusion of high-risk patients, the SMR for cardiovascular disease mortality remained increased in women. The reason for this sex difference remained unclear, although undertreatment of gonadal deficiency in premenopausal women has been suggested to be of influence (Isley, 2002). Mortality due to malignancies was not elevated. Interestingly, recent data in patients treated for nonfunctioning pituitary adenomas showed that higher glucocorticoid replacement doses were associated with increased overall mortality (Zueger et al., 2012). Thus, hypopituitarism in general is associated with an approximately twofold increased mortality in the presence of significant heterogeneity between the historically treated patient cohorts in terms of differences in pituitary disease, hormone replacement, and radiation therapies.

CARDIOVASCULAR MORBIDITY AND PITUITARY DISEASE

Patients with pituitary disease have an increased prevalence of cardiometabolic complications. This is evident in patients exposed to GH excess (acromegaly) and cortisol excess (Cushing's disease), but also the presence of hypopituitarism negatively affects the cardiometabolic phenotype. Patients with acromegaly have an increased prevalence of cardiovascular risk factors (more hypertension, diabetes mellitus, and more dyslipidemia), which, when compared with controls, is also reflected in an increased Framingham risk score (FRS). Treatment of acromegaly reduces FRS after 1 year, but only in those with controlled disease, indicating that long-term effective control of disease reduces the risk for coronary heart disease (Berg et al., 2010). Active acromegaly is also associated with a specific cardiomyopathy, characterized by biventricular hypertrophy, left ventricular systolic and diastolic dysfunction, valvular insufficiency, aortic root dilatation, and arrthymias (Colao et al., 2004). The prevalence of valvular insufficiency is dependent on the duration of exposure to increased GH concentrations, with a 19% increase in odds per year. Intriguingly, whereas the effects on LV function and mass appear to be reversible, (occult) valvular insufficiencies and arrhythmias may persist even after long-term remission (Pereira et al., 2004). This indicates that appropriate follow-up care and monitoring (for instance antibiotic prophylaxis for any nonsterile procedures) is required, especially for patients with inadequate control of GH overproduction.

Patients with Cushing's disease not only have a phenotypical resemblance to the metabolic syndrome, but also fulfill some of the criteria (Newell-Price et al., 2006). Carotid intima media thickness (IMT) is increased and vessel wall plaques are more common in patients with Cushing's (Colao et al., 1999; Faggiano et al., 2003). In conformity with this, patients with Cushing's disease have markedly increased cardiovascular morbidity and mortality (Dekkers et al., 2007). Intriguingly, some of the features of the metabolic syndrome in these patients persisted even after long-term successful correction of cortisol excess. Indeed, 1 year after remission, patients treated for Cushing's still demonstrated impaired glucose tolerance and increased insulin levels after oral glucose tolerance test (Faggiano et al., 2003; Giordano et al., 2011). Furthermore, waist circumference was persistently increased after 1 year of remission, and even after long-term remission, visceral fat mass was higher without affecting body mass index (Barahona et al., 2009). A prospective study in only 14 patients with Cushing's disease using whole body magnetic resonance imaging (MRI) demonstrated that remission reduced fat depots and reverted fat distribution to a more favorable cardiovascular risk, but also decreased skeletal muscle (Geer et al., 2012). Finally, the adverse metabolic profile in Cushing's disease also includes a hypercoagulable state, reflected by a high risk of venous thromboembolism (Stuijver et al., 2011), and is characterized by increased production of procoagulant factors and impaired fibrinolysis, which also persists, at least in the presence of biochemical remission induced by medical therapy (van der Pas et al., 2012).

A higher prevalence of (features of) the metabolic syndrome is also found in patients with hypopituitarism *per se* (Abs et al., 2006; Verhelst et al., 2011). Type 2 diabetes mellitus and cardiovascular risk factors such as dyslipidemia (46% versus 19%), hypertension (66% versus 36%), and abdominal obesity (38% versus 23%) were more prevalent in patients when compared with controls (van der Klaauw et al., 2007). Consequently, the prevalence of the metabolic syndrome was more than double in patients (38% versus 16%). Interestingly, treatment with recombinant human GH did not affect the prevalence of CVR factors (as evaluated by the metabolic syndrome (NCEP III criteria)) in adult patients with hypopituitarism after 1 and 5 years of treatment (van der Klaauw et al., 2007; Attanasio et al., 2010). These findings have recently been replicated in a study with a much longer duration of follow-up. In this study, in which patients with hypopituitarism were followed for up to 15 years after optimal hormone substitution therapy, including rGH therapy, the increase in the prevalence of the metabolic syndrome during the follow-up period was higher than expected as a consequence of aging alone in adults without hypopituitarism (Claessen et al., 2013). Interestingly, it appears that long-term supraphysiologic cortisol replacement, as was given in the majority of historically treated cohorts of patients with hypopituitarism, contributes to this phenotype, accounting for at least part of the adverse metabolic profile. When the glucocorticoid replacement dose was reduced from 20–30 mg/day to 15 mg/day, beneficial effects were found both on lipids and on waist circumference (Filipsson et al, 2006). This indicates that apparently appropriate hormonal substitution in adult patients with hypopituitarism may improve the adverse cardiovascular risk profile. Furthermore, the cause for the observed increased cardiovascular morbidity and mortality in hypopituitarism cannot be explained by hormonal deficiencies alone. In addition, these observations strongly suggest that patients with hypopituitarism might benefit from additional treatment with stringent criteria to reduce cardiovascular risk, an issue that has not yet routinely been incorporated into clinical practice. The question then arises as to the mechanism(s) involved in the explanation for the increased prevalence of cardiometabolic complications in patients with pituitary disease. It is likely that, apart from hypopituitarism and hormone excess syndromes, the failure to mimic physiologic hormone secretion with substitution and irreversible hypothalamic dysfunction play an important role.

FAILURE TO MIMIC PHYSIOLOGIC HORMONE SECRETION WITH SUBSTITUTION

Even when hormone substitution is complete, hypopituitarism is associated with intrinsic imperfections of endocrine replacement, i.e., the failure to mimic physiologic hormone secretion with substitution. Physiologic hormone secretion is characterized by strong adaptation to the environment (e.g., stress factors, food availability), and by pulsatile secretion with circadian variation, enabling intermittent signal exchange (Veldhuis et al., 2010). In addition, hormone-specific binding proteins and hormone-specific enzymes regulate the biological activity of the hormone via (de)activation at the tissue level, for instance deiodinases for thyroid hormone and 11-betahydroxysteroid dehydrogenases for cortisol. It is evident that such a fine-tuned homeostatic ensemble cannot be mimicked by oral or subcutaneous hormone replacement therapy. Moreover, there are no tools in clinical practice to measure the biological effects of hormones at the tissue level. This is illustrated by the following examples: first, the adequacy of thyroid hormone supplementation. Titration of thyroid hormone dose is only possible by measuring serum TSH concentrations, and this can be used only in primary hypothyroidism. However, even in primary hypothyroidism, when pituitary hormone secretion is intact but is only temporarily stimulated, it is not possible to obtain an optimal dose titration based on serum TSH concentrations alone. This is because TSH secretion varies widely between individuals, and consequently the normal reference range for TSH varies approximately tenfold (Jensen et al., 2004). In the case of pituitary disease and TSH deficiency, there is no single parameter available for titration of the dose of thyroid hormone, which poses an extra dilemma (see Ch. 27 in this volume). The thyroid gland secretes both thyroxine and triiodothyronine; however, only the latter is considered as the biologically active form of the hormone because to date thyroxine-specific receptors have not been identified. Animal data have shown that neither thyroxine nor triiodothyronine is able to normalize thyroid hormone concentrations in thyroidectomized mice in all tissues simultaneously (Escobar-Morreale et al., 1996). A second example is the adequacy of growth hormone (GH) replacement therapy in adult patients with GH deficiency (for an extensive discussion of this issue the reader is referred to Ch. 28 in this volume). This is only possible by measuring IGF-1 concentration; this, unlike GH, is not subject to pulsatile secretion and reflects many, but not all of the biological effects of GH. Nevertheless, IGF-1 is the parameter upon which the dose of rGH dose is titrated. However, approximately 50% of adult patients over 40 years of age with GH deficiency have IGF-1 levels within the normal reference range (Hilding et al., 1999), precluding correct titration of the dose of rGH. Intriguingly, current clinical guidelines recommend that rGH dosing regimens should be individualized and titrated on individual IGF-1 SDS scores (preferably between 0 and 2 SDS) corrected for age and gender (Molitch et al., 2011). This implicates that, by definition, 50% of patients will be oversubstituted, which hampers a reliable evaluation of the effects of physiologic rGH replacement. Consequently, titration of endocrine replacement therapy is possible only within certain limits. It is very likely that these intrinsic imperfections in endocrine replacement therapies result in subtle physiologic derangements (Romijn et al., 2003). In agreement with this hypothesis are recent findings with a dual-release hydrocortisone preparation that provides a more circadian-based serum cortisol profile. When compared to a standard treatment with oral hydrocortisone taken three times per day, a once-daily administration of the dual-release hydrocortisone tablet reduced both body weight and blood pressure, and improved glucose metabolism within 3 months (Johannsson et al., 2012).

HYPOTHALAMIC DYSFUNCTION

Another factor that may affect the prevalence of cardiometabolic morbidity is irreversible hypothalamic dysfunction. The hypothalamus, situated in close vicinity to the pituitary, is easily involved in suprasellar tumor extension, which may compromise hypothalamic function even after surgical treatment. In addition, it is well documented that the hypothalamus is more radiosensitive than the pituitary. Radiotherapy for pituitary adenomas therefore frequently affects hypothalamic neurosecretory function and precedes pituitary insufficiencies (Darzy and Shalet, 2009). Hypothalamic neurosecretory dysfunction is reflected by the incidence of pituitary insufficiencies that develop after radiotherapy: radiation doses below 30 Gy result in isolated growth hormone deficiency in only 30% of cases, while doses of 30–50 Gy not only result in growth hormone deficiency in the majority of the patients but also in multiple pituitary hormone deficits in 30–60% during long-term follow-up (Darzy and Shalet, 2009). However, the hypothesis that radiation-induced growth hormone deficiency is primarily related to hypothalamic dysfunction, based on relative preservation of GH responses to exogenous GHRH and arginine stimulation and blunted responses to an insulin tolerance test (ITT), is an oversimplification. Irradiated patients with normal individual responses to ITT and GHRH + arginine had preserved GH pulsatility and diurnal variation but overall group responses were reduced by 50% compared to matched controls, arguing for dual damage to both the hypothalamus and the pituitary (Darzy et al., 2007). The relatively preserved GH responses to GHRH + arginine might be explained by endogenous hyperactivation of the hypothalamus–pituitary axis resulting in (near) normalization of spontaneous GH secretion. Theoretically, prolactin is a sensitive tool to estimate the likelihood of radiation-induced damage to the hypothalamus. However, because hyperprolactinemia might be a consequence of decreased hypothalamic dopamine secretion, hyperprolactinemia is variable in severity and often subclinical, and therefore is not always detected. In time, prolactin levels diminish and may even normalize due to slowly evolving radiation-induced damage of lactotrophs. If hyperprolactinemia is to be considered as an indicator of disturbed hypothalamic dopamine secretion, it is plausible to assume that the prolactin level will be high in the first years after radiation and normal after several years when lactrotroph function has declined. This has been documented in several anecdotal cases and small patient series, though mainly in young women. It should be noted that a transient mild increase within the reference ranges in men and older individuals would not be easily detected (Darzy and Shalet, 2009).

Besides regulation of pituitary hormone secretion, the hypothalamus synchronizes activity and rest to the day/night cycle by means of biological clock mechanisms, which enables the individual to keep a stable internal environment. Although the hypothalamus is a relatively small structure of only 4 mL, it contains many groups of nerve cell bodies forming distinct nuclei, which have highly diverse structural, molecular, and functional organizations (Swaab et al., 1992). These hypothalamic nuclei are crucial to integrating and conveying the different signals, informing the brain of the internal and external environment. As a consequence, hypothalamic dysfunction may result in altered regulation of central clock mechanisms, which predisposes to alterations in metabolism (Kalsbeek et al., 2001). In addition, damage to other hypothalamic nuclei involved in the integrative physiology of metabolism will accentuate these effects. Considering the large proportion of patients with damage to the optic nerves, it is possible that the effects of treatment may result in damage to the suprachiasmatic hypothalamic nucleus. Hypothalamic damage thus may contribute to the observed morbidity, but convincing clinical associations were not available until recently. A case-control study in patients with pituitary insufficiency and a history of optic chiasm compression demonstrated that proximal skin temperature and its relation to sleep were affected in contrast to those without chiasm compression (Romeijn et al., 2012). In addition, another study from the same group reported that a history of cranial radiotherapy was associated with a higher visceral to subcutaneous fat ratio in men treated for (supra)sellar tumors with pituitary insufficiency (Borgers et al., 2012). Finally, a postmortem study from the same group of investigators found reduced arginine-vasopressine immunoreactivity in the suprachiasmatic nucleus of the hypothalamus in patients with suprasellar tumors, leading to permanent visual field defects when compared to controls (Borgers et al., 2013). Thus, although these preliminary data merely reflect associations and do not prove causality, it is tempting to speculate that hypothalamic dysfunction, as a result of hypothalamic damage induced by either irradiation or compression, may explain, at least in part, hypopituitarism-related morbidity. Thus, hypothalamic dysfunction can affect the regulation of body fat distribution and contribute to sleep–wake disturbances, which in turn facilitate the development of an adverse cardiometabolic phenotype, such as the metabolic syndrome.

QUALITY OF LIFE

Quality of life (QoL) estimation has emerged as an important tool in the evaluation of the success of treatment. QoL has been evaluated in patients with pituitary adenomas using general health-related questionnaires in both untreated and treated disease. These reports demonstrated that QoL generally improves after treatment (Koltowska-Häggström and Mattsson, 2006), but also that QoL remains impaired even after successful treatment, with disease-specific features (van der Klaauw et al., 2008). However, quality of life fails to normalize in the vast majority of studies in the long-term, and it is not yet known exactly why this is the case.

When QoL was evaluated using general health questionnaires, patients treated for acromegaly were more severely impaired when compared to patients treated for Cushing's disease, prolactinoma, or nonfunctioning adenoma (van der Klaauw et al., 2008). Specifically, patients treated for acromegaly predominantly reported both impairment in physical performance and an increase in bodily pain (Biermasz et al., 2005), whereas patients treated for Cushing's disease reported impairment in physical functioning (van Aken et al., 2005). However, these results were obtained using general health questionnaires and not disease-specific ones (See also Table 24.1).

In addition, these QoL studies revealed psychological impairments on various quality of life questionnaires, both in general health and disease-specific questionnaires. As stated above, the QoL questionnaires are not designed for an in-depth assessment of psychological functioning. While biological effects of cortisol and GH excess on psychological functioning have been

Table 24.1

Functional domains affecting general well-being, cognition, mood, and behavior after long-term remission for pituitary disease

	Quality of Life	Cognitive Function	Psychopathology	Coping Strategies	Illness Perceptions
Hypopituitarism	Impaired*	Impaired	Increased tenseness and impaired self-confidence	Not studied	Not studied
Non-functioning macroadenoma (NFMA)	Impaired	Normal or impaired	Apathy, anxiety, depression, and affective lability #	Less effective coping	Not studied
Acromegaly	Impaired* (Arthropathy)	Normal or impaired	Maladaptive personality traits, especially: negative affect, somatic arrousal, oppositionality	Less effective coping	Strong correlation with quality of life. Lack of personal control$
Prolactinoma	Impaired	Not studied	Reduced extraversion and increased shyness with strangers, tending to be more neurotic**	Not studied	Not studied
Cushing's disease	Impaired*	Impaired	Maladaptive personality traits, especially: Apathy, irritability, anxiety, negative affect and lack of positive affect	Less effective coping	Strong correlation with quality of life. More negative illness perceptions ##

Quality of Life: evaluated using general health related questionaires, and disease specific questionaires* (for adult growth hormone deficiency in case of hypopituitarism; in acromegaly the presence of arthropathy was the most frequently reported contributor that negatively affected quality of life).

Cognitive function: various tests, which evaluated global cognitive functioning, memory, and executive functioning.

Psychopathology: evaluated using questionaires, focussing on frequently occurring psychiatric symptoms in somatic illness. #: NFMA vs controls. **: prolactinoma vs NFMA.

Coping strategies: compared with the normal population. No differences between patients in remission for Cushing's disease, acromegaly, and NFMA, besides more social support seeking in patients with Cushing's disease than in patients treated for NFMA.

Illness perceptions: evaluated using the Illness Perception Questionnaire (IPQ) Revised. Illness perceptions showed a strong correlation with quality of life. $: compared with patients with acute and chronic pain; ##: compared with patients with other acute or chronic conditions.

For literature references: see text in corresponding sections.

reported in several studies in untreated Cushing's disease and in acromegaly, and in some studies after short-term remission (Pereira et al., 2010), it was unknown if, and to what extent, cognitive dysfunction and psychopathology were present in these patients in the long term.

COGNITIVE FUNCTION AND PSYCHOPATHOLOGY

Cognition, mood, and personality may be affected by pituitary disease, but also by its treatment, by disrupting the connections between the prefrontal cortex and other limbic structures, thereby impairing the behavioral control exerted by the prefrontal cortex on the limbic system (Weitzner, 1998).

Specifically, limbic structures such as the hippocampus and the prefrontal cortex are rich in glucocorticoid, GH, and IGF-1 receptors (Lai et al., 1993; de Kloet et al., 2005). This suggests that these structures are particularly vulnerable to cortisol excess, as is present in Cushing's disease (Forget et al., 2000), and possibly also to the GH excess found in acromegaly. Accordingly, impaired cognitive function, psychopathology, and maladaptive personality have all been documented in patients with active acromegaly and Cushing's disease (Richert et al., 1987; Flitsch et al., 2000; Starkman et al., 2001). After successful surgery physical and psychiatric symptoms improve significantly within the first year (Pereira et al., 2010).

In addition, substitution of GH-deficient patients with recombinant human GH resulted in a rapid and sustained amelioration of cognitive functioning (Arwert et al., 2006).

An increasing number of data indicate that neuropsychological dysfunction may persist even after many years of remission (See also Table 24.1). For instance, 36% of the patients that were considered cured from acromegaly showed elevated scores for anxiety and depression (Biermasz et al., 2004). Recent studies have demonstrated that patients cured of Cushing's disease have persisting impairments in cognitive functioning (Tiemensma et al., 2010a). Cognitive function was not different between patients successfully treated for acromegaly and nonfunctioning adenomas (Tiemensma et al., 2010b; Brummelman et al., 2011). Intriguingly, in the study by Tiemensma et al. cognitive function was not different from controls, whereas Brummelman and colleagues found worse scores on memory and executive functioning in patients treated for acromegaly and for nonfunctioning adenomas when compared to controls (Brummelman et al., 2012). In addition, recent reports indicate that behavior and mood are affected despite long-term remission. In a study that included both uncontrolled and controlled patients with acromegaly and nonfunctioning adenomas, increased anxiety-related personality traits were found in both patient groups when compared to healthy controls. In addition, acromegaly was associated with reduced impulsivity and novelty seeking behavior (Sievers et al., 2009). In the Leiden series, patients considered to be cured from Cushing's disease and acromegaly had more psychopathology and maladaptive personality traits when compared to both patients treated for a nonfunctioning pituitary tumors and healthy controls. These differences were most prominent in the patients treated for Cushing's with disease-specific features (Tiemensma et al., 2010b, c), although altered personality profiles have also been found in patients with prolactinomas (Athanasoulia et al., 2012). Finally, it appears that other factors, such as coping strategies and illness perceptions, are related to psychopathology, and consequently may affect quality of life after long-term remission. When compared to controls, patients treated for Cushing's disease, for acromegaly, and for nonfunctioning pituitary adenomas displayed different and less effective coping strategies (Tiemensma et al., 2011a) (Fig. 24.2). Illness perceptions were negatively affected in patients after remission for both acromegaly and Cushing's disease and were strongly related to quality of life (Tiemensma et al., 2011b, c).

ACROMEGALIC ARTHROPATHY AS A MODEL FOR DISEASE-SPECIFIC PERSISTENT MORBIDITY

At present, acromegalic arthropathy most probably is the best model representing disease-specific persistent morbidity. Patients with acromegaly have a high prevalence of joint-related morbidity. Joint complaints

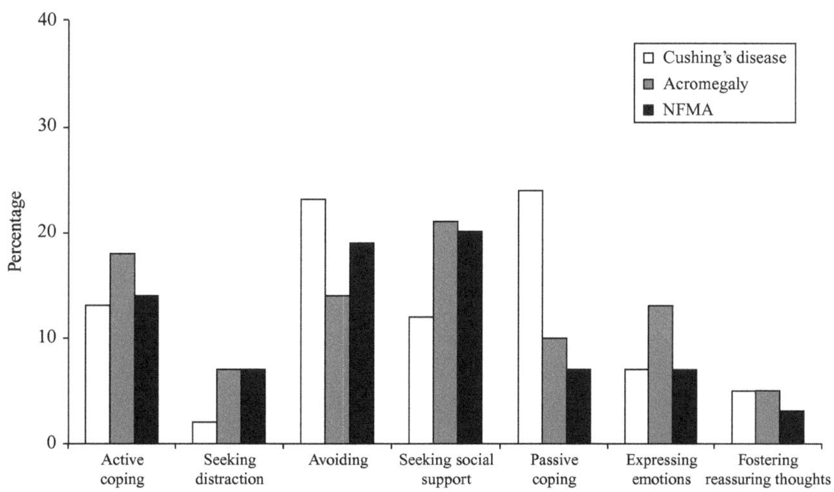

Fig. 24.2. Distribution of scores of patients with long-term cure of pituitary adenomas and the reference groups. Means are displayed in this figure. The coping strategies Avoiding and Passive coping are negative strategies, which means that a higher score indicates a more maladaptive coping strategy. The coping strategies Active coping, Seeking distraction, Seeking social support, Expressing emotions, and Fostering reassuring thoughts are positive strategies, which means that lower scores indicate a more maladaptive coping strategy.

(arthropathy) were the most important contributor for the persistent impairment in quality of life in patients with acromegaly despite long-term remission (Biermasz et al., 2004). In a case-control study, patients with acromegaly had a four- to 12-fold increased risk of developing osteoarthritis (especially in the hand and spine) in comparison to the general population despite long-term remission, and at a significantly younger age (Wassenaar et al., 2009). Radiologic characteristics of acromegalic osteoarthritis typically include widening of joint spaces and severe osteophytosis. Increased insulin-like growth factor-1 (IGF-1) levels most likely play a key role in the pathophysiology that leads to secondary osteoarthritis in acromegaly. IGF-1 is highly expressed in osteophytes in murine experimental osteoarthritis models (Okazaki et al., 1999) and in proliferative chondrocytes in the growth plate, and is directly regulated by growth hormone (Nilsson et al., 1986). A recent meta-analysis demonstrated that a functional growth hormone receptor polymorphism that enhances GH signal transduction, exon 3 deleted GHR, was associated with radiographic osteoarthritis in females (Claessen et al., 2014). For the development of acromegalic arthropathy, it is hypothesized that the first stage, in which elevated GH and IGF-1 levels stimulate the growth of articular cartilage and periarticular ligaments, is (at least partly) reversible. At this stage cartilage hypertrophy develops, limiting range of movement. The second (irreversible) stage is characterized by mechanical changes where the acromegalic joints acquire the characteristics of degenerative joint disease (Barkan, 2001). In a follow-up study of 67 patients with long-term disease control of almost 13 years, the severity of GH/IGF-1 excess at diagnosis was related to the prevalence of radiographic osteoarthritis (Biermasz et al., 2009). It is noteworthy that the distribution of radiologic abnormalities in acromegaly remains different from regular degenerative joint disease, i.e., primary osteoarthritis. The radiographic phenotype in acromegaly is predominantly characterized by severe osteophytosis, frequently in combination with preserved normal or even widened joint spaces.

Thus, acromegaly arthropathy represents a valuable clinical model of the impact of transient overexposure of GH IGF-1 excess on joints and can be used to delineate the role of GH and IGF-1 in the regulation of degenerative joint disease.

IMPLICATIONS FOR TREATMENT AND FOLLOW-UP

Patients in long-term remission after effective treatment for pituitary adenomas frequently encounter persistent morbidity in the face of near normalization of mortality. The fact that long-term remission is accompanied by persistent multiple physical and psychological impairments in many patients with reduced quality of life and that, by inference, these patients cannot be considered really cured has gained much attention, especially during the last decade. Hypopituitarism, intrinsic imperfections of endocrine replacement therapy, and persistent effects of previous hormone excess on the central nervous system may all affect long-term morbidity, general well-being, and mortality. This suggests that treatment goals for patients with pituitary adenomas will shift from long-term cure to long-term care. Further research is therefore needed to get further insight into these factors of influence, such as the extent of reversibility of hormone excess syndromes on cardiovascular risk and behavior. The fact that coping strategies are altered despite long-term remission, and illness perceptions are affected, strongly suggests that long-term care should incorporate self-management interventions that might help to improve quality of life for these patients.

References

Abs R, Feldt-Rasmussen U, Mattsson AF et al. (2006). Determinants of cardiovascular risk in 2589 hypopituitary GH-deficient adults – a KIMS database analysis. Eur J Endocrinol 155: 79–90.

Albarel F, Castinetti F, Morange I et al. (2013). Outcome of multimodal therapy in operated acromegalic patients, a study in 115 patients. Clin Endocrinol (Oxf) 78: 263–270, 2013.

Arwert LI, Veltman DJ, Deijen JB et al. (2006). Effects of growth hormone substitution therapy on cognitive functioning in growth hormone deficient patients: a functional MRI study. Neuroendocrinology 83: 12–19.

Athanasoulia AP, Ising M, Pfister H et al. (2012). Distinct dopaminergic personality patterns in patients with prolactinomas: a comparison with nonfunctioning pituitary adenoma patients and age- and gender-matched controls. Neuroendocrinology 96: 204–211.

Attanasio AF, Mo D, Erfurth EM et al. (2010). Prevalence of metabolic syndrome in adult hypopituitary growth hormone (GH)-deficient patients before and after GH replacement. J Clin Endocrinol Metab 95: 74–81.

Barahona MJ, Sucunza N, Resmini E et al. (2009). Persistent body fat mass and inflammatory marker increases after long-term cure of Cushing's syndrome. J Clin Endocrinol Metab 94: 3365–3371.

Barkan AL (2001). Acromegalic arthropathy. Pituitary 4: 263–264.

Berg C, Petersenn S, Lahner H et al. (2010). Cardiovascular risk factors in patients with uncontrolled and long-term acromegaly: comparison with matched data from the general population and the effect of disease control. J Clin Endocrinol Metab 95: 3648–3656.

Biermasz NR, van Thiel SW, Pereira AM et al. (2004). Decreased quality of life in patients with acromegaly despite long-term cure of growth hormone excess. J Clin Endocrinol Metab 89: 5369–5376.

Biermasz NR, Pereira AM, Smit JW et al. (2005). Morbidity after long-term remission for acromegaly: persisting joint-related complaints cause reduced quality of life. J Clin Endocrinol Metab 90: 2731–2739.

Biermasz NR, Wassenaar MJ, van der Klaauw AA et al. (2009). Pretreatment insulin-like growth factor-I concentrations predict radiographic osteoarthritis in acromegalic patients with long-term cured disease. J Clin Endocrinol Metab 94: 2374–2379.

Borgers AJ, Alkemade A, Venema HW et al. (2012). A history of cranial radiotherapy is associated with a higher visceral to subcutaneous fat ratio in men with pituitary insufficiency. Eur J Endocrinol 166: 619–624.

Borgers AJ, Fliers E, Siljee JE et al. (2013). Arginine vasopressin immunoreactivity is decreased in the hypothalamic suprachiasmatic nucleus of subjects with suprasellar tumors. Brain Pathol 23: 440–444.

Brummelman P, Elderson MF, Dullaart RP et al. (2011). Cognitive functioning in patients treated for nonfunctioning pituitary macroadenoma and the effects of pituitary radiotherapy. Clin Endocrinol (Oxf) 74: 481–487.

Brummelman P, Koerts J, Dullaart RP et al. (2012). Effects of previous growth hormone excess and current medical treatment for acromegaly on cognition. Eur J Clin Invest 42: 1317–1324.

Burgers AM, Kokshoorn NE, Pereira AM et al. (2011). Low incidence of adrenal insufficiency after transsphenoidal surgery in patients with acromegaly: a long-term follow-up study. J Clin Endocrinol Metab 96: E1163–E1170.

Colao A, Pivonello R, Spiezia S et al. (1999). Persistence of increased cardiovascular risk in patients with Cushing's disease after five years of successful cure. J Clin Endocrinol Metab 84: 2664–2672.

Colao A, Ferone D, Marzullo P et al. (2004). Systemic complications of acromegaly: epidemiology, pathogenesis, and management. Review. Endocr Rev 25: 102–152.

Colao A, Petersenn S, Newell-Price J et al. (2012). Pasireotide B2305 study group. A 12-month phase 3 study of pasireotide in Cushing's disease. N Engl J Med 366: 914–924.

Claessen KM, Appelman-Dijkstra NM, Adoptie DM et al. (2013). Metabolic profile in growth hormone deficient (GHD) adults after long-term recombinant human growth hormone (rhGH) therapy. J Clin Endocrinol Metab 98: 352–361.

Claessen KM, Kloppenburg M, Kroon HM et al. (2014). Relationship between the functional exon 3 deleted growth hormone receptor polymorphism and symptomatic osteoarthritis in women. Ann Rheum Dis 73: 433–436.

Darzy KH, Shalet SM (2009). Hypopituitarism following radiotherapy. Review. Pituitary 12: 40–50.

Darzy KH, Pezzoli SS, Thorner MO et al. (2007). Cranial irradiation and growth hormone neurosecretory dysfunction: a critical appraisal. J Clin Endocrinol Metab 92: 1666–1672.

de Kloet ER, Joëls M, Holsboer F (2005). Stress and the brain: from adaptation to disease. Nat Rev Neurosci 6: 463–475.

Dekkers OM, Pereira AM, Roelfsema F et al. (2006). Observation alone after transsphenoidal surgery for nonfunctioning pituitary macroadenoma. J Clin Endocrinol Metab 91: 1796–1801.

Dekkers OM, Biermasz NR, Pereira AM et al. (2007). Mortality in patients treated for Cushing's disease is increased, compared with patients treated for nonfunctioning pituitary macroadenoma. J Clin Endocrinol Metab 92: 976–981.

Dekkers OM, Biermasz NR, Pereira AM et al. (2008). Mortality in acromegaly: a metaanalysis. J Clin Endocrinol Metab 93: 61–67.

Escobar-Morreale HF, del Rey FE, Obregon MJ et al. (1996). Only the combined treatment with thyroxine and triiodothyronine ensures euthyroidism in all tissues of the thyroidectomized rat. Endocrinology 137: 2490–2502.

Faggiano A, Pivonello R, Spiezia S et al. (2003). Cardiovascular risk factors and common carotid artery caliber and stiffness in patients with Cushing's disease during active disease and 1 year after disease remission. J Clin Endocrinol Metab 88: 2527–2533.

Feelders RA, de Bruin C, Pereira AM et al. (2010). Pasireotide alone or with cabergoline and ketoconazole in Cushing's disease. N Engl J Med 362: 1846–1848.

Feelders RA, Pulgar SJ, Kempel A et al. (2012). The burden of Cushing's disease (CD): clinical and health-related quality of life aspects. Eur J Endocrinol 167: 311–326.

Filipsson H, Monson JP, Koltowska-Häggström M et al. (2006). The impact of glucocorticoid replacement regimens on metabolic outcome and comorbidity in hypopituitary patients. J Clin Endocrinol Metab 91: 3954–3961.

Flitsch J, Spitzner S, Ludecke DK et al. (2000). Emotional disorders in patients with different types of pituitary adenomas and factors affecting the diagnostic process. Exp Clin Endocrinol Diabetes 108: 480–485.

Fleseriu M, Biller BM, Findling JW et al. (2012). Mifepristone, a glucocorticoid receptor antagonist, produces clinical and metabolic benefits in patients with Cushing's syndrome. J Clin Endocrinol Metab 97: 2039–2049.

Forget H, Lacroix A, Somma M et al. (2000). Cognitive decline in patients with Cushing's syndrome. J Int Neuropsychol Soc 6: 20–29.

Geer EB, Shen W, Strohmayer E et al. (2012). Body composition and cardiovascular risk markers after remission of Cushing's disease: a prospective study using whole-body MRI. J Clin Endocrinol Metab 97: 1702–1711.

Giordano R, Picu A, Marinazzo E et al. (2011). Metabolic and cardiovascular outcomes in patients with Cushing's syndrome of different aetiologies during active disease and 1 year after remission. Clin Endocrinol (Oxf) 75: 354–360.

Hammer GD, Tyrrell JB, Lamborn KR et al. (2004). Transsphenoidal microsurgery for Cushing's disease: initial outcome and long-term results. J Clin Endocrinol Metab 89: 6348–6357.

Hilding A, Hall K, Wivall-Helleryd IL et al. (1999). Serum levels of insulin-like growth factor I in 152 patients with growth hormone deficiency, aged 19–82 years, in relation to those in healthy subjects. J Clin Endocrinol Metab 84: 2013–2019.

Honegger J, Schmalisch K, Beuschlein F et al. (2012). Contemporary microsurgical concept for the treatment of Cushing's disease: endocrine outcome in 83 consecutive patients. Clin Endocrinol (Oxf) 76: 560–567.

Isley WL (2002). Growth hormone therapy for adults: not ready for prime time? Review. Ann Intern Med 137: 190–196.

Jensen E, Hyltoft Petersen P, Blaabjerg O et al. (2004). Establishment of a serum thyroid stimulating hormone (TSH) reference interval in healthy adults. The importance of environmental factors, including thyroid antibodies. Clin Chem Lab Med 42: 824–832.

Johannsson G, Nilsson AG, Bergthorsdottir R et al. (2012). Improved cortisol exposure-time profile and outcome in patients with adrenal insufficiency: a prospective randomized trial of a novel hydrocortisone dual-release formulation. J Clin Endocrinol Metab 97: 473–481.

Kalsbeek A, Fliers E, Romijn JA et al. (2001). The suprachiasmatic nucleus generates the diurnal changes in plasma leptin levels. Endocrinology 142: 2677–2685.

Karatsoreos IN, Bhagat SM, Bowles NP et al. (2010). Endocrine and physiological changes in response to chronic corticosterone: a potential model of the metabolic syndrome in mouse. Endocrinology 151: 2117–2127.

Kars M, Pereira AM, Smit JW et al. (2008). Long-term outcome of patients with macroprolactinomas initially treated with dopamine agonists. Eur J Intern Med 20: 387–393.

Koltowska-Häggström M, Mattsson AF (2006). Does long-term GH replacement therapy in hypopituitary adults with GH deficiency normalise quality of life? Eur J Endocrinol 155: 109–119.

Lai Z, Roos P, Zhai O et al. (1993). Age-related reduction of human growth hormone-binding sites in the human brain. Brain Res 621: 260–266.

Lee RS, Tamashiro KL, Yang X et al. (2010). Chronic corticosterone exposure increases expression and decreases deoxyribonucleic acid methylation of Fkbp5 in mice. Endocrinology 151: 4332–4343.

Melmed S (2008). Update in pituitary disease. Review. J Clin Endocrinol Metab 93: 331–338.

Molitch ME, Clemmons DR, Malozowski S et al. (2011). Evaluation and treatment of adult growth hormone deficiency: an Endocrine Society clinical practice guideline. J Clin Endocrinol Metab 96: 1587–1609.

Newell-Price J, Bertagna X, Grossman AB et al. (2006). Cushing's syndrome. Lancet 367: 1605–1617.

Nilsson A, Isgaard J, Lindahl A et al. (1986). Regulation by growth hormone of number of chondrocytes containing IGF-I in rat growth plate. Science 233: 571–574.

Okazaki K, Jingushi S, Ikenoue T et al. (1999). Expression of insulin-like growth factor I messenger ribonucleic acid in developing osteophytes in murine experimental osteoarthritis and in rats inoculated with growth hormone-secreting tumor. Endocrinology 140: 4821–4830.

Pereira AM, van Aken MO, van Dulken H et al. (2003). Long-term predictive value of postsurgical cortisol concentrations for cure and risk of recurrence in Cushing's disease. J Clin Endocrinol Metab 88: 5858–6584.

Pereira AM, van Thiel SW, Lindner JR et al. (2004). Increased prevalence of regurgitant valvular heart disease in acromegaly. J Clin Endocrinol Metab 89: 71–75.

Pereira AM, Tiemensma J, Romijn JA (2010). Neuropsychiatric disorders in Cushing's syndrome. Neuroendocrinology 92 (Suppl 1): 65–70.

Richert S, Strauss A, Fahlbusch R et al. (1987). Psychopathologic symptoms and personality traits in patients with florid acromegaly. Schweiz Arch Neurol Psychiatr 138: 61–86.

Roelfsema F, Biermasz NR, Pereira AM (2012). Clinical factors involved in the recurrence of pituitary adenomas after surgical remission: a structured review and meta-analysis. Pituitary 15: 71–83.

Romeijn N, Borgers AJ, Fliers E et al. (2012). Medical history of optic chiasm compression in patients with pituitary insufficiency affects skin temperature and its relation to sleep. Chronobiol Int 29: 1098–1108.

Romijn JA, Smit JWA, Lamberts SWJ (2003). Intrinsic imperfections of endocrine replacement therapy. Eur J Endocrinol 149: 91–97.

Shimon I, Melmed S (1998). Management of pituitary tumors. Review. Ann Intern Med 129: 472–483.

Sievers C, Ising M, Pfister H et al. (2009). Personality in patients with pituitary adenomas is characterized by increased anxiety-related traits: comparison of 70 acromegalic patients with patients with non-functioning pituitary adenomas and age- and gender-matched controls. Eur J Endocrinol 160: 367–373.

Starkman MN, Giordani B, Berent S et al. (2001). Elevated cortisol levels in Cushing's disease are associated with cognitive decrements. Psychosom Med 63: 985–993.

Stuijver DJ, van Zaane B, Feelders RA et al. (2011). Incidence of venous thromboembolism in patients with Cushing's syndrome: a multicenter cohort study. J Clin Endocrinol Metab 96: 3525–3532.

Swaab DF, Gooren LJ, Hofman MA (1992). The human hypothalamus in relation to gender and sexual orientation. Prog Brain Res 93: 205–217.

Swearingen B, Biller BM, Barker FG et al. (1999). Long-term mortality after transsphenoidal surgery for Cushing disease. Ann Intern Med 130: 821–824.

Tiemensma J, Kokshoorn NE, Biermasz NR et al. (2010a). Subtle cognitive impairments in patients with long-term cure of Cushing's disease. J Clin Endocrinol Metab 95: 2699–2714.

Tiemensma J, Biermasz NR, van der Mast RC et al. (2010b). Increased psychopathology and maladaptive personality traits, but normal cognitive functioning, in patients after long-term cure of acromegaly. J Clin Endocrinol Metab 95: E392–E402.

Tiemensma J, Biermasz NR, Middelkoop HAM et al. (2010c). Increased prevalence of psychopathology and maladaptive personality traits after long-term cure of Cushing's disease. J Clin Endocrinol Metab 95: E129–E141.

Tiemensma J, Kaptein AA, Pereira AM et al. (2011a). Coping strategies in patients after treatment for functioning or non-functioning pituitary adenomas. J Clin Endocrinol Metab 96: 964–971.

Tiemensma J, Kaptein AA, Pereira AM (2011b). Affected illness perceptions and the association with impaired quality of life in patients with long-term remission of acromegaly. J Clin Endocrinol Metab 96: 3550–3558.

Tiemensma J, Kaptein AA, Pereira AM (2011c). Negative illness perceptions are associated with impaired quality of life in patients after long-term remission of Cushing's syndrome. Eur J Endocrinol 165: 527–535.

Tomlinson JW, Holden N, Hills RK et al. (2001). Association between premature mortality and hypopituitarism. West Midlands Prospective Hypopituitary Study Group. Lancet 357: 425–431.

Wassenaar MJ, Biermasz NR, van Duinen N et al. (2009). High prevalence of arthropathy, according to the definitions of radiological and clinical osteoarthritis, in patients with long-term cure of acromegaly: a case-control study. Eur J Endocrinol 160: 357–365.

van Aken MO, Pereira AM, Biermasz NR et al. (2005). Quality of life in patients after long-term biochemical cure of Cushing's disease. J Clin Endocrinol Metab 90: 3279–3286.

van Bunderen CC, van Nieuwpoort IC, Arwert LI et al. (2011). Does growth hormone replacement therapy reduce mortality in adults with growth hormone deficiency? Data from the Dutch National Registry of Growth Hormone Treatment in adults. J Clin Endocrinol Metab 96: 3151–3159.

van der Klaauw AA, Biermasz NR, Feskens EJ et al. (2007). The prevalence of the metabolic syndrome is increased in patients with GH deficiency, irrespective of long-term substitution with recombinant human GH. Eur J Endocrinol 156: 455–462.

van der Klaauw AA, Kars M, Biermasz NR et al. (2008). Disease-specific impairments in quality of life during long-term follow-up of patients with different pituitary adenomas. Clin Endocrinol (Oxf) 69: 775–784.

van der Pas R, de Bruin C, Leebeek FW et al. (2012). The hypercoagulable state in Cushing's disease is associated with increased levels of procoagulant factors and impaired fibrinolysis, but is not reversible after short-term biochemical remission induced by medical therapy. J Clin Endocrinol Metabol 97: 1303–1310.

Veldhuis JD, Keenan DM, Pincus SM (2010). Regulation of complex pulsatile and rhythmic neuroendocrine systems: the male gonadal axis as a prototype. Review. Prog Brain Res 181: 79–110.

Verhelst J, Mattsson AF, Luger A et al. (2011). Prevalence and characteristics of the metabolic syndrome in 2479 hypopituitary patients with adult-onset GH deficiency before GH replacement: a KIMS analysis. Eur J Endocrinol 165: 881–889.

Webb SM (2011). Pituitary tumors: coping with "cured" pituitary tumors. Nat Rev Endocrinol 7: 251–252.

Weitzner MA (1998). Neuropsychiatry and pituitary disease: an overview. Review. Psychother Psychosom 67: 125–132.

Zueger T, Kirchner P, Herren C et al. (2012). Glucocorticoid replacement and mortality in patients with nonfunctioning pituitary adenoma. J Clin Endocrinol Metab 97: E1938–E1942.

Handbook of Clinical Neurology, Vol. 124 (3rd series)
Clinical Neuroendocrinology
E. Fliers, M. Korbonits, and J.A. Romijn, Editors

Chapter 25

Neuroendocrine mechanisms in athletes

MADHUSMITA MISRA*

Pediatric Endocrine and Neuroendocrine Units, Massachusetts General Hospital and Harvard Medical School, Boston, MA, USA

INTRODUCTION

Increased athletic activity is characterized by significant energy expenditure. When energy intake is sufficient to allow for a balanced caloric state, the beneficial effects of physical activity are evident. In addition to a positive body image, improved self-esteem, and increased likelihood of academic success, bone health is optimized. However, when energy intake is insufficient to meet individual needs, several adaptive neuroendocrine mechanisms come into play either to conserve energy for the most essential functions, or to allow the body to draw on its reserves to meet energy needs. These adaptive or physiologic neuroendocrine changes then lead to changes in body composition and weight, which can eventually impact bone metabolism. Low bone density is an important pathologic consequence in low energy states, notwithstanding the beneficial effects of exercise and mechanical loading on bone. Subsequent sections will discuss neuroendocrine changes that occur in athletes based on whether or not they are energy-replete, and the effects of these neuroendocrine changes on bone.

NEUROENDOCRINE ALTERATIONS IN ATHLETES

Hypothalamic–pituitary–gonadal axis

SPECTRUM OF MENSTRUAL FUNCTION IN THE FEMALE ATHLETE

The functionality of the hypothalamic–pituitary–gonadal (HPG) axis in athletes depends on their state of energy availability, although a contribution of genetic factors appears likely, based on recent data (Caronia et al., 2011). Low energy availability from excessive energy expenditure (from exercise), and/or a decrease in energy intake, presumably leads to varying degrees of perturbation of the HPG axis, particularly in female athletes (Loucks and Thuma, 2003). There is, in fact, a spectrum of menstrual function reported in the female athlete that ranges from ovulatory eumenorrhea to subclinical menstrual dysfunction (luteal phase defects and anovulatory eumenorrhea) to functional hypothalamic oligoamenorrhea (Nattiv et al., 2007).

One study reported that amongst normal cycling (eumenorrheic) recreational runners, 45% of the women have luteal phase defects and 12% have anovulatory cycles, compared to a 90% prevalence of ovulatory cycles amongst sedentary controls in the study (De Souza et al., 1998). Menstrual irregularity occurs in up to 24% of adolescent athletes (Nichols et al., 2006, 2007b) and is variably reported in 3–66% of adult athletes. The prevalence of oligoamenorrhea (the most extreme form of menstrual dysfunction in the female athlete) depends on several factors, which include the nature of exercise, the intensity of training, and the nutritional status of the athlete. A higher prevalence of oligoamenorrhea is reported in endurance sports (such as running, cycling, and swimming), gymnastics and in ballet dancers, and the prevalence increases with increasing intensity of training, and with decreasing nutritional status of the athlete (Sanborn et al., 1982).

Disordered eating behavior is common in the female athlete, and is reported in up to 18–20% of adolescent athletes (Nichols et al., 2006, 2007b), and 25% of college athletes (Beals and Hill, 2006). This contributes to a low energy state from decreased caloric intake, and the triad of low energy availability, menstrual dysfunction, and low bone density is commonly referred to as the "female athlete triad." Disordered eating is particularly common in sports that emphasize leanness or appearance (such as gymnastics, diving, and figure skating), increases with the level of competition, and is more

*Correspondence to: Madhusmita Misra, MD, MPH, BUL 457, Neuroendocrine Unit, Massachusetts General Hospital, Boston, MA 02114, USA. Tel: 617-726-3870, Fax: 617-726-5072, E-mail: mmisra@partners.org

prevalent in judged compared with refereed sports (Rosen and Hough, 1988). It is also more common in amenorrheic than eumenorrheic athletes, and in one study, 62% of adolescent amenorrheic athletes reported disordered eating behaviors compared with 11% of eumenorrheic athletes (Christo et al., 2008b). Another study reported eating disorders in 68% of amenorrheic compared with 38% of oligomenorrheic girls visiting a youth clinic (Loucks, 2007). Additionally, high school athletes who have a history of disordered eating are more than twice as likely to report oligoamenorrhea as those without this history (Nichols et al., 2007b).

Low energy availability in turn may impact GnRH pulsatility (as reflected by luteinizing hormone (LH) pulsatility). This has been demonstrated in acute settings where energy availability was restricted to varying degrees in eumenorrheic healthy women in a clinical research setting (Loucks and Thuma, 2003). The study reported a change in the pattern of LH pulsatility when energy availability decreased below 30 kcal/kg lean body mass/day (45 kcal/kg lean body mass/day being a state of energy balance), with a decrease in pulse frequency and an increase in pulse amplitude (Loucks and Thuma, 2003). There are, however, limited data regarding net energy availability in chronically exercising women.

One study reported lower net energy balance (standardized for body weight) in normally cycling anovulatory recreational runners compared with runners with luteal phase defects or ovulatory runners (De Souza et al., 1998). However, in chronically exercising women, studies have not been able to demonstrate a clear or consistent cut-off below which women are more likely to become oligoamenorrheic. This may be consequent to other contributory factors that impact the HPG axis, such as genetic factors. Women with hypothalamic amenorrhea are more likely to have heterozygous mutations in genes that impact GnRH neuron migration or function (Caronia et al., 2011). Presence of these mutations may predispose some women to menstrual dysfunction even when energy availability is above 30 kcal/kg lean body mass. Additionally, some women may be more capable than others of implementing adaptive changes that allow them to continue to have normal cycles despite very low energy availability.

Luteinizing hormone and follicle-stimulating hormone secretion in female athletes

Pulsatility patterns of luteinizing hormone and follicle-stimulating hormone in athletes

Data from adult and adolescent athletes regarding LH pulsatility are limited, but do indicate differences in oligoamenorrheic athletes compared with eumenorrheic athletes and nonathletes. Reported results are not always consistent, possibly because available studies differ with respect to (1) timing of frequent blood sampling in relation to menses (days 1–3 (Loucks et al., 1989), days 1–5 (Rickenlund et al., 2004), days 2–5 (Berga et al., 1989), days 3–6 (Laughlin and Yen, 1996), and days 1–10 (Ackerman et al., 2012c)); (2) the frequency of blood sampling (every 10 minutes during the sleep cycle and every 20 minutes during the wake cycle (Loucks et al., 1989), every 10 minutes for 8 hours overnight (Ackerman et al., 2012c), or every 10 (Laughlin and Yen, 1996), 15 (Berga et al., 1989) or 20 minutes for 24 hours (Rickenlund et al., 2004)); (3) methods used to analyze data from frequent sampling (Cluster (Berga et al., 1989; Loucks et al., 1989; Laughlin and Yen, 1996), Pulsar (Rickenlund et al., 2004) or Deconvolution (Ackerman et al., 2012c)).

Overall, studies in adults report lowest LH pulse frequency in amenorrheic athletes, intermediate pulse frequency in eumenorrheic athletes, and highest pulse frequency in nonathletes (Loucks et al., 1989; Laughlin and Yen, 1996; Rickenlund et al., 2004), particularly during the waking hours (Loucks et al., 1989), with either no differences in pulse amplitude (Laughlin and Yen, 1996; Rickenlund et al., 2004) or higher amplitude in eumenorrheic athletes compared with nonathletes (Loucks et al., 1989). One adolescent study that examined nighttime LH pulsatility reported no differences in LH pulse frequency among groups; however, amenorrheic girls had lower pulse amplitude and total pulsatile secretion than nonathletes (Ackerman et al., 2012c). The 24 h mean LH concentration or area under the curve (AUC) have been variably reported to be low (Loucks et al., 1989; Ackerman et al., 2012c) or unchanged (Laughlin and Yen, 1996; Rickenlund et al., 2004) in amenorrheic athletes compared with eumenorrheic athletes and nonathletes. In contrast, most studies report no differences in mean follicle-stimulating hormone (FSH) or FSH AUC in amenorrheic athletes compared with eumenorrheic athletes and nonathletes (Loucks et al., 1989; Laughlin and Yen, 1996).

Determinants of altered luteinizing hormone pulsatility in athletes

If indeed low energy availability determines why some (but not all athletes) develop oligoamenorrhea (Loucks and Thuma, 2003), there are likely metabolic and presumably hormonal links between decreased energy availability and impaired LH pulsatility. Although studies have not clearly identified low energy availability in chronically exercising women with amenorrhea in an ambulatory setting (Ackerman et al., 2013a), this may relate to inaccuracies inherent in the methodologies used to assess energy intake and expenditure, such as food records, exercise questionnaires, and accelerometers (Trost et al., 2006; Pietilainen et al., 2010). Fat mass and percent body fat are surrogate and objective measures of energy status as fat is a source of fuel in long-standing energy restricted states. Amenorrheic athletes do have lower fat mass and percent body fat than eumenorrheic athletes and nonathletes, even when body mass index (BMI) and lean body mass are in the normal range (Christo et al., 2008a; Ackerman et al., 2013a). Of note, in anorexia nervosa, a severely energy restricted state, increases in fat mass are the strongest predictor of resumption of menses (Misra et al., 2006). In addition to serving as a measure of energy stores, fat mass may be a link between low energy availability and altered pulsatile GnRH secretion through secretion of adipokines, such as leptin and adiponectin, that have a known impact on the HPG axis (Welt et al., 2004; Lu et al., 2008; Wen et al., 2008). Fat mass is also an important determinant of hormones that are altered in low energy states such as ghrelin, peptide YY (PYY), cortisol, insulin, and insulin-like growth factor 1 (IGF-1) (Misra et al., 2007; Christo et al., 2008a, b; Ackerman et al., 2012c, 2013a), which may in turn impact the HPG axis (Vulliemoz et al., 2004; Fernandez-Fernandez et al., 2005; Pinilla et al., 2006; Kluge et al., 2007; Sadler et al., 2010). Subsequent sections will discuss alterations in these hormones and hormonal axes in athletes, and their potential to impact GnRH pulsatility.

Genetic factors may also explain why some excessively exercising women are more likely than others to develop hypothalamic amenorrhea. In one study of 55 women with functional hypothalamic amenorrhea, seven women (two of whom were exercisers) had functional heterozygous mutations in the FGFR-1, PROKR2, GnRHR, or KAL-I genes, compared with none of 422 eumenorrheic control women and 47 menstruating exercisers (Caronia et al., 2011). Further studies are necessary to determine whether a genetic predisposition to hypogonadotropic hypogonadism explains which exercisers are more likely to become amenorrheic.

Finally, in girls with anorexia nervosa, one study has implicated coexisting polycystic ovarian syndrome (PCOS) as a predisposing factor for developing amenorrhea at a relatively higher weight (Pinhas-Hamiel et al., 2006). Rickenlund et al. reported higher 24 hour testosterone concentrations in oligomenorrheic athletes compared with both amenorrheic and regularly menstruating athletes and nonathletes (Rickenlund et al., 2004). The same study also reported a positive association between 24 hour testosterone concentration and menarchal age. Other studies have reported either lower levels of testosterone in amenorrheic compared with eumenorrheic athletes (Loucks et al., 1989; Christo et al., 2008a), or no differences among groups (Laughlin and Yen, 1996). More studies are necessary to determine whether an underlying predisposition to PCOS explains early development of amenorrhea in some athletes. Estradiol levels are generally lower in amenorrheic athletes compared with nonathletes (Loucks et al., 1989; Laughlin and Yen, 1996) or eumenorrheic athletes (Christo et al., 2008a).

Kisspeptin in athletes and nonathletes

Kisspeptin is an important driver of normal GnRH pulsatility and gonadotropin secretion (Navarro et al., 2005); however, data are limited regarding kisspeptin levels in athletes versus nonathletes. In prepubertal rats, a 3 day fast leads to decreased Kiss-1 expression in the hypothalami in both males and females, and LH responses to administered kisspeptin were potentiated in a fasting state (Navarro et al., 2004). These findings suggest that a decrease in Kiss-1 expression in conditions of low energy availability may lead to decreased pulsatile GnRH secretion and decreased levels of gonadotropins. In women with hypothalamic amenorrhea, twice daily kisspeptin administration leads to an acute increase in LH, FSH, and estradiol secretion, although subsequently, desensitization occurs (Jayasena et al., 2009). More studies are necessary to determine the impact of changes in kisspeptin levels on the HPG axis in athletes.

Prolactin and oxytocin in athletes and nonathletes

Levels of prolactin are typically lower in amenorrheic athletes compared with normally cycling women (Laughlin and Yen, 1996; Rickenlund et al., 2004). Similarly, pooled overnight oxytocin levels are lower in amenorrheic athletes compared with normally cycling women, even after controlling for estradiol levels (Lawson et al., 2013).

Appetite regulating and gut peptides that may regulate energy homeostasis and impact the hypothalamic–pituitary–gonadal axis

Leptin

Leptin is an adipokine that correlates very strongly with absolute fat mass, and levels of leptin decrease both acutely and chronically with energy deprivation (Grinspoon et al., 1996, 1997) and in conditions associated with low fat mass (Misra et al., 2005b). Leptin is also anorexigenic, and a decrease in leptin in conditions of low energy availability is an appropriate adaptive response to prevent further suppression of food intake. Both fasting (Christo et al., 2008a) and integrated measures of overnight leptin secretion (Ackerman et al., 2012c) are markedly lower in adolescent and young adult amenorrheic athletes compared with eumenorrheic athletes and nonathletes, subsequent to decreases in pulse amplitude, pulse mass, and total pulsatile secretion. Leptin is also lower in adult ovulatory exercisers, exercisers with luteal phase deficiency, and anovulatory eumenorrheic exercisers compared with sedentary controls (De Souza et al., 2003, 2007), and lower in amenorrheic exercisers compared with ovulating exercisers (Corr et al., 2011).

Leptin has a known impact on the HPG axis. Leptin-deficient mice and humans are hypogonadal, and replacing leptin reverses this state of hypogonadotropic hypogonadism (Chehab et al., 1996; Strobel et al., 1998). Leptin is an important determinant of pulsatile LH secretion, as demonstrated in amenorrheic athletes and in a combined group of athletes and nonathletes, independent of BMI (Ackerman et al., 2012c). Administration of leptin to women with hypothalamic amenorrhea (not specifically athletes) normalizes gonadotropin secretion and induces ovulation in 30–40% of these women (Welt et al., 2004; Chou et al., 2011).

Ghrelin

Ghrelin is an orexigenic hormone and GH secretagogue that also increases secretion of ACTH, FSH and LH (Arvat et al., 2001; Vulliemoz et al., 2004; Kluge et al., 2007). Levels of ghrelin reflect overall energy status, and are inversely associated with fat mass (Ackerman et al., 2012c). High ghrelin levels are common in severely energy-restricted states (Misra et al., 2005a), likely an adaptive response. Adolescent normal-weight amenorrheic athletes have higher fasting acylated ghrelin levels than eumenorrheic athletes and nonathletes (Christo et al., 2008a). Additionally, amenorrheic normal-weight adolescent and young adult athletes have higher overnight total ghrelin secretion than nonathletes subsequent to increased pulse amplitude and pulse mass (Ackerman et al., 2012c). In adult women, fasting ghrelin levels are higher in exercising amenorrheic women compared with exercising eumenorrheic and sedentary women (De Souza et al., 2004, 2007).

Ghrelin administration suppresses LH and FSH pulsatility in animals and humans (Vulliemoz et al., 2004; Kluge et al., 2007), and consistent with these findings, overnight ghrelin secretion is a strong inverse determinant of overnight LH secretion in adolescent and young adult amenorrheic athletes and in a combined group of athletes and nonathletes (Ackerman et al., 2012c). Interestingly, these effects persist after controlling for fat mass.

Peptide YY

Peptide YY (PYY) is an anorexigenic hormone secreted by the L cells of the gut, and levels of PYY are increased in conditions of energy deficiency (Misra et al., 2005c) and decreased in conditions of energy excess (le Roux et al., 2006). PYY levels correlate inversely with BMI and fat mass (Misra et al., 2005c; Russell et al., 2009). Amenorrheic athletes and exercisers have higher PYY than eumenorrheic athletes and exercisers and sedentary controls (Russell et al., 2009; Scheid et al., 2009). Importantly, PYY has been shown to attenuate GnRH secretion in *in vitro* models and gonadotropin secretion in prepubertal rats. In one study, PYY was an inverse predictor of testosterone levels in adolescent athletes and nonathletes (Russell et al., 2009).

Insulin

Insulin and glucose levels are typically decreased in low energy states, and low insulin may impact the HPG axis. Insulin increases GnRH promoter activity in GnRH neurons (Kim et al., 2005) through the transcription factor Egr-1 (DiVall et al., 2007), and increases aromatase activity in granulosa cells (Paul et al., 2010). Additionally, administration of insulin to healthy women increases LH pulse frequency while also decreasing pulse interval (Moret et al., 2009). Within normally cycling exercisers, those with luteal phase defects have lower insulin levels than ovulatory exercisers with normal luteal phase length and sedentary ovulatory women (De Souza et al., 2003, 2007). Insulin levels are lower in amenorrheic (Laughlin and Yen, 1996) or oligomenorrheic (Rickenlund et al., 2004) athletes compared with eumenorrheic athletes and nonathletes, and insulin AUC is positively correlated with LH pulse frequency (Laughlin and Yen, 1996; Rickenlund et al., 2004).

Adiponectin

Adiponectin levels may be increased in energy restricted states, although this is not a consistent finding. *In vitro* data indicate that adiponectin may suppress GnRH and gonadotropin secretion (Lu et al., 2008; Wen et al., 2008). Although absolute adiponectin levels do not differ in adolescent amenorrheic versus eumenorrheic athletes and nonathletes (Russell et al., 2009), adult amenorrheic athletes have been reported to have higher adiponectin relative to fat mass than eumenorrheic athletes and nonathletes (O'Donnell and De Souza, 2011).

Reproductive function in male athletes

In contrast to female athletes, there are limited data regarding the impact of exercise on the reproductive axis in male athletes. Overall, data suggest that sexual and hormonal maturity during the childhood and adolescent years does not change with increased athletic activity (Moore et al., 2010), and baseline testosterone levels do not differ significantly in sprinters (Grandys et al.,

2011) and other athletes (Lucia et al., 1996) compared with untrained subjects. Testosterone levels may decrease with increased intensity of training within sprinters (Grandys et al., 2011) and with energy restriction in bodybuilders (Maestu et al., 2010), although another study reported no change in hormonal profiles over different intensities of training in three groups of athletes (Lucia et al., 1996). One study of elite cyclists reported higher levels of testosterone compared with recreational athletes (Fitzgerald et al., 2012), and another reported lower sperm mobility in cyclists during the training period (Lucia et al., 1996). More studies are necessary to better understand these findings. Significant energy restriction in young men and boys associated with low weight, as in anorexia nervosa, is associated with lower testosterone levels than in normal-weight controls (Misra et al., 2008).

With respect to other hormones that may impact the HPG axis, IGF-1 and insulin levels decrease with energy restriction in athletes (Maestu et al., 2010). Low baseline leptin levels have been reported in elite gymnasts (Weimann et al., 1999) and marathon runners (Bobbert et al., 2012), and a decrease in leptin and an increase in ghrelin occurs after an acute training session in male rowers (Jurimae et al., 2007, 2011).

Hypothalamic–pituitary–adrenal axis

Data are conflicting regarding the impact of athletic activity on the hypothalamic–pituitary–adrenal (HPA) axis. Loucks et al. reported no differences in ACTH secretory patterns and cortisol secretory pulse frequency in amenorrheic athletes compared with eumenorrheic athletes and nonathletes (Loucks et al., 1989). However, overnight cortisol levels were higher in both groups of athletes compared with nonathletes, and whereas pulse amplitude remained high over the course of the day in amenorrheic athletes, pulse amplitude decreased in eumenorrheic athletes. As a consequence, 24 hour urinary cortisol measurements were higher in amenorrheic athletes than the other two groups. Rickenlund et al. have reported higher baseline cortisol secretion in amenorrheic athletes compared with eumenorrheic athletes and nonathletes (Rickenlund et al., 2004). In contrast, another study reported higher 24 hour serum cortisol in both groups of athletes (amenorrheic and eumenorrheic) than nonathletes (Laughlin and Yen, 1996). In an adolescent and young adult population, Ackerman et al. reported higher cortisol pulse amplitude, mass, half-life, and AUC in amenorrheic athletes than eumenorrheic athletes and nonathletes (Ackerman et al., 2013a). As with studies of LH pulsatility, differences in reports of cortisol pulsatile secretion in athletes and nonathletes may be attributable to differences in sample size, frequency of blood draws, and methods used to assess secretory hormone dynamics. Overall, data suggest higher cortisol levels in athletes who are amenorrheic and presumably in a state of low energy availability, compared with normally menstruating athletes and nonathletes. The diurnal rhythm of cortisol secretion is maintained in athletes.

Cortisol is a gluconeogenic hormone, and one proposed reason for elevated cortisol in conditions of low energy availability (as in heavily exercising athletes) is an adaptive response to maintain euglycemia. This is consistent with an inverse association of cortisol with fat mass (Rickenlund et al., 2004; Ackerman et al., 2013a). However, Laughlin et al. found no association of cortisol with fasting glucose (Laughlin and Yen, 1996).

An inverse relationship exists between integrated measures of cortisol secretion (such as 24 hour mean serum cortisol) and LH pulse parameters. Several studies have reported an inverse association between 24 hour mean serum cortisol or baseline cortisol secretion and LH pulse frequency (Loucks et al., 1989; Laughlin and Yen, 1996; Rickenlund et al., 2004), and between cortisol levels and the number of menses in the preceding year in amenorrheic athletes (Rickenlund et al., 2004). In a study of adolescent and young adult athletes, overnight cortisol AUC correlated inversely with LH AUC even after controlling for fat mass and other hormones such as ghrelin and leptin (Ackerman et al., 2013a). These data are consistent with stress-like elevations in cortisol causing a decrease in GnRH secretion, and an impaired gonadotrope responsiveness to GnRH in animal models (Breen et al., 2008; Oakley et al., 2009). This effect appears to be mediated by the type II glucocorticoid receptor within the pituitary (Breen et al., 2008). Additionally, a CRH receptor-1 antagonist can mitigate the deleterious effects of stress and cortisol on LH pulse frequency (Herod et al., 2011). Of interest, ghrelin can stimulate ACTH and therefore cortisol secretion (Arvat et al., 2001), and leptin inhibits cortisol secretion (Glasow et al., 1998). This may explain why associations of integrated measures of ghrelin and leptin secretion with LH are lost when cortisol measures are added to the regression model, while inverse associations of cortisol with LH persist (Ackerman et al., 2013a).

Growth hormone–insulin-like growth factor 1 (IGF-1) axis

Rickenlund et al. reported higher baseline GH secretion in amenorrheic (but not oligomenorrheic) athletes compared with regularly menstruating athletes and nonathletes (Rickenlund et al., 2004), with no differences between groups for pulse frequency or amplitude, or for 24 hour AUC. Another study reported higher 24 hour GH concentrations in both amenorrheic and eumenorrheic athletes compared with nonathletes, with higher pulse frequency in the amenorrheic women, and higher pulse amplitude in the eumenorrheic athletes compared with the other groups (Laughlin and Yen, 1996).

In conditions of extreme energy deprivation, as in anorexia nervosa, there is an increase in basal and pulsatile GH secretion and AUC compared with normal-weight women, associated with low systemic IGF-1 levels, suggestive of a nutritionally acquired hepatic resistance to GH (Misra et al., 2003). This is further corroborated by a decrease in GH binding protein (GHBP), the extracellular portion of the GH receptor, in anorexia nervosa (Counts et al., 1992), and a failure to increase IGF-1 levels with administration of supraphysiologic doses of rhGH (Fazeli et al., 2010).

In athletes, who are typically not as severely energy deprived as women with anorexia nervosa, GHBP levels are lower in amenorrheic women compared with eumenorrheic athletes and nonathletes, and correlate inversely with GH levels (Laughlin and Yen, 1996). However, IGF-1 levels are not as low as in anorexia nervosa. A study in adolescent athletes reported lower IGF-1 levels in the amenorrheic exercisers compared with nonathletes (Christo et al., 2008b); however, studies in adult athletes have reported no differences among groups for IGF-1, IGFBP-3 or the ratio of IGF-1/IGFBP-3 (Laughlin and Yen, 1996). In contrast, IGFBP-1 is higher in the amenorrheic group, leading to lower ratios of IGF-1/IGFBP-1 in amenorrheic athletes (Laughlin and Yen, 1996). Within normally cycling exercisers (with normal ovulatory cycles or with luteal phase deficiency), levels of GH, IGF-1, IGFBP3, and ratios of IGF-1/IGFBP-1 and IGF-1/IGFBP-3 do not differ from sedentary controls (De Souza et al., 2003).

GH levels correlate inversely with percent body fat (Laughlin and Yen, 1996; Rickenlund et al., 2004), whereas GHBP levels correlate positively with body fat mass and 24 hour integrated glucose concentrations (Laughlin and Yen, 1996), consistent with net energy status being an important driver of GH secretion.

Importantly, alterations in the GH-IGF-1 axis in athletes have the potential to impact the HPG axis. GnRH neurons express both IGF-1 and the IGF-1 receptor, and central IGF-1 has a direct effect on GnRH neurons (Daftary and Gore, 2005). IGF-1 also stimulates kisspeptin gene expression (Hiney et al., 2010) and therefore GnRH secretion, and potentiates gonadal steroidogenesis and oocyte maturation *in vitro* (Sadler et al., 2010). Additionally, IGF-1 increases aromatase activity in granulosa cells (Paul et al., 2010). Of note, GH and IGFBP-1 levels are inversely associated with LH secretion in athletes and nonathletes (Laughlin and Yen, 1996; Rickenlund et al., 2004), whereas the ratio of IGF-1/IGFBP-1 is positively associated with 24 hour LH pulse frequency (Laughlin and Yen, 1996). One study reported that IGF-1 levels were a positive determinant of estradiol levels in adolescent athletes and nonathletes after controlling for potential confounders (Russell et al., 2009).

Hypothalamic–pituitary–thyroid axis

There are few data regarding the HPT axis in athletes compared with nonathletes. In thin women with functional hypothalamic amenorrhea (not specifically athletes), the circadian rhythm of TSH secretion is maintained with a nadir between 12:00 and 17:00 hours and a peak between 21:00 and 02:00 hours (Berga et al., 1989). These women, however, have lower total T_3 and T_4 levels than normally cycling controls (Berga et al., 1989), a finding also observed in anorexia nervosa, a condition of amenorrhea and extreme energy deprivation (Misra et al., 2004), although free T_4 levels only trend lower in anorexia nervosa.

No differences were noted in TSH, free T_4 or total T_4 levels among sedentary ovulatory women, exercising ovulatory women, and exercising women with luteal phase deficiency, though total T_3 levels were lower in both groups of exercising women compared with the sedentary group (De Souza et al., 2003). Another study reported lower total T_3 levels in exercising women who were amenorrheic, anovulatory eumenorrheic, or eumenorrheic with short luteal phases compared with sedentary women (De Souza et al., 2007).

IMPACT ON BONE METABOLISM OF ATHLETIC ACTIVITY AND ASSOCIATED NEUROENDOCRINE CHANGES

The various neuroendocrine changes that occur in athletes, particularly in heavy exercisers, are mostly physiologic and in response to a low energy state. These adaptive changes (1) conserve energy for more important and vital functions (by suppression of the HPG axis and thyroid hormones), and (2) increase substrate availability (by increases in cortisol and GH, both of which are gluconeogenic, as well as decreases in leptin and increases in ghrelin, which should stimulate caloric intake). However, if persistent, these physiologic changes can subsequently lead to the major pathologic consequence of energy deficiency, namely poor bone accrual over time in adolescents, and low bone density in adults.

Areal bone density in athletes

The impact of athletic activity on bone depends of various factors. These include the nature of athletic activity, weight-bearing patterns, intensity of training, and associated neuroendocrine hormonal alterations. Bone metabolism is adversely affected in athletes with restricted energy availability (namely amenorrheic athletes), despite the beneficial effects of physical exercise on bone (Cobb et al., 2003; Christo et al., 2008b; Ackerman et al., 2011).

Impact of physical activity and the nature of the sport

In states of *adequate energy availability and eumenorrhea*, bone density is not only preserved, but may be increased compared with a sedentary population. Thus, eumenorrheic weight-bearing endurance athletes have higher bone density than eumenorrheic nonathletes, particularly at sites of predominantly cortical bone such as the hip (Ackerman et al., 2011). This is attributable to the known beneficial effects of weight-bearing and mechanical loading on bone. Eumenorrhea is protective to bone in exercising women even when subclinical menstrual dysfunction exists, and normally cycling exercising women with luteal phase defects do not differ for bone density at the spine, hip, or femoral neck compared with cycling and ovulatory exercising women (De Souza et al., 1997).

Of importance, the kind of impact experienced with any specific sport determines the direction and nature of mechanical loading on bone, and can differentially affect bone metabolism. Thus, eumenorrheic high school athletes involved in high/odd impact sports such as soccer, softball, volleyball, tennis, lacrosse, sprinting, and jumping have higher bone density at the total hip than eumenorrheic athletes involved in repetitive/nonimpact sports such as cross-country, track, and swimming (Nichols et al., 2007a).

Impact of energy deficiency and/or hypogonadism

Both energy status and gonadal status determine the impact of athletic activity on bone. *Reduced energy intake* in runners, as indicated by increased dietary restraint, is associated with lower bone density (Barrack et al., 2008). In both adult and adolescent amenorrheic athletes, who are presumably *energy deficient and hypoestrogenemic*, bone density is deleteriously affected at the lumbar spine, a site of predominantly trabecular bone, compared with eumenorrheic athletes and nonathletes (Cobb et al., 2003; Christo et al., 2008b; Ackerman et al., 2011), and at the total hip compared with eumenorrheic athletes (Cobb et al., 2003; Ackerman et al., 2011). Similarly, oligo/amenorrheic high school athletes participating in repetitive/nonimpact sports have lower spine bone density than eumenorrheic athletes who participate in high/odd impact sports (Nichols et al., 2007a). In one study, 38% of adolescent amenorrheic weight-bearing endurance athletes had lumbar spine bone density Z-scores of < -1 compared with only 11% each of eumenorrheic athletes and nonathletes (Christo et al., 2008b). In an energy replete state, one study reported that hypogonadism in exercisers was not associated with perturbations in bone formation or resorption (as indicated by surrogate markers of bone turnover); however, bone density Z-scores at the femoral neck were lower in the hypogonadal group (De Souza et al., 2008). In male athletes, lower estradiol levels are associated with lower bone density (Ackerman et al., 2012b).

Modifying effect of the nature of impact

It is interesting to note that even in athletes with estrogen deficiency, bone density may be preserved or be higher than in nonathletes and other athletes, depending on the nature of the sport. Robinson et al. have reported higher bone density at the femoral neck and preservation of bone density at the spine in gymnasts compared with sedentary controls, whereas runners have lower bone density at both sites compared with gymnasts and controls, despite a similar prevalence of amenorrhea in gymnasts and runners (Robinson et al., 1995). This has been attributed to the impact loading effect of gymnastics versus the repetitive impact associated with running, and higher lean body mass, muscle strength and IGF-1 levels in the gymnasts.

Bone turnover in athletes

Adolescent athletes with amenorrhea have a coupled decrease in bone turnover compared with eumenorrheic athletes and nonathletes (Christo et al., 2008b). This is in contrast to normal adolescence, which is characterized by increased bone turnover in early adolescence, and a decrease in bone turnover to approach adult levels in late adolescence (Mora et al., 1999). Adult exercising amenorrheic women, however, have an uncoupling of bone turnover, with a decrease in bone formation and an increase in bone resorption markers (De Souza et al., 2008).

Determinants of bone density in athletes

As discussed above, important determinants of bone density in athletes are the nature of physical activity, the nutritional status of the athlete (driven by net energy availability) and associated neuroendocrine changes. Nutritional status is reflected by BMI, fat mass, lean (or fat-free) mass, IGF-1, and leptin levels. Although lean mass is preserved in normal-weight athletes, it decreases when weight decreases below 90% of what is ideal for age and height. As indicated earlier, important endocrine alterations occur in low energy states (in addition to low IGF-1 and leptin levels), and include hypothalamic amenorrhea with associated hypogonadism, high ghrelin, PYY and cortisol, and low total T_3. Many of these changes impact bone density in energy-deficient athletes.

In one study, important determinants of bone density in young elite rhythmic gymnasts included BMI, fat-free mass, and levels of gonadal steroids and IGF-1 (Maimoun et al., 2010). These are also important determinants of bone density in young weight-bearing endurance athletes (Christo et al., 2008b). Additionally, relatively higher cortisol and PYY levels and lower leptin contribute to lower bone density (Russell et al., 2009; Scheid et al., 2011). PYY acts primarily through the Y2 receptor, and the Y2 receptor knockout mice have increased bone mass (Baldock et al., 2006), suggesting that high PYY should be associated with lower bone density. Higher PYY levels are associated with lower levels of bone formation markers and lower lumbar bone mineral apparent density Z-scores in adolescent athletes (Russell et al., 2009), and with lower hip and whole body bone density Z-scores in adult exercising women (Scheid et al., 2011). Leptin is a bone anabolic hormone, and low leptin levels in amenorrheic athletes are also associated with lower bone density measures (Christo et al., 2008b; Russell et al., 2009).

Limitations of areal bone density assessment

A limitation of bone density measured by dual energy X-ray absorptiometry (DXA) is that it assesses areal (rather than volumetric) bone density, which is affected by stature and body composition (Bachrach, 2006, 2007). Therefore, adjustment for body size is important at the extremes of stature, and this is particularly important in certain kinds of sports, and inherent to the nature of the sport and its demands. For example, adjustment for height may be important in certain gymnasts and basketball players. Whether or not adjustment for bone age is necessary in adolescents is a matter of controversy, but may need to be factored into the evaluation of bone density in adolescent amenorrheic athletes.

Also, areal bone density cannot provide measures of bone geometry or differentiate between trabecular and cortical bone at different weight-bearing and non-weight-bearing sites (Burrows et al., 2010). High resolution peripheral quantitative tomography (HRpQCT) is a novel technique that allows assessment of bone structure (cortical, trabecular, and total cross-sectional area), cortical and trabecular microarchitecture (cortical porosity, trabecular number, thickness, and separation) and volumetric measures of cortical, trabecular, and total bone density. Using finite element models, HRpQCT data can also be used to derive estimated measures of bone strength (stiffness and failure load). Although studies have shown that bone microarchitecture provides additional estimates of bone strength not provided by DXA measures of bone density (Boutroy et al., 2008; Bredella et al., 2008), there are limited data regarding bone microarchitecture in adolescent and adult athletes.

Bone microarchitecture, volumetric bone density, and estimates of bone strength in athletes

Existing studies have attempted to differentiate changes consequent to mechanical loading (evident in all athletes within a study) from those consequent to energy deficiency and hypogonadism (evident in amenorrheic athletes).

Bone structural changes

In adolescent and young adult weight-bearing endurance athletes, athletic activity is associated with an expansion outwards of weight-bearing bone (distal tibia), as evidenced by an increase in trabecular and total cross-sectional area in both eumenorrheic and amenorrheic athletes. These changes are not evident at the non-weight-bearing distal radius (Ackerman et al., 2011). Amenorrheic normal-weight athletes, however, have lower percent cortical area at weight-bearing bone than nonathletes, attributable to increased endosteal bone resorption in a hypogonadal state.

Cortical and trabecular microarchitectural changes

Adult elite fencers have higher trabecular number and decreased trabecular separation in the femur (assessed using MRI) (Chang et al., 2008). At weight-bearing sites, adolescent amenorrheic normal-weight athletes have lower trabecular number and increased trabecular separation (but no differences in trabecular thickness) compared with eumenorrheic athletes and nonathletes (Ackerman et al., 2011). Cortical pore diameter is highest in amenorrheic athletes (Ackerman et al., 2012a).

Volumetric bone density

Elite fencers have higher trabecular bone density than controls in the femur (Chang et al., 2008). However, cortical density is lower at weight-bearing sites in adolescent athletes (amenorrheic and eumenorrheic) compared with nonathletes, believed to be consequent to delayed mineralization of an outwardly expanding cortex. In addition, total density at weight-bearing bone, and trabecular density at non-weight-bearing bone is lower in amenorrheic compared with eumenorrheic athletes and nonathletes (Ackerman et al., 2011).

Estimated bone strength and fractures

Amenorrheic athletes have lower measures of estimated bone strength (stiffness and failure load) at non-weight-bearing bone compared with nonathletes, and lose the beneficial effect of exercise at weight-bearing sites (Ackerman et al., 2012a). Similar data have been reported for the hip in adolescent and young adults athletes and nonathletes using hip structural analysis (Ackerman et al., 2013b). A study in adult athletes compared women with a history of lower extremity stress fractures with those without a history of such fractures. This study reported that women with a history of fractures had thinner tibia after adjusting for body size, and a higher load was carried by the cortex in these women. In addition, they had lower trabecular bone mineral density in the posterior region of the tibia and less cortical area (Schnackenburg et al., 2011).

Determinants of bone structure, microarchitecture, volumetric bone density, and estimated strength

Hypogonadism has significant and deleterious effects on bone microarchitecture in athletes. In addition, lean mass has a positive impact, and older age at menarche has a negative impact on bone microarchitecture and strength (Ackerman et al., 2011, 2012a). Interestingly, oxytocin, now known to have bone anabolic effects (Colaianni et al., 2014), is associated positively with bone microarchitectural parameters in amenorrheic athletes (Lawson et al., 2013).

Strategies to optimize bone health in athletes

It is essential to increase energy availability in amenorrheic athletes. This may be achieved by increasing nutritional intake, reducing the intensity of training, or both (Nattiv et al., 2007). Resumption of menses, an indicator of increasing levels of gonadal steroids, is critical to optimizing bone health in athletes with functional hypothalamic amenorrhea. In addition, normalizing energy balance is likely to cause a normalization of other hormones that impact bone such as IGF-1, leptin, PYY, and cortisol.

Studies in adult amenorrheic women (exercisers or nonexercisers) are conflicting regarding the effect of hormone replacement on bone density. Some report no change (Gibson et al., 1999), others report an increase (Hergenroeder et al., 1997; Castelo-Branco et al., 2001) and some a decrease (Burr et al., 2000; Hartard et al., 2004) in bone density following estrogen administration. Additionally, the studies are usually small (Hergenroeder et al., 1997), often retrospective (Hartard et al., 2004), not randomized (Burr et al., 2000; Castelo-Branco et al., 2001), and do not control for weight changes (Hergenroeder et al., 1997; Castelo-Branco et al., 2001). A study in adolescents with anorexia nervosa, an extreme energy restricted state, has recently reported an increase in bone density and maintenance of bone density Z-scores at the lumbar spine and the hip using the transdermal estrogen patch with cyclic progesterone (Misra et al., 2011). However, catch up of bone density did not occur, and this strategy has not been studied in an exclusively athlete population. Another study in adolescent and young adult women with anorexia nervosa has reported stabilization of lumbar spine bone density Z-scores using a combination of oral estrogen–progesterone with micronized oral dehyroepiandrosterone, but without catch-up (Divasta et al., 2012).

Calcium and vitamin D intake is usually sufficient in most athletes, but should be optimized if not sufficient, with the aim to maintain vitamin D levels between 80 and 125 nmol/L. There are no studies of recombinant human (rh) IGF-1 administration in athletes, although a beneficial effect of rhIGF-1 when given with estrogen has been reported in adult women with anorexia nervosa (Grinspoon et al., 2002). Administration of metreleptin (a commercial form of leptin) to women with functional hypothalamic amenorrhea (not necessarily exercisers) caused resumption of menses in 60–70% of the women, resumption of ovulatory cycles in 30–40%, and was associated with an increase in bone formation markers and in bone mineral content (Welt et al., 2004; Chou et al., 2011; Sienkiewicz et al., 2011). However, a significant decrease in weight and fat mass occurred with leptin administration. No effective strategies currently exist to reduce cortisol or PYY levels or their effects in this population.

IMPACT ON NEUROCOGNITIVE FUNCTION

Although participation in athletic activities is associated with a positive body image, improved self-esteem and increased likelihood of academic success, there are limited data regarding the impact of amenorrhea on neurocognitive function in athletes. Hypogonadism in other conditions, such as anorexia nervosa and Turner syndrome, is associated with specific changes in neurocognitive parameters that improve with hormone replacement (Collaer et al., 2002; Chui et al., 2008). Additionally, testosterone levels have been associated with mood in women with anorexia nervosa (Miller et al., 2007), while physiologic estrogen replacement causes a reduction in anxiety scores in adolescents with anorexia nervosa, and prevents increases in anxiety and body dissatisfaction following weight gain (Misra et al., 2013). Further investigations are necessary to determine whether amenorrhea leads to a blunting of the beneficial effects of athletic participation on neurocognition and mood.

IMPACT ON FERTILITY

Infertility in athletes with anovulatory cycles and in amenorrheic athletes can be treated with puslatile GnRH or gonadotropin combinations. In other conditions of hypogonadotropic hypogonadism, ovulatory rates of up to 96% are reported (Martin et al., 1993; Borges et al., 2007). Low-weight women with functional hypothalamic amenorrhea have a higher prevalence of low birth weight offspring than normal-weight women (Gordon, 2010).

CONCLUSION

Athletic activity is characterized by changes in many neuroendocrine axes, and the degree of affectation of these axes depends on the state of energy availability in the athlete, although genetic factors and an underlying predisposition to PCOS may play an important role in determining which athletes become amenorrheic early. Figure 25.1 summarizes adaptive changes in neuroendocrine axes that result from a state of low energy availability. These adaptive changes in neuroendocrine axes in athletes then lead to alterations in bone metabolism, and may affect neurocognition and mood. Figure 25.2 summarizes the impact of

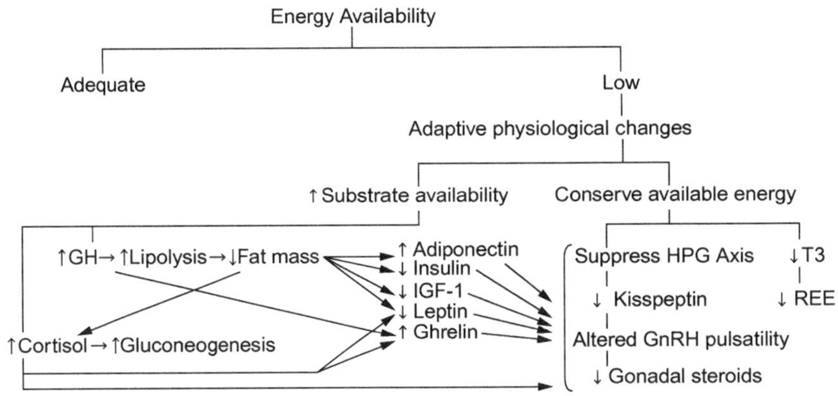

Fig. 25.1. Pathways demonstrating physiological and adaptive mechanisms that result from a state of low energy availability, and resulting alterations in various hormonal axes.

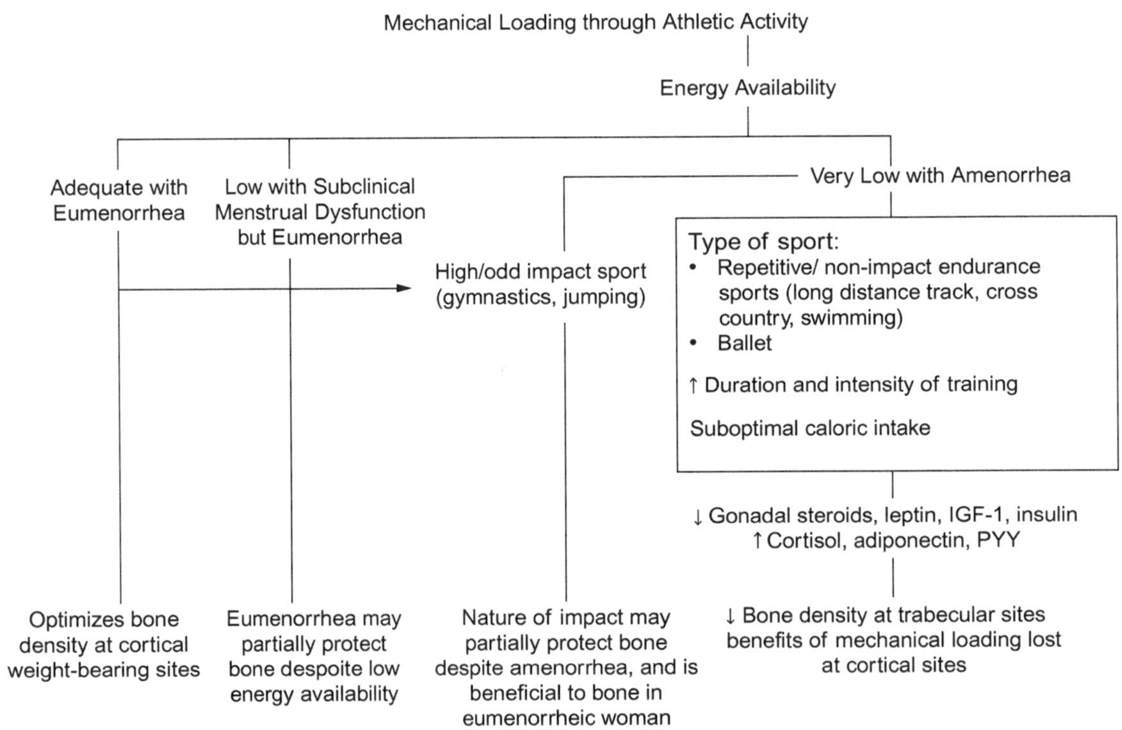

Fig. 25.2. Impact of energy availability, menstrual status, and nature of mechanical loading on bone density.

energy and menstrual status, and the nature of mechanical loading on bone. The American College of Sports Medicine (ACSM) in its position stand (Nattiv et al., 2007) has emphasized the importance of educating athletes, parents, coaches, trainers, and judges about the consequences of overexercising and the female athlete triad. The ACSM recommends screening for triad components during preparticipation examinations, during the annual health screening examination and as needed. It is critical to discourage unhealthy weight loss practices amongst athletes. If an athlete develops oligoamenorrhea, a multidisciplinary treatment team needs to be established to include a healthcare professional, a registered dietician, a mental health professional (if eating issues are evident), and additional members (such as a certified athletic trainer, exercise physiologist, athlete's coach, and family members). Nutritional counseling and monitoring may suffice for some athletes, whereas others may require a reduction in exercise activity (a 10–20% decrease in activity per week is recommended). If eating disorders are identified, psychotherapy becomes essential, with clear criteria for continuation of exercise. When necessary, training and competition intensity may need to be modified. The position stand emphasizes that no pharmacologic agent at this time is known to adequately treat functional hypothalamic amenorrhea or low bone density in athletes.

ACKNOWLEDGMENTS

The author has no conflicts to disclose.

This manuscript was supported by NIH grant 1 R01 HD060827-01A1.

References

Ackerman KE, Nazem T, Chapko DR et al. (2011). Bone microarchitecture is impaired in adolescent amenorrheic athletes compared with eumenorrheic athletes and nonathletic controls. J Clin Endocrinol Metab 96: 3123–3133.

Ackerman KE, Putman M, Guereca G et al. (2012a). Cortical microstructure and estimated bone strength in young amenorrheic athletes eumenorrheic athletes and non-athletes. Bone 51: 680–687.

Ackerman KE, Skrinar GS, Medvedova E et al. (2012b). Estradiol levels predict bone mineral density in male collegiate athletes: a pilot study. Clin Endocrinol (Oxf) 76: 339–345.

Ackerman KE, Slusarz K, Guereca G et al. (2012c). Higher ghrelin and lower leptin secretion are associated with lower LH secretion in young amenorrheic athletes compared with eumenorrheic athletes and controls. Am J Physiol Endocrinol Metab 302: E800–E806.

Ackerman KE, Patel KT, Guereca G et al. (2013a). Cortisol secretory parameters in young exercisers in relation to LH secretion and bone parameters. Clin Endocrinol (Oxf) 78: 114–119.

Ackerman KE, Pierce L, Guereca G et al. (2013b). Hip structural analysis in adolescent and young adult oligoamenorrheic and eumenorrheic athletes and nonathletes. J Clin Endocrinol Metab 98: 1742–1749.

Arvat E, Maccario M, Di Vito L et al. (2001). Endocrine activities of ghrelin a natural growth hormone secretagogue (GHS) in humans: comparison and interactions with hexarelin a nonnatural peptidyl GHS and GH-releasing hormone. J Clin Endocrinol Metab 86: 1169–1174.

Bachrach LK (2006). Measuring bone mass in children: can we really do it? Horm Res 65 (Suppl 2): 11–16.

Bachrach LK (2007). Consensus and controversy regarding osteoporosis in the pediatric population. Endocr Pract 13: 513–520.

Baldock PA, Allison S, McDonald MM et al. (2006). Hypothalamic regulation of cortical bone mass: opposing activity of Y2 receptor and leptin pathways. J Bone Miner Res 21: 1600–1607.

Barrack MT, Rauh MJ, Barkai HS et al. (2008). Dietary restraint and low bone mass in female adolescent endurance runners. Am J Clin Nutr 87: 36–43.

Beals KA, Hill AK (2006). The prevalence of disordered eating menstrual dysfunction and low bone mineral density among US collegiate athletes. Int J Sport Nutr Exerc Metab 16: 1–23.

Berga SL, Mortola JF, Girton L et al. (1989). Neuroendocrine aberrations in women with functional hypothalamic amenorrhea. J Clin Endocrinol Metab 68: 301–308.

Bobbert T, Mai K, Brechtel L et al. (2012). Leptin and endocrine parameters in marathon runners. Int J Sports Med 33: 244–248.

Borges Jr E, Rossi LM, Locambo de Freitas CV et al. (2007). Fertilization and pregnancy outcome after intracytoplasmic injection with fresh or cryopreserved ejaculated spermatozoa. Fertil Steril 87: 316–320.

Boutroy S, van Rietbergen B, Sornay-Rendu E et al. (2008). Finite element analysis based on in vivo HR-pQCT images of the distal radius is associated with wrist fracture in postmenopausal women. J Bone Miner Res 23: 392–399.

Bredella MA, Misra M, Miller KK et al. (2008). Distal radius in adolescent girls with anorexia nervosa: trabecular structure analysis with high-resolution flat-panel volume. CT. Radiology 249: 938–946.

Breen KM, Davis TL, Doro LC et al. (2008). Insight into the neuroendocrine site and cellular mechanism by which cortisol suppresses pituitary responsiveness to gonadotropin-releasing hormone. Endocrinology 149: 767–773.

Burr DB, Yoshikawa T, Teegarden D et al. (2000). Exercise and oral contraceptive use suppress the normal age-related increase in bone mass and strength of the femoral neck in women 18–31 years of age. Bone 27: 855–863.

Burrows M, Liu D, Perdios A et al. (2010). Assessing bone microstructure at the distal radius in children and adolescents using HR-pQCT: a methodological pilot study. J Clin Densitom 13: 451–455.

Caronia LM, Martin C, Welt CK et al. (2011). A genetic basis for functional hypothalamic amenorrhea. N Engl J Med 364: 215–225.

Castelo-Branco C, Vicente JJ, Pons F et al. (2001). Bone mineral density in young hypothalamic oligoamenorrheic women treated with oral contraceptives. J Reprod Med 46: 875–879.

Chang G, Pakin SK, Schweitzer ME et al. (2008). Adaptations in trabecular bone microarchitecture in Olympic athletes determined by 7 T MRI. J Magn Reson Imaging 27: 1089–1095.

Chehab FF, Lim ME, Lu R (1996). Correction of the sterility defect in homozygous obese female mice by treatment with the human recombinant leptin. Nat Genet 12: 318–320.

Chou SH, Chamberland JP, Liu X et al. (2011). Leptin is an effective treatment for hypothalamic amenorrhea. Proc Natl Acad Sci U S A 108: 6585–6590.

Christo K, Cord J, Mendes N et al. (2008a). Acylated ghrelin and leptin in adolescent athletes with amenorrhea eumenorrheic athletes and controls: a cross-sectional study. Clin Endocrinol (Oxf) 69: 628–633.

Christo K, Prabhakaran R, Lamparello BC et al. (2008b). Bone metabolism in adolescent amenorrheic and eumenorrheic athletes and controls. Pediatrics 121: 1127–1136.

Chui HT, Christensen BK, Zipursky RB et al. (2008). Cognitive function and brain structure in females with a history of adolescent-onset anorexia nervosa. Pediatrics 122: e426–e437.

Cobb KL, Bachrach LK, Greendale G et al. (2003). Disordered eating menstrual irregularity and bone mineral density in female runners. Med Sci Sports Exerc 35: 711–719.

Colaianni G, Tamma R, di Benedetto A et al. (2014). The oxytocin-bone axis. J Neuroendocrinol 26: 53–57.

Collaer ML, Geffner ME, Kaufman FR et al. (2002). Cognitive and behavioral characteristics of Turner syndrome: exploring a role for ovarian hormones in female sexual differentiation. Horm Behav 41: 139–155.

Corr M, De Souza MJ, Toombs RJ et al. (2011). Circulating leptin concentrations do not distinguish menstrual status in exercising women. Hum Reprod 26: 685–694.

Counts D, Gwirtsman H, Carlsson L et al. (1992). The effect of anorexia nervosa and refeeding on growth hormone-binding protein the insulin-like growth factors (IGFs) and the IGF-binding proteins. J Clin Endocrinol Metab 75: 762–767.

Daftary SS, Gore AC (2005). IGF-1 in the brain as a regulator of reproductive neuroendocrine function. Exp Biol Med (Maywood) 230: 292–306.

De Souza MJ, Miller BE, Sequenzia LC et al. (1997). Bone health is not affected by luteal phase abnormalities and decreased ovarian progesterone production in female runners. J Clin Endocrinol Metab 82: 2867–2876.

De Souza MJ, Miller BE, Loucks AB et al. (1998). High frequency of luteal phase deficiency and anovulation in recreational women runners: blunted elevation in follicle-stimulating hormone observed during luteal-follicular transition. J Clin Endocrinol Metab 83: 4220–4232.

De Souza MJ, van Heest J, Demers LM et al. (2003). Luteal phase deficiency in recreational runners: evidence for a hypometabolic state. J Clin Endocrinol Metab 88: 337–346.

De Souza MJ, Leidy HJ, O'Donnell E et al. (2004). Fasting ghrelin levels in physically active women: relationship with menstrual disturbances and metabolic hormones. J Clin Endocrinol Metab 89: 3536–3542.

De Souza MJ, Lee DK, Vanheest JL et al. (2007). Severity of energy-related menstrual disturbances increases in proportion to indices of energy conservation in exercising women. Fertil Steril 88: 971–975.

De Souza MJ, West SL, Jamal SA et al. (2008). The presence of both an energy deficiency and estrogen deficiency exacerbate alterations of bone metabolism in exercising women. Bone 43: 140–148.

Divall SA, Radovick S, Wolfe A (2007). Egr-1 binds the GnRH promoter to mediate the increase in gene expression by insulin. Mol Cell Endocrinol 270: 64–72.

Divasta AD, Feldman HA, Giancaterino C et al. (2012). The effect of gonadal and adrenal steroid therapy on skeletal health in adolescents and young women with anorexia nervosa. Metabolism 61: 1010–1020.

Fazeli PK, Lawson EA, Prabhakaran R et al. (2010). Effects of recombinant human growth hormone in anorexia nervosa: a randomized placebo-controlled study. J Clin Endocrinol Metab 95: 4889–4897.

Fernandez-Fernandez R, Aguilar E, Tena-Sempere M et al. (2005). Effects of polypeptide YY(3-36) upon luteinizing hormone-releasing hormone and gonadotropin secretion in prepubertal rats: in vivo and in vitro studies. Endocrinology 146: 1403–1410.

Fitzgerald LZ, Robbins WA, Kesner JS et al. (2012). Reproductive hormones and interleukin-6 in serious leisure male athletes. Eur J Appl Physiol 112: 3765–3773.

Gibson J, Mitchell A, Reeve J et al. (1999). Treatment of reduced bone mineral density in athletic amenorrhea: a pilot study. Osteoporos Int 10: 284–289.

Glasow A, Haidan A, Hilbers U et al. (1998). Expression of Ob receptor in normal human adrenals: differential regulation of adrenocortical and adrenomedullary function by leptin. J Clin Endocrinol Metab 83: 4459–4466.

Gordon CM (2010). Clinical practice. Functional hypothalamic amenorrhea. N Engl J Med 363: 365–371.

Grandys M, Majerczak J, Zapart-Bukowska J et al. (2011). Gonadal hormone status in highly trained sprinters and in untrained men. J Strength Cond Res 25: 1079–1084.

Grinspoon S, Gulick T, Askari H et al. (1996). Serum leptin levels in women with anorexia nervosa. J Clin Endocrinol Metab 81: 3861–3863.

Grinspoon SK, Askari H, Landt ML et al. (1997). Effects of fasting and glucose infusion on basal and overnight leptin concentrations in normal-weight women. Am J Clin Nutr 66: 1352–1356.

Grinspoon S, Thomas L, Miller K et al. (2002). Effects of recombinant human IGF-I and oral contraceptive administration on bone density in anorexia nervosa. J Clin Endocrinol Metab 87: 2883–2891.

Hartard M, Kleinmond C, Kirchbichler A et al. (2004). Age at first oral contraceptive use as a major determinant of vertebral bone mass in female endurance athletes. Bone 35: 836–841.

Hergenroeder AC, Smith EO, Shypailo R et al. (1997). Bone mineral changes in young women with hypothalamic amenorrhea treated with oral contraceptives medroxyprogesterone or placebo over 12 months. Am J Obstet Gynecol 176: 1017–1025.

Herod SM, Pohl CR, Cameron JL (2011). Treatment with a CRH-R1 antagonist prevents stress-induced suppression of the central neural drive to the reproductive axis in female macaques. Am J Physiol Endocrinol Metab 300: E19–E27.

Hiney JK, Srivastava VK, les Dees W (2010). Insulin-like growth factor-1 stimulation of hypothalamic KiSS-1 gene expression is mediated by Akt: effect of alcohol. Neuroscience 166: 625–632.

Jayasena CN, Nijher GM, Chaudhri OB et al. (2009). Subcutaneous injection of kisspeptin-54 acutely stimulates gonadotropin secretion in women with hypothalamic amenorrhea but chronic administration causes tachyphylaxis. J Clin Endocrinol Metab 94: 4315–4323.

Jurimae J, Jurimae T, Purge P (2007). Plasma ghrelin is altered after maximal exercise in elite male rowers. Exp Biol Med (Maywood) 232: 904–909.

Jurimae J, Ramson R, Maestu J et al. (2011). Interactions between adipose bone and muscle tissue markers during acute negative energy balance in male rowers. J Sports Med Phys Fitness 51: 347–354.

Kim HH, Divall SA, Deneau RM et al. (2005). Insulin regulation of GnRH gene expression through MAP kinase signaling pathways. Mol Cell Endocrinol 242: 42–49.

Kluge M, Schussler P, Uhr M et al. (2007). Ghrelin suppresses secretion of luteinizing hormone in humans. J Clin Endocrinol Metab 92: 3202–3205.

Laughlin G, Yen S (1996). Nutritional and endocrine-metabolic aberrations in amenorrheic athletes. J Clin Endocrinol Metab 81: 4301–4309.

Lawson EA, Ackerman KE, Estella NM et al. (2013). Nocturnal oxytocin secretion is lower in amenorrheic athletes than nonathletes and associated with bone microarchitecture and finite element analysis parameters. Eur J Endocrinol 168: 457–464.

le Roux CW, Batterham RL, Aylwin SJ et al. (2006). Attenuated peptide YY release in obese subjects is associated with reduced satiety. Endocrinology 147: 3–8.

Loucks AB (2007). Energy availability and infertility. Curr Opin Endocrinol Diabetes Obes 14: 470–474.

Loucks AB, Thuma JR (2003). Luteinizing hormone pulsatility is disrupted at a threshold of energy availability in regularly menstruating women. J Clin Endocrinol Metab 88: 297–311.

Loucks A, Mortola J, Girton L et al. (1989). Alterations in the hypothalamic–pituitary–ovarian and the hypothalamic–pituitary–adrenal axes in athletic women. J Clin Endocrinol Metab 68: 402–411.

Lu M, Tang Q, Olefsky JM et al. (2008). Adiponectin activates adenosine monophosphate-activated protein kinase and decreases luteinizing hormone secretion in LbetaT2 gonadotropes. Mol Endocrinol 22: 760–771.

Lucia A, Chicharro JL, Perez M et al. (1996). Reproductive function in male endurance athletes: sperm analysis and hormonal profile. J Appl Physiol 81: 2627–2636.

Maestu J, Eliakim A, Jurimae J et al. (2010). Anabolic and catabolic hormones and energy balance of the male bodybuilders during the preparation for the competition. J Strength Cond Res 24: 1074–1081.

Maimoun L, Coste O, Jaussent A et al. (2010). Bone mass acquisition in female rhythmic gymnasts during puberty: no direct role for leptin. Clin Endocrinol (Oxf) 72: 604–611.

Martin KA, Hall JE, Adams JM et al. (1993). Comparison of exogenous gonadotropins and pulsatile gonadotropin-releasing hormone for induction of ovulation in hypogonadotropic amenorrhea. J Clin Endocrinol Metab 77: 125–129.

Miller KK, Wexler TL, Zha AM et al. (2007). Androgen deficiency: association with increased anxiety and depression symptom severity in anorexia nervosa. J Clin Psychiatry 68: 959–965.

Misra M, Miller K, Bjornson J et al. (2003). Alterations in growth hormone secretory dynamics in adolescent girls with anorexia nervosa and effects on bone metabolism. J Clin Endocrinol Metab 88: 5615–5623.

Misra M, Miller KK, Almazan C et al. (2004). Alterations in cortisol secretory dynamics in adolescent girls with anorexia nervosa and effects on bone metabolism. J Clin Endocrinol Metab 89: 4972–4980.

Misra M, Miller K, Kuo K et al. (2005a). Secretory dynamics of ghrelin in adolescent girls with anorexia nervosa and healthy adolescents. Am J Physiol Endocrinol Metab 289: E347–E356.

Misra M, Miller KK, Kuo K et al. (2005b). Secretory dynamics of leptin in adolescent girls with anorexia nervosa and healthy adolescents. Am J Physiol Endocrinol Metab 289: E373–E381.

Misra M, Miller KK, Tsai P et al. (2005c). Elevated peptide YY levels in adolescent girls with anorexia nervosa. J Clin Endocrinol Metab 91: 1027–1033.

Misra M, Prabhakaran R, Miller KK et al. (2006). Role of cortisol in menstrual recovery in adolescent girls with anorexia nervosa. Pediatr Res 59: 598–603.

Misra M, Miller KK, Cord J et al. (2007). Relationships between serum adipokines insulin levels and bone density in girls with anorexia nervosa. J Clin Endocrinol Metab 92: 2046–2052.

Misra M, Katzman DK, Cord J et al. (2008). Bone metabolism in adolescent boys with anorexia nervosa. J Clin Endocrinol Metab 93: 3029–3036.

Misra M, Katzman D, Miller KK et al. (2011). Physiologic estrogen replacement increases bone density in adolescent girls with anorexia nervosa. J Bone Miner Res 26: 2430–2438.

Misra M, Katzman DK, Estella NM et al. (2013). Impact of physiologic estrogen replacement on anxiety symptoms body shape perception and eating attitudes in adolescent girls with anorexia nervosa: data from a randomized controlled trial. J Clin Psychiatry 74: e765–e771.

Moore SA, Moore M, Klentrou P et al. (2010). Maturity status in male child and adolescent athletes. J Sports Med Phys Fitness 50: 486–493.

Mora S, Pitukcheewanont P, Kaufman FR et al. (1999). Biochemical markers of bone turnover and the volume and the density of bone in children at different stages of sexual development. J Bone Miner Res 14: 1664–1671.

Moret M, Stettler R, Rodieux F et al. (2009). Insulin modulation of luteinizing hormone secretion in normal female volunteers and lean polycystic ovary syndrome patients. Neuroendocrinology 89: 131–139.

Nattiv A, Loucks AB, Manore MM et al. (2007). American College of Sports Medicine position stand. The female athlete triad. Med Sci Sports Exerc 39: 1867–1882.

Navarro VM, Fernandez-Fernandez R, Castellano JM et al. (2004). Advanced vaginal opening and precocious activation of the reproductive axis by KiSS-1 peptide the endogenous ligand of GPR54. J Physiol 561: 379–386.

Navarro VM, Castellano JM, Fernandez-Fernandez RT et al. (2005). Characterization of the potent luteinizing hormone-releasing activity of KiSS-1 peptide the natural ligand of GPR54. Endocrinology 146: 156–163.

Nichols JF, Rauh MJ, Lawson MJ et al. (2006). Prevalence of the female athlete triad syndrome among high school athletes. Arch Pediatr Adolesc Med 160: 137–142.

Nichols JF, Rauh MJ, Barrack MT et al. (2007a). Bone mineral density in female high school athletes: interactions of menstrual function and type of mechanical loading. Bone 41: 371–377.

Nichols JF, Rauh MJ, Barrack MT et al. (2007b). Disordered eating and menstrual irregularity in high school athletes in lean-build and nonlean-build sports. Int J Sport Nutr Exerc Metab 17: 364–377.

O'Donnell E, De Souza MJ (2011). Increased serum adiponectin concentrations in amenorrheic physically active women are associated with impaired bone health but not with estrogen exposure. Bone 48: 760–767.

Oakley AE, Breen KM, Clarke IJ et al. (2009). Cortisol reduces gonadotropin-releasing hormone pulse frequency in follicular phase ewes: influence of ovarian steroids. Endocrinology 150: 341–349.

Paul S, Pramanick K, Kundu S et al. (2010). Regulation of ovarian steroidogenesis in vitro by IGF-I and insulin in common carp *Cyprinus carpio*: stimulation of aromatase activity and P450arom gene expression. Mol Cell Endocrinol 315: 95–103.

Pietilainen KH, Korkeila M, Bogl LH et al. (2010). Inaccuracies in food and physical activity diaries of obese subjects: complementary evidence from doubly labeled water and co-twin assessments. Int J Obes (Lond) 34: 437–445.

Pinhas-Hamiel O, Pilpel N, Carel C et al. (2006). Clinical and laboratory characteristics of adolescents with both polycystic ovary disease and anorexia nervosa. Fertil Steril 85: 1849–1851.

Pinilla L, Fernandez-Fernandez R, Vigo E et al. (2006). Stimulatory effect of PYY-(3-36) on gonadotropin secretion is potentiated in fasted rats. Am J Physiol Endocrinol Metab 290: E1162–E1171.

Rickenlund A, Thoren M, Carlstrom K et al. (2004). Diurnal profiles of testosterone and pituitary hormones suggest different mechanisms for menstrual disturbances in endurance athletes. J Clin Endocrinol Metab 89: 702–707.

Robinson TL, Snow-Harter C, Taaffe DR et al. (1995). Gymnasts exhibit higher bone mass than runners despite similar prevalence of amenorrhea and oligomenorrhea. J Bone Miner Res 10: 26–35.

Rosen L, Hough D (1988). Pathogenic weight-control behaviors of female college gymnasts. Phys Sportsmed 16: 141–146.

Russell M, Stark J, Nayak S et al. (2009). Peptide YY in adolescent athletes with amenorrhea eumenorrheic athletes and non-athletic controls. Bone 45: 104–109.

Sadler SE, Angleson JK, Dsouza M (2010). IGF-1 receptors in *Xenopus laevis* ovarian follicle cells support the oocyte maturation response. Biol Reprod 82: 591–598.

Sanborn C, Martin B, Wagner WJ (1982). Is athletic amenorrhea specific to runners? Am J Obstet Gynecol 143: 859–861.

Scheid JL, Williams NI, West SL et al. (2009). Elevated PYY is associated with energy deficiency and indices of subclinical disordered eating in exercising women with hypothalamic amenorrhea. Appetite 52: 184–192.

Scheid JL, Toombs RJ, Ducher G et al. (2011). Estrogen and peptide YY are associated with bone mineral density in premenopausal exercising women. Bone 49: 194–201.

Schnackenburg KE, Macdonald HM, Ferber R et al. (2011). Bone quality and muscle strength in female athletes with lower limb stress fractures. Med Sci Sports Exerc 43: 2110–2119.

Sienkiewicz E, Magkos F, Aronis KN et al. (2011). Long-term metreleptin treatment increases bone mineral density and content at the lumbar spine of lean hypoleptinemic women. Metabolism 60: 1211–1221.

Strobel A, Issad T, Camoin L et al. (1998). A leptin missense mutation associated with hypogonadism and morbid obesity. Nat Genet 18: 213–215.

Trost SG, Way R, Okely AD (2006). Predictive validity of three ActiGraph energy expenditure equations for children. Med Sci Sports Exerc 38: 380–387.

Vulliemoz NR, Xiao E, Xia-Zhang L et al. (2004). Decrease in luteinizing hormone pulse frequency during a five-hour peripheral ghrelin infusion in the ovariectomized rhesus monkey. J Clin Endocrinol Metab 89: 5718–5723.

Weimann E, Blum WF, Witzel C et al. (1999). Hypoleptinemia in female and male elite gymnasts. Eur J Clin Invest 29: 853–860.

Welt CK, Chan JL, Bullen J et al. (2004). Recombinant human leptin in women with hypothalamic amenorrhea. N Engl J Med 351: 987–997.

Wen JP, Lv WS, Yang J et al. (2008). Globular adiponectin inhibits GnRH secretion from GT1-7 hypothalamic GnRH neurons by induction of hyperpolarization of membrane potential. Biochem Biophys Res Commun 371: 756–761.

Handbook of Clinical Neurology, Vol. 124 (3rd series)
Clinical Neuroendocrinology
E. Fliers, M. Korbonits, and J.A. Romijn, Editors

Chapter 26

Uncertainties in endocrine substitution therapy for central hypocortisolism

FRANCESCA M. SWORDS*
Norwich Medical School and Directorate of Endocrinology, Norfolk and Norwich University Hospital NHS Foundation Trust, Norwich, UK

INTRODUCTION

Central hypocortisolism is common. In adults, this is most frequently encountered in patients with pituitary tumors or other intrasellar pathologies such as Rathke's cleft cysts or pituitary apoplexy, and following their treatment: surgery or radiotherapy. The hypothalamopituitary region is also frequently included in the field of radiotherapy given for other intracranial malignancies leading to late onset hypopituitarism. Rarer causes of hypopituitarism include pituitary metastases, granulomatous hypophysisits, hemachromatosis, internal carotid artery aneurysm, and congenital or genetic conditions including the empty sella syndrome.

Isolated adrenocorticotropic hormone (ACTH) deficiency can also occur with no obvious cause and may also follow traumatic brain injury in which the usual pattern of sequential loss of pituitary hormones is not observed (Agha and Thompson, 2006).

Finally, the commonest cause of apparent isolated ACTH deficiency is iatrogenic suppression of the hypothalamic–pituitary–adrenal axis due to long-term glucocorticoid exposure. This group of patients may develop longstanding "central" hypocortisolism, which requires conventional treatment and patient education in exactly the same manner as patients with true "central" hypocortisolism, although full recovery of the axis is expected in time.

Whatever the cause of central hypocortisolism, the treatment aims remain the same: to maximize quality of life while minimizing treatment-related adverse effects. The majority of patients with central hypocortisolism now receive hydrocortisone as their replacement therapy. However, various areas of controversy remain: how to assess the patient with suspected hypocortisolism, which agent to use for optimal effect, what is the optimal total daily dose, how to administer divided daily doses, how to monitor therapy and to individually tailor doses, whether to replace other adrenal androgens, how to approach the patient with adrenal suppression, and how best to educate patients with hypocortisolism and treat them in emergency situations. This chapter will discuss the evidence behind each of these controversial areas in turn and reflect on future solutions for this major clinical problem.

CLINICAL ASSESSMENT FOR HYPOCORTISOLISM

Hypocortisolism classically presents with typical features of nausea, weight loss, and dizziness. Laboratory findings suggestive of the diagnosis include hyponatremia, hyperkalemia, normochromic normocytic anemia, and eosinophilia. A useful initial screen for hypocortisolism in the face of clinical suspicion is to perform an early morning serum cortisol. A result over 500 nmol/L clearly excludes hypocortisolism in the absence of steroid therapy, completely obviating the need for further tests. More recently, this cut-off has been refined downward, for example, to 450 nmol/L in one adult series and to 381 nmol/L in a pediatric series, and so pragmatically, most units now consider a 9 a.m. cortisol > 450 nmol/L as normal, requiring no further dynamic testing, although this is not fully established in international clinical practice (Courtney et al., 2000; Maguire et al., 2008).

*Correspondence to: Dr Francesca M. Swords, Consultant Endocrinologist and Honorary Senior Lecturer, Norwich Medical School, Directorate of Endocrinology, Norfolk and Norwich University Hospital NHS Foundation Trust, Colney Lane, Norwich NR4 7UY, UK. Tel: + 44-1603-286-286, Fax: +44-1603-287320, E-mail: Francesca.swords@nnuh.nhs.uk

Patients with a high clinical suspicion of hypocortisolism with an early morning cortisol result below 450 nmo/L will require dynamic testing to confirm or refute the diagnosis of central hypocortisolism. However, the optimal choice of dynamic test is not always clear-cut as discussed below.

Patients with abnormal results on dynamic testing but with early morning (9 a.m.) cortisol levels over 250 nmol/L may be completely asymptomatic. This leads to a second area of controversy, since these patients do have a proven inability to mount an adequate response to stressful stimuli, but do not necessarily require daily treatment for symptom control. Some units continue to recommend that all such patients receive life-long glucocorticoid replacement therapy, though others pragmatically suggest treatment at times of intercurrent illness only. What is clear is that these patients require the full education provided to patients with profound glucocorticoid deficiency: to wear appropriate emergency medical identification, take supplementary steroid therapy during intercurrent illness, and have access to an intramuscular injection of hydrocortisone in the event of vomiting. However, the decision on whether to institute daily replacement therapy should be made on an individual case basis.

DYNAMIC TESTS OF HYPOCORTISOLISM

The gold standard test of ACTH reserve is the insulin stress test (Landon et al., 1963; Plumpton and Besser, 1969). This is the optimum test to be performed in patients suspected of having central hypocortisolism as it tests the entire hypothalamo–pituitary–adrenal axis. Insulin induced hypoglycemia should stimulate a stress response leading to corticotropin-releasing hormone (CRH), adrenocorticotropic hormone (ACTH), and subsequently cortisol release. A suboptimal cortisol response therefore confirms the axis to be insufficient.

The test is unnecessary in patients with a high clinical suspicion of central hypocortisolism with a documented 9 a.m. cortisol below 100 nmol/L (Courtney et al., 2000). For example, a patient with a documented undetectable early morning serum cortisol following surgery for Cushing's disease is clearly ACTH-deficient and no dynamic test is required to confirm this. Furthermore, prolonged refractory hypoglycemia may follow the test in this group of patients, making the test dangerous as well as unnecessary to perform.

The insulin stress test is also unreliable in patients with untreated thyroid-stimulating hormone (TSH) deficiency, as adequate free thyroid hormone is necessary for the maintenance of normal ACTH (and GH) secretion. The test should therefore be deferred until the serum free thyroxine level has been normalized for 3 months. In practice, this may require the patient to be treated empirically with glucocorticoid-replacement therapy as well as levothyroxine for 3 months before testing can be reliably performed.

The insulin stress test should be performed after a 12 hour fast. In this test 0.15 units rapid-acting insulin is administered as a bolus intravenously to provoke symptomatic hypoglycemia < 2.2 mmol/L. A repeat dose should be administered if this is not achieved by 45 minutes, and a double dose (0.3 units) is usually required for patients with active acromegaly. The full protocol for this and all other tests listed is available at www.endobible.com/investigations/. Cortisol is expected to rise to above 550 nmol/L and lesser rises indicate suboptimal ACTH release with central hypocortisolism (Dökmetaş et al., 2000).

However, the test has multiple absolute contraindications. These include epilepsy, cerebrovascular disease, ischemic heart disease or an abnormal resting electrocardiogram. Furthermore, the test requires highly trained and experienced staff, with continuous secure intravenous access, and glucose, fluid, and hydrocortisone resuscitation facilities available in case of severe or prolonged hypoglycemia and hypotension. For these reasons, many centers routinely use alternative tests for central hypocortisolism.

The glucagon test is reserved for second-line use as its specificity and sensitivity are lower than those of the insulin stress test (Mitchell et al., 1970). This test is performed by monitoring serum cortisol levels after a single intramuscular dose of 1 mg glucagon (1.5 mg in subjects over 90 kg). Although many people feel nauseated and unwell during this test, it has few cautions or contraindications, and again does test the entire hypothalamo–pituitary–adrenal axis, with the same cut-off ranges as the insulin stress test above.

The Synacthen test, on the other hand, is quick, simple to perform, and has no contraindications or cautions (Wood et al., 1965). This test involves checking serum cortisol response at 30 and 60 minutes following a single intramuscular dose of 250 μg synthetic ACTH (Synacthen®). A rise in cortisol to above 500 nmol/L indicates a normal adrenal response to ACTH stimulation (Agwu et al., 1999). This makes it a useful test in patients with possible adrenal dysfunction, or with suspected HPA axis suppression due to exogenous glucocorticoids, but makes it a poor test for hypocortisolism due to central causes. Normal responses have been redefined in recent years with a 5th percentile defined as above 510 nmol/L (Clark et al., 1998). In practice, this test can be useful for pituitary patients in some cases, but a "normal" result ($>$500 nmol/L) must be treated with caution, particularly in patients with a relatively short history in whom higher cut-off values

(>600 nmol/L) have been suggested by some groups (Bangar and Clayton, 1998; Agha et al., 2006). In practice, this is not a problem in patients with long-standing symptoms but means the test should not be used to assess patients immediately following pituitary surgery, for example. In these cases, and if an insulin stress test is unavailable or contraindicated, and the 9 a.m. cortisol does not conclusively indicate adequate ACTH reserve (i.e., not > 450 nmol/L), it may be safer to treat the patient for presumed hypocortisolism in the first instance and then perform a Synacthen test at least 6 weeks later. By this time, if the hypothalamo–pituitary axis is no longer able to produce ACTH, the adrenal glands should theoretically have lost responsitivity to exogenous ACTH. Failure to respond to ACTH at this stage then indicates likely central hypocortisolism. However, normal results should still be interpreted with caution and an insulin stress test remains the gold standard test.

Finally, the metyrapone test is also used in some units. This test is performed by administering 2 g metyrapone at midnight, then taking a 9 a.m. blood sample for ACTH, cortisol, and 11 deoxycortisol. The test relies on a rise in pituitary ACTH release following blockade of cortisol production by high-dose metyrapone, and as such should test the entire HPA axis. A rise in 11 deoxycortisol to > 200 nmol/L demarcates a normal response although the sensitivity of this test is poor at 47% (Berneis et al., 2002). The sensitivity is increased by using the sum of 11 deoxycortisol and cortisol (normal range > 450 nmol/L), and by measuring ACTH responses. However, this test is not widely used. It requires an 11 deoxycortisol assay which is not performed routinely in many laboratories, and which is known to cross-react with cortisol. Furthermore, it relies on multiple sequential steps which may be affected unpredictably in patients with suspected central hypocortisolism, further limiting its usefulness.

STANDARD TREATMENTS FOR HYPOCORTISOLISM

What is the optimal agent?

In most countries, hydrocortisone (HC) is the drug of choice and standard replacement therapy for hypocortisolism. This was used in 76% patients in one major international patient survey (Forss et al., 2012).

Cortisone acetate (CA) remains the most commonly used alternative in Scandinavia and Australia (6% worldwide usage). CA is biologically inactive, requiring conversion to cortisol by the enzyme 11BHSD1 for activity, and has an approximate dose equivalence of 20 mg HC = 25 mg CA. This leads to a slight delay and reduction in achieved peak serum cortisol levels with this agent, making it less popular internationally (Swords et al., 2003).

However, examination of urinary steroid ratios in patients converting between CA and HC reveals significantly lower, and therefore more physiologic, 11BHSD1 activity when patients are taking CA compared to when they are taking HC. This suggests that bioequivalent doses of CA lead to lower and hence more physiologic tissue exposure to glucocorticoid (Swords et al., 2003).

Almost any glucocorticoid can be used to replace hypocortisolism, and so prednisolone and dexamethasone are also used (in 11% and 4% respectively, as reported by Forss et al., 2012). Both of these agents offer the advantage of once daily dosing due to their prolonged half-life in serum. However, both these agents offer significantly less physiologic replacement when compared to HC or CA. The prolonged half-life itself is problematic, leading to glucocorticoid exposure overnight, which may be responsible for the significantly worse side-effect profiles as compared to CA or HC. Prednisolone and in particular dexamethasone are associated with more weight gain, adverse metabolic profile, and worsened bone mineral metabolism (Koetz et al., 2012). Furthermore, dexamethasone is not measurable on standard cortisol assays, and prednisolone tends to cross over by only 30% in standard assays, again complicating the assessment of hypocortisolemic patients, particularly in the presence of limited endogenous cortisol production.

Total daily glucocorticoid dosing

Traditionally, hydrocortisone doses between 20 and 30 mg daily (or cortisone acetate 25–60 mg) have been administered to patients with hypocortisolism of any cause, and relatively high doses may still be required to maintain symptom control in patients with primary adrenal disease. However, work in the 1990s has convincingly demonstrated that daily cortisol production rates in healthy individuals are significantly lower than previously thought. Esteban et al. (1991) calculated daily production rates to be approximately 5.7 mg/m^2/day, less than half of the previously accepted values. Other modern estimates are similar: 6–11 mg/m^2/day (Linder et al., 1990; Kraan et al., 1998), equating to approximately 10–20 mg total daily dose allowing for incomplete oral absorption (Ten et al., 2001). Since then, multiple series have reported improved bone markers, more physiologic biochemical profiles, and no diminution of quality of life with lower total daily doses (Wichers et al., 1999; Mah et al., 2004; Behan et al., 2011).

Furthermore, Bergthorsdottir et al. (2006) confirmed that patients requiring life-long glucocorticoid replacement therapy (for Addison's disease) still have a twofold higher risk of death compared to normal controls, despite modern dosing regimens, with this excess mortality largely due to cardiovascular, malignant, and infectious diseases. This confirms multiple similar historic observations of increased mortality with long-term glucocorticoid exposure in Cushing's disease and hypopituitarism (Plotz et al., 1952; Rosen and Bengtsson, 1990; Tomlinson et al., 2001). More recently, significant evidence suggests that a large proportion of the increased standardized mortality ratio in pituitary patients is in fact due to supraphysiologic glucocorticoid replacement therapy (and untreated growth hormone deficiency). High tissue exposure to glucocorticoids carries multiple potential detrimental effects, on glucose tolerance, lipid profile, body mass index, bone mineral density, and metabolic and cardiovascular risk (Pupo et al., 1966; Lukert and Raisz, 1990; Beshyah et al., 1994; Canalis, 1996), and some reports now relate this excess morbidity directly to total daily glucocorticoid dose (Filipsson et al., 2006).

It seems logical to assume that a reduced total daily dose will reduce tissue exposure. However, the pharmacokinetics of orally administered hydrocortisone demand multiple daily dosing, with subsequent peaks and troughs despite lower total daily doses. Even very low total daily doses of hydrocortisone, given in multiple divided doses to mimic mean endogenous serum cortisol levels, are probably still associated with glucocorticoid overexposure at the tissue level.

Although cortisol is carried bound to CBG, it is largely metabolized by 11betahydroxysteroid dehydrogenase (11BHSD). This enzyme exists in two isoforms with distinct tissue distributions and roles. 11BHSD1 is largely expressed in visceral adipose tissue and is bidirectional but largely acts as a reductase, responsible for the conversion of inactive cortisone (E) to active cortisol (F). The activity of this enzyme can be assessed by examining urine steroid profiles of E and F metabolites (Em and Fm): the higher the ratio of Fm:Em, the higher the type 1 activity. 11BHSD2 is predominantly expressed in the kidneys to dehydrogenate or inactivate F to E to protect the renal collecting tubules from the mineralocorticoid actions of cortisol. Another urine ratio urinary free cortisol:urinary free cortisone indicates type 2 activity: the lower the ratio, the higher the 11BHSD2 activity. These urinary steroid profile ratios indicate that the 11BHSD1 enzyme, which greatly increases cortisol availability, is less active in females (Raven and Taylor, 1996; Weaver et al., 1998), is less active with increasing adiposity and with increasing insulin resistance (Walker and Best, 1995), and is less active in the presence of increasing GH and IGF-1 (Gelding et al., 1998; Moore et al., 1999; Toogood et al., 2000; Trainer et al., 2001).

Examination of these ratios in patients with central hypocortisolism and growth hormone deficiency reveals a dramatic increase in 11BHSD1 activity compared to normal subjects (Fm:Em ratios 59% increased compared to normal subjects) (Fig. 26.1). Increased 11BHSD1 activity increases cortisol bioavailability at tissue level, perhaps explaining the reduced doses required for central versus adrenal hypocortisolism. However, this also implies that tissue exposure to glucocorticoid remains elevated even in hypopituitary patients receiving "low dose" (median daily dose 20 mg) hydrocortisone replacement therapy compared to eupituitary subjects (Swords et al., 2003).

Interestingly, this overexposure is abrogated when growth hormone replacement therapy is initiated. This GH related fall in 11BHSD1 activity towards normal therefore appears to reduce tissue exposure to glucocorticoid, and is also paralleled by a measurable fall in serum cortisol levels (14.5% reduction in area under the curve, $p < 0.05$) without an associated change in symptoms. Therefore, in the presence of untreated growth hormone deficiency, it is even more important to use the minimum total daily dose associated with an acceptable quality of life for the individual patient.

Some patients do report increased subjective well-being in the short term on higher doses of glucocorticoids, indeed courses of glucocorticoids are frequently used in the palliative setting to improve quality of life in people with an intact hypothalamo–pituitary–adrenal axis. However, controlled studies with repeated quality of life assessments have failed to show a significant difference with lower total daily doses: bodily pain scores may improve with increased total daily dose, but all other measures, for example, of emotional well-being, mental

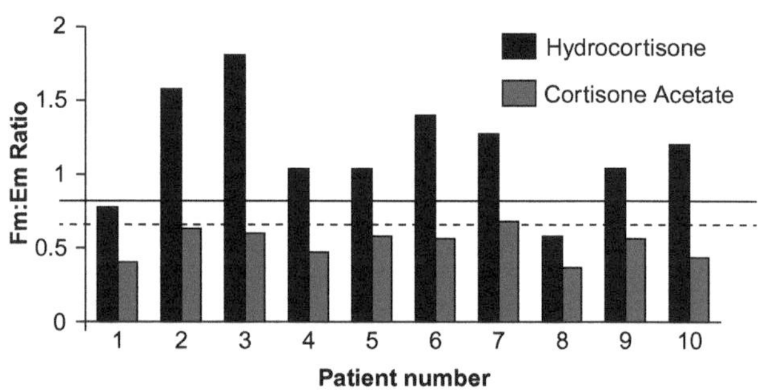

Fig. 26.1. Supraphysiologic Fm:Em ratios on hydrocortisone replacement therapy. This figure illustrates the elevated cortisol (Fm) to cortisone (Em) metabolite ratios observed in hypopituitary patients taking hydrocortisone replacement therapy (black bars), and the fall towards normal when the same individuals take cortisone acetate replacement therapy (shaded bars). Values for normal subjects are indicated with the horizontal lines (unbroken line = males, broken line = females). (Reproduced from Swords et al., 2003.)

health, and vitality, fail to improve with increased glucocorticoiod exposure (Peacey et al., 1997; Wichers et al., 1999; Forss et al., 2012).

Recommended total daily doses have therefore come down and are typically 15–20 mg hydrocortisone daily. However, even such low total daily doses are probably still associated with supraphysiologic tissue glucocorticoid exposure and so weight-based dosing regimens typically equating to 10–15 mg hydrocortisone daily are increasingly being used, as discussed below (Mah et al., 2004).

Multiple daily dosing and monitoring of glucocorticoid replacement

Endogenous cortisol production is known to have a dramatic circadian rhythm. Many patients also report fatigue and headache between once or twice daily doses, and there is now good objective evidence of improved quality of life with multiple daily dosing, so a tailored approach with multiple daily doses (typically three) according to symptom control is now the norm (Riedel et al., 1993; Hahner et al., 2007; Ekman et al., 2012).

The majority of endogenous circulating cortisol is bound to corticosteroid-binding globulin (CBG). Unfortunately, CBG becomes saturated following oral hydrocortisone administration leading to supraphysiologic urine free cortisol (UFC) excretion. UFC is therefore not a useful means of monitoring cortisol replacement. Newell-Price et al. (2008) performed detailed analysis of serum cortisol levels in eupituitary subjects which elegantly demonstrates the diurnal variation in endogenous serum cortisol (Fig. 26.2).

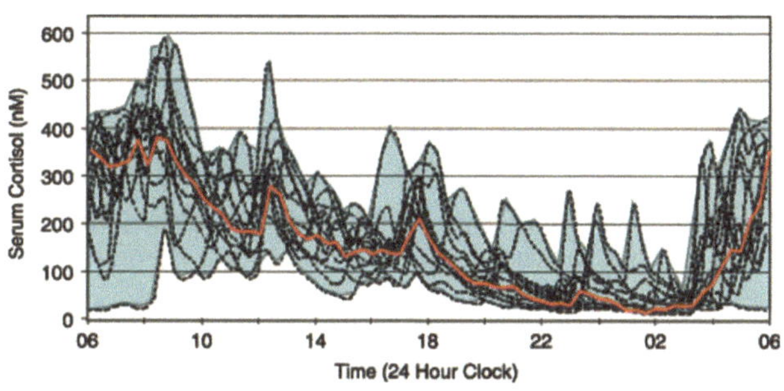

Fig. 26.2. Diurnal variation of serum cortisol in normal individuals. Serum cortisol–time profiles in normal subjects. (To convert to μg/dL divide by 27.59.) The normal range for the 15 BMI-matched volunteers was constructed by recording the minimum and maximum reported concentrations of cortisol at each time in the normal controls, and calculating the median values. The normal range is indicated by the light blue area; values for individual normal subjects are shown as broken black lines; the median value is indicated by the thick red line. (Reproduced from Newell-Price et al., 2008.)

Experimentally, serum sampling of serum cortisol levels before and after hydrocortisone doses has usefully confirmed that with three times daily dosing: typically of 10 mg on waking, 5 mg after 5 hours, and 5 mg after 9 hours, oral hydrocortisone mimics the peaks and troughs of endogenous cortisol production fairly closely (Fig. 26.3) (Swords et al., 2003; Behan et al., 2011). Furthermore, the metabolic effects of cortisol are different at different times of day. This is particularly obvious in patients with diabetes who note a larger rise in blood glucose following the same dose of hydrocortisone if this is administered later in the day. Therefore, both excessive total daily doses, but also larger doses of hydrocortisone later in the day, have been implicated in the poor metabolic outcomes in patients with hypocortisolism (Plat et al., 1999).

Serial sampling of serum cortisol levels at 0, 1, 3, 5, and 9 hours post initial hydrocortisone dose (or more

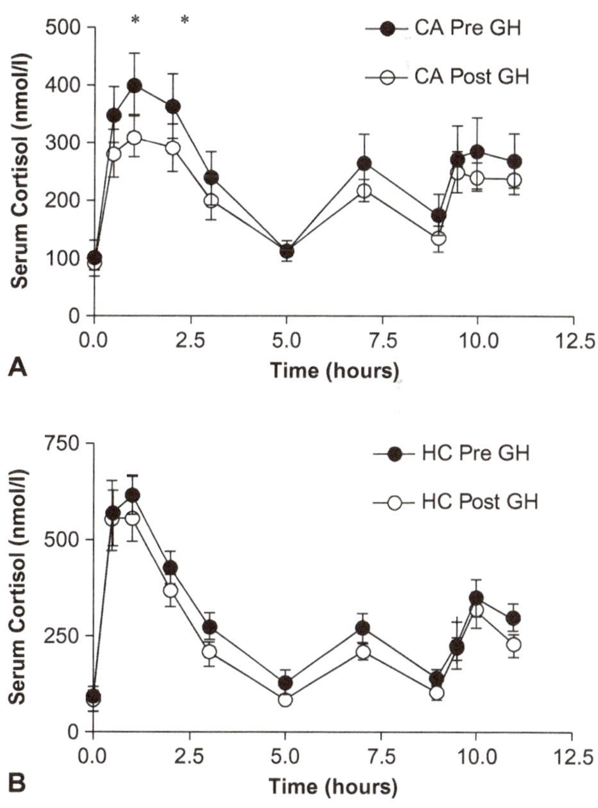

Fig. 26.3. Diurnal variability in cortisol levels during hydrocortisone and cortisone acetate replacement therapy. Mean cortisol values with standard errors are shown for 10 patients studied while on cortisone acetate (**A**) and while on hydrocortisone (**B**). In both cases, the closed circles indicate values recorded before growth hormone therapy and the open circles indicate the values recorded during growth hormone therapy. *Indicate values at which there is a significant difference between the two groups ($p < 0.05$). (Reproduced from Swords et al., 2003.)

frequently in some institutions) can be useful to rule out excessive glucocorticoid levels, particularly late in the day – aiming for no troughs below 100 nmol/L and no peaks above 700 nmol/L. However, careful clinical assessment of the patient's symptoms, as well as their postural blood pressure, will usually yield similarly useful information. If the aim of treatment is to optimize patient well-being, with the minimum exposure to glucocorticoids, the patients themselves will often be able to recognize how low each dose can be titrated down to without adversely affecting symptoms or inducing postural hypotension. This means that such detailed circadian profiling is generally unnecessary for routine clinical use, although concerns remain regarding potentially increasing the risk of hypocortisolemic crisis with lower doses, with potentially very low troughs between doses.

Furthermore, there is now good evidence to support the use of weight-based dosing regimens. Typically, a dose of 8.1 mg/m^2/day is recommended with 50% total dose (approximately 0.12 mg/kg) given on waking prior to food, and 25% given at 5 and 9 hours (Mah et al., 2004). This group have confirmed that cortisol absorbtion is highly predictable according to weight. They recommend a test dose of 0.12 mg/kg should be administered in the fasting state. If a single 4 hour postdose serum cortisol level falls within the predicted range, then a standard dosing regimen can be given, with no further dose adjustment or monitoring required. Levels falling outside the predicted range suggest atypical absorption and should prompt dose adjustment accordingly. This has led to further refinement and reduction in the total daily exposure of patients to hydrocortisone. Conventional tablet sizes of 10 or 20 mg only can lead to practical issues when patients need to take a quarter of a tablet, but these can usually be overcome with detailed education, with doses such as 7.5 mg on waking, 5 mg with the midday meal, and 2.5 mg late afternoon for a 70 kg patient now becoming increasingly common. Long-term data are not yet available on whether this leads to the predicted improvement in metablic parameters, though short-term safety data with no increase in hypocortisolemic crises or adverse patient quality of life are encouraging.

Modified-release hydrocortisone

The pharmacodynamics of standard oral hydrocortisone leads to a peak in serum cortisol levels 30–60 minutes after oral administration, with levels gradually falling over the next few hours depending on the initial peak achieved and dose given (Fig. 26.3). For example, a dose of 10 mg given to a hypopituitary 70 kg male would be expected to achieve a peak serum cortisol of 500–600 nmol/L, gradually falling to ~100 nmol/L by 5 hours, and becoming undetectable by 7 hours (Swords et al., 2003). Multiple studies have identified the loss of circadian rhythm during ill health in normal individuals, and so it is assumed that mimicking the normal rhythm would be beneficial (Schuetz and Müller, 2006). Furthermore, quality of life studies reveal that well-being is at its lowest in hypocortisolemic patients just before their morning dose, improving by midday, and then falling again prior to their next dose of glucocorticoid, compelling evidence that mimicking the circadian profile of cortisol would be of benefit (Groves et al., 1988). This necessitates either very large doses given in the morning to delay the trough level, with a consequent highly supraphysiologic peak in the morning, or repeated daily dosing with the consequent inconvenience for the patient as discussed above. In one study 38% patients surveyed were dissatisfied with multiple daily dosing (Forss et al., 2012), and quality of life measures were improved with four times daily dosing of standard hydrocortisone preparations, though this may cause difficulties with concordance long term (Ekman et al., 2012). These observations have led to the development and recent launch of the modified-release hydrocortisone preparation Plenadren (Johannsson et al., 2009). This preparation comprises an immediate-release coating of hydrocortisone surrounding an extended release core.

This is administered once daily on waking and should be taken fasting and with water alone. This leads to a similar peak in cortisol within 30–60 minutes, but with a sustained level falling more gradually over 12 hours. Overall, the bioavailability is 20% less than with the equivalent total daily dose of standard hydrocortisone. A crossover study comparing once daily 20 mg Plenadren with hydrocortisone 20 mg in divided doses (10/5/5 mg) was associated with a fall in weight of 0.7 kg, a fall in systolic blood pressure of 5 mmHg and a fall in glycosylated hemaglobin of 0.1% in normal subjects, and 0.6% in diabetic individuals (Johannsson et al., 2012). Whether this represents a clinically important improvement in care, or whether similar results would have been achieved by a 20% reduction in hydrocortisone is not yet clear. Plenadren does offer the benefit of once-daily dosing, with potential benefits for compliance and possible but unproven improved quality of life with the smoother serum profile. Data so far indicate a good safety profile with no increase in hypocortisolemic crises or hospital admissions. However, there is limited experience with this agent, particularly in acutely unwell patients, and its role is yet to be established fully.

The circadian profile of endogenous cortisol also highlights another issue with hydrocortisone replacement therapy. Figure 26.2 demonstrates the rise in serum cortisol which occurs prior to waking, and this is extremely difficult to replicate with oral replacement therapy. In some centers (notably in Germany) it is common practice to recommend patients set an alarm clock to take their hydrocortisone at least 30 minutes before the patient plans to rise for the day. In practice, this is difficult to achieve. However, other work has focused on this issue and an alternative modified-release agent has been developed for nocturnal administration (Debono et al., 2009). This agent was designed to replicate the very low serum levels overnight, with a rise in serum cortisol approximately 5 hours after administration at 23.00 but prior to waking. Modeling studies suggest that a second dose would be required on waking, to lead to a further slow fall in levels during the day. This agent has yet to be marketed but safety and quality of life studies will be interesting.

Future glucocorticoid treatments

Serial serum cortisol measurements from patients on divided daily dosing and once-daily dosing with a modified-release hydrocortisone are able to demonstrate a good circadian pattern – with the largest peak in cortisol early in the morning, smaller peaks associated with meal times, and low trough levels overnight. These roughly approximate to the mean levels observed in eupituitary patients (Figs. 26.2 and 26.3). However, all of these treatments fail to replicate the dynamic ultradian rhythm comprising discrete pulses of ACTH and endogenous cortisol production in eupituitary subjects (Carnes et al., 1989; Windle et al., 1998). This pulsatility is not an accident, but rather attributable to an oscillatory feedforward–feedback relationship between the pituitary and adrenal gland (Walker et al., 2010). More recent work by Russell et al. (2010) confirms that this pulsatility, as well as the overall trend in cortisol levels, is important. Large doses of long-acting potent steroids such as dexamethasone are well known to suppress the HPA axis, and this forms the basis of dynamic tests of glucocorticoid excess. However, Russell et al. have now demonstrated that continuous infusion of prednisolone also leads to almost immediate loss of CRH responsiveness and that this is mediated through the glucocorticoid receptor in humans. Extrapolation from these data would suggest that the continuous low-level elevation of serum cortisol levels seen with oral hydrocortisone administration is supraphysiologic and would lead to HPA axis suppression, whereas pulsatile delivery of cortisol would be more physiologic.

A possible explanation for this lies in the way in which cortisol is transported: 90% serum cortisol is bound to CBG, which saturates at levels of approximately 400 nmol/L (Tunn et al., 1992).

It is therefore likely that most tissues are only usually exposed to serum-derived free cortisol during the short pulses of cortisol release in which serum levels transiently rise over and above this individual saturation level, which itself has a circadian fluctuation. The resultant episodic activation of the glucocorticoid receptor is thought to be critical for normal regulation of gene activity and physiology and the blunting of endogenous pulsatility which is seen with aging and with chronic inflammatory states might then in part contribute to the adverse metabolic milieu in these states.

Circadian fluctuations in the binding capacity of CBG are lost in patients on chronic replacement (Angeli et al., 1978). Furthermore, with oral hydrocortisone administration, serum levels are only likely to fall much below that saturation point just prior to the second and third doses of the day, leading to near continuous exposure of tissue to plasma-derived rather than locally generated glucocorticoid.

For this reason, the ultimate glucocorticoid replacement therapy to mimic endogenous production as closely as possible would require very short duration pulsatile delivery system such as via an infusion pump. Continuous intravenous hydrocortisone infusions have been used to good effect experimentally, but to date no such pulsatile delivery system has been used in clinical practice (Merza et al., 2006).

It is possible, therefore, that the focus on ever smoother exogenous cortisol levels may not in fact lead to the expected improvements in adverse effects as minute-to-minute tissue glucocorticoid exposure will remain supraphysiologic due to the lack of pulsatility.

Other adrenal androgens

Even in primary adrenal failure, the replacement of adrenal androgens remains contentious. In ACTH deficiency or central hypocortisolism, adrenal androgens have now also been confirmed to be reduced (Isidori et al., 2003). Furthermore, dehydroepiandrostenedione (DHEA) replacement has been shown to reduce growth hormone requirements in panhypopituitary subjects. However, hard evidence of benefit in this group of patients is still lacking, with recent work failing to show any significant improvement in quality of life or insulin sensitivity in hypopituitary patients as opposed to patients with primary adrenal failure (Dhatariya and Nair, 2003; McHenry et al., 2012). The role of other adrenal androgens with central hypocortisolism therefore remains unclear.

Adrenal suppression

Long-term glucocorticoid therapy for any indication is now well recognized as a cause of hypothalamo–pituitary–adrenal axis suppression by endocrinologists.

However, recognition of this phenomenon remains patchy and controversial within other fields. For example, most healthcare professionals now recognize that treatment with prednisolone doses > 5 mg daily for greater than 3 months puts patients at risk of loss of bone mineral density and so are proactive at minimizing bone risk, e.g., by considering calcium and vitamin D supplementation and bisphosphonate therapy (reviewed in Weinstein, 2012). However, it is not yet universal for these patients to be considered at risk of hypothalamo–pituitary–adrenal axis suppression. These patients, as well as others taking long-term inhaled, intranasal, or topical long-term glucocorticoids, should be assessed for possible hypocortisolism as this and iatrogenic Cushing's syndrome are increasingly common. More and more patients now survive conditions requiring long-term steroid use such as organ transplantation. High risk patients such as these and those with confirmed suboptimal cortisol responsiveness should therefore receive exactly the same education and advice as those with other causes of hypocortisolism as above (Wass and Arlt, 2012; further information available via the Addison's disease weblink below).

Patient education

As with all chronic diseases, patient education is a key part of the management of hypocortisolism of any kind. Patients are at risk of life-threatening hypoadrenal crises in the event of intercurrent illness or omitted doses, and so patients and their relatives require dedicated education to help them understand the importance of their treatment, understand the need for specific timing of their doses, and understand how to deal with intercurrent illness, minor surgical procedures, etc.

In the UK, the Addison's disease self-help group offers excellent practical and detailed guidance on this: namely to double the usual doses of glucocorticoids during intercurrent illness, and to take an immediate dose of 100 mg intramuscular hydrocortisone in the event of vomiting or the inability to take a standard oral dose (www.addisons.org.uk).

However, this remains controversial with some patients and healthcare professionals. There remains resistance in some communities to the self-administration of an intramuscular dose, and the lack of a commercially available emergency hydrocortisone injection device complicates this further. Furthermore, although the requirement for depot intramuscular doses is fully understood and accepted within the endocrine community, the practice of administering intermittent boluses of intravenous hydrocortisone remains widespread for other indications, and many healthcare professionals from other disciplines continue to resist 6 hourly depot intramuscular doses or the other safe alternative of a low-dose continuous infusion of hydrocortisone despite clear evidence in support of its superiority (Wass and Arlt, 2012).

The UK Addison's society also offers practical guidance on dental procedures, gastrointestinal endoscopy, general and local anesthesia, etc., though not all relevant professionals are fully versant with this (see weblink below at end of reference list).

Patients need to understand the importance of emergency identification to facilitate this treatment during unpredictable emergency situations. Standard medical advice is therefore for all patients with hypocortisolemia, of any cause, to carry and wear recognizable emergency steroid identification at all times, though again this is not acceptable to all affected individuals.

CONCLUSIONS

Hydrocortisone in two to three divided doses with a total daily dose between 10 and 20 mg, or a weight-based regimen of 8.1 mg/m^2/day, remains the standard treatment for the replacement of central hypocortisolism throughout most of the world.

Once-daily regimens using dexamethasone, prednisolone are simple and convenient to take, but associated with a significantly higher incidence of adverse effects than HC or cortisone acetate. Cortisone acetate requires conversion to cortisol to be active, and has a delayed absorption, making this less popular internationally, though it is probably associated with a nearer physiologic tissue exposure than hydrocortisone. However, even the most detailed dose titration according to weight, symptom control, lifestyle, and serum cortisol monitoring has been shown to result in overexposure to glucocorticoids at a tissue level.

Newer prolonged-release agents have recently been launched that mimic the overall circadian profile of cortisol in a once-daily dose, and may aid compliance for some patients. However, these are also likely to lead to overexposure due to the lack of pulsatility of exposure at tissue level.

Other agents in development have sought to mimic the dawn phenomenon of a rise in serum cortisol prior to waking, though their role remains uncertain and they are not yet commercially available.

The holy grail of cortisol replacement associated with pulsatile delivery of cortisol of varying magnitude throughout the day therefore remains elusive. This can currently only be achieved in experimental situations using continuous subcutaneous infusion pumps, but as a standard treatment option, this remains a long way off.

References

Agha A, Thompson CJ (2006). Anterior pituitary dysfunction following traumatic brain injury (TBI). Clin Endocrinol (Oxf) 64: 481–488.

Agha A, Tomlinson JW, Clark PM et al. (2006). The long-term predictive accuracy of the short Synacthen (corticotropin) stimulation test for assessment of the hypothalamic–pituitary–adrenal axis. J Clin Endocrinol Metab 91: 43–47.

Agwu J, Spoudeas H, Hindmarsh P et al. (1999). Tests of adrenal insufficiency. Arch Dis Child 80: 330–333.

Angeli A, Frajria R, De Paoli R et al. (1978). Diurnal variation of prednisolone binding to serum corticosteroid-binding globulin in man. Clin Pharmacol Ther 23: 47–53.

Bangar V, Clayton RN (1998). How reliable is the short Synacthen test for the investigation of the hypothalamic–pituitary–adrenal axis? Eur J Endocrinol 139: 580–583.

Behan LA, Rogers B, Hannon MJ et al. (2011). Optimizing glucocorticoid replacement therapy in severely adrenocorticotropin-deficient hypopituitary male patients. Clin Endocrinol (Oxf) 75: 505–513.

Bergthorsdottir R, Leonsson-Zachrisson M, Oden A et al. (2006). Premature mortality in patients with Addison's disease: a population-based study. J Clin Endocrinol Metab 91: 4849–4853.

Berneis K, Staub JJ, Gessler A et al. (2002). Combined stimulation of adrenocorticotropin and compound-S by single dose metyrapone test as an outpatient procedure to assess hypothalamic–pituitary–adrenal function. J Clin Endocrinol Metab 87: 5470–5475.

Beshyah SA, Shahi M, Skinner E et al. (1994). Cardiovascular effects of growth hormone replacement therapy in hypopituitary adults. Eur J Endocrinol 130: 451–458.

Canalis E (1996). Mechanisms of glucocorticoid action in bone: implications to glucocorticoid-induced osteoporosis. J Clin Endocrinol Metab 81: 3441–3447.

Carnes M, Lent S, Feyzi J et al. (1989). Plasma adrenocorticotropic hormone in the rat demonstrates three different rhythms within 24 h. Neuroendocrinology 50: 17–25.

Clark PM, Neylon I, Raggatt PR et al. (1998). Defining the normal cortisol response to the short Synacthen test: implications for the investigation of hypothalamic–pituitary disorders. Clin Endocrinol (Oxf) 49: 287–292.

Courtney CH, McAllister AS, McCance DR et al. (2000). Comparison of one week 0900 h serum cortisol, low and standard dose Synacthen tests with a 4 to 6 week insulin hypoglycaemia test after pituitary surgery in assessing HPA axis. Clin Endocrinol (Oxf) 53: 431–436.

Debono M, Ghobadi C, Rostami-Hodjegan A et al. (2009). Modified-release hydrocortisone to provide circadian cortisol profiles. J Clin Endocrinol Metab 94: 1548–1554.

Dhatariya KK, Nair KS (2003). Dehydroepiandrosterone: is there a role for replacement? Mayo Clin Proc 78: 1257–1273.

Dökmetaş HS, Colak R, Keleştimur F et al. (2000). A comparison between the 1-microg adrenocorticotropin (ACTH) test, the short ACTH (250 microg) test, and the insulin tolerance test in the assessment of hypothalamo–pituitary–adrenal axis immediately after pituitary surgery. J Clin Endocrinol Metab 85: 3713.

Ekman B, Bachrach-Lindström M, Lindström T et al. (2012). A randomized, double-blind, crossover study comparing two- and four-dose hydrocortisone regimen with regard to quality of life, cortisol and ACTH profiles in patients with primary adrenal insufficiency. Clin Endocrinol (Oxf) 77: 18–25.

Esteban NV, Loughlin T, Yergey AL et al. (1991). Daily cortisol production rate in man determined by stable isotope dilution/mass spectrometry. J Clin Endocrinol Metab 72: 39–45.

Filipsson H, Monson JP, Koltowska-Häggström M et al. (2006). The impact of glucocorticoid replacement regimens on metabolic outcome and comorbidity in hypopituitary patients. J Clin Endocrinol Metab 91: 3954–3961.

Forss M, Batcheller G, Skrtic S et al. (2012). Current practice of glucocorticoid replacement therapy and patient-perceived health outcomes in adrenal insufficiency – a worldwide patient survey. BMC Endocr Disord 12: 8.

Gelding SV, Taylor NF, Wood PJ et al. (1998). The effect of growth hormone replacement therapy on cortisol-cortisone interconversion in hypopituitary adults: evidence for growth hormone modulation of extrarenal 11 beta-hydroxysteroid dehydrogenase activity. Clin Endocrinol (Oxf) 48: 153–162.

Groves RW, Toms GC, Houghton BJ et al. (1988). Corticosteroid replacement therapy: twice or thrice daily? J R Soc Med 81: 514–516.

Hahner S, Loeffler M, Fassnacht M et al. (2007). Impaired subjective health status in 256 patients with adrenal insufficiency on standard therapy based on cross-sectional analysis. J Clin Endocrinol Metab 92: 3912–3922.

Isidori AM, Kaltsas GA, Perry L et al. (2003). The effect of growth hormone replacement therapy on adrenal androgen secretion in adult onset hypopituitarism. Clin Endocrinol (Oxf) 58: 601–611.

Johannsson G, Bergthorsdottir R, Nilsson AG et al. (2009). Improving glucocorticoid replacement therapy using a novel modified-release hydrocortisone tablet: a pharmacokinetic study. Eur J Endocrinol 161: 119–130.

Johannsson G, Nilsson AG, Bergthorsdottir R et al. (2012). Improved cortisol exposure-time profile and outcome in patients with adrenal insufficiency: a prospective randomized trial of a novel hydrocortisone dual-release formulation. J Clin Endocrinol Metab 97: 473–481.

Koetz KR, Ventz M, Diederich S et al. (2012). Bone mineral density is not significantly reduced in adult patients on low-dose glucocorticoid replacementtherapy. J Clin Endocrinol Metab 97: 85–92.

Kraan GP, Dullaart RP, Pratt JJ et al. (1998). The daily cortisol production reinvestigated in healthy men. The serum and urinary cortisol production rates are not significantly different. J Clin Endocrinol Metab 83: 1247–1252.

Landon J, Wynn V, James VH (1963). The adrenocortical response to insulin-induced hypoglycaemia. J Endocrinol 27: 183–192.

Linder BL, Esteban NV, Yergey AL et al. (1990). Cortisol production rate in childhood and adolescence. J Pediatr 117: 892–896.

Lukert BP, Raisz LG (1990). Glucocorticoid-induced osteoporosis: pathogenesis and management. Ann Intern Med 112: 352–364.
Maguire AM, Biesheuvel CJ, Ambler GR et al. (2008). Evaluation of adrenal function using the human corticotrophin-releasing hormone test, low dose Synacthen test and 9 am cortisol level in children and adolescents with central adrenal insufficiency. Clin Endocrinol (Oxf) 68: 683–691.
Mah PM, Jenkins RC, Rostami-Hodjegan A et al. (2004). Weight-related dosing, timing and monitoring hydrocortisone replacement therapy in patients with adrenal insufficiency. Clin Endocrinol (Oxf) 61: 367–375.
McHenry CM, Bell PM, Hunter SJ et al. (2012). Effects of dehydroepiandrosterone sulphate (DHEAS) replacement on insulin action and quality of life in hypopituitary females: a double-blind, placebo-controlled study. Clin Endocrinol (Oxf) 77: 423–429.
Merza Z, Rostami-Hodjegan A, Memmott A et al. (2006). Circadian hydrocortisone infusions in patients with adrenal insufficiency and congenital adrenal hyperplasia. Clin Endocrinol (Oxf) 65: 45–50.
Mitchell ML, Byrne MJ, Sanchez Y et al. (1970). Detection of growth-hormone deficiency. N Engl J Med 282: 539–541.
Moore JS, Monson JP, Kaltsas G et al. (1999). Modulation of 11beta-hydroxysteroid dehydrogenase isozymes by growth hormone and insulin-like growth factor: in vivo and in vitro studies. J Clin Endocrinol Metab 84: 4172–4177.
Newell-Price J, Whiteman M, Rostami-Hodjegan A et al. (2008). Modified-release hydrocortisone for circadian therapy: a proof-of-principle study in dexamethasone-suppressed normal volunteers. Clin Endocrinol (Oxf) 68: 130–135.
Peacey SR, Guo CY, Robinson AM et al. (1997). Glucocorticoid replacement therapy: are patients over treated and does it matter? Clin Endocrinol (Oxf) 46: 255–261.
Plat L, Leproult R, L'Hermite-Baleriaux M et al. (1999). Metabolic effects of short-term elevations of plasma cortisol are more pronounced in the evening than in the morning. J Clin Endocrinol Metab 84: 3082–3092.
Plotz CM, Knowlton AI, Ragan C (1952). The natural history of Cushing's syndrome. Am J Med 13: 597–614.
Plumpton FS, Besser GM (1969). The adrenocortical response to surgery and insulin-induced hypoglycaemia in corticosteroid-treated and normal subjects. Br J Surg 56: 216–219.
Pupo AA, Wajchenberg BL, Schnaider J (1966). Carbohydrate metabolism in hyperadrenocortisolism. Diabetes 15: 24–29.
Raven P, Taylor NF (1996). Sex differences in the human metabolism of cortisol. Endocr Res 22: 751–755.
Riedel M, Wiese A, Schürmeyer TH et al. (1993). Quality of life in patients with Addison's disease: effects of different cortisol replacement modes. Exp Clin Endocrinol 101: 106–111.
Rosen T, Bengtsson BA (1990). Premature mortality due to cardiovascular disease in hypopituitarism. Lancet 336: 285–288.
Russell GM, Henley DE, Leendertz J et al. (2010). Rapid glucocorticoid receptor-mediated inhibition of hypothalamic–pituitary–adrenal ultradian activity in healthy males. J Neurosci 30: 6106–6115.
Schuetz P, Muller B (2006). The hypothalamic–pituitary–adrenal axis in critical illness. Endocrinol Metab Clin North Am 35: 823–838.
Swords FM, Carroll PV, Kisalu J et al. (2003). The effects of growth hormone deficiency and replacement on glucocorticoid exposure in hypopituitary patients on cortisone acetate and hydrocortisone replacement. Clin Endocrinol (Oxf) 59: 613–620.
Ten S, New M, Maclaren N (2001). Addison's disease 2001. J Clin Endocrinol Metab 86: 2909–2922.
Tomlinson JW, Holden N, Hills RK et al. (2001). Association between premature mortality and hypopituitarism. West Midlands Prospective Hypopituitary Study Group. Lancet 357: 425–431.
Toogood AA, Taylor NF, Shalet SM et al. (2000). Modulation of cortisol metabolism by low-dose growth hormone replacement in elderly hypopituitary patients. J Clin Endocrinol Metab 85: 1727–1730.
Trainer PJ, Drake WM, Perry LA et al. (2001). Modulation of cortisol metabolism by the growth hormone receptor antagonist pegvisomant in patients with acromegaly. J Clin Endocrinol Metab 86: 2989–2992.
Tunn S, Möllmann H, Barth J et al. (1992). Simultaneous measurement of cortisol in serum and saliva after different forms of cortisol administration. Clin Chem 38: 1491–1494.
Walker BR, Best R (1995). Clinical investigation of 11 beta-hydroxysteroid dehydrogenase. Endocr Res 21: 379–387.
Walker JJ, Terry JR, Lightman SL (2010). Origin of ultradian pulsatility in the hypothalamic–pituitary–adrenal axis. Proc Biol Sci 277: 1627–1633.
Wass JAH, Arlt W (2012). How to avoid precipitating an acute adrenal crisis. BMJ 345: e6333.
Weaver JU, Taylor NF, Monson JP et al. (1998). Sexual dimorphism in 11 beta hydroxysteroid dehydrogenase activity and its relation to fat distribution and insulin sensitivity; a study in hypopituitary subjects. Clin Endocrinol (Oxf) 49 (1): 13–20.
Weinstein RS (2012). Glucocorticoid-induced osteoporosis and osteonecrosis. Endocrinol Metab Clin North Am 41: 595–611.
Wichers M, Springer W, Bidlingmaier F et al. (1999). The influence of hydrocortisone substitution on the quality of life and parameters of bone metabolism in patients with secondary hypocortisolism. Clin Endocrinol (Oxf) 50: 759–765.
Windle RJ, Wood SA, Lightman SL et al. (1998). The pulsatile characteristics of hypothalamo–pituitary–adrenal activity in female Lewis and Fischer 344 rats and its relationship to differential stress responses. Endocrinology 139: 4044–4052.
Wood JB, Frankland AW, Landon J (1965). A rapid test of adrenocortical function. Lancet 1: 243–245.

Web links

http://www.addisons.org.uk
http://www.addisons.org.uk/comms/publications/surgicalguidelines-bw.pdf
www.endobible.com/investigations/

Handbook of Clinical Neurology, Vol. 124 (3rd series)
Clinical Neuroendocrinology
E. Fliers, M. Korbonits, and J.A. Romijn, Editors

Chapter 27

Uncertainties in endocrine substitution therapy for central endocrine insufficiencies: hypothyroidism

LUCA PERSANI[1,2]* AND MARCO BONOMI[1,2]

[1]*Department of Clinical Sciences and Community Health, University of Milan, Milan, Italy*

[2]*Division of Endocrine and Metabolic Diseases, San Luca Hospital, Istituto Auxologico Italiano, Milan, Italy*

INTRODUCTION

Central hypothyroidism (CeH) is a disease characterized by a defect of thyroid hormone production due to insufficient stimulation by thyrotropin (TSH) of an otherwise normal thyroid gland. This condition is the consequence of anatomic or functional disorders of the pituitary gland or the hypothalamus causing variable alterations of TSH secretion (Persani, 2012; Persani and Beck-Peccoz, 2012).

Although an isolated failure of thyrotrope cells can sometimes be observed, the defective TSH secretion is more frequently part of combined pituitary hormone deficiencies (CPHDs), a condition complicating both diagnosis and clinical management of CeH. Diagnosis is usually made biochemically with low circulating free T_4 concentrations associated with low/normal serum TSH levels. Therefore, CeH represent a major false-negative result of the "reflex TSH strategy", which is a worldwide diffuse method used to screen thyroid function by first-line TSH measurement (Price and Weetman, 2001). CeH itself does not severely reduce life expectancy but quality of life can be severely affected at all ages by the hypothyroid state. Therefore, the existence of mild forms of CeH should always be suspected in patients with hypothalamic–pituitary disorders.

CeH most frequently occurs as a sporadic form of hypothyroidism. CeH can affect patients of all ages and, unlike what is observed in primary hypothyroidism, there is no female preponderance. CeH apparently accounts for about 1 out 1000 hypothyroid patients as its prevalence was estimated to range from 1:16 000 to about 1:100 000 in the general adult or neonatal populations (Price and Weetman, 2001; Kempers et al., 2006; Nebesio et al., 2010; Adachi et al., 2012). This variability probably depends on several factors, including ethnicity, but also on the differences in sensitivity of the various diagnostic strategies used.

The mechanisms underlying CeH pathogenesis variably involve both hypothalamic and pituitary cells but they remain as yet undetermined in several cases. The major causes of CeH are listed in Table 27.1. In this condition, pituitary TSH reserve and/or the intrinsic bioactivity of secreted TSH molecules is impaired, thus contributing to the generation of CeH (Faglia et al., 1979; Beck-Peccoz et al., 1985; Persani et al., 1993, 2000; Horimoto et al., 1995; Persani, 1998).

FACTS AND UNCERTAINTIES IN CENTRAL HYPOTHYROIDISM DIAGNOSIS

Inheritable central hypothyroidism

Inheritable forms of CeH due to biallelic *TSHβ* mutations are frequently associated with severe neonatal onset and characterized by the typical manifestations of congenital primary hypothyroidism (jaundice, macroglossia, coarse cry, failure to thrive and retarded growth, umbilical hernia, hypotonia, etc.). If untreated with L-T4 within the first 6 weeks of life, these patients may develop cretinism (Bonomi et al., 2001; Miyai, 2007). In neonates, CeH can be identified only by screening programs based on concomitant TSH and total or free T_4 measurements on dried blood spot testing (Kempers et al., 2006; Baquedano et al., 2010; Nebesio et al., 2010; LaFranchi, 2010; Ramos et al., 2010; Adachi et al., 2012). CeH confirmation by serum free T_4

*Correspondence to: Luca Persani, MD, PhD, San Luca Hospital, IRCCS Istituto Auxologico Italiano, Piazzale Brescia 20. 20149 Milan, Italy. Tel: +39-02-619112738, Fax: +39-02-619112777, E-mail: luca.persani@unimi.it

Table 27.1

Known causes of central hypothyroidism (CeH)

Invasive or compressive lesions	Pituitary macroadenomas Craniopharyngiomas Meningiomas or gliomas Rathke's cleft cysts Empty sella Metastatic seeding Carotid aneurysm
Iatrogenic causes	Cranial surgery or irradiation Drugs
Injuries	Head traumas Traumatic delivery
Vascular accidents	Pituitary infarction Sheehan syndrome Subarachnoid hemorrhage
Autoimmune diseases	Postpartum hypophysitis Lymphocytic hypophysitis Polyglandular autoimmune diseases
Infiltrative lesions	Iron overload Sarcoidosis Histiocytosis X
Inheritable defects	CPHDs – pituitary transcription factor defects – *LEPR* mutations Isolated CeH – *TSHβ, TRHR* or *IGSF1* mutations
Infective diseases	Tuberculosis Mycoses Syphilis

CPHDs, combined pituitary hormone deficiencies.

and abnormal TSH response to thyrotropin-releasing hormone (TRH) testing may reveal the risk of CPHDs and impending adrenal crisis. In pituitary transcription factor defects, CeH can also have a delayed onset and be associated with hypoglycemia, typical craniofacial abnormalities, and severity of growth retardation (Miyai, 2007; Yamada and Mori, 2008; Pfäffle and Klammt, 2011). The association of CeH with high α-GSU levels in an infant is invariably indicative of a TSHβ gene defect (Bonomi et al., 2000).

No *TRH* gene defect has been documented so far in humans, but defective TRH action due to natural mutations in the *TRHR* gene has been so far described in two families (Collu et al., 1997; Bonomi et al., 2009). We recently reported a family with a homozygous nonsense mutation causing an early stop codon in the TRH receptor (Bonomi et al., 2009). Complete TRH resistance does not cause severe neonatal hypothyroidism and the early development of patients with complete TRH resistance appeared uneventful as the diagnosis in the male proband with homozygous *TRHR* mutations was reached because of delayed growth accompanied by lethargy and fatigue at 11 years of age (Bonomi et al., 2009). The presence of this defect can be suggested by the blunted TSH and PRL responses to TRH testing. Unexpectedly, the same diagnosis was reached in the sister in this family by genetic testing during her second pregnancy, when she was 33 years old. This woman, with complete TRH resistance, had reached her target height and had a normal IQ, and has so far delivered three heterozygous babies with normal pre- and postnatal growth. With none of these children has she experienced any defect of lactation. This study showed that in the absence of hypothalamic stimulation the pituitary feedback mechanism is set at a level inadequate to maintain free thyroxine levels in the normal range. Nevertheless, hypothyroidism was not severe, showing that the bioactivity of circulating TSH is not completely damaged in the absence of TRH activity. In addition, the nocturnal TSH surge was blunted but still significant in this condition, indicating that TRH activity influences amplitude, but additional sleep-related factors account for the determination of circadian oscillation (Bonomi et al., 2009).

Very recently, several familial cases of X-linked CeH from the Netherlands, the UK, and Italy have been reported to be associated with genetic defects in the *IGSF1* gene (Sun et al., 2012). This gene encodes a membrane protein with as yet unclear biological functions that is expressed in the pituitary and testes. IGSF1 defects are associated with a novel syndrome including CeH and macrorchidism, and seldom GH deficiency. CeH in these cases is associated with blunted TSH and PRL responses to TRH testing consistent with the finding of a reduced Trh-r expression in the pituitaries of IGSF1 knockout mice. In addition, Igsf1-deficient male mice show diminished pituitary and serum TSH concentrations, decreased triiodothyronine concentration, and increased body mass (Sun et al., 2012).

Acquired forms of central hypothyroidism

The hypothyroid state is mild to moderate in most patients with acquired CeH, as the pituitary TSH reserve is rarely depleted and a residual thyroid hormone secretion may occur via the constitutive activity of the TSH receptor (Neumann et al., 2010; Barbesino et al., 2012). Although manifestations of CeH are similar to those of primary hypothyroidism, they can be masked by the coexisting CPHDs (Ferretti et al., 1999; Alexopoulou et al., 2004; Persani, 2012). CeH represents a major false-negative result of the reflex TSH strategy for the diagnosis of thyroid dysfunction (Price and Weetman, 2001; Wardle et al., 2001; Demers and Spencer, 2003;

Baquedano et al., 2010). Therefore, the diagnosis of acquired CeH is generally made biochemically in patients being evaluated for hypothalamic/pituitary disease and is usually suggested by the low free T_4 concentrations, associated with low/normal TSH levels (Ferretti et al., 1999; Alexopoulou et al., 2004). Nevertheless, some CeH patients with a predominant hypothalamic defect have high serum immunoreactive TSH levels, but devoid of full biological activity. In these cases, TSH elevations are similar to those generally found in subclinical or mild primary hypothyroidism and may lead to the misdiagnosis (Faglia et al., 1979; Beck-Peccoz et al., 1985; Horimoto et al., 1995; Lee et al., 1995; Persani et al., 2000).

Table 27.2 lists the conditions that should be considered in the differential diagnosis of subjects with low/normal TSH and low free T_4 levels. When a low free T_4 is combined with a normal TSH value, the existence of an interference in free T_4 or TSH measurement should always be considered (Demers and Spencer, 2003; Gurnell et al., 2011; Persani, 2012). Among the variables of thyroid function, the one that provides the greatest accuracy in the diagnosis of CeH is indeed the measurement of free T_4 (Ferretti et al., 1999), as total T_4 levels are influenced by the variations of serum binding proteins (Demers and Spencer, 2003; Gurnell et al., 2011). Free T_4 is more stable throughout these conditions but the absolute values are dependent upon the assay used. In general, automated free T_4 assays are less reliable than equilibrium dialysis, which, however, is not compatible with the routine work carried out (Gurnell et al., 2011). Nevertheless, free T_4 assays involving two-step immunoextraction (back-titration) are less affected by thyroid autoantibodies or abnormal binding proteins than are one-step assays. If interference is suspected, this should be explored by using a two-step assay or by mass spectrometry. If the problem persists, hormone measurement following equilibrium dialysis remains the gold-standard for eliminating free T_4 assay interference. Less frequently, TSH immunometric measurement can be interfered with by the presence of antianimal antibodies (heterophile antibodies) in a patient's serum if directed against the same species as the assay antibodies: thus, a heterophile antibody that blocks TSH binding to either capture or detection antibodies will result in negative interference in the immunoassay, causing a falsely low TSH readout and potentially indicating a central instead of a primary hypothyroidism. Though most of the manufacturers nowadays provide reagents including the preimmune serum from the source animal, heterophile antibodies may still interfere with the TSH determination in some instances. If interference is suspected, the discordant TSH concentration should be checked: (1) by means of an alternative immunoassay using a different antibody pair, (2) after immunosubtraction by treatment with polyethylene glycol (PEG) or protein G, or (3) by dilution or recovery tests (Gurnell et al., 2011; Persani, 2012).

Table 27.2

Conditions associated with low/normal TSH and low free T_4 serum levels

Central hypothyroidism (hypothalamic hypothyroidism may be associated with TSH values above the upper limit of normal range)
Severe forms of nonthyroidal illness or sick euthyroid syndrome
L-T4 withdrawal syndrome
Prolonged TSH suppression after recovery from thyrotoxicosis
Intermittent thyrotoxicosis during autoimmune thyroditis
Drugs inhibiting TSH secretion
Allan–Herndon–Dudley syndrome (MCT8 mutations)
TRα1 mutations

TSH, thyrotropin.

In the absence of any interference, the finding of low free T_4 combined with an abnormally low TSH accurately delineates the diagnosis of overt forms of CeH, but the diagnosis of milder defects, characterized by free T_4 levels still within the normal range (mild or hidden CeH), remains unsolved. For instance, cranial irradiation can cause hypothalamic defects with TRH secretory abnormalities resulting in either hidden CeH (CeH with free T_4 values included in the normal range that can be recognized only by the demonstration of abnormal circadian or stimulated TSH secretory kinetics) or manifest CeH (most frequently associated with low free T_4) (Rose, 2001; Persani, 2012). Since mild CeH may be associated with a decreased growth velocity in children surviving cancer, several groups investigated possible solutions for the diagnosis of this condition. Rose et al. (1999) reported the frequent diagnosis of hidden CeH in survivors of childhood cancer by evaluating the nocturnal TSH surge and showed that many patients with mild thyrotrope insufficiency were not diagnosed on the basis of basal thyroid function screening. Though abnormalities in circadian TSH secretion may not correlate with free T_4 levels (Darzy and Shalet, 2005), the lack of a nocturnal TSH rise may therefore be useful in the diagnosis of CeH patients (Rose et al., 1999; Yamakita et al., 2001), but can be evaluated only in hospitalized patients. The TRH testing is not available in the US, but it may confirm the suspicion of mild CeH and may be of help in the differential diagnosis between tertiary (hypothalamic) and secondary (pituitary) hypothyroidism, as the two defects may be associated with exaggerated/delayed/prolonged or blunted TSH responses, respectively (Costom et al., 1971; Faglia, 1998; Rose, 2001; Yamakita et al., 2001;

Hartoft-Nielsen et al., 2004; Atmaca et al., 2007). However, it must be underscored that a significant proportion of patients with CeH may still have a normal TSH increase after TRH stimulation (Mehta et al., 2003; Darzy and Shalet, 2005), and a clear distinction between the two forms of CeH may be difficult, as both sites are affected in most patients (Persani et al., 2000). The practical utility of TRH testing is therefore limited to the patients with uncertain diagnosis, in whom the abnormal TSH response to TRH may confirm the CeH. In addition, the lack of a thyroid hormone rise despite an exaggerated and prolonged TSH increase may be an indirect estimate of the poor bioactivity of circulating TSH (Beck-Peccoz and Persani, 1994; Horimoto et al., 1995; Persani, 1998).

Interestingly, time-related decreases in circulating free T_4 concentrations larger than 20% versus the initial free T_4 determination were reported to support the diagnosis of CeH in patients with different pituitary diseases followed up for several years (Alexopoulou et al., 2004). This cut-off value was set on the basis of a 10% variation over time of T_4 levels in normal individuals (Andersen et al., 2002). Provided that free T_4 determination is repeatedly performed in the same laboratory, this approach would then allow the diagnosis and treatment of mild or hidden hypothyroid states of central origin.

The indices of peripheral thyroid hormone action, such as sex hormone-binding globulin (SHBG), bone markers, serum lipids and others, lack sufficient sensitivity and specificity for the diagnosis of mild or subclinical hypothyroidism, especially in patients who present with CPHDs, which may *per se* affect the levels of these indices (Ferretti et al., 1999; Alexopoulou et al., 2004). Very recently, the determination of parameters of Doppler echocardiography including the isovolumic contraction time, isovolumic contraction time/ejection time, and myocardial performance index were, however, demonstrated to correlate with the presence of CeH in a cohort of patients with hypothalamic–pituitary diseases, thus suggesting a potential use in the diagnosis of hidden CeH (Doin et al., 2012). Since abnormalities in cardiac parameters were reverted during L-T4 replacement, these findings may also indicate the requirement for L-T4 treatment even in milder forms of the disease, as previously claimed in subclinical primary hypothyroidism (Cooper and Biondi, 2012).

In the presence of low thyroid hormone levels, the exclusion of a primary thyroid defect may be required either because CeH may sometimes result from an intermittent thyrotoxic state (Table 27.2) or because hypothalamic hypothyroidism may be associated with slightly raised TSH concentrations at immunoassay. Indeed, the exclusion of primary thyroid disease by biochemical testing and/or ultrasound examination is the main objective in this differential diagnosis and the repetition of thyroid function tests may sometimes discriminate. Conversely, a family history of CeH or the clinical history (e.g., head trauma) or manifestations (e.g., headaches or visual field defects) may be suggestive of the presence of hypothalamic–pituitary lesions, and the MRI imaging generally confirms the central origin of hypothyroidism (Persani and Beck-Peccoz, 2012; Persani, 2012).

Patients with severe and chronic nonthyroidal illness (NTI) or sick euthyroid syndrome have values of thyroid function tests that largely overlap with those of CeH patients (Burman and Wartofski, 2001; DeGroot, 2003; Demers and Spencer, 2003). Since NTI is in most cases considered an adaptation of the organism to a sick condition aimed at preventing an excessive metabolic stimulation, thyroid hormone treatment is generally avoided. Therefore, the presence of concomitant diseases at the time of blood sampling should always be excluded before suspecting "true" CeH.

Patients with Allan–Herndon–Dudley syndrome, an X-linked form of mental retardation associated with tissue-specific resistance to thyroid hormones, can have low free T_4 and normal or slightly elevated TSH levels (Friesema et al., 2010). This disease is caused by mutations in the *MCT8* gene encoding a membrane thyroid hormone transporter. These patients can be distinguished from those with CeH by the severe clinical phenotype, including cognitive and psychomotor retardation, and the typical elevation of T_3 circulating levels, that are usually two- to threefold higher than in normal subjects. Similar biochemical findings can be found also in patients with heterozygous mutations in the *THRA* gene, encoding the thyroid hormone receptor $\alpha 1$ (TRα1) (van Mullem et al., 2012). Severe constipation, growth and mental retardation, and delayed bone development appear as distinct features of this disease (Bochukova et al., 2012; van Mullem et al., 2012).

FACTS AND UNCERTAINTIES IN CENTRAL HYPOTHYROIDISM REPLACEMENT THERAPY

As in primary hypothyroidism, treatment of CeH should restore appropriate serum concentrations of thyroid hormones (Beck-Peccoz, 2011; Persani, 2012). There is insufficient evidence that L-T4 + L-T3 combination therapy is better than L-T4 monotherapy (Cassio et al., 2003; Slawik et al., 2007), and it is recommended that L-T4 monotherapy remain the standard treatment of hypothyroidism. L-T4 + L-T3 combination therapy might be considered as an experimental approach in compliant L-T4-treated hypothyroid patients who have persistent complaints despite adequate thyroid function tests. In such cases, the guidelines of the European Thyroid Association (Wiersinga et al., 2012) suggest starting combination therapy in an L-T4/L-T3 dose ratio between 13:1 and 20:1 by weight (L-T4 once daily, and the daily L-T3 dose in two doses). Currently available combined preparations with an L-T4/L-T3 dose ratio of less than 13:1 would not be recommended.

A full replacement dose can generally be advised from the start of L-T4 treatment. In the elderly or in patients with long-standing hypothyroidism who are at risk of untoward effects mainly due to concomitant heart diseases, L-T4 treatment could be started at low daily dosage and then progressively increased during the following weeks. As in primary hypothyroidism (Helfand and Crapo, 1990), younger CeH patients require higher doses than older ones (Ferretti et al., 1999; Alexopoulou et al., 2004).

In normal infants and children, thyroid hormone levels are higher than in adults (Hollowell et al., 2002; LaFranchi 2010). Therefore, distinctly higher L-T4 doses are recommended in hypothyroid pediatric patients and treatment should be started at full-replacement doses, especially in patients with neonatal onset, in order to rapidly reach adequate circulating free T_4 levels and promptly support neurological development (Selva et al., 2002). Guidelines recommend initiating treatment of neonatal disorders with 10–15 μg/kg of L-T4 and adjusting doses on the basis of free T_4 measurements every 2–4 weeks (Rose et al., 2006). The target range should be that observed in normal children. L-T4 treatment has been reported to promote an acceleration of growth velocity, allowing patients to reach their target height (Collu et al., 1997; Rose, 2001; Bonomi et al., 2009). Progressively lower doses are required in childhood and in transition to adulthood (Koch and Sarlis, 2001).

In patients at risk of CPHDs, concomitant central adrenal insufficiency must be assessed before starting L-T4 therapy, because of the risk of triggering an adrenal crisis. If adrenal function cannot be excluded prior to the start of L-T4, a prophylactic treatment with corticosteroids is recommended and assessment of corticotrope function can be postponed (Beck-Peccoz, 2011; Persani, 2012).

The L-T4 replacement is easily tuned in primary hypothyroidism by evaluating circulating TSH levels, but this index has a different significance in CeH patients. In particular, the finding of unsuppressed serum TSH levels during L-T4 treatment strongly indicates undertreatment in CeH. Indeed, our group (Ferretti et al., 1999) reported that about half of the final L-T4 substitutive dose is sufficient to suppress TSH secretion in about 80% of CeH patients, despite serum free T_4 levels still in the hypothyroid range in most (Fig. 27.1). Similarly, the large majority of 135 CeH patients was reported to have subnormal serum TSH concentrations during apparently adequate L-T4 treatment (Carrozza et al., 1999). Afterwards, TSH levels above 1.0 mU/L were reported to reflect an insufficient L-T4 replacement (Shimon et al., 2002).

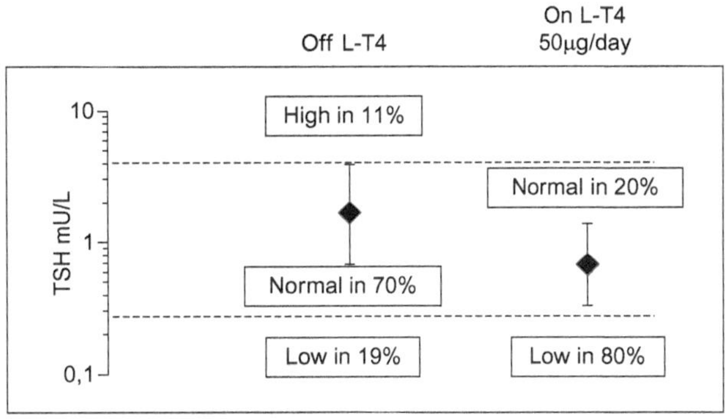

Fig. 27.1. Percentage distribution of serum thyrotropin (TSH) concentrations in patients with central hypothyroidism (CeH) during withdrawal of L-T4 replacement and 2 weeks after the start of the administration of 50 μg/day of L-T4. The diamonds indicate mean (±standard deviation). Note that many CeH patients have TSH values below the lower limit of normal soon after the start of such a low L-T4 dose (data from Ferretti et al., 1999; Persani, 2004).

The determination of circulating free thyroid hormone levels is of major significance in monitoring L-T4 treatment in CeH patients (Ferretti et al., 1999; Alexopoulou et al., 2004; Slawik et al., 2007; Iverson and Mariash, 2008; Koulouri et al., 2011; Beck-Peccoz, 2011; Persani, 2012). If blood is withdrawn before the morning administration of L-T4, low free T_4 values may suggest undertreatment and high free T_3 values are more sensitive for disclosing overtreatment. In certain cases, the evaluation of biochemical indices of thyroid hormone action at the tissue level (e.g., SHBG, bone GLA protein, or cholesterol) may become useful in monitoring the efficacy of L-T4 treatment. However, the evaluation of these indices should take into account the possible interference by alterations in somatotrope, gonadal, or adrenal functions (Ferretti et al., 1999; Alexopoulou et al., 2004) and, with the exception of cholesterol, are generally more effective in documenting thyrotoxicosis (Demers and Spencer, 2003). Mainly due to this lack of sensitivity and specificity, only longitudinal evaluation of such indices may be potentially helpful in CeH patients (Persani, 2004, 2012).

Very recently, Koulouri et al. (2011) compared free T_4 values in a large series of patients with hypothalamic–pituitary lesions with those of patients with primary thyroid disease adequately treated with L-T4, i.e., those with normal levels of circulating TSH during replacement therapy. The conclusions were that CeH patients are frequently undertreated and it was suggested that levels of free T_4 in the middle of the normal range might represent an appropriate target in most treated CeH patients. Although the individual free T_4 values are quite widely distributed in normal subjects and genetically determined (Medici et al., 2011), this conclusion is similar to the one reached in the past by other authors, who even suggested targeting free T_4 values in the upper part of the normal range (Slawik et al., 2007; Iverson and Mariash, 2008). Indeed, mean L-T4 daily doses of 1.5 ± 0.3 or 1.6 ± 0.5 μg/kg body weight/day, similar to those reported for primary hypothyroidism (Oppenheimer et al., 1995), were judged sufficient in the large majority of treated Italian and Belgian CeH patients, respectively (Ferretti et al., 1999; Alexopoulou et al., 2004).

Importantly, significant differences in L-T4 doses also depend on the concomitant treatment for CPHDs. Estrogens had been reported to increase L-T4 requirements in hypothyroid patients (Arafah, 2001; Alexopoulou et al., 2004). As this is a consequence of an increase in thyroxine-binding globulin (TBG) levels, the required adjustment of L-T4 should be carried out to saturate the increased T4-binding capacity of plasma proteins. Moreover, patients on recombinant hGH (rhGH) replacement therapy also require significantly higher L-T4 doses (Jorgensen et al., 1989; Portes et al., 2000; Porretti et al., 2002; Persani, 2004; Agha et al., 2007; Losa et al., 2008). Accordingly, GH deficiency may mask subclinical forms of CeH that achieve subnormal values only after institution of rhGH replacement therapy. The effects of rhGH on thyroid hormone metabolism and the activity of the hypothalamic–pituitary–thyroid axis are not transient (Portes et al., 2000; Losa et al., 2008), but are biologically relevant only in patients with CPHDs who already have a partial impairment of thyrotrope function (Giavoli et al., 2003). The conditions suggesting under- or overtreatment in CeH are summarized in Table 27.3.

Table 27.3

Events and findings that may suggest the need of a revision of L-T4 regimen in central hypothyroidism

Possible undertreatment

- Serum TSH > 0.5 mU/L, in particular if associated with serum free T_4 values below the lower tertile of normal range
- Fall of serum free T_4 values below the lower tertile of normal range
- Introduction of GH replacement therapy in the context of combined pituitary hormone deficiencies
- Introduction of oral contraceptives or estrogen replacement therapy
- Introduction of treatments impacting L-T4 absorption or thyroid hormone metabolism
- Variations of biochemical indices of thyroid hormone action in the hypothyroid range
- In the presence of clinical manifestations suggestive of hypothyroidism, in particular when associated with one of the above events or findings

Possible overtreatment

- Serum values of free T_4 and/or free T_3 above the upper tertile of normal range
- Withdrawal of GH or estrogen replacement therapy
- Withdrawal of oral contraceptives or transition into menopause
- Withdrawal of treatments impacting L-T4 absorption or thyroid hormone metabolism
- Variations of biochemical indices of thyroid hormone action in the hyperthyroid range
- In the presence of clinical manifestations suggestive of thyrotoxicosis, in particular when associated with one of the above events or findings

GH, growth hormone.

NOVEL PERSPECTIVES FOR THERAPY OF CENTRAL HYPOTHYROIDISM

The recent paper by Adachi et al. (2012) showing the possibility of determining free T_4 levels on dried blood spot testing, which in principle should be more sensitive than the total T_4 determination, may open novel perspectives for the early diagnosis and treatment of patients with CeH.

Several studies in children and adult primary hypothyroid patients did not find a superiority of combined L-T4 plus triiodothyronine (L-T3) treatment (Cassio et al., 2003; Wiersinga et al., 2012). An improved understanding of thyroid hormone metabolism and action at the tissue level may probably provide novel markers for a tailored replacement therapy, as nowadays can be more easily achieved in primary thyroid failure by the determination of serum TSH concentrations.

The discovery of polymorphisms in the genes of the deiodinases and thyroid hormone transporters led to the obvious hypothesis that subgroups of hypothyroid patients respond differently to monotherapy versus combined therapy depending on their genetic background (Bianco and Casula, 2012). This hypothesis was tested in 552 individuals in the WATTS study and the results suggest that the combined therapy is associated with a positive clinical outcome in patients exhibiting the Thr92Ala polymorphism in the DIO2 gene (Panicker et al., 2009). In the WATTS study, the thyroid hormone formulation used included the simultaneous reduction in the dose of L-T4 by 50 μg/day and the introduction of 10 μg/day of L-T3. Since, this D2 polymorphism is present in a relatively small proportion of the population taking L-T4, previous studies are likely to have been underpowered to see this effect (Appelhof et al., 2005). No preference for the combined therapy were reported by a relatively small sample of patients carrying polymorphisms in an efficient T_4 membrane transporter such as OATP1C1 (van der Deure et al., 2008).

Perhaps, in the near future, we will be able to predict the optimal free T_4 values in the various patients with CeH under replacement therapy by testing the genetic loci determining the individual setpoint of hypothalamus–pituitary–thyroid axis and the interindividual variations in TSH and free T_4 levels (Medici et al., 2011).

As for other pituitary defects, the cure for central hypothyroidism may finally be given by the future advent of regenerative medicine (Castinetti et al., 2011; Antonica et al., 2012), as recently suggested by the generation of functional pituitary cell lineages in three-dimensional culture (Suga et al., 2011).

References

Adachi M, Soneda A, Asakura Y et al. (2012). Mass screening of newborns for congenital hypothyroidism of central origin by free thyroxine measurement of blood samples on filter paper. Eur J Endocrinol 166: 829–838.

Agha A, Walker D, Perry L et al. (2007). Unmasking of central hypothyroidism following growth hormone replacement in adult hypopituitary patients. Clin Endocrinol (Oxf) 66: 72–77.

Alexopoulou O, Beguin C, De Nayer P et al. (2004). Clinical and hormonal characteristics of central hypothyroidism at diagnosis and during follow-up in adult patients. Eur J Endocrinol 150: 1–8.

Andersen S, Pedersen KM, Bruun NH et al. (2002). Narrow individual variations in serum T4 and T3 in normal subjects: a clue to understanding of subclinical thyroid disease. J Clin Endocrinol Metab 87: 1068–1072.

Antonica F, Figini-Kasprzyk D, Opitz R et al. (2012). Generation of functional thyroid from embryonic stem cells. Nature 491: 66–71.

Appelhof BC, Peeters RP, Wiersinga WM et al. (2005). Polymorphisms in type 2 deiodinase are not associated with wellbeing, neurocognitive functioning, and preference for combined thyroxine/3,5,3′-triiodothyronine therapy. J Clin Endocrinol Metab 90: 6296–6299.

Arafah BM (2001). Increased need for thyroxine in women with hypothyroidism during estrogen therapy. N Engl J Med 344: 1743–1749.

Atmaca H, Tanriverdi F, Gokce C et al. (2007). Do we still need the TRH stimulation test? Thyroid 17: 529–533.

Baquedano MS, Ciaccio M, Dujovne N et al. (2010). Two novel mutations of the TSH-beta subunit gene underlying congenital central hypothyroidism undetectable in neonatal TSH screening. J Clin Endocrinol Metab 95: E98–E103.

Barbesino G, Sluss PM, Caturegli P (2012). Central hypothyroidism in a patient with pituitary autoimmunity: evidence for TSH-independent thyroid hormone synthesis. J Clin Endocrinol Metab 97: 345–350.

Beck-Peccoz P (2011). Treatment of central hypothyroidism. Clin Endocrinol (Oxf) 74: 671–672.

Beck-Peccoz P, Persani L (1994). Variable biological activity of circulating thyrotropin (TSH). Eur J Endocrinol 131: 331–340.

Beck-Peccoz P, Amr S, Menezes-Ferreira MM et al. (1985). Decreased receptor binding of biologically inactive thyrotropin central hypothyroidism: effect of treatment with thyrotropin-releasing hormone. N Engl J Med 312: 1085–1090.

Bianco AC, Casula S (2012). Thyroid hormone replacement therapy: three "simple" questions, complex answers. Eur Thyroid J 1: 88–98.

Bochukova E, Schoenmakers N, Agostini M et al. (2012). A mutation in the thyroid hormone receptor α gene. N Engl J Med 366: 243–249.

Bonomi M, Proverbio MC, Weber G et al. (2001). Hyperplastic pituitary gland, high serum glycoprotein hormone α-subunit, and variable circulating thyrotropin (TSH) levels as hallmark of central hypothyroidism due to mutations of the TSHβ gene. J Clin Endocrinol Metab 86: 1600–1604.

Bonomi M, Busnelli M, Beck-Peccoz P et al. (2009). A family with complete resistance to thyrotropin-releasing hormone. N Engl J Med 360: 731–734.

Burman KD, Wartofsky L (2001). Thyroid function in the intensive care unit setting. Crit Care Clin 17: 43–57.

Carrozza V, Csako G, Yanovski JA et al. (1999). Levothyroxine replacement therapy in central hypothyroidism: a practice report. Pharmacotherapy 19: 349–355.

Cassio A, Cacciari E, Cicognani A et al. (2003). Treatment for congenital hypothyroidism: thyroxine alone or thyroxine plus triiodothyronine? Pediatrics 111: 1055–1060.

Castinetti F, Davis SW, Brue T et al. (2011). Pituitary stem cell update and potential implications for treating hypopituitarism. Endocr Rev 32: 453–471.

Collu R, Tang J, Castagnè J et al. (1997). A novel mechanism for isolated central hypothyroidism: inactivating mutations in the thyrotropin-releasing hormone receptor gene. J Clin Endocrinol Metab 82: 1561–1565.

Cooper DS, Biondi B (2012). Subclinical thyroid disease. Lancet 379: 1142–1154.

Costom BH, Grumbach MM, Kaplan SL (1971). Effect of thyrotropin-releasing factor on serum thyroid-stimulating hormone. An approach to distinguishing hypothalamic from pituitary forms of idiopathic hypopituitary dwarfism. J Clin Invest 50: 2219–2225.

Darzy KH, Shalet SM (2005). Circadian and stimulated thyrotropin secretion in cranially irradiated adult cancer survivors. J Clin Endocrinol Metab 90: 6490–6497.

DeGroot LJ (2003). Nonthyroidal illness syndrome is functional central hypothyroidism, and if severe, hormone replacement therapy is appropriate in light of present knowledge. J Endocrinol Invest 26: 1163–1170.

Demers LM, Spencer CA (2003). Laboratory medicine practice guidelines for the diagnosis and monitoring of thyroid disease. Monograph "Consensus guidelines for thyroid testing in the new millennium". Thyroid 13: 2–125.

Doin FC, Rosa-Borges M, Martins MA et al. (2012). Diagnosis of subclinical central hypothyroidism in patients with hypothalamic-pituitary disease by Doppler echocardiography. Eur J Endocrinol 166: 631–640.

Faglia G (1998). The clinical impact of the thyrotropin-releasing hormone test. Thyroid 8: 903–908.

Faglia G, Bitenski L, Pinchera A et al. (1979). Thyrotropin secretion in patients with central hypothyroidism: evidence for reduced biological activity of immunoreactive thyrotropin. J Clin Endocrinol Metab 48: 989–998.

Ferretti E, Persani L, Jaffrain-Rea ML et al. (1999). Evaluation of the adequacy of l-T4 replacement therapy in patients with central hypothyroidism. J Clin Endocrinol Metab 84: 924–929.

Friesema EC, Visser WE, Visser TJ (2010). Genetics and phenomics of thyroid hormone transport by MCT8. Mol Cell Endocrinol 322: 107–113.

Giavoli C, Porretti S, Ferrante E et al. (2003). Recombinant hGH replacement therapy and the hypothalamus–pituitary–thyroid axis in children with GH deficiency: when should we be concerned about the occurrence of central hypothyroidism? Clin Endocrinol (Oxf) 59: 806–810.

Gurnell M, Halsall DJ, Chatterjee VK (2011). What should be done when thyroid function tests do not make sense? Clin Endocrinol (Oxf) 74: 673–678.

Hartoft-Nielsen ML, Lange M, Rasmussen AK et al. (2004). Thyrotropin-releasing hormone stimulation test in patients with pituitary pathology. Horm Res 61: 53–57.

Helfand M, Crapo LM (1990). Monitoring therapy in patients taking levothyroxine. Ann Intern Med 113: 450–454.

Hollowell JG, Staehling NW, Flanders WD et al. (2002). Serum TSH, T(4), and thyroid antibodies in the United States population (1988 to 1994): National Health and Nutrition Examination Survey (NHANES III). J Clin Endocrinol Metab 87: 489–499.

Horimoto M, Nishikawa M, Ishihara T et al. (1995). Bioactivity of thyrotropin (TSH) in patients with central hypothyroidism: comparison between in vivo 3,5,3-triiodothyronine response to TSH and in vitro bioactivity of TSH. J Clin Endocrinol Metab 80: 1124–1128.

Iverson JF, Mariash CN (2008). Optimal free thyroxine levels for thyroid hormone replacement in hypothyroidism. Endocr Pract 14: 550–555.

Jorgensen JO, Pedersen SA, Laurberg P et al. (1989). Effects of growth hormone therapy on thyroid function of growth hormone-deficient adults with and without concomitant thyroxine-substituted central hypothyroidism. J Clin Endocrinol Metab 69: 1127–1132.

Kempers MJ, Lanting CI, van Heijst AF et al. (2006). Neonatal screening for congenital hypothyroidism based on thyroxine, thyrotropin, and thyroxine-binding globulin measurement: potentials and pitfalls. J Clin Endocrinol Metab 91: 3370–3376.

Koch CA, Sarlis NJ (2001). The spectrum of thyroid diseases in childhood and its evolution during transition to adulthood: natural history, diagnosis, differential diagnosis and management. J Endocrinol Invest 24: 659–675.

Koulouri O, Auldin MA, Agarwal R et al. (2011). Diagnosis and treatment of hypothyroidism in TSH deficiency compared to primary thyroid disease: pituitary patients are at risk of underreplacement with levothyroxine. Clin Endocrinol (Oxf) 74: 744–749.

LaFranchi SH (2010). Newborn screening strategies for congenital hypothyroidism: an update. J Inherit Metab Dis 33 (Suppl 2): S225–S233.

Lee KO, Persani L, Tan M et al. (1995). Thyrotropin with decreased bioactivity, a delayed consequence of cranial irradiation for nasopharyngeal carcinoma. J Endocrinol Invest 18: 800–805.

Losa M, Scavini M, Gatti E et al. (2008). Long-term effects of growth hormone replacement therapy on thyroid function in adults with growth hormone deficiency. Thyroid 18: 1249–1254.

Medici M, van der Deure WM, Verbiest M et al. (2011). A large-scale association analysis of 68 thyroid hormone pathway genes with serum TSH and FT4 levels. Eur J Endocrinol 164: 781–788.

Mehta A, Hindmarsh PC, Stanhope RG et al. (2003). Is the thyrotropin-releasing hormone test necessary in the diagnosis of central hypothyroidism in children? J Clin Endocrinol Metab 88: 5696–5703.

Miyai K (2007). Congenital thyrotropin deficiency – from discovery to molecular biology, postgenome and preventive medicine. Endocr J 54: 191–203.

Nebesio TD, McKenna MP, Nabhan ZM et al. (2010). Newborn screening results in children with central hypothyroidism. J Pediatr 156: 990–993.

Neumann S, Raaka BM, Gershengorn MC (2010). Constitutively active thyrotropin and thyrotropin-releasing hormone receptors and their inverse agonists. Methods Enzymol 485: 147–160.

Oppenheimer JH, Braverman LE, Toft A et al. (1995). A therapeutic controversy. Thyroid hormone treatment: when and what? J Clin Endocrinol Metab 80: 2873–2883.

Panicker V, Saravanan P, Vaidya B et al. (2009). Common variation in the DIO2 gene predicts baseline psychological well-being and response to combination thyroxine plus triiodothyronine therapy in hypothyroid patients. J Clin Endocrinol Metab 94: 1623–1629.

Persani L (1998). Hypothalamic thyrotropin-releasing hormone and thyrotropin biological activity. Thyroid 8: 941–946.

Persani L (2004). Thyroid hormone replacement in central hypothyroidism. In: R Ross, EM Erfurth (Eds.), Pituitary Function through the Ages. Bioscientifica Ltd, Bristol, UK, pp. 111–124.

Persani L (2012). Central hypothyroidism: pathogenic, diagnostic and therapeutic challenges. J Clin Endocrinol Metab 97: 3068–3078.

Persani L, Beck-Peccoz P (2012). Central hypothyroidism. In: LE Braverman, D Cooper (Eds.), Werner and Ingbar's The Thyroid: A Fundamental and Clinical Text, 10th edn. Lippincott Williams and Wilkins/Wolters Kluwer Health, Philadelphia, pp. 560–568, ch. 38.

Persani L, Tonacchera M, Beck-Peccoz P et al. (1993). Measurement of cAMP accumulation in Chinese hamster ovary cells transfected with the recombinant human TSH receptor (CHO-R): a new bioassay for human thyrotropin. J Endocrinol Invest 16: 511–519.

Persani L, Ferretti E, Borgato S et al. (2000). Circulating TSH bioactivity in sporadic central hypothyroidism. J Clin Endocrinol Metab 85: 3631–3635.

Pfäffle R, Klammt J (2011). Pituitary transcription factors in the aetiology of combined pituitary hormone deficiency. Best Pract Res Clin Endocrinol Metab 25: 43–60.

Porretti S, Giavoli C, Ronchi C et al. (2002). Recombinant human GH replacement therapy and thyroid function in a large group of adult GH-deficient patients: when does L-T4 therapy become mandatory? J Clin Endocrinol Metab 87: 2042–2045.

Portes ES, Oliveira JH, Maccagnan P et al. (2000). Changes in serum thyroid hormones levels and their mechanisms during long-term growth hormone (GH) replacement therapy in GH deficient children. Clin Endocrinol (Oxf) 53: 183–189.

Price A, Weetman AP (2001). Screening for central hypothyroidism is unjustified. BMJ 322: 798.

Ramos HE, Labedan I, Carré A et al. (2010). New cases of isolated congenital central hypothyroidism due to homozygous thyrotropin beta gene mutations: a pitfall to neonatal screening. Thyroid 20: 639–645.

Rose SR (2001). Cranial irradiation and central hypothyroidism. Trends Endocrinol Metab 12: 97–104.

Rose SR, Lustig RH, Pitukcheewanont P et al. (1999). Diagnosis of hidden central hypothyroidism in survivors of childhood cancer. J Clin Endocrinol Metab 84: 4472–4479.

Rose SR, Brown RS, Foley T et al. (2006). Update of newborn screening and therapy for congenital hypothyroidism. Pediatrics 117: 2290–2303.

Selva KA, Mandel SH, Rien L et al. (2002). Initial treatment dose of L-T4 in congenital hypothyroidism. J Pediatr 141: 786–792.

Shimon I, Cohen O, Lubetsky A et al. (2002). Thyrotropin suppression by thyroid hormone replacement is correlated with thyroxine level normalization in central hypothyroidism. Thyroid 12: 823–827.

Slawik M, Klawitter B, Meiser E et al. (2007). Thyroid hormone replacement for central hypothyroidism: a randomized controlled trial comparing two doses of thyroxine (T4) with a combination of T4 and triiodothyronine. J Clin Endocrinol Metab 92: 4115–4122.

Suga H, Kadoshima T, Minaguchi M et al. (2011). Self-formation of functional adenohypophysis in three-dimensional culture. Nature 480: 57–62.

Sun Y, Bak B, Schoenmakers N et al. (2012). Loss-of-function mutations in IGSF1 cause a novel X-linked syndrome of central hypothyroidism and testicular enlargement. Nat Genet 44: 1375–1381.

van der Deure W, Appelhof BC, Peeters RP et al. (2008). Polymorphism in the brain-specific thyroid hormone transporter OATP-C1 are associated with fatigue and depression in hypothyroid patients. Clin Endocrinol (Oxf) 69: 8.

van Mullem A, van Heerebeek R, Chrysis D et al. (2012). Clinical phenotype and mutant TRα1. N Engl J Med 366: 1451–1453.

Wardle CA, Fraser WD, Squire CR (2001). Pitfalls in the use of thyrotropin concentration as a first-line thyroid-function test. Lancet 357: 1013–1014.

Wiersinga WM, Duntas L, Fadeyev V et al. (2012). 2012 ETA Guidelines: the use of L-T4 + L-T3 in the treatment of hypothyroidism. Eur Thyroid J 1: 55–71.

Yamada M, Mori M (2008). Mechanisms related to the pathophysiology and management of central hypothyroidism. Nat Clin Pract Endocrinol Metab 4: 683–694.

Yamakita N, Komaki T, Takao T et al. (2001). Usefulness of thyrotropin (TSH)-releasing hormone test and nocturnal surge of TSH for diagnosis of isolated deficit of TSH secretion. J Clin Endocrinol Metab 86: 1054–1060.

Handbook of Clinical Neurology, Vol. 124 (3rd series)
Clinical Neuroendocrinology
E. Fliers, M. Korbonits, and J.A. Romijn, Editors

Chapter 28

Uncertainties in endocrine substitution therapy for central endocrine insufficiencies: growth hormone deficiency

EVA-MARIE ERFURTH*
Department of Endocrinology, Lund University, Lund, Sweden

INTRODUCTION

The growth hormone (GH) deficiency syndrome is associated with several metabolic abnormalities (Cuneo et al., 1992; Carroll and Christ, 1998; Maison et al., 2004), and it has been postulated that the increased cardiovascular morbidity and mortality recorded in GH-deficient (GHD) patients may be related to the missing metabolic effects of GH (Rosén and Bengtsson, 1990; Bülow et al., 1997). In addition, it has been suggested that the adult GHD phenotype shares features of the metabolic syndrome, such as abdominal obesity, dyslipidemia, and insulin resistance, thus a cluster of risk factors for cardiovascular disease (CVD) and type 2 diabetes mellitus (Ford, 2005; Cornier et al., 2008). GH has a positive effect on body composition, with increased lean mass and reduced fat mass, and with a reduction particularly in waist measurements and a favorable effect on lipid levels (Maison et al., 2004). Many of the traditional CVD risk factors show improvements after GH therapy, which might decrease CVD risk in these patients and improve longevity. In particular, if patients with newly developed GHD received early GH replacement, the negative effects of GHD could possibly be avoided.

GH affects linear growth and is important for skeletal development in young adults and for bone turnover in adult skeletal tissue (Ohlsson et al., 1998). Reduced bone mineral density (BMD) has been recorded both in patients with isolated GHD and in those with multiple pituitary deficiencies, indicating that GHD *per se* is, at least in part, responsible for the low BMD in both types (De Boer et al., 1994; Holmes et al., 1994). There are, however, contradictory results from studies in adult onset GHD showing both normal BMD, particularly in older subjects (Toogood et al., 1997; Fernholm et al., 2000), and low BMD (Rosén et al., 1993; Degerblad et al., 1995). In contrast, in patients with childhood-onset GHD (Kaufman et al., 1992; O'Halloran et al., 1993; De Boer et al., 1994) there seems to be a clear reduction in BMD, which may illustrate the potential role of GH in the acquisition of peak bone mass (De Boer et al., 1994).

These matters are, however, more complicated, as hypopituitary patients with GHD appear in several phenotypes due to different background diagnoses. Furthermore, GHD is not the only risk factor for CVD or low BMD in these patients. This notion is based on the observation that the effect of complete hormone substitutions, including GH therapy, differs depending on the phenotypes of hypopituitary patients. However, these phenotypes are not clear-cut and not absolute in individual patients as patients may transit between the different phenotypes over time due to extension of the pathology and/or the effects of treatment (surgery and/or radiotherapy). Moreover, in clinical practice patients with similar pituitary diseases may fall into different categories of GHD and may even transit from one category (e. g., GHD only) to a second category (e.g., GHD with multiple hormone deficiencies).

The purpose of this chapter is to discuss three different phenotypes of hypopituitary patients with GHD, with a focus on CVD risk and bone health:

1. Patients with isolated GHD, e.g., caused by prophylactic cranial radiotherapy for lymphoblastic leukemia in childhood
2. Patients with GHD and multiple hormone deficiencies, caused by a pituitary macroadenoma, treated by surgery

*Correspondence to: Eva-Marie Erfurth, Department of Endocrinology, Lasarettsgatan 15, SE 221 85 Lund, Sweden. Tel: +46-46-172-36-3, E-mail: Eva_Marie.Erfurth@med.lu.se

3. Patients with GHD and multiple hormone deficiencies and with hypothalamic involvement caused by a craniopharyngioma.

PATIENTS WITH ISOLATED GROWTH HORMONE DEFICIENCY (E.G., CAUSED BY TREATMENT WITH PROPHYLACTIC CRANIAL RADIOTHERAPY FOR LYMPHOBLASTIC LEUKEMIA IN CHILDHOOD)

Cardiovascular risk after acute lymphoblastic leukemia

Acute lymphoblastic leukemia (ALL) is the most common childhood malignancy and constitutes 25% of all childhood cancers. The peak incidence is at the age of 2–4 years. The survival after ALL has improved dramatically during recent decades, and is currently about 85%, which emphasises the importance of long-term treatment complications (Oeffinger et al., 2000). GHD is common after long-term follow-up in survivors of childhood ALL treated with prophylactic cranial radiotherapy (Brennan et al., 1998; Link et al., 2004). Even if cranial radiotherapy is no longer routinely used in Sweden and other countries, we estimate that about 500 survivors of ALL have been subjected to this therapy in Sweden (9 million inhabitants), with corresponding numbers in other countries. The longevity is hampered in ALL survivors, as they are at elevated risk for obesity (Oeffinger et al., 2003), CVD (Talvensaari et al., 1996; Link et al., 2004), and late occurring stroke (Bowers et al., 2006).

After proper evaluation, we recorded GHD in 90% of former ALL survivors 20 years (range 8–27 years) after cranial radiotherapy (Link et al., 2004). Indeed, the same pattern of increased prevalence of CVD risk factors, e.g., lipid abnormalities, increased fat mass, decrease in lean mass, was shown in this population (Talvensaari et al., 1996; Link et al., 2004). In a long-term study of GH therapy in former ALL patients, 5 years with GH therapy, compared to 8 years without GH therapy, resulted in a significant positive change in plasma glucose, ApoB/ApoA1, and HDL-cholesterol levels (Fig. 28.1). In addition, a reduction in fat mass and increase in lean mass was recorded with a significant improvement in insulin sensitivity after 5 years of GH therapy (Follin et al., 2010). Thus, this study showed that GH therapy reduced the prevalence of CVD risk factors in this young ALL population (Follin et al., 2010). However, there are data that reveal that GHD is not the only factor explaining increased CVD risk, which is the leading nonmalignant cause of death, among survivors of childhood cancers (Mertens et al., 2008).

Cranial radiotherapy, particularly when given to patients at a young age, causes alterations to the hypothalamus, and it has been shown that young adult survivors of ALL have "leptin resistance" (Brennan et al., 1999). Leptin is produced by adipocytes and inhibits food intake via interaction with receptors at the hypothalamus (Zhang et al., 1994). Thus, leptin, like GH, plays an important role in the regulation of body composition and carbohydrate metabolism. In an investigation of GH-deficient ALL survivors previously treated with prophylactic cranial radiotherapy for ALL and sex-, age-, and body mass index (BMI)-matched controls, an increased serum leptin/kg fat mass was recorded (Bülow et al., 2004). Twelve months of GH treatment increased serum insulin-like growth factor 1 (IGF-1), but did not result in a significant change in serum leptin/kg fat mass, which argues against the assumption that elevated leptin levels in this group of patients are entirely an effect of GHD. In addition, a study comparing irradiated to nonirradiated ALL patients 3–15 years after cranial radiotherapy showed that irradiated patients not only had GHD, but also a significantly lower level of resting metabolic rate, measured by indirect calorimetry (Mayer et al., 2000). Caloric intake was adequate, but physical activity was reduced in the group with cranial radiotherapy. Thus, the higher risk of obesity in this ALL population was not only associated with GHD but also with leptin resistance, lower resting metabolic rate, and lower physical activity. The lower resting metabolic rate was most likely caused by hypothalamic involvement by cranial radiotherapy.

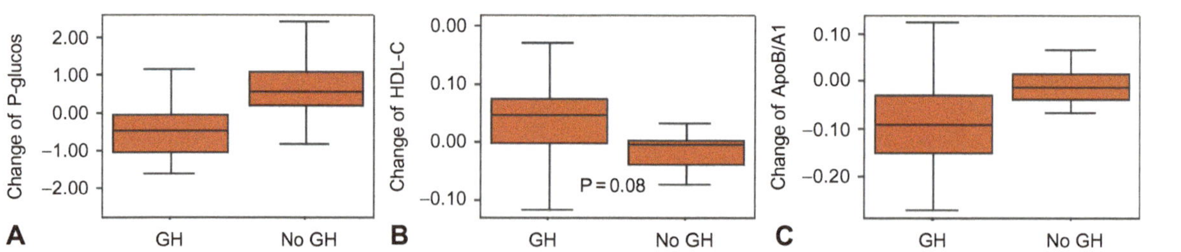

Fig. 28.1. Differences in the change of cardiovascular (CV) risk factors between 16 GH- and 13 non-GH-treated acute lymphoblastic leukemia (ALL) patients. Data are presented as median (range = min–max). (Reproduced from Follin et al., 2010.)

Bone health after acute lymphoblastic leukemia

ALL and its treatment include many known risk factors for low BMD. High doses of glucocorticosteroids and methotrexate, included in nearly all ALL treatment regimens, can, at least temporarily, affect bone formation (Robson et al., 1998; Sala and Barr, 2007). Furthermore, boys with ALL treated with irradiation of the testes need adequate testosterone supplementation and in addition they are at risk for GHD (Brennan et al., 1998; Link et al., 2004). There are conflicting results regarding long-term BMD in former ALL patients after cranial radiotherapy, with both low BMD (Gilsanz et al., 1990; Nussey et al., 1994) and normal BMD (Mandel et al., 2004; Jarfelt et al., 2006). Compared to baseline, a significantly lower BMD Z-score at femoral neck and at L2–L4 was recorded after 8 years without GH therapy (Follin et al., 2011). With GH therapy for 5 years (0.5 mg/day) only female ALL patients had a significantly lower femoral neck Z-score (Fig. 28.2). However, these female ALL patients reached a serum IGF-1 level of −0.7 SD, but in men the level was + 0.05 SD. If Z-scores continue to decrease, there is a premature risk for osteoporosis in this young ALL population.

PATIENTS WITH GROWTH HORMONE DEFICIENCY AND MULTIPLE HORMONE DEFICIENCIES CAUSED BY NONSECRETING PITUITARY MACROADENOMAS TREATED BY SURGERY

Cardiovascular risk in hypopituitary patients with nonsecreting pituitary macroadenomas

Epidemiologic studies have revealed increased cerebral and cardiovascular mortality in patients with hypopituitarism on conventional hormone treatment but without GH therapy. The greatest increase was seen in cerebrovascular disease (Bülow et al., 1997; Tomlinson et al., 2001), with a more pronounced risk in women (Bülow et al., 1997; Tomlinson et al., 2001), but without gender difference in cardiac mortality (Bülow et al., 1997). These epidemiologic studies included particularly patients with nonfunctioning pituitary macroadenomas, whereas patients with Cushing's disease and acromegaly were excluded. The cause of this increased mortality is possibly multifactorial, but GHD is one of the reasons. This statement is based on the results from a meta-analysis of 37 clinical trials, which documented that GH therapy had positive effects on lean mass, fat mass,

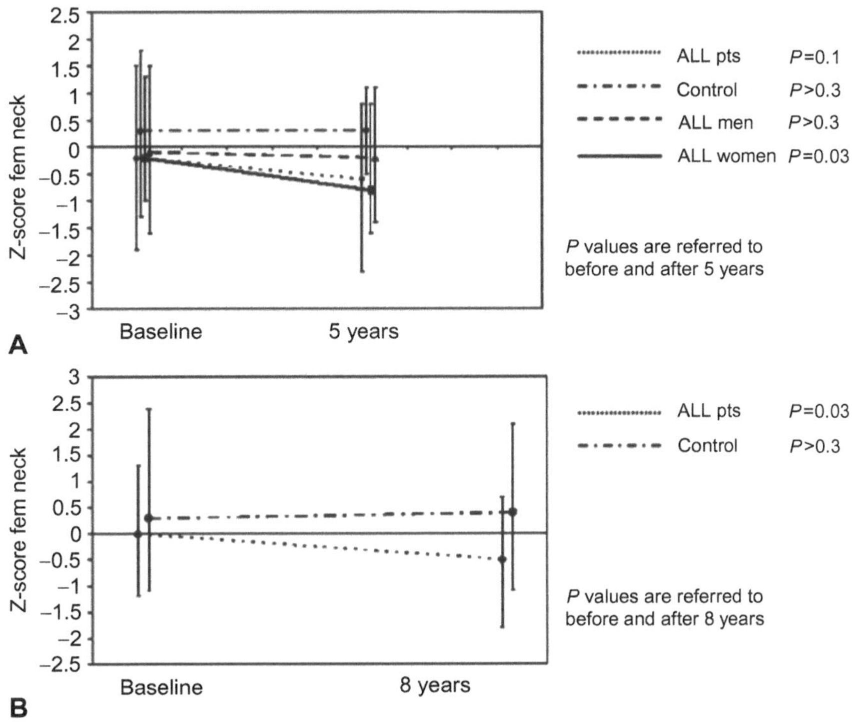

Fig. 28.2. (**A**) Bone mineral density (BMD) Z-score at femoral neck in 15 acute lymphoblastic leukemia (ALL) patients before and after 5 years of GH therapy, in 15 controls before and after 5 years, (**B**) in 13 ALL patients before and after 8 years without GH therapy and in 13 controls before and after 8 years. Data are presented as median and range. *P* values are referred to before and after 5 and 8 years, respectively. (Reproduced from Follin et al., 2011.)

HDL-cholesterol, LDL-cholesterol, total cholesterol, diastolic blood pressure, but also resulted in increments of insulin and glucose levels (Maison et al., 2004). During GH therapy increased insulin resistance has been shown, which, however, was improved by longer duration of GH therapy (Svensson et al., 2002). There is also a GH dose-response dependence, with no change in insulin sensitivity with the use of low GH doses for 6–12 months (Bülow and Erfurth, 1999; Segerlantz et al., 2003; Yuen et al., 2005), and even an improvement in insulin sensitivity with low GH doses after 5 years (Follin et al., 2010). Thus, many of the traditional CVD risk factors show improvements after GH therapy, which might possibly decrease CVD risk and improve longevity in these patients.

There is some hope that this supposition will prove to be correct. A Swedish nationwide study included 750 adult-onset GHD patients, on replacement therapy, with 6 years of GH therapy (Holmer et al., 2007a). Patients and matched population controls responded to a questionnaire on stroke and cardiac events. The vast majority of these patients had been operated for nonfunctioning pituitary adenomas. No increase in the incidence of nonfatal stroke and cardiac events was recorded in patients of either gender. Further, there was a significant decrease in myocardial infarction in men with hypopituitarism (Holmer et al., 2007). An additional explanation for these CVD improvements in GHD patients was the fact that the women in the study had a significantly higher intake of lipid-lowering drugs and the men a higher intake of antihypertensive drugs. A recent study from the KIMS (Pfizer International Metabolic Database) on 13 983 GHD patients treated for 4.9 years with GH therapy showed a modest increase in standardized mortality rate (SMR) of 1.13 (95% confidence interval (CI) 1.04–1.24) (Gaillard et al., 2012). There was an increased risk among females, younger age, diagnosis of Cushing's disease, craniopharyngioma, and aggressive tumors and the presence of diabetes insipidus. A study from the Netherlands, which included patients from the Dutch National Registry of Growth Hormone Treatments, showed only a slight increase in mortality (SMR 1.27 95% CI 1.04–1.56) (van Bunderen et al., 2011). However, interestingly, men receiving GH therapy had no increased mortality. Women had an increased risk, which was lowered after exclusion of high-risk patients, which again were craniopharyngiomas, or malignant causes of hypopitutiarism, i.e., previous cancer treatment. Thus, interestingly, craniopharyngioma and aggressive tumors, diabetes insipidus, and previous cancer treatment have in common that these conditions often involve the hypothalamus. Thus, it seems that men with a pituitary adenoma on GH therapy but without hypothalamic involvement have normal longevity. The reason for the increased mortality among women with a pituitary adenoma is unknown, but unsubstituted hypogonadism before diagnosis (Erfurth et al., 2002), but also unfortunate exposure to sex hormones after diagnosis, have been suggested (Barrett-Connor, 2003; Anderson et al., 2004).

Bone health in patients with growth hormone deficiency due to pituitary macroadenomas

Three studies (Wüster et al., 1991, 2001; Rosén et al., 1997) have reported increased fracture risk among GHD patients without GH therapy. In the first fracture incidence study, 832 patients with GHD and 2581 matched population controls answered a questionnaire about fractures and other background information (Holmer et al., 2007b). A more than doubled incidence rate ratio (IRR) of 2.29 (95% CI 1.23–4.28) for nonosteoporotic fractures was recorded in women with childhood-onset GHD, whereas no increase in risk was observed among childhood-onset GHD men (IRR 0.61) and adult-onset GHD women (IRR 1.08). A significantly decreased incidence of fractures (IRR 0.54 95% CI 0.34–0.86) was recorded in adult-onset GHD men. Increased fracture risk in childhood onset GHD women was most likely explained by the interaction between oral estrogen and GH–IGF-1 axis. The median age at the start of sex steroid substitution was 17 years for both men and women, and late-onset puberty may have been related to low peak bone mass in both genders (Rivera-Woll et al., 2004; Guo et al., 2005), though apparently more deleterious to the female skeleton. It is noteworthy that IGF-1 production in the liver is impaired by oral estrogens, and enhanced by testosterone (Gibney et al., 2005; Mah et al., 2005). Thus, a more likely explanation of the increased fracture risk in childhood-onset GHD women is estrogen insufficiency and an interaction between oral estrogen and the GH–IGF-1 axis during a time window of relevance for bone development and maturation. The adequate substitution rate of testosterone (90%) and GH (94%) that was shown in Holmer et al.'s study may have resulted in significantly lower fracture risk in adult onset GHD men (Holmer et al., 2007).

PATIENTS WITH GROWTH HORMONE DEFICIENCY AND MULTIPLE HORMONE DEFICIENCIES AND WITH HYPOTHALAMIC INVOLVEMENT CAUSED BY A CRANIOPHARYNGIOMA

Cardiovascular risk in patients with a craniopharyngioma

In cohort studies of patients with craniopharyngiomas (CP) the SMR varies from 2.88 to 9.28 (Bülow et al., 1998; Tomlinson et al., 2001; Pereira et al., 2005).

In addition, CP patients have a particularly high CVD mortality in comparison to the general population (3–19-fold higher) (Bülow et al., 1998; Tomlinson et al., 2001; Pereira et al., 2005). Women with CP have an even higher risk (Bülow et al., 1998; Pereira et al., 2005).

Patients with hypothalamic involvement by a CP have significantly more operations and were more often treated with cranial irradiation than patients without hypothalamic involvement (Holmer et al., 2009). On the other hand, the number of pituitary deficiencies seems not to differ between these groups (Holmer et al., 2009). In 54–100% of CP patients at least three pituitary hormone deficiencies have been reported (Karavitaki et al., 2006). The prevalence of GHD is about 90% (Crowley et al., 2010). This pituitary deficiency *per se* might through various metabolic effects contribute to the enhanced cardiovascular morbidity and mortality seen in epidemiologic studies. In addition, in comparison to patients with nonfunctioning pituitary macroadenomas, a higher prevalence of pituitary deficiencies and more obesity and dyslipidemia were recorded in CP patients (Verhelst et al., 2005). Two years of GH replacement showed the same effect on fat-free mass and lipids, but CP patients were less likely to lose body fat, which is possibly due to a hypothalamic effect of the CP. Holmer et al. (2009) evaluated the prevalence of CVD risk factors after long-term GH therapy in adults with childhood-onset CP and recorded increased CVD risk factors in particularly CP women, and in patients with hypothalamic involvement by the tumor (Holmer et al., 2009). Patients with hypothalamic involvement had significantly higher weight, BMI, waist, fat mass, and lower fat free mass compared to patients without such involvement (Table 28.1). Indeed, the increase in CVD risk factors resulted in higher CVD morbidity, as patients with

Table 28.1

Anthropometric measurements, cardiovascular disease risk, and risk factors in patients with and without tumor growth into the third ventricle

	TGTV (n = 25, (women = 11))	Non-TGTV (n = 17, (women = 9))	
	Median (range)	Median (range)	*p*
Anthropometric measurements and body composition (BIA)			
Weight (kg)	96 (60–149)	80 (53–110)	0.001
BMI (kg/m^2)	32 (21–41)	25 (19–35)	0.001
Waist (cm)	105 (77–131)	89 (72–109)	0.002
WHR	0.94 (0.83–1.05)	0.88 (0.76–0.99)	0.02
Fat mass (kg)	40 (9–64)	20 (13–56)	0.003
Fat mass (%)	39 (16–51)	30 (19–52)	0.04
Muscle mass (kg)	44 (19–59)	25 (21–60)	0.4
Muscle mass (%)	35 (12–44)	28 (15–44)	0.8
Fat free mass (kg)	57 (41–86)	51 (40–68)	0.015
Fat free mass (%)	61 (49–84)	70 (48–81)	0.04
CVD risk and risk factors			
Increased CVD risk (HDL-C + hs-CRP)	16/25 (64%)	4/17 (24%)	<0.006
CVD treatment/MetS (IDF)	10	1	<0.02
P-leptin (ng/mL)	28 (4–215)	16 (3–67)	0.07
Leptin/kg fat mass (BIA)	0.74 (0.26–4.51)	0.95 (0.16–1.56)	>0.3
S-insulin (mIU/L)	7 (2–38)	4 (1–8)	<0.001
Insulin/kg fat mass (BIA)	0.21 (0.05–0.59)	0.15 (0.06–0.35)	0.047
P-glucose (mmol/L)	4.6 (3.6–11.0)	4.4 (3.7–5.9)	>0.3
P-LDL (mmol/L)	2.9 (0.5–5.4)	3.2 (2.2–4.7)	0.23
P-HDL (mmol/L)	1.2 (0.7–8.5)	1.3 (0.9–2.4)	0.23
ApoB/ApoA-I ratio	0.68 (0.35–1.40)	0.66 (0.37–1.09)	>0.3
P-fibrinogen (g/L)	3 (2–5)	3 (2–4)	>0.3
P-hs-CRP (mg/L)	2.8 (0.4–23)	0.4 (0.4–16)	0.07

CVD, cardiovascular disease; TGTV, tumor growth into the third ventricle; BMI, body mass index; P, plasma; S, serum; ApoA-I, apolipoprotein A-I; ApoB, apolipoprotein B; BIA, bioelectric impedance analysis; HDL-C, high-density lipoprotein cholesterol; hs-CRP, high sensitivity C-reactive protein; IGF-1, insulin-like growth factor 1; LDL-C, low-density lipoprotein cholesterol; MetS, metabolic syndrome; WHR, waist/hip circumference ratio.

hypothalamic involvement were more often on treatment with antihypertensive and antidiabetes and lipid-lowering drugs (Holmer et al., 2009).

Hypothalamic damage and obesity in craniopharyngioma patients

Hypothalamic damage is the most devastating consequence of a CP. Approximately 50% of children treated for CP are obese at follow-up and hypothalamic damage is a major cause (Holmer et al., 2009; Trivin et al., 2009). Destruction of the ventromedial hypothalamus may cause hyperphagia, and later uncontrolled obesity (Hetelekidis et al., 1993; Duff et al., 2000; Karavitaki et al., 2006), but overt hyperphagia is conspicuously absent in the long-term follow-up of these patients (Holmer et al., 2010). Elevated leptin levels in relation to BMI, thus leptin resistance, have been recorded in both children (Roth et al., 1998) and adults with CP (Holmer et al., 2010). Vagally mediated hyperinsulinemia and autonomic imbalance (Bray et al., 1981; Lustig, 2002), with insulin resistance at the intracellular signaling levels, causes a relative resistance to leptin, by enabling the body to tolerate new adiposity. There is also cross-talk between signaling events downstream of insulin receptors and leptin receptors and thus cross-desensitization between insulin and leptin pathways can occur.

The primary pathogenesis of hypothalamic damage is associated with hyperinsulinemia, associated with damage to the ventromedial hypothalamic nuclei causing imbalance of autonomic nervous system, resulting in suppression of the sympathetic nervous system and stimulation of the vagus (Bray and Gallagher, 1975). Indeed, CP patients have hyperinsulinemia, with increased levels in relation to tumor growth (Holmer et al., 2009) (Fig. 28.3). Hyperinsulinemia increases lipogenesis in liver and adipose tissue and also lipoprotein lipase activity accelerating endogenous (very low density lipoproteins and triglycerides) lipid production (Bray et al., 1981). Fat mass is accumulated with an increase in weight and BMI (Bray et al., 1981).

Adipose tissue is richly innervated by sympathetic nerve fibers that control lipolyses. Lipogenesis is also controlled by parasympathetic innervations of adipose tissue originating from separate sympathetic and parasympathetic neurons in the periventricular nucleus and suprachiasmatic nucleus (Kreier et al., 2002) of the hypothalamus. Thus, the suprachiasmatic nucleus has a unique position to balance both branches of the autonomic nervous system. A large proportion of CP patients have suprasellar involvement and thus damage to the suprachiasmatic nucleus with an altered regulation of central clock mechanism, which predispose to alterations in metabolism (Kreier et al., 2002; Stripp et al., 2004).

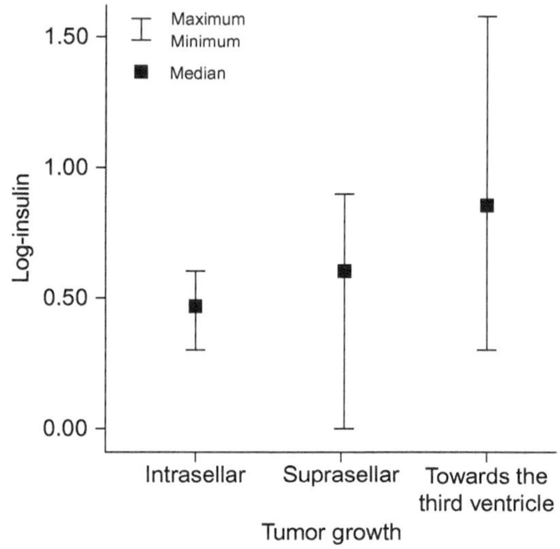

Fig. 28.3. Association between tumor growth and serum insulin in 42 craniopharyngioma patients ($r=0.57$, $p<0.001$). Plasma insulin increased in relation to tumor growth. (Reproduced from Holmer et al., 2009.)

A reduction in energy expenditure was recorded in a study of children with hypothalamic obesity suffering from tumors in the hypothalamic region (Shaikh et al., 2008). Furthermore, Holmer et al. (2010) showed that adults with childhood-onset CP on complete hormone supplementation including GH therapy had significantly lower resting metabolic rate compared to controls. In addition, the total caloric intake was not significantly different and similar dietary macronutrient composition was found in patients and controls, and only significantly higher scales in restricting food intake were recorded in patients. Thus, a reduction in resting metabolic rate due to hypothalamic involvement but not higher caloric intake, together with lower physical activity, was the most likely cause of obesity in these patients (Holmer et al., 2010).

Bone health in craniopharyngioma patients

Overweight is protective against low BMD in healthy subjects (Felson et al., 1993), possibly by the action of increased leptin levels (Hamrick and Ferrari, 2008). However, in CP patients leptin levels were shown to be higher in relation to BMI, indicating leptin resistance (Roth et al., 1998), and among adolescents leptin resistance was shown to be negatively correlated to BMD (Do Prado et al., 2009). In a study by Holmer et al. (2011) in childhood onset CP on complete hormone treatment including long-term GH therapy, 31% had Z-scores ≤ -2.0 SD, and the majority of them had tumor growth into the hypothalamus. It seems reasonable to believe that the hypothalamic damage, together with leptin resistance, recorded in many of the CP patients, is of paramount importance for low BMD in this population. Thus, high BMI was not protective against low BMD in this patient population.

The hypothalamus regulates bone and adipose tissue via a complex and finely tuned interplay of endocrine mechanisms (of which neuropeptide Y is a key regulator) together with the sympathetic nervous system. Whereas many of the effects occur via direct actions on osteoblasts or adipocytes, sex hormones can also mediate effects on bone and adipose tissue via interaction with neuronal pathways. Thus, as was shown in previous studies (Holmer et al., 2007b, 2011), both early and persistent hypothalamic dysfunction, and insufficient sex hormone replacement at the time of disease onset are likely to have contributed to the observed long-term effects on bone in these patients. Thus, the interaction between the adipose tissue-related endocrine system and bone seems to be complex and may be modulated by central and peripheral mechanisms as well as local resistance to the putative protective effects of insulin and leptin on bone.

GENERAL UNCERTAINTIES OF GROWTH HORMONE THERAPY

GH therapy of adults has generally been regarded as quite a safe treatment, although concerns remain regarding the potential for cancer risk, as IGF-1 has mitogenic effects on both normal and cancerous cells (Maculty, 1992). In general, GH treatment is contraindicated in the presence of active malignancy. Although GH treatment reduces insulin sensitivity, the worsening of glycemic control has been minimal and transient, and low GH doses may even improve insulin sensitivity. However, GH treatment in patients with diabetes mellitus may require adjustment in antidiabetic medication. It is also of importance to carefully monitor thyroid and adrenal function during GH therapy.

There is no clear suggestion as to how to establish the exact GH dose in adults with GHD. It is clear that the GH dosing should be individualized, should start with a low dose and be titrated according to clinical responses and IGF-1 levels. Side-effects with peripheral tissue swelling and joint pain may occur, but is mostly transient. If not, these symptoms will disappear after lowering the GH dose. In general, a serum IGF-1 response into the mid-ranges of normal is recommended. The GH dosing should take age, gender, and estrogen status into consideration. Older age, above 70–80 years, is no contraindication, but the GH dose should be minimal and carefully adjusted (around 0.10–0.2 mg/day subcutaneously), often aiming at a low normal serum IGF-1 level. However, there is little evidence that GH therapy in these older GHD subjects is beneficial, since old age has been an exclusion criterion in most studies and since GH levels decrease considerably during old age in general. GH therapy should initially be monitored at intervals of 1–2 months during dose titration with clinical assessment of adverse effects, and semiannually with measurers of body composition (e.g., Bio Impedance analyses).

CONCLUSION

If GHD was an isolated entity the metabolic effects of GH therapy could be easily predicted. However, based on different background diagnoses, there are several phenotypes of the hypopituitary patient with GHD. In clinical practice these patients may even transit between the different phenotypes over time due to extension of the pathology and/or the effects of treatments (surgery and/or radiotherapy). Furthermore, hypothalamic involvement, but also hormone substitutions of other pituitary hormone deficiencies, is of paramount importance for the resulting phenotype. Thus, the effects of GH therapy are rendered less clear as a result all these factors. This can be clearly seen among CP patients with or without hypothalamic involvement, in whom the devastating effects of hypothalamic damage dominate over the positive metabolic effects of GH therapy. In most studies on the effects of GH treatment in GHD patients, these differential effects of the underlying pituitary diseases on the manifestations of GHD and on the effects of GH treatment are not taken into consideration.

References

Anderson GL, Limacher M, Assaf AR et al. (2004). Effects of conjugated equine estrogen in postmenopausal women with hysterectomy: the Women's Health Initiative randomized controlled trial. JAMA 291: 1701–1712.

Barrett-Connor E (2003). Clinical review 162: Cardiovascular Endocrinology 3. An epidemiologist looks at hormones and heart disease in women. J Clin Endocrinol Metab 88: 4031–4042.

Bowers DC, Liu Y, Leisenring W et al. (2006). Late-occurring stroke among long-term survivors of childhood leukemia and brain tumors: a report from the Childhood Cancer Survivor Study. J Clin Oncol 24: 5277–5282.

Bray GA, Gallagher Jr TF (1975). Manifestations of hypothalamic obesity in man: a comprehensive investigation of eight patients and a review of the literature. Medicine (Baltimore) 54: 301–330.

Bray GA, Inoue S, Nishizawa Y (1981). Hypothalamic obesity. The autonomic hypothesis and the lateral hypothalamus. Diabetologia 20 (Suppl): 366–377.

Brennan BMD, Rahim A, Mackie EM et al. (1998). Growth hormone status in adults treated for acute lymphoblastic leukaemia in childhood. Clin Endocrinol (Oxf) 48: 777–783.

Brennan BM, Rahim A, Blum WF et al. (1999). Hyperleptinaemia in young adults following cranial irradiation in childhood: growth hormone deficiency or leptin insensitivity? Clin Endocrinol (Oxf) 50: 163–169.

Bülow B, Erfurth EM (1999). A low individualized GH dose in young patients with childhood onset GH deficiency normalized serum IGF-I without significant deterioration in glucose tolerance. Clin Endocrinol (Oxf) 50: 45–55.

Bülow B, Hagmar L, Mikoczy Z et al. (1997). Increased cerebrovascular mortality in patients with hypopituitarism. Clin Endocrinol (Oxf) 46: 75–81.

Bülow B, Attewell R, Hagmar L et al. (1998). Postoperative prognosis in craniopharyngioma with respect to cardiovascular mortality, survival, and tumor recurrence. J Clin Endocrinol Metab 83: 3897–3904.

Bülow B, Link K, Ahrén B et al. (2004). Survivors of childhood acute lymphoblastic leukaemia, with radiation-induced GH deficiency, exhibit hyperleptinaemia and impaired insulin sensitivity, unaffected by 12 months of GH treatment. Clin Endocrinol (Oxf) 61: 683–691.

Carroll PV, Christ ER (1998). Growth hormone deficiency in adulthood and the effects of growth hormone replacement: a review. J Clin Endocrinol Metab 3: 382–395.

Cornier MA, Dabalea A, Hernandez TL et al. (2008). The metabolic syndrome. Endocr Rev 29: 777–822.

Crowley RK, Hamnvik OP, O'Sullivan EP et al. (2010). Morbidity and mortality in patients with craniopharyngioma after surgery. Clin Endocrinol (Oxf) 73: 516–521.

Cuneo RC, Salomon F, McGauley GA et al. (1992). The growth hormone deficiency syndrome in adults. Clin Endocrinol (Oxf) 37: 387–397.

De Boer H, Blok GJ, van Lingen A et al. (1994). The consequences of childhood-onset growth hormone deficiency for adult bone mass. J Bone Miner Res 9: 1319–1326.

Degerblad M, Bengtsson BÅ, Bramnert M et al. (1995). Reduced bone mineral density in adults with growth hormone (GH) deficiency: increased bone turnover during 12 months of GH substitution therapy. Eur J Endocrinol 133: 180–188.

Do Prado WL, de Piano A, Lazaretti-Castro M et al. (2009). Relationship between bone mineral density, leptin and insulin concentration in Brazilian obese adolscents. J Bone Miner Metab 27: 613–619.

Duff J, Meyer FB, Ilstrup DM et al. (2000). Long-term outcomes for surgically resected craniopharyngiomas. Neurosurgery 46: 291–302.

Erfurth EM, Bülow B, Svahn-Tapper G et al. (2002). Riskfactors for cerebrovascular deaths in patients operated and irradiated for pituitary tumors. J Clin Endocrinol Metab 87: 4892–4899.

Felson DT, Zhang Y, Hannan MT et al. (1993). Effects of weight and body mass index on bone mineral density in men and women: the Framingham study. J Bone Miner Res 8: 567–573.

Fernholm R, Bramnert M, Hägg E et al. (2000). Growth hormone replacement therapy improves body composition and increases bone metabolism in elderly patients with pituitary disease. J Clin Endocrinol Metab 85: 4104–4112.

Follin C, Thilén U, Österberg K et al. (2010). Cardiovascular risk, cardiac function, physical activity and quality of life with and without long-term GH therapy in adult survivors of childhood acute lymphoblastic leukaemia. J Clin Endocrinol Metab 95: 3726–3735.

Follin C, Link K, Wiebe T et al. (2011). Bone loss after childhood acute lymphoblastic leukaemia: an observational study with and without GH therapy. Eur J Endocrinol 164: 695–703.

Ford ES (2005). Risks for all-cause mortality, cardiovascular disease, and diabetes associated with the metabolic syndrome: a summary of the evidence. Diabetes Care 28: 1769–1778.

Gaillard RC, Mattsson AF, Akerblad AC et al. (2012). Overall and cause-specific mortality in GH-deficient adults on GH replacement. Eur J Endocrinol 166: 1069–1077.

Gibney J, Wolthers T, Johannsson G et al. (2005). Growth hormone and testosterone interact positively to enhance protein and energy metabolism in hypopituitary men. Am J Physiol Endocrinol Metab 2: 266–271.

Gilsanz V, Carlsson M, Roe T et al. (1990). Osteoporosis after cranial irradiation for acute lymphoblastic leukemia. J Pediatr 117: 238–244.

Guo Y, Zhao LJ, Shen H et al. (2005). Genetic and environmental correlations between age at menarche and bone mineral density at different skeletal sites. Calcif Tissue Int 77: 356–360.

Hamrick MW, Ferrari SL (2008). Leptin and the sympathetic connection of fat to bone. Review. Osteoporos Int 19: 905–912.

Hetelekidis S, Barnes PD, Tao ML et al. (1993). 20-year experience in childhood craniopharyngioma. Int J Radiat Oncol Biol Phys 27: 189–195.

Holmer H, Svensson J, Rylander L et al. (2007a). Non-fatal stroke, cardiac disease, and diabetes mellitus in hypopituitary patients on complete hormone replacement including GH. J Clin Endocrinol Metab 92: 3560–3567.

Holmer H, Svensson J, Rylander L et al. (2007b). Fracture incidence in GH deficient patients on complete hormone replacement including GH. J Bone Miner Res 22: 1842–1850.

Holmer H, Ekman B, Björk J et al. (2009). Hypothalamic involvement predicts cardiovascular risk in adults with childhood onset craniopharyngioma on long-term GH therapy. Eur J Endocrinol 161: 671–679.

Holmer H, Pozarek G, Wirfält E et al. (2010). Reduced energy expenditure and impaired feeding-related signals but not high energy intake reinforces hypothalamic obesity in adults with childhood onset craniopharyngioma. J Clin Endocrinol Metab 95: 5395–5402.

Holmer H, Popovic V, Ekman B et al. (2011). Hypothalamic involvement and insufficient sex steroid supplementation are associated with low bone mineral density in women with childhood onset craniopharyngioma. Eur J Endocrinol 165: 25–31, April.

Holmes SJ, Economou G, Whitehouse RW et al. (1994). Reduced bone mineral density in patients with adult onset growth hormone deficiency. J Clin Endocrinol Metab 78: 669–674.

Jarfelt M, Fors H, Lannering B et al. (2006). Bone mineral density and bone turnover in young adult survivors of childhood acute lymphoblastic leukaemia. Eur J Endocrinol 154: 303–309.

Karavitaki N, Cudlip S, Adams CB et al. (2006). Craniopharyngiomas. Endocr Rev 27: 371–397.

Kaufman JM, Taelman P, Vermeulen A et al. (1992). Bone mineral status growth hormone-deficient males with isolated and multiple pituitary deficiencies of childhood onset. J Clin Endocrinol Metab 74: 118–123.

Kreier F, Fliers E, Voshol PJ et al. (2002). Selective parasympathetic innervation of subcutaneous and intra-abdominal fat – functional implications. J Clin Invest 110: 1243–1250.

Link K, Moëll C, Garwicz S et al. (2004). Growth hormone deficiency predicts cardiovascular risk in young adults treated for acute lymphoblastic leukaemia in childhood. J Clin Endocrinol Metab 89: 5005–5012.

Lustig RH (2002). Hypothalamic obesity: the sixth cranial. Endocrinopathy Endocrinologist 12: 210–217.

Maculty VM (1992). Insulin-like growth factors and cancer. Br J Cancer 65: 311–320.

Mah PM, Webster J, Jönsson P et al. (2005). Estrogen replacement in women of fertile years with hypopituitarism. J Clin Endocrinol Metab 90: 5964–5969.

Mandel K, Atkinson S, Barr R et al. (2004). Skeletal morbidity in childhood acute lymphoblastic leukaemia. J Clin Oncol 22: 2015–2021.

Maison P, Griffin S, Nicoue-Beglah M et al. (2004). Impact of growth hormone (GH) treatment on cardiovascular risk factors in GH-deficient adults: a metaanalysis of blinded, randomized, placebo-controlled trials. J Clin Endocrinol Metab 89: 2192–2199.

Mayer EI, Reuter M, Dopfer RE et al. (2000). Energy expenditure, energy intake and prevalence of obesity after therapy for acute lymphoblastic leukemia during childhood. Horm Res 53: 193–199.

Mertens AC, Liu Q, Neglia JP et al. (2008). Cause-specific late mortality among 5-year survivors of childhood cancer: the Childhood Cancer Survivor Study. J Natl Cancer Inst 100: 1368–1379.

Nussey S, Hyer S, Brada M et al. (1994). Bone mineralization after treatment of growth hormone deficiency in survivors of childhood malignancy. Acta Paediatr Suppl 399: 9–14.

Oeffinger KC, Mertens AC, Sklar CA et al. (2003). Obesity in adult survivors of childhood acute lymphoblastic leukemia: a report from the Childhood Cancer Survivor Study. J Clin Oncol 21: 1359–1365.

Oeffinger KC, Eshelman DA, Tomlinson GE et al. (2000). Providing primary care for long-term survivors of childhood acute lymphoblastic leukemia. J Fam Pract 49: 1133–1146.

O'Halloran D, Tsatosoulis A, Whitehouse R et al. (1993). Increased bone density after recombinant human growth hormone (GH) therapy in adults with isolated GH deficiency. J Clin Endocrinol Metab 76: 1344–1348.

Ohlsson C, Bengtsson BÅ, Isaksson OG et al. (1998). Growth hormone and bone. Endocr Rev 19: 55–79.

Pereira AM, Schmid EM, Schutte PJ et al. (2005). High prevalence of long-term cardiovascular, neurological and psychosocial morbidity after treatment for craniopharyngioma. Clin Endocrinol (Oxf) 62: 197–204.

Robson H, Anderson E, Eden O et al. (1998). Chemotherapeutic agents used in the treatment of childhood malignancies have a direct effects on growth plate chondrocyte proliferation. J Endocrinol 157: 225–235.

Rivera-Woll LM, Papalia M, Davis SR et al. (2004). Androgen insufficiency in women: diagnostic and therapeutic implications. Hum Reprod Update 10: 421–432.

Rosén T, Bengtsson BÅ (1990). Premature mortality due to cardiovascular disease in hypopituitarism. Lancet 336: 285–288.

Rosén T, Hansson T, Granhed H et al. (1993). Reduced bone mineral content in adult patients with growth hormone deficiency. Acta Endocrinol (Copenh) 129: 201–206.

Rosén T, Wilhelmsen L, Landin-Whilhelmsen K et al. (1997). Increased fracture frequency in adult patients with hypopituitarism and GH deficiency. Eur J Endocrinol 137: 240–245.

Roth C, Wilken B, Hanefels F et al. (1998). Hyperphagia in children with craniopharyngioma is associated with hyperleptinaemia and a failure in the downregulation of appetite. Eur J Endocrinol 138: 89–91.

Sala A, Barr R (2007). Osteopenia and cancer in children and adolescents. Cancer 109: 1420–1431.

Segerlantz M, Bramnert M, Manhem P et al. (2003). Inhibition of lipolysis during acute GH exposure increases insulin sensitivity in previously untreated GH-deficient adults. Eur J Endocrinol 149: 511–519.

Shaikh MG, Grundy RG, Kirk JM (2008). Reductions in basal metabolic rate and physical activity contribute to hypothalamic obesity. J Clin Endocrinol Metab 93: 2588–2593.

Stripp DC, Maity A, Janss AJ et al. (2004). Surgery with or without radiation therapy in the management of craniopharyngiomas in children and young adults. Int J Radiat Oncol Biol Phys 58: 714–720.

Svensson J, Fowelin J, Landin K et al. (2002). Effects of seven years of growth hormone (GH) replacement therapy on insulin sensitivity in GH deficient adults. J Clin Endocrinol Metab 87: 2121–2127.

Talvensaari KK, Lanning M, Tapanainen P et al. (1996). Long-term survivors of childhood cancer have an increased risk of manifesting the metabolic syndrome. J Clin Endocrinol Metab 81: 3051–3055.

Tomlinson JW, Holden N, Hills RK et al. (2001). Association between premature mortality and hypopituitarism. Lancet 357: 425–431.

Toogood A, Adams J, O'Neill P et al. (1997). Elderly patients with adult-onset growth hormone deficiency are not osteopenic. J Clin Endocrinol Metab 82: 1462–1466.

Trivin C, Busiah K, Mahlaoui N et al. (2009). Childhood craniopharyngioma: greater hypothalamic involvement before surgery is associated with higher homeostasis model insulin resistance index. BMC Pediatr 9: 24.

van Bunderen CC, van Nieuwpoort IC, Arwert LI et al. (2011). Does growth hormone replacement therapy reduce mortality in adults with growth hormone deficiency? Data from the Dutch National Registry of Growth Hormone Treatment in adults. J Clin Endocrinol Metab 96: 3151–3159.

Verhelst J, Kendall-Taylor P, Erfurth EM et al. (2005). Baseline characteristics and response to 2 years of growth hormone (GH) replacement of hypopituitary patients with GH deficiency due to adult-onset craniopharyngioma in comparison with patients with nonfunctioning pituitary adenoma: data from KIMS (Pfizer International Metabolic Database). J Clin Endocrinol Metab 90: 4636–4643.

Wüster C, Slenczka E, Ziegler R (1991). Increased prevalence of osteoporosis and arteriosclerosis in conventionally substituted anterior pituitary insufficiency: is there a need for additional growth hormone substitution? [In German]. Klin Wochenschr 69: 769–773.

Wüster C, Abs R, Bengtsson BÅ et al. (2001). The influence of growth hormone deficiency, growth hormone replacement therapy and other aspects of hypopituitarism on fracture rate and bone mineral density. J Bone Miner Res 2: 398–405.

Yuen K, Frystyk J, White D et al. (2005). Improvement in insulin sensitivity without concomitant changes in body composition and cardiovascular risk markers following fixed administration of a very low growth hormone (GH) dose in adults with severe GH deficiency. Clin Endocrinol (Oxf) 63: 428–436.

Zhang Y, Proenca R, Maffei M et al. (1994). Positional cloning of the mouse obese gene and its human homologue. Nature 1: 372.

Handbook of Clinical Neurology, Vol. 124 (3rd series)
Clinical Neuroendocrinology
E. Fliers, M. Korbonits, and J.A. Romijn, Editors

Chapter 29

Autoimmune hypophysitis: new developments

YUTAKA TAKAHASHI*
Division of Diabetes and Endocrinology, Department of Internal Medicine, Kobe University Graduate School of Medicine, Kobe, Japan

INTRODUCTION

Autoimmune hypophysitis (AH), often referred to as lymphocytic hypophysitis, is defined as an inflammatory condition of the pituitary gland of autoimmune etiology that leads to pituitary dysfunction (Caturegli et al., 2005; Hamnvik et al., 2010); however, its pathogenesis is still incompletely defined. It is classified into three subtypes based on the affected anatomic region: lymphocytic adenohypophysitis (LAH), lymphocytic infundibuloneurophypophysitis (LINH), and lymphocytic panhypophysitis (LPH) (Caturegli et al., 2005). Although AH is a rare disease, its incidence has increased as physicians become increasingly aware of the entity. As the prevalence, five cases of AH were detected in 619 consecutive pituitary surgeries (0.8%) (Buxton and Robertson, 2001), and Leung et al. reported 13 cases among 2000 patients who underwent transsphenoidal surgery (0.65%) (Leung et al., 2004).

LAH is the most common subtype, with clinical and histologic involvement primarily of the anterior pituitary. LAH was first described in 1962 by Goudie and Pinkerton (Goudie and Pinkerton, 1962). They reported a 22-year-old woman who died 14 months after her second delivery, presumably because of adrenal insufficiency. LAH primarily affects women between 30 and 40 years of age and is associated with pregnancy. In 20–50% of patients, LAH is associated with other autoimmune disorders, such as Hashimoto's thyroiditis, Graves' disease, autoimmune adrenalitis, and pernicious anemia, as well as type 1 diabetes, in which it is associated with autoimmune polyglandular syndrome (APS). The pathologic and serologic features of LAH are consistent with an autoimmune etiology. Lymphocytic infiltrate is observed in the pituitary, and several autoantibodies directed against pituitary antigens, e.g., growth hormone (GH) (Tanaka et al., 2002), α-enolase (Crock, 1998), and secretogranin-2 (Bensing et al., 2007b) have been detected.

LINH was first described in 1970 by Saito et al., who observed a 66-year-old woman with a 1 month history of central diabetes insipidus (Saito et al., 1970). LINH affects the posterior pituitary and the pituitary stalk, with diabetes insipidus as its main clinical feature. LPH was first reported in 1991 in a 40-year-old male with a 3 month history of headaches, impotence, polyuria, and polydipsia. A histologic examination revealed extensive infiltration of the adenohypophysis and neurohypophysis by lymphocytes, plasma cells, and histiocytes (Nussbaum et al., 1991).

While the etiology of AH is unknown, it exhibits several epidemiologic features of autoimmune mechanisms. Moreover, few triggers of AH – apart from pregnancy – have been described. One intriguing exception, which also suggests an autoimmune etiology, is exposure to the cytotoxic T lymphocyte protein 4 (CTLA4)-blocker tremelimumab (Shaw et al., 2007). CTLA4 blockade – used in the treatment of malignant melanoma – has also been implicated in the development of several other tissue-specific autoimmune endocrinopathies. Although the mechanisms underlying this association have not yet been elucidated, given that single nucleotide polymorphisms in the CTLA4 locus are associated with Graves' disease and type 1 diabetes (Ueda et al., 2003), CTLA4 may be closely related to the tissue-specific autoimmunity, including the pituitary. The definition and precise mechanisms of AH have not yet been clarified; however, novel clinical entities (IgG4-related hypophysitis and anti-PIT1 antibody syndrome) related to the unique autoantibodies found in AH have recently been reported. Therefore, these

*Correspondence to: Dr Yutaka Takahashi, Division of Diabetes and Endocrinology, Department of Internal Medicine, Kobe University Graduate School of Medicine, 7-5-1, Kusunoki-cho, Chuo-ku, Kobe 650-0017, Japan. Tel: +81-78-382-5861, Fax: +81-78-382-2080, E-mail: takahash@med.kobe-u.ac.jp

clinical entities will be instrumental in understanding the pathophysiology of AH.

PITUITARY AUTOANTIBODIES

Although T lymphocytes are more likely to be critical for the development of AH because of the predominant infiltrate of CD4+ T cells, only the function of B lymphocytes has been assessed by measuring autoantibodies against the pituitary, due to the technical limitations. Antibodies against pituitary antigens have been mainly detected by indirect immunofluorescence or immunoblotting (Caturegli et al., 2005). Depending on the specific method used for screening, 6–57% of patients with confirmed hypophysitis have positive autoantibodies (Hamnvik et al., 2010). The specificity of pituitary antibodies is low, as they have also been found in various nonautoimmune pituitary diseases, including pituitary adenomas, empty sella syndrome, and Sheehan syndrome, as well as in other autoimmune diseases, such as type 1 diabetes and autoimmune thyroid diseases. The pituitary antigens reported include GH1 or placental GH2 (Takao et al., 2001), α-enolase (Crock, 1998), pituitary gland-specific factors 1a and 2 (PGSF1a, 2) (Tanaka et al., 2002), secretogranin-2 (Bensing et al., 2007b), CGI99, chorionic somatomammotropin (Lupi et al., 2008), and TDRD6 (Bensing et al., 2007a). In general, these autoantibodies are used as a marker with limitations because of the specificity and unclarified pathophysiologic significance in hypophysitis. In fact, many autoantigens are located within the cytosol, sequestered from the blood; thus it is unlikely that their function is directly affected by antibodies.

Although pathogenic autoantibodies in AH have not yet been reported, it has been suggested that some antibodies may be closely related to pathogenesis, e.g., antipituitary antibodies against gonadotropin-secreting cells in adult male patients with idiopathic hypogonadotropic hypogonadism (De Bellis et al., 2007). Furthermore, De Bellis et al. reported that antipituitary antibodies were present at high titers in 4 of 27 patients with ACTH deficiency (15%), 4 of 20 patients with GH deficiency (20%), and 5 of 19 with hypogonadotropic hypogonadism (21%), and that these antibodies targeted corticotrophs, somatotrophs, and gonadotrophs, respectively (De Bellis et al., 2011). Recently, Smith et al. screened a pituitary cDNA expression library by using sera from patients with lymphocytic hypophysitis and identified an autoantigen, Tpit, as well as PGSF1a, PGSF2, and neuron-specific enolase (NSE), which were reported previously (Smith et al., 2012). Significantly positive autoantibody reactivity against Tpit was found in 9 of 86 hypophysitis patients (11%) compared with 1 of 90 control subjects. Autoantibodies were also detected against chromodomain helicase DNA-binding protein 8 (CHD8), presynaptic cytomatrix protein (piccolo), Ca2+-dependent secretion activator (CADPS), PGSF2, and NSE; however, such antibodies were detected at a frequency that did not differ from that of the healthy controls. Intriguingly, Tpit is a transcription factor that is essential for the differentiation of corticotrophs and mutations in the *TPIT* gene cause congenital isolated ACTH deficiency (Lamolet et al., 2001; Pulichino et al., 2003a, b). Given that corticotrophs are often the first cell type to be affected in lymphocytic hypophysitis and that some cases of isolated ACTH deficiency are related to autoimmune diseases including APS (Crock, 1998; Kasperlik-Zaluska et al., 2003), a causal involvement of anti-Tpit antibody in the development of lymphocytic hypophysitis has been speculated. However, this antibody is not specific to lymphocytic hypophysitis, as it has also been detected in other autoimmune endocrine diseases. Further investigation is necessary to elucidate the significance of anti-Tpit antibody.

Molecular mimicry has been proposed to mediate the underlying mechanism responsible for pathogenesis in AH, which is often associated with pregnancy. GH2, chorionic somatomammotropin, and γ-enolase proteins are expressed in the placenta, and it is possible that immune recognition of these proteins spreads during pregnancy to GH1 and α-enolase (O'Dwyer et al., 2002; Wegner et al., 2009). A number of studies have investigated the significance of pituitary antibodies. In the patients with APS, organ-specific antibodies are frequently observed (Michels and Gottlieb, 2010). These autoantibodies are usually detected in more than 90% of affected patients at the onset of the clinical phase, but they may sometimes also be detected in the preclinical phase of the disease (Falorni et al., 2004; Maghnie et al., 2006). Bellastella et al. conducted a longitudinal study for 5 years observing 149 antipituitary antibody (APA)-positive and 50 APA-negative patients with APS and normal pituitary function (Bellastella et al., 2010). APA was evaluated in this study by an indirect immunofluorescence method using prepubertal baboon pituitary glands. During the 5 years of the study, hypopituitarism occurred in 28 of 149 (19%) APA-positive patients, but in none of the 50 APA-negative patients. All patients that developed pituitary dysfunction throughout the study span exhibited a characteristic immunostaining pattern, in which a part of pituitary cells was positive for the immunofluorescence. These data suggest that measurement of APA may help to predict the occurrence of hypopituitarism.

IGG4-RELATED HYPOPHYSITIS

Immunoglobulin G4 (IgG4)-related disease is a newly recognized clinical entity that was proposed following the close observation of patients with autoimmune pancreatitis (Hamano et al., 2001). It is characterized by elevated serum IgG4 concentration and tumefaction or tissue infiltration by IgG4-positive plasma cells. IgG4-related disease involves many tissues and is associated with Mikulicz's disease, autoimmune pancreatitis, hypophysitis, Riedel's thyroiditis, interstitial nephritis, prostatitis, lymphadenopathy, retroperitoneal fibrosis, inflammatory aortic aneurysm, and inflammatory pseudotumor (Umehara et al., 2012). IgG4-related hypophysitis has been reported during the last few years; it was initially diagnosed on clinical grounds in 2004 (van der Vliet and Perenboom, 2004), and then by pathology in 2007 (Wong et al., 2007). The first case reported was a 66-year-old woman with multiple pseudotumors in the salivary glands, pancreas, and retroperineum. The first pathologically proven case presented pituitary expansion with blurred vision and hypogonadism and a history of autoimmune pancreatitis. Histopathology of the pituitary mass showed a dense lymphoplasmacytic infiltrate among residual nests of adenohypophysial cells and fibrosis. Immunohistochemical staining for IgG4 and κ/λlight chains highlighted the presence of numerous polyclonal plasma cells in pituitary, pancreas, and gall bladder specimens. Shimatsu et al. reviewed 22 Japanese patients with IgG4-related hypophysitis (Shimatsu et al., 2009). The majority of cases involved middle-aged to elderly men presenting various degrees of hypopituitarism and diabetes insipidus who also demonstrated a thickened pituitary stalk and/or pituitary mass. These structures shrank significantly in response to glucocorticoid therapy, even in low doses. The presence of IgG4-related systemic disease and an elevated level of IgG4 before glucocorticoid therapy were the main clues that led to correct diagnosis. Leporati et al. reported the first Caucasian patient with IgG4-related hypophysitis, who manifested severe headache with panhypopituitarism and symmetric enlargement of pituitary and thickened stalk shown by magnetic resonance imaging (MRI), reviewed the published literature, and proposed diagnostic criteria (Leporati et al., 2011). These criteria are as follows: pituitary histopathology of mononuclear infiltration of the pituitary gland, rich in lymphocytes and plasma cells, with IgG4-positive cells; sellar mass and/or thickened pituitary stalk; biopsy-proven involvement of other organs; increased serum IgG4 concentrations of more than 140 mg/dL; and shrinkage of the pituitary mass and symptom improvement with steroids. At present, the pathogenic mechanism and the underlying immunologic abnormalities of IgG4-related disease remain unclear. IgG4 is thought to protect against type I allergy (IgE-mediated type I hypersensitivity) by inhibiting the functions of IgE, and it may also prevent type II (cytotoxic and antibody-dependent hypersensitivity) and III (immune complex disease) allergy by blocking the Fc-mediated effector functions of IgG1 and by inhibiting the formation of large immune complexes. Moreover, IgG4 is predominantly expressed under conditions of chronic antigen exposure, which may be related to the chronic features of IgG4-related disease (Umehara et al., 2012). Aberrant immunologic findings have been observed in patients with IgG4-related disease. The numbers of regulatory T cells (Treg) expressing CD4+CD25+Foxp3 are significantly higher in the affected tissues and peripheral blood of patients (Zen and Nakanuma, 2011). Furthermore, a Th2-dominant immune response and increased production of Th2-type cytokines such as IL-4, IL-5, IL-10, and IL-13 are observed (Akitake et al., 2010; Suzuki et al., 2010). Indeed, overexpression of the regulatory cytokines, namely IL-10 and transforming growth factor (TGF-β), has been reported in patients with IgG4-related disease (Nakashima et al., 2010). IL-10 and TGF-β have potent activities in directing B cells to produce IgG4 or to induce fibroplasias, respectively. IL-4, IL-5, and IL-13 are important for class switching to IgE production and eosinophil migration. Thus, the data suggest that abnormalities in the production of these cytokines may be involved in the pathogenesis of IgG4-related disease.

ANTI-PIT-1 ANTIBODY SYNDROME

The pituitary-specific transcriptional factor 1 (PIT-1; also known as POU1F1) plays a crucial role in regulating the expression of GH, prolactin (PRL), and TSH-β in the anterior pituitary. PIT-1 is essential for the differentiation, proliferation, and maintenance of somatotrophs, lactotrophs, and thyrotrophs in the pituitary (Ingraham et al., 1988; Cohen et al., 1996). As a result, abnormalities in the *PIT-1* gene result in short stature and congenital combined pituitary hormone deficiency (CPHD), which is characterized by GH, PRL, and TSH deficiencies (Pfaffle et al., 1992; Tatsumi et al., 1992). In contrast, acquired CPHD is generally caused by various types of damage to the hypothalamic–pituitary region, resulting in impaired hormone secretion in a nonspecific pattern.

Recently, Yamamoto et al. reported three patients with acquired deficiency of GH, PRL, and TSH (Yamamoto et al., 2011). Mutations previously associated with CPHD were excluded in these patients, who otherwise exhibited relatively normal ACTH and gonadotropin function. Moreover, all three patients showed normal growth and displayed no signs of pituitary deficiencies earlier in life. Intriguingly, anti-PIT-1 antibody and various other autoantibodies were detected in the patients' sera. An ELISA-based screening revealed that anti-PIT-1 antibody was highly specific to the disease and

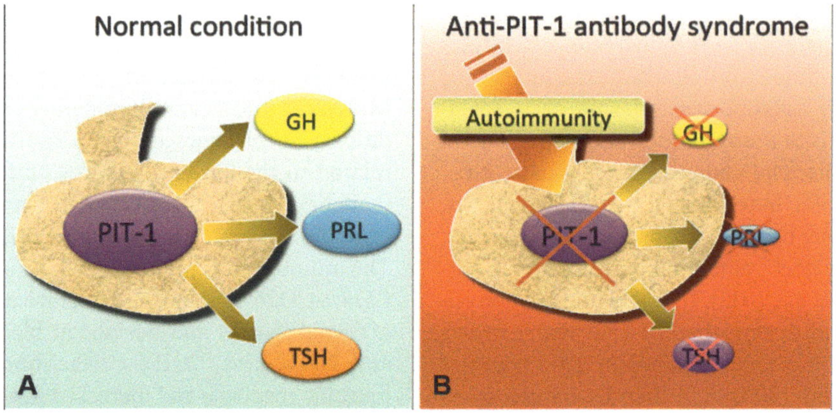

Fig. 29.1. In a normal condition, the transcriptional factor PIT-1 maintains the GH-, PRL-, and TSH-secreting cells (**A**); however, autoimmunity against PIT-1 impaired these cells and caused a specific defect in these hormones (**B**).

absent in control subjects. Immunohistochemical analysis demonstrated that PIT-1-, GH-, PRL-, and TSH-positive cells were absent in the pituitary of patient, who exhibited a range of autoimmune endocrinopathies. These clinical manifestations were compatible with the definition of APS. However, the main manifestations of APS-1 – hypoparathyroidism and *Candida* infection – were not observed. Moreover, the pituitary abnormalities clearly differed from those of common hypophysitis associated with APS, in which pituitary hormone secretion is impaired nonspecifically. Therefore, a novel clinical entity termed "anti-PIT-1 antibody syndrome" was proposed, which is related to APS. Anti-PIT-1 antibodies were present in these patients' sera but not in the sera of control subjects or patients with various other autoimmune-related diseases, indicating that the anti-PIT-1 antibody is highly specific to the disease. Furthermore, other autoantibodies previously reported in common hypophysitis, such as anti-GH, -α-enolase, or -TDRD6 antibodies, were not detected by immunoblotting. In addition, any obvious pituitary abnormalities such as an enlargement, which is frequently observed in hypophysitis, were not detected by MRI, demonstrating the difference of anti-PIT-1 antibody syndrome from common hypophysitis.

Pathogenic autoantibodies such as anti-TSH receptor antibody in Graves' disease and antiacetylcholine receptor antibody in myasthenia gravis are generally directed against cell surface antigens. Thus, it is unlikely that PIT-1 protein – a nuclear transcription factor – is a direct target for anti-PIT-1 antibodies. It is possible that immune intolerance to PIT-1 occurred by an as yet unknown mechanism, thereby provoking the attack of PIT-1-expressing cells by cytotoxic T cells through recognition of PIT-1 epitopes exposed with MHC (HLA) antigen on the cell surface. As a result, anti-PIT-1 antibodies would be produced. Regarding abnormal T cell function, lymphocytic infiltration of the pituitary, thyroid, adrenal gland, liver, and pancreas was observed (Yamamoto et al., 2011). The combination of the impaired GH, PRL, and TSH hormones; the antigen, PIT-1; and the specificity of the autoantibody to this disease strongly suggest that the anti-PIT-1 antibody is closely related to pathogenesis (Fig. 29.1). However, further investigation is required in order to elucidate the underlying mechanisms by transferring of antibody or T cells from the patients to animals to reconstitute the disease. There are several interesting implications of this novel syndrome. It demonstrated a new cause of cell-restricted pituitary hormone deficiencies that is very similar to PIT-1-dependent congenital CPHD. The striking specificity of the pituitary cell death observed in the patient may warrant the re-evaluation of patients with CPHD without detectable mutations for the presence of autoimmune antibodies against the specific transcriptional factors (Drouin and Takayasu, 2011). Indeed, most patients with CPHD do not display the gene abnormalities in the expected transcription factors. Another interesting hypothesis raised by this syndrome is that if autoimmunity to nuclear proteins – particularly transcription factors – is related to their functional impairment, various acquired diseases may be caused by autoimmunity to transcription factors. Although few cases demonstrating the presence of autoantibody against transcription factors have been reported thus far (Hedstrand et al., 2001), this concept may apply to various acquired conditions.

AUTOIMMUNITY AND METABOLIC DISEASE

Recently, Winer et al. reported that B cells promote insulin resistance through modulation of T cells and production of pathogenic IgG antibodies (Winer et al., 2011).

They demonstrated that B cells play a pivotal role in inducing insulin resistance in diet-induced obese (DIO) mice. The transfer of IgG from DIO mice rapidly induced insulin resistance and glucose intolerance, indicating that pathogenic IgG was produced in DIO mice. IgG autoantibodies that segregate with insulin sensitivity were also detected in insulin-resistant human subjects. Moreover, the antigens targeted by such IgG autoantibodies are mostly intracellular proteins, many of which are expressed in tissues including immune cells, the pancreas, nervous tissue, muscle, or fat. Taken together, these results imply that the pathologic relevance of autoantibodies needs to be re-evaluated in order to understand the pathogenesis of endocrine and metabolic diseases with unknown etiology.

CONCLUSION

Novel clinical entities associated with hypophysitis that have recently been reported demonstrate the heterogeneity of this disease and provide an important clues for understanding pathogenesis and definition of hypophysitis, as well as the significance of antipituitary antibodies.

ABBREVIATIONS

AH, autoimmune hypophysitis; LAH, lymphocytic adenohypophysitis; LINH, lymphocytic infundibuloneurophypophysitis; LPH, lymphocytic panhypophysitis; APS, autoimmune polyglandular syndrome; GH, growth hormone; CTLA4, cytotoxic T-lymphocyte protein 4; PGSFla, 2, pituitary gland specific factors la and 2; NSE, neuron-specific enolase; CHD8, chromodomain helicase DNA-binding protein 8; CADPS, Ca2+-dependent secretion activator; APA, antipituitary antibody; IgG4, immunoglobulin G4; PIT-1, POU1F1, pituitary-specific transcriptional factor-1; CPHD, combined pituitary hormone deficiency; DIO, diet-induced obese.

References

Akitake R, Watanabe T, Zaima C et al. (2010). Possible involvement of T helper type 2 responses to Toll-like receptor ligands in IgG4-related sclerosing disease. Gut 59: 542–545.

Bellastella G, Rotondi M, Pane E et al. (2010). Predictive role of the immunostaining pattern of immunofluorescence and the titers of antipituitary antibodies at presentation for the occurrence of autoimmune hypopituitarism in patients with autoimmune polyendocrine syndromes over a five-year follow-up. J Clin Endocrinol Metab 95: 3750–3757.

Bensing S, Fetissov SO, Mulder J et al. (2007a). Pituitary autoantibodies in autoimmune polyendocrine syndrome type 1. Proc Natl Acad Sci U S A 104: 949–954.

Bensing S, Hulting AL, Hoog A et al. (2007b). Lymphocytic hypophysitis: report of two biopsy-proven cases and one suspected case with pituitary autoantibodies. J Endocrinol Invest 30: 153–162.

Buxton N, Robertson I (2001). Lymphocytic and granulocytic hypophysitis: a single centre experience. Br J Neurosurg 15: 242–245, discussion 245–246.

Caturegli P, Newschaffer C, Olivi A et al. (2005). Autoimmune hypophysitis. Endocr Rev 26: 599–614.

Cohen LE, Wondisford FE, Radovick S (1996). Role of Pit-1 in the gene expression of growth hormone, prolactin, and thyrotropin. Endocrinol Metab Clin North Am 25: 523–540.

Crock PA (1998). Cytosolic autoantigens in lymphocytic hypophysitis. J Clin Endocrinol Metab 83: 609–618.

De Bellis A, Sinisi AA, Conte M et al. (2007). Antipituitary antibodies against gonadotropin-secreting cells in adult male patients with apparently idiopathic hypogonadotropic hypogonadism. J Clin Endocrinol Metab 92: 604–607.

De Bellis A, Pane E, Bellastella G et al. (2011). Detection of antipituitary and antihypothalamus antibodies to investigate the role of pituitary or hypothalamic autoimmunity in patients with selective idiopathic hypopituitarism. Clin Endocrinol (Oxf) 75: 361–366.

Drouin J, Takayasu S (2011). Autoimmunity: acquired versus inherited pituitary deficiency – same difference? Nat Rev Endocrinol 7: 255–256.

Falorni A, Laureti S, De Bellis A et al. (2004). Italian addison network study: update of diagnostic criteria for the etiological classification of primary adrenal insufficiency. J Clin Endocrinol Metab 89: 1598–1604.

Goudie RB, Pinkerton PH (1962). Anterior hypophysitis and Hashimoto's disease in a young woman. J Pathol Bacteriol 83: 584–585.

Hamano H, Kawa S, Horiuchi A et al. (2001). High serum IgG4 concentrations in patients with sclerosing pancreatitis. N Engl J Med 344: 732–738.

Hamnvik OP, Laury AR, Laws Jr ER et al. (2010). Lymphocytic hypophysitis with diabetes insipidus in a young man. Nat Rev Endocrinol 6: 464–470.

Hedstrand H, Ekwall O, Olsson MJ et al. (2001). The transcription factors SOX9 and SOX10 are vitiligo autoantigens in autoimmune polyendocrine syndrome type I. J Biol Chem 276: 35390–35395.

Ingraham HA, Chen RP, Mangalam HJ et al. (1988). A tissue-specific transcription factor containing a homeodomain specifies a pituitary phenotype. Cell 55: 519–529.

Kasperlik-Zaluska AA, Czarnocka B, Czech W (2003). Autoimmunity as the most frequent cause of idiopathic secondary adrenal insufficiency: report of 111 cases. Autoimmunity 36: 155–159.

Lamolet B, Pulichino AM, Lamonerie T et al. (2001). A pituitary cell-restricted T box factor, Tpit, activates POMC transcription in cooperation with Pitx homeoproteins. Cell 104: 849–859.

Leporati P, Landek-Salgado MA, Lupi I et al. (2011). IgG4-related hypophysitis: a new addition to the hypophysitis spectrum. J Clin Endocrinol Metab 96: 1971–1980.

Leung GK, Lopes MB, Thorner MO et al. (2004). Primary hypophysitis: a single-center experience in 16 cases. J Neurosurg 101: 262–271.

Lupi I, Broman KW, Tzou SC et al. (2008). Novel autoantigens in autoimmune hypophysitis. Clin Endocrinol (Oxf) 69: 269–278.

Maghnie M, Ghirardello S, De Bellis A et al. (2006). Idiopathic central diabetes insipidus in children and young adults is commonly associated with vasopressin-cell antibodies and markers of autoimmunity. Clin Endocrinol (Oxf) 65: 470–478.

Michels AW, Gottlieb PA (2010). Autoimmune polyglandular syndromes. Nat Rev Endocrinol 6: 270–277.

Nakashima H, Miyake K, Moriyama M et al. (2010). An amplification of IL-10 and TGF-beta in patients with IgG4-related tubulointerstitial nephritis. Clin Nephrol 73: 385–391.

Nussbaum CE, Okawara SH, Jacobs LS (1991). Lymphocytic hypophysitis with involvement of the cavernous sinus and hypothalamus. Neurosurgery 28: 440–444.

O'Dwyer DT, Clifton V, Hall A et al. (2002). Pituitary autoantibodies in lymphocytic hypophysitis target both gamma- and alpha-Enolase – a link with pregnancy? Arch Physiol Biochem 110: 94–98.

Pfaffle RW, DiMattia GE, Parks JS et al. (1992). Mutation of the POU-specific domain of Pit-1 and hypopituitarism without pituitary hypoplasia. Science 257: 1118–1121.

Pulichino AM, Vallette-Kasic S, Couture C et al. (2003a). Human and mouse TPIT gene mutations cause early onset pituitary ACTH deficiency. Genes Dev 17: 711–716.

Pulichino AM, Vallette-Kasic S, Tsai JP et al. (2003b). Tpit determines alternate fates during pituitary cell differentiation. Genes Dev 17: 738–747.

Saito T, Yoshida S, Nakao K et al. (1970). Chronic hypernatremia associated with inflammation of the neurohypophysis. J Clin Endocrinol Metab 31: 391–396.

Shaw SA, Camacho LH, McCutcheon IE et al. (2007). Transient hypophysitis after cytotoxic T lymphocyte-associated antigen 4 (CTLA4) blockade. J Clin Endocrinol Metab 92: 1201–1202.

Shimatsu A, Oki Y, Fujisawa I et al. (2009). Pituitary and stalk lesions (infundibulo-hypophysitis) associated with immunoglobulin G4-related systemic disease: an emerging clinical entity. Endocr J 56: 1033–1041.

Smith CJ, Bensing S, Burns C et al. (2012). Identification of Tpit and other novel autoantigens in lymphocytic hypophysitis; immunoscreening of a pituitary cDNA library and development of immunoprecipitation assays. Eur J Endocrinol 166: 391–398.

Suzuki K, Tamaru J, Okuyama A et al. (2010). IgG4-positive multi-organ lymphoproliferative syndrome manifesting as chronic symmetrical sclerosing dacryo-sialadenitis with subsequent secondary portal hypertension and remarkable IgG4-linked IL-4 elevation. Rheumatology (Oxford) 49: 1789–1791.

Takao T, Nanamiya W, Matsumoto R et al. (2001). Antipituitary antibodies in patients with lymphocytic hypophysitis. Horm Res 55: 288–292.

Tanaka S, Tatsumi KI, Kimura M et al. (2002). Detection of autoantibodies against the pituitary-specific proteins in patients with lymphocytic hypophysitis. Eur J Endocrinol 147: 767–775.

Tatsumi K, Miyai K, Notomi T et al. (1992). Cretinism with combined hormone deficiency caused by a mutation in the PIT1 gene. Nat Genet 1: 56–58.

Ueda H, Howson JM, Esposito L et al. (2003). Association of the T-cell regulatory gene CTLA4 with susceptibility to autoimmune disease. Nature 423: 506–511.

Umehara H, Okazaki K, Masaki Y et al. (2012). A novel clinical entity, IgG4-related disease (IgG4RD): general concept and details. Mod Rheumatol 22: 1–14.

van der Vliet HJ, Perenboom RM (2004). Multiple pseudotumors in IgG4-associated multifocal systemic fibrosis. Ann Intern Med 141: 896–897.

Wegner N, Wait R, Venables PJ (2009). Evolutionarily conserved antigens in autoimmune disease: implications for an infective aetiology. Int J Biochem Cell Biol 41: 390–397.

Winer DA, Winer S, Shen L et al. (2011). B cells promote insulin resistance through modulation of T cells and production of pathogenic IgG antibodies. Nat Med 17: 610–617.

Wong S, Lam WY, Wong WK et al. (2007). Hypophysitis presented as inflammatory pseudotumor in immunoglobulin G4-related systemic disease. Hum Pathol 38: 1720–1723.

Yamamoto M, Iguchi G, Takeno R et al. (2011). Adult combined GH, prolactin, and TSH deficiency associated with circulating PIT-1 antibody in humans. J Clin Invest 121: 113–119.

Zen Y, Nakanuma Y (2011). Pathogenesis of IgG4-related disease. Curr Opin Rheumatol 23: 114–118.

Index

NB: Page numbers in *italics* refer to figures and tables.

D

E

I

J

Q

R

FSC
www.fsc.org
MIX
Papier | Fördert
gute Waldnutzung
FSC® C083411

Kolibri 360 GmbH
Ferdinand-Jühlke-Straße 7
99095 Erfurt, Deutschland
produktsicherheit@kolibri360.de